disha
Publication Inc

AF646083

34 YEARS

NTA NEET (UG)

PHYSICS

CHAPTERWISE & TOPICWISE SOLVED PAPERS (2021-1988)

with Value Added Notes

Corporate Office

DISHA PUBLICATION
45, 2nd Floor, Maharishi Dayanand Marg,
Corner Market, Malviya Nagar, New Delhi - 110017
Tel : 49842349 / 49842350

By:
Sanjeev Kumar Jha

Typeset by Disha DTP Team

Write to us at feedback_disha@aiets.co.in

Contents

NEET FAQs

<table>
<tr><th>Query</th><th>Response</th></tr>
<tr><td rowspan="5">Any changes in the NEET exam pattern for 2022?</td><td>Up till now, there has been no intimation on changes in the exam pattern for 2022. However, changes will be disclosed with the information brochure of NEET 2022</td></tr>
<tr><td>Subjects – Physics, Chemistry, Botany & Zoology</td></tr>
<tr><td>Number of questions – 200</td></tr>
<tr><td>Maximum marks – 720 marks</td></tr>
<tr><td>Exam medium – 13 languages</td></tr>
<tr><td>How many questions will appear from each subject?</td><td><table>
<tr><th>Subject</th><th>Section-A</th><th>Section-B</th></tr>
<tr><td>Physics</td><td>35 MCQs</td><td>15 MCQs</td></tr>
<tr><td>Chemistry</td><td>35 MCQs</td><td>15 MCQs</td></tr>
<tr><td>Botany</td><td>35 MCQs</td><td>15 MCQs</td></tr>
<tr><td>Zoology</td><td>35 MCQs</td><td>15 MCQs</td></tr>
</table>
Attempt all 35 questions of section-A & any 10 questions of section-B in each subject.</td></tr>
<tr><td>What will the syllabus for NEET be?</td><td>NEET Syllabus covers NCERT syllabus from classes 11 & 12 of all the subjects</td></tr>
<tr><td>Which type of questions will be asked?</td><td>Objective (MCQs) type questions will be asked</td></tr>
<tr><td>What is the marking scheme for NEET 2020?</td><td>4 marks will be allotted for every right response. 1 mark will be deducted for every incorrect response. No marks will be awarded if left unanswered</td></tr>
</table>

ELIGIBILITY CRITERIA

In this section, we have mentioned some of the important NEET FAQs related to the eligibility criteria:

Q1. I have crossed more than 24 years of my age. Can I attempt for NEET 2022?

Ans: Yes, you can appear for NEET 2022. There is a minimum age criterion of 17 years but there are no such stipulated criteria of maximum age.

Q2. I am a candidate from open school education. Can I apply for NEET 2022?

Ans: No, candidates who have passed 10+2 as the private candidates or from open schools are not eligible to take the NEET exam.

Q3. What is the basic eligibility criteria to appear for NEET 2022?

Ans: Candidates have to make sure that they meet the following criteria. Failing to meet these specifications will lead to rejection:

- Must have attained 17 years of age as on December 31, 2022.
- Must have completed class 12 or equivalent exams from a centrally recognized or state board with major subjects as Physics, Chemistry, Biology, and English.
- Must have qualifying marks of 50% if they belong to general category. For SC/ST/OBC, it is 40%.
- OCIs, PIOs, NRIs and foreign nationals can apply for the exam.

Q4. Can I appear for NEET exams even if I have a gap of one year between my class 11 & 12?

Ans: Yes, you can appear for the NEET if you have a year gap. Earlier there was no rule regarding the year gaps, but the amendments were made by the Delhi High Court.

Mode of Counselling: Online, Offline

As per the information brochure and the latest guidelines, NEET counselling is broadly of two types. One is the All India counselling conducted by DGHS for 15% All India Quota seats in goverment medical colleges and all seats in Central Institutions including AIIMS, JIPMER, AFMC, ESI, Delhi University (DU), BHU, AMU and Deemed Universities. The second is the state counselling held by each individual state medical council for the rest of the 85% of the seats in the government colleges, state run universities and the private medical colleges of the state. Only students who qualify in NEET will be eligible for the counselling subject to certain regulations.

NEET counselling is generally held in two rounds followed by a mop-up round for any vacant seats left after the first two rounds. Candidates must apply for the counselling as per their eligibility. An important point to note is that all candidates who qualify in NEET 2022 will be eligible to apply for the All India counselling. However, for the respective state counselling processes, admissions will be based on factors like domicile, residency in state etc. Candidates must check whether they meet the state admission criteria before applying for the state counselling.

NEET seat allotment will be based on the All India Rank of the applicants, choice of colleges entered, seat availability, reservation criteria etc.

NEET COUNSELLING GUIDELINES

Type of Counselling	Counselling Body
Counselling for 15% AIQ Seats (except in Jammu & Kashmir) in Government medical and dental colleges	DGHS, on behalf of MCC
Counselling for all MBBS seats in AIIMs and JIPMER	
Counselling for all the seats in Central and Deemed Universities including BDS at Jamia Millia Islamia, New Delhi	
Counselling for 85% State Quota seats in DU colleges (MAMC, LHMC, UCMS)	
Counselling for IP Quota seats in ESIC Colleges	
Counselling for Armed Forces Medical College (AFMC)	DGHS, on behalf of MCC and AFMC Pune
Counselling for 85% State Quota seats and 100% state private colleges.	Respective State Counselling Authorities
In case of Jammu & Kashmir - 100% of all government and private colleges	

NEET SEAT INTAKE: (No. may increase in NEET 2022)

- AIIMS MBBS (15 institutes) - 1205 seats
- JIPMER MBBS (2 institutes) - 200 seats
- MBBS seats in 532 colleges - 76,928
- BDS seats in 313 colleges - 26,949
- Ayush seats in 914 colleges - 52,720 seats

NEET CUT OFF FOR TOP MEDICAL COLLEGES

Students appearing in NEET 2022 can check Previous year cut off of some well-known medical institutes in India from the table given below:

Name of the College	Number of Seats	Closing Rank
Maulana Azad Medical College	250	32
VMMC and Safdarjung Hospital	150	157
University College of Medical Science	150	171
Govt. Medical College	250	360
Seth G.S Medical College	180	638
Lady Hardinge Medical College	150	489
CSM Medical University	250	3692
S. M. S Medical College	250	760
Grant Medical College & Sir J. J. Hospital	200	1329
Pt. B. D. Sharma PGIMS	200	1825
King George Medical College	NA	908
Sardar Patel Medical College	250	2875
Madras Medical College	250	501
Dr S.N Medical College	250	1128
T. D. Medical College	150	747
M. G. M. Medical College	100	8265
Rural Institute of Med Science & Research	150	1953

Note :- This is a rough estimate & many vary from year to year

NEET Result - Marks vs Rank 2020, 2019, 2018 and 2017

(will be updated with 2021 once results are out)

Marks Range	Rank Range (2020)	Rank Range (2019)	Rank Range (2018)	Rank Range (2017)
701-720	1-14	-	-	-
691-700	15-20	1	1	1 - 9
681-690	21-80	20 - 88	2 - 7	10 - 25
671-680	81-200	99 - 214	8 - 31	26 - 83
661-670	201-450	223 - 476	32 - 63	84 - 163
651-660	451-950	568 - 930	64 - 122	164 - 301
641-650	951-1700	946 - 1714	123 - 232	302 - 535
631-640	1701-2500	1809 - 2570	233 - 398	534 - 870
621-630	2501-4000	2788 - 3956	399 - 639	871 - 1308
611-620	4001-5500	4074 - 5630	640 - 994	1309 - 1962
601-610	5501-7500	5692 - 7580	995 - 1505	1963 - 2786
591-600	7501-10000	7784 - 10036	1506 - 2169	2787 - 3874
581-590	10001-12800	10248 - 12898	2170 - 3084	3875 - 5229
571-580	12801-13000	13064 - 16008	3085 - 4202	5230 - 6788
561-570	13001-16000	16173 - 19478	4203 - 5615	6789 - 8736
551-560	16001-19000	19967 - 23501	5616 - 7433	8737 - 10851
541-550	19001-23000	23695 - 27650	7434 - 9493	10851 - 13353
531-540	23001-27000	27994 - 32317	9494 - 11885	13354 - 16163
521-530	27001-32000	32796 - 37464	11886 - 14629	16163 - 18876
511-520	32001-37000	37780 - 38736	14630 - 17816	18876 - 22372
501-510	37001-44000	38822 - 44553	17817 - 21337	22372 - 25842
491-500	44001-51000	45023 - 51086	21338 - 25229	25843 - 29557
481-490	51001-55000	51498 - 58114	25230 - 29528	29558 - 33893
471-480	55001-58000	58214 - 65316	29529 - 34037	33894 - 38152
461-470	58001-68000	65801 - 73197	34038 - 38947	38153 - 43019
451-460	68001-73000	73337 - 81607	38948 - 44227	43020 - 47809
441-450	73001-80000	82216 - 89872	44228 - 49907	47810 - 53184
431-440	80001-82000	90825 - 99323	49908 - 55928	53185 - 59177
421-430	82001-89000	99914 - 109429	55929 - 62506	59178 - 65280

1 Physical World, Units and Measurements

Trend Analysis with Important Topics & Sub-Topics

Topic Name	Sub-Topic	2021 QNS.	2021 LOD	2020 QNS.	2020 LOD	2019 QNS.	2019 LOD	2018 QNS.	2018 LOD	2017 QNS.	2017 LOD	2016 QNS.	2016 LOD
Units of Physical Quantities	Units of Physical Quantities					1	E						
Dimensions of Physical Quantities	Dimensions of Physical Q.	1	A	1	E								
	Application of Dimensions	1	A							1	A	1	A
Errors in Measurements	Relative & % Error					1	E						
	Diameter = MSR + CSR (L.C.) – Zero error	1	A	1	A			1	A				
	Significant figures			1	E								

LOD - Level of Difficulty | E - Easy | A - Average | D - Difficult | Qns - No. of Questions

Topic 1 Units of Physical Quantities

1. The unit of thermal conductivity is : *[2019]*
   (a) $J\,m\,K^{-1}$ (b) $J\,m^{-1}\,K^{-1}$
   (c) $W\,m\,K^{-1}$ (d) $W\,m^{-1}K^{-1}$
2. The density of material in CGS system of units is $4g/cm^3$. In a system of units in which unit of length is 10 cm and unit of mass is 100 g, the value of density of material will be *[2011M]*
   (a) 0.4 (b) 40
   (c) 400 (d) 0.04
3. The unit of permittivity of free space, ε_o is *[2004]*
   (a) $Coulomb^2/(Newton\text{-}metre)^2$
   (b) Coulomb/Newton-metre
   (c) $Newton\text{-}meter^2/Coulomb^2$
   (d) $Coulomb^2/Newton\text{-}meter^2$
4. The unit of the Stefan-Boltzmann's constant is *[2002]*
   (a) W/m^2K^4 (b) W/m^2
   (c) W/m^2K (d) W/m^2K^2
5. In a particular system, the unit of length, mass and time are chosen to be 10 cm, 10 g and 0.1 s respectively. The unit of force in this system will be equivalent to *[1994]*
   (a) 0.1 N (b) 1 N
   (c) 10 N (d) 100 N
6. If $x = at + bt^2$, where x is the distance travelled by the body in kilometers while t is the time in seconds, then the unit of b is *[1989]*
   (a) km/s (b) kms
   (c) km/s^2 (d) kms^2

Topic 2 Dimensions of Physical Quantities

7. If E and G respectively denote energy and gravitational constant, then $\frac{E}{G}$ has the dimensions of *[2021]*
   (a) $[M^2][L^{-2}][T^{-1}]$ (b) $[M^2][L^{-1}][T^0]$
   (c) $[M][L^{-1}][T^{-1}]$ (d) $[M][L^0][T^0]$
8. If force $[F]$, acceleration $[A]$ and time $[T]$ are chosen as the fundamental physical quantities. Find the dimensions of energy. *[2021]*
   (a) $[F][A^{-1}][T]$ (b) $[F][A][T]$
   (c) $[F][A][T^2]$ (d) $[F][A][T^{-1}]$
9. Dimensions of stress are : *[2020]*
   (a) $[ML^2T^{-2}]$ (b) $[ML^0T^{-2}]$
   (c) $[ML^{-1}T^{-2}]$ (d) $[MLT^{-2}]$

10. A physical quantity of the dimensions of length that can be formed out of c, G and $\frac{e^2}{4\pi\varepsilon_0}$ is [c is velocity of light, G is universal constant of gravitation and e is charge] *[2017]*

(a) $c^2\left[G\frac{e^2}{4\pi\varepsilon_0}\right]^{1/2}$ (b) $\frac{1}{c^2}\left[\frac{e^2}{G4\pi\varepsilon_0}\right]^{1/2}$

(c) $\frac{1}{c}G\frac{e^2}{4\pi\varepsilon_0}$ (d) $\frac{1}{c^2}\left[G\frac{e^2}{4\pi\varepsilon_0}\right]^{1/2}$

11. If energy (E), velocity (V) and time (T) are chosen as the fundamental quantities, the dimensional formula of surface tension will be: *[2015]*

(a) $[EV^{-1}T^{-2}]$ (b) $[EV^{-2}T^{-2}]$
(c) $[E^{-2}V^{-1}T^{-3}]$ (d) $[EV^{-2}T^{-1}]$

12. If dimensions of critical velocity υ_c of a liquid flowing through a tube are expressed as $[\eta^x\rho^y r^z]$, where η, ρ and r are the coefficient of viscosity of liquid, density of liquid and radius of the tube respectively, then the values of x, y and z are given by: *[2015 RS]*

(a) –1, –1, 1 (b) –1, –1, –1
(c) 1, 1, 1 (d) 1, –1, –1

13. If force (F), velocity (V) and time (T) are taken as fundamental units, then the dimensions of mass are: *[2014]*

(a) $[F\,V\,T^{-1}]$ (b) $[F\,V\,T^{-2}]$
(c) $[F\,V^{-1}\,T^{-1}]$ (d) $[F\,V^{-1}\,T]$

14. The pair of quantities having same dimensions is *[NEET Kar. 2013]*

(a) Young's modulus and energy
(b) impulse and surface tension
(c) angular momentum and work
(d) work and torque

15. The dimensions of $(\mu_0 \epsilon_0)^{\frac{-1}{2}}$ are *[2012M, 2011]*

(a) $[L^{1/2}\,T^{-1/2}]$ (b) $[L^{-1}\,T]$
(c) $[L\,T^{-1}]$ (d) $[L^{-1/2}\,T^{1/2}]$

16. The dimension of $\frac{1}{2}\varepsilon_0 E^2$, where ε_0 is permittivity of free space and E is electric field, is: *[2010]*

(a) $[ML^2\,T^{-2}]$ (b) $[ML^{-1}T^{-2}]$
(c) $[ML^2T^{-1}]$ (d) $[MLT^{-1}]$

17. If the dimensions of a physical quantity are given by $M^aL^bT^c$, then the physical quantity will be: *[2009]*

(a) Velocity if $a = 1, b = 0, c = -1$
(b) Acceleration if $a = 1, b = 1, c = -2$
(c) Force if $a = 0, b = -1, c = -2$
(d) Pressure if $a = 1, b = -1, c = -2$

18. Which two of the following five physical parameters have the same dimensions? *[2008]*

(A) Energy density
(B) Refractive index
(C) Dielectric constant
(D) Young's modulus
(E) Magnetic field

(a) (B) and (D) (b) (C) and (E)
(c) (A) and (D) (d) (A) and (E)

19. Dimensions of resistance in an electrical circuit, in terms of dimension of mass [M], of length [L], of time [T] and of current [I], would be *[2007]*

(a) $[ML^2T^{-2}]$ (b) $[ML^2T^{-1}I^{-1}]$
(c) $[ML^2T^{-3}I^{-2}]$ (d) $[ML^2T^{-3}I^{-1}]$

20. The velocity v of a particle at time t is given by $v = at + \frac{b}{t+c}$, where a, b and c are constant. The dimensions of a, b and c are respectively *[2006]*

(a) $[L^2$, T and $LT^2]$ (b) $[LT^2$, LT and L]
(c) [L, LT and $T^2]$ (d) $[LT^{-2}$, L and T]

21. The ratio of the dimension of Planck's constant and that of the moment of inertia is the dimension of *[2005]*

(a) time
(b) frequency
(c) angular momentum
(d) velocity

22. The dimensions of universal gravitational constant is *[2004]*

(a) $[M^{-2}\,L^2\,T^{-1}]$ (b) $[M^{-1}\,L^3\,T^{-2}]$
(c) $[M\,L^2\,T^{-1}]$ (d) $[M^{-2}\,L^3\,T^{-2}]$

23. The dimensions of Planck's constant are same as *[2001]*

(a) energy (b) power
(c) momentum (d) angular momentum

24. Which one of the following groups have quantities that do not have the same dimensions? *[2000]*
(a) pressure, stress (b) velocity, speed
(c) force, impulse (d) work, energy

25. The dimensional formula for magnetic flux is *[1999]*
(a) $[ML^2T^{-2}A^{-1}]$ (b) $[ML^3T^{-2}A^{-2}]$
(c) $[M^0L^{-2}T^2A^{-2}]$ (d) $[ML^2T^{-1}A^2]$

26. The force F on a sphere of radius '*a*' is moving in a medium with velocity *v* is given by F = 6πηav. The dimensions of η are *[1997]*
(a) $[ML^{-3}]$ (b) $[ML^{-2}]$
(c) $[ML^{-1}]$ (d) $[ML^{-1}T^{-1}]$

27. An equation is given as : $\left(P+\frac{a}{V^2}\right)=b\frac{\theta}{V}$
where P = Pressure, V = Volume & θ = Absolute temperature. If *a* and *b* are constants, then dimensions of *a* will be *[1996]*
(a) $[ML^5T^{-2}]$ (b) $[M^{-1}L^5T^2]$
(c) $[ML^{-5}T^{-1}]$ (d) $[ML^5T^1]$

28. Which of the following will have the dimensions of time *[1996]*
(a) LC (b) $\frac{R}{L}$ (c) $\frac{L}{R}$ (d) $\frac{C}{L}$

29. Which of the following is a dimensional constant? *[1995]*
(a) Refractive index (b) Poissons ratio
(c) Relative density
(d) Gravitational constant

30. The time dependence of a physical quantity p is given by $p=p_0\exp(-\alpha t^2)$, where α is a constant and t is the time. The constant α *[1993]*
(a) is dimensionless (b) has dimensions T^{-2}
(c) has dimensions T^2 (d) has dimensions of p

31. Turpentine oil is flowing through a tube of length ℓ and radius r. The pressure difference between the two ends of the tube is p. The viscosity of oil is given by

$$\eta=\frac{p\left(r^2-x^2\right)}{4vl}$$

where v is the velocity of oil at a distance x from the axis of the tube. The dimensions of η are *[1993]*
(a) $[M^0L^0T^0]$ (b) $[MLT^{-1}]$
(c) $[ML^2T^{-2}]$ (d) $[ML^{-1}T^{-1}]$

32. P represents radiation pressure, c represents speed of light and S represents radiation energy striking unit area per sec. The non zero integers x, y, z such that $P^x S^y c^z$ is dimensionless are *[1992]*
(a) $x=1, y=1, z=1$ (b) $x=-1, y=1, z=1$
(c) $x=1, y=-1, z=1$ (d) $x=1, y=1, z=-1$

33. The dimensional formula for permeability μ is given by *[1991]*
(a) $[MLT^{-2}A^{-2}]$ (b) $[M^0L^1T]$
(c) $[M^0L^2T^{-1}A^2]$ (d) None of the above

34. According to Newton, the viscous force acting between liquid layers of area A and velocity gradient ΔV/ΔZ is given by $F=-\eta A\frac{\Delta V}{\Delta Z}$ where η is constant called coefficient of viscosity. The dimensional formula of η is *[1990]*
(a) $ML^{-2}T^{-2}$ (b) $M^0L^0T^0$
(c) ML^2T^{-2} (d) $ML^{-1}T^{-1}$

35. The frequency of vibration f of a mass m suspended from a spring of spring constant k is given by a relation of the type $f=c\,m^x k^y$, where c is a dimensionless constant. The values of x and y are *[1990]*
(a) $x=\frac{1}{2}, y=\frac{1}{2}$ (b) $x=-\frac{1}{2}, y=-\frac{1}{2}$
(c) $x=\frac{1}{2}, y=-\frac{1}{2}$ (d) $x=-\frac{1}{2}, y=\frac{1}{2}$

36. The dimensional formula of pressure is *[1990]*
(a) $[MLT^{-2}]$ (b) $[ML^{-1}T^2]$
(c) $[ML^{-1}T^{-2}]$ (d) $[MLT^2]$

37. The dimensional formula of torque is *[1989]*
(a) $[ML^2T^{-2}]$ (b) $[MLT^{-2}]$
(c) $[ML^{-1}T^{-2}]$ (d) $[ML^{-2}T^{-2}]$

38. Dimensional formula of self inductance is *[1989]*
(a) $[MLT^{-2}A^{-2}]$ (b) $[ML^2T^{-1}A^{-2}]$
(c) $[ML^2T^{-2}A^{-2}]$ (d) $[ML^2T^{-2}A^{-1}]$

39. Of the following quantities, which one has dimension different from the remaining three? *[1989]*
(a) Energy per unit volume
(b) Force per unit area
(c) Product of voltage and charge per unit volume
(d) Angular momentum.

40. The dimensional formula for angular momentum is [1988]
(a) $[M^0L^2T^{-2}]$ (b) $[ML^2T^{-1}]$
(c) $[MLT^{-1}]$ (d) $[ML^2T^{-2}]$

41. If C and R denote capacitance and resistance, the dimensional formula of CR is [1988]
(a) $[M^0L^0T^1]$ (b) $[M^0L^0T^0]$
(c) $[M^0L^0T^{-1}]$
(d) not expressible in terms of M,L,T.

Topic 3 Errors in Measurements

42. A screw gauge gives the following readings when used to measure the diameter of a wire
Main scale reading : 0 mm
Circular scale reading : 52 divisions
Given that 1 mm on main scale corresponds to 100 divisions on the circular scale. The diameter of the wire from the above data is [2021]
(a) 0.052 cm (b) 0.52 cm
(c) 0.026 cm (d) 0.26 cm

43. Taking into account of the significant figures, what is the value of 9.99 m – 0.0099 m? [2020]
(a) 9.98 m (b) 9.980 m
(c) 9.9 m (d) 9.9801 m

44. A screw gauge has least count of 0.01 mm and there are 50 divisions in its circular scale. The pitch of the screw gauge is : [2020]
(a) 0.25 mm (b) 0.5 mm
(c) 1.0 mm (d) 0.01 mm

45. In an experiment, the percentage of error occurred in the measurement of physical quantities A, B, C an D are 1%, 2%, 3% and 4% respectively. Then the maximum percentage of error in the measurement X, where $X = \frac{A^2 B^{1/2}}{C^{1/3} D^3}$ will be : [2019]
(a) $\left(\frac{3}{13}\right)\%$ (b) 16%
(c) –10% (d) 10%

46. The main scale of a vernier callipers has n divisions/cm. n divisions of the vernier scale coincide with $(n - 1)$ divisions of main scale. The least count of the vernier callipers is, [NEET Odisha, 2019]
(a) $\frac{1}{n(n+1)}$ cm (b) $\frac{1}{(n+1)(n-1)}$ cm
(c) $\frac{1}{n}$ cm (d) $\frac{1}{n^2}$ cm

47. A student measured the diameter of a small steel ball using a screw gauge of least count 0.001 cm. The main scale reading is 5 mm and zero of circular scale division coincides with 25 divisions above the reference level. If screw gauge has a zero error of –0.004 cm, the correct diameter of the ball is [2018]
(a) 0.521 cm (b) 0.525 cm
(c) 0.529 cm (d) 0.053 cm

48. In an experiment four quantities a, b, c and d are measured with percentage error 1%, 2%, 3% and 4% respectively. Quantity P is calculated as follows $P = \frac{a^3 b^2}{cd}$ % error in P is: [2013]
(a) 10% (b) 7% (c) 4% (d) 14%

49. If the error in the measurement of radius of a sphere is 2%, then the error in the determination of volume of the sphere will be: [2008]
(a) 4% (b) 6% (c) 8% (d) 2%

50. The density of a cube is measured by measuring its mass and length of its sides. If the maximum error in the measurement of mass and length are 4% and 3% respectively, the maximum error in the measurement of density will be [1996]
(a) 7% (b) 9% (c) 12% (d) 13%

51. The percentage errors in the measurement of mass and speed are 2% and 3% respectively. The error in kinetic energy obtained by measuring mass and speed will be [1995]
(a) 12% (b) 10% (c) 8% (d) 2%

52. In a vernier calliper N divisions of vernier scale coincides with (N – 1) divisions of main scale (in which length of one division is 1 mm). The least count of the instrument should be [1994]
(a) N (b) N – 1
(c) 1/10 N (d) 1/N – 1

53. A certain body weighs 22.42 gm and has a measured volume of 4.7 cc. The possible error in the measurement of mass and volume are 0.01 gm and 0.1 cc. Then maximum error in the density will be [1991]
(a) 22% (b) 2% (c) 0.2% (d) 0.02%

ANSWER KEY

1	(d)	7	(b)	13	(d)	19	(c)	25	(a)	31	(d)	37	(a)	43	(a)	49	(b)
2	(b)	8	(c)	14	(d)	20	(d)	26	(d)	32	(c)	38	(c)	44	(b)	50	(d)
3	(d)	9	(c)	15	(c)	21	(b)	27	(a)	33	(a)	39	(d)	45	(b)	51	(c)
4	(a)	10	(d)	16	(b)	22	(b)	28	(c)	34	(d)	40	(b)	46	(d)	52	(c)
5	(a)	11	(b)	17	(d)	23	(d)	29	(d)	35	(d)	41	(a)	47	(c)	53	(b)
6	(c)	12	(d)	18	(c)	24	(c)	30	(b)	36	(c)	42	(a)	48	(d)		

Hints & Solutions

1. (d) In steady state, the amount of heat flowing from one face to the other face in time dt is given by

$$H = \frac{kA(T_1 - T_2)dt}{\ell}$$

$\Rightarrow \frac{dH}{dt} = \frac{kA}{\ell}\Delta T$ (k = coefficient of thermal conductivity)

$$\therefore\ k = \frac{\ell dH}{A\, dt\, \Delta T}$$

Unit of $k = Wm^{-1} K^{-1}$

2. (b) In CGS system, density $d = 4\frac{g}{cm^3}$

unit of length = 1 cm
unit of mass = 1 g
And in another system of units
The unit of mass is 100g and unit of length is 10 cm, so substitute these values

$$\text{Density} = \frac{4\left(\frac{100g}{100}\right)}{\left(\frac{10}{10}cm\right)^3} = \frac{\left(\frac{4}{100}\right)}{\left(\frac{1}{10}\right)^3}\frac{(100g)}{(10cm)^3}$$

$$= \frac{4}{100}\times(10)^3\cdot\frac{100g}{(10cm)^3} = 40 \text{ unit}$$

> **NOTES**: Apart from fundamental and derived units, we have also used practical units *e.g.*, horse power (h.p.) is a practical units of power. Practical units may or may not belong to a system but can be expressed in any system of units.

3. (d) $\varepsilon_o = \frac{q^2}{(r^2)4\pi F}$

$\Rightarrow$ unit of ε_o is (coulomb)^2/ newton-metre^2

4. (a) According to Stefan's law, $E = \sigma AT^4$
where, E is energy dissipated per second,
A = surface area
T = absolute temperature

$$\sigma = \frac{E}{AT^4} = \frac{W}{m^2K^4}$$

5. (a) As we know force = Mass × Acceleration
= Mass × length × time^{-2} = (10g)(10 cm)(0.1s)$^{-2}$
= (10^{-2} kg) (10^{-1}m) (10^{-1}s)$^{-2}$ = 10^{-1}N.

6. (c) Given, $x = at + bt^2$
where, x = distance in km
t = time in sec
By the principle of homogencity, the dimensions of each terms of a dimensional equation on both sides are the same.
So, from equation, $x = at + bt^2$
dimension of left side = $[L]$
dimension of right side should be = [L]
According to homogeneity's law,

$$[L] = [b][T^2]$$

$$[b] = \frac{[L]}{[T^{+2}]} = \text{dimension of acceleration}$$

so unit of b should be km/sec^2.

7. (b) Dimensional formula of energy $E = \frac{1}{2}mv^2$

$[E] = [M^1 L^2 T^{-2}]$ ...(I)
Dimensional formula of gravitational constant

$$G = \frac{Fr^2}{m_1 m_2}$$

$[G] = [M^{-1} L^3 T^{-2}]$ ...(II)
From eq. (I) & (II)

$$\therefore \frac{E}{G} = \frac{[M^1 L^2 T^{-2}]}{[M^{-1} L^3 T^{-2}]} = [M^2 L^{-1} T^0]$$

NOTES Numerical constant like, $\frac{1}{2}$, 1, 2 or 2π has no dimension.

8. **(c)** Energy, $E \propto F^a A^b T^c$

$[E] = [F^a][A^b][T^c]$

$\Rightarrow [ML^2T^{-2}] = [MLT^{-2}]^a [LT^{-2}]^b [T]^c$

$[ML^2T^{-2}] = [M^a L^{a+b} T^{-2a-2b+c}]$

Comparing dimensions on both sides.

$a = 1$; $a + b = 2$ and $-2 = -2a - 2b + c$

$\therefore b = 1$ and $-2 = -2 - 2 + c$ or, $c = 2$

$\therefore$ Dimensions of energy $= [FAT^2]$

9. **(c)** Stress $= \frac{\text{Force}}{\text{Area}}$

Dimensions of force $= [MLT^{-2}]$

Dimensions of area $= [L^2]$

$\therefore$ Stress $= \frac{[MLT^{-2}]}{[L^2]} = [ML^{-1}T^{-2}]$

10. **(d)** Let dimensions of length is related as,

$$[L] = [c]^x [G]^y \left[\frac{e^2}{4\pi\varepsilon_0}\right]^z$$

$$\Rightarrow \frac{e^2}{4\pi\varepsilon_0} = [ML^3T^{-2}]$$

$[L] = [LT^{-1}]^x [M^{-1}L^3T^{-2}]^y [ML^3T^{-2}]^z$

$[L] = [L^{x+3y+3z} M^{-y+z} T^{-x-2y-2z}]$

Comparing both sides

$-y + z = 0 \Rightarrow y = z$ ...(i)

$x + 3y + 3z = 1$ ...(ii)

$-x - 4z = 0$ $(\because y = z)$ ...(iii)

From (i), (ii) and (iii)

$z = y = \frac{1}{2}$, $x = -2$

Hence, $[L] = c^{-2}\left[G \cdot \frac{e^2}{4\pi\varepsilon_0}\right]^{1/2}$

11. **(b)** As we know that, surface tension(s) $= \frac{\text{Force}[F]}{\text{Length}[L]}$

So, $[S] = \frac{[MLT^{-2}]}{[L]} = [MT^{-2}]$

Energy, $(E) =$ Force $\times$ Displacement,

$[E] = [ML^2T^{-2}]$

Velocity (V) $= \frac{\text{displacement}}{\text{time}}$

$[V] = [LT^{-1}]$

Let surface tension expressed as,

$s = E^a V^b T^c$ where a, b, c are constant.

Put the value

$$\frac{[MLT^{-2}]}{[L]} = [ML^2T^{-2}]^a \left[\frac{L}{T}\right]^b [T]^c$$

From the principle of homogeneity,

Equating the dimension of LHS and RHS

$[ML^0T^{-2}] = [M^a L^{2a+b} T^{-2a-b+c}]$

$\Rightarrow a = 1, 2a + b = 0, -2a - b + c = -2$

$\Rightarrow a = 1, b = -2, c = -2$

Hence, the dimensions of surface tension are $[E V^{-2} T^{-2}]$

NOTES Length, mass and time are arbitrarily chosen as fundamental quantities in mechanics. In fact any three quantities in mechanics can be termed as fundamental as all other quantities can be expressed in terms of these. If force (F) and acceleration (a) are taken as fundamental quantities, then mass will be defined as force (F)/acceleration (a) and will be termed as derived quantity.

12. **(d)** Applying dimensional method :

$v_c = \eta^x \rho^y r^z$

here,

dimension of critical velocity, $V_0 = [LT^{-1}]$

co-efficient of viscosity, $\eta = \frac{F}{6\pi rv}$

so dimension of $\eta = \frac{[MLT^{-2}]}{[L][LT^{-1}]} = [ML^{-1}T^{-1}]$

dimension of density, $\rho = \frac{[M]}{[L^3]} = [ML^{-3}]$

dimension of radius, $r = [L]$

Put these values in equation (i),

$[M^0LT^{-1}] = [ML^{-1}T^{-1}]^x [ML^{-3}T^0]^y [M^0LT^0]^z$

Equating powers both sides

$x + y = 0$; $-x = -1 \therefore x = 1$

$1 + y = 0 \therefore y = -1$

$-x - 3y + z = 1$

$-1 - 3(-1) + z = 1$

$-1 + 3 + z = 1$

$\therefore z = -1$

13. (d) Force = mass × acceleration

$\Rightarrow$ Mass $= \dfrac{\text{force}}{\text{acceleration}}$

$= \dfrac{\text{force}}{\text{velocity / time}} = [\text{F V}^{-1}\text{T}]$

14. (d) Work = Force × displacement
$= [\text{MLT}^{-2}][\text{L}]$
$= [\text{ML}^2\text{T}^{-2}]$
Torque = Force × force arm
= mass × acceleration × length
$= [\text{M}] \times [\text{LT}^{-2}] \times [\text{L}] = [\text{M L}^2\text{T}^{-2}]$

NOTES: If dimensions are given, physical quantity may not be unique as many physical quantities have same dimensions. *e.g.*, If the dimensional formula of a physical quantity is $[\text{ML}^2\text{T}^{-2}]$ it may be work or energy or torque.

15. (c) $(\mu_0\varepsilon_0)^{-1/2} = \dfrac{1}{\sqrt{\mu_0\varepsilon_0}} \Rightarrow c$ = speed of light

where ε_0 = permittivity of free space
μ_0 = permeability of free space
So dimensions of $(\mu_0\varepsilon_0)^{-1/2}$ will be $[\text{LT}^{-1}]$

16. (b) $\dfrac{1}{2}\varepsilon_0 E^2$ represents energy density *i.e.*, energy per unit volume.

$$\Rightarrow \left[\frac{1}{2}\varepsilon_0 E^2\right] = \frac{[ML^2T^{-2}]}{[L^3]} = [\text{ML}^{-1}\text{T}^{-2}]$$

17. (d) Pressure $= \dfrac{\text{Force}}{\text{Area}}$

$$\Rightarrow \frac{[\text{MLT}^{-2}]}{[\text{L}^2]} = [\text{ML}^{-1}\text{T}^{-2}]$$

$\Rightarrow a = 1, b = -1, c = -2.$

18. (c) Energy density $= \dfrac{\text{Energy}}{\text{Volume}}$

$$\Rightarrow \frac{[\text{ML}^2\text{T}^{-2}]}{[\text{L}^3]} = [\text{ML}^{-1}\text{T}^{-2}]$$

Refractive index and dielectric constant have no dimensions.

Young's Modulus $= \dfrac{F}{A} \times \dfrac{l}{\Delta l}$

$$\Rightarrow \frac{[\text{MLT}^{-2}]}{[\text{L}^2[} \cdot \frac{[\text{L}]}{[\text{L}]} = [\text{ML}^{-1}\text{T}^{-2}].$$

Magnetic field, $B = \dfrac{F}{i\ell} = \dfrac{[\text{MLT}^{-2}]}{[\text{A}][\text{L}]}$
$= [\text{MT}^{-2}\text{A}^{-1}].$

19. (c) As we know that

$$\text{R} = \frac{[\text{V}]}{[\text{I}]} = \frac{w}{q \cdot i} = \frac{w}{i \cdot t \cdot i}$$

Dimension of Resistance

$$= \frac{[\text{ML}^2\text{T}^{-3}\text{I}^{-1}]}{[\text{T}][\text{I}]} = [\text{ML}^2\text{T}^{-3}\text{I}^{-2}]$$

20. (d) Dimension of $a.t$ = dimension of velocity
$a.t = [\text{LT}^{-1}] \;\Rightarrow\; [a = \text{LT}^{-2}]$
Dimension of c = dimension of t
(two physical quantity of same dimension can only be added)
So, dimension of c = [T]

Dimension of $\dfrac{b}{t+c}$ = Dimension of velocity

$$\frac{b}{\text{T}+\text{T}} = [\text{LT}^{-1}][\text{LT}^{-1}] \Rightarrow [b \cdot \text{T}^{-1}] = [\text{LT}^{-1}]$$

$\Rightarrow b = [\text{L}]$
So, answer is $[\text{LT}^{-2}]$, [L] and [T]

NOTES: A dimensionally correct equation may or may not be physically correct. In a dimensionally correct equation, the dimensions of each term on both sides of an equation must be the same.

21. (b) Dimension formula for the planck's constant, $h = [\text{ML}^2\text{T}^1]$
Dimension formula for the moment of inertia, $\text{I} = [\text{ML}^2]$
So, the ratio between the plank's constant and moment of inertia is

$$\Rightarrow \frac{h}{I} = \frac{[\text{ML}^2\text{T}^{-1}]}{[\text{ML}^2]} \Rightarrow [\text{T}^{-1}]$$

$$\Rightarrow \frac{\text{h}}{\text{I}} = [\text{T}^{-1}] \Rightarrow \text{dimension of frequency}$$

22. (b) $F = \dfrac{G M_1 m_2}{r^2} \Rightarrow G = \dfrac{F r^2}{M_1 m_2}$

$\therefore$ dimension of G is $\dfrac{[\text{MLT}^{-2}][\text{L}^2]}{[\text{M}][\text{M}]}$
$= [\text{M}^{-1}\text{L}^3\text{T}^{-2}]$

23. (d) We know that $\text{E} = h\nu$

$$h = \frac{\text{E}}{\nu} = \frac{[\text{ML}^2\text{T}^{-2}]}{[\text{T}^{-1}]} = [\text{ML}^2\text{T}^{-1}]$$

Angular momentum = Iω
$= [\text{ML}^2][\text{T}^{-1}] = [\text{ML}^2\text{T}^{-1}]$

24. (c) Force has dimension $[MLT^{-2}]$ while impulse has dimension $[MLT^{-1}]$, both have different dimensions.

25. (a) Dimension of magnetic flux
= Dimension of magnetic field × Dimension of area
$[ML^0T^{-2}A^{-1}][L^2] = [ML^2T^{-2}A^{-1}]$

26. (d) $F = 6\pi\eta av$

$$\eta = \frac{F}{6\pi av} = \frac{[MLT^{-2}]}{[L][LT^{-1}]} = [ML^{-1}T^{-1}]$$

27. (a) $\left(P + \frac{a}{V^2}\right) = b\frac{\theta}{V}$

According to the principle of homogeneity quantity with same dimension can be added or subtracted.

Hence, Dimension of P = Dimension of $\frac{a}{V^2}$

$\Rightarrow$ Dimension of $\frac{\text{Force}}{\text{Area}}$ = Dimension of $\frac{a}{V^2}$

$$\Rightarrow \left[\frac{MLT^{-2}}{L^2}\right] = \frac{a}{[L^3]^2} \Rightarrow a = [ML^5T^{-2}]$$

NOTES: To get the dimensions of physical constant, we write any formula or equation incorporating the given constant and then by substituting the dimensional formula of all other quantities, we can find the dimensions of the required constant or coefficients.

28. (c) $\varepsilon = -L\frac{di}{dt}$ ....(1)

$\varepsilon = iR$ ....(2)

From (1) & (2), $iR = -L\frac{di}{dt}$

$\therefore$ Dimension of L.H.S. = Dimension of R.H.S.

$[A]R = L[AT^{-1}] \Rightarrow \frac{L}{R} = [T]$

NOTES: $\frac{L}{R}$ is time constant of R-L circuit so, dimensions of $\frac{L}{R}$ is same as that of time.

29. (d) A quantity which has dimensions and a constant value is called dimensional constant. Therefore, gravitational constant (G) is a dimensional constant. Value of $G = 6.67 \times 10^{-11}$ $m^2/kg/sec^2$
dimension of $G = [M^{-1}L^3T^{-2}]$

NOTES: Relative density, refractive index and poisson ratio all the three are ratios, therefore they are dimensionless constants.
Angle is an exceptional physical quantity, which though is a ratio of two similar physical quantities $\left(\text{angle} = \frac{\text{arc}}{\text{radius}}\right)$ but still requires a unit but no dimensions.

30. (b) In $p = p_0 \exp(-\alpha t^2)$, where αt^2 where is dimensionless

$$\therefore \alpha = \frac{1}{t^2} = \frac{1}{[T^2]} = [T^{-2}]$$

31. (d) $\eta = \frac{p(r^2 - x^2)}{4vl} = \frac{[ML^{-1}T^{-2}][L^2]}{[LT^{-1}][L]}$
$= [ML^{-1}T^{-1}]$

NOTES: According to the principle of homogeneity, the dimensions of each term on the L.H.S. must be equal to the dimensions of the terms on the R.H.S. Only then dimensional equation or formula is dimensionally correct.

32. (c) Let the expression, $\alpha = P^x S^y c^z$ ...(i)
and given that dimension of $\alpha = [M^0L^0T^0]$ ...(ii)
= dimensionless

Dimension of radiation pressure $P = \frac{\text{Force}}{\text{Area}}$

$$= \frac{[MLT^{-2}]}{[L^2]} = [ML^{-1}T^{-2}]$$

Dimension of radiation energy/unit area unit time

$$S = \frac{\text{Energy}}{\text{Area} \times \text{Time}} = \frac{[ML^2T^{-2}]}{[L^2][T]} = [MT^{-3}]$$

Dimension of speed of light, $c = [LT^{-1}]$
By equation (i) we get,
So, the dimension of $\alpha = [ML^{-1}T^{-2}]^x [MT^{-3}]^y [LT^{-1}]^z$
According to equation (ii),
$\Rightarrow [M^0L^0T^0] = [ML^{-1}T^{-2}]^x [MT^{-3}]^y [LT^{-1}]^z$
$\Rightarrow [M^0L^0T^0] = [M^{x+y} L^{-x+z} T^{-2x-3y-z}]$

Applying the principle of homegenity of dimension we get,
$x + y = 0$ ...(iii)

$-x + z = 0$ ...(iv)

$-2x - 3y - z = 0$ ...(v)

After solving above three equation we get,

$x = 1;\ y = -1;\ z = 1$

NOTES Try out the given alternatives.

When $x = 1, y = -1, z = 1$

$$P^x S^y c^z = P^1 S^{-1} c^1 = \frac{Pc}{S}$$

$$= \frac{[ML^{-1}T^{-2}][LT^{-1}]}{[ML^2T^{-2}/L^2T]} = [M^0L^0T^0]$$

33. (a) Permeability of free space,

$$\mu_0 = \frac{2\pi \times \text{force} \times \text{distance}}{(\text{current})^2 \times \text{length}}$$

So, dimensional formula

$$\mu_0 = \frac{[MLT^{-2}][L]}{[A^2][L]} = [MLT^{-2}A^{-2}]$$

NOTES Also find the dimensional formula by using the relation,

Speed of light, $c = \dfrac{1}{\sqrt{\mu_0 \varepsilon_0}}$

34. (d) $F = -\eta A \dfrac{\Delta V}{\Delta Z}$

$$\Rightarrow \eta = (-1)\frac{F\Delta Z}{A\Delta V}$$

So dimensional formula of η

$$\Rightarrow \frac{[MLT^{-2}][L]}{[L^2][LT^{-1}]}$$

$$\Rightarrow [ML^{-1}T^{-1}]$$

35. (d) $f = c\, m^x k^y$;

Spring constant k = force/length.

$[M^0L^0T^{-1}] = [M^x][MT^{-2}]^y = [M^{x+y}T^{-2y}]$

$\Rightarrow x + y = 0, -2y = -1$ or $y = \dfrac{1}{2}$

Therefore, $x = -\dfrac{1}{2}$

NOTES The method of dimensions cannot be used to derive relations other than product of power functions.

36. (c) Pressure = Force / Area

So dimensional formula

$$= \frac{[MLT^{-2}]}{[L^2]} = [ML^{-1}T^{-2}]$$

37. (a) τ = Force × distance

So dimensional formula,

$$= [MLT^{-2}][L] = [ML^2T^{-2}]$$

38. (c) $L = \varepsilon\left(\dfrac{dt}{dI}\right) = \dfrac{W}{q}\left[\dfrac{di}{dt}\right] = \dfrac{W}{i \cdot t}\left[\dfrac{di}{dt}\right]$

$$\text{or, } [L] = \frac{[ML^2T^{-2}][T]}{[AT][A]} = [ML^2T^{-2}A^{-2}]$$

39. (d) For angular momentum, the dimensional formula is $[ML^2T^{-1}]$. For other three, it is $[ML^{-1}T^{-2}]$.

40. (b) Angular momentum

= Momentum of inertia × Angular velocity

So dimensional formula,

$= [ML^2] \times [T^{-1}]$

$= [ML^2T^{-1}]$

41. (a) $CR = \left(\dfrac{q}{V}\right)\left(\dfrac{V}{i}\right) \Rightarrow \left(\dfrac{i \cdot t}{i}\right) \Rightarrow t = \text{time}$

$$= [T] = [M^0L^0T^1]$$

RC is the time constant of the circuit.

42. (a) Least count LC $= \dfrac{\text{Pitch}}{\text{CSD}} = \dfrac{1}{100} = 0.01\,\text{mm}$

$= 0.001$ cm

$\therefore$ Diameter of the wire = MSR + (CSR × LC)

$= 0 + (52 \times 0.001\text{ cm})$

$= 0.052$ cm

43. (a) In subtraction the number of decimal places in the result should be equal to the number of decimal places of that term in the operation which contain lesser number of decimal places.

$$\begin{array}{r} 9.99 \\ -\,0.0099 \\ \hline 9.9801 \end{array}$$

As the least number of decimal places is 3. So, answer should be 9.98 m.

44. (b) Least count of screw gauge = 0.01 mm

Least count

$$= \frac{\text{Pitch}}{\text{No. of divisions on circular scale}}$$

$$\Rightarrow 0.01\text{ mm} = \frac{\text{Pitch}}{50}$$

$\Rightarrow$ Pitch = 0.5 mm

45. **(b)** Given, $x = \frac{A^2B^{1/2}}{C^{1/3}D^3}$

% error, $\frac{\Delta x}{x}\times 100 = 2\frac{\Delta A}{A}\times 100 + \frac{1}{2}\frac{\Delta B}{B}\times 100 + \frac{1}{3}\frac{\Delta C}{C}\times 100 + 3\frac{\Delta D}{D}\times 100$

$= 2\times 1\% + \frac{1}{2}\times 2\% + \frac{1}{3}\times 3\% + 3\times 4\%$

$= 2\% + 1\% + 1\% + 12\% = 16\%$

46. **(d)** n VSD = $(n-1)$ MSD

$1\text{ VSD} = \frac{(n-1)}{n}\text{MSD}$

$\text{L.C.} = 1\text{ MSD} - 1\text{ VSD} = 1\text{ MSD} - \frac{(n-1)}{n}\text{MSD}$

$= \frac{1}{n}\text{MSD} = \frac{1}{n}\times\frac{1}{n} = \frac{1}{n^2}$

47. **(c)** Diameter of the ball

= MSR + CSR × (ℓeast count) – zero error

= 0.5 cm + 25 × 0.001 – (–0.004)

= 0.5 + 0.025 + 0.004 = 0.529 cm

48. **(d)** Given, $P = \frac{a^3b^2}{cd}$

Therefore, $\frac{\Delta P}{P}\times 100\% = 3\frac{\Delta a}{a}\times 100\% + 2\frac{\Delta b}{b}\times 100\% + \frac{\Delta c}{c}\times 100\% + \frac{\Delta d}{d}\times 100\%$

$= 3\times 1\% + 2\times 2\% + 3\% + 4\% = 14\%$

NOTES When we multiply or divide two measured quantities, the relative error in the final result is equal to the sum of the relative errors in the measured quantities. And when we add or subtract two measured quantities the absolute error in the final result is equal to the sum of the absolute error in the measured quantities.

49. **(b)** Given, error in the measurement of radius of a sphere $\frac{\Delta r}{r}\times 100 = 2\%$

Volume of the sphere $V = \frac{4}{3}\pi r^3$

$\therefore$ Error in the volume $\frac{\Delta V}{V}\times 100$

$= \pm 3\cdot\frac{\Delta r}{r}\times 100 = 3\times 2\% = \pm 16\%$

50. **(d)** As we know, density = $\frac{\text{mass}}{\text{volume}}$

Maximum error in the measurement of density

$\rho = \frac{M}{L^3}$

$\therefore \frac{\Delta\rho}{\rho} = \frac{\Delta M}{M} + 3\frac{\Delta L}{L}$

% error in density = % error in Mass +3 (% error in length)

= 4 + 3(3) = 13%

51. **(c)** Percentage error in mass $\left(\frac{\Delta m}{m}\times 100\right) = 2$

and percentage error in speed $\left(\frac{\Delta v}{v}\times 100\right) = 3$.

Kinetic energy, $k = \frac{1}{2}mv^2$.

$\therefore$ Error in measurement of kinetic energy

$\frac{\Delta K}{K} = \frac{\Delta m}{m} + 2\left(\frac{\Delta v}{v}\right)$

$= \left(\frac{2}{100}\right) + \left(2\times\frac{3}{100}\right) = \frac{8}{100} = 8\%$

$\therefore$ %age error = 8%.

52. **(c)** Least count = 1MSD – 1 VSD

$= 1\text{MSD} - \left(\frac{N-1}{N}\right)\text{MSD}$

$= \frac{1}{N}\text{MSD} = \frac{1}{N}\times\frac{1}{10}\text{cm} = \frac{1}{10N}$

NOTES Smaller value of the least count, higher is the accuracy of measurement. Accuracy of measurement is higher when number of significant figure after the decimal in measurement is larger.

53. **(b)** Density, $D = \frac{\text{Mass(M)}}{\text{Volume(V)}}$

$\therefore \frac{\Delta D}{D} = \frac{\Delta M}{M} + \frac{\Delta V}{V} = \left(\frac{0.01}{22.42} + \frac{0.1}{4.7}\right)\times 100 = 2\%$

2 Motion in a Straight Line

Trend Analysis with Important Topics & Sub-Topics

Topic Name	Sub-Topic	2021 QNS.	2021 LOD	2020 QNS.	2020 LOD	2019 QNS.	2019 LOD	2018 QNS.	2018 LOD	2017 QNS.	2017 LOD	2016 QNS.	2016 LOD
Distance, Displacement & Uniform motion	Average Speed												
	Integration & Differentiation											1	E·
Non-uniform motion	Acceleration down the inclined plane & eqn of motion	1	A										
Relative Velocity	Use of eqn. of motion			1	E								
Motion Under Gravity	Velocity-time graph									1	A		
	Rever-Man problem					1	A						

LOD - Level of Difficulty | E - Easy | A - Average | D - Difficult | Qns - No. of Questions

Topic 1 Distance, Displacement & Uniform motion

1. A person travelling in a straight line moves with a constant velocity v_1 for certain distance 'x' and with a constant velocity v_2 for next equal distance. The average velocity v is given by the relation *[NEET Odisha 2019]*

(a) $v = \sqrt{v_1 v_2}$ (b) $\frac{1}{v} = \frac{1}{v_1} + \frac{1}{v_2}$

(c) $\frac{2}{v} = \frac{1}{v_1} + \frac{1}{v_2}$ (d) $\frac{v}{2} = \frac{v_1 + v_2}{2}$

2. Preeti reached the metro station and found that the escalator was not working. She walked up the stationary escalator in time t_1. On other days, if she remains stationary on the moving escalator, then the escalator takes her up in time t_2. The time taken by her to walk up on the moving escalator will be: *[2017]*

(a) $\frac{t_1 t_2}{t_2 - t_1}$ (b) $\frac{t_1 t_2}{t_2 + t_1}$

(c) $t_1 - t_2$ (d) $\frac{t_1 + t_2}{2}$

3. A particle covers half of its total distance with speed v_1 and the rest half distance with speed v_2. Its average speed during the complete journey is *[2011M]*

(a) $\frac{v_1 v_2}{v_1 + v_2}$ (b) $\frac{2v_1 v_2}{v_1 + v_2}$

(c) $\frac{2v_1^2 v_2^2}{v_1^2 + v_2^2}$ (d) $\frac{v_1 + v_2}{2}$

4. A car moves from X to Y with a uniform speed v_u and returns to Y with a uniform speed v_d. The average speed for this round trip is *[2007]*

(a) $\sqrt{v_u v_d}$ (b) $\frac{v_d v_u}{v_d + v_u}$

(c) $\frac{v_u + v_d}{2}$ (d) $\frac{2v_d v_u}{v_d + v_u}$

5. If a car at rest accelerates uniformly to a speed of 144 km/h in 20 s, it covers a distance of *[1997]*

(a) 2880 m (b) 1440 m

(c) 400 m (d) 20 m

6. A bus travelling the first one third distance at a speed of 10 km/h, the next one third at 20 km/h and the last one-third at 60 km/h. The average speed of the bus is [1991]
(a) 9 km/h (b) 16 km/h
(c) 18 km/h (d) 48 km/h

7. A car moves a distance of 200 m. It covers the first half of the distance at speed 40 km/h and the second half of distance at speed v. The average speed is 48 km/h. Find the value of v [1991]
(a) 56 km/h (b) 60 km/h
(c) 50 km/h (d) 48 km/h

8. A car covers the first half of the distance between two places at 40 km/h and other half at 60 km/h. The average speed of the car is [1990]
(a) 40 km/h (b) 48 km/h
(c) 50 km/h (d) 60 km/h

Topic 2 Non-uniform motion

9. A small block slides down on a smooth inclined plane, starting from rest at time $t = 0$. Let S_n be the distance travelled by the block in the interval $t = n-1$ to $t = n$. Then, the ratio $\frac{S_n}{S_{n+1}}$ is [2021]
(a) $\frac{2n}{2n-1}$ (b) $\frac{2n-1}{2n}$
(c) $\frac{2n-1}{2n+1}$ (d) $\frac{2n+1}{2n-1}$

10. If the velocity of a particle is $v = At + Bt^2$, where A and B are constants, then the distance travelled by it between 1s and 2s is : [2016]
(a) $\frac{3}{2}A + 4B$ (b) $3A + 7B$
(c) $\frac{3}{2}A + \frac{7}{3}B$ (d) $\frac{A}{2} + \frac{B}{3}$

11. A particle of unit mass undergoes one-dimensional motion such that its velocity varies according to $v(x) = bx^{-2n}$ where b and n are constants and x is the position of the particle. The acceleration of the particle as the function of x, is given by: [2015]
(a) $-2nb^2x^{-4n-1}$ (b) $-2b^2x^{-2n+1}$
(c) $-2nb^2e^{-4n+1}$ (d) $-2nb^2x^{-2n-1}$

12. The displacement 'x' (in meter) of a particle of mass 'm' (in kg) moving in one dimension under the action of a force, is related to time 't' (in sec) by $t = \sqrt{x} + 3$. The displacement of the particle when its velocity is zero, will be [NEET Kar. 2013]
(a) 2m (b) 4m
(c) zero (d) 6m

13. A particle has initial velocity $(2\vec{i} + 3\vec{j})$ and acceleration $(0.3\vec{i} + 0.2\vec{j})$. The magnitude of velocity after 10 seconds will be : [2012]
(a) $9\sqrt{2}$ units (b) $5\sqrt{2}$ units
(c) 5 units (d) 9 units

14. The motion of a particle along a straight line is described by equation :
$x = 8 + 12t - t^3$
where x is in metre and t in second. The retardation of the particle when its velocity becomes zero, is : [2012]
(a) 24 ms^{-2} (b) zero
(c) 6 ms^{-2} (d) 12 ms^{-2}

15. A body is moving with velocity 30 m/s towards east. After 10 seconds its velocity becomes 40 m/s towards north. The average acceleration of the body is [2011]
(a) 1 m/s^2 (b) 7 m/s^2
(c) $\sqrt{7}$ m/s^2 (d) 5 m/s^2

16. A particle has initial velocity $(3\hat{i} + 4\hat{j})$ and has acceleration $(0.4\hat{i} + 0.3\hat{j})$. It's speed after 10 s is: [2010]
(a) 7 units (b) $7\sqrt{2}$ units
(c) 8.5 units (d) 10 units

17. A particle moves a distance x in time t according to equation $x = (t + 5)^{-1}$. The acceleration of particle is proportional to: [2010]
(a) $(\text{velocity})^{3/2}$ (b) $(\text{distance})^2$
(c) $(\text{distance})^{-2}$ (d) $(\text{velocity})^{2/3}$

18. A particle starts its motion from rest under the action of a constant force. If the distance covered in first 10 seconds is S_1 and that covered in the first 20 seconds is S_2, then: [2009]
(a) $S_2 = 3S_1$ (b) $S_2 = 4S_1$
(c) $S_2 = S_1$ (d) $S_2 = 2S_1$

19. The distance travelled by a particle starting from rest and moving with an acceleration $\frac{4}{3}\text{ms}^{-2}$, in the third second is: *[2008]*

(a) 6m (b) 4m

(c) $\frac{10}{3}$ m (d) $\frac{19}{3}$ m

20.

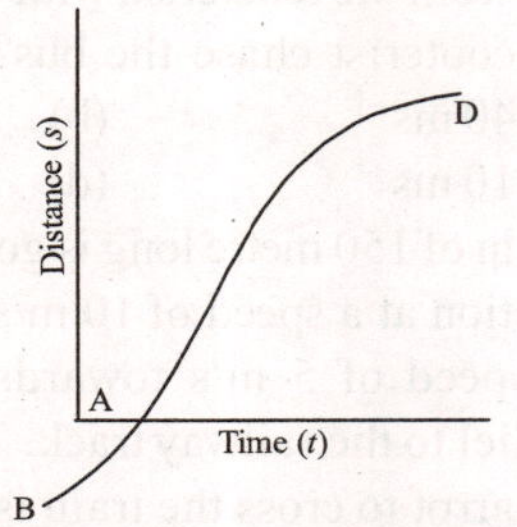

A particle shows distance-time curve as given in this figure. The maximum instantaneous velocity of the particle is around the point: *[2008]*

(a) B (b) C

(c) D (d) A

21. A particle moves in a straight line with a constant acceleration. It changes its velocity from 10 ms^{-1} to 20 ms^{-1} while passing through a distance 135 m in t second. The value of t is: *[2008]*

(a) 10 (b) 1.8

(c) 12 (d) 9

22. The position x of a particle with respect to time t along x-axis is given by $x = 9t^2 - t^3$ where x is in metres and t in second. What will be the position of this particle when it achieves maximum speed along the +ve x direction? *[2007]*

(a) 54m (b) 81 m

(c) 24m (d) 32m.

23. A particle moving along x-axis has acceleration f, at time t, given by $f = f_0\left(1 - \frac{t}{T}\right)$, where f_0 and T are constants. The particle at $t = 0$ has zero velocity. In the time interval between $t = 0$ and the instant when $f = 0$, the particle's velocity (v_x) is *[2007]*

(a) $\frac{1}{2}f_0T^2$ (b) f_0T^2

(c) $\frac{1}{2}f_0T$ (d) f_0T

24. A particle moves along a straight line OX. At a time t (in seconds) the distance x (in metres) of the particle from O is given by $x = 40 + 12t - t^3$. How long would the particle travel before coming to rest? *[2006]*

(a) 40m (b) 56m

(c) 16m (d) 24m

25. The displacement x of a particle varies with time t as $x = ae^{-\alpha t} + be^{\beta t}$, where a, b, α and β are positive constants. The velocity of the particle will *[2005]*

(a) be independent of α and β

(b) drop to zero when $\alpha = \beta$

(c) go on decreasing with time

(d) go on increasing with time

26. The displacement of a particle is represented by the following equation : $s = 3t^3 + 7t^2 + 5t + 8$ where s is in metre and t in second. The acceleration of the particle at $t = 1$s is *[2000]*

(a) 14 m/s^2 (b) 18 m/s^2

(c) 32 m/s^2 (d) zero

27. A car moving with a speed of 40 km/h can be stopped by applying brakes at least after 2 m. If the same car is moving with a speed of 80 km/h, what is the minimum stopping distance? *[1998]*

(a) 8m (b) 6m

(c) 4m (d) 2m

28. The displacement of a particle varies with time (t) as: $s = at^2 - bt^3$. The acceleration of the particle will be zero at time t equal to *[1997]*

(a) $\frac{a}{b}$ (b) $\frac{a}{3b}$

(c) $\frac{3b}{a}$ (d) $\frac{2a}{3b}$

29. A car accelerates from rest at a constant rate α for some time, after which it decelerates at a constant rate β and comes to rest. If the total time elapsed is t, then the maximum velocity acquired by the car is *[1994]*

(a) $\left(\frac{\alpha^2+\beta^2}{\alpha\beta}\right)t$ (b) $\left(\frac{\alpha^2-\beta^2}{\alpha\beta}\right)t$

(c) $\frac{(\alpha+\beta)t}{\alpha\beta}$ (d) $\frac{\alpha\beta t}{\alpha+\beta}$

30. The displacement time graph of a moving particle is shown below

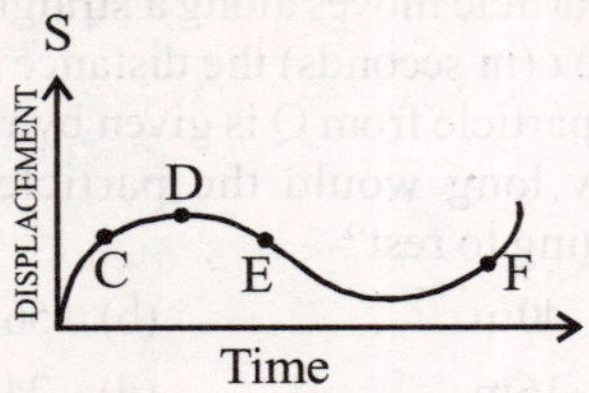

The instantaneous velocity of the particle is negative at the point *[1994]*

(a) D (b) F
(c) C (d) E

31. A particle moves along a straight line such that its displacement at any time t is given by $s = (t^3 - 6t^2 + 3t + 4)$ metres
The velocity when the acceleration is zero is *[1994]*

(a) $3\ ms^{-1}$ (b) $-12\ ms^{-1}$
(c) $42\ ms^{-2}$ (d) $-9\ ms^{-1}$

32. A body starts from rest, what is the ratio of the distance travelled by the body during the 4th and 3rd seconds? *[1993]*

(a) $\frac{7}{5}$ (b) $\frac{5}{7}$
(c) $\frac{7}{3}$ (d) $\frac{3}{7}$

33. Which of the following curve does not represent motion in one dimension? *[1992]*

(a)
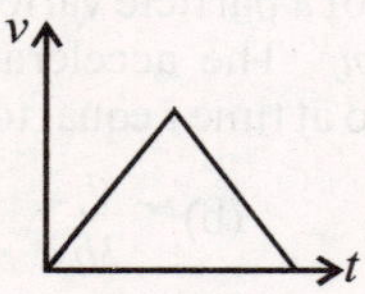

(b)
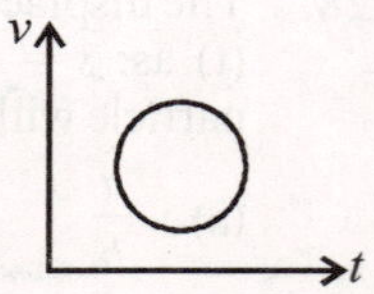

(c)
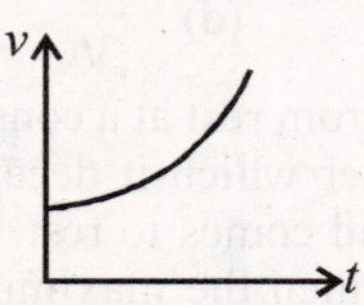

(d)
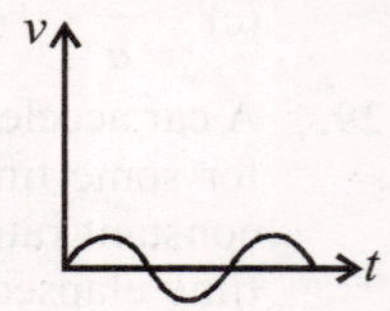

34. A car is moving along a straight road with a uniform acceleration. It passes through two points P and Q separated by a distance with velocity 30 km/h and 40 km/h respectively. The velocity of the car midway between P and Q is *[1988]*

(a) 33.3 km/h (b) $20\sqrt{2}$ km/h
(c) $25\sqrt{2}$ km/h (d) 35 km/h

Topic 3 Relative Velocity

35. A bus is moving with a speed of $10\ ms^{-1}$ on a straight road. A scooterist wishes to overtake the bus in 100 s. If the bus is at a distance of 1 km from the scooterist, with what speed should the scooterist chase the bus? *[2009]*

(a) $40\ ms^{-1}$ (b) $25\ ms^{-1}$
(c) $10\ ms^{-1}$ (d) $20\ ms^{-1}$

36. A train of 150 metre long is going towards north direction at a speed of 10 m/s. A parrot flies at the speed of 5 m/s towards south direction parallel to the railway track. The time taken by the parrot to cross the train is *[1988]*

(a) 12 sec (b) 8 sec
(c) 15 sec (d) 10 sec

Topic 4 Motion Under Gravity

37. A ball is thrown vertically downward with a velocity of 20 m/s from the top of a tower. It hits the ground after some time with a velocity of 80 m/s. The height of the tower is : ($g = 10\ m/s^2$) *[2020]*

(a) 340 m (b) 320 m
(c) 300 m (d) 360 m

38. A stone falls freely under gravity. It covers distances h_1, h_2 and h_3 in the first 5 seconds, the next 5 seconds and the next 5 seconds respectively. The relation between h_1, h_2 and h_3 is *[2013]*

(a) $h_1 = \frac{h_2}{3} = \frac{h_3}{5}$
(b) $h_2 = 3h_1$ and $h_3 = 3h_2$
(c) $h_1 = h_2 = h_3$
(d) $h_1 = 2h_2 = 3h_3$

39. A boy standing at the top of a tower of 20m height drops a stone. Assuming $g = 10\ ms^{-2}$, the velocity with which it hits the ground is *[2011]*

(a) 10.0 m/s (b) 20.0 m/s
(c) 40.0 m/s (d) 5.0 m/s

40. A ball is dropped from a high rise platform at $t = 0$ starting from rest. After 6 seconds another ball is thrown downwards from the same platform with a speed v. The two balls meet at $t = 18$s. What is the value of v? *[2010]*
(take $g = 10\ m/s^2$)
(a) 75 m/s (b) 55 m/s (c) 40 m/s (d) 60 m/s

41. A man of 50 kg mass is standing in a gravity free space at a height of 10 m above the floor. He throws a stone of 0.5 kg mass downwards with a speed 2 m/s. When the stone reaches the floor, the distance of the man above the floor will be: *[2010]*
(a) 9.9m (b) 10.1m
(c) 10m (d) 20m

42. Two bodies, A (of mass 1 kg) and B (of mass 3 kg), are dropped from heights of 16m and 25m, respectively. The ratio of the time taken by them to reach the ground is *[2006]*
(a) 12/5 (b) 5/12
(c) 4/5 (d) 5/4

43. A ball is thrown vertically upward. It has a speed of 10 m/sec when it has reached one half of its maximum height. How high does the ball rise? *[2005, 2001]*
Take $g = 10\ m/s^2$.
(a) 10m (b) 5m
(c) 15m (d) 20m

44. If a ball is thrown vertically upwards with speed u, the distance covered during the last t seconds of its ascent is *[2003]*
(a) $(u + gt)\,t$ (b) ut
(c) $\frac{1}{2}gt^2$ (d) $ut - \frac{1}{2}gt^2$

45. A man throws balls with the same speed vertically upwards one after the other at an interval of 2 seconds. What should be the speed of the throw so that more than two balls are in the sky at any time? [Given $g = 9.8\ m/s^2$] *[2003]*
(a) Only with speed 19.6 m/s
(b) More than 19.6 m/s
(c) At least 9.8 m/s
(d) Any speed less than 19.6 m/s

46. If a ball is thrown vertically upwards with a velocity of 40 m/s, then velocity of the ball after two seconds will be ($g = 10\ m/s^2$) *[1996]*
(a) 15 m/s (b) 20 m/s
(c) 25 m/s (d) 28 m/s

47. A body is thrown vertically upward from the ground. It reaches a maximum height of 20 m in 5 sec. After what time, it will reach the ground from its maximum height position? *[1995]*
(a) 2.5 sec (b) 5 sec
(c) 10 sec (d) 25 sec

48. A stone released with zero velocity from the top of a tower, reaches the ground in 4 sec. The height of the tower is ($g = 10 m/s^2$) *[1995]*
(a) 20m (b) 40m
(c) 80m (d) 160m

49. Three different objects of masses m_1, m_2 and m_3 are allowed to fall from rest and from the same point O along three different frictionless paths. The speeds of the three objects on reaching the ground will be in the ratio of *[1995]*
(a) $m_1 : m_2 : m_3$ (b) $m_1 : 2m_2 : 3m_3$
(c) 1:1:1 (d) $\frac{1}{m_1} : \frac{1}{m_2} : \frac{1}{m_3}$

50. The water drops fall at regular intervals from a tap 5 m above the ground. The third drop is leaving the tap at an instant when the first drop touches the ground. How far above the ground is the second drop at that instant? (Take $g = 10\ m/s^2$) *[1995]*
(a) 1.25m (b) 2.50m
(c) 3.75m (d) 5.00m

51. A body dropped from top of a tower fall through 40 m during the last two seconds of its fall. The height of tower is ($g = 10\ m/s^2$) *[1991]*
(a) 60m (b) 45m
(c) 80m (d) 50m

52. What will be the ratio of the distances moved by a freely falling body from rest in 4th and 5th seconds of journey? *[1989]*
(a) 4 : 5 (b) 7 : 9
(c) 16 : 25 (d) 1 : 1

ANSWER KEY

1	(c)	7	(b)	13	(b)	19	(c)	25	(d)	31	(d)	37	(c)	43	(a)	49	(c)
2	(b)	8	(b)	14	(d)	20	(b)	26	(c)	32	(a)	38	(a)	44	(c)	50	(c)
3	(b)	9	(c)	15	(d)	21	(d)	27	(a)	33	(b)	39	(b)	45	(b)	51	(b)
4	(d)	10	(c)	16	(b)	22	(a)	28	(b)	34	(c)	40	(a)	46	(b)	52	(b)
5	(c)	11	(a)	17	(a)	23	(c)	29	(d)	35	(d)	41	(b)	47	(b)		
6	(c)	12	(c)	18	(b)	24	(b)	30	(d)	36	(d)	42	(c)	48	(c)		

Hints & Solutions

1. (c) As $t_1 = \frac{x}{v_1}$ and $t_2 = \frac{x}{v_2}$

$$\therefore \quad v = \frac{x+x}{t_1+t_2} = \frac{2x}{\frac{x}{v_1}+\frac{x}{v_2}} = \frac{2v_1v_2}{v_1+v_2}$$

$$\therefore \quad \frac{2}{v} = \frac{1}{v_1} + \frac{1}{v_2}$$

2. (b) Let the distance be 'd' time taken by preeti to travel up the stationary escalator $= t_1$ Velocity of preeti w.r.t. elevator $v_1 = \frac{d}{t_1}$

Since the distance is same let the time taken when preeti stands on the moving escalator $= t_2$.

Velocity of elevator w.r.t. ground $v_2 = \frac{d}{t_2}$

Then net velocity of preeti w.r.t. ground

$$v = v_1 + v_2$$

$$\frac{d}{t} = \frac{d}{t_1} + \frac{d}{t_2}$$

$$\frac{1}{t} = \frac{1}{t_1} + \frac{1}{t_2}$$

$\therefore \; t = \frac{t_1t_2}{(t_1+t_2)}$ (time taken by preeti to walk up on the moving escalator)

3. (b) Let the total distance covered by the particle be 2s. Then

$$v_{av} = \frac{2s}{\frac{s}{v_1}+\frac{s}{v_2}} = \frac{2v_1v_2}{v_1+v_2}$$

> **NOTES** The average speed of an object is the total distance travelled by the object divides by the elapsed time to cover that distance. It's a scalar quantity which means it is defined only by magnitude. A related concept, average velocity, is a vector quantity. A vector quantity is defined by magnitude and direction both.

4. (d) Average speed

$$= \frac{\text{total distance travelled}}{\text{total time taken}}$$

Let s be the distance from X to Y.

$$\therefore \text{ Average speed } = \frac{s+s}{t_1+t_2} = \frac{2s}{\frac{s}{v_u}+\frac{s}{v_d}}$$

$$= \frac{2v_uv_d}{v_d+v_u}$$

5. (c) Initial velocity of car $(u) = 0$

Final velocity of car $(v) = 144$ km/hr $= 40$ m/s

Time taken $= 20$ s

We know that, $v = u + at$

$40 = a \times 20 \Rightarrow a = 2\ \text{m/s}^2$

Also, $v^2 - u^2 = 2as$

$$\Rightarrow s = \frac{v^2-u^2}{2a}$$

$$\Rightarrow s = \frac{(40)^2-(0)^2}{2\times 2} = \frac{1600}{4} = 400 \text{ m.}$$

6. (c) Average speed $= \frac{s}{\frac{s/3}{10}+\frac{s/3}{20}+\frac{s/3}{60}}$

$$= \frac{s}{s/18} = 18 \text{km/h}$$

> **NOTES** In case speed is continuously changing with time, then, $Vav = \frac{\int vdt}{\int dt}$

7. (b) Given, Total distance = 200 m speed in first half distance = 40 km/hr speed in second half distance = $v$km/hr.

$$\text{So, } v_{av} = \frac{\text{Total distance covered}}{\text{Total time elapsed}}$$

$$48 = \frac{200\times 10^{-3}}{\left(\frac{100\times 10^{-3}}{40}\right)+\left(\frac{100\times 10^{-3}}{v}\right)}$$

or $\frac{1}{40}+\frac{1}{v}=\frac{2}{48}=\frac{1}{24}$

or $\frac{1}{v}=\frac{1}{24}-\frac{1}{40}=\frac{2}{120}=\frac{1}{60}$

or $v=60$ km/h

NOTES
$v_{av}=\frac{2v_1v_2}{v_1+v_2}\Rightarrow 48=\frac{2\times 40\times v}{40+v}$

$\Rightarrow v=60$ km/h

8. (b) Total distance = s;
Total time taken

$=\frac{s/2}{40}+\frac{s/2}{60}=\frac{5s}{240}=\frac{s}{48}$

$\therefore$ Average speed $=\frac{\text{total distance}}{\text{total time}}$

$=\frac{s}{s/48}=48\text{km/h}$

$v_{av}=\frac{2v_1v_2}{v_1+v_2}=\frac{2\times 40\times 60}{40+60}=48$ km/h

9. (c) Acceleration along inclined plane,
$a = g\sin\theta$
S_n = distance travelled by object during n^{th} second.
Initial speed u = 0

$$S_n=u+\frac{a}{2}(2n-1)$$

$$S_n=0+\frac{g\sin\theta}{2}(2n-1)=\frac{g\sin\theta}{2}(2n-1)\ \text{...(i)}$$

Distance travelled during $(n+1)^{th}$ second.

$$S_{n+1}=0+\frac{g\sin\theta}{2}[2(n+1)-1]$$

$$=\frac{g\sin\theta}{2}(2n+1)\ \text{...(ii)}$$

Dividing equations (i) and (ii)

$$\frac{S_n}{S_{n+1}}=\frac{\frac{g\sin\theta}{2}(2n-1)}{\frac{g\sin\theta}{2}(2n+1)}=\frac{(2n-1)}{(2n+1)}$$

NOTES
The distance covered by a body in n^{th} second i.e., $S_n\propto(2n-1)$ if body starts from rest and moves with uniform acceleration.

10. (c) **Given :** Velocity $v=At+Bt^2$

$$\Rightarrow \frac{dx}{dt}=At+Bt^2$$

By integrating we get distance travelled by the particle between 1s and 2s,

$$\Rightarrow \int_0^x dx=\int_1^2(At+Bt^2)dt$$

$$x=\frac{A}{2}(2^2-1^2)+\frac{B}{3}(2^3-1^3)=\frac{3A}{2}+\frac{7B}{3}$$

11. (a) Given, $v(x)=bx^{-2n}$

$$a=\frac{dv}{dt}=\frac{dv}{dx}\cdot\frac{dx}{dt}$$

$$=v\cdot\frac{dv}{dx}$$

So, $\frac{dv}{dx}=-2\,nb\,x^{-2n-1}$

Acceleration of the particle as function of x,

$$a=v\frac{dv}{dx}=bx^{-2n}\{b(-2n)x^{-2n-1}\}$$

$$=-2nb^2x^{-4n-1}$$

NOTES
For one dimensional motin, the angle between velocity and acceleration is either 0° or 180° and it does not change with time.

12. (c) $\because t=\sqrt{x}+3$

$\Rightarrow \sqrt{x}=t-3\Rightarrow x=(t-3)^2$ ...(i)

$v=\frac{dx}{dt}=2(t-3)=0$

$\Rightarrow t=3$

From equation (i)

$\therefore x=(3-3)^2$

$\Rightarrow x=0$

13. (b) $\vec{v}=\vec{u}+\vec{a}t$

$v=(2\hat{i}+3\hat{j})+(0.3\hat{i}+0.2\hat{j})\times 10=5\hat{i}+5\hat{j}$

$|\vec{v}| = \sqrt{5^2 + 5^2}$

$|\vec{v}| = 5\sqrt{2}$

14. **(d)** $x = 8 + 12t - t^3$

The final velocity of the particle will be zero, because it retarded.

$\therefore v = 0 + 12 - 3t^2 = 0$

$3t^2 = 12$

$t = 2$ sec

Now the retardation

$a = \frac{dv}{dt} = 0 - 6t$

$a[t = 2] = -12 \text{ m/s}^2$

retardation = 12 m/s^2

15. **(d)** Average acceleration

$<a> = \frac{\text{Change in velocity}}{\text{Total Time}}$

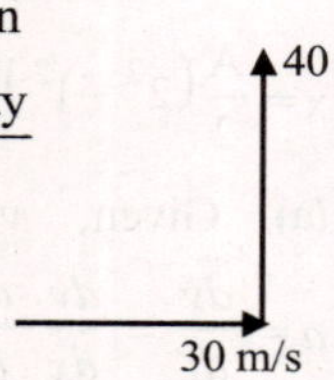

$<a> = \frac{|40\hat{j} - 30\hat{i}|}{10}$

$<a> = \sqrt{4^2 + (-3)^2}$

$<a> = 5 \text{ m/s}^2$

16. **(b)** Given, $\vec{u} = 3\hat{i} + 4\hat{j}$ and $\vec{a} = 0.4\hat{i} + 0.3\hat{j}$

$\Rightarrow u_x = 3$ units, $u_y = 4$ units

$\Rightarrow a_x = 0.4$ units, $a_y = 0.3$ units

Along x-axis,

$\therefore v_x = u_x + a_x \times 10 = 3 + 4 = 7$ units

Along y-axis,

and $v_y = 4 + 0.3 \times 10 = 4 + 3 = 7$ units

Net final velocity

$\therefore v = \sqrt{v_x^2 + v_y^2} = 7\sqrt{2}$ units

$\vec{v}_i = 3\hat{i} + 4\hat{j}$ and $\vec{a} = 0.4\hat{i} + 0.3\hat{j}$

time, $t = 10$ sec.

Final velocity $\vec{v}_f$ after time $t = 10$ sec, $\vec{v}_f = \vec{v}_i + \vec{a}t$

$\vec{v}_f = (3\hat{i} + 4\hat{j}) + (0.4\hat{i} + 0.3\hat{j})(10) = 7\hat{i} + 7\hat{j}$

NOTES The particle speeds up i.e., the speed of the particle increases when the angle between $\vec{a}$ and $\vec{v}$ lies between 0° + 90°. The particle speeds down i.e., the speed of the particle decreases when the angle between $\vec{a}$ and $\vec{v}$ lies between + 90° and 180°.

17. **(a)** distance $x = \frac{1}{t+5}$

$\therefore$ velocity $v = \frac{dx}{dt} = \frac{-1}{(t+5)^2}$

$\therefore$ acceleration $a = \frac{d^2x}{dt^2} = \frac{2}{(t+5)^3} = 2x^3$

Therefore, $v^{3/2} = -(t+5)^{-3}$

So, $a \propto v^{3/2}$

18. **(b)** $u = 0, t_1 = 10\text{s}, t_2 = 20s$

Using the relation, $S = ut + \frac{1}{2}at^2$

Acceleration being the same in two cases,

$S_1 = \frac{1}{2}a \times t_1^2, S_2 = \frac{1}{2}a \times t_2^2$

$\therefore \frac{S_1}{S_2} = \left(\frac{t_1}{t_2}\right)^2 = \left(\frac{10}{20}\right)^2 = \frac{1}{4}$

$S_2 = 4S_1$

19. **(c)** Distance travelled in the nth second is given by $S_n = u + \frac{a}{2}(2n - 1)$

Given $u = 0$, $a = \frac{4}{3}\text{ms}^{-2}$, $n = 3$

$\therefore S_n = 0 + \frac{4}{3 \times 2}(2 \times 3 - 1) = \frac{4}{6} \times 5 = \frac{10}{3}$ m

20. **(b)** The slope of the graph $\frac{ds}{dt}$ is maximum at C and hence the instantaneous velocity is maximum at C.

Instantaneous velocity,

$\vec{v}_{ins} = \lim_{\Delta t \to 0} \frac{\Delta\vec{s}}{\Delta t} = \frac{d\vec{s}}{dt}$

NOTES The instantaneous velocity of an object at a given instant is first derivative of displacement with respect to time.
The slope of displacement-time graph at any instant of time gives the measure of instantaneous velocity of an object at that instant.

21. **(d)** Initial velocity, $u = 10 \text{ ms}^{-1}$

Final velocity, $v = 20 \text{ ms}^{-1}$

Distance, $s = 135$ m

Let, acceleration = a

Using the formula, $v^2 = u^2 + 2as$

$a = \frac{v^2 - u^2}{2s}$

$= \frac{(20)^2 - (10)^2}{2 \times 135} = \frac{400 - 100}{2 \times 135}$

or, $a = \frac{300}{2\times 135} = \frac{150}{135} = \frac{10}{9}\,\text{ms}^{-2}$

Now, using the relation, $v = u + at$

$$t = \frac{v-u}{a} = \frac{20-10}{10/9} = \frac{10}{10}\times 9\,\text{sec} = 9\text{ sec.}$$

22. **(a)** Speed $v = \frac{dx}{dt} = \frac{d}{dt}(9t^2 - t^3) = 18t - 3t^2$

For maximum speed its acceleration should be zero,

Acceleration $\frac{dv}{dt} = 0 \Rightarrow 18 - 6t = 0 \Rightarrow t = 3$

At $t = 3$ its speed is max

$\Rightarrow \quad x_{max} = 81 - 27 = 54$ m

> **NOTES**
> Position of any point is completely expressed by two factors its distance from the observer and its direction with respect to observer.
> That is why position is characterised by a vector known as position vector.

23. **(c)** Here, $f = f_0\left(1 - \frac{t}{T}\right)$

or, $\frac{dv}{dt} = f_0\left(1 - \frac{t}{T}\right)$

If $f = 0$, then

$$0 = f_0\left(1 - \frac{t}{T}\right) \Rightarrow t = T$$

Hence, particle's velocity in the time interval $t = 0$ and $t = T$ is given by

$$v_x = \int_{v=0}^{v=V_z} dv = \int_{t=0}^{T}\left[f_0\left(1 - \frac{t}{T}\right)\right]dt$$

$$= f_0\left[\left(t - \frac{t^2}{2T}\right)\right]_0^T$$

$$= f_0\left(T - \frac{T^2}{2T}\right) = f_0\left(T - \frac{T}{2}\right) = \frac{1}{2}f_0 T.$$

24. **(b)** $x = 40 + 12t - t^3$

$$v = \frac{dx}{dt} = 12 - 3t^2$$

For $v = 0$; $t = \sqrt{\frac{12}{3}} = 2$ sec

So, after 2 seconds velocity becomes zero.

Value of x in 2 secs $= 40 + 12\times 2 - 2^3$

$= 40 + 24 - 8 = 56$ m

25. **(d)**

26. **(c)** Displacement

$s = 3t^3 + 7t^2 + 5t + 8;$

Velocity $= \frac{ds}{dt} = 9t^2 + 14t + 5$

Acceleration $= \frac{d^2s}{dt^2} = 18t + 14$

Acceleration at ($t = 1$s)

$= 18\times 1 + 14 = 18 + 14 = 32\text{ m/s}^2$

27. **(a)** $v^2 - u^2 = 2as$

$$\Rightarrow a = \frac{v^2 - u^2}{2s}$$

$$= -\frac{u_1^2}{2s},$$

For same retarding force $s \propto u^2$

$$\therefore \frac{s_2}{s_1} = \frac{u_2^2}{u_1^2} \Rightarrow \frac{s_2}{s_1} = \left(\frac{80}{40}\right)^2 = 4$$

$\therefore s_2 = 4s_1 = 8$ m

If F is retarding force and s the stopping distance, then $\frac{1}{2}mv^2 = Fs$

For same retarding force, $s \propto v^2$

$$\therefore \frac{s_2}{s_1} = \left(\frac{v_2}{v_1}\right)^2 = \left(\frac{80\text{km/h}}{40\text{km/h}}\right)^2 = 4$$

$\therefore \quad s_2 = 4s_1 = 4\times 2 = 8$ m

28. **(b)** distance, $s = at^2 - bt^3$

velocity, $v = \frac{ds}{dt} = 2at - 3bt^2$

acceleration $a = \frac{dv}{dt} = 2a - 6bt$

Acceleration is zero at

$$2a - 6bt = 0 \Rightarrow t = \frac{a}{3b}$$

29. **(d)**

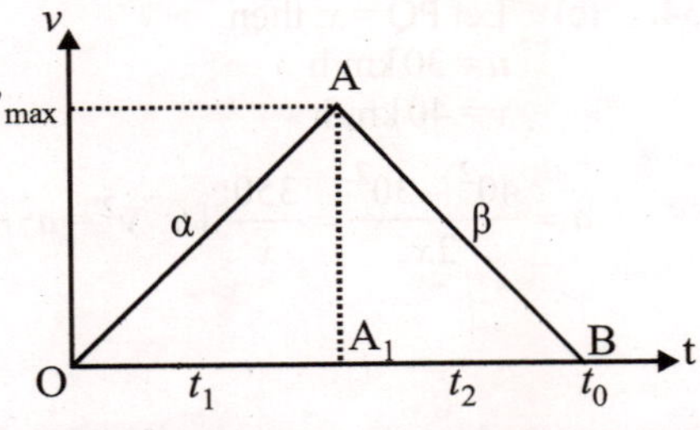

In Fig.

$AA_1 = v_{max.} = \alpha t_1 = \beta t_2$

But $t = t_1 + t_2 = \frac{v_{max}}{\alpha} + \frac{v_{max}}{\beta}$

$$= v_{max}\left(\frac{1}{\alpha} + \frac{1}{\beta}\right) = v_{max}\left(\frac{\alpha+\beta}{\alpha\beta}\right)$$

or, $v_{max} = t\left(\frac{\alpha\beta}{\alpha+\beta}\right)$

> **NOTES** If a body starting from rest accelerates at a constant rate α for certain time and then retards at constant β and comes to rest after t. second from the starting point, then
>
> Distance travelled by the body $= \frac{\alpha\beta t^2}{(2\alpha + 2\beta)}$

30. **(d)** At E, the slope of the curve is negative.

> **NOTES** The line bending towards time axis represents decreasing velocity or negative velocity of the particle. It means the particle possesses retardation.

31. **(d)** Velocity, $v = \frac{ds}{dt} = 3t^2 - 12t + 3$

Acceleration, $a = \frac{dv}{dt} = 6t - 12$; For $a = 0$, we have, $0 = 6t - 12$ or $t = 2$s. Hence, at $t = 2$ s the velocity will be

$v = 3 \times 2^2 - 12 \times 2 + 3 = -9 \text{ ms}^{-1}$

32. **(a)** $\frac{S_4}{S_3} = \frac{0 + \frac{a}{2}(2\times4-1)}{0 + \frac{a}{2}(2\times3-1)} = \frac{7}{5}$

> **NOTES** Equation of the motion of uniformly accelerated motion, the distance travelled in n^{th} sec is given by,
>
> $S_n = u + \frac{a}{2}(2n-1)$

33. **(b)** In one dimensional motion, the body can have at a time one velocity but not two values of velocities.

34. **(c)** Let PQ = x, then

$u = 30$ km/h

$v = 40$ km/h

$$a = \frac{40^2 - 30^2}{2x} = \frac{350}{x} \quad [\because v^2 = u^2 + 2as]$$

Also, velocity at mid point is given by v_m,

$\Rightarrow v_m = u^2 + 20(7)$

$\Rightarrow v_m^2 - 30^2 = 2 \times \frac{350}{x} \times \frac{x}{2}$

This gives $v = 25\sqrt{2}$ km/h

35. **(d)** Let v be the relative velocity of scooter w.r.t bus as $v = v_S - v_B$

$v = \frac{1000}{100} = 10 \text{ ms}^{-1}$ $v_B = 10\text{ms}^{-1}$

$\therefore \quad v_S = v + v_B$,

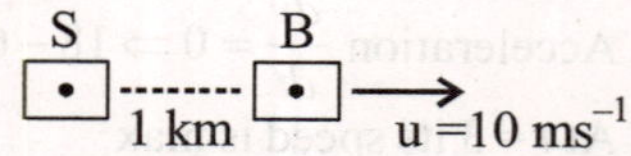

$= 10 + 10 = 20 \text{ ms}^{-1}$

$\therefore$ velocity of scooter = 20 ms^{-1}

> **NOTES** When two bodies are moving in same directions, relative velocity is between difference individual velocities.
>
> $\vec{V}_{AB} = \vec{V}_A - \vec{V}_B$

36. **(d)** Relative velocity of parrot w.r.t the train $= 10 - (-5) = 15 \text{ ms}^{-1}$.

Time taken by parrot to cross the train

$= \frac{150}{15} = 10$ s

> **NOTES** When two bodies are moving in direction relative velocity is sum of individual velocities.
>
> $\vec{V}_{AB} = \vec{V}_A - \vec{V}_B$
>
> The distance which bird has to travel to cross the train is equal to the length of train. Now since the train and the bird are travelling in opposite direction, therefore the speed will be added up.

37. **(c)**

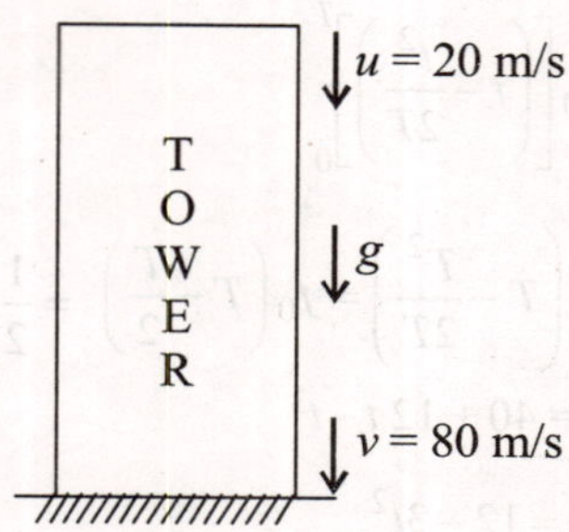

Using $v^2 = u^2 + 2gh$

Height, $h = \frac{v^2 - u^2}{2g} = \frac{(80)^2 - (20)^2}{2 \times 10}$

$$= \frac{6400-400}{20} = 300 \text{ m}$$

38. **(a)** $\because \quad h = \frac{1}{2}gt^2$

$\therefore \quad h_1 = \frac{1}{2}g(5)^2 = 125$

$h_1 + h_2 = \frac{1}{2}g(10)^2 = 500$

$\Rightarrow \quad h_2 = 375$

$h_1 + h_2 + h_3 = \frac{1}{2}g(15)^2 = 1125$

$\Rightarrow \quad h_3 = 625$

$\Rightarrow \quad h_2 = 3h_1,\ h_3 = 5h_1$

or $\quad h_1 = \frac{h_2}{3} = \frac{h_3}{5}$

> **NOTES** The distance covered in time t, $2t$, $3t$, etc. will be in the ratio of $1^2 : 2^2 : 3^2$ i.e., square of integers i.e., h $\propto t^2$.

39. **(b)** Here, $u = 0$

We have, $v^2 = u^2 + 2gh$

$\Rightarrow v = \sqrt{2gh} = \sqrt{2 \times 10 \times 20} = 20$ m/s

40. **(a)** Clearly distance moved by 1st ball in 18s = distance moved by 2nd ball in 12s. ...(i)

Now, distance moved in 18 s by 1st ball $= \frac{1}{2} \times 10 \times 18^2 = 90 \times 18 = 1620$ m

Distance moved in 12 s by 2nd ball

$= ut + \frac{1}{2}gt^2 \quad \Rightarrow \quad 12v + \frac{1}{2} \times 10 \times (12)^2$

$\Rightarrow \quad 12\,v + 720$

From equation (i)

$\therefore \quad 1620 = 12\,v + 5 \times 144$

$\Rightarrow \quad v = 135 - 60 = 75 \text{ ms}^{-1}$

41. **(b)** No external force is acting, therefore, momentum is conserved.

By momentum conservation,

$50\,u + 0.5 \times 2 = 0$

where u is the velocity of man.

$u = -\frac{1}{50} \text{ ms}^{-1}$

Negative sign of u shows that man moves upward.

Time taken by the stone to reach the ground

$= \frac{10}{2} = 5$ sec

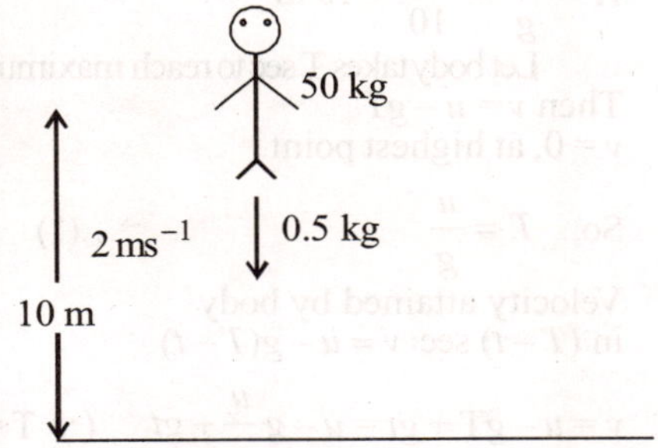

Distance moved by the man $= 5 \times \frac{1}{50} = 0.1$ m

$\therefore$ when the stone reaches the floor, the distance of the man above floor = 10.1 *m*

42. **(c)** Let t_1 & t_2 be the time taken by A and B respectively to reach the ground then from the formula,

$h = \frac{1}{2}gt^2,$

For first body, $\quad 16 = \frac{1}{2}gt_1^2$

For second body, $\; 25 = \frac{1}{2}gt_2^2$

$\therefore \quad \frac{16}{25} = \frac{t_1^2}{t_2^2} \Rightarrow \frac{t_1}{t_2} = \frac{4}{5}.$

> **NOTES** If a body is a dropped from some height, the motion is independent of mass of the body. The time taken to reach the ground $t = \sqrt{2h/g}$ and final velocity $V = \sqrt{2gh}$ and initial velocity, $u = 0$.

43. **(a)** For part *AB*

From 3rd equation of motion

$v^2 = u^2 - 2gH$

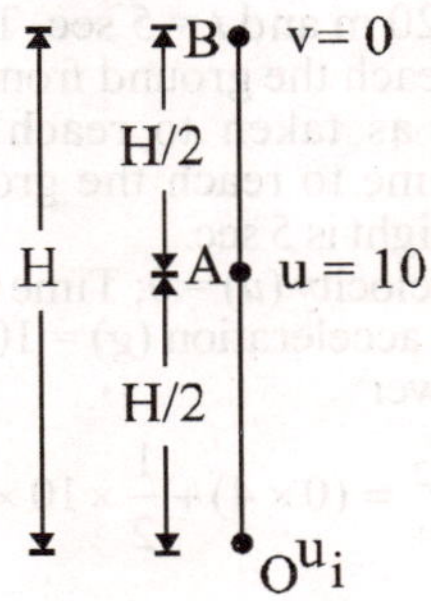

$0 = u^2 - 2g(H/2) = u^2 - gH$

$$H = \frac{u^2}{g} = \frac{10^2}{10} = 10 \text{ m}$$

44. **(c)** Let body takes T sec to reach maximum height. Then $v = u - gT$
$v = 0$, at highest point.

$$\text{So, } T = \frac{u}{g} \quad \ldots(1)$$

Velocity attained by body in $(T - t)$ sec $v = u - g(T - t)$

$$v = u - gT + gt = u - g\frac{u}{g} + gt \quad (\because T = u/g)$$

or $v = gt$ ...(2)

$\therefore$ Distance travelled in last t sec of its ascent

$$s = vt - \frac{1}{2}gt^2$$

$$s = (gt)t - \frac{1}{2}gt^2 = \frac{1}{2}gt^2$$

NOTES If a body is projected vertically upward, then, final velocity, $v = 0$; intial speed, $u = gt$ and height attained, $h = \frac{u^2}{2g} = \frac{g^2t^2}{2g} = \frac{1}{2}gt^2$

45. **(b)** Let the required speed of throw be u ms^{-1}. Then time taken to reach maximum height,

$$t = \frac{u}{g}$$

For two balls to remain in air at any time, t must be greater than 2.

$$\therefore \quad \frac{u}{g} > 2 \Rightarrow u > 19.6 \text{ m/s}$$

46. **(b)** Initial velocity $(u) = 40$ m/s
Acceleration $a = -g \text{ m/s}^2 = -10 \text{ m/s}^2$
Time = 2 seconds
By Ist equation of motion,
$v = u + at$
$v = 40 - 10(2) = 20$ m/s

47. **(b)** $h_{max} = 20$ m and $t = 5$ sec. Time taken by the body to reach the ground from some height is the same as taken to reach that height. Therefore, time to reach the ground from its maximum height is 5 sec.

48. **(c)** Initial velocity $(u) = 0$; Time $(t) = 4$ sec and gravitational acceleration $(g) = 10 \text{ m/s}^2$.
Height of tower

$$h = ut + \frac{1}{2}gt^2 = (0 \times 4) + \frac{1}{2} \times 10 \times (4)^2. = 80 \text{m.}$$

49. **(c)** The speed of an object, falling freely due to gravity, depends only on its height and not on its mass $V = \sqrt{2gh}$. Since the paths are frictionless and all the objects fall through the same height, therefore, their speeds on reaching the ground will be in the ratio of 1 : 1 : 1.

50. **(c)** Height of tap = 5m and $(g) = 10 \text{ m/sec}^2$.

For the first drop, $S = ut + \frac{1}{2}gt^2$

$$5 = (0 \times t) + \frac{1}{2} \times 10t^2 = 5t^2 \text{ or } t^2 = 1 \text{ or } t = 1 \text{ sec.}$$

It means that the third drop leaves after one second of the first drop. Or, each drop leaves after every 0.5 sec.
Distance covered by the second drop in 0.5 sec

$$= ut + \frac{1}{2}gt^2 = (0 \times 0.5) + \frac{1}{2} \times 10 \times (0.5)^2$$

$= 1.25$ m.
Therefore, distance of the second drop above the ground $= 5 - 1.25 = 3.75$ m.

51. **(b)** Let the body fall through the height of tower in nth seconds. From,

$D_n = u + \frac{a}{2}(2n - 1)$ we have, total distance travelled in last 2 seconds of fall is
$D = D_t + D_{(t-1)}$

$$= \left[0 + \frac{g}{2}(2n-1)\right] + \left[0 + \frac{g}{2}\{2(n-1) - 1\}\right]$$

$$= \frac{g}{2}(2n-1) + \frac{g}{2}(2n-3) = \frac{g}{2}(4n-4)$$

$$= \frac{10}{2} \times 4(n-1)$$

or, $40 = 20(n-1)$ or $n = 2 + 1 = 3$s
Distance travelled in t seconds is
where, $t = 3$ sec

$$s = ut + \frac{1}{2}at^2 = 0 + \frac{1}{2} \times 10 \times 3^2 = 45 \text{ m}$$

52. **(b)**

$$\frac{x(4)}{x(5)} = \frac{\frac{g}{2}(2 \times 4 - 1)}{\frac{g}{2}(2 \times 5 - 1)} = \frac{7}{9}$$

$$\left[\because S_{n^{\text{th}}} = u + \frac{a}{2}(2n-1) \text{ and } u = 0, a = g\right]$$

NOTES In the absence of air resistance, all bodies irrespective of the size, weight or composition fall with the same acceleration near the surface of the earth. This motion of a body falling towards the earth from a small altitude (h<<R) is called free fall.

Distance travelled in nth second $h_n = \frac{g}{2}(2n-1)$

3 Motion in a Plane

Trend Analysis with Important Topics & Sub-Topics

Topic Name	Sub-Topic	2021 QNS.	2021 LOD	2020 QNS	2020 LOD	2019 QNS.	2019 LOD	2018 QNS.	2018 LOD	2017 QNS.	2017 LOD	2016 QNS.	2016 LOD
Vectors	Vectors Product							1	E				
Motion in a Plane with Constant acceleration	Addition, Subtraction & Mod of Vectors											1	E
	Motion in 2-dimension free fall	1	E							1	E		
Projectile Motion	Projectile motion & circular motion	1	A									1	A
Relative Velocity in Two Dimensions & Uniform Circular Motion	River-Man Problem in 2D							1	A				
	Uniform Circular Motion					1	E						

LOD - Level of Difficulty | E - Easy | A - Average | D - Difficult | Qns - No. of Questions

Topic 1 Vectors

1. The moment of the force, $\vec{F} = 4\hat{i} + 5\hat{j} - 6\hat{k}$ at (2, 0, –3), about the point (2, –2, –2), is given by *[2018]*
   (a) $-8\hat{i} - 4\hat{j} - 7\hat{k}$ (b) $-4\hat{i} - \hat{j} - 8\hat{k}$
   (c) $-7\hat{i} - 4\hat{j} - 8\hat{k}$ (d) $-7\hat{i} - 8\hat{j} - 4\hat{k}$

2. If the magnitude of sum of two vectors is equal to the magnitude of difference of the two vectors, the angle between these vectors is : *[2016]*
   (a) 0° (b) 90°
   (c) 45° (d) 180°

3. If vectors $\vec{A} = \cos\omega t\hat{i} + \sin\omega t\hat{j}$ and $\vec{B} = \cos\frac{\omega t}{2}\hat{i} + \sin\frac{\omega t}{2}\hat{j}$ are functions of time, then the value of t at which they are orthogonal to each other is : *[2015 RS]*
   (a) $t = \frac{\pi}{2\omega}$ (b) $t = \frac{\pi}{\omega}$
   (c) $t = 0$ (d) $t = \frac{\pi}{4\omega}$

4. A particle is moving such that its position coordinate (x, y) are
   (2m, 3m) at time $t = 0$
   (6m, 7m) at time $t = 2$ s and
   (13m, 14m) at time $t = 5$s.
   Average velocity vector $(\vec{V}_{av})$ from $t = 0$ to $t = 5$s is : *[2014]*
   (a) $\frac{1}{5}(13\hat{i} + 14\hat{j})$ (b) $\frac{7}{3}(\hat{i} + \hat{j})$
   (c) $2(\hat{i} + \hat{j})$ (d) $\frac{11}{5}(\hat{i} + \hat{j})$

5. Vectors $\vec{A}, \vec{B}$ and $\vec{C}$ are such that $\vec{A} \cdot \vec{B} = 0$ and $\vec{A} \cdot \vec{C} = 0$. Then the vector parallel to $\vec{A}$ is *[NEET Kar. 2013]*
   (a) $\vec{B}$ and $\vec{C}$ (b) $\vec{A} \times \vec{B}$
   (c) $\vec{B} + \vec{C}$ (d) $\vec{B} \times \vec{C}$

6. Six vectors, $\vec{a}$ through $\vec{f}$ have the magnitudes and directions indicated in the figure. Which of the following statements is true? *[2010]*

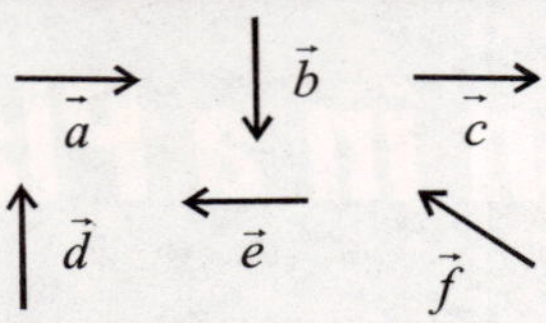

(a) $\vec{b}+\vec{c}=\vec{f}$ (b) $\vec{d}+\vec{c}=\vec{f}$
(c) $\vec{d}+\vec{e}=\vec{f}$ (d) $\vec{b}+\vec{e}=\vec{f}$

7. $\vec{A}$ and $\vec{B}$ are two vectors and θ is the angle between them, if $|\vec{A}\times\vec{B}|=\sqrt{3}(\vec{A}.\vec{B})$, the value of θ is *[2007]*
(a) 45° (b) 30°
(c) 90° (d) 60°

8. The vectors $\vec{A}$ and $\vec{B}$ are such that
$|\vec{A}+\vec{B}|=|\vec{A}-\vec{B}|$
The angle between the two vectors is *[2006, 2001, 1996, 1991]*
(a) 60° (b) 75°
(c) 45° (d) 90°

9. If a vector $2\hat{i}+3\hat{j}+8\hat{k}$ is perpendicular to the vector $4\hat{j}-4\hat{i}+\alpha\hat{k}$, then the value of α is *[2005]*
(a) 1/2 (b) –1/2
(c) 1 (d) –1

10. If the angle between the vectors $\vec{A}$ and $\vec{B}$ is θ, the value of the product $(\vec{B}\times\vec{A})\cdot\vec{A}$ is equal to *[2005]*
(a) $BA^2\sin\theta$ (b) $BA^2\cos\theta$
(c) $BA^2\sin\theta\cos\theta$ (d) zero

11. If $|\vec{A}\times\vec{B}|=\sqrt{3}\vec{A}\cdot\vec{B}$ then the value of $|\vec{A}+\vec{B}|$ is *[2004]*
(a) $(A^2+B^2+\sqrt{3}AB)^{1/2}$
(b) $(A^2+B^2+AB)^{1/2}$
(c) $\left(A^2+B^2+\frac{AB}{\sqrt{3}}\right)^{1/2}$
(d) $A+B$

12. The vector sum of two forces is perpendicular to their vector differences. In that case, the forces *[2003]*
(a) cannot be predicted
(b) are equal to each other
(c) are equal to each other in magnitude
(d) are not equal to each other in magnitude

13. The angle between the two vectors $\vec{A}=3\hat{i}+4\hat{j}+5\hat{k}$ and $\vec{B}=3\hat{i}+4\hat{j}-5\hat{k}$ will be *[2001, 1994]*
(a) zero (b) 45°
(c) 90° (d) 180°

14. A particle moves with a velocity $\vec{v}=6\hat{i}-4\hat{j}+3\hat{k}$ m/s under the influence of a constant force $\vec{F}=20\hat{i}+15\hat{j}-5\hat{k}$ N. The instantaneous power applied to the particle is *[2000]*
(a) 45 J/s (b) 35 J/s
(c) 25 J/s (d) 195 J/s

15. What is the linear velocity if angular velocity vector $\vec{\omega}=3\hat{i}-4\hat{j}+\hat{k}$ and position vector $\vec{r}=5\hat{i}-6\hat{j}+6\hat{k}$ *[1999]*
(a) $6\hat{i}+2\hat{j}-3\hat{k}$ (b) $-18\hat{i}-13\hat{j}+2\hat{k}$
(c) $18\hat{i}+13\hat{j}-2\hat{k}$ (d) $6\hat{i}-2\hat{j}+8\hat{k}$

16. The angle between two vectors of magnitude 12 and 18 units when their resultant is 24 units, is *[1999]*
(a) 63°51′ (b) 75°52′
(c) 82°31′ (d) 89°16′

17. If a unit vector is represented by $0.5\hat{i}+0.8\hat{j}+c\hat{k}$, the value of c is *[1999]*
(a) 1 (b) $\sqrt{0.11}$
(c) $\sqrt{0.01}$ (d) 0.39

18. Find the torque of a force $\vec{F}=-3\hat{i}+\hat{j}+5\hat{k}$ acting at the point $\vec{r}=7\hat{i}+3\hat{j}+\hat{k}$. *[1997]*
(a) $-21\hat{i}+3\hat{j}+5\hat{k}$ (b) $-14\hat{i}+3\hat{j}+16\hat{k}$
(c) $4\hat{i}+4\hat{j}+6\hat{k}$ (d) $14\hat{i}-38\hat{j}+16\hat{k}$

19. Which of the following is not a vector quantity? *[1996]*
(a) displacement (b) electric field
(c) work (d) acceleration

20. Which of the following is not a vector quantity? *[1995]*
(a) speed (b) velocity
(c) torque (d) displacement

21. A body constrained to move in y-direction, is subjected to a force given by $\vec{F} = (-2\hat{i} + 15\hat{j} + 6\hat{k})N$. What is the workdone by this force in moving the body through a distance of 10m along y-axis? *[1994]*

(a) 190 J (b) 160 J
(c) 150 J (d) 20 J

22. The resultant of ($\vec{A} \times 0$) will be equal to *[1992]*

(a) zero (b) A
(c) zero vector (d) unit vector

23. The angle between $\vec{A}$ and $\vec{B}$ is θ. The value of the triple product $\vec{A} \cdot (\vec{B} \times \vec{A})$ is *[1989]*

(a) A^2B (b) zero
(c) $A^2B \sin\theta$ (d) $A^2B \cos\theta$

24. The magnitudes of vectors $\vec{A}$, $\vec{B}$ and $\vec{C}$ are 3, 4 and 5 units respectively. If $\vec{A} + \vec{B} = \vec{C}$, then the angle between $\vec{A}$ and $\vec{B}$ is *[1988]*

(a) $\pi/2$ (b) $\cos^{-1} 0.6$
(c) $\tan^{-1} 7/5$ (d) $\pi/4$

Topic 2 Motion in a Plane with Constant acceleration

25. A car starts from rest and accelerates at 5 m/s^2. At t = 4 s, a ball is dropped out of a window by a person sitting in the car. What is the velocity and acceleration of the ball at t = 6 s? (Take g = 10 m/s^2) *[2021]*

(a) $20\sqrt{2}$ m/s, 10 m/s^2
(b) 20 m/s, 5 m/s^2
(c) 20 m/s, 0
(d) $20\sqrt{2}$ m/s, 0

26. The x and y coordinates of the particle at any time are $x = 5t - 2t^2$ and $y = 10t$ respectively, where x and y are in meters and t in seconds. The acceleration of the particle at $t = 2$s is *[2017]*

(a) 5 m/s^2 (b) –4 m/s^2
(c) –8 m/s^2 (d) 0

27. A particle starting from the origin (0, 0) moves in the (x, y) plane. Its coordinates at a later time are $(\sqrt{3}, 3)$. The path of the particle makes with the x-axis an angle of *[2007]*

(a) 45° (b) 60°
(c) 0° (d) 30°

28. A particle moves in a plane with constant acceleration in a direction different from the initial velocity. The path of the particle is *[2005]*

(a) an ellipse (b) a parabola
(c) an arc of a circle (d) a straight line

29. A body of 3 kg moves in the XY plane under the action of a force given by $6t^2\hat{i} + 4t\hat{j}$. Assuming that the body is at rest at time $t = 0$, the velocity of the body at $t = 3$s is *[2002]*

(a) $6\hat{i} + 6\hat{j}$ (b) $18\hat{i} + 6\hat{j}$
(c) $18\hat{i} + 12\hat{j}$ (d) $12\hat{i} + 18\hat{j}$

30. Two particles A and B are connected by a rigid rod AB. The rod slides along perpendicular rails as shown here. The velocity of A to the left is 10 m/s. What is the velocity of B when angle α = 60°? *[1998]*

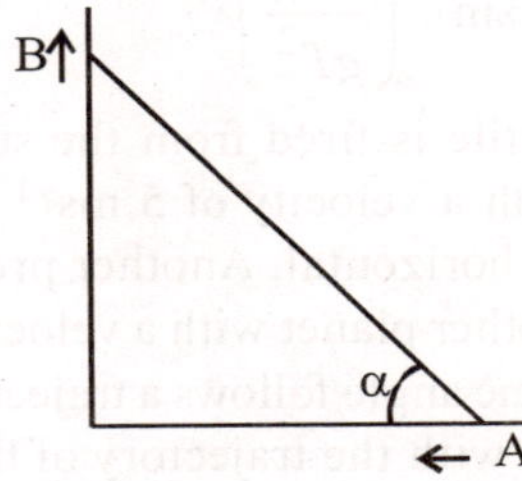

(a) 5.8 m/s (b) 9.8 m/s
(c) 10 m/s (d) 17.3 m/s

31. The position vector of a particle is $\vec{r} = (a\cos\omega t)\hat{i} + (a\sin\omega t)\hat{j}$. The velocity of the particle is *[1995]*

(a) directed towards the origin
(b) directed away from the origin
(c) parallel to the position vector
(d) perpendicular to the position vector

32. A bullet is fired from a gun with a speed of 1000 m/s in order to hit a target 100 m away. At what height above the target should the gun be aimed? (The resistance of air is negligible and g = 10 m/s^2) *[1995]*

(a) 5 cm (b) 10 cm
(c) 15 cm (d) 20 cm

Topic 3 Projectile Motion

33. A particle moving in a circle of radius R with a uniform speed takes a time T to complete one revolution. If this particle were projected with the same speed at an angle 'θ' to the horizontal, the maximum height attained by it equals 4R. The angle of projection, θ, is then given by : [2021]

(a) $\theta = \sin^{-1}\left(\frac{2gT^2}{\pi^2 R}\right)^{1/2}$

(b) $\theta = \cos^{-1}\left(\frac{gT^2}{\pi^2 R}\right)^{1/2}$

(c) $\theta = \cos^{-1}\left(\frac{\pi^2 R}{gT^2}\right)^{1/2}$

(d) $\theta = \sin^{-1}\left(\frac{\pi^2 R}{gT^2}\right)^{1/2}$

34. A projectile is fired from the surface of the earth with a velocity of 5 ms^{-1} and angle θ with the horizontal. Another projectile fired from another planet with a velocity of 3 ms^{-1} at the same angle follows a trajectory which is identical with the trajectory of the projectile fired from the earth. The value of the acceleration due to gravity on the planet is (in ms^{-2}) given g = 9.8 m/s^2 [2014]

(a) 3.5 (b) 5.9
(c) 16.3 (d) 110.8

35. The velocity of a projectile at the initial point A is $\left(2\hat{i}+3\hat{j}\right)$ m/s. It's velocity (in m/s) at point B is [2013]

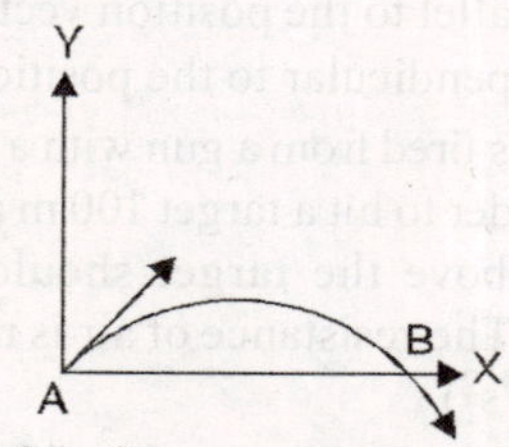

(a) $-2\hat{i}+3\hat{j}$ (b) $2\hat{i}-3\hat{j}$
(c) $2\hat{i}+3\hat{j}$ (d) $-2\hat{i}-3\hat{j}$

36. The horizontal range and the maximum height of a projectile are equal. The angle of projection of the projectiles is : [2012]

(a) $\theta = \tan^{-1}\left(\frac{1}{4}\right)$ (b) $\theta = \tan^{-1}(4)$
(c) $\theta = \tan^{-1}(2)$ (d) $\theta = 45°$

37. A missile is fired for maximum range with an initial velocity of 20 m/s. If $g = 10$ m/s^2, the range of the missile is [2011]

(a) 40m (b) 50m
(c) 60m (d) 20m

38. A projectile is fired at an angle of 45° with the horizontal. Elevation angle of the projectile at its highest point as seen from the point of projection is [2011M]

(a) 60° (b) $\tan^{-1}\left(\frac{1}{2}\right)$
(c) $\tan^{-1}\left(\frac{\sqrt{3}}{2}\right)$ (d) 45°

39. A particle of mass m is projected with velocity v making an angle of 45° with the horizontal. When the particle lands on the level ground the magnitude of the change in its momentum will be: [2008]

(a) 2mv (b) $mv/\sqrt{2}$
(c) $mv\sqrt{2}$ (d) zero

40. For angles of projection of a projectile (45° – θ) and (45° + θ), the horizontal ranges described by the projectile are in the ratio of [2006]

(a) 1:3 (b) 1:2
(c) 2:1 (d) 1:1

41. Two projectiles are fired from the same point with the same speed at angles of projection 60° and 30° respectively. Which one of the following is true? [2000]

(a) Their maximum height will be same
(b) Their range will be same
(c) Their landing velocity will be same
(d) Their time of flight will be same

42. If a body A of mass M is thrown with velocity v at an angle of 30° to the horizontal and another body B of the same mass is thrown with the same speed at an angle of 60° to the horizontal, the ratio of horizontal range of A to B will be [1992]

(a) 1:3 (b) 1:1
(c) $1:\sqrt{3}$ (d) $\sqrt{3}:1$

43. The maximum range of a gun of horizontal terrain is 16 km. If $g = 10\ ms^{-2}$, then muzzle velocity of a shell must be **[1990]**

(a) $160\ ms^{-1}$ (b) $200\sqrt{2}\ ms^{-1}$
(c) $400\ ms^{-1}$ (d) $800\ ms^{-1}$

44. Two bodies of same mass are projected with the same velocity at an angle 30° and 60° respectively. The ratio of their horizontal ranges will be **[1990]**

(a) 1 : 1 (b) 1 : 2
(c) 1 : 3 (d) $2 : \sqrt{2}$

Topic 4 Relative Velocity in Two Dimensions & Uniform Circular Motion

45. The speed of a swimmer in still water is 20 m/s. The speed of river water is 10 m/s and is flowing due east. If he is standing on the south bank and wishes to cross the river along the shortest path, the angle at which he should make his strokes w.r.t. north is given by : **[2019]**

(a) 30°west (b) 0°
(c) 60° west (d) 45° west

46. A particle moves so that its position vector is given by $\vec{r} = \cos\omega t\hat{x} + \sin\omega t\hat{y}$. Where ω is a constant. Which of the following is true? **[2016]**

(a) Velocity and acceleration both are perpendicular to $\vec{r}$
(b) Velocity and acceleration both are parallel to $\vec{r}$
(c) Velocity is perpendicular to $\vec{r}$ and acceleration is directed towards the origin
(d) Velocity is perpendicular to $\vec{r}$ and acceleration is directed away from the origin

47. A ship A is moving Westwards with a speed of $10\ km\ h^{-1}$ and a ship B 100 km South of A, is moving Northwards with a speed of $10\ km\ h^{-1}$. The time after which the distance between them becomes shortest, is : **[2015]**

(a) 5 h (b) $5\sqrt{2}$ h
(c) $10\sqrt{2}$ h (d) 0 h

48. The position vector of a particle $\vec{R}$ as a function of time is given by:

$\vec{R} = 4\sin(2\pi t)\hat{i} + 4\cos(2\pi t)\hat{j}$

Where R is in meter, t in seconds and $\hat{i}$ and $\hat{j}$ denote unit vectors along x-and y-directions, respectively.

Which one of the following statements is wrong for the motion of particle? **[2015 RS]**

(a) Magnitude of acceleration vector is $\frac{v^2}{R}$, where v is the velocity of particle
(b) Magnitude of the velocity of particle is 8 meter/second
(c) path of the particle is a circle of radius 4 meter.
(d) Acceleration vector is along $\vec{R}$

49. A particle moves in a circle of radius 5 cm with constant speed and time period 0.2 πs. The acceleration of the particle is **[2011]**

(a) $15\ m/s^2$ (b) $25\ m/s^2$
(c) $36\ m/s^2$ (d) $5\ m/s^2$

50. A car runs at a constant speed on a circular track of radius 100 m, taking 62.8 seconds in every circular loop. The average velocity and average speed for each circular loop respectively, is **[2006]**

(a) 0, 10 m/s (b) 10 m/s, 10 m/s
(c) 10 m/s, 0 (d) 0, 0

51. Two boys are standing at the ends A and B of a ground where $AB = a$. The boy at B starts running in a direction perpendicular to AB with velocity v_1. The boy at A starts running simultaneously with velocity v and catches the other boy in a time t, where t is **[2005]**

(a) $a/\sqrt{v^2 + v_1^2}$ (b) $a/(v + v_1)$
(c) $a/(v - v_1)$ (d) $\sqrt{a^2/(v^2 - v_1^2)}$

52. A stone tied to the end of a string of 1 m long is whirled in a horizontal circle with a constant speed. If the stone makes 22 revolutions in 44 seconds, what is the magnitude and direction of acceleration of the stone? **[2005]**

(a) $\pi^2\ m\ s^{-2}$ and direction along the radius towards the centre
(b) $\pi^2\ m\ s^{-2}$ and direction along the radius away from the centre
(c) $\pi^2\ m\ s^{-2}$ and direction along the tangent to the circle
(d) $\pi^2/4\ m\ s^{-2}$ and direction along the radius towards the centre

53. The circular motion of a particle with constant speed is *[2005]*
(a) periodic but not simple harmonic
(b) simple harmonic but not periodic
(c) periodic and simple harmonic
(d) neither periodic nor simple harmonic

54. A stone is tied to a string of length ℓ and is whirled in a vertical circle with the other end of the string as the centre. At a certain instant of time, the stone is at its lowest position and has a speed u. The magnitude of the change in velocity as it reaches a position where the string is horizontal (g being acceleration due to gravity) is *[2004]*
(a) $\sqrt{2g\ell}$ (b) $\sqrt{2(u^2 - g\ell)}$
(c) $\sqrt{u^2 - g\ell}$ (d) $u - \sqrt{u^2 - 2g\ell}$

55. A particle moves along a circle of radius $\left(\frac{20}{\pi}\right)$ m with constant tangential acceleration. If the velocity of the particle is 80 m/s at the end of the second revolution after motion has begun, the tangential acceleration is *[2003]*
(a) $40\pi\,m/s^2$ (b) $40\,m/s^2$
(c) $640\pi\,m/s^2$ (d) $160\pi\,m/s^2$

56. Two particles of mass *M* and *m* are moving in a circle of radii *R* and *r*. If their time-periods are same, what will be the ratio of their linear velocities? *[2001]*
(a) $MR : mr$ (b) $M : m$
(c) $R : r$ (d) 1 : 1

57. A small sphere is attached to a cord and rotates in a vertical circle about a point *O*. If the average speed of the sphere is increased, the cord is most likely to break at the orientation when the mass is at *[2000]*

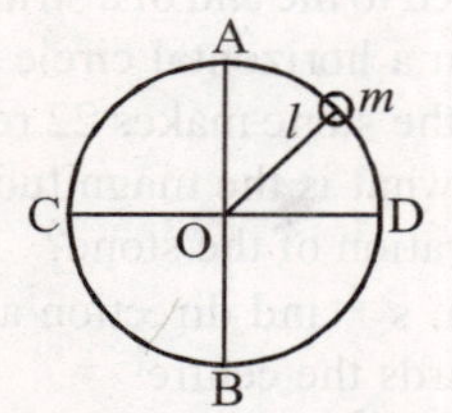

(a) bottom point B (b) the point C
(c) the point D (d) top point A

58. A boat which has a speed of 5 km/hr in still water crosses a river of width km along the shortest possible path in 15 minutes. The velocity of the river water in km/hr is *[2000, 1998]*
(a) 3 (b) 4
(c) $\sqrt{21}$ (d) 1

59. A stone tied with a string, is rotated in a vertical circle. The minimum speed with which the string has to be rotated *[1999]*
(a) is independent of the mass of the stone
(b) is independent of the length of the string
(c) decreases with increasing mass of the stone
(d) decreases with increasing length of the string

60. A person swims in a river aiming to reach exactly opposite point on the bank of a river. His speed of swimming is 0.5 m/s at an angle 120° with the direction of flow of water. The speed of water in stream is *[1999]*
(a) 1.0 m/s (b) 0.5 m/s
(c) 0.25 m/s (d) 0.43 m/s.

61. A ball of mass 0.25 kg attached to the end of a string of length 1.96 m is moving in a horizontal circle. The string will break if the tension is more than 25 N. What is the maximum speed with which the ball can be moved? *[1998]*
(a) 14 m/s (b) 3 m/s
(c) 5 m/s (d) 3.92 m/s

62. A body is whirled in a horizontal circle of radius 20 cm. It has an angular velocity of 10 rad/s. What is its linear velocity at any point on circular path *[1996]*
(a) $\sqrt{2}$ m/s (b) 2 m/s
(c) 10 m/s (d) 20 m/s

63. When a body moves with a constant speed along a circle *[1994]*
(a) its velocity remains constant
(b) no force acts on it
(c) no work is done on it
(d) no acceleration is produced in it

64. A boat is sent across a river with a velocity of 8 km h^{-1}. If the resultant velocity of boat is 10 km h^{-1}, then the velocity of the river is *[1993]*
(a) 12.8 km h^{-1} (b) 6 km h^{-1}
(c) 8 km h^{-1} (d) 10 km h^{-1}

65. An electric fan has blades of length 30 cm measured from the axis of rotation. If the fan is rotating at 120 rpm, the acceleration of a point on the tip of the blade is *[1990]*
(a) 1600 ms^{-2} (b) 47.4 ms^{-2}
(c) 23.7 ms^{-2} (d) 50.55 ms^{-2}

ANSWER KEY																	
1	(c)	9	(b)	17	(b)	25	(a)	33	(a)	41	(b)	49	(d)	57	(a)	65	(c)
2	(b)	10	(d)	18	(d)	26	(b)	34	(a)	42	(b)	50	(a)	58	(a)		
3	(b)	11	(b)	19	(c)	27	(b)	35	(b)	43	(c)	51	(d)	59	(a)		
4	(d)	12	(c)	20	(a)	28	(b)	36	(b)	44	(a)	52	(a)	60	(c)		
5	(d)	13	(c)	21	(c)	29	(b)	37	(a)	45	(a)	53	(a)	61	(a)		
6	(c)	14	(a)	22	(c)	30	(a)	38	(b)	46	(c)	54	(b)	62	(b)		
7	(d)	15	(b)	23	(b)	31	(d)	39	(c)	47	(a)	55	(b)	63	(c)		
8	(d)	16	(b)	24	(a)	32	(a)	40	(d)	48	(b)	56	(c)	64	(b)		

Hints & Solutions

1. **(c)** Moment of force, $\vec{\tau} = \vec{r} \times \vec{F}$

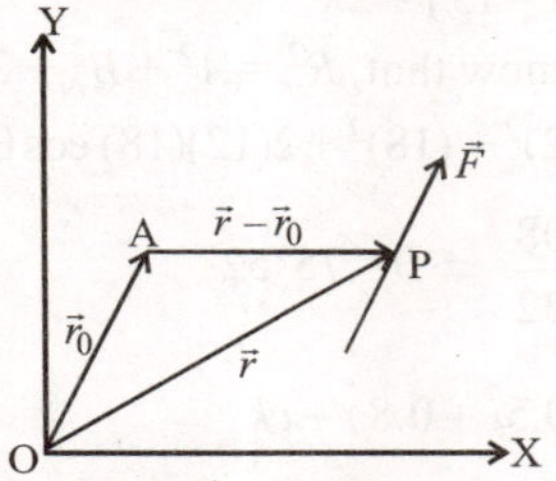

$$\vec{\tau} = (\vec{r} - \vec{r}_0) \times \vec{F}$$

$$\vec{r} - \vec{r}_0 = (2\hat{i} + 0\hat{j} - 3\hat{k}) - (2\hat{i} - 2\hat{j} - 2\hat{k})$$

$$= 0\hat{i} + 2\hat{j} - \hat{k}$$

$$\vec{\tau} = (0\hat{i} + 2\hat{j} - \hat{k})(4\hat{i} + 5\hat{j} - 6\hat{k})$$

$$\vec{\tau} = \begin{vmatrix} \hat{i} & \hat{j} & \hat{k} \\ 0 & 2 & -1 \\ 4 & 5 & -6 \end{vmatrix} = -7\hat{i} - 4\hat{j} - 8\hat{k}$$

2. **(b)** $|\vec{A} + \vec{B}| = |\vec{A} - \vec{B}|$

Squaring on both sides

$$|\vec{A} + \vec{B}|^2 = |\vec{A} - \vec{B}|^2$$

$$\Rightarrow \vec{A} \cdot \vec{A} + 2\vec{A} \cdot \vec{B} + \vec{B} \cdot \vec{B}$$

$$= \vec{A} \cdot \vec{A} - 2\vec{A} \cdot \vec{B} + \vec{B} \cdot \vec{B}$$

$$\Rightarrow 4\vec{A} \cdot \vec{B} = 0 \Rightarrow 4AB\cos\theta = 0$$

$$\Rightarrow \cos\theta = 0 \Rightarrow \theta = 90°$$

3. **(b)** Two vectors are

$$\vec{A} = \cos\omega t\hat{i} + \sin\omega t\hat{j}$$

$$\vec{B} = \cos\frac{\omega t}{2}\hat{i} + \sin\frac{\omega t}{2}\hat{j}$$

For two vectors $\vec{A}$ and $\vec{B}$ to be orthogonal

$$\vec{A}.\vec{B} = 0$$

$$\vec{A}.\vec{B} = 0 = \cos\omega t.\cos\frac{\omega t}{2} + \sin\omega t.\sin\frac{\omega t}{2}$$

$$= \cos\left(\omega t - \frac{\omega t}{2}\right) = \cos\left(\frac{\omega t}{2}\right)$$

So, $\frac{\omega t}{2} = \frac{\pi}{2}$ $\therefore$ $t = \frac{\pi}{\omega}$

4. **(d)** $\vec{v}_{av} = \frac{\Delta\vec{r}\text{ (displacement)}}{\Delta t\text{ (time taken)}}$

$$= \frac{(13-2)\hat{i} + (14-3)\hat{j}}{5-0} = \frac{11}{5}(\hat{i} + \hat{j})$$

> **NOTES**
> When a point have coordinate (x, y) then its position vector $= x\hat{i} + y\hat{j}$
> When a particle moves from point (x_1, y_1) to (x_2, y_2) then its displacement vector
> $\vec{r} = (x_2 - x_1)\hat{i} + (y_2 - y_1)\hat{j}$

5. **(d)** Vector triple product

$$\vec{A} \times (\vec{B} \times \vec{C}) = \vec{B}(\vec{A} \cdot \vec{C}) - \vec{C}(\vec{A} \cdot \vec{B}) = 0$$

$$\Rightarrow \vec{A} \,||\, (\vec{B} \times \vec{C})$$

6. **(c)** Using the law of vector addition, $(\vec{d} + \vec{e})$ is as shown in the fig.

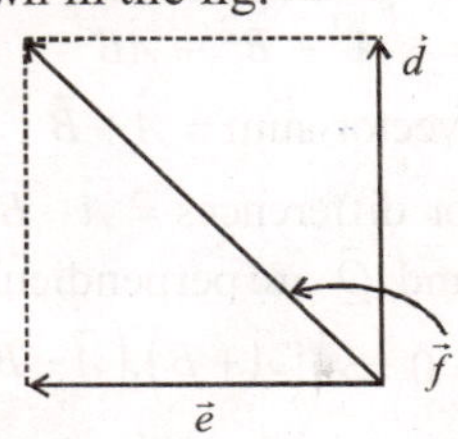

$$\therefore \quad \vec{d} + \vec{e} = \vec{f}$$

7. **(d)** $\left|\vec{A}\times\vec{B}\right| = \sqrt{3}(\vec{A}.\vec{B})$

$\Rightarrow AB\sin\theta = \sqrt{3}AB\cos\theta$

$\Rightarrow \tan\theta = \sqrt{3} \Rightarrow \theta = 60°$

8. **(d)** $|\vec{A}+\vec{B}|^2 = |\vec{A}-\vec{B}|^2$

$= |\vec{A}|^2 + |\vec{B}|^2 + 2\vec{A}.\vec{B} = A^2 + B^2 + 2AB\cos\theta$

$= |\vec{A}-\vec{B}|^2 = |\vec{A}|^2 + |\vec{B}|^2 - 2\vec{A}.\vec{B}$

$= A^2 + B^2 - 2AB\cos\theta$

So, $A^2 + B^2 + 2AB\cos\theta$

$= A^2 + B^2 - 2AB\cos\theta$

$4AB\cos\theta = 0 \Rightarrow \cos\theta = 0$

$\theta = 90°$

So, angle between A & B is 90º.

9. **(b)** For two vectors to be perpendicular to each other

$\vec{A}\cdot\vec{B} = 0$

$(2\hat{i}+3\hat{j}+8\hat{k})\cdot(4\hat{j}-4\hat{i}+\alpha\hat{k}) = 0$

$-8+12+8\alpha = 0$

$\alpha = -\frac{4}{8} = -\frac{1}{2}$

10. **(d)** $(\vec{B}\times\vec{A})\cdot\vec{A} = \vec{C}\cdot\vec{A} = CA\cos 90° = 0.$

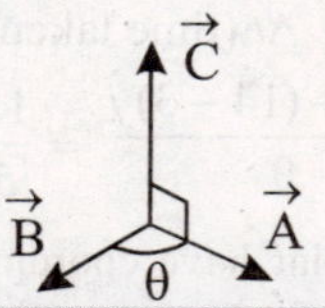

> **NOTES** Dot product of any two perpendicular vector is zero.

11. **(b)** $|\vec{A}\times\vec{B}| = AB\sin\theta$

$\vec{A}.\vec{B} = AB\cos\theta$

$|\vec{A}\times\vec{B}| = \sqrt{3}\ \vec{A}.\vec{B} \Rightarrow AB\sin\theta = \sqrt{3}AB\cos\theta$

or, $\tan\theta = \sqrt{3}$, $\quad\therefore \theta = 60º$

$\therefore |\vec{A}+\vec{B}| = \sqrt{A^2+B^2+2AB\cos 60º}$

$= \sqrt{A^2+B^2+AB}$

12. **(c)** $\vec{P}$ = vector sum = $\vec{A}+\vec{B}$

$\vec{Q}$ = Vector differences = $\vec{A}-\vec{B}$

Since $\vec{P}$ and $\vec{Q}$ are perpendicular

$\therefore \vec{P}.\vec{Q} = 0 \Rightarrow (\vec{A}+\vec{B}).(\vec{A}-\vec{B}) = 0$

$\Rightarrow A^2 = B^2 \Rightarrow |\vec{A}| = |\vec{B}|$

13. **(c)** $\vec{A} = 3\hat{i}+4\hat{j}+5\hat{k}$, $\vec{B} = 3\hat{i}+4\hat{j}-5\hat{k}$

$\vec{A}.\vec{B} = (3\hat{i}+4\hat{j}+5\hat{k}).(3\hat{i}+4\hat{j}-5\hat{k})$

$|\vec{A}|\,|\vec{B}|\cos\theta = 9+16-25 = 0$

$|\vec{A}| \neq 0$, $|\vec{B}| \neq 0$, hence, $\cos\theta = 0, \theta = 90°$

14. **(a)** As we know,

$P = \vec{F}.\vec{v} = (6\hat{i}-4\hat{j}+3\hat{k}).(20\hat{i}+15\hat{j}-5\hat{k})$

$= 6\times 20 - 4\times 15 - 3\times 5 = 45$ J/s

> **NOTES** Instantanesous power can be calculated by
>
> $P_{ins} = F\cdot\cos\theta v = \vec{F}\cdot\vec{v}$
>
> which is the scalar product of force and velocity vector.

15. **(b)** As we know that

$\vec{v} = \vec{\omega}\times\vec{r} = (3\hat{i}-4\hat{j}+\hat{k})\times(5\hat{i}-6\hat{j}+6\hat{k})$

$= -18\hat{i}-13\hat{j}+2\hat{k}$

16. **(b)** We know that, $R^2 = A^2+B^2+2AB\cos\theta$

$(24)^2 = (12)^2+(18)^2+2(12)(18)\cos\theta$

$\cos\theta = \frac{108}{432} \Rightarrow \theta = 75°52'$

17. **(b)** $\hat{r} = 0.5\hat{i}+0.8\hat{j}+c\hat{k}$

$|\hat{r}| = 1 = \sqrt{(0.5)^2+(0.8)^2+c^2}$

$(0.5)^2+(0.8)^2+c^2 = 1$

$c^2 = 0.11 \Rightarrow c = \sqrt{0.11}$

> **NOTES** Unit vector is a vector which has a magnitude of one. It is a vector divided by its magnitude. Unit vector for
>
> $\vec{A}$ is $\hat{A}: \hat{A} = \frac{\vec{A}}{A}$
>
> Unit vector gives direction.

18. **(d)** $\vec{F} = -3\hat{i}+\hat{j}+5\hat{k}$; $\vec{r} = 7\hat{i}+3\hat{j}+\hat{k}$

Torque $(\vec{\tau}) = \vec{r}\times\vec{F}$

$= (7\hat{i}+3\hat{j}+\hat{k})\times(-3\hat{i}+\hat{j}+5\hat{k})$

$= 7\hat{k}+35(-\hat{j})-9(-\hat{k})+15\hat{i}-3\hat{j}+(-\hat{i})$

$= 14\hat{i}-38\hat{j}+16\hat{k}$

19. **(c)**

20. **(a)** A vector quantity has both magnitude and direction. In the given options, speed has only magnitude, therefore, it is non-vector or scalar quantity.

NOTES: If a physical quantity has magnitude and direction both, then it does not always imply that it is a vector. for it to be a vector, laws of vector algebra has to be satisfied.

21. **(c)** Since displacement is along the y-direction, hence displacement $\vec{s} = 10\hat{j}$.

Work done $= \vec{F} \cdot \vec{s}$

$= (-2\hat{i} + 15\hat{j} + 6\hat{k}) \cdot 10\hat{j} = 150J$

22. **(c)** When a vector is multiplied with a scalar, the result is a vector.

NOTES: The cross product of $\vec{A}$ and $\vec{B}$ is a vector with its direction perpendicular to both vector $\vec{A}$ and $\vec{B}$.

Vector $\vec{A} \times 0$ is a zero vector because 0 is a scalar then also the product is zero but a (scalar × vector) is also vector.

23. **(b)** Note that $(\vec{B} \times \vec{A}) \perp \vec{A}$. Hence their dot product is zero.

NOTES: Dot product of any two perpendicular vector is zero

24. **(a)** $(\vec{A} + \vec{B})^2 = (\vec{C})^2$

$\Rightarrow \quad A^2 + B^2 + 2\vec{A} \cdot \vec{B} = C^2$

$\Rightarrow \quad 3^2 + 4^2 + 2\vec{A} \cdot \vec{B} = 5^2$

$\Rightarrow \quad 2\vec{A} \cdot \vec{B} = 0 \quad \text{or} \Rightarrow \vec{A} \cdot \vec{B} = 0 \quad \therefore \vec{A} \perp \vec{B}$

Here $A^2 + B^2 = C^2$. Hence, $\vec{A} \perp \vec{B}$

25. **(a)** Velocity of car at t = 4 s = velocity the ball dropped out of a window v = u + at

$\Rightarrow v = 0 + 5 \times 4 = 20 \text{ ms}^{-1}$

∴ Velocity of ball along the horizontal and is 20ms^{-1}.

After 2 seconds of motion :

Vertical velocity of ball $(v_y) = u_y + a_y t$

$v_y = 0 + 10 \times 2 = 20 \text{ ms}^{-1}$ ($\because a_y = g = 10 \text{ m/s}^2$)

∴ Magnitude of velocity of ball

$$(v) = \sqrt{v_x^2 + v_y^2} = 20\sqrt{2} \text{ m/s}$$

Since the ball is under free fall.

Acceleration of ball (a) at t = 6s a = g = 10 m/s²

NOTES: All bodies, in the absence of air resistance, irrespective of the size, weight or composition fall with same acceleration near the surface of the earth is called free fall.

26. **(b)** Given:

$x = 5t - 2t^2 \qquad y = 10t$

$v_x = \dfrac{dx}{dt} = 5 - 4t \qquad v_y = \dfrac{dy}{dt} = 10$

$a_x = \dfrac{dv_x}{dt} = -4 \qquad a_y = \dfrac{dv_y}{dt} = 0$

$\vec{a} = a_x i + a_y j \qquad \vec{a} = -4i \text{ m/s}^2$

Hence, acceleration of particle at $(t = 2 \text{ s}) = -4 \text{m/s}^2$

27. **(b)**

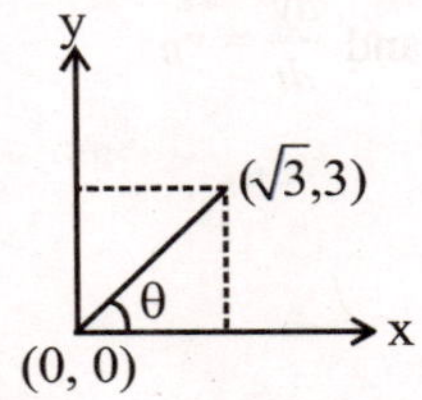

Let θ be the angle which the particle makes with x-axis.

From figure, $\tan\theta = \dfrac{3}{\sqrt{3}} = \sqrt{3}$

$\Rightarrow \theta = \tan^{-1}(\sqrt{3}) = 60°$

NOTES: If a vector $\vec{R}$ in X–Y plane then its orthogonal vector $R_X = R \cos\theta$ and $R_Y = R \sin\theta$.

And $\dfrac{R\sin\theta}{R\cos\theta} = \tan\theta = \dfrac{R_Y}{R_X}$

28. **(b)**

29. **(b)** $\vec{F} = 6t\,\hat{i} + 4t\,\hat{j}$

$F_x = 6t, F_y = 4t$

$a_x = \dfrac{6t}{3} = 2t, a_y = \dfrac{4t}{3}$

$\dfrac{dv_x}{dt} = 2t^2$

$\int_0^{v_x} dV_x = \int_0^t 2t^2 dt$

$\Rightarrow \quad v_x = \dfrac{2}{3}t^3 = \dfrac{2}{3} \cdot 3^3 = 18$

$\dfrac{dV_y}{dt} = \dfrac{4}{3}t \Rightarrow \int_0^{V_y} dV_y = \dfrac{4}{3}\int_0^t t\,dt$

$\Rightarrow \quad V_y = \dfrac{4}{3}\dfrac{t^2}{2} = \dfrac{4}{3} \cdot \dfrac{3^2}{2} = 6$

$\Rightarrow \quad V = 18\hat{i} + 6\hat{j}$

30. **(a)** Given, $v_A = 10 \text{m/sec}$

$\alpha = 60°$

let length of the rod = L

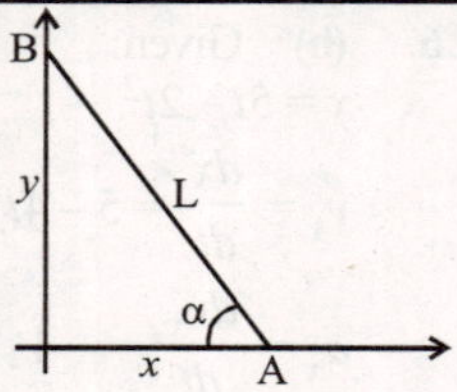

From figure,

$\Rightarrow \quad x^2+y^2=L$

differentiation with respect to time 't'

$$\Rightarrow 2x\frac{dx}{dt}+2y\frac{dy}{dt}=0$$

$$\Rightarrow x\frac{dx}{dt}+y\frac{dy}{dt}=0$$

where, $\frac{dx}{dt}=v_A$ and $\frac{dy}{dt}=v_B$

So, $xv_A+yv_B=0$

$$v_B=\frac{-x}{y}v_A$$

where $\frac{x}{y}=\cot\alpha$

So, $v_B=-v_A\cot\alpha$

$v_B=10\cot 60°$

$$v_B=\frac{-10}{\sqrt{3}}=-5.773\simeq 5.8\text{ m/sec}$$

31. **(d)** Position vector,

$\vec{r}=(a\cos\omega t)\hat{i}+(a\sin\omega t)\hat{j}$

Velocity vector,

$$\vec{v}=\frac{d(\vec{r})}{dt}=\frac{d}{dt}\{(a\cos\omega t)\hat{i}+(a\sin\omega t)\hat{j}\}$$

$=(-a\omega\sin\omega t)\hat{i}+(a\omega\cos\omega t)\hat{j}$

$=\omega[(-a\sin\omega t)\hat{i}+(a\cos\omega t)\hat{j}]$

Slope of position vector $=\frac{a\sin\omega t}{a\cos\omega t}=\tan\omega t$

Slope of velocity vector, $=\frac{-a\cos\omega t}{a\sin\omega t}=\frac{-1}{\tan\omega t}$

$\therefore$ velocity is perpendicular to the displacement.

32. **(a)** Speed of the bullet (v) = 1000 m/s and horizontal distance of the target (s) = 100 m. Time taken to cover the horizontal distance (t)

$$=\frac{100}{1000}=0.1\text{ sec}.$$

During this time, the bullet will fall down vertically due to gravitational acceleration.

$$\therefore \text{height } (h)=ut+\frac{1}{2}gt^2$$

$$=(0\times 0.1)+\frac{1}{2}\times 10(0.1)^2$$

$$=0.05\text{m}=5\text{cm}$$

33. **(a)** Speed of particle to complete one revolution, $(u)=\frac{2\pi R}{T}$ ... (i)

Particle is projected with same speed (U) at an angle θ to the horizontal

$$\therefore \text{Maximum Height}(H)=\frac{u^2\sin^2\theta}{2g}$$

$$\text{or, } \frac{u^2\sin^2\theta}{2g}=4R \quad [\because H=4R \text{ given}]$$

$$\Rightarrow \sin^2\theta=\frac{8gR}{u^2}$$

$$\text{or, } \sin^2\theta=\frac{8gRT^2}{4\pi^2R^2}=\frac{2gT^2}{\pi^2R} \text{ (using equation (i))}$$

$$\therefore \theta=\sin^{-1}\left(\frac{2gT^2}{\pi^2R}\right)^{1/2}$$

34. **(a)** Trajectory is identical for both horizontal range $=\frac{u^2\sin 2\theta}{g}$

$$\text{So that, } \frac{g_{\text{planet}}}{g_{\text{earth}}}=\frac{(u_{\text{planet}})^2}{(u_{\text{earth}})^2}$$

$$\text{Therefore, } g_{\text{planet}}=\left(\frac{3}{5}\right)^2(9.8\text{ m/s}^2)$$

$=3.5\text{ m/s}^2$

35. **(b)** At point B the direction of velocity component of the projectile along y-component reverses, while the x-component remains unchanged.

Hence, $\vec{V}_B=2\hat{i}-3\hat{j}$

36. **(b)** Horizontal range,

$$R=\frac{u^2\sin 2\theta}{g} \quad \text{....(1)}$$

Maximum height,

$$H=\frac{u^2\sin^2\theta}{2g} \quad \text{....(2)}$$

According to the condition,

$R=H$

$$\Rightarrow \frac{u^2\sin 2\theta}{g}=\frac{u^2\sin^2\theta}{2g}$$

$$\Rightarrow 2\sin\theta\cos\theta=\frac{\sin^2\theta}{2}$$

$$2\cos\theta = \frac{\sin\theta}{2}$$

$$\Rightarrow \quad \cot\theta = \frac{1}{4}$$

$$\Rightarrow \quad \tan\theta = 4$$

$$\Rightarrow \quad \theta = [\tan^{-1}(4)]$$

> **NOTES**
> In case of projective motion range R is n times the maximum height H
>
> i.e., $R = nH \Rightarrow \frac{u^2\sin 2\theta}{g} = n\frac{u^2\sin^2\theta}{2g}$
>
> $\Rightarrow \text{Tan}\,\theta = \frac{4}{n}$ Angle of projection $\theta = \tan^{-1}\left(\frac{4}{n}\right)$

37. **(a)** For maximum range, the angle of projection, $\theta = 45°$.

$$\therefore \quad R = \frac{u^2\sin 2\theta}{g}$$

$$= \frac{(20)^2\sin(2\times 45°)}{10} = \frac{400\times 1}{10} = 40.$$

38. **(b)** Maximum height,

$$H = \frac{u^2\sin^2 45°}{2g} = \frac{u^2}{4g} \quad \text{... (1)}$$

$$\text{Range, } R = \frac{u^2\sin 90°}{g} = \frac{u^2}{g}$$

$$\therefore \quad \frac{R}{2} = \frac{u^2}{2g} \quad \text{... (2)}$$

According to the required condition,

$$\therefore \quad \tan\alpha = \frac{H}{R/2}$$

Put the values from equation (1) and (2)

$$\Rightarrow \quad \frac{\left(\frac{u^2}{4g}\right)}{\left(\frac{u^2}{2g}\right)} = \tan\alpha \qquad \therefore \alpha = \tan^{-1}\left(\frac{1}{2}\right)$$

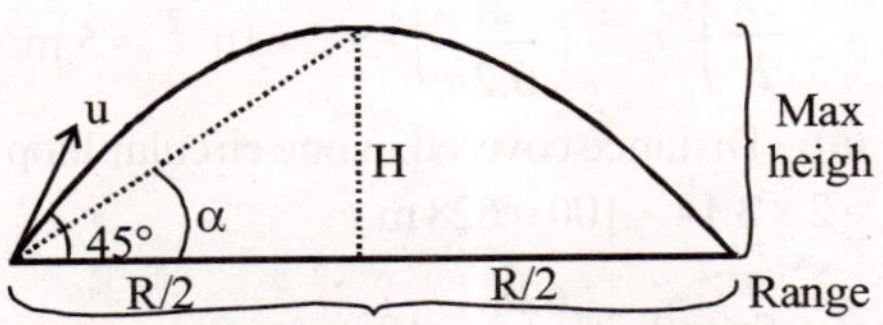

39. **(c)** The momentum along x-axis remains unchanged

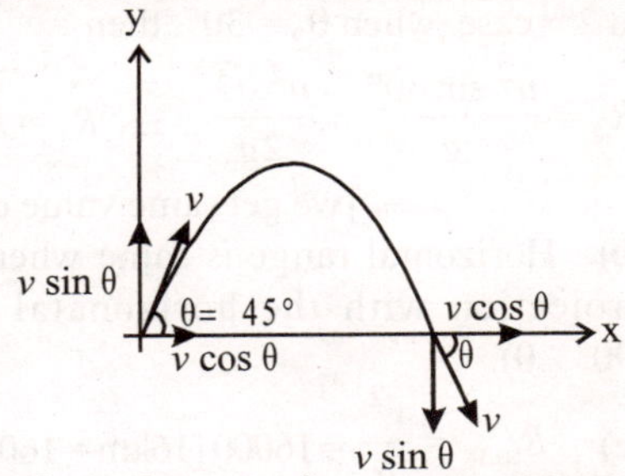

Clearly, change in momentum along x-axis $= mv\cos\theta - mv\cos\theta = 0$

Momentum changed only in vertical direction or y-axis.

So, $\Delta P = \Delta P_{vertical}$

$$\Rightarrow \quad P_{final} - P_{initial}$$

$$= mv\sin\theta - (-mv\sin\theta)$$

$$= 2\,mv\sin\theta = 2mv \times \sin 45°$$

$$= 2mv \times \frac{1}{\sqrt{2}} = \sqrt{2}mv$$

Hence, resultant change in momentum $= \sqrt{2}mv$

40. **(d)** Horizontal range for projection angle $(45° - \theta)$ is, $R_1 = \frac{u^2\sin 2(45-\theta)}{g}$

Horizontal range projection angle $(45° + \theta)$ is, $R_2 = \frac{u^2\sin 2(45+\theta)}{g}$

According to the condition,

$$\Rightarrow \quad \frac{R_1}{R_2} = \frac{u^2\sin 2(45-\theta)}{u^2\sin 2(45+\theta)} = \frac{\sin(90-2\theta)}{\sin(90+2\theta)}$$

$$\Rightarrow \quad \frac{R_1}{R_2} = \frac{\cos 2\theta}{\cos 2\theta} = \frac{1}{1}$$

So, $R_1 : R_2 : 1 : 1$

> **NOTES**
> The angle of elevation (ϕ) of the highest point of the projectile and the angle of projection θ are related to each other as $\tan\phi = \frac{1}{2}\tan\theta$

41. **(b)** Given, $u_1 = u_2 = u$, $\theta_1 = 60°$, $\theta_2 = 30°$

In 1st case, we know that range

$$R_1 = \frac{u^2\sin 2(60°)}{g} = \frac{u^2\sin 120°}{g}$$

$$= \frac{u^2\sin(90° + 30°)}{g}$$

$$= \frac{u^2(\cos 30°)}{g} = \frac{\sqrt{3}u^2}{2g}$$

In 2nd case, when $\theta_2 = 30°$, then

$$R_2 = \frac{u^2 \sin 60°}{g} = \frac{u^2\sqrt{3}}{2g} \Rightarrow R_1 = R_2$$

[we get same value of ranges].

42. **(b)** Horizontal range is same when angle of projection with the horizonatal is θ and (90° – θ).

43. **(c)** $R_{max} = \frac{v^2}{g} = 16000$ [16km = 16000m]

or $v = (16000g)^{1/2} = (16000 \times 10)^{1/2} = 400\ ms^{-1}$

44. **(a)** Horizontal range is same when angle of projection is θ or (90° – θ).

45. **(a)** Velocity of swimmer w.r.t. river $V_{SR} = 20$ m/s

Velocity of river w.r.t. ground $V_{RG} = 10$ m/s

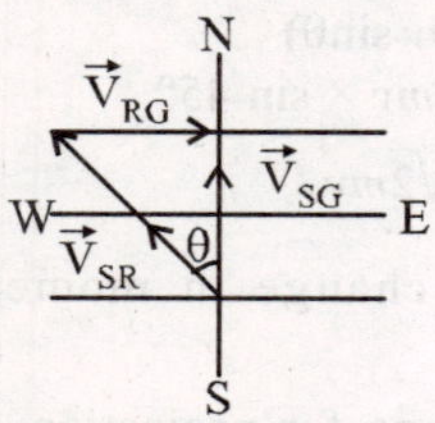

$$\vec{V}_{SG} = \vec{V}_{SR} + \vec{V}_{RG}$$

$$\sin\theta = \left|\frac{\vec{V}_{RG}}{\vec{V}_{SR}}\right| \Rightarrow \sin\theta = \frac{10}{20}$$

$$\Rightarrow \sin\theta = \frac{1}{2} \quad \therefore \theta = 30° \text{ west}$$

i.e.,to cross the river along the shortest path, swimmer should make his strokes 30° west.

46. **(c)** **Given:** Position vector

$\vec{r} = \cos\omega t\ \hat{x} + \sin\omega t\ \hat{y}$

∴ Velocity vector, $\vec{v} = -\omega\sin\omega t\ \hat{x} + \omega\cos\omega t\ \hat{y}$ acceleration vector,

$\vec{a} = -\omega^2\cos\omega t\ \hat{x} - \omega^2\sin\omega t\ \hat{y} = -\omega^2\vec{r}$

$\vec{r}\cdot\vec{v} = 0$ hence $\vec{r} \perp \vec{v}$ and $\vec{a}$ is directed towards the origin.

> **NOTES** Nothing actually moves in the direction of the angular velocity vector $\vec{\omega}$. The direction of $\vec{\omega}$ simply represents that the circular motion is taking place in a plane perpendicular to it.

47. **(a)** Let *t* be the time passed and *x* be distance covered by the two ships after which the distance between them becomes shortest.

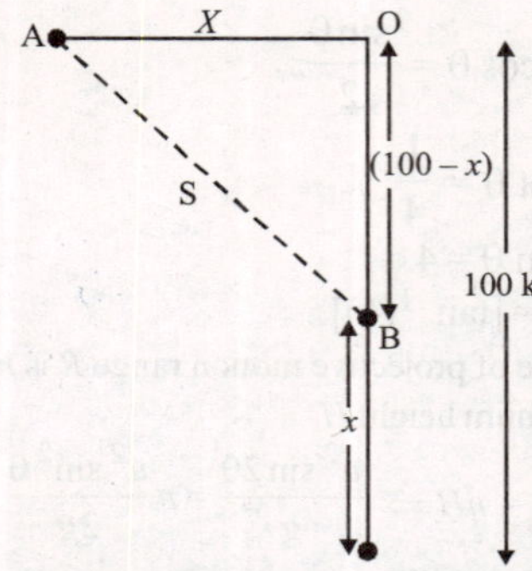

distance(s) between the ship is given by,

$$S = \sqrt{(100-x)^2 + x^2}$$

For *S* to be minimum, $\frac{ds}{dt} = 0$

$$\Rightarrow \frac{ds}{dt} = \frac{1}{9[(100-x)^2 + x^2]^{-1/2}}$$

$$[-2(100-x) + 2x] = 0$$

$$\Rightarrow 4x - 200 = 0 \Rightarrow x = 50\text{ m}$$

So, after both the ships have covered 50 m, distance between them becomes shortest. Time taken for it, will

$$t = \frac{x}{10} = \frac{50}{10} = 5\text{hr.}$$

48. **(b)** Here, $x = 4\sin(2\pi t)$ ...(*i*)

$y = 4\cos(2\pi t)$ ...(*ii*)

Squaring and adding equation (*i*) and (*ii*)

$x^2 + y^2 = 4^2 \Rightarrow R = 4$

Motion of the particle is circular motion, acceleration vector is along $-\vec{R}$ and its magnitude $= \frac{V^2}{R}$

Velocity of particle, $V = \omega R = (2\pi)(4) = 8\pi$

49. **(d)** Centripetal acceleration $a_c = \omega^2 r$

$$= \left(\frac{2\pi}{T}\right)^2 r = \left(\frac{2\pi}{0.2\pi}\right)^2 \times 5 \times 10^{-2} = 5\text{ m/s}^2$$

50. **(a)** Distance covered in one circular loop $= 2\pi r$
$= 2 \times 3.14 \times 100 = 628$ m

avg Speed $= \frac{628}{62.8} = 10$ m/sec

Displacement in one circular loop = 0

avg Velocity $= \frac{0}{\text{time}} = 0$

51. (d) According to the given condition,

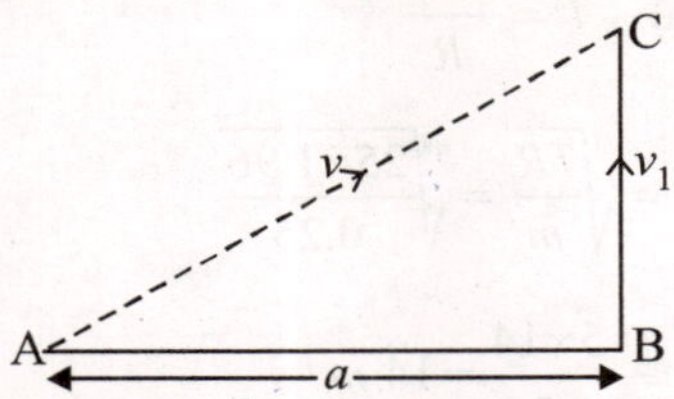

Let both y boy meet at c: in time t then,

AC = vt and BC = $v_1 t$

From pythagorus theorem,

$AC^2 = AB^2 + BC^2$

$v^2t^2 = a^2 + {v_1}^2t^2$

$t^2(v^2 - {v_1}^2) = a^2$

$$t^2 = \frac{a^2}{(v^2 - v_1^2)}$$

$$t = \sqrt{\frac{a}{(v^2 - v_1^2)}}$$

52. (a) $a_r = \omega^2 R$ & $a_t = \frac{dv}{dt} = 0$

or, $a_r = (2\pi n)^2 R = 4\pi^2 n^2 R^2 = 4\pi^2\left(\frac{22}{44}\right)^2 (1)^2$

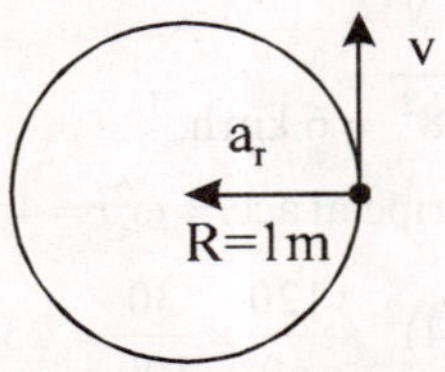

$a_{net} = a_r = \pi^2 ms^{-2}$ and direction along the radius towards the centre.

NOTES If the force acting on a particle is always perpendicular to the velocity of the particle, then the path of the particle is a circle. The centripetal force (F_c) is always perpendicular to the velocity of the particle, i.e., $F_c \perp v$

53. (a) In circular motion of a particle with constant speed, particle repeats its motion after a regular interval of time but does not oscillate about a fixed point. So, motion of particle is periodic but not simple harmonic.

54. (b) $W_{mg} = \Delta K$

$\Rightarrow -mg\,\ell = ½\,mv^2 - ½\,mu^2$

or, $mv^2 = m(u^2 - 2\,g\ell]$

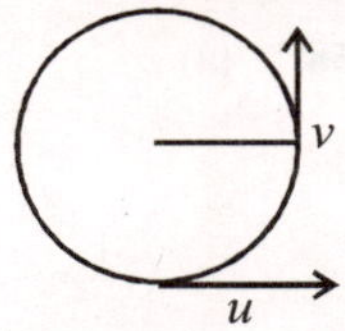

or, $v = \sqrt{u^2 - 2g\ell}\ \ \hat{j}$

$\vec{u} = u\hat{i}$

$\therefore \vec{v} - \vec{u} = \sqrt{u^2 - 2g\ell}\ \hat{j} - u\hat{i}$

$\therefore |\vec{v} - \vec{u}| = [(u^2 - 2g\ell) + u^2]^{½} = \sqrt{2(u^2 - g\ell)}$

55. (b) Given, $r = \frac{20}{\pi}$ m

$v_j = 80$ m/sec $\Rightarrow w_f = \frac{80\pi}{2\theta} \Rightarrow 4\pi$

$\theta = 2reN = 4\pi$ radian

From equation,

$\omega_f^2 = \omega_0^2 + 2\alpha\theta \quad [\because \omega_0 = 0]$

$(4\pi)^2 = 0 + 2.\alpha.4\pi$

$\alpha = 2\pi$

tangential acceleration

$a_t = \alpha.r$

$at = 2\pi \cdot \frac{20}{\pi} = 40$ m/sec^2

56. (c) Linear velocity $v = r\omega$

$v_1 = \omega r_1, v_2 = \omega r_2$

[ω is same in both cases because time period is same]

$$\frac{v_1}{v_2} = \frac{r_1}{r_2} = \frac{R}{r}$$

57. (a) In the case of a body describing a vertical circle,

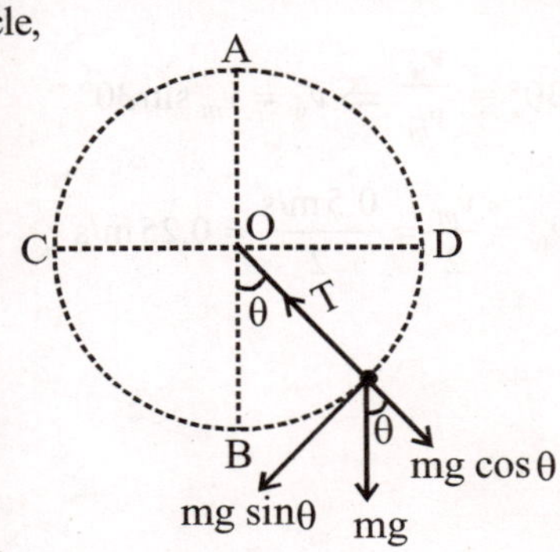

$$T - mg\cos\theta = \frac{mv^2}{l}\ ;\ T = mg\cos\theta + \frac{mv^2}{l}$$

Tension is maximum when cos θ = +1 and velocity is maximum

Both conditions are satisfied at θ = 0° (*i.e.*, at lowest point *B*]

58. **(a)**

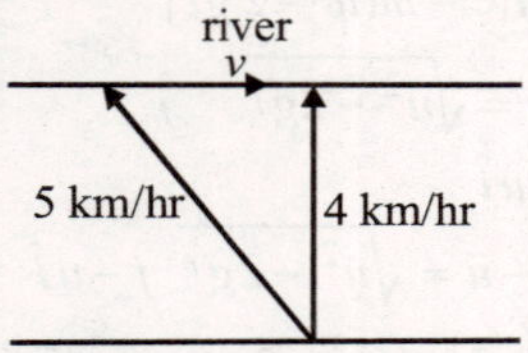

Speed along the shortest path

$=\frac{1}{15/60}=4\text{km/hr}$

Speed of river $v=\sqrt{5^2-4^2}=3\text{ km/hr}$

59. **(a)** Minimum speed with which the string is rotating in a vertical circle (v) = $\sqrt{gr}$

The minimum speed of stone is independent of mass of stone.

60. **(c)**

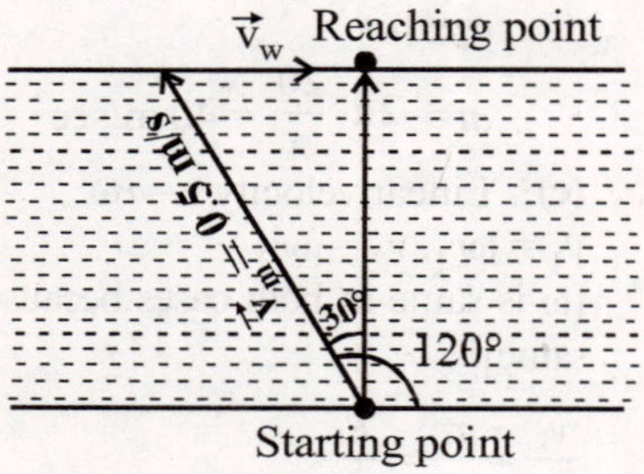

Velocity of person, $\vec{v}_m=0.5\text{ m/s}$

$\sin 30°=\frac{v_w}{v_m}\Rightarrow v_w=v_m\sin 30°$

$\Rightarrow v_w=\frac{v_m}{2}=\frac{0.5\text{m/s}}{2}=0.25\text{m/s}$

61. **(a)** $T=\frac{mv^2}{R}$

$v=\sqrt{\frac{TR}{m}}=\sqrt{\frac{25\times 1.96}{0.25}}$

$=\frac{5\times 14}{5}=14\ m/s$

NOTES

In the horizontal circular motion of the ball tension in the string is balanced by the centrifugal force $\left(\frac{mv^2}{\ell}\right)$ and hence the maximum tension in the string will be for the maximum speed of the ball.

62. **(b)** Radius of circular path = 20 cm = $\frac{2}{10}$ m

Angular speed of body = 10 rad/s

Linear velocity = radius × Angular speed

$=\frac{2}{10}\times 10=2\text{ m/s}$

63. **(c)** On circular motion, the force acts along the radius and displacement at a location is perpendicular to the radius *i.e.*, $\theta=90°$

As work done = $\vec{F}.\vec{S}=FS\cos 90°=0$

64. **(b)** $v_r=\sqrt{v_R{}^2-v_B^2}$

$=\sqrt{10^2-8^2}=6\text{ kmh}^{-1}$

65. **(c)** Centripetal acc. $=\omega^2 r=4\pi^2\nu^2 r$

$=4\times(3.14)^2\times\frac{120}{60}\times\frac{30}{100}=23.7\text{ ms}^{-2}$

$[\because \omega=2\pi\nu]$

4 Laws of Motion

Trend Analysis with Important Topics & Sub-Topics

Topic Name	Sub-Topic	2021		2020		2019		2018		2017		2016	
		QNS.	LOD	QNS.	LOD	QNS.	LOD	QNS.	LOD	QNS.	LOD	QNS.	LOD
Ist, IInd & IIIrd Laws of Motion	Impulse-momentum Theorem	1	A										
Motion of Connected Bodies, Pulley & Equilibrium of Forces	Acceleration of Motion of Connected Bodies			1	A								
	FBD							1	E				
Friction	Friction							1	A				
Circular Motion, Banking of Road	Centripetal Force					1	A			1	A		
	Maximum Safe Velocity											1	E
	Vel. Body Enter Vertical Loop											1	E

LOD - Level of Difficulty E - Easy A - Average D - Difficult Qns - No. of Questions

Topic 1 Ist, IInd & IIIrd Laws of Motion

1. A ball of mass 0.15 kg is dropped from a height 10 m, strikes the ground and rebounds to the same height. The magnitude of impulse imparted to the ball is ($g = 10 \text{ m/s}^2$) nearly *[2021]*
(a) 1.4 kg m/s (b) 0 kg m/s
(c) 4.2 kg m/s (d) 2.1 kg m/s

2. A particle moving with velocity is acted by three forces shown by the vector triangle PQR. The velocity of the particle will : *[2019]*

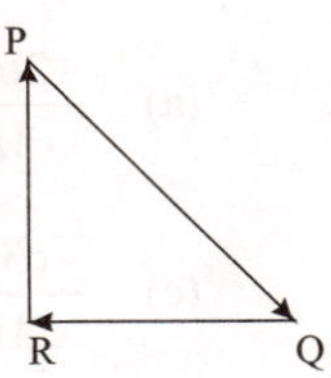

(a) increase
(b) decrease
(c) remain constant
(d) change according to the smallest force

3. The force 'F' acting on a particle of mass 'm' is indicated by the force-time graph shown below. The change in momentum of the particle over the time interval from zero to 8 s is : *[2014]*

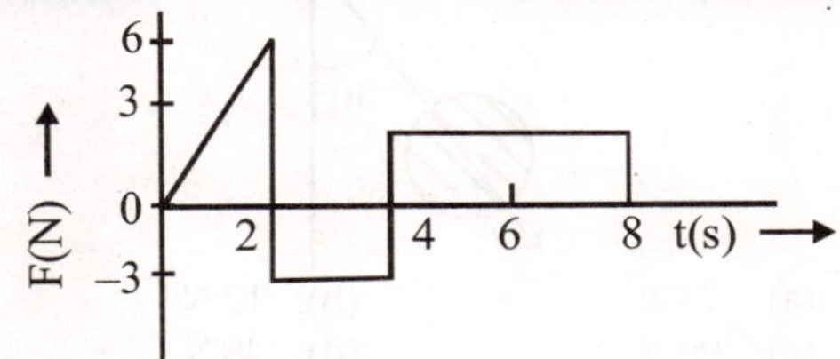

(a) 24 Ns (b) 20 Ns
(c) 12 Ns (d) 6 Ns

4. A stone is dropped from a height h. It hits the ground with a certain momentum P. If the same stone is dropped from a height 100% more than the previous height, the momentum when it hits the ground will change by : *[2012M]*
(a) 68% (b) 41%
(c) 200% (d) 100%

5. A body of mass M hits normally a rigid wall with velocity V and bounces back with the same velocity. The impulse experienced by the body is *[2011]*
(a) MV (b) 1.5 MV
(c) 2 MV (d) zero

6. A body under the action of a force $\vec{F} = 6\hat{i} - 8\hat{j} + 10\hat{k}$, acquires an acceleration of $1\ m/s^2$. The mass of this body must be *[2009]*
(a) 10 kg (b) 20 kg
(c) $10\sqrt{2}$ kg (d) $2\sqrt{10}$ kg

7. Sand is being dropped on a conveyor belt at the rate of M kg/s. The force necessary to keep the belt moving with a constant velocity of v m/s will be: *[2008]*
(a) Mv newton (b) 2 Mv newton
(c) $\frac{Mv}{2}$ newton (d) zero

8. A 0.5 kg ball moving with speed of 12 m/s strikes a hard wall at an angle of 30° with the wall. It is reflected with the same speed and at the same angle. If the ball is in contact with the wall for 0.25 seconds, the average force acting on the wall is *[2006]*

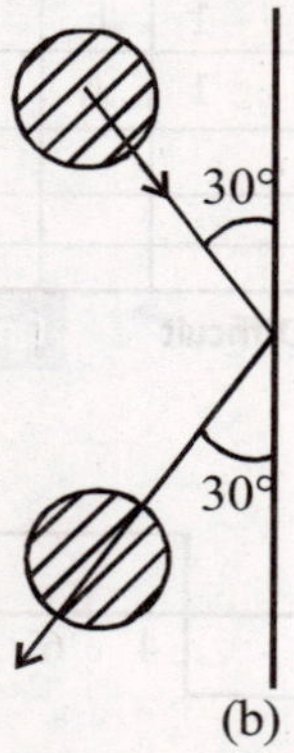

(a) 24 N (b) 12 N
(c) 96 N (d) 48 N

9. If *a* cricketer catches a ball of mass 150 gm moving with a velocity of 20 m/s, then he experiences a force of (Time taken to complete the catch is 0.1 sec.) *[2001]*
(a) 300 N (b) 30 N
(c) 3 N (d) 0.3 N

10. A 3 kg ball strikes a heavy rigid wall with a speed of 10 m/s at an angle of 60°. It gets reflected with the same speed and angle as shown here. If the ball is in contact with the wall for 0.20s, what is the average force exerted on the ball by the wall? *[2000]*

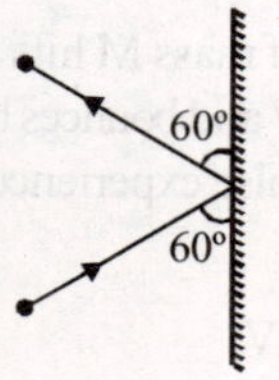

(a) 150 N
(b) zero
(c) $150\sqrt{3}$ N
(d) 300N

11. A bullet is fired from a gun. The force on the bullet is given by $F = 600 - 2 \times 10^5\ t$ where, F is in newton and t in second. The force on the bullet becomes zero as soon as it leaves the barrel. What is the average impulse imparted to the bullet? *[1998]*
(a) 1.8 N-s (b) zero
(c) 9 N-s (d) 0.9 N-s

12. A 5000 kg rocket is set for vertical firing. The exhaust speed is $800\ ms^{-1}$. To give an initial upward acceleration of $20\ ms^{-2}$, the amount of gas ejected per second to supply the needed thrust will be ($g = 10\ ms^{-2}$) *[1998]*
(a) $127.5\ kg\ s^{-1}$ (b) $187.5\ kg\ s^{-1}$
(c) $185.5\ kg\ s^{-1}$ (d) $137.5\ kg\ s^{-1}$

13. A 10 N force is applied on a body produces an acceleration of $1\ m/s^2$. The mass of the body is
(a) 5 kg (b) 10 kg *[1996]*
(c) 15 kg (d) 20 kg

14. A ball of mass 150 g, moving with an acceleration $20\ m/s^2$, is hit by a force, which acts on it for 0.1 sec. The impulsive force is *[1996]*
(a) 0.5 N (b) 0.1 N
(c) 0.3 N (d) 1.2 N

15. If the force on a rocket moving with a velocity of 300 m/sec is 345 N, then the rate of combustion of the fuel, is *[1995]*
(a) 0.55 kg/sec (b) 0.75 kg/sec
(c) 1.15 kg/sec (d) 2.25 kg/sec

16. A satellite in a force free space sweeps stationary interplanetary dust at a rate (dM/dt) = αv. The acceleration of satellite is *[1994]*
(a) $\frac{-2\alpha v^2}{M}$ (b) $\frac{-\alpha v^2}{M}$
(c) $\frac{-\alpha v^2}{2M}$ (d) $-\alpha v^2$

17. Physical independence of force is a consequence of *[1991]*
(a) third law of motion
(b) second law of motion
(c) first law of motion
(d) all of these laws

18. A 600 kg rocket is set for a vertical firing. If the exhaust speed is 1000 ms^{-1}, the mass of the gas ejected per second to supply the thrust needed to overcome the weight of rocket is *[1990]*
(a) 117.6 kg s^{-1} (b) 58.6 kg s^{-1}
(c) 6 kg s^{-1} (d) 76.4 kg s^{-1}

19. A particle of mass m is moving with a uniform velocity v_1. It is given an impulse such that its velocity becomes v_2. The impulse is equal to *[1990]*
(a) $m[|v_2|-|v_1|]$ (b) $\frac{1}{2}m[v_2^2 - v_1^2]$
(c) $m[v_1 + v_2]$ (d) $m[v_2 - v_1]$

Topic 2 Motion of Connected Bodies, Pulley & Equilibrium of Forces

20. Two bodies of mass 4 kg and 6 kg are tied to the ends of a massless string. The string passes over a pulley which is frictionless (see figure). The acceleration of the system in terms of acceleration due to gravity (g) is: *[2020]*

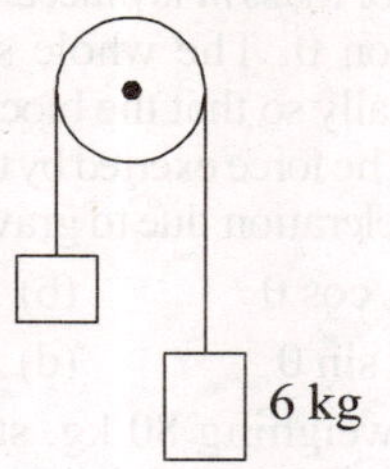

(a) g/2 (b) g/5
(c) g/10 (d) g

21. Two blocks A and B of masses 3 m and m respectively are connected by a massless and inextensible string. The whole system is suspended by a massless spring as shown in figure. The magnitudes of acceleration of A and B immediately after the string is cut, are respectively : *[2017]*

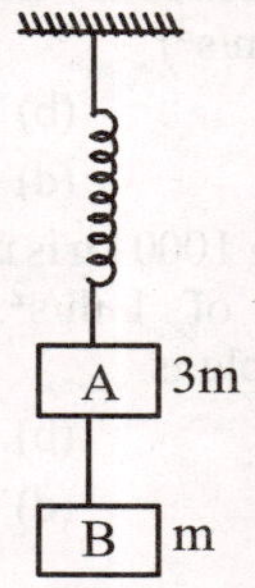

(a) $\frac{g}{3}, g$ (b) g, g
(c) $\frac{g}{3}, \frac{g}{3}$ (d) $g, \frac{g}{3}$

22. One end of string of length l is connected to a particle of mass 'm' and the other end is connected to a small peg on a smooth horizontal table. If the particle moves in circle with speed 'v' the net force on the particle (directed towards centre) will be (T represents the tension in the string) : *[2017]*
(a) $T + \frac{mv^2}{l}$ (b) $T - \frac{mv^2}{l}$
(c) Zero (d) T

23. Three blocks A, B and C of masses 4 kg, 2 kg and 1 kg respectively, are in contact on a frictionless surface, as shown. If a force of 14 N is applied on the 4 kg block then the contact force between A and B is *[2015]*

(a) 6 N (b) 8 N
(c) 18 N (d) 2 N

24. A system consists of three masses m_1, m_2 and m_3 connected by a string passing over a pulley P. The mass m_1 hangs freely and m_2 and m_3 are on a rough horizontal table (the coefficient of friction = μ). The pulley is frictionless and of negligible mass. The downward acceleration of mass m_1 is : (Assume $m_1 = m_2 = m_3 = m$) *[2014]*

(a) $\frac{g(1 - g\mu)}{g}$

(b) $\frac{2g\mu}{3}$

(c) $\frac{g(1 - 2\mu)}{3}$

(d) $\frac{g(1 - 2\mu)}{2}$

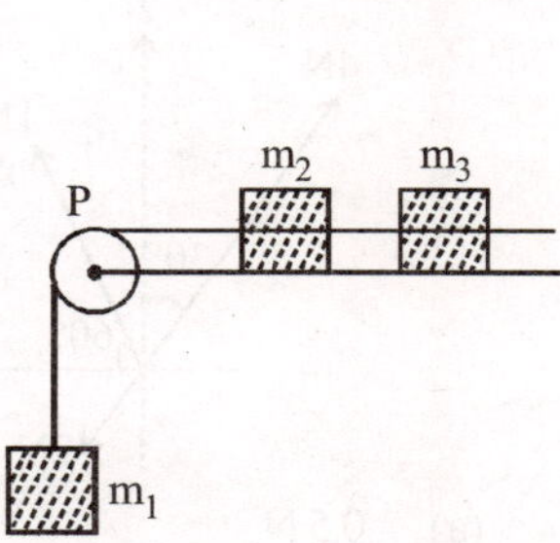

25. A balloon with mass 'm' is descending down with an acceleration 'a' (where a < g). How much mass should be removed from it so that it starts moving up with an acceleration 'a'? *[2014]*

(a) $\frac{2ma}{g+a}$ (b) $\frac{2ma}{g-a}$

(c) $\frac{ma}{g+a}$ (d) $\frac{ma}{g-a}$

26. Three blocks with masses m, 2 m and 3 m are connected by strings as shown in the figure. After an upward force F is applied on block m, the masses move upward at constant speed v. What is the net force on the block of mass 2m? (g is the acceleration due to gravity) *[2013]*

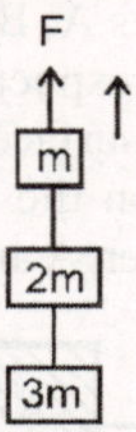

(a) 2 mg

(b) 3 mg

(c) 6 mg

(d) zero

27. A person of mass 60 kg is inside a lift of mass 940 kg and presses the button on control panel. The lift starts moving upwards with an acceleration 1.0 m/s^2. If $g = 10 \text{ ms}^{-2}$, the tension in the supporting cable is *[2011]*

(a) 8600 N (b) 9680 N

(c) 11000 N (d) 1200 N

28. Three forces acting on a body are shown in the figure. To have the resultant force only along the y- direction, the magnitude of the minimum additional force needed is: *[2008]*

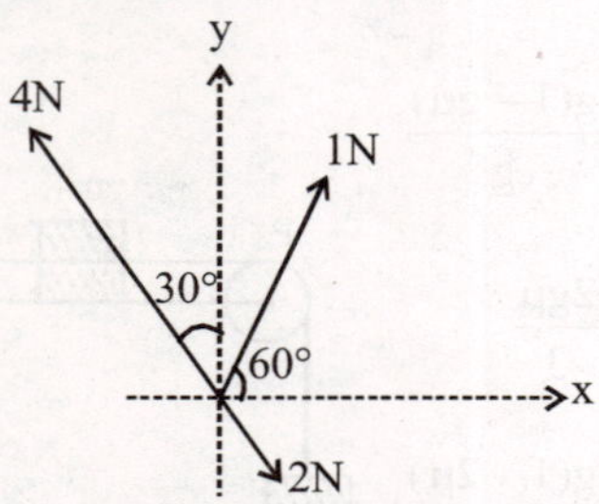

(a) 0.5 N (b) 1.5 N

(c) $\frac{\sqrt{3}}{4}$ N (d) $\sqrt{3}$N

29. The mass of a lift is 2000 kg. When the tension in the supporting cable is 28000 N, then its acceleration is: *[2009]*

(a) 4 ms^{-2} upwards

(b) 4 ms^{-2} downwards

(c) 14 ms^{-2} upwards

(d) 30 ms^{-2} downwards

30. The coefficient of static friction, μ_s, between block A of mass 2 kg and the table as shown in the figure is 0.2. What would be the maximum mass value of block B so that the two blocks do not move? The string and the pulley are assumed to be smooth and massless. ($g = 10 \text{ m/s}^2$) *[2004]*

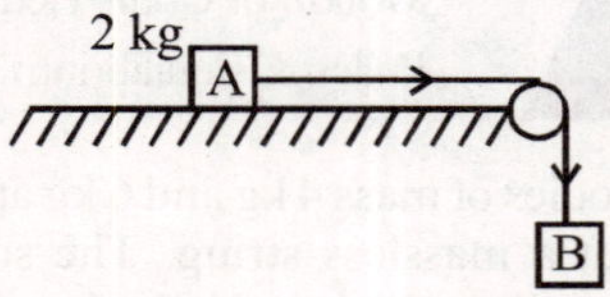

(a) 0.4 kg (b) 2.0 kg

(c) 4.0 kg (d) 0.2 kg

31. A block of mass *m* is placed on a smooth wedge of inclination θ. The whole system is accelerated horizontally so that the block does not slip on the wedge. The force exerted by the wedge on the block (g is acceleration due to gravity) will be *[2004]*

(a) mg/cos θ (b) mg cos θ

(c) mg sin θ (d) mg

32. A man weighing 80 kg, stands on a weighing scale in a lift which is moving upwards with a uniform acceleration of 5m/s^2. What would be the reading on the scale? ($g = 10 \text{ m/s}^2$) *[2003]*

(a) 1200 N (b) zero

(c) 400 N (d) 800 N

33. A monkey of mass 20 kg is holding a vertical rope. The rope will not break when a mass of 25 kg is suspended from it but will break if the mass exceeds 25 kg. What is the maximum acceleration with which the monkey can climb up along the rope? ($g = 10 \text{ m/s}^2$) *[2003]*

(a) 2.5 m/s^2 (b) 5 m/s^2

(c) 10 m/s^2 (d) 25 m/s^2

34. A lift weighing 1000 kg is moving upwards with an accelertion of 1 m/s^2. The tension in the supporting cable is *[2002]*

(a) 980 N (b) 10800 N

(c) 9800 N (d) 8800 N

35. Two blocks $m_1 = 5$ gm and $m_2 = 10$ gm are hung vertically over a light frictionless pulley as shown here. What is the acceleration of the masses when they are left free? *[2000]*

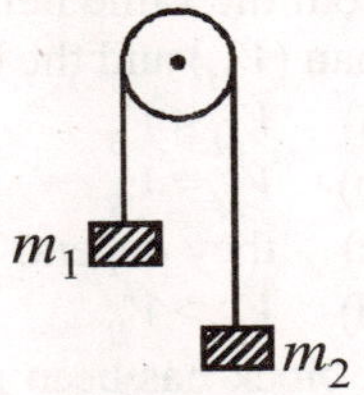

(a) g / 3

(b) g / 2

(c) g

(d) g / 5

(where g is acceleration due to gravity)

36. A mass of 1 kg is suspended by a thread. It is
(i) lifted up with an acceleration 4.9 m/s^2,
(ii) lowered with an acceleration 4.9 m/s^2.
The ratio of the tensions is *[1998]*

(a) 3 : 1 (b) 1 : 2
(c) 1 : 3 (d) 2 : 1

37. A monkey is decending from the branch of a tree with constant acceleration. If the breaking strength is 75% of the weight of the monkey, the minimum acceleration with which monkey can slide down without breaking the branch is *[1993]*

(a) g (b) $\frac{3g}{4}$
(c) $\frac{g}{4}$ (d) $\frac{g}{2}$

Topic 3 Friction

38. A body of mass m is kept on a rough horizontal surface (coefficient of friction = $\propto$). A horizontal force is applied on the body, but it does not move. The resultant of normal reaction and the frictional force acting on the object is given by F, where F is, *[NEET Odisha 2019]*

(a) $|\vec{F}| = mg$ (b) $|\vec{F}| = mg + \propto mg$
(c) $|\vec{F}| = \propto mg$ (d) $|\vec{F}| \le mg\sqrt{1+\mu^2}$

39. Which one of the following statements is incorrect? *[2018]*

(a) Rolling friction is smaller than sliding friction.
(b) Limiting value of static friction is directly proportional to normal reaction.
(c) Coefficient of sliding friction has dimensions of length.
(d) Frictional force opposes the relative motion.

40. A block of mass m is placed on a smooth inclined wedge ABC of inclination θ as shown in the figure. The wedge is given an acceleration 'a' towards the right. The relation between a and θ for the block to remain stationary on the wedge is *[2018]*

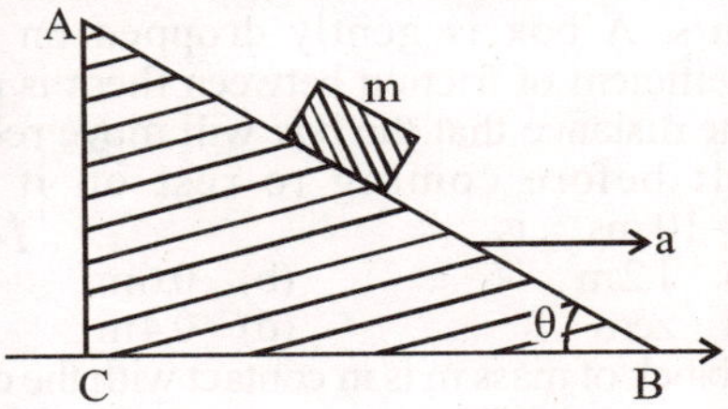

(a) $a = \frac{g}{\text{cosec}\,\theta}$ (b) $a = \frac{g}{\sin\theta}$
(c) $a = g\tan\theta$ (d) $a = g\cos\theta$

41. A block A of mass m_1 rests on a horizontal table. A light string connected to it passes over a frictionless pulley at the edge of table and from its other end another block B of mass m_2 is suspended. The coefficient of kinetic friction between the block and the table is μ_k. When the block A is sliding on the table, the tension in the string is *[2015]*

(a) $\frac{(m_2 - \mu k\, m_1)g}{(m_1 + m_2)}$ (b) $\frac{m_1 m_2 (1+\mu_k)g}{(m_1 + m_2)}$

(c) $\frac{m_1 m_2 (1-\mu_k)g}{(m_1 + m_2)}$ (d) $\frac{(m_2 + \mu_k\, m_1)g}{(m_1 + m_2)}$

42. A plank with a box on it at one end is gradually raised about the other end. As the angle of inclination with the horizontal reaches 30° the box starts to slip and slides 4.0 m down the plank in 4.0s. The coefficients of static and kinetic friction between the box and the plank will be, respectively : *[2015 RS]*

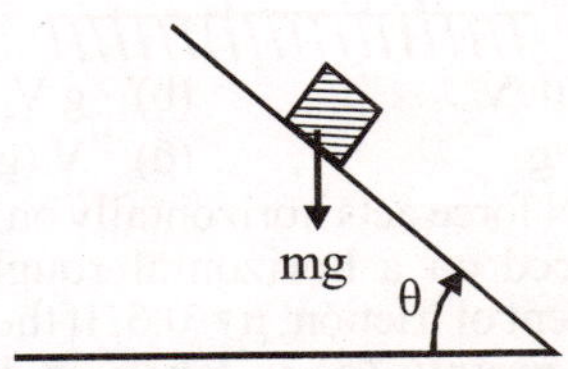

(a) 0.6 and 0.5 (b) 0.5 and 0.6
(c) 0.4 and 0.3 (d) 0.6 and 0.6

43. The upper half of an inclined plane of inclination θ is perfectly smooth while lower half is rough. A block starting from rest at the top of the plane will again come to rest at the bottom, if the coefficient of friction between the block and lower half of the plane is given by *[2013]*

(a) $\mu = \frac{2}{\tan\theta}$ (b) $\mu = 2\tan\theta$

(c) $\mu = \tan\theta$ (d) $\mu = \frac{1}{\tan\theta}$

44. A conveyor belt is moving at a constant speed of 2m/s. A box is gently dropped on it. The coefficient of friction between them is $\mu = 0.5$. The distance that the box will move relative to belt before coming to rest on it taking $g = 10\ ms^{-2}$, is *[2011M]*

(a) 1.2m (b) 0.6m

(c) zero (d) 0.4m

45. A block of mass m is in contact with the cart C as shown in the **Figure**. *[2010]*

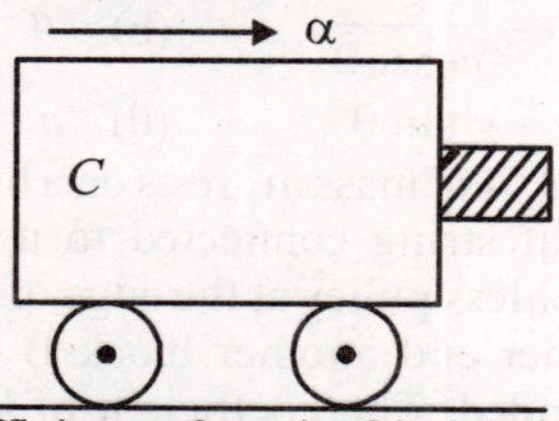

The coefficient of static friction between the block and the cart is μ. The acceleration α of the cart that will prevent the block from falling satisfies:

(a) $\alpha > \frac{mg}{\mu}$ (b) $\alpha > \frac{g}{\mu m}$

(c) $\alpha \geq \frac{g}{\mu}$ (d) $\alpha < \frac{g}{\mu}$

46. A block B is pushed momentarily along a horizontal surface with an initial velocity V. If μ is the coefficient of sliding friction between B and the surface, block B will come to rest after a time *[2007]*

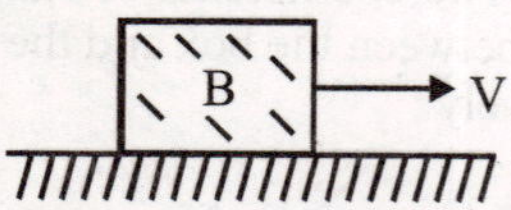

(a) $g\mu/V$ (b) g/V

(c) V/g (d) $V/(g\mu)$.

47. A 100 N force acts horizontally on a block of 10 kg placed on a horizontal rough surface of coefficient of friction $\mu = 0.5$. If the acceleration due to gravity (g) is taken as 10 ms^{-2}, the acceleration of the block (in ms^{-2}) is *[2002]*

(a) 2.5 (b) 10

(c) 5 (d) 7.5

48. A block of mass 1 kg is placed on a truck which accelerates with acceleration $5 m/s^2$. The coefficient of static friction between the block and truck is 0.6. The frictional force acting on the block is *[2001]*

(a) 5N (b) 6N

(c) 5.88N (d) 4.6N

49. A person slides freely down a frictionless inclined plane while his bag falls down vertically from the same height. The final speeds of the man (V_M) and the bag (V_B) should be such that *[2000]*

(a) $V_M < V_B$

(b) $V_M = V_B$

(c) they depend on the masses

(d) $V_M > V_B$

50. A block has been placed on an inclined plane with the slope angle θ, block slides down the plane at constant speed. The coefficient of kinetic friction is equal to *[1993]*

(a) $\sin\theta$ (b) $\cos\theta$

(c) g (d) $\tan\theta$

51. Consider a car moving along a straight horizontal road with a speed of 72 km/h. If the coefficient of static friction between the tyres and the road is 0.5, the shortest distance in which the car can be stopped is (taking $g = 10\ m/s^2$) *[1992]*.

(a) 30m (b) 40m

(c) 72m (d) 20m

52. A heavy uniform chain lies on horizontal table top. If the coefficient of friction between the chain and the table surface is 0.25, then the maximum fraction of the length of the chain that can hang over one edge of the table is **[1991]**

(a) 20% (b) 25%

(c) 35% (d) 15%

53. Starting from rest, a body slides down a 45° inclined plane in twice the time it takes to slide down the same distance in the absence of friction. The coefficient of friction between the body and the inclined plane is *[1988]*

(a) 0.80 (b) 0.75

(c) 0.25 (d) 0.33

Topic 4 Circular Motion, Banking of Road Ird Laws of Motion

54. A block of mass 10 kg is in contact against the inner wall of a hollow cylindrical drum of radius 1 m. The coefficient of friction between the block and the inner wall of the cylinder is 0.1. The minimum angular velocity needed for the cylinder to keep the block stationary when the cylinder is vertical and rotating about its axis, will be : ($g = 10\ m/s^2$) *[2019]*

(a) $\sqrt{10}$ rad/s (b) $\frac{10}{2\pi}$ rad/s

(c) 10 rad/s (d) 10 π rad/s

55. A mass m is attached to a thin wire and whirled in a vertical circle. The wire is most likely to break when: *[2014]*

(a) the mass is at the highest point

(b) the wire is horizontal

(c) the mass is at the lowest point

(d) inclined at an angle of 60° from vertical

56. A car is negotiating a curved road of radius R. The road is banked at an angle θ. the coefficient of friction between the tyres of the car and the road is μ_s. The maximum safe velocity on this road is : *[2016]*

(a) $\sqrt{gR^2\left(\frac{\mu_s+\tan\theta}{1-\mu_s\tan\theta}\right)}$

(b) $\sqrt{gR\left(\frac{\mu_s+\tan\theta}{1-\mu_s\tan\theta}\right)}$

(c) $\sqrt{\frac{g}{R}\left(\frac{\mu_s+\tan\theta}{1-\mu_2\tan\theta}\right)}$

(d) $\sqrt{\frac{g}{R^2}\left(\frac{\mu_s+\tan\theta}{1-\mu_s\tan\theta}\right)}$

57. What is the minimum velocity with which a body of mass m must enter a vertical loop of radius R so that it can complete the loop? *[2016]*

(a) $\sqrt{gR}$ (b) $\sqrt{2gR}$

(c) $\sqrt{3gR}$ (d) $\sqrt{5gR}$

58. Two stones of masses m and 2 m are whirled in horizontal circles, the heavier one in radius $\frac{r}{2}$ and the lighter one in radius r. The tangential speed of lighter stone is n times that of the value of heavier stone when they experience same centripetal forces. The value of n is : *[2015 RS]*

(a) 3 (b) 4

(c) 1 (d) 2

59. A car is moving in a circular horizontal track of radius 10 m with a constant speed of 10 m/s. A bob is suspended from the roof of the car by a light wire of length 1.0 m. The angle made by the wire with the vertical is *[NEET Kar. 2013]*

(a) 0° (b) $\frac{\pi}{3}$

(c) $\frac{\pi}{6}$ (d) $\frac{\pi}{4}$

60. A car of mass 1000 kg negotiates a banked curve of radius 90 m on a frictionless road. If the banking angle is 45°, the speed of the car is : *[2012]*

(a) 20 ms^{-1} (b) 30 ms^{-1}

(c) 5 ms^{-1} (d) 10 ms^{-1}

61. A car of mass m is moving on a level circular track of radius R. If μ_s represents the static friction between the road and tyres of the car, the maximum speed of the car in circular motion is given by : *[2012M]*

(a) $\sqrt{\mu_s mRg}$ (b) $\sqrt{Rg/\mu_s}$

(c) $\sqrt{mRg/\mu_s}$ (d) $\sqrt{\mu_s Rg}$

62. A 500 kg car takes a round turn of radius 50 m with a velocity of 36 km/h. The centripetal force is *[1999]*

(a) 250 N (b) 750 N

(c) 1000 N (d) 1200 N

63. A body of mass 0.4 kg is whirled in a vertical circle making 2 rev/sec. If the radius of the circle is 1.2 m, then tension in the string when the body is at the top of the circle, is *[1999]*

(a) 41.56 N (b) 89.86 N

(c) 109.86 N (d) 115.86 N

64. What will be the maximum speed of a car on a road turn of radius 30 m if the coefficient of friction between the tyres and the road is 0.4 (Take $g = 9.8\ m/s^2$) *[1995]*

(a) 10.84 m/s (b) 9.84 m/s

(c) 8.84 m/s (d) 6.84 m/s

65. A particle of mass M is moving in a horizontal circle of radius R with uniform speed V. When it moves from one point to a diametrically opposite point, its *[1992]*

(a) kinetic energy changes by $MV^2/4$

(b) momentum does not change

(c) momentum changes by 2 MV

(d) kinetic energy changes by MV^2

66. When milk is churned, cream gets separated due to *[1991]*

(a) centripetal force

(b) centrifugal force

(c) frictional force

(d) gravitational force

ANSWER KEY																	
1	(c)	9	(b)	17	(c)	25	(a)	33	(a)	41	(b)	48	(a)	55	(c)	62	(c)
2	(c)	10	(c)	18	(c)	26	(d)	34	(b)	42	(a)	49	(b)	56	(b)	63	(a)
3	(c)	11	(d)	19	(d)	27	(c)	35	(a)	43	(b)	50	(d)	57	(d)	64	(a)
4	(b)	12	(b)	20	(b)	28	(a)	36	(a)	44	(d)	51	(b)	58	(d)	65	(c)
5	(c)	13	(b)	21	(a)	29	(a)	37	(c)	45	(c)	52	(a)	59	(d)	66	(b)
6	(c)	14	(c)	22	(d)	30	(a)	38	(d)	46	(d)	53	(b)	60	(c)		
7	(a)	15	(c)	23	(a)	31	(a)	39	(c)	47	(c)	54	(c)	61	(d)		
8	(a)	16	(b)	24	(c)	32	(a)	40	(c)								

Hints & Solutions

1. **(c)** Impulse, $\vec{I}$ = change in linear momentum

or, $I = \Delta\vec{P} = \vec{P}_f - \vec{P}_i$

or, $I = m_2v_2 - m_1v_1$

Velocity of ball at ground

$(v) = \sqrt{2gh}\left[\because v^2 - u^2 = 2gh\right]$

$= \sqrt{2\times10\times10} = 10\sqrt{2}\ m/s$

$\therefore \vec{I} = 0.15\times10\sqrt{2}\left(-\hat{j}\right) - 0.15\times10\sqrt{2}\left(\hat{j}\right)$

$\vec{I} = 2\times0.15\times10\sqrt{2}\left(-\hat{j}\right) = 4.2\left(\hat{j}\right) = 4.2$ kg m/s

NOTES Impulse $I = \overrightarrow{\Delta P} = \vec{P}_f - \vec{P}_i$ = change in momentum is called impulse momentum theorem. Impulse of a force is a measure of total effect of force.

2. **(c)** As three forces are forming closed loop in same order, so net force is zero.

i.e., $\vec{F}_{net} = 0$

or $m\dfrac{d\vec{v}}{dt} = 0$

$\therefore$ Velocity of the particle

$\vec{v}$ = Constant

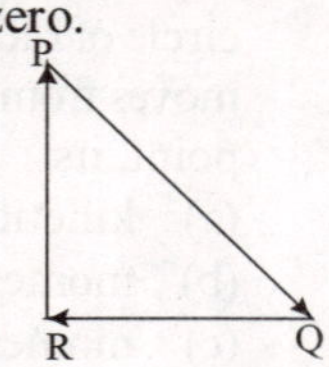

3. **(c)** Change in momentum,

$\Delta p = \int F dt$

= Area of F-t graph

= ar of Δ − *ar* of ▭ + *ar* of ▭

$= \dfrac{1}{2}\times2\times6 - 3\times2 + 4\times3 = 12$ N-s

NOTES According to Newton's 2nd law. $F \propto \dfrac{d\vec{p}}{dt} \Rightarrow F$

$= K\dfrac{d\vec{p}}{dt}$ or, $F = \dfrac{d\vec{p}}{dt}$ [$\because k = 1$ in S.I & CGS system]

And $d\vec{p} = \vec{F}dt$

4. **(b)** Momentum $P = mv = m\sqrt{2gh}$

($\because v^2 = u^2 + 2gh$; Here $u = 0$)

When stone hits the ground momentum,

$$P = m\sqrt{2gh}$$

when same stone dropped from $2h$ (100% of initial) then momentum,

$$P' = m\sqrt{2g(2h)} = \sqrt{2}P$$

50% change in momentum, $\dfrac{(P'-P)}{P}\times100\%$

$\Rightarrow \left(\dfrac{\sqrt{2}P-P}{P}\right)\times100\% = \left(\sqrt{2}-1\right)\times100\%$

$\Rightarrow$ (1.414 − 1) × 100%

$\Rightarrow$ 414 × 100% = 41.4%

Which is changed by 41% of initial.

5. **(c)** Impulse experienced by the body

= change in momentum

= MV − (−MV)

= 2MV

6. **(c)** $\vec{F} = 6\hat{i} - 8\hat{j} + 10\hat{k}$,

$|\vec{F}| = \sqrt{36+64+100} = 10\sqrt{2}$ N

$\left(\because F = \sqrt{F_x^2 + F_y^2 + F_z^2}\right)$

Given, $a = 1\ ms^{-2}$

$\because F = ma \Rightarrow m = \dfrac{10\sqrt{2}}{1} = 10\sqrt{2}$ kg

7. (a) $F = \frac{d(Mv)}{dt} = M\frac{dv}{dt} + v\frac{dM}{dt}$

$\therefore$ v is constant,

$\therefore F = v\frac{dM}{dt}$ But $\frac{dM}{dt} = Mkg/s$

$\therefore$ F = vM newton.

8. (a)

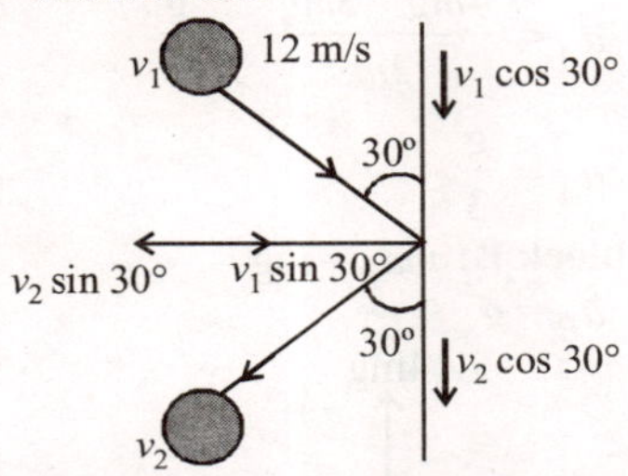

The magnitude of both velocity v_1 and v_2 are equal. Components of velocity along the wall are equal and are in same direction.

But the component of velocity perpendicular to wall are equal but opposite in direction.

So, net change in momentum in perpendicular direction,

$\Delta P = mv_1 \sin 30° - (-mv_2 \sin 30°)$

$\Rightarrow 2mv \sin 30°$ [As $v_1 = v_2 = v$]

Average force acting on the wall will be given by

$\Rightarrow Ft = 2mv \sin 30°$

$\Rightarrow F = \frac{2mv\sin 30°}{t}$

$\Rightarrow F = \frac{2 \times 0.5^2 \times 12 \times 1}{2 \times 0.25} = 24$

$F = 24N$

9. (b) Net force experienced $= \frac{\text{Total Impulse}}{\text{Time taken}}$

$= \frac{m\Delta v}{t} = 0.15 \times \frac{20}{0.1} = 30N$

10. (c) Change in momentum along the wall

$= mv \cos 60° - mv \cos 60° = 0$

Change in momentum perpendicular to the wall

$= mv \sin 60° - (-mv \sin 60°) = 2mv \sin 60°$

$\therefore$ Applied force $= \frac{\text{Change in momentum}}{\text{Time}}$

$= \frac{2\, mv \sin 60°}{0.20}$

$= \frac{2 \times 3 \times 10 \times \sqrt{3}}{2 \times 20} = 50 \times 3\sqrt{3}$

$= 150\sqrt{3}$ newton

11. (d) Given $F = 600 - (2 \times 10^5 t)$

The force is zero at time t, given by

$0 = 600 - 2 \times 10^5 t$

$\Rightarrow t = \frac{600}{2 \times 10^5} = 3 \times 10^{-3}$ seconds

$\therefore$ Impulse $= \int_0^t F dt = \int_0^{3\times10^{-3}} (600 - 2\times10^5 t)\, dt$

$= \left[600t - \frac{2 \times 10^5 t^2}{2}\right]_0^{3\times10^{-3}}$

$= 600 \times 3 \times 10^{-3} - 10^5 (3 \times 10^{-3})^2$

$= 1.8 - 0.9 = 0.9$ Ns

NOTES In case of impulse or motion of a charged particle in an alternating electric field, force is time dependent.

12. (b) Given : Mass of rocket (m) = 5000 kg

Exhaust speed (v) = 800 m/s

Acceleration of rocket (a) = 20 m/s^2

Gravitational acceleration (g) = 10 m/s^2

Thrust, $\Rightarrow \frac{vdm}{dt}$

We know that upward force,

$F = m(g + a) = 5000(10 + 20)$

$= 5000 \times 30 = 150000$ N

This thrust gives upward force, $F = \frac{vdm}{dt}$

We also know that amount of gas ejected

$\Rightarrow \left(\frac{dm}{dt}\right) = \frac{F}{v} = \frac{150000}{800} = 187.5$ kg/s

13. (b) By Newton's IInd law of motion, $F = ma$

$\Rightarrow 10 = m(1) \Rightarrow m = 10$ kg.

14. (c) Mass = 150 gm $= \frac{150}{1000}$ kg

Force = Mass × acceleration

$= \frac{150}{1000} \times 20N = 3N$

Impulsive force $= F \cdot \Delta t = 3 \times 0.1 = 0.3$ N

15. (c) Velocity of the rocket (u) = 300 m/s and force (F) = 345N. According to Newton's second law, thrust (force) = Rate of change of linear momentum.

$\frac{F}{u} = \left(\frac{dm}{dt}\right) = 1.15 kg/sec$

16. **(b)** Thrust on the satellite,

$$F = \frac{-vdM}{dt} = -v(\alpha v) = -\alpha v^2$$

$$\text{Acceleration} = \frac{F}{M} = \frac{-\alpha v^2}{M}$$

17. **(c)** Newton's first law of motion is related to physical independence of force.

> **NOTES** If no net force acts on a body, then the velocity of the body cannot change *i.e.*, the body cannot accelerate.

18. **(c)** Thrust = weight

$$\Rightarrow \frac{udM}{dt} = mg \Rightarrow \frac{dM}{dt} = \frac{mg}{u}$$

$$= \frac{600 \times 10}{1000} = 6\ \text{kg s}^{-1}$$

19. **(d)** Impulse = final momentum − initial momentum = $m(v_2 - v_1)$

20. **(b)** Given : Mass $M_1 = 4$ kg and $M_2 = 6$ kg.

Acceleration of the system,

$$a = \frac{(m_1 - m_2)g}{(m_1 + m_2)} \text{ where } m_1 > m_2$$

$$\therefore a = \frac{(6-4)g}{6+4} = \frac{g}{5}$$

> **NOTES** Here no option is given according to acceleration of COM of the system.

21. **(a)** Before cutting the string

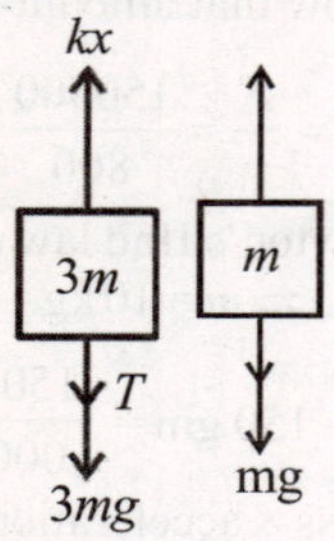

$kx = T + 3\,mg$ ...(i)

$T = mg$ ...(ii)

Force an string,

$\Rightarrow \quad kx = 4mg$

After cutting the string T = 0

For block A: $3ma_A = 4mg - 3mg$

$$a_A = \frac{4mg - 3mg}{3m} = \frac{g}{3}$$

$$a_A = \frac{g}{3}$$

For block B: $ma_B = mg$

$$a_B = g$$

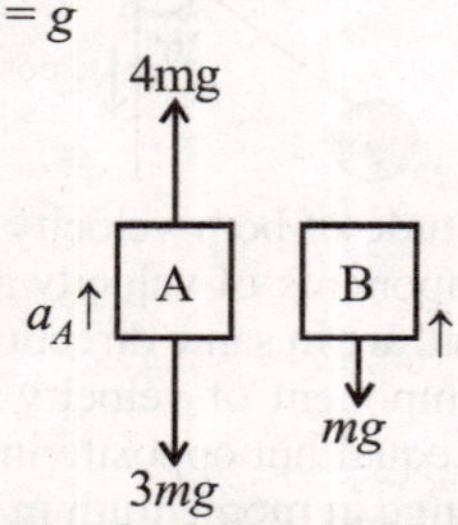

22. **(d)** Net force on particle in uniform circular motion is centripetal force $\left(\frac{mv^2}{\ell}\right)$ which is provided by tension in string so the net force will be equal to tension *i.e.*, T.

23. **(a)** Acceleration of system, $a = \frac{F_{net}}{M_{total}}$

$$= \frac{14}{4+2+1} = \frac{14}{7} = 2\ \text{m/s}^2$$

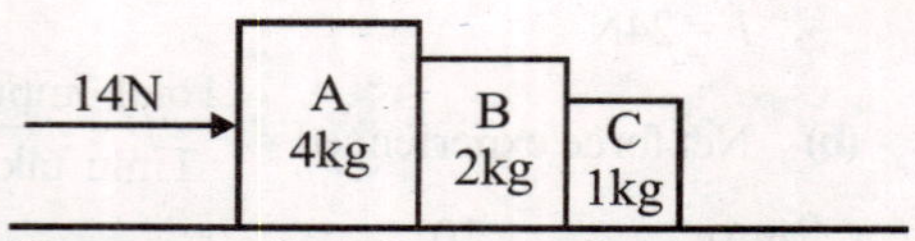

The contact force between A and B

$$= (m_B + m_C) \times a = (2+1) \times 2 = 6\text{N}$$

> **NOTES**

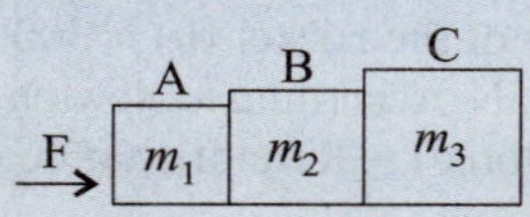

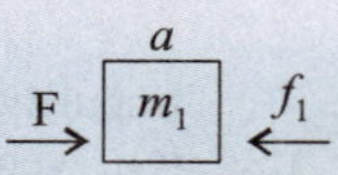	$F - f_1 = m_2 a$	$a = \frac{F}{m_1 + m_2 + m_3}$
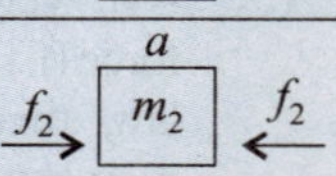	$f_1 - f_2 = m_2 a$	$f_1 = \frac{(m_2 + m_3)F}{m_1 + m_2 + m_3}$
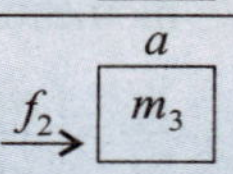	$f_2 = m_2 a$	$f_2 = \frac{m_3 F}{m_1 + m_2 + m_3}$

24. **(c)** Acceleration

$$= \frac{\text{Net force in the direction of motion}}{\text{Total mass of system}}$$

$$= \frac{m_1g - \mu(m_2 + m_3)g}{(m_1 + m_2 + m_3)} = \frac{g}{3}(1 - 2\mu)$$

$(\because m_1 = m_2 = m_3 = m \text{ given})$

25. **(a)** Let upthrust of air be F_a then

For downward motion of balloon,

$F_a = mg - ma \Rightarrow Fa = m(g - a)$ ...(i)

For upward motion,

$F_a - (m - \Delta m)g = (m - \Delta m)a$ ...(ii)

From equation (i) and (ii),

Therefore, $\Delta m = \dfrac{2ma}{(g + a)}$

26. **(d)**

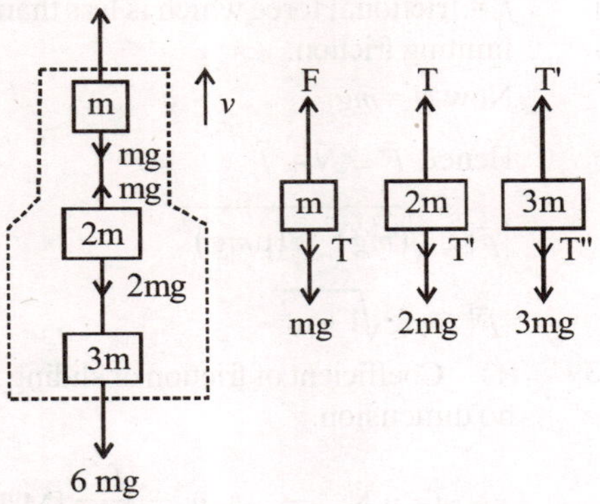

From figure

$F = 6$ mg,

As speed is constant, acceleration $a = 0$

$\therefore$ 6 mg = 6ma = 0, F = 6 mg

$\therefore$ T = 5 mg, T' = 3 mg

T'' = 0

F_{net} on block of mass 2 m

= T – T' – 2 mg = 0

NOTES: As all blocks are moving with constant speed, then acceleration is zero, so force is zero.

$\because v$ = constant

so, $a = 0$, Hence, $F_{net} = ma = 0$

27. **(c)**

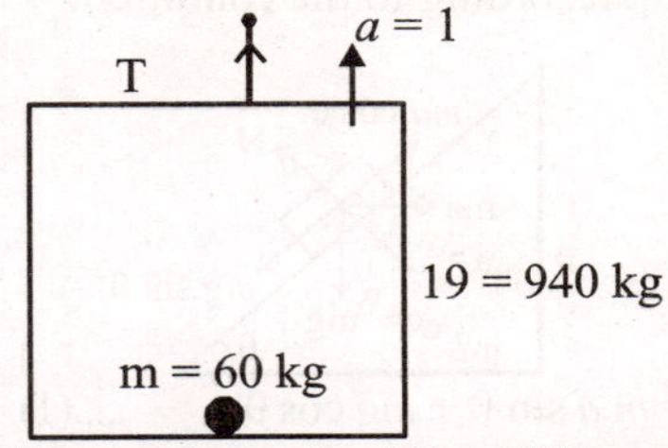

Total mass = (m + m) = (60 + 940) kg = 1000 kg

Let T be the tension in the supporting cable, then

T – 1000g = 1000 × 1 [$\because a = 1$ m/sec]

$\Rightarrow$ T = 11000 N

28. **(a)** The components of 1 N and 2N forces along + x-axis = 1 cos 60° + 2 sin 30°

$$= 1 \times \frac{1}{2} + 2 \times \frac{1}{2} = \frac{1}{2} + 1 = \frac{3}{2} = 1.5\text{N}$$

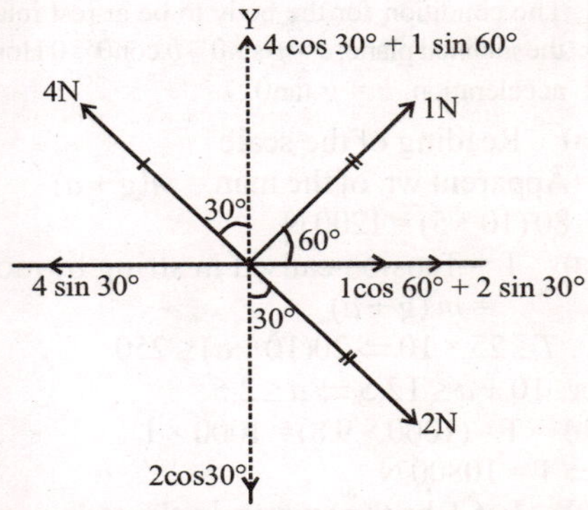

The component of 4 N force along –x-axis

$= 4 \sin 30° = 4 \times \dfrac{1}{2} = 2\text{N}$.

Therefore, if a force of 0.5N is applied along + x-axis, the resultant force along x-axis will become zero and the resultant force will be obtained only along y-axis.

29. **(a)** Net force, F = T – mg

ma = T – mg

2000 a = 28000 – 20000 = 8000

$$a = \frac{8000}{2000} = 4\,\text{ms}^{-2} \uparrow$$

30. **(a)** Condition for the max value of mass of block B so that two blocks do not move is,

$m_Bg = \mu_s m_Ag$

$\Rightarrow m_B = \mu_s m_A$

or, $m_B = 0.2 \times 2 = 0.4$ kg

31. **(a)** According to the condition,

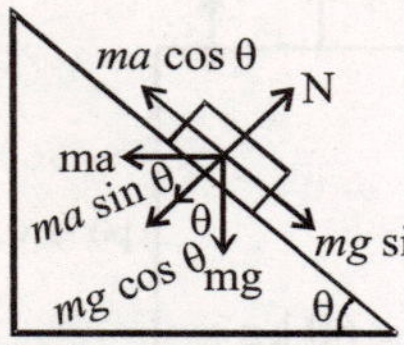

$N = m\,a\sin\theta + mg\cos\theta$ .....(1)

Also, $m\,g\sin\theta = m\,a\cos\theta$ ....(2)

From (1) & (2), $a = g\tan\theta$

$\therefore\ N = mg\dfrac{\sin^2\theta}{\cos\theta} + mg\cos\theta\,.$

$= \dfrac{mg}{\cos\theta}(\sin^2\theta + \cos^2\theta) = \dfrac{mg}{\cos\theta}$

or, $N = \dfrac{mg}{\cos\theta}$

NOTES The condition for the body to be at rest relative to the inclined plane, $a = g\sin\theta - b\,\mathrm{con}\theta = 0$ Horizontal acceleration, $b = g\tan\theta$.

32. **(a)** Reading of the scale

$=$ Apparent wt. of the man $= m(g + a)$

$= 80\,(10 + 5) = 1200$ N

33. **(a)** T = Tension caused in string by monkey

$= m\,(g + a)$

$\therefore\ T \le 25 \times 10 \Rightarrow 20(10 + a) \le 250$

or, $10 + a \le 12.5 \Rightarrow a \le 2.5$

34. **(b)** $T - (1000 \times 9.8) = 1000 \times 1$

$\Rightarrow T = 10800$ N

35. **(a)** Let T be the tension in the string.

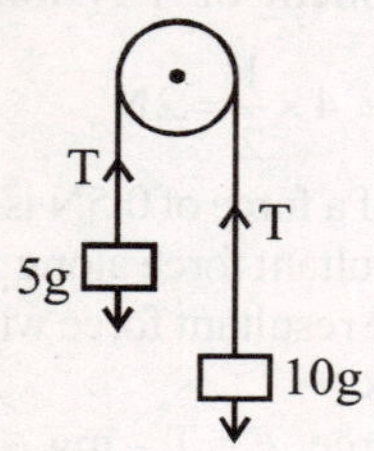

$\therefore\ 10g - T = 10a$ ....(i)

$T - 5g = 5a$ ....(ii)

Adding (i) and (ii),

$5g = 15a \Rightarrow a = \dfrac{g}{3}$ m/s^2

NOTES Whe pulley have a finite mass m then acceleration,

$$a = \frac{m_1 - m_2}{m_1 + m_2 + \dfrac{m}{2}}$$

36. **(a)** In case (i) we have

$T_1 - (1 \times g) = 1 \times 4.9$

$\Rightarrow T_1 = 9.8 + 4.9 = 14.7$ N

In case (ii), $1 \times g - T_2 = 1 \times 4.9$

$\Rightarrow T_2 = 9.8 - 4.9 = 4.9$ N

$\therefore\ \dfrac{T_1}{T_2} = \dfrac{14.7}{4.9} = \dfrac{3}{1}$

37. **(c)** Let T be the tension in the branch of a tree when monkey is descending with acceleration a. Then mg – $T = ma$; and $T = 75\%$ of weight of monkey,

$\therefore\ ma = mg - \left(\dfrac{75}{100}\right)mg = \left(\dfrac{1}{4}\right)mg$

or $a = \dfrac{g}{4}$.

38. **(d)** Since body does not move hence it is in equilibrium.

f_r = frictional force which is less than or equal to limiting friction.

Now $N = mg$

Hence $\vec{F} = \vec{N} + \vec{f}_r$

$|\vec{F}| \le \sqrt{(mg)^2 + (\mu mg)^2}$

$|\vec{F}| \le mg\sqrt{1 + \mu^2}$

39. **(c)** Coefficient of friction or sliding friction has no dimension.

$f = \mu_s N \ \Rightarrow\ \mu_s = \dfrac{f}{N} = [M^0L^0T^0]$

40. **(c)** Let the mass of block is m. It will remains stationary if forces acting on it are in equilibrium. *i.e.*, ma cos θ = mg sin θ ⇒ a = g tan θ

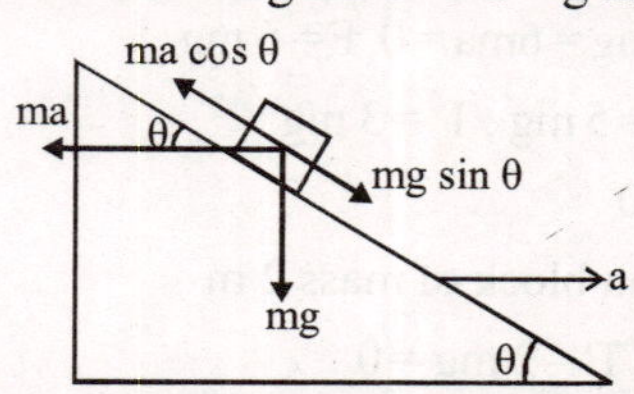

Here ma = Pseudo force on block, mg = weight.

41. **(b)** For the motion of both the blocks

$m_1a = T - \mu_k m_1 g$

$m_2 g - T = m_2 a$

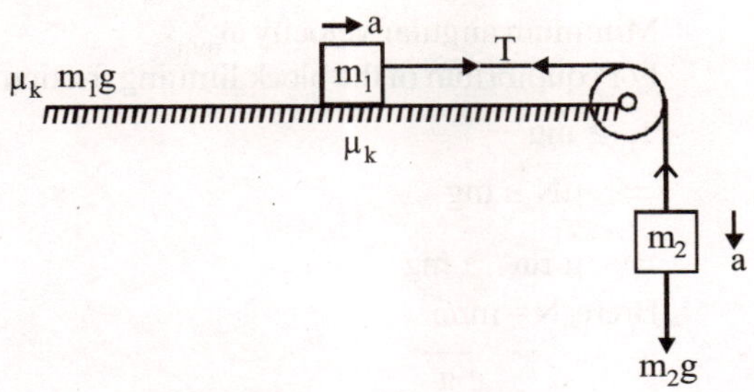

$$a = \frac{m_2g - \mu_k\, m_1g}{m_1 + m_2}$$

$$m_2g - T = (m_2)\left(\frac{m_2g - \mu_k\, m_1g}{m_1 + m_2}\right)$$

solving we get tension in the string

$$T = \frac{m_1m_2\,(1 + \mu_k)\,g}{m_1 + m_2}$$

42. **(a)** Coefficient of static friction, $\mu_s = \tan\theta$

$$\Rightarrow \quad \mu_s = \tan 30° = \frac{1}{\sqrt{3}} = 0.577 \cong 0.6$$

So, distance covered by plank,

$$S = ut + \frac{1}{2}at^2$$

where, $u = 0$ and $a = g\sin\theta - \mu_k(g)\cos\theta$

$$4 = \frac{1}{2}a(4)^2 \Rightarrow a = \frac{1}{2} = 0.5$$

$[\because\ \theta = 30°,\ s = 4\text{m and } t = 4\text{s given}]$

$$\Rightarrow \mu_k = \frac{0.9}{\sqrt{3}} = 0.5$$

43. **(b)**

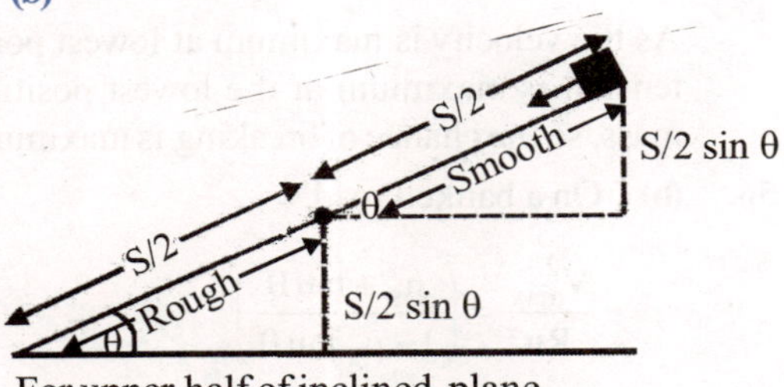

For upper half of inclined plane

$v^2 = u^2 + 2a\,S/2 = 2\,(g\sin\theta)\,S/2 = gS\sin\theta$

For lower half of inclined plane

$0 = u^2 + 2\,g\,(\sin\theta - \mu\cos\theta)\,S/2$

$\Rightarrow \quad -gS\sin\theta = gS\,(\sin\theta - \mu\cos\theta)$

$\Rightarrow \quad 2\sin\theta = \mu\cos\theta$

$$\Rightarrow \quad \mu = \frac{2\sin\theta}{\cos\theta} = 2\tan\theta$$

44. **(d)** Frictional force on the box $f = \mu mg$

$\therefore$ Acceleration in the box, $a = \dfrac{f}{m} = \dfrac{\mu mg}{m}$

$a = \mu g = 5\ \text{ms}^{-2}$

$v^2 = u^2 + 2as$

$\Rightarrow \quad 0 = 2^2 + 2 \times (5)\,s$

$\Rightarrow \quad s = -\dfrac{2}{5}$ w.r.t. belt

$\Rightarrow$ distance $= 0.4$ m

45. **(c)** Forces acting on the block are as shown in the fig. Normal reaction N equals to ma

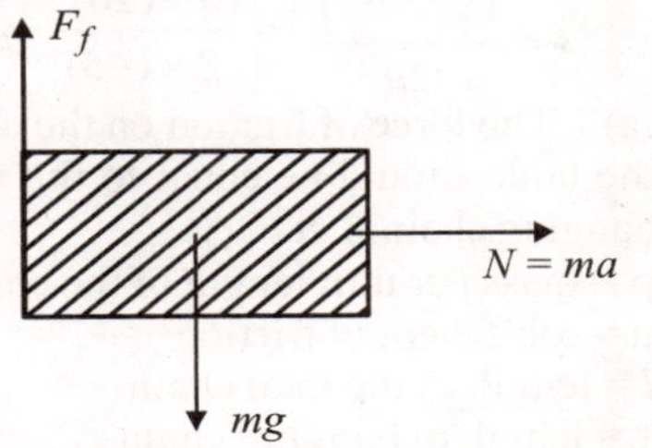

$\therefore \quad N = ma$

For the block not to fall, frictional force, $F_f \geq mg$

$\Rightarrow \quad \mu N \geq mg$

$\Rightarrow \quad \mu ma \geq mg$

$\Rightarrow \quad a \geq g/\mu$

NOTES When a cart moves with some acceleration towards right then a psendo force (ma) acts on block towards left. This force ma is action force by a block on cart.

46. **(d)** Friction is the retarding force for the block

$F = ma = \mu R = \mu mg$

Therefore, from the first equation of motion

$v = u - at$

$$0 = V - \mu g \times t \Rightarrow \frac{V}{\mu g} = t$$

47. **(c)** $a = \dfrac{F - \mu R}{m} = \dfrac{100 - 0.5 \times (10 \times 10)}{10} = 5\ \text{ms}^{-2}$

48. **(a)** Maximum friction force $= \mu mg$

$= .6 \times 1 \times 9.8 = 5.88$ N

But here required friction force

$= ma = 1 \times 5 = 5$ N

NOTES For a body placed on a moving body and Ist body not to slide - applied force $F <$ limiting friction ($= Ms\ mg$) combined system will move together with common acceleration $a_1 = a_2 = \dfrac{F}{M + m}$.

49. (b) As there is only gravitational field which works. We know it is conservative field and depends only on the end points. So, $V_M = V_B$

50. (d) When the block slides down the plane with a constant speed, then the inclination of the plane is equal to angle of repose (θ).
Coeff. of friction = tan of the angle of repose
$= \tan\theta$.

51. (b) Here u = 72 km/h = 20 m/s; v = 0;
$a = -\mu g = -0.5 \times 10 = -5\ m/s^2$
As $v^2 = u^2 + 2as$,

$$\therefore\ s = \frac{(v^2 - u^2)}{2a} = \frac{(0-(20)^2}{2\times(-5)} = 40\,m$$

52. (a) The force of friction on the chain lying on the table should be equal to the weight of the hanging chain. Let
ρ = mass per unit length of the chain
μ = coefficient of friction
l = length of the total chain
x = length of hanging chain
Now, $\mu(l - x)\rho g = x\rho g$ or $\mu(l - x) = x$
or $\mu l = (\mu + 1)x$ or $x = \mu l/(\mu + 1)$

$$\therefore\ x = \frac{0.25l}{(0.25+1)} = \frac{0.25l}{1.25} = 0.2l$$

$$\frac{x}{l} = 0.2 \Rightarrow \frac{x}{\ell}\times 100 = 20\%$$

53. (b) In presence of friction a = (g sinθ – μg cos θ)
∴ Time taken to slide down the plane

$$t_1 = \sqrt{\frac{2s}{a}} = \sqrt{\frac{2s}{g(\sin\theta - \mu\cos\theta)}}$$

In absence of friction, $t_2 = \sqrt{\frac{2s}{g\sin\theta}}$

According to the condition,

$$t_1 = 2t_2 \quad \therefore\ t_1^2 = 4t_2^2$$

$$\text{or}\ ,\ \frac{2s}{g(\sin\theta - \mu\cos\theta)} = \frac{2s\times 4}{g\sin\theta}$$

$$\sin\theta = 4\sin\theta - 4\mu\cos\theta$$

$$\mu = \frac{3}{4}\tan\theta = \frac{3}{4} = 0.75$$

54. (c) Given mass of block, m = 10 kg; radius of cylindrical drum, r = 1m; coefficient of friction between the block and the inner wall of the cylinder μ = 0.1;

Minimum angular velocity ω_{min}
For equilibrium of the block limiting friction
$f_L \geq mg$

$\Rightarrow\ \mu N \geq mg$

$\Rightarrow\ \mu r\omega^2 \geq mg$
Hrere, $N = mr\omega^2$

$$\text{or,}\quad m \geq \sqrt{\frac{g}{r\mu}}$$

$$\text{or,}\quad \omega_{min} = \sqrt{\frac{g}{r\mu}}$$

$$\therefore\ \omega_{min} = \sqrt{\frac{10}{0.1\times 1}} = 10\ rad/s$$

55. (c) $T - mg = \frac{mv^2}{r}$ [centripetal force] $= \frac{mv^2}{r}$

$$\Rightarrow\ T = mg + \frac{mv^2}{r}$$

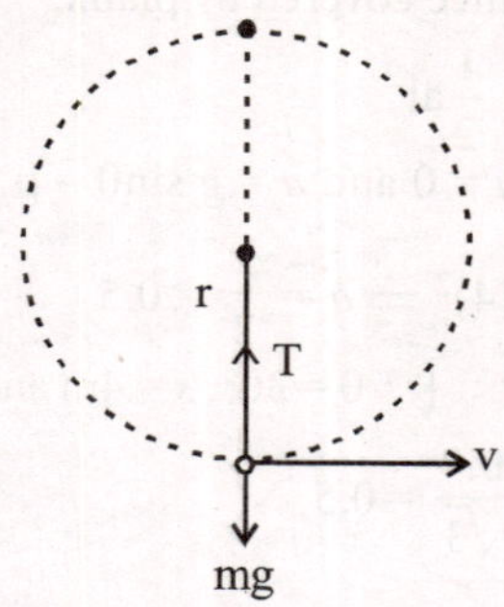

As the velocity is maximum at lowest point so tension is maximum at the lowest position of mass, so the chance of breaking is maximum.

56. (b) On a banked road,

$$\frac{V_{max}^2}{Rg} = \left(\frac{\mu_s + \tan\theta}{1 - \mu_s\tan\theta}\right)$$

Maximum safe velocity of a car on the banked road

$$V_{max} = \sqrt{Rg\left[\frac{\mu_s + \tan\theta}{1 - \mu_s\tan\theta}\right]}$$

57. (d) $\sqrt{5gR}$ is the minimum velocity which the body must possess at the bottom of the circle so as to go round the circle completely.

58. **(d)** According to question, two stones experience same centripetal force

i.e., $F_{C_1} = F_{C_2}$

or, $\frac{mv_1^2}{r} = \frac{2mv_2^2}{(r/2)}$ or, $V_1^2 = 4V_2^2$

So, $V_1 = 2V_2$ *i.e.*, $n = 2$

59. **(d)** Given; speed = 10 m/s; radius $r = 10$ m

Angle made by the wire with the vertical is θ.

Then, $T\cos\theta = mg$

$$T\sin\theta = \frac{mv^2}{r}$$

So,

$$\tan\theta = \frac{v^2}{rg} = \frac{10^2}{10\times10} = 1$$

$$\Rightarrow \theta = 45^\circ = \frac{\pi}{4}$$

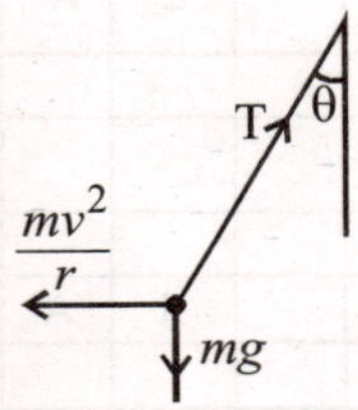

60. **(c)** For banking, $\tan\theta = \frac{V^2}{Rg}$

$$\tan 45^\circ = \frac{V^2}{90\times10} = 1$$

$V = 30$ m/s

> **NOTES** If friction is present then maximum safe speed, on a banked frictional road, $V = \sqrt{\frac{Rg(\mu + \tan\theta)}{1 - \mu\tan\theta}}$

61. **(d)** For smooth driving maximum speed of car v then,

$$\frac{mv^2}{R} = \mu_s mg$$

$$v = \sqrt{\mu_s Rg}$$

62. **(c)** Centripetal force $= \frac{mv^2}{r} = \frac{500\times(10)^2}{50}$

$= 1000$ N [∵ 36 km/hr = 10 m/s]

63. **(a)** Given : Mass $(m) = 0.4$ kg

Its frequency $(n) = 2$ rev/sec

Radius $(r) = 1.2$ m. We know that linear velocity of the body $(v) = \omega r = (2\pi n)r$

$= 2\times3.14\times2\times1.2 = 15.08$ m/s.

Therefore, tension in the string when the body is at the top of the circle (T)

$$= \frac{mv^2}{r} - mg = \frac{0.4\times(15.08)^2}{2} - (0.4\times9.8)$$

$= 45.78 - 3.92 = 41.56$ N

64. **(a)** $r = 30\,m$ and $\mu = 0.4$.

$$v_{max} = \sqrt{\mu rg} = \sqrt{0.4\times30\times9.8} = 10.84 \text{ m/s}$$

65. **(c)** On the diametrically opposite points, the velocities have same magnitude but opposite directions. Therefore, change in momentum is MV − (−MV) = 2MV

66. **(b)** Cream gets separated from a churned milk due to centrifugal force.

> **NOTES** Centrifugal force is a pseudo force that is equal and opposite to the centripetal force. Centripetal force can be mechanical, electrical or magnetic force.

5 Work, Energy and Power

Trend Analysis with Important Topics & Sub-Topics

Topic Name	Sub-Topic	2021 QNS.	2021 LOD	2020 QNS	2020 LOD	2019 QNS.	2019 LOD	2018 QNS.	2018 LOD	2017 QNS.	2017 LOD	2016 QNS.	2016 LOD
Work	Work done, $w = \int_{y_1}^{y} F \cdot dy$					1	A						
	Work-energy Theorem									1	A		
Energy	Conservation of Mechanical energy	1	E									1	A
Power	Power, P = F.v = w/t	1	E									1	E
Collisions	Elastic Collision					1	A						
	Coefficient of Restitution							1	A				

LOD - Level of Difficulty | E - Easy | A - Average | D - Difficult | Qns - No. of Questions

Topic 1 Work

1. A force F = 20 + 10*y* acts on a particle in *y*-direction where F is in newton and *y* in meter. Work done by this force to move the particle from *y* = 0 to *y* = 1 m is : *[2019]*
(a) 30 J (b) 5 J (c) 25 J (d) 20 J

2. An object of mass 500 g, initially at rest, is acted upon by a variable force whose X-component varies with *X* in the manner shown. The velocities of the object at the points *X* = 8 m and *X* = 12 m, would have the respective values of (nearly) *[NEET Odisha 2019]*

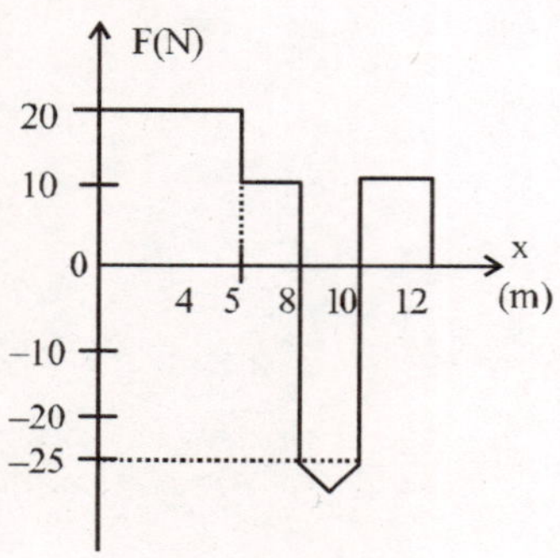

(a) 18 m/s and 20.6 m/s
(b) 18 m/s and 24.4 m/s
(c) 23 m/s and 24.4 m/s
(d) 23 m/s and 20.6 m/s

3. Consider a drop of rain water having mass 1 g falling from a height of 1 km. It hits the ground with a speed of 50 m/s. Take 'g' constant with a value 10 m/s^2. The work done by the (i) gravitational force and the (ii) resistive force of air is *[2017]*
(a) (i) 1.25 J (ii) –8.25 J
(b) (i) 100 J (ii) 8.75 J
(c) (i) 10 J (ii) – 8.75 J
(d) (i) – 10 J (ii) –8.25 J

4. Two similar springs P and Q have spring constants K_P and K_Q, such that $K_P > K_Q$. They are stretched, first by the same amount (case a,) then by the same force (case b). The work done by the springs W_P and W_Q are related as, in case (a) and case (b), respectively *[2015]*
(a) $W_P = W_Q$; $W_P = W_Q$
(b) $W_P > W_Q$; $W_Q > W_P$
(c) $W_P < W_Q$; $W_Q < W_P$
(d) $W_P = W_Q$; $W_P > W_Q$

5. A uniform force of $(3\hat{i}+\hat{j})$ newton acts on a particle of mass 2 kg. The particle is displaced from position $(2\hat{i}+\hat{k})$ meter to position $(4\hat{i}+3\hat{j}-\hat{k})$ meter. The work done by the force on the particle is *[2013]*
(a) 6 J (b) 13 J
(c) 15 J (d) 9 J

6. Force F on a particle moving in a straight line varies with distance d as shown in the figure. The work done on the particle during its displacement of 12 m is *[2011]*

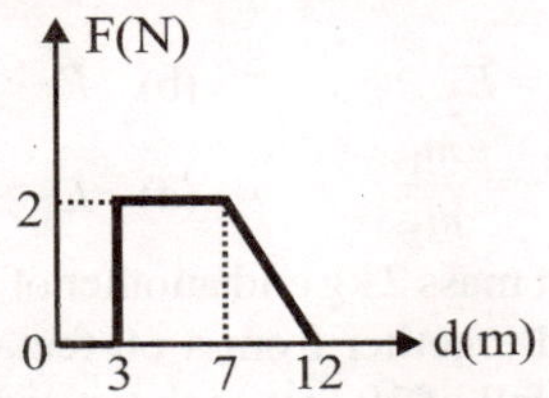

(a) 18 J (b) 21 J
(c) 26 J (d) 13 J

7. A vertical spring with force constant k is fixed on a table. A ball of mass m at a height h above the free upper end of the spring falls vertically on the spring so that the spring is compressed by a distance d. The net work done in the process is *[2007]*
(a) $mg(h+d)-\frac{1}{2}kd^2$ (b) $mg(h-d)-\frac{1}{2}kd^2$
(c) $mg(h-d)+\frac{1}{2}kd^2$ (d) $mg(h+d)+\frac{1}{2}kd^2$

8. A body of mass 3 kg is under a constant force which causes a displacement s in metres in it, given by the relation $s=\frac{1}{3}t^2$, where t is in seconds. Work done by the force in 2 seconds is *[2006]*
(a) $\frac{3}{8}$ J (b) $\frac{8}{3}$ J (c) $\frac{19}{5}$ J (d) $\frac{5}{19}$ J

9. 300 J of work is done in sliding a 2 kg block up an inclined plane of height 10 m. Taking g = 10 m/s^2, work done against friction is *[2006]*
(a) 100 J (b) zero
(c) 1000 J (d) 200 J

10. A force F acting on an object varies with distance x as shown here. The force is in N and x in m. The work done by the force in moving the object from $x = 0$ to $x = 6$ m is *[2005]*

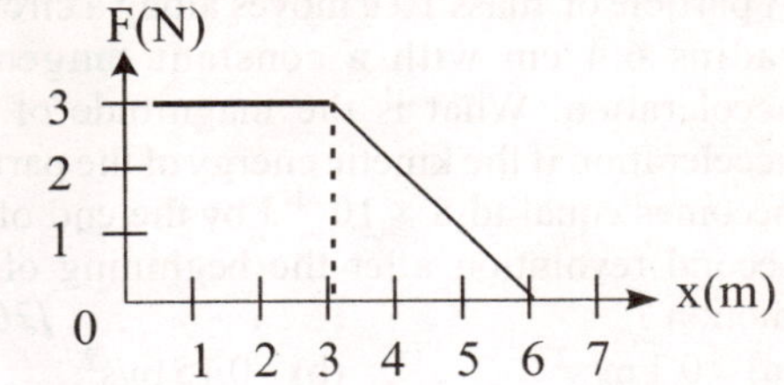

(a) 18.0 J (b) 13.5 J
(c) 9.0 J (d) 4.5 J

11. A force of 250 N is required to lift a 75 kg mass through a pulley system. In order to lift the mass through 3 m, the rope has to be pulled through 12m. The efficiency of system is *[2001]*
(a) 50% (b) 75%
(c) 33% (d) 90%

12. A force acts on a 30 gm particle in such a way that the position of the particle as a function of time is given by $x = 3t - 4t^2 + t^3$, where x is in metres and t is in seconds. The work done during the first 4 seconds is *[1998]*
(a) 576mJ (b) 450mJ
(c) 490mJ (d) 530mJ

13. A position dependent force, $F = (7 - 2x + 3x^2)$ N acts on a small body of mass 2 kg and displaces it from x = 0 to $x = 5\ m$. Work done in joule is *[1992]*
(a) 35 (b) 70
(c) 135 (d) 270

14. A bullet of mass 10g leaves a rifle at an initial velocity of 1000 m/s and strikes the earth at the same level with a velocity of 500 m/s. The work done in joules overcoming the resistance of air will be *[1989]*
(a) 375 (b) 3750
(c) 5000 (d) 500

Topic 2 Energy

15. A particle is released from height S from the surface of the Earth. At a certain height its kinetic energy is three times its potential energy. The height from the surface of earth and the speed of the particle at that instant are respectively *[2021]*
(a) $\frac{S}{4}, \sqrt{\frac{3gS}{2}}$ (b) $\frac{S}{4}, \frac{3gS}{2}$
(c) $\frac{S}{4}, \frac{\sqrt{3gS}}{2}$ (d) $\frac{S}{2}, \frac{\sqrt{3gS}}{2}$

16. A particle of mass 10 g moves along a circle of radius 6.4 cm with a constant tangential acceleration. What is the magnitude of this acceleration if the kinetic energy of the particle becomes equal to 8×10^{-4} J by the end of the second revolution after the beginning of the motion ? *[2016]*

(a) $0.1\ m/s^2$ (b) $0.15\ m/s^2$

(c) $0.18\ m/s^2$ (d) $0.2\ m/s^2$

17. A block of mass 10 kg, moving in x direction with a constant speed of $10\ ms^{-1}$, is subject to a retarding force $F = 0.1x$ J/m during its travel from $x = 20$ m to 30 m. Its final KE will be: *[2015]*

(a) 450 J (b) 275 J

(c) 250 J (d) 475 J

18. A person holding a rifle (mass of person and rifle together is 100 kg) stands on a smooth surface and fires 10 shots horizontally, in 5 s. Each bullet has a mass of 10 g with a muzzle velocity of $800\ ms^{-1}$. The final velocity acquired by the person and the average force exerted on the person are *[NEET Kar. 2013]*

(a) $-1.6\ ms^{-1}$; 8 N (b) $-0.08\ ms^{-1}$; 16 N

(c) $-0.8\ ms^{-1}$; 8 N (d) $-1.6\ ms^{-1}$; 16 N

19. A particle with total energy E is moving in a potential energy region $U(x)$. Motion of the particle is restricted to the region when *[NEET Kar. 2013]*

(a) $U(x) > E$ (b) $U(x) < E$

(c) $U(x) = O$ (d) $U(x) \leq E$

20. The potential energy of a system increases if work is done *[2011]*

(a) upon the system by a non conservative force

(b) by the system against a conservative force

(c) by the system against a non conservative force

(d) upon the system by a conservative force

21. An engine pumps water continuously through a hose. Water leaves the hose with a velocity v and m is the mass per unit length of the water jet. What is the rate at which kinetic energy is imparted to water? *[2009]*

(a) mv^2 (b) $\frac{1}{2}mv^2$

(c) $\frac{1}{2}m^2v^2$ (d) $\frac{1}{2}mv^3$

22. A body of mass 1 kg is thrown upwards with a velocity 20 m/s. It momentarily comes to rest after attaining a height of 18 m. How much energy is lost due to air friction? (g = 10 m/s^2) *[2009]*

(a) 30 J (b) 40 J

(c) 10 J (d) 20 J

23. A particle of mass m_1 is moving with a velocity v_1 and another particle of mass m_2 is moving with a velocity v_2. Both of them have the same momentum but their different kinetic energies are E_1 and E_2 respectively. If $m_1 > m_2$ then *[2004]*

(a) $E_1 = E_2$ (b) $E_1 < E_2$

(c) $\frac{E_1}{E_2} = \frac{m_1}{m_2}$ (d) $E_1 > E_2$

24. A ball of mass 2 kg and another of mass 4 kg are dropped together from a 60 feet tall building. After a fall of 30 feet each towards earth, their respective kinetic energies will be in the ratio of *[2004]*

(a) $1 : \sqrt{2}$ (b) $\sqrt{2} : 1$

(c) 1 : 4 (d) 1 : 2

25. When a long spring is stretched by 2 cm, its potential energy is U. If the spring is stretched by 10 cm, the potential energy stored in it will be *[2003]*

(a) 25U (b) U/5

(c) 5U (d) 10U

26. If the kinetic energy of a particle is increased by 300%, the momentum of the particle will increase by *[2002]*

(a) 20% (b) 200%

(c) 100% (d) 50%

27. In a simple pendulum of length l the bob is pulled aside from its equilibrium position through an angle θ and then released. The bob passes through the equilibrium position with speed *[2000]*

(a) $\sqrt{2g\ell(1+\cos\theta)}$ (b) $\sqrt{2g\ell\sin\theta}$

(c) $\sqrt{2g\ell}$ (d) $\sqrt{2g\ell(1-\cos\theta)}$

28. Two bodies with kinetic energies in the ratio 4:1 are moving with equal linear momentum. The ratio of their masses is *[1999]*

(a) 1 : 2 (b) 1 : 1

(c) 4 : 1 (d) 1 : 4

29. A rubber ball is dropped from a height of 5m on a plane, where the acceleration due to gravity is not shown. On bouncing it rises to 1.8m. The ball loses its velocity on bouncing by a factor of [1998]

(a) $\frac{16}{25}$ (b) $\frac{2}{5}$

(c) $\frac{3}{5}$ (d) $\frac{9}{25}$

30. Two bodies of masses m and 4 m are moving with equal K.E. The ratio of their linear momenta is [1997]

(a) 4 : 1 (b) 1 : 1

(c) 1 : 2 (d) 1 : 4

31. The kinetic energy acquired by a mass (m) in travelling distance (s) starting from rest under the action of a constant force is directly proportional to [1996, 1994]

(a) $1/\sqrt{m}$ (b) 1/m

(c) $\sqrt{m}$ (d) m^0

32. If the momentum of a body is increased by 50%, then the percentage increase in its kinetic energy is [1995]

(a) 50% (b) 100%

(c) 125% (d) 200%

33. Consider a car moving along a straight horizantal road with a speed of 72 km/h. If the coefficient of static friction between road and tyres is 0.5, the shortest distance in which the car can be stopped is [1994]

(a) 30m (b) 40m

(c) 72m (d) 20m

34. Two masses of 1g and 9g are moving with equal kinetic energies. The ratio of the magnitudes of their respective linear momenta is [1993]

(a) 1 : 9 (b) 9 : 1

(c) 1 : 3 (d) 3 : 1

35. A 4 kg mass and 1 kg are moving with equal kinetic energies. The ratio of the magnitudes of their linear momenta is [1989]

(a) 1 : 2 (b) 1 : 1

(c) 2 : 1 (d) 4 : 1

36. Two bodies of masses m and 4m are moving with equal kinetic energies. The ratio of their linear momenta will be [1988]

(a) 1 : 4 (b) 4 : 1

(c) 1 : 2 (d) 2 : 1

Topic 3 Power

37. Water falls from a height of 60 m at the rate of 15 kg/s to operate a turbine. The losses due to frictional force are 10% of the input energy. How much power is generated by the turbine? ($g = 10\ m/s^2$) [2021]

(a) 7.0 kW (b) 10.2 kW

(c) 8.1 kW (d) 12.3 kW

38. A body of mass 1 kg begins to move under the action of a time dependent force $\vec{F} = (2t\hat{i} + 3t^2\hat{j})$ N, where $\hat{i}$ and $\hat{j}$ are unit vectors alogn *x* and *y*-axis. What power will be developed by the force at the time *t*? [2016]

(a) $(2t^2 + 3t^3)$W (b) $(2t^2 + 4t^4)$W

(c) $(2t^3 + 3t^4)$ W (d) $(2t^3 + 3t^5)$W

39. A particle of mass m is driven by a machine that delivers a constant power of *k* watts. If the particle starts from rest the force on the particle at time *t* is [2015]

(a) $\sqrt{mk}\ t^{-1/2}$ (b) $\sqrt{2mk}\ t^{-1/2}$

(c) $\frac{1}{2}\sqrt{mk}\ t^{-1/2}$ (d) $\sqrt{\frac{mk}{2}}t^{-1/2}$

40. The heart of man pumps 5 litres of blood through the arteries per minute at a pressure of 150 mm of mercury. If the density of mercury be 13.6 $\times 10^3\ kg/m^3$ and $g = 10 m/s^2$ then the power of heart in watt is : [2015 RS]

(a) 2.35 (b) 3.0

(c) 1.50 (d) 1.70

41. One coolie takes 1 minute to raise a suitcase through a height of 2 m but the second coolie takes 30 s to raise the same suitcase to the same height. The powers of two coolies are in the ratio of [NEET Kar. 2013]

(a) 1 : 2 (b) 1 : 3

(c) 2 : 1 (d) 3 : 1

42. A car of mass *m* starts from rest and accelerates so that the instantaneous power delivered to the car has a constant magnitude P_0. The instantaneous velocity of this car is proportional to : [2012M]

(a) $t^2 P_0$ (b) $t^{1/2}$

(c) $t^{-1/2}$ (d) $\frac{t}{\sqrt{m}}$

43. A body projected vertically from the earth reaches a height equal to earth's radius before returning to the earth. The power exerted by the gravitational force is greatest *[2011]*
(a) at the highest position of the body
(b) at the instant just before the body hits the earth
(c) it remains constant all through
(d) at the instant just after the body is projected

44. An engine pumps water through a hose pipe. Water passes through the pipe and leaves it with a velocity of 2 m/s. The mass per unit length of water in the pipe is 100 kg/m. What is the power of the engine? *[2010]*
(a) 400 *W* (b) 200 *W*
(c) 100 *W* (d) 800 *W*

45. Water falls from a height of 60 m at the rate of 15 kg/s to operate a turbine. The losses due to frictional force are 10% of energy. How much power is generated by the turbine? *[2008]* ($g = 10\ m/s^2$)
(a) 8.1 kW (b) 10.2 kW
(c) 12.3 kW (d) 7.0 kW

46. How much water, a pump of 2 kW can raise in one minute to a height of 10 m, take g = 10 m/s^2? *[1990]*
(a) 1000 litres (b) 1200 litres
(c) 100 litres (d) 2000 litres

Topic 4 Collisions

47. Body A of mass 4m moving with speed *u* collides with another body B of mass 2m, at rest. The collision is head on and elastic in nature. After the collision the fraction of energy lost by the colliding body A is : *[2019]*
(a) $\frac{1}{9}$ (b) $\frac{8}{9}$
(c) $\frac{4}{9}$ (d) $\frac{5}{9}$

48. An object flying in air with velocity $(20\hat{i} + 25\hat{j} - 12\hat{k})$ suddenly breaks into two pieces whose masses are in the ratio 1 : 5. The smaller mass flies off with a velocity $(100\hat{i} + 35\hat{j} + 8\hat{k})$. The velocity of the larger piece will be *[NEET Odisha 2019]*
(a) $-20\hat{i} - 15\hat{j} - 80\hat{k}$ (b) $4\hat{i} + 23\hat{j} - 16\hat{k}$
(c) $-100\hat{i} - 35\hat{j} - 8\hat{k}$ (d) $20\hat{i} + 15\hat{j} - 80\hat{k}$

49. Two bullets are fired horizontally and simultaneously towards each other from roof tops of two buildings 100 m apart of same height of 200 m, with the same velocity of 25 m/s. When and where will the two bullets collide? (g = 10 m/s^2) *[NEET Odisha 2019]*
(a) They will not collide
(b) After 2 s at a height of 180 m
(c) After 2 s at a height of 20 m
(d) After 4 s at a height of 120 m

50. A particle of mass 5 m at rest suddenly breaks on its own into three fragments. Two fragments of mass m each move along mutually perpendicular direction with speed *v* each. The energy released during the process is, *[NEET Odisha 2019]*
(a) $\frac{4}{3}mv^2$ (b) $\frac{3}{5}mv^2$
(c) $\frac{5}{3}mv^2$ (d) $\frac{3}{2}mv^2$

51. A moving block having mass m, collides with another stationary block having mass 4m. The lighter block comes to rest after collision. When the initial velocity of the lighter block is *v*, then the value of coefficient of restitution (e) will be *[2018]*
(a) 0.5 (b) 0.25
(c) 0.4 (d) 0.8

52. Two particles A and B, move with constant velocities $\vec{v}_1$ and $\vec{v}_2$. At the initial moment their position vectors are $\vec{r}_1$ and $\vec{r}_2$ respectively. The condition for particles A and B for their collision is: *[2015 RS]*
(a) $\vec{r}_1 \cdot \vec{v}_1 = \vec{r}_2 \cdot \vec{v}_2$
(b) $\vec{r}_1 \times \vec{v}_1 = \vec{r}_2 \times \vec{v}_2$
(c) $\vec{r}_1 - \vec{r}_2 = \vec{v}_1 - \vec{v}_2$
(d) $\frac{\vec{r}_1 - \vec{r}_2}{|\vec{r}_1 - \vec{r}_2|} = \frac{\vec{v}_2 - \vec{v}_1}{|\vec{v}_2 - \vec{v}_1|}$

53. On a frictionless surface a block of mass M moving at speed *v* collides elastically with another block of same mass M which is initially at rest. After collision the first block moves at an angle θ to its initial direction and has a speed $\frac{v}{3}$. The second block's speed after the collision is : *[2015 RS]*

(a) $\frac{3}{4}v$ (b) $\frac{3}{\sqrt{2}}v$

(c) $\frac{\sqrt{3}}{2}v$ (d) $\frac{2\sqrt{2}}{3}v$

54. A ball is thrown vertically downwards from a height of 20 m with an initial velocity v_0. It collides with the ground loses 50 percent of its energy in collision and rebounds to the same height. The initial velocity v_0 is : *[2015 RS]* (Take $g = 10\ ms^{-2}$)

(a) $20\ ms^{-1}$ (b) $28\ ms^{-1}$

(c) $10\ ms^{-1}$ (d) $14\ ms^{-1}$

55. A body of mass (4m) is lying in *x-y* plane at rest. It suddenly explodes into three pieces. Two pieces, each of mass (m) move perpendicular to each other with equal speeds (*v*). The total kinetic energy generated due to explosion is : *[2014]*

(a) mv^2 (b) $\frac{3}{2}mv^2$

(c) $2\ mv^2$ (d) $4\ mv^2$

56. An explosion breaks a rock into three parts in a horizontal plane. Two of them go off at right angles to each other. The first part of mass 1 kg moves with a speed of $12\ ms^{-1}$ and the second part of mass 2 kg moves with speed $8\ ms^{-1}$. If the third part flies off with speed $4\ ms^{-1}$ then its mass is *[2013]*

(a) 5 kg (b) 7 kg

(c) 17 kg (d) 3 kg

57. A solid cylinder of mass 3 kg is rolling on a horizontal surface with velocity $4\ ms^{-1}$. It collides with a horizontal spring of force constant 200 Nm^{-1}. The maximum compression produced in the spring will be : *[2012]*

(a) 0.5 m (b) 0.6 m

(c) 0.7 m (d) 0.2 m

58. Two spheres *A* and *B* of masses m_1 and m_2 respectively collide. *A* is at rest initially and *B* is moving with velocity *v* along *x*-axis. After collision *B* has a velocity $\frac{v}{2}$ in a direction perpendicular to the original direction. The mass *A* moves after collision in the direction. *[2012]*

(a) Same as that of *B*

(b) Opposite to that of *B*

(c) $\theta = \tan^{-1}(1/2)$ to the *x*-axis

(d) $\theta = \tan^{-1}(-1/2)$ to the *x*-axis

59. A mass m moving horizontally (along the *x*-axis) with velocity v collides and sticks to mass of 3m moving vertically upward (along the *y*-axis) with velocity 2*v*. The final velocity of the combination is *[2011M]*

(a) $\frac{1}{4}v\hat{i} + \frac{3}{2}v\hat{j}$ (b) $\frac{1}{3}v\hat{i} + \frac{2}{3}v\hat{j}$

(c) $\frac{2}{3}v\hat{i} + \frac{1}{3}v\hat{j}$ (d) $\frac{3}{2}v\hat{i} + \frac{1}{4}v\hat{j}$

60. A ball moving with velocity 2 m/s collides head on with another stationary ball of double the mass. If the coefficient of restitution is 0.5, then their velocities (in m/s) after collision will be: *[2010]*

(a) 0, 1 (b) 1, 1

(c) 1, 0.5 (d) 0, 2

61. An explosion blows a rock into three parts. Two parts go off at right angles to each other. These two are, 1 kg first part moving with a velocity of $12\ ms^{-1}$ and 2 kg second part moving with a velocity of $8\ ms^{-1}$. If the third part flies off with a velocity of $4\ ms^{-1}$, its mass would be: *[2009]*

(a) 7 kg (b) 17 kg

(c) 3 kg (d) 5 kg

62. A shell of mass 200 gm is ejected from a gun of mass 4 kg by an explosion that generates 1.05 kJ of energy. The initial velocity of the shell is: *[2008]*

(a) $100\ ms^{-1}$ (b) $80\ ms^{-1}$

(c) $40\ ms^{-1}$ (d) $120\ ms^{-1}$

63. A bomb of mass 30 kg at rest explodes into two pieces of masses 18 kg and 12 kg. The velocity of 18 kg mass is $6\ ms^{-1}$. The kinetic energy of the other mass is *[2005]*

(a) 324 J (b) 486 J

(c) 256 J (d) 524 J

64. A mass of 0.5 kg moving with a speed of 1.5 m/s on a horizontal smooth surface, collides with a nearly weightless spring of force constant $k = 50$ N/m. The maximum compression of the spring would be *[2004]*

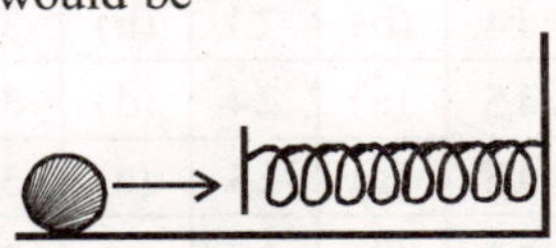

(a) 0.5 m (b) 0.15 m

(c) 0.12 m (d) 1.5 m

65. A stationary particle explodes into two particles of masses m_1 and m_2 which move in opposite directions with velocities v_1 and v_2. The ratio of their kinetic energies E_1/E_2 is *[2003]*

(a) m_1v_2/m_2v_1 (b) m_2/m_1
(c) m_1/m_2 (d) 1

66. A bomb of mass 1 kg is thrown vertically upwards with a speed of 100 m/s. After 5 seconds it explodes into two fragments. One fragment of mass 400 gm is found to go down with a speed of 25 m/s. What will happen to the second fragment just after the explosion? ($g = 10\,m/s^2$) *[2000]*

(a) It will go upward with speed 40 m/s
(b) It will go upward with speed 100 m/s
(c) It will go upward with speed 60 m/s
(d) It will also go downward with speed 40m/s

67. Two equal masses m_1 and m_2 moving along the same straight line with velocities + 3 m/s and – 5m/s respectively, collide elastically. Their velocities after the collision will be respectively. *[1998]*

(a) –3 m/s & +5 m/s (b) +4 m/s for both
(c) –4 m/s & +4 m/s (d) –5m/s & +3 m/s

68. A molecule of mass m of an ideal gas collides with the wall of a vessel with a velocity v and returns back with the same velocity. The change in linear momentum of molecule is *[1997]*

(a) 2 mv (b) 4 mv
(c) 8 mv (d) 10 mv

69. A metal ball of mass 2 kg moving with a velocity of 36 km/h has a head on collision with a stationary ball of mass 3 kg. If after the collision, the two balls move together, the loss in kinetic energy due to collision is *[1997]*

(a) 140 J (b) 100 J
(c) 60 J (d) 40 J

70. A body of mass m moving with velocity 3 km/h collides with a body of mass 2 m at rest. Now the coalesced mass starts to move with a velocity *[1996]*

(a) 1 km/h (b) 2 km/h
(c) 3 km/h (d) 4 km/h

71. A shell is fired from a cannon, it explodes in mid air, its total *[1994]*

(a) momentum increases
(b) momentum decreases
(c) K.E. increases
(d) K.E. decreases

72. Two identical balls A and B moving with velocities +0.5 m/s and –0.3 m/s respectively, collide head on elastically. The velocities of the balls *A* and *B* after collision, will be, respectively *[1991]*

(a) +0.5 m/s and +0.3 m/s
(b) – 0.3 m/s and +0.5 m/s
(c) +0.3 m/s and 0.5 m/s
(d) –0.5 m/s and +0.3 m/s

73. A body of mass 5 kg explodes at rest into three fragments with masses in the ratio 1 : 1 : 3. The fragments with equal masses fly in mutually perpendicular directions with speeds of 21 m/s. The velocity of heaviest fragment in m/s will be *[1989]*

(a) $7\sqrt{2}$ (b) $5\sqrt{2}$
(c) $3\sqrt{2}$ (d) $\sqrt{2}$

74. The co-efficient of restitution e for a perfectly elastic collision is *[1988]*

(a) 1 (b) 0
(c) ∞ (d) –1

ANSWER KEY																	
1	(c)	10	(b)	19	(b)	28	(d)	37	(c)	46	(b)	55	(b)	64	(b)	73	(a)
2	(d)	11	(b)	20	(d)	29	(b)	38	(d)	47	(b)	56	(a)	65	(b)	74	(a)
3	(c)	12	(a)	21	(d)	30	(c)	39	(d)	48	(b)	57	(b)	66	(b)		
4	(b)	13	(c)	22	(d)	31	(d)	40	(d)	49	(b)	58	(d)	67	(d)		
5	(d)	14	(b)	23	(b)	32	(c)	41	(a)	50	(a)	59	(a)	68	(a)		
6	(d)	15	(a)	24	(d)	33	(b)	42	(b)	51	(b)	60	(a)	69	(c)		
7	(a)	16	(a)	25	(a)	34	(c)	43	(b)	52	(d)	61	(d)	70	(a)		
8	(b)	17	(d)	26	(c)	35	(c)	44	(d)	53	(d)	62	(a)	71	(c)		
9	(a)	18	(c)	27	(d)	36	(c)	45	(a)	54	(a)	63	(b)	72	(b)		

Hints & Solutions

1. **(c)** Work done by variable force

Work done, $W = \int_{y_i}^{y_f} F dy \Rightarrow \int_{y=0}^{y_f=1} F \cdot dy$

where, $F = 20 + 10y$

$\therefore W = \int_0^1 (20+10y)dy = \left[20y + \frac{10y^2}{2}\right]_0^1 = 25\ J$

2. **(d)** Using work-energy theorem, ΔK = work = area under $F - x$ graph

From $x = 0$ to $x = 8$ m

$\frac{1}{2}mv^2 = 100 + 30$

$\therefore v = \sqrt{520} = 23$ m/s

From $x = 0$ to $x = 12$ m

$\frac{1}{2}mv^2 = 100 + 30 - 47.5 + 20$

$\therefore v = \sqrt{410} = 20.6$ m/s

3. **(c)** From work-energy theorem,

$W_g + W_a = \Delta K.E.$

or, $mgh + W_a = \frac{1}{2}mv^2 - 0$

$10^{-3} \times 10 \times 10^3 + W_a = \frac{1}{2} \times 10^{-3} \times (50)^2$

$\Rightarrow W_a = -8.75$ J

which is the work done due to air resistance

Work done due to gravity = mgh

$= 10^{-3} \times 10 \times 10^3 = 10$ J

4. **(b)** Case (a): Suppose the two springs are stretched by the same distance x. Then

$$\frac{W_P}{W_Q} = \frac{\frac{1}{2}k_P x^2}{\frac{1}{2}k_Q x^2} = \frac{k_P}{k_Q}$$

As $k_p > k_Q$ so, $W_p > W_Q$

Case (b): Suppose the two springs stretched by distance x_P and x_Q by the same force F.

Then, $F = k_p x_p = k_Q x_Q$

$$\frac{W_p}{W_Q} = \frac{\frac{1}{2}k_P x_P^2}{\frac{1}{2}k_Q x_Q^2} = \frac{k_P \cdot x_P \cdot x_P}{k_Q \cdot x_Q \cdot x_Q} = \frac{Fx_P}{Fx_Q} = \frac{k_Q}{k_p}$$

As $k_P > k_Q$ $\therefore$ $W_Q > W_P$

5. **(d)** Given : $\vec{F} = 3\hat{i} + \hat{j}$

$\vec{r}_1 = (2\hat{i} + \hat{k})$, $\vec{r}_2 = (4\hat{i} + 3\hat{j} - \hat{k})$

$\vec{r} = \vec{r}_2 - \vec{r}_1 = (4\hat{i} + 3\hat{j} - \hat{k}) - (2\hat{i} + \hat{k})$

or $\vec{r} = 2\hat{i} + 3\hat{j} - 2\hat{k}$

So work done by the given force, $w = \vec{f} \cdot \vec{r}$

$= (3\hat{i} + \hat{j}) \cdot (2\hat{i} + 3\hat{j} - 2\hat{k}) = 6 + 3 = 9$J

6. **(d)** Work done = area under F-d graph

$= [2 \times (7-3)] + \left[\frac{1}{2} \times 2 \times (12-7)\right]$

$= 8 + 5$

$= 13$ J.

7. **(a)**

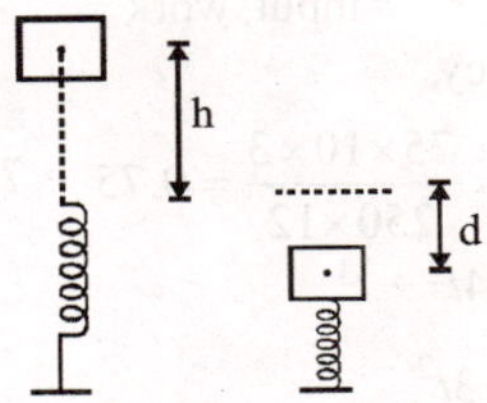

W = Potential energy stored in the spring + Loss of potential energy of mass

$W = mg(h + d) - \frac{1}{2}kd^2$

NOTES

Gravitational potential energy of ball gets converted into elastic potential energy of the spring.

$mg(h + d) = \frac{1}{2}kd^2$

Net work done = $mg(h + d) - \frac{1}{2}kd^2 = 0$

8. **(b)** Acceleration $= \frac{d^2s}{dt^2} = \frac{2}{3}$ m/s^2

Force acting on the body

$= 3 \times \frac{2}{3} = 2$ newton

Displacement in 2 secs $= \frac{1}{3} \times 2 \times 2 = \frac{4}{3}$m

Work done $= 2 \times \frac{4}{3} = \frac{8}{3}$ J

9. (a) Work done against gravity = mg sin θ × *d*

$(d \sin\theta = 10)$

$= 2 \times 10 \times 10$

$= 200$ J

Actual work done = 300 J

Work done against friction = 300 − 200 = 100 J

10. (b)

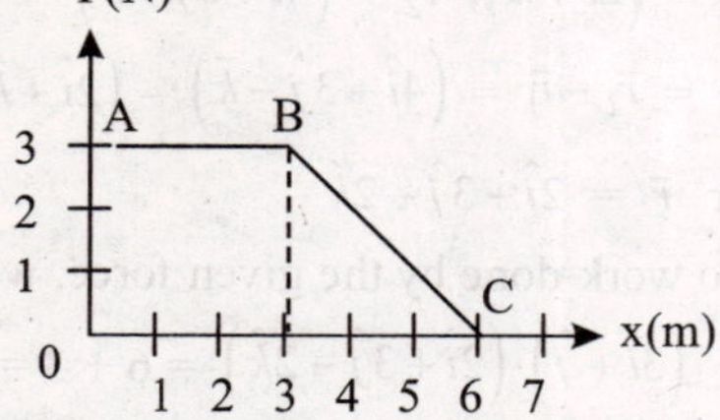

Work done = area under F-*x* graph

= area of trapezium OABC

$= \frac{1}{2}(3+6)(3) = 13.5$ J

11. (b) $\text{Efficiency} = \frac{\text{output work}}{\text{input work}}$

i.e., Efficiency,

$= \frac{75g \times 3}{250 \times 12} = \frac{75 \times 10 \times 3}{250 \times 12} = 0.75 = 75\%$

12. (a) $x = 3t - 4t^2 + t^3$

$\frac{dx}{dt} = 3 - 8t + 3t^2$

Acceleration $= \frac{d^2x}{dt^2} = -8 + 6t$

Acceleration after 4 sec

$= -8 + 6 \times 4 = 16 \text{ ms}^{-2}$

Displacement in 4 sec

$= 3 \times 4 - 4 \times 4^2 + 4^3 = 12$ m

∴ Work = Force × displacement

= Mass × acc. × disp.

$= 3 \times 10^{-3} \times 16 \times 12 = 576$ mJ

13. (c) $W = \int_0^5 F dx = \int_0^5 (7 - 2x + 3x^2) dx$

$= [7x - x^2 + x^3]_0^5 = 135 J$

14. (b) $W = \Delta E = \frac{1}{2} m(v_1^2 - v_2^2)$

$= \frac{1}{2} \times 0.01[(1000)^2 - (500)^2] = 3750\text{J}.$

15. (a) Let the height from the surface of the earth be x.

When body from rest falls through height (S − x)

Then from $v^2 = u^2 + 2gs$

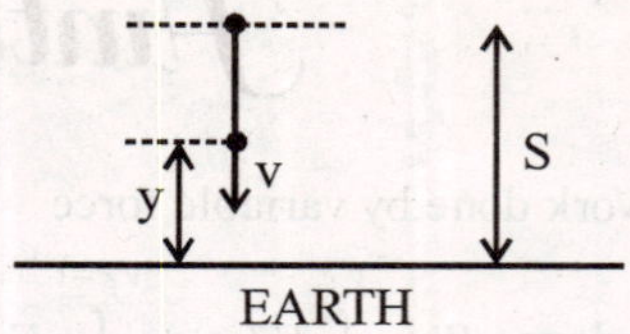

$v^2 = 0^2 + 2g(S - x)$

$v = \sqrt{2g(S-x)}$ ...(i)

At this point, U = mgx

From question, kinetic energy, K = 3U

$\therefore \frac{1}{2} m(v)^2 = 3 \times mg(x)$

$\frac{1}{2} \times m \times 2g(S - x) = 3 \times mgx$ (using (i))

$S - x = 3x$

$\therefore x = \frac{S}{4}$ ... (ii)

$\therefore v = \sqrt{2 \times g\left(S - \frac{S}{4}\right)} = \sqrt{\frac{3gS}{2}}$ ... (iii)

16. (a) Given: Mass of particle, $M = 10\text{g} = \frac{10}{1000}$ kg

radius of circle R = 6.4 cm

Kinetic energy E of particle $= 8 \times 10^{-4}$ J

acceleration $a_t = ?$

$\frac{1}{2}mv^2 = E \Rightarrow \frac{1}{2}\left(\frac{10}{1000}\right)v^2 = 8 \times 10^{-4}$

$\Rightarrow v^2 = 16 \times 10^{-2}$

$\Rightarrow v = 4 \times 10^{-1} = 0.4$ m/s

Now using

$v^2 = u^2 + 2a_t s$ $(s = 4\pi R)$

$(0.4)^2 = 0^2 + 2a_t\left(4 \times \frac{22}{7} \times \frac{6.4}{100}\right)$

$\Rightarrow a_t = (0.4)^2 \times \frac{7 \times 100}{8 \times 22 \times 6.4} = 0.1 \text{ m/s}^2$

17. (d) Given, m = 10 kg

$u = 10$ m/sec

Retarding force, F = 0.1*x* J/m

Acceleration, $a = \frac{F}{m}$

$a = \frac{0.1x}{10} = 0.01x$

From, $v^2 = u^2 - 2as$

$$\Rightarrow \quad v^2 = (10)^2 - 2 \times 0.01 \int_{20}^{30} x\,dx$$

$$\Rightarrow \quad v^2 = 100 - 0.02 \left[\frac{x^2}{2} \right]_{20}^{30}$$

$$\Rightarrow \quad v^2 = 100 - \frac{0.02}{2} [(30)^2 - (20)^2]$$

$$\Rightarrow \quad v^2 = 100 - 0.01[900 - 400]$$

$$\Rightarrow \quad v^2 = 100 - 0.01[500]$$

$$\Rightarrow v^2 = 95 \text{ m/sec}$$

$$\text{Final K.E.} = \frac{1}{2} mv^2$$

$$= \frac{1}{2} \times 10 \times 95 \Rightarrow 475 \text{ J}$$

NOTES

From, $F = ma$

$$a = \frac{F}{m} = \frac{0.1x}{10} = 0.01x = V\frac{dV}{dx}$$

$$\text{So, } \int_{v_1}^{v_2} v\,dV = \int_{20}^{30} \frac{x}{100} dx$$

$$-\frac{V^2}{2}\bigg|_{V_1}^{V_2} = \frac{x^2}{200}\bigg|_{20}^{30} = \frac{30 \times 30}{200} - \frac{20 \times 20}{200}$$

$$= 4.5 - 2 = 2.5$$

$$\frac{1}{2} m\left(V_2^2 - V_1^2\right) = 10 \times 2.5\, J = -25 J$$

Final K.E.

$$= \frac{1}{2} mv_2^2 = \frac{1}{2} mv_1^2 - 25 = \frac{1}{2} \times 10 \times 10 \times 10 - 25$$

$$= 500 - 25 = 475 \text{ J}$$

18. **(c)** According to law of conservation of momentum

$$MV + mnv = 0$$

$$\Rightarrow V = \frac{-mNv}{M} = \frac{-0.01 \text{ kg} \times 10 \times 800 \text{ m/s}}{100}$$

$$\Rightarrow -0.8 \text{ m/s}$$

According to work energy theorem,

Average work done = Change in average kinetic energy

$$i.e., F_{av} \times S_{av} = \frac{1}{2} mV^2$$

$$\Rightarrow \frac{F_{av} V_{max} t}{2} = \frac{1}{2} m \frac{V^2}{2}$$

$$\Rightarrow F_{av} = 8 \text{ N}$$

19. **(b)** As the particle is moving in a potential energy region.

$\therefore$ Kinetic energy > 0

And, total energy E = K.E. + P.E.

$\Rightarrow U(x) < E$

20. **(d)** When work is done upon a system by a conservative force then its potential energy increases.

21. **(d)** m = mass per unit length

$\therefore$ Mass flowing per second $= mv$

$$\text{Rate of K.E.} = \frac{1}{2}(mv)v^2 = \frac{1}{2}mv^3$$

22. **(d)** When the body is thrown upwards. its K.E is converted into P.E.

The loss of energy due to air friction is the difference of K.E and P.E.

$$\frac{1}{2}mv^2 - mgh = \frac{1}{2} \times 1 \times 400 - 1 \times 18 \times 10$$

$$= 200 - 180 = 20 \text{ J}$$

23. **(b)** $E = \dfrac{p^2}{2m}$

$$\text{or, } E_1 = \frac{p_1^2}{2m_1}, E_2 = \frac{p_2^2}{2m_2}$$

$$\text{or, } m_1 = \frac{p_1^2}{2E_1}, m_2 = \frac{p_2^2}{2E_2}$$

$$m_1 > m_2 \Rightarrow \frac{m_1}{m_2} > 1$$

$$\therefore \frac{p_1^2 E_2}{E_1 P_2^2} > 1 \Rightarrow \frac{E_2}{E_1} > 1 \qquad [\because p_1 = p_2]$$

$$\text{or, } E_2 > E_1$$

NOTES

Kinetic energy of body of mass m moving with velocity v is

$$\text{K.E.} = \frac{1}{2}mv^2$$

$$\Rightarrow \text{K.E.} = \frac{1}{2}\frac{mv^2 \times m}{m}$$

[Multiplying both numerator and denominator by m]

$$\Rightarrow \text{K.E.} = \frac{1}{2}\frac{m^2v^2}{m}$$

$$\Rightarrow \text{K.E.} = \frac{1}{2}\frac{p^2}{m} \qquad [\because \text{Momentum, } P = mv]$$

$$\Rightarrow P^2 = 2m\,\text{K.E.}$$

$$\Rightarrow P = \sqrt{2mK.E.}$$

24. **(d)**

> **NOTES** Since height is same for both balls, their velocities on reaching the ground will be same.
>
> $$\therefore \frac{K.E_1}{K.E_2} = \frac{\frac{1}{2} m_1 v_0^2}{\frac{1}{2} m_2 v_0^2} = \frac{m_1}{m_2} = \frac{2}{4} = \frac{1}{2}$$

25. **(a)** If k be the spring constant, then

$$U = \frac{1}{2} \times k \times (2)^2 = 2k$$

$$U_{\text{final}} = \frac{1}{2} \times k \times (10)^2 = 50k$$

$$\Rightarrow \frac{U}{U_{\text{final}}} = \frac{2k}{50k} = \frac{1}{25}$$

$\Rightarrow U_{\text{final}} = 25U$

26. **(c)** New K.E., $E' = E + 3E = 4E$

$p = \sqrt{2mE}$ and $p' = \sqrt{2mE'}$

$$\frac{p'}{p} = \sqrt{\frac{2m \times 4E}{2mE}} = 2$$

$\frac{p'}{p} - 1 = 2 - 1$ [on substrating 1 in both sides.]

$$\frac{p' - p}{p} \times 100 = (2-1) \times 100 = 100\%$$

27. **(d)** If l is length of pendulum and θ be angular amplitude then height

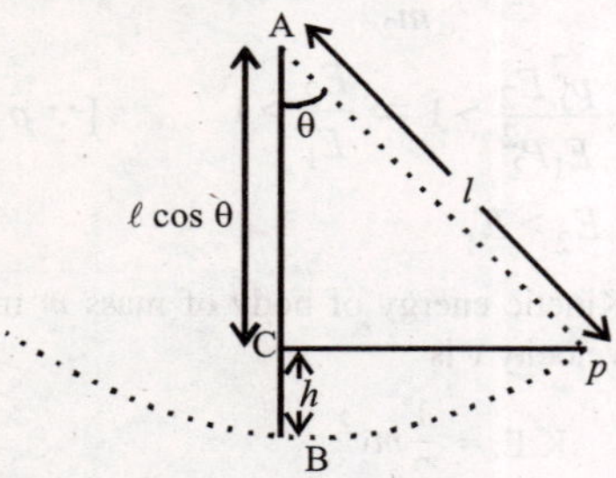

h = AB – AC = l – l cos θ = l(1 – cos θ)

At extreme position, potential energy is maximum and kinetic energy is zero; At mean (equilibrium) position potential energy is zero and kinetic energy is maximum, so from principle of conservation of energy.

(KE + PE) at P = (KE + PE) at B

$$0 + mgh = \frac{1}{2} mv^2 + 0$$

$$\Rightarrow v = \sqrt{2gh} = \sqrt{2g\ell(1 - \cos\theta)}$$

28. **(d)** $\frac{(K.E)_1}{(K.E)_2} = \frac{\frac{1}{2} m_1 v_1^2}{\frac{1}{2} m_2 v_2^2} = \frac{4}{1} \Rightarrow \frac{m_1 v_1^2}{m_2 v_2^2} = \frac{4}{1}$

$$\Rightarrow \frac{(m_1 v_1)^2}{(m_2 v_2)^2} \frac{m_2}{m_1} = \frac{4}{1} \Rightarrow \frac{p_1^2}{p_2^2} \times \frac{m_2}{m_1} = \frac{4}{1}$$

[Given: $p_1 = p_2$]

$$\frac{m_2}{m_1} = \frac{4 \times p_2^2}{1 \times p_1^2} = \frac{4}{1} \times 1 \Rightarrow \frac{m_1}{m_2} = \frac{1}{4}$$

29. **(b)** According to principle of conservation of energy,

Loss in potential energy = Gain in kinetic energy

$$\Rightarrow mgh = \frac{1}{2} mv^2 \Rightarrow v = \sqrt{2gh}$$

If h_1 and h_2 are initial and final heights, then

$v_1 = \sqrt{2gh_1}$ and $v_2 = \sqrt{2gh_2}$

Loss in velocity,

$$\Delta v = v_1 - v_2 = \sqrt{2gh_1} - \sqrt{2gh_2}$$

∴ Fractional loss in velocity,

$$= \frac{\Delta v}{v_1} = \frac{\sqrt{2gh_1} - \sqrt{2gh_2}}{\sqrt{2gh_1}} = 1 - \sqrt{\frac{h_2}{h_1}}$$

$$= 1 - \sqrt{\frac{1.8}{5}} = 1 - \sqrt{0.36} = 1 - 0.6 = 0.4 = \frac{2}{5}$$

30. **(c)** $m_1 = m, m_2 = 4\text{ m}$

$K.E_1 = K.E_2$

$$\frac{1}{2} m_1 v_1^2 = \frac{1}{2} m_2 v_2^2 ; \quad \frac{1}{2} m v_1^2 = \frac{1}{2} 4m v_2^2$$

$$\frac{v_1}{v_2} = 2 \Rightarrow v_1 = 2v_2$$

$$\frac{\text{Linear momentum of first body}}{\text{Linear momentum of second body}} = \frac{m_1 v_1}{m_2 v_2} = \frac{m.2v_2}{4mv_2} = \frac{1}{2}$$

31. **(d)** $K.E. = \frac{1}{2} mv^2$

Further, $v^2 = u^2 + 2as = 0 + 2ad = 2ad$
$= 2(F/m)d$

Hence, $K.E. = \frac{1}{2} m \times 2(F/m) d = Fd$

or, K.E. acquired = Work done
$= F \times d$ = constant.

i.e., it is independent of mass m.

32. **(c)** Initial momentum $(p_1) = p$; Final momentum $(p_2) = 1.5\,p$ and initial kinetic energy $(K_1) = K$.

Kinetic energy $(K) = \frac{p^2}{2m} \propto p^2$

or, $\frac{K_1}{K_2} = \left(\frac{p_1}{p_2}\right)^2 = \left(\frac{p}{1.5p}\right)^2 = \frac{1}{2.25}$

or, $K_2 = 2.25$ K.

Therefore, increase in kinetic energy is 2.25 K − K = 1.25 K or 125%.

33. **(b)** Force due to friction = kinetic energy

$\mu\, mg\, s = \frac{1}{2} mv^2$

[Here, $v = 72$ km/h $= \frac{72000}{60 \times 60} = 20$m / s]

or , $s = \frac{v^2}{2\mu g} = \frac{20 \times 20}{2 \times 0.5 \times 10} = 40$m

NOTES

If a body of mass m moves with velocity u on a rough surface and stops after travelling a distance S due to friction. Then, frictional force, F = ma = μR ⇒ ma = μmg ⇒ $a = \mu g$

Using $v^2 = u^2 - 2as$

$\Rightarrow 0 = u^2 - 2\mu g S$

$\Rightarrow S = \frac{u^2}{2\mu g}$

34. **(c)** $\frac{1}{2}(1)v_1^2 = \frac{1}{2}(9)v_2^2$

$\Rightarrow \frac{v_1^2}{v_2^2} = 9$ or $\frac{v_1}{v_2} = 3$

Ratio of their linear momenta

$= \frac{m_1 v_1}{m_2 v_2} = \frac{1}{9} \times (3) = \frac{1}{3}$

35. **(c)** $E = \frac{1}{2} mv^2$. Hence, $mv = (2mE)^{1/2}$.

NOTES

For same KE, momentum $\propto \sqrt{m}$.

Hence, the ratio is 2 : 1.

36. **(c)** $\frac{K_1}{K_2} = \frac{p_1^2}{m_1} \times \frac{m_2}{p_2^2}$ $[\because p = mv \Rightarrow K = \frac{p^2}{2m}]$

Hence, $\frac{p_1}{p_2} = \sqrt{\frac{M_1}{M_2}} = \sqrt{\frac{1}{4}} = \frac{1}{2}$

37. **(c)** Power on turbine $P = \frac{d(mgh)}{dt}$

$= gh\frac{dm}{dt} = 10 \times 60 \times 15 = 9000$W

Due to frictional force, lossses are 10%

∴ Power generated

$= \left(1 - \frac{10}{100}\right) \times 9000 = 8.1$ kW

38. **(d)** Given force $\vec{F} = 2t\hat{i} + 3t^2\hat{j}$

According to Newton's second law of motion,

$m\frac{d\vec{v}}{dt} = 2t\hat{i} + 3t^2\hat{j}$ (m = 1 kg)

$\Rightarrow \int_0^{\vec{v}} d\vec{v} = \int_0^t \left(2t\hat{i} + 3t^2\hat{j}\right)dt$

$\Rightarrow \vec{v} = t^2\hat{i} + t^3\hat{j}$

Power $P = \vec{F} \cdot \vec{v}$ $(2t\hat{i} + 3t^2\hat{j}) \cdot (t^2\hat{i} + t^3\hat{j})$

$= (2t^3 + 3t^5)$W

39. **(d)** As we know power $P = \frac{dw}{dt}$

$\Rightarrow w = Pt = \frac{1}{2} mV^2$

So, $v = \sqrt{\frac{2Pt}{m}}$

Hence, acceleration $a = \frac{dV}{dt} = \sqrt{\frac{2P}{m}} \cdot \frac{1}{2\sqrt{t}}$

Therefore, force on the particle at time 't'

$= ma = \sqrt{\frac{2Km^2}{m}} \cdot \frac{1}{2\sqrt{t}} = \sqrt{\frac{Km}{2t}} = \sqrt{\frac{mK}{2}}\, t^{-1/2}$

40. **(d)** Work done by the heart, W = PΔV

where, pressure $P = \rho g h$

and $h = 150$ mm − 0.15 m

$g = 10$ m/sec^2

$\rho = 13.6 \times 10^3$ kg/m^3

So, Power $= \frac{W}{t} = \frac{P \cdot \Delta V}{t}$

Power $= \frac{\rho g h \times 5 \times 10^{-3}}{60}$

$\Rightarrow \frac{13.6 \times 10^3 \times 10 \times 0.15 \times 5 \times 10^{-3}}{60}$

$\Rightarrow$ 1.7 watt

NOTES

Power $\vec{F}\cdot\vec{V} = PA\vec{V} = \rho ghAV$

$\left[\because P = \frac{F}{A} \text{ and } P = \rho gh\right]$

$\therefore$ P = 13.6 × 10³ × 10 × 150 × 10⁻³ × 0.5 × 10⁻³/60

$= \frac{102}{60} = 1.70$ watt

41. (a) $\because$ Power $P = \frac{W}{t}$

$\Rightarrow \frac{P_1}{P_2} = \frac{t_2}{t_1} = \frac{30s}{1\text{ minute}} = \frac{30s}{60s} = \frac{1}{2}$

($t_1 = 1$ minute; $t_2 = 30$ second given)

42. (b) Constant power of car $P_0 = F.V = ma.v$

$P_0 = m\frac{dv}{dt}\cdot v$

$P_0 dt = mvdv$ Integrating

$P_0 \cdot t = \frac{mv^2}{2}$

$v = \sqrt{\frac{2P_0 t}{m}}$

$\because$ P_0, m and 2 are constant

$\therefore \quad v \propto \sqrt{t}$

At constant power, P = constant

NOTES

$P = Fv = \frac{mdv}{dt}v = P \quad \left(\because F = \frac{mdv}{dt}\right)$

$\Rightarrow vdv = \frac{P}{m}dt$

Integrating both sides, we get

$\int vdv = \int \frac{P}{m}t$

$\Rightarrow \frac{v^2}{2} = \frac{P}{m}t + C_1$ ...(i)

At $t = 0, v = 0$

$\therefore C_1 = 0$

From equation (i) we get

$V^2 = \frac{2Pt}{m} \quad \Rightarrow \quad V = \left(\frac{2Pt}{m}\right)^{1/2}$

43. (b) Power exerted by a force is given by P = F.v

When the body is just above the earth's surface, its velocity is greatest. At this instant, gravitational force is also maximum. Hence, the power exerted by the gravitational force is greatest at the instant just before the body hits the earth.

44. (d) Amount of water flowing per second from the pipe

$= \frac{m}{\text{time}} = \frac{m}{\ell}\cdot\frac{\ell}{t} = \left(\frac{m}{\ell}\right)v$

Power = K.E. of water flowing per second

$= \frac{1}{2}\left(\frac{m}{\ell}\right)v\cdot v^2$

$= \frac{1}{2}\left(\frac{m}{\ell}\right)v^3$

$= \frac{1}{2}\times 100 \times 8 = 400$ W

45. (a) Given, $h = 60$m, $g = 10$ ms^{-2},

Rate of flow of water = 15 kg/s

$\therefore$ Power of the falling water

= 15 kgs^{-1} × 10 ms^{-2} × 60 m = 9000 watt.

Loss in energy due to friction

$= 9000 \times \frac{10}{100} = 900$ watt.

$\therefore$ Power generated by the turbine

= (9000 – 900) watt = 8100 watt = 8.1 kW

46. (b) $P = \frac{W}{t}$. Here, P = 2kW = 2000 W.

W = Mgh = M × 10 × 10 = 100 M and t = 60 s.

This gives, M = 1200 kg

Its volume = 1200 litre as 1 litre of water contains 1 kg of its mass.

47. (b) $m_1u_1 + m_2u_2 = m_1v_1 + m_2v_2$

Also, $\frac{1}{2}m_1u_1^2 + \frac{1}{2}m_2u_2^2$

$= \frac{1}{2}m_1v_1^2 + \frac{1}{2}m_2v_2^2$

For elastic collision

$v_2 = \frac{2m_1u_1}{m_1 + m_2}$

as $u_2 = 0; u_1 = u$

$m_1 = 4$m and $m_2 = 2$m

so, $v_2 = \frac{4}{3}u$

$\frac{\Delta KE}{KE} = \frac{1}{2}\times 2m \times \left(\frac{4}{3}u\right)^2$

$= \frac{1}{2}\times 2 \times \frac{16}{9}$

48. (b) $\vec{P}_i = \vec{P}_f$

$$m\vec{v}_i = \left(\frac{m}{6}\vec{v}_1 + \frac{5m}{6}\vec{v}_2\right)$$

$$\vec{v}_i = \left(\frac{\vec{v}_1}{6} + \frac{5}{6}\vec{v}_2\right)$$

$$20\hat{i} + 25\hat{j} - 12\hat{k} = \frac{(100\hat{i} + 35\hat{j} + 8\hat{k})}{6} + \frac{5\vec{v}_2}{6}$$

$$\vec{v}_2 = \frac{20\hat{i} + 115\hat{j} - 80\hat{k}}{5}$$

$$\therefore \quad \vec{v}_2 = 4\hat{i} + 23\hat{j} - 16\hat{k}$$

49. (b)

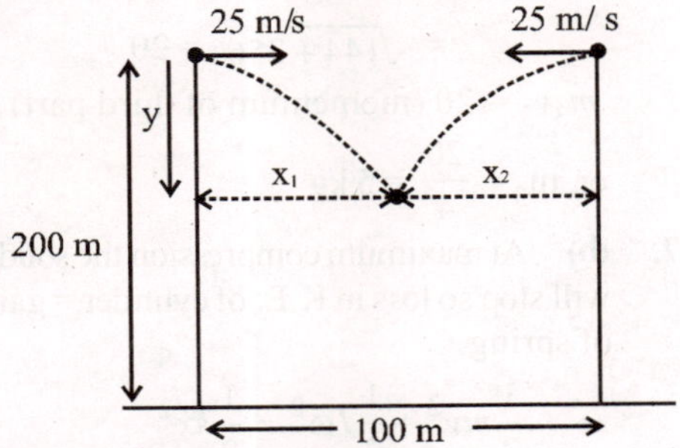

Bullets collide at time t

$x_1 + x_2 = 100$ m

$25t + 25t = 100$

$t = 2s$

$$y = \frac{1}{2}gt^2 = \frac{1}{2} \times 10 \times 2^2 = 20\text{ m}$$

$h = 200 - 20 = 180$ m

50. (a) From conservation of linear momentum.

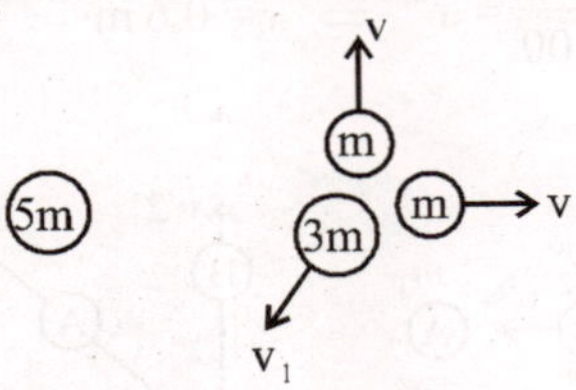

$$0 = mv\hat{j} + mv\hat{i} + 3m\vec{v}_1$$

$$\vec{v}_1 = -\frac{v}{3}(\hat{i} + \hat{j})$$

$$v_1 = \frac{\sqrt{2}}{3}v$$

$$K_f = \frac{1}{2}mv^2 + \frac{1}{2}mv^2 + \frac{1}{2}(3m)\left(\frac{\sqrt{2}}{3}\right)^2 v^2$$

$$= mv^2 + \frac{mv^2}{3} = \frac{4}{3}mv^2$$

$$\Delta KE = K_f - K_i = \frac{4}{3}mv^2$$

51. (b)

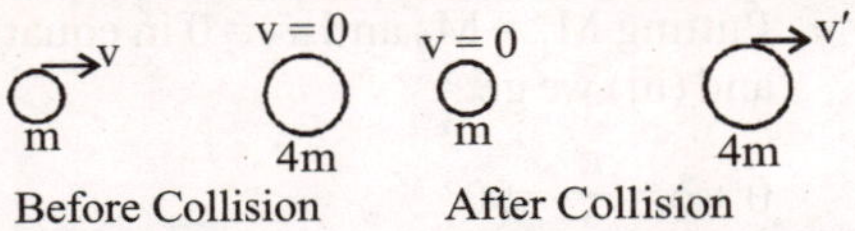

According to law of conservation of linear momentum,

$$mv + 4m \times 0 = 4\,mv' + 0 \Rightarrow \quad v' = \frac{v}{4}$$

Coefficient of restitution,

$$e = \frac{\text{Relative velocity of separation}}{\text{Relative velocity of approach}}$$

$$= \frac{\frac{V}{4}}{V}$$

or, $$e = \frac{1}{4} = 0.25$$

52. (d) For collision $\vec{V}_{B/A}$ should be along

$B \rightarrow A(\vec{r}_{A/B})$

So, $$\frac{\vec{V}_2 - \vec{V}_1}{|V_2 - V_1|} = \frac{\vec{r}_1 - \vec{r}_2}{|r_1 - r_2|}$$

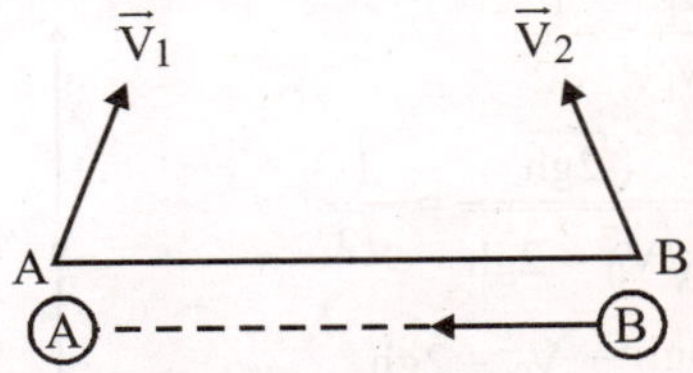

53. (d) Here, $M_1 = M_2$ and $u_2 = 0$

$$u_1 = V, \quad V_1 = \frac{V}{3}; \quad V_2 = ?$$

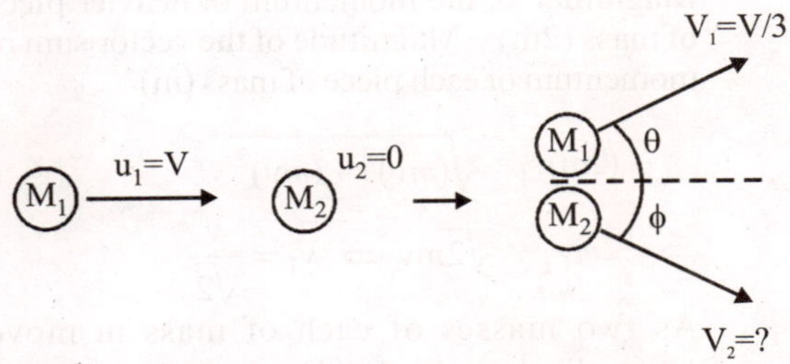

From figure, along x-axis,

$M_1u_1 + M_2u_2 = M_1V_1\cos\theta + M_2V_2\cos\phi$ ...(i)

Along y-axis

$0 = M_1V_1\sin\theta - M_2V_s\sin\phi$ ...(ii)

By law of conservation of kinetic energy,

$$\frac{1}{2}M_1u_1^2 + \frac{1}{2}M_2u_2^2 = \frac{1}{2}M_1V_1^2 + \frac{1}{2}M_2V_2^2 \quad \text{...(iii)}$$

Putting $M_1 = M_2$ and $u_2 = 0$ in equation (i), (ii) and (iii) we get,

$$\theta + \phi = \frac{\pi}{2} = 90°$$

and $u_1^2 = V_1^2 + V_2^2$

$$V^2 = \left(\frac{V}{3}\right)^2 + V_2^2 \quad \left[\because u_1 = V \text{ and } V_1 = \frac{V}{3}\right]$$

$$\text{or, } V^2 - \left(\frac{V}{3}\right)^2 = V_2^2$$

$$V^2 - \frac{V^2}{9} = V_2^2$$

$$\text{or } V_2^2 = \frac{8}{9}V^2 \Rightarrow V_2 = \frac{2\sqrt{2}}{3}V$$

54. (a) When ball collides with the ground it loses its 50% of energy

$$\therefore \frac{KE_f}{KE_i} = \frac{1}{2} \Rightarrow \frac{\frac{1}{2}mV_f^2}{\frac{1}{2}mV_i^2} = \frac{1}{2}$$

$$\text{or } \frac{V_f}{V_i} = \frac{1}{\sqrt{2}}$$

$$\text{or, } \frac{\sqrt{2gh}}{\sqrt{V_0^2 + 2gh}} = \frac{1}{\sqrt{2}}$$

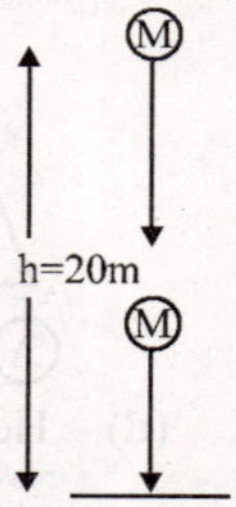

$$\text{or, } 4gh = V_0^2 + 2gh$$

$$\therefore V_0 = 20\ ms^{-1}$$

55. (b) By conservation of linear momentum Magnitude of the momentum of heavier piece of mass (2m) = Magnitude of the vector sum of momentum of each piece of mass (m)

$$(2m)v_1 = \sqrt{(mv)^2 + (mv)^2}$$

$$\Rightarrow 2mv_1 = \sqrt{2}mv \Rightarrow v_1 = \frac{v}{\sqrt{2}}$$

As two masses of each of mass m move perpendicular to each other.

Total KE generated

$$= \frac{1}{2}mv^2 + \frac{1}{2}mv^2 + \frac{1}{2}(2m)v_1^2$$

$$= mv^2 + \frac{mv^2}{2} = \frac{3}{2}mv^2$$

56. (a)

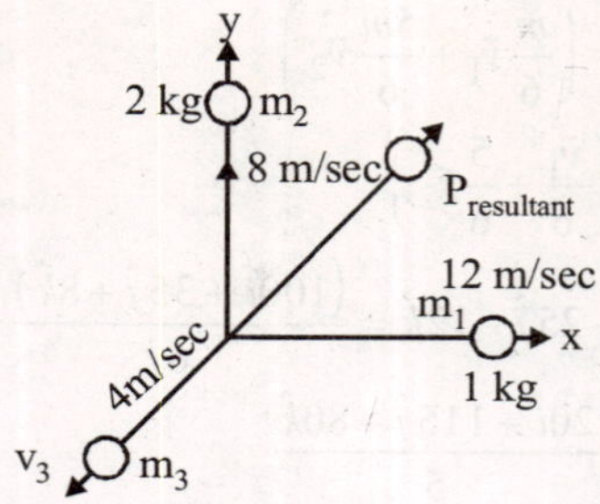

$$P_{resultant} = \sqrt{12^2 + 16^2}$$

$$= \sqrt{144 + 256} = 20$$

$m_3 v_3 = 20$ (momentum of third part)

$$\text{or, } m_3 = \frac{20}{4} = 5\ kg$$

57. (b) At maximum compression the solid cylinder will stop so loss in K.E. of cylinder = gain in P.E. of spring

$$\Rightarrow \frac{1}{2}mv^2 + \frac{1}{2}I\omega^2 = \frac{1}{2}kx^2$$

$$\Rightarrow \frac{1}{2}mv^2 + \frac{1}{2}\frac{mR^2}{2}\left(\frac{v}{R}\right)^2 = \frac{1}{2}kx^2$$

$$\Rightarrow \frac{3}{4}mv^2 = \frac{1}{2}kx^2$$

$$\Rightarrow \frac{3}{4} \times 3 \times (4)^2 = \frac{1}{2} \times 200\,x^2$$

$$\Rightarrow \frac{36}{100} = x^2 \Rightarrow x = 0.6\ m$$

58. (d)

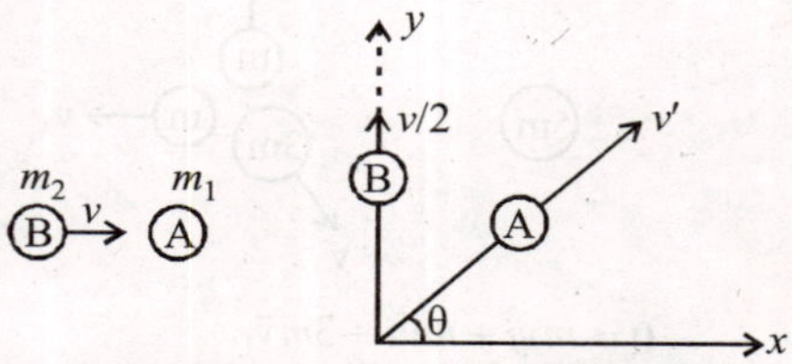

[Before collision] [After collision]

According to law of conservation of linear momentum along x-axis, we get

$$\Rightarrow m_1(0) + m_2 v = m_1 v' \cos\theta$$

$$\Rightarrow \cos\theta = \frac{m_2 v}{m_1 v'} \quad ...(i)$$

Similarly, law of conservation of linear momentum along y-axis, we get

$$\Rightarrow m_1(0) + m_2(0) = m_1 v' \sin\theta + m_2\left(\frac{v}{2}\right)$$

$$\Rightarrow \sin\theta = \frac{m_2 v}{2m_1 v'} \quad ...(ii)$$

From equations, (i) and (ii),

$$\tan\theta = -\left(\frac{m_2 v}{2mv'}\right)\times\left(\frac{m_1 v'}{m_2 v}\right) = -\frac{1}{2}$$

$$\theta = \tan^{-1}\left(\frac{-1}{2}\right) \text{ to the } x\text{-axis.}$$

NOTES

Let A moves in the direction, which makes an angle θ with initial direction *i.e.*,

$$\tan\theta = \frac{v_y}{v_x} = \frac{m_2 v}{2m_1}\bigg/\frac{m_2 v}{2m_1}$$

$$\tan\theta = \frac{1}{2}$$

$$\Rightarrow \theta = \tan^{-1}\left(\frac{1}{2}\right) \text{ to the } x\text{-axis.}$$

59. **(a)** As the two masses stick together after collision, hence it is inelastic collision. Therefore, only momentum is conserved.

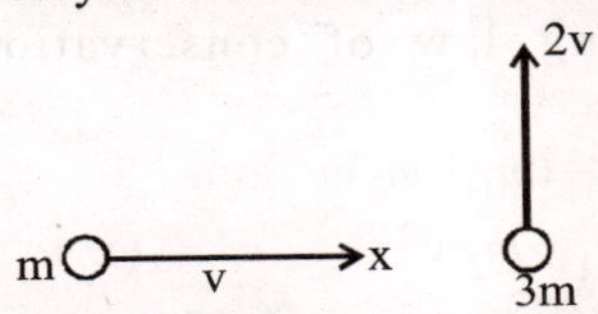

$$\therefore\ mv\hat{i} + 3m(2v)\hat{j} = (4m)\vec{v}$$

$$\vec{v} = \frac{v}{4}\hat{i} + \frac{6}{4}v\hat{j}$$

$$= \frac{v}{4}\hat{i} + \frac{3}{2}v\hat{j}$$

60. **(a)** Clearly $v_1 = 2\ \text{ms}^{-1}$, $v_2 = 0$

$m_1 = m$ (say), $m_2 = 2m$

$v_1' = ?$, $v_2' = ?$

$$e = \frac{v_1' - v_2'}{v_2 - v_1} \quad \text{....(i)}$$

By conservation of momentum,

$$2m = mv_1' + 2mv_2' \quad \text{....(ii)}$$

From (i),

$$0.5 = \frac{v_2' - v_1'}{2}$$

$$\therefore\ v_2' = 1 + v_1'$$

From (ii),

$$2 = v_1' + 2 + 2v_1'$$

$$\Rightarrow\ v_1' = 0 \text{ and } v_2' = 1\ \text{ms}^{-1}$$

61. **(d)** Momentum of first part

$\Rightarrow\ 1 \times 12 = 12$

Momentum of second part

$\Rightarrow\ 2 \times 8 = 16$

Resultant momentum,

$$= \sqrt{(12)^2 + (16)^2} = 20$$

Then M_1 be the mass of third part,

$$M \cdot 4 = 20$$

$$M = \frac{20}{4} = 5\ \text{kg}$$

NOTES

Let two parts of the rock move along x-axis and y-axis respectively.

If M and v be the mass and velocities of third part then

$Mv\cos\theta = 12$

$Mv\sin\theta = 16$

$$\tan\theta = \frac{16}{12} = \frac{4}{3}$$

$$\cos\theta = \frac{3}{5}$$

$v = 4$ m/s

$$M = \frac{12}{v\cos\theta}$$

$$M = \frac{12\times 5}{4\times 3} = \frac{60}{12} = 5\text{kg}$$

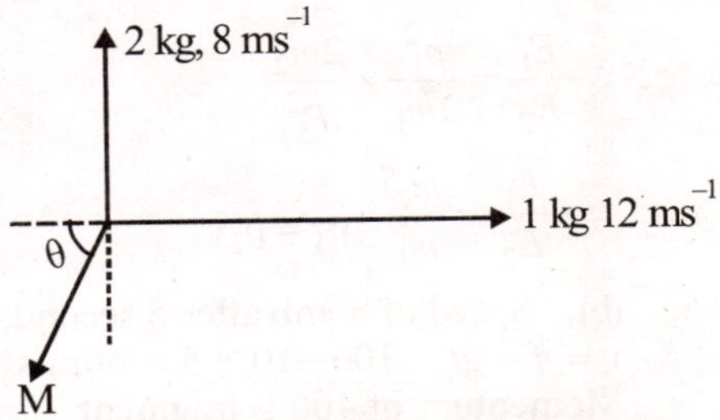

62. **(a)** Let the initial velocity of the shell be v, then by the conservation of momentum $mv = Mv'$ where v' = velocity of gun.

$$\therefore\ v' = \left(\frac{m}{M}\right)v$$

$$\text{Now, total K.E.} = \frac{1}{2}mv^2 + \frac{1}{2}Mv'^2$$

$$= \frac{1}{2}mv^2 + \frac{1}{2}M\left(\frac{m}{M}\right)^2 v^2$$

$$= \frac{1}{2}mv^2\left[1 + \frac{m}{M}\right]$$

$$= \left(\frac{1}{2}\times 0.2\right)\left(1 + \frac{0.2}{4}\right)v^2 = (0.1\times 1.05)v^2$$

But total K.E. = 1.05 kJ = 1.05×10^3 J

$$\therefore\ 1.05\times 10^3 = 0.1 \times 1.05 \times v^2$$

$$\Rightarrow v^2 = \frac{1.05\times 10^3}{0.1\times 1.05} = 10^4$$

$$\therefore\ v = 10^2 = 100\ \text{ms}^{-1}.$$

63. **(b)** From conservation of linear momentum

$m_1v_1 + m_2v_2 = 0$

$$v_2 = \left(\frac{-m_1}{m_2}\right)v_1 = \left(\frac{-18}{12}\right)6 = -9\text{ms}^{-1}$$

$$\text{K.E.} = \frac{1}{2}m_2v_2^2 = \frac{1}{2}\times 12\times 9^2 = 486\text{ J}$$

64. **(b)** $\frac{1}{2}mv^2 = \frac{1}{2}kx^2$

$\Rightarrow mv^2 = kx^2$

or $0.5 \times (1.5)^2 = 50 \times x^2$

$\therefore x = 0.15\text{ m}$

65. **(b)** From conservation law of momentum, before collision and after collision linear momentum (p) will be same. That is, initial momentum = final momentum.

$\Rightarrow 0 = m_1v_1 - m_2v_2 \Rightarrow m_1v_1 = m_2v_2$

$p_1 = p_2$

Now, $E = \frac{p^2}{2m}$

$$\therefore \frac{E_1}{E_2} = \frac{p_1^2}{2m_1}\times\frac{2m_2}{P_2^2}$$

$$\Rightarrow \frac{E_1}{E_2} = \frac{m_2}{m_1}\quad [p_1 = p_2]$$

66. **(b)** Speed of bomb after 5 second,

$v = u - gt = 100 - 10 \times 5 = 50\text{m/s}$

Momentum of 400 g fragment

$= \frac{400}{1000}\times(-25)$ [downward]

Momentum of 600g fragment $= \frac{600}{1000}v$

Momentum of bomb $= 1 \times 50 = 50$

From conservation of momentum

Total momentum before splitting = total momentum after splitting.

$$\Rightarrow 50 = -\frac{400}{1000}\times 25 + \frac{600}{1000}v$$

$\Rightarrow v = 100\text{ m/s}$ [upward]

67. **(d)** In elastic collision, the velocities get inter changed if the colliding objects have equal masses.

68. **(a)**

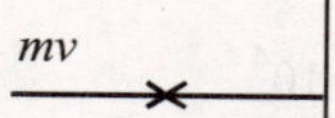

Initial momentum = mv

Final momentum = $m(-v)$

change in momentum

$= mv - m(-v) = 2mv$

69. **(c)** Applying conservation of momentum,

$m_1v_1 = (m_1 + m_2)v$

$$v = \frac{m_1v_1}{(m_1 + m_2)}$$

Here, $v_1 = 36\text{ km/hr} = 10\text{ m/s}$

$m_1 = 2\text{ kg}, m_2 = 3\text{ kg}$

$$v = \frac{10\times 2}{5} = 4\text{ m/s}$$

$$\text{K.E. (initial)} = \frac{1}{2}\times 2\times(10)^2 = 100\text{ J}$$

$$\text{K.E. (Final)} = \frac{1}{2}\times(3+2)\times(4)^2 = 40\text{ J}$$

Loss in K.E. = 100 − 40 = 60 J

70. **(a)** Applying law of conservation of momentum,

$m_1u_1 + m_2u_2 = (m_1 + m_2)v$

or, $m_1u_1 = (m_1 + m_2)v \qquad (\because u_2 = 0)$

$$\Rightarrow m\frac{(3\times 1000)}{3600} = 3m(v)$$

$$\Rightarrow v = \frac{1000}{3600}\text{ m/s} = 1\text{ km/hr}$$

71. **(c)** When shell explodes in mid air its chemical energy is partly converted into mechanical energy, hence K.E. increases.

72. **(b)** When the identical balls collide head-on, their velocities are exchanged.

73. **(a)** Momentum of Ist part = m × 21 = 21m

Momentum of 2nd part = m × 21 = 21m

Resultant momentum, $= \sqrt{(m21)^2 + (m21)^2}$

Resultant momentum = momentum of 3rd part

$\Rightarrow \quad 21\sqrt{2}\text{ m} = 3mv$

$\Rightarrow \quad v = 7\sqrt{2}\text{ m/sec}$

NOTES

Masses of the pieces are 1, 1, 3 kg. Hence $(1 \times 21)^2 + (1 \times 21)^2 = (3 \times V)^2$

That is, $V = 7\sqrt{2}$ m/s

74. **(a)** $e = |v_1 - v_2| / |u_1 - u_2|$ which is 1 for a perfectly elastic collision.

6 System of Particles and Rotational Motion

Trend Analysis with Important Topics & Sub-Topics

Topic Name	Sub-Topic	2021 QNS.	2021 LOD	2020 QNS	2020 LOD	2019 QNS.	2019 LOD	2018 QNS.	2018 LOD	2017 QNS.	2017 LOD	2016 QNS.	2016 LOD
Centre of Mass, Centre of Gravity & Principle of Moments	Centre of mass of two particle			1	E								
	Centre of mass definition									1	E		
Angular Displacement, Velocity and Acceleration	$\tau = F \times R = I\alpha$					1	A						
Torque, Couple and Angular Momentum	Torque, $\vec{\tau} = \vec{r} \times \vec{F}$	1	A	1	E								
Moment of Inertia, Rotational K.E. and Power	Moment of inertia	1	A					1	E	1	A	1	D
	Work Energy theorem in rotation							2	A				
Rolling Motion	Rolling Motion					1	A					1	E

LOD - Level of Difficulty | E - Easy | A - Average | D - Difficult | Qns - No. of Questions

Topic 1 Centre of Mass, Centre of Gravity & Principle of Moments

1. Two particles of mass 5 kg and 10 kg respectively are attached to the two ends of a rigid rod of length 1 m with negligible mass. The centre of mass of the system from the 5 kg particle is nearly at a distance of : *[2020]*

(a) 50 cm (b) 67 cm
(c) 80 cm (d) 33 cm

2. Which of the following statements are correct?

(A) Centre of mass of a body always coincides with the centre of gravity of the body
(B) Centre of mass of a body is the point at which the total gravitational torque on the body is zero
(C) A couple on a body produce both translational and rotation motion in a body
(D) Mechanical advantage greater than one means that small effort can be used to lift a large load *[2017]*

(a) (A) and (B) (b) (B) and (C)
(c) (C) and (D) (d) (B) and (D)

3. Two persons of masses 55 kg and 65 kg respectively, are at the opposite ends of a boat. The length of the boat is 3.0 m and weighs 100 kg. The 55 kg man walks up to the 65 kg man and sits with him. If the boat is in still water the centre of mass of the system shifts by : *[2012]*

(a) 3.0 m (b) 2.3 m
(c) zero (d) 0.75 m

4. Three masses are placed on the x-axis : 300 g at origin, 500 g at $x = 40$ cm and 400 g at $x = 70$ cm. The distance of the centre of mass from the origin is : *[2012M]*

(a) 40 cm (b) 45 cm
(c) 50 cm (d) 30 cm

5. Two particles which are initially at rest, move towards each other under the action of their internal attraction. If their speeds are v and $2v$ at any instant, then the speed of centre of mass of the system will be: *[2010]*

(a) $2v$ (b) zero
(c) 1.5 (d) v

6. Two bodies of mass 1 kg and 3 kg have position vectors $\hat{i}+2\hat{j}+\hat{k}$ and $-3\hat{i}-2\hat{j}+\hat{k}$ respectively. The centre of mass of this system has a position vector: *[2009]*
(a) $-2\hat{i}-\hat{j}+\hat{k}$ (b) $2\hat{i}-\hat{j}-2\hat{k}$
(c) $-\hat{i}+\hat{j}+\hat{k}$ (d) $-2\hat{i}+2\hat{k}$

7. If the linear density (mass per unit length) of a rod of length 3m is proportional to x, where x is the distance from one end of the rod, the distance of the centre of gravity of the rod from this end is *[2002]*
(a) 2.5m (b) 1m
(c) 1.5m (d) 2m

8. A solid sphere of radius R is placed on a smooth horizontal surface. A horizontal force F is applied at height h from the lowest point. For the maximum acceleration of the centre of mass, *[2002]*
(a) $h=R$
(b) $h=2R$
(c) $h=0$
(d) The acceleration will be same whatever h may be

9. The centre of mass of a system of particles does not depend upon *[1997]*
(a) masses of the particles
(b) forces acting on the particles
(c) position of the particles
(d) relative distances between the particles

10. In carbon monoxide molecule, the carbon and the oxygen atoms are separated by a distance 1.12×10^{-10} m. The distance of the centre of mass, from the carbon atom is *[1997]*
(a) 0.64×10^{-10}m (b) 0.56×10^{-10}m
(c) 0.51×10^{-10}m (d) 0.48×10^{-10}m

Topic 2 Angular Displacement, Velocity and Acceleration

11. Two particles A and B are moving in uniform circular motion in concentric circles of radii r_A and r_B with speed υ_A and υ_B respectively. Their time period of rotation is the same. The ratio of angular speed of A to that of B will be : *[2019]*
(a) $r_A : r_B$ (b) $\upsilon_A : \upsilon_B$
(c) $r_B : r_A$ (d) 1 : 1

12. A particle starting from rest, moves in a circle of radius 'r'. It attains a velocity of V_0 m/s in the nth round. Its angular acceleration will be, *[NEET Odisha 2019]*
(a) $\frac{V_0^2}{4\pi nr}$ rad/s² (b) $\frac{V_0}{n}$ rad/s²
(c) $\frac{V_0^2}{2\pi nr^2}$ rad/s² (d) $\frac{V_0^2}{4\pi nr^2}$ rad/s²

13. A wheel has angular acceleration of 3.0 rad/sec^2 and an initial angular speed of 2.00 rad/sec. In a time of 2 sec it has rotated through an angle (in radian) of *[2007]*
(a) 10 (b) 12
(c) 4 (d) 6

14. A wheel of radius 1m rolls forward half a revolution on a horizontal ground. The magnitude of the displacement of the point of the wheel initially in contact with the ground is *[2002]*
(a) π (b) 2π
(c) $\sqrt{2}\pi$ (d) $\sqrt{\pi^2+4}$

15. Two racing cars of masses m_1 and m_2 are moving in circles of radii r_1 and r_2 respectively. Their speeds are such that each makes a complete circle in the same time t. The ratio of the angular speeds of the first to the second car is *[1999]*
(a) 1 : 1 (b) $m_1 : m_2$
(c) $r_1 : r_2$ (d) $m_1 m_2 : r_1 r_2$

16. If a flywheel makes 120 revolutions/minute, then its angular speed will be *[1996]*
(a) 8 π rad/sec (b) 6 π rad/sec
(c) 4 π rad/sec (d) 2 π rad/s

17. The angular speed of an engine wheel making 90 revolutions per minute is *[1995]*
(a) 1.5 π rad/s (b) 3π rad/s
(c) 4.5 π rad/s (d) 6π rad/s

18. Two racing cars of masses m and 4m are moving in circles of radii r and 2r respectively. If their speeds are such that each makes a complete circle in the same time, then the ratio of the angular speeds of the first to the second car is *[1995]*
(a) 8 : 1 (b) 4 : 1
(c) 2 : 1 (d) 1 : 1

Topic 3 Torque, Couple and Angular Momentum

19. A uniform rod of length 200 cm and mass 500 g is balanced on a wedge placed at 40 cm mark. A mass of 2 kg is suspended from the rod at 20 cm and another unknown mass 'm' is suspended from the rod at 160 cm mark as shown in the figure. Find the value of 'm' such that the rod is in equilibrium. ($g = 10\ m/s^2$) *[2021]*

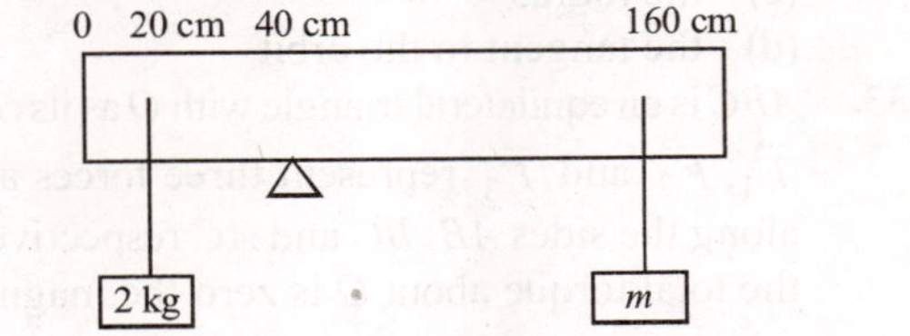

(a) $\frac{1}{12}$ kg (b) $\frac{1}{2}$ kg

(c) $\frac{1}{3}$ kg (d) $\frac{1}{6}$ kg

20. Find the torque about the origin when a force of $3\hat{j}$ N acts on a particle whose position vector is $2\hat{k}$ m. *[2020]*

(a) $6\hat{j}$ Nm (b) $-6\hat{i}$ Nm

(c) $6\hat{k}$ Nm (d) $6\hat{i}$ Nm

21. A solid cylinder of mass 2 kg and radius 4 cm is rotating about its axis at the rate of 3 rpm. The torque required to stop after 2π revolutions is : *[2019]*

(a) 2×10^{-6} N m (b) 2×10^{-3} N m

(c) 12×10^{-4} N m (d) 2×10^{6} N m

22. A rope is wound around a hollow cylinder of mass 3 kg and radius 40 cm. What is the angular acceleration of the cylinder if the rope is pulled with a force of 30 N? *[2017]*

(a) 0.25 rad/s^2 (b) 25 rad/s^2

(c) 5 m/s^2 (d) 25 m/s^2

23. A uniform circular disc of radius 50 cm at rest is free to turn about an axis which is perpendicular to its plane and passes through its centre. It is subjected to a torque which produces a constant angular acceleration of 2.0 rad s^{-2}. Its net acceleration in ms^{-2} at the end of 2.0s is approximately: *[2016]*

(a) 8.0 (b) 7.0

(c) 6.0 (d) 3.0

24. A mass m moves in a circle on a smooth horizontal plane with velocity v_0 at a radius R_0. The mass is attached to string which passes through a smooth hole in the plane as shown.

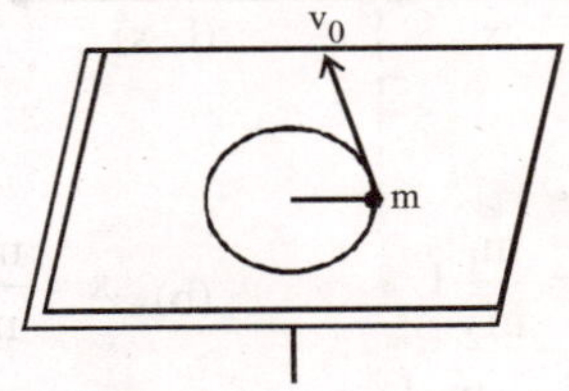

The tension in the string is increased gradually and finally m moves in a circle of radius $\frac{R_0}{2}$. The final value of the kinetic energy is *[2015]*

(a) $\frac{1}{4}mv_0^2$ (b) $2mv_0^2$

(c) $\frac{1}{2}mv_0^2$ (d) mv_0^2

25. A rod of weight W is supported by two parallel knife edges A and B and is in equilibrium in a horizontal position. The knives are at a distance d from each other. The centre of mass of the rod is at distance x from A. The normal reaction on A is *[2015]*

(a) $\frac{Wd}{x}$ (b) $\frac{W(d-x)}{x}$

(c) $\frac{W(d-x)}{d}$ (d) $\frac{Wx}{d}$

26. An automobile moves on a road with a speed of 54 km h^{-1}. The radius of its wheels is 0.45 m and the moment of inertia of the wheel about its axis of rotation is 3 kg m^2. If the vehicle is brought to rest in 15s, the magnitude of average torque transmitted by its brakes to the wheel is: *[2015 RS]*

(a) 8.58 kg m^2 s^{-2} (b) 10.86 kg m^2 s^{-2}

(c) 2.86 kg m^2 s^{-2} (d) 6.66 kg m^2 s^{-2}

27. Point masses m_1 and m_2 are placed at the opposite ends of a rigid rod of length L, and negligible mass. The rod is to be set rotating about an axis perpendicular to it. The position of point P on this rod through which the axis should pass so that the work required to set the rod rotating with angular velocity ω_0 is minimum, is given by : *[2015 RS]*

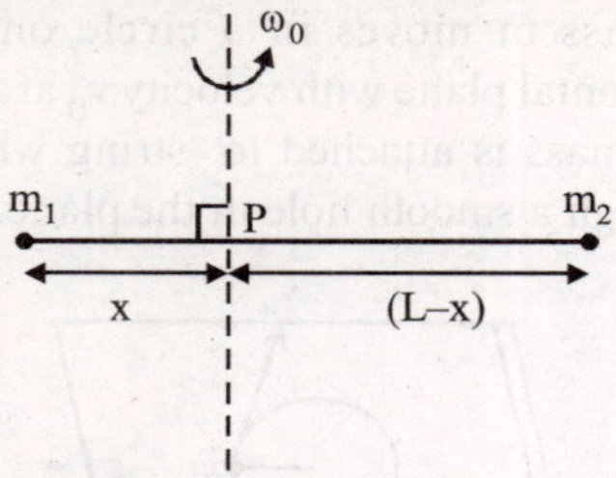

(a) $x = \frac{m_1}{m_2}L$ (b) $x = \frac{m_2}{m_1}L$

(c) $x = \frac{m_2 L}{m_1 + m_2}$ (d) $x = \frac{m_1 L}{m_1 + m_2}$

28. A force $\vec{F} = \alpha\hat{i} + 3\hat{j} + 6\hat{k}$ is acting at a point $\vec{r} = 2\hat{i} - 6\hat{j} - 12\hat{k}$. The value of α for which angular momentum about origin is conserved is : *[2015 RS]*

(a) 2 (b) zero

(c) 1 (d) –1

29. A solid cylinder of mass 50 kg and radius 0.5 m is free to rotate about the horizontal axis. A massless string is wound round the cylinder with one end attached to it and other hanging freely. Tension in the string required to produce an angular acceleration of 2 revolutions s^{-2} is :

(a) 25 N (b) 50 N *[2014]*

(c) 78.5 N (d) 157 N

30. A rod PQ of mass M and length L is hinged at end P. The rod is kept horizontal by a massless string tied to point Q as shown in figure. When string is cut, the initial angular acceleration of the rod is *[2013]*

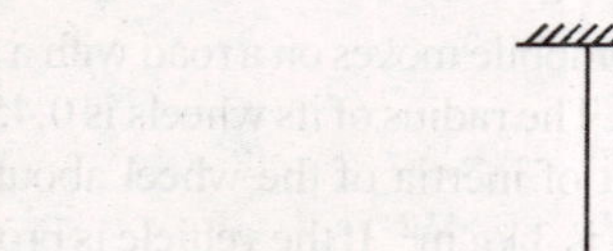

(a) g /L (b) 2g/L

(c) $\frac{2g}{3L}$ (d) $\frac{3g}{2L}$

31. Two discs are rotating about their axes, normal to the discs and passing through the centres of the discs. Disc D_1 has 2 kg mass and 0.2 m radius and initial angular velocity of 50 rad s^{-1}. Disc D_2 has 4kg mass, 0.1 m radius and initial angular velocity of 200 rad s^{-1}. The two discs are brought in contact face to face, with their axes of rotation coincident. The final angular velocity (in rad s^{-1}) of the system is *[NEET Kar. 2013]*

(a) 40 (b) 60

(c) 100 (d) 120

32. When a mass is rotating in a plane about a fixed point, its angular momentum is directed along : *[2012]*

(a) a line perpendicular to the plane of rotation

(b) the line making an angle of 45° to the plane of rotation

(c) the radius

(d) the tangent to the orbit

33. *ABC* is an equilateral triangle with *O* as its centre. $\vec{F}_1, \vec{F}_2$ and $\vec{F}_3$ represent three forces acting along the sides *AB*, *BC* and *AC* respectively. If the total torque about *O* is zero the magnitude of $\vec{F}_3$ is : *[2012, 1998]*

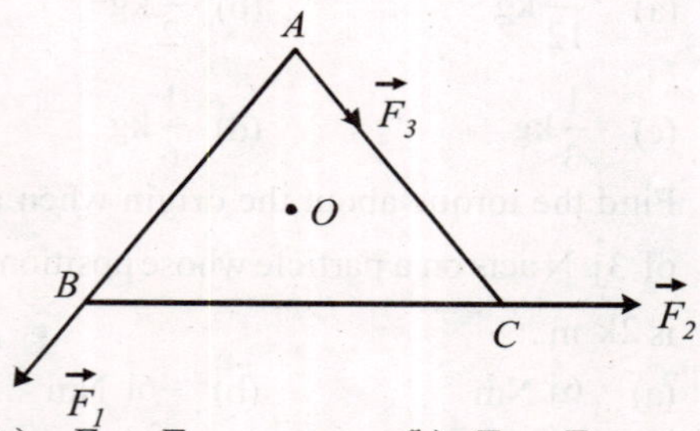

(a) $F_1 + F_2$ (b) $F_1 - F_2$

(c) $\frac{F_1 + F_2}{2}$ (d) $2(F_1 + F_2)$

34. A circular platform is mounted on a frictionless vertical axle. Its radius $R = 2$ m and its moment of inertia about the axle is 200 kgm^2. It is initially at rest. A 50 kg man stands on the edge of the platform and begins to walk along the edge at the speed of 1 ms^{-1} relative to the ground. Time taken by the man to complete one revolution is

(a) πs (b) $\frac{3\pi}{2}$s *[2012M]*

(c) 2πs (d) $\frac{\pi}{2}$s

35. The instantaneous angular position of a point on a rotating wheel is given by the equation $\theta(t) = 2t^3 - 6t^2$. The torque on the wheel becomes zero at *[2011]*

(a) t = 1 s (b) t = 0.5 s

(c) t = 0.25 s (d) t = 2s

36. A circular disk of moment of inertia I_t is rotating in a horizontal plane, its symmetry axis, with a constant angular speed ω_i. Another disk of moment of inertia I_b is dropped coaxially onto the rotating disk. Initially the second disk has zero angular speed. Eventually both the disks rotate with a constant angular speed ω_f. The energy lost by the initially rotating disk to friction is: **[2010]**

(a) $\frac{1}{2}\frac{I_b^2}{(I_t+I_b)}\omega_i^2$ (b) $\frac{I_t^2}{(I_t+I_b)}\omega_i^2$

(c) $\frac{I_b-I_t}{(I_t+I_b)}\omega_i^2$ (d) $\frac{1}{2}\frac{I_b I_t}{(I_t+I_b)}\omega_i^2$

37. A thin circular ring of mass M and radius R is rotating in a horizontal plane about an axis vertical to its plane with a constant angular velocity ω. If two objects each of mass m be attached gently to the opposite ends of a diameter of the ring, the ring will then rotate with an angular velocity: **[2009, 1998]**

(a) $\frac{\omega M}{M+2m}$ (b) $\frac{\omega(M+2m)}{M}$

(c) $\frac{\omega M}{M+m}$ (d) $\frac{\omega(M-2m)}{M+2m}$

38. If $\vec{F}$ is the force acting on a particle having position vector $\vec{r}$ and $\vec{\tau}$ be the torque of this force about the origin, then: **[2009]**

(a) $\vec{r}\cdot\vec{\tau}>0$ and $\vec{F}\cdot\vec{\tau}<0$

(b) $\vec{r}\cdot\vec{\tau}=0$ and $\vec{F}\cdot\vec{\tau}=0$

(c) $\vec{r}\cdot\vec{\tau}=0$ and $\vec{F}\cdot\vec{\tau}\neq 0$

(d) $\vec{r}\cdot\vec{\tau}\neq 0$ and $\vec{F}\cdot\vec{\tau}=0$

39. A uniform rod AB of length ℓ, and mass m is free to rotate about point A. The rod is released from rest in the horizontal position. Given that the moment of inertia of the rod about A is $\frac{m\ell^2}{3}$, the initial angular acceleration of the rod will be **[2007]**

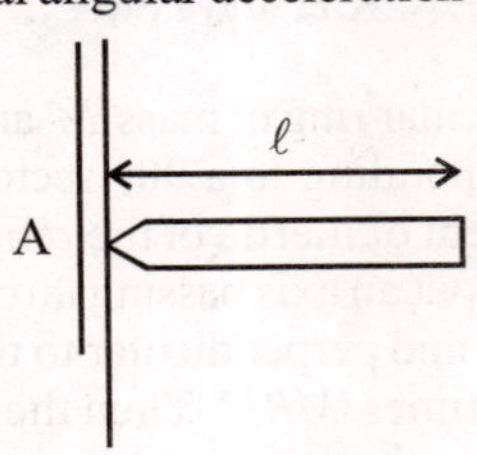

(a) $\frac{mg\ell}{2}$ (b) $\frac{3}{2}g\ell$

(c) $\frac{3g}{2\ell}$ (d) $\frac{2g}{3\ell}$

40. A particle of mass m moves in the XY plane with a velocity v along the straight line AB. If the angular momentum of the particle with respect to origin O is L_A when it is at A and L_B when it is at B, then **[2007]**

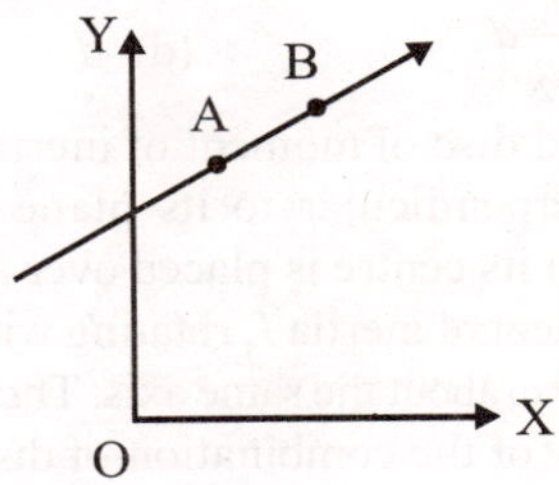

(a) $L_A = L_B$

(b) the relationship between L_A and L_B depends upon the slope of the line AB

(c) $L_A < L_B$

(d) $L_A > L_B$

41. A tube of length L is filled completely with an incompressible liquid of mass M and closed at both the ends. The tube is then rotated in a horizontal plane about one of its ends with a uniform angular velocity ω. The force exerted by the liquid at the other end is **[2006]**

(a) $\frac{ML^2\omega}{2}$ (b) $ML\omega^2$

(c) $\frac{ML^2\omega^2}{2}$ (d) $\frac{ML\omega^2}{2}$

42. Two bodies have their moments of inertia I and 2I respectively about their axis of rotation. If their kinetic energies of rotation are equal, their angular momenta will be in the ratio **[2005]**

(a) 2 : 1 (b) 1 : 2

(c) $\sqrt{2}:1$ (d) $1:\sqrt{2}$

43. A wheel having moment of inertia 2 kg-m² about its vertical axis, rotates at the rate of 60 rpm about this axis. The torque which can stop the wheel's rotation in one minute would be **[2004]**

(a) $\frac{\pi}{18}$ N-m (b) $\frac{2\pi}{15}$ N-m

(c) $\frac{\pi}{12}$ N-m (d) $\frac{\pi}{15}$ N-m

44. Consider a system of two particles having masses m_1 and m_2. If the particle of mass m_1 is pushed towards the centre of mass of particles through a distance d, by what distance would the particle of mass m_2 move so as to keep the centre of mass of particles at the original position? [2004]

(a) $\frac{m_2}{m_1}d$ (b) $\frac{m_1}{m_1+m_2}d$

(c) $\frac{m_1}{m_2}d$ (d) d

45. A round disc of moment of inertia I_2 about its axis perpendicular to its plane and passing through its centre is placed over another disc of moment of inertia I_1 rotating with an angular velocity ω about the same axis. The final angular velocity of the combination of discs is [2004]

(a) $\frac{(I_1+I_2)\omega}{I_1}$ (b) $\frac{I_2\omega}{I_1+I_2}$

(c) ω (d) $\frac{I_1\omega}{I_1+I_2}$

46. A thin circular ring of mass M and radius r is rotating about its axis with a constant angular velocity ω. Four objects each of mass m, are kept gently to the opposite ends of two perpendicular diameters of the ring. The angular velocity of the ring will be [2003]

(a) $\frac{(M-4m)\omega}{M+4m}$ (b) $\frac{M\omega}{4m}$

(c) $\frac{M\omega}{M+4m}$ (d) $\frac{(M+4m)\omega}{M}$

47. A disc is rotating with angular velocity ω. If a child sits on it, what is conserved? [2002]

(a) Linear momentum
(b) Angular momentum
(c) Kinetic energy
(d) Moment of inertia

48. A boy suddenly comes and sits on a circular rotating table. What will remain conserved? [2002]

(a) Angular velocity
(b) Angular momentum
(c) Linear momentum
(d) Kinetic energy

49. A constant torque of 1000 N-m turns a wheel of moment of inertia 200 kg-m^2 about an axis through its centre. Its angular velocity after 3 seconds is [2001]

(a) 1 rad/s (b) 5 rad/s
(c) 10 rad/s (d) 15 rad/s

50. A weightless ladder 20 ft long rests against a frictionless wall at an angle of 60° from the horizontal. A 150 pound man is 4 ft from the top of the ladder. A horizontal force is needed to keep it from slipping. Choose the correct magnitude of the force from the following [1998]

(a) 175 lb (b) 100 lb
(c) 120 lb (d) 69.2 lb

51. A couple produces [1997]

(a) no motion
(b) purely linear motion
(c) purely rotational motion
(d) linear and rotational

52. The angular momentum of a body with mass (m), moment of inertia (I) and angular velocity (ω) rad/sec is equal to [1996]

(a) Iω (b) Iω^2 (c) $\frac{I}{\omega}$ (d) $\frac{I}{\omega^2}$

53. Angular momentum is [1994]

(a) vector (axial) (b) vector (polar)
(c) scalar (d) none of the above

54. A particle of mass m = 5 is moving with a uniform speed $v = 3\sqrt{2}$ in the XOY plane along the line $y = x + 4$. The magnitude of the angular momentum of the particle about the origin is [1991]

(a) 60 units (b) $40\sqrt{2}$ units
(c) zero (d) 7.5 units

55. A solid homogeneous sphere of mass M and radius R is moving on a rough horizontal surface, partly rolling and partly sliding. During this kind of motion of the sphere [1988]

(a) total kinetic energy is conserved
(b) the angular momentum of the sphere about the point of contact with the plane is conserved
(c) only the rotational kinetic energy about the centre of mass is conserved
(d) angular momentum about the centre of mass is conserved.

Topic 4 Moment of Inertia, Rotational K.E. and Power

56. From a circular ring of mass 'M' and radius 'R' an arc corresponding to a 90° sector is removed. The moment of inertia of the remaining part of the ring about an axis passing through the centre of the ring and perpendicular to the plane of the ring is 'K' times 'MR^2'. Then the value of 'K' is

(a) $\frac{1}{8}$ (b) $\frac{3}{4}$ *[2021]*
(c) $\frac{7}{8}$ (d) $\frac{1}{4}$

57. A solid sphere is rotating freely about its symmetry axis in free space. The radius of the sphere is increased keeping its mass same. Which of the following physical quantities would remain constant for the sphere? *[2018]*
(a) Angular velocity
(b) Moment of inertia
(c) Angular momentum
(d) Rotational kinetic energy

58. Two discs of same moment of inertia rotating about their regular axis passing through centre and perpendicular to the plane of disc with angular velocities ω_1 and ω_2. They are brought into contact face to face coinciding the axis of rotation. The expression for loss of energy during this process is:- *[2017]*
(a) $\frac{1}{4}I(\omega_1-\omega_2)^2$ (b) $I(\omega_1-\omega_2)^2$
(c) $\frac{1}{8}(\omega_1-\omega_2)^2$ (d) $\frac{1}{2}I(\omega_1+\omega_2)^2$

59. From a disc of radius R and mass M, a circular hole of diameter R, whose rim passes through the centre is cut. What is the moment of inertia of the remaining part of the disc about a perpendicular axis, passing through the centre ? *[2016]*
(a) $15\,MR^2/32$ (b) $13\,MR^2/32$
(c) $11\,MR^2/32$ (d) $9\,MR^2/32$

60. Three identical spherical shells, each of mass m and radius r are placed as shown in figure. Consider an axis XX' which is touching to two shells and passing through diameter of third shell. Moment of inertia of the system consisting of these three spherical shells about XX' axis is *[2015]*

(a) $3mr^2$
(b) $\frac{16}{5}mr^2$
(c) $4mr^2$
(d) $\frac{11}{5}mr^2$

61. The ratio of radii of gyration of a circular ring and a circular disc, of the same mass and radius, about an axis passing through their centres and perpendicular to their planes are *[NEET Kar. 2013, 2008]*
(a) $\sqrt{2}:1$ (b) $1:\sqrt{2}$
(c) $3:2$ (d) $2:1$

62. The moment of inertia of a uniform circular disc is maximum about an axis perpendicular to the disc and passing through : *[2012M]*

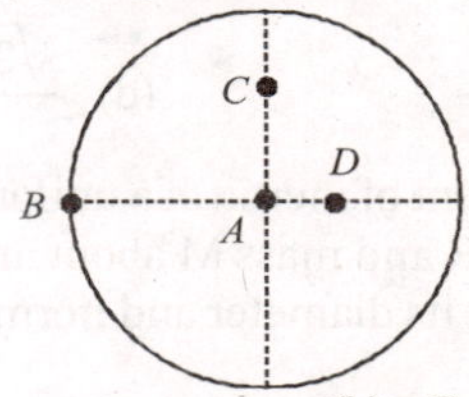

(a) B (b) C
(c) D (d) A

63. The moment of inertia of a thin uniform rod of mass M and length L about an axis passing through its midpoint and perpendicular to its length is I_0. Its moment of inertia about an axis passing through one of its ends and perpendicular to its length is *[2011]*
(a) $I_0+ML^2/2$ (b) $I_0+ML^2/4$
(c) I_0+2ML^2 (d) I_0+ML^2

64. A small mass attached to a string rotates on frictionless table top as shown. If the tension in the string is increased by pulling the string causing the radius of the circular motion to decrease by a factor of 2, the kinetic energy of the mass will *[2011M]*

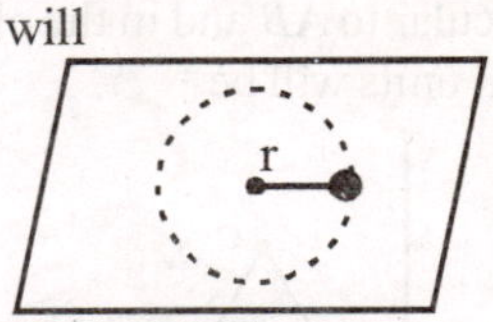

(a) remain constant
(b) increase by a factor of 2
(c) increase by a factor of 4
(d) decrease by a factor of 2

65. Four identical thin rods each of mass M and length l, form a square frame. Moment of inertia of this frame about an axis through the centre of the square and perpendicular to its plane is : *[2009]*
(a) $\frac{2}{3}Ml^2$ (b) $\frac{13}{3}Ml^2$
(c) $\frac{1}{3}Ml^2$ (d) $\frac{4}{3}Ml^2$

66. A thin rod of length L and mass M is bent at its midpoint into two halves so that the angle between them is 90°. The moment of inertia of the bent rod about an axis passing through the bending point and perpendicular to the plane defined by the two halves of the rod is: *[2008]*

(a) $\frac{ML^2}{24}$ (b) $\frac{ML^2}{12}$

(c) $\frac{ML^2}{6}$ (d) $\frac{\sqrt{2}ML^2}{24}$

67. The moment of inertia of a uniform circular disc of radius R and mass M about an axis touching the disc at its diameter and normal to the disc is *[2006, 2005]*

(a) $\frac{2}{5}MR^2$ (b) $\frac{3}{2}MR^2$

(c) $\frac{1}{2}MR^2$ (d) MR^2

68. The ratio of the radii of gyration of a circular disc about a tangential axis in the plane of the disc and of a circular ring of the same radius about a tangential axis in the plane of the ring is *[2004]*

(a) $1:\sqrt{2}$ (b) $1:3$

(c) $2:1$ (d) $\sqrt{5}:\sqrt{6}$

69. Three particles, each of mass m gram, are situated at the vertices of an equilateral triangle *ABC* of side *l* cm (as shown in the figure). The moment of inertia of the system about a line *AX* perpendicular to A*B* and in the plane of *ABC*, in gram-cm^2 units will be *[2004]*

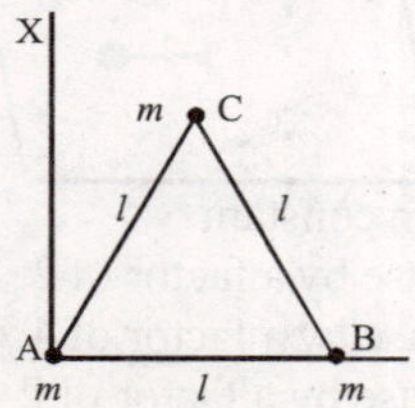

(a) $\frac{3}{2}m\ell^2$ (b) $\frac{3}{4}m\ell^2$

(c) $2m\ell^2$ (d) $\frac{5}{4}m\ell^2$

70. A ball rolls without slipping. The radius of gyration of the ball about an axis passing through its centre of mass is K. If radius of the ball be R, then the fraction of total energy associated with its rotational energy will be *[2003]*

(a) $\frac{R^2}{K^2+R^2}$ (b) $\frac{K^2+R^2}{R^2}$

(c) $\frac{K^2}{R^2}$ (d) $\frac{K^2}{K^2+R^2}$

71. A composite disc is to be made using equal masses of aluminium and iron so that it has as high a moment of inertia as possible. This is possible when *[2002]*

(a) the surfaces of the discs are made of iron with aluminium inside

(b) the whole of aluminium is kept in the core and the iron at the outer rim of the disc

(c) the whole of the iron is kept in the core and the aluminium at the outer rim of the disc

(d) the whole disc is made with thin alternate sheets of iron and aluminium

72. There is a flat uniform triangular plate *ABC* such that *AB* = 4 cm, *BC* = 3 cm and angle *ABC* = 90º. The moment of inertia of the plate about *AB*, *BC* and *CA* as axis is respectively I_1, I_2 and I_3. Which one of the following is true? *[2000]*

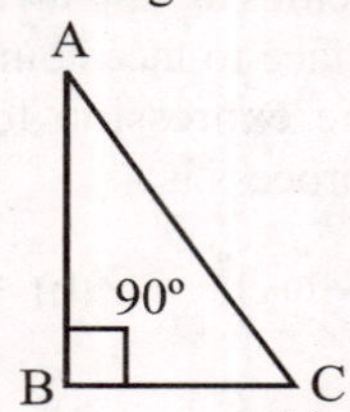

(a) $I_3 > I_2$ (b) $I_2 > I_1$

(c) $I_3 > I_1$ (d) $I_1 > I_2$

73. The moment of inertia of a disc of mass M and radius R about an axis, which is tangential to the circumference of the disc and parallel to its diameter, is *[1999]*

(a) $\frac{3}{2}MR^2$ (b) $\frac{2}{3}MR^2$

(c) $\frac{5}{4}MR^2$ (d) $\frac{4}{5}MR^2$

74. ABC is a triangular plate of uniform thickness. The sides are in the ratio shown in the figure. I_{AB}, I_{BC} and I_{CA} are the moments of inertia of the plate about *AB*, *BC* and *CA* as axes respectively. Which one of the following relations is correct? *[1995]*

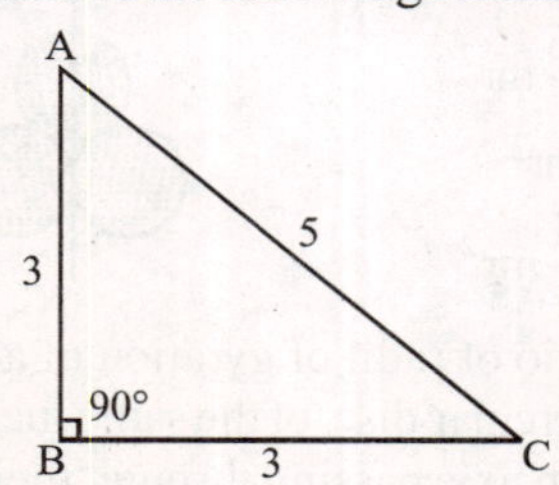

(a) $I_{AB} > I_{BC}$ (b) $I_{BC} > I_{AB}$

(c) $I_{AB} + I_{BC} = I_{CA}$ (d) I_{CA} is maximum

75. In a rectangle $ABCD$ ($BC = 2\,AB$). The moment of inertia is minimum along axis through *[1993]*

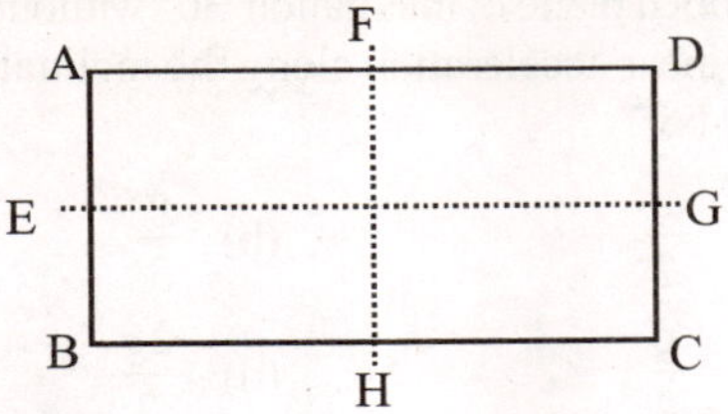

(a) BC (b) BD
(c) HF (d) EG

76. Moment of inertia of a uniform circular disc about a diameter is I. Its moment of inertia about an axis ⊥ to its plane and passing through a point on its rim will be *[1990]*
(a) 5 I (b) 3 I
(c) 6 I (d) 4 I

77. The moment of inertia of a body about a given axis is 1.2 kg m^2. Initially, the body is at rest. In order to produce a rotational kinetic energy of 1500 joule, an angular acceleration of 25 radian/sec^2 must be applied about that axis for a duration of *[1990]*
(a) 4 seconds (b) 2 seconds
(c) 8 seconds (d) 10 seconds

78. A fly wheel rotating about a fixed axis has a kinetic energy of 360 joule when its angular speed is 30 radian/sec. The moment of inertia of the wheel about the axis of rotation is *[1990]*
(a) 0.6 kg/m^2 (b) 0.15 kg m^2
(c) 0.8 kg m^2 (d) 0.75 kg m^2

79. A ring of mass m and radius r rotates about an axis passing through its centre and perpendicular to its plane with angular velocity ω. Its kinetic energy is *[1988]*
(a) $\frac{1}{2}mr^2\omega^2$ (b) $mr\omega^2$
(c) $mr^2\omega^2$ (d) $\frac{1}{2}mr\omega^2$

Topic 5 Rolling Motion

80. A disc of radius 2 m and mass 100 kg rolls on a horizontal floor. Its centre of mass has speed of 20 cm/s. How much work is needed to stop it? *[2019]*
(a) 3 J (b) 30 kJ
(c) 2 J (d) 1J

81. A solid cylinder of mass 2 kg and radius 50 cm rolls up an inclined plane of angle of inclination 30°. The centre of mass of the cylinder has speed of 4 m/s. The distance travelled by the cylinder on the inclined surface will be, [take g = 10 m/s^2] *[NEET Odisha 2019]*
(a) 2.4m (b) 2.2m
(c) 1.6m (d) 1.2m

82. A solid sphere is in rolling motion. In rolling motion a body possesses translational kinetic energy (K_t) as well as rotational kinetic energy (K_r) simultaneously. The ratio $K_t : (K_t + K_r)$ for the sphere is *[2018]*
(a) 7 : 10 (b) 5 : 7
(c) 2 : 5 (d) 10 : 7

83. Three objects, A : (a solid sphere), B : (a thin circular disk) and C : (a circular ring), each have the same mass M and radius R. They all spin with the same angular speed ω about their own symmetry axes. The amounts of work (W) required to bring them to rest, would satisfy the relation *[2018]*
(a) $W_C > W_B > W_A$ (b) $W_A > W_B > W_C$
(c) $W_A > W_C > W_B$ (d) $W_B > W_A > W_C$

84. A disk and a sphere of same radius but different masses roll off on two inclined planes of the same altitude and length. Which one of the two objects gets to the bottom of the plane first? *[2016]*
(a) Disk
(b) Sphere
(c) Both reach at the same time
(d) Depends on their masses

85. The ratio of the accelerations for a solid sphere (mass 'm' and radius 'R') rolling down an incline of angle 'θ' without slipping and slipping down the incline without rolling is: *[2014]*
(a) 5 : 7 (b) 2 : 3
(c) 2 : 5 (d) 7 : 5

86. A small object of uniform density rolls up a curved surface with an initial velocity 'v'. It reaches upto a maximum height of $\frac{3v^2}{4g}$ with respect to the initial position. The object is *[2013]*
(a) solid sphere (b) hollow sphere
(c) disc (d) ring

87. A drum of radius R and mass M, rolls down without slipping along an inclined plane of angle θ. The frictional force *[2005]*
(a) dissipates energy as heat
(b) decreases the rotational motion
(c) decreases the rotational and translational motion
(d) converts translational energy to rotational energy

88. A solid cylinder of mass m and radius R rolls down an inclined plane of height h without slipping. The speed of its centre of mass when it reaches the bottom is *[2003, 1989]*
(a) $\sqrt{(2gh)}$ (b) $\sqrt{4gh/3}$
(c) $\sqrt{3gh/4}$ (d) $\sqrt{4g/h}$

89. A solid cylinder and a hollow cylinder both of the same mass and same external diameter are released from the same height at the same time on an inclined plane. Both roll down without slipping. Which one will reach the bottom first? *[2000]*
(a) Both together
(b) Solid cylinder
(c) One with higher density
(d) Hollow cylinder

90. A spherical ball rolls on a table without slipping. Then the fraction of its total energy associated with rotation is *[1994]*
(a) 2/5 (b) 2/7
(c) 3/5 (d) 3/7

91. A thin uniform circular ring is rolling down an inclined plane of inclination 30° without slipping. Its linear acceleration along the inclination plane will be *[1994]*
(a) $\frac{g}{2}$ (b) $\frac{g}{3}$
(c) $\frac{g}{4}$ (d) $\frac{2g}{3}$

92. A solid sphere, disc and solid cylinder all of the same mass and made of the same material are allowed to roll down (from rest) on the inclined plane, then *[1993]*
(a) solid sphere reaches the bottom first
(b) solid sphere reaches the bottom last
(c) disc will reach the bottom first
(d) all reach the bottom at the same time

93. The speed of a homogenous solid sphere after rolling down an inclined plane of vertical height h from rest without sliding is *[1992]*
(a) $\sqrt{\frac{10}{7}gh}$ (b) $\sqrt{gh}$
(c) $\sqrt{\frac{6}{5}gh}$ (d) $\sqrt{\frac{4}{3}gh}$

94. If a sphere is rolling, the ratio of the translational energy to total kinetic energy is given by *[1991]*
(a) 7:10 (b) 2:5
(c) 10:7 (d) 5:7

ANSWER KEY

1	(b)	12	(d)	23	(a)	34	(c)	45	(d)	56	(b)	67	68	78	70	89	72
2	(d)	13	(a)	24	(b)	35	(a)	46	(c)	57	(c)	66	(d)	79	(a)	90	(b)
3	(c)	14	(d)	25	(c)	36	(d)	47	(b)	58	(a)	69	(d)	80	(a)	91	(c)
4	(a)	15	(a)	26	(d)	37	(a)	48	(b)	59	(b)	70	(d)	81	(a)	92	(a)
5	(b)	16	(c)	27	(c)	38	(b)	49	(d)	60	(c)	71	(b)	82	(b)	93	(a)
6	(a)	17	(b)	28	(d)	39	(c)	50	(d)	61	(a)	72	(b)	83	(a)	94	(d)
7	(d)	18	(d)	29	(d)	40	(a)	51	(c)	62	(a)	73	(c)	84	(b)		
8	(d)	19	(a)	30	(d)	41	(d)	52	(a)	63	(b)	74	(b)	85	(a)		
9	(b)	20	(b)	31	(c)	42	(d)	53	(a)	64	(c)	75	(d)	86	(c)		
10	(a)	21	(a)	32	(a)	43	(d)	54	(a)	65	(d)	76	(c)	87	(d)		
11	(d)	22	(b)	33	(a)	44	(c)	55	(b)	66	(b)	77	(b)	88	(b)		

Hints & Solutions

1. (b)

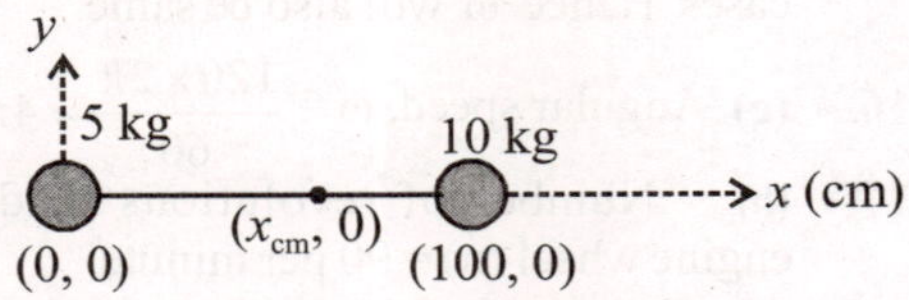

Centre of mass of a system of two particles of masses m_1 and m_2 having position vectors $\overrightarrow{x_1}$ and $\overrightarrow{x_2}$ is

$$x_{cm} = \frac{m_1x_1 + m_2x_2}{m_1 + m_2}$$

Taking particle of mass 5 kg at origin.

$$\Rightarrow x_{cm} = \frac{5\times 0 + 100\times 10}{5+10} = \frac{200}{3} = 66.66 \text{ cm}$$

$$\Rightarrow x_{cm} = 67 \text{ cm}$$

2. (d) Mechanical advantage, M.A.= $\frac{\text{Load}}{\text{Effort}}$

If M.A. > 1 $\Rightarrow$ Load > Effort

> **NOTES**
> Centre of mass may or may not coincide with centre of gravity. Net torque of gravitational pull is zero about centre of mass.
>
> $\tau_g = \Sigma\tau_i = \Sigma r_i \times m_i g = 0$

3. (c)

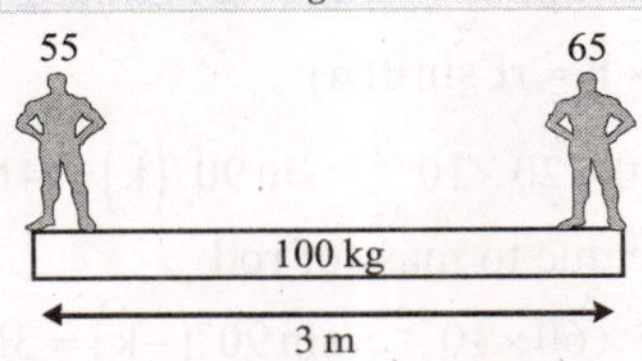

There is no external force so centre of mass of the system will not shift

4. (a) $X_{cm} = \frac{m_1x_1 + m_2x_2 + m_3x_3}{m_1 + m_2 + m_3}$

$$X_{cm} = \frac{300\times(0) + 500(40) + 400\times 70}{300 + 500 + 400}$$

$$X_{cm} = \frac{500\times 40 + 400\times 70}{1200}$$

$$X_{cm} = \frac{50 + 70}{3} = \frac{120}{3} = 40 \text{ cm}$$

5. (b) If no external force action a system of particles, the centre of mass remains at rest. So, speed of centre of mass is zero.

> **NOTES**
> Force acting on the rigid body,
>
> $\vec{F} = M\vec{a}_{cm} = \frac{Md^2\vec{r}}{dt^2}$
>
> When external force acting on the rigid body is zero.
>
> $\vec{F} = M\frac{d}{dt}(\vec{V}_{cm}) = 0 \Rightarrow \vec{V}_{cm} = \text{Constant}$
>
> So, when no external force acts on a body its centre of mass will remain at rest or move with constant velocity.

6. (a) The position vector of the centre of mass of two particle system is given by

$$\vec{R}_{cm} = \frac{m_1\vec{r}_1 + m_2\vec{r}_2}{(m_1 + m_2)}$$

$$= \frac{1}{4}[-8\hat{i} - 4\hat{j} + 4\hat{k}] = -2\hat{i} - \hat{j} + \hat{k}$$

7. (d) Consider an element of length dx at a distance x from end A.

Here, mass per unit length λ of rod

$\lambda \propto x \Rightarrow \lambda = kx$

$\therefore \; dm = \lambda dx = kx\,dx$

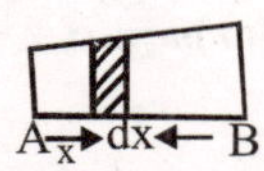

Position of centre of gravity of rod from end A.

$$x_{CG} = \frac{\int_0^L x\,dm}{\int_0^L dm}$$

$$\therefore \; x_{CG} = \frac{\int_0^3 x(kx\,dx)}{\int_0^3 kx\,dx} = \frac{\left[\frac{x^3}{3}\right]_0^3}{\left[\frac{x^2}{2}\right]_0^3} = \frac{\frac{(3)^3}{3}}{\frac{(3)^3}{2}} = 2m$$

8. (d) As friction is absent at the point of contact,

$$\text{Acceleration} = \frac{\text{Force}}{\text{Mass}}$$

It is independent of h

9. **(b)** Centre of mass of system depends upon position and masses of particle. Also, it depends upon relative distance between particles.

10. **(a)**

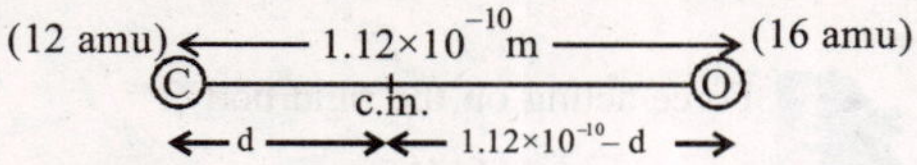

From definition of centre of mass.

$$d = \frac{16 \times 1.12 \times 10^{-10} + 12 \times 0}{16 + 12}$$

$$= \frac{16 \times 1.12 \times 10^{-10}}{28} = 0.64 \times 10^{-10} \text{ m.}$$

11. **(d)** Let T_A and T_B are the time periods of particle A and B respectively. According to question,

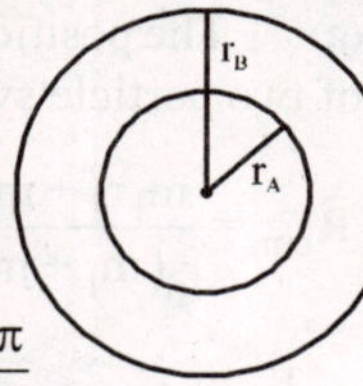

$T_A = T_B = T$

If ω_A and ω_B are their angular speeds, then

$$\omega_A = \frac{2\pi}{T_A} \text{ and } \omega_B = \frac{2\pi}{T_B}$$

$$\therefore \frac{\omega_A}{\omega_B} = \frac{T_B}{TA} = \frac{T}{T} = 1:1$$

12. **(d)** $V^2 = u^2 + 2as$

$$V_0^2 = 0 + 2r\alpha(2\pi r)n \Rightarrow \alpha = \frac{V_0^2}{4\pi n r^2}$$

13. **(a)** Since, $\theta = \omega_0 t + \frac{1}{2}\alpha t^2$

Where α is angular aceleration, ω_0 is the initial angular speed.

$t = 2$s

$$\theta = 2 \times 2 + \frac{1}{2} \times 3(2)^2 = 4 + 6 = 10 \text{ rad}$$

14. **(d)**

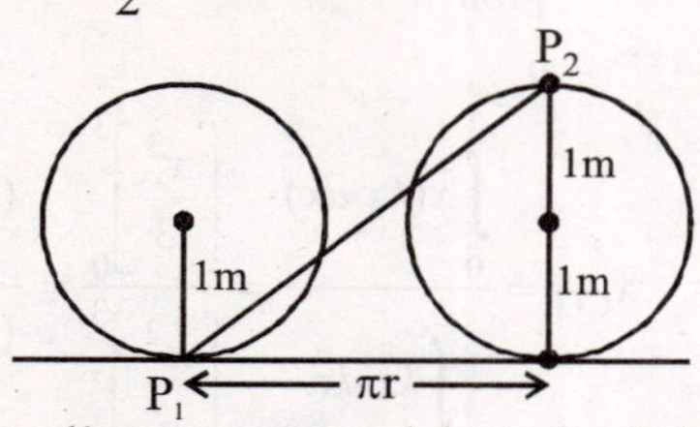

Linear distance moved by wheel in half revolution $= \pi r$. Point P_1 after half revolution reaches at P_2 vertically 2m above the ground.

$\therefore$ Displacement P_1P_2

$$= \sqrt{\pi^2 r^2 + 2^2} = \sqrt{\pi^2 + 4} \quad [\because r = 1m]$$

15. **(a)** As time taken by both car to complete one revolution is same.

As $\omega = \frac{2\pi}{T} \Rightarrow \omega \propto \frac{1}{T}$, as T is same in both cases. Hence 'ω' will also be same.

16. **(c)** Angular speed, $\omega = \frac{120 \times 2\pi}{60} = 4\pi$ rad/sec

17. **(b)** Number of revolutions made by the engine wheel (n) = 90 per minute.

Angular speed of the engine wheel

$$(\omega) = \frac{2\pi n}{60} = \frac{2\pi \times 90}{60} = 3\pi \text{ rad/s.}$$

18. **(d)** We know that both the cars take the same time to complete the circle, therefore ratio of angular speeds of the cars will be 1 : 1.

19. **(a)**

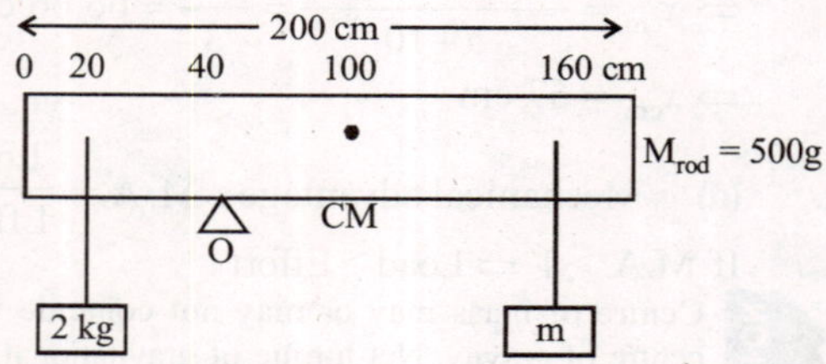

Rod will be in equilibrium, if $\vec{\tau}_{net} = 0$ about point O.

Torque at point O due to 2 kg mass

$$\vec{\tau} = \vec{r} \times \vec{F} = rF\sin\theta(\hat{n})$$

$$\tau_1 = 20 \times 20 \times 10^{-2} \times \sin 90°(\hat{k}) = 4\text{Nm}(\hat{k})$$

Torque due to mass of rod

$$\tau_2 = 5 \times 60 \times 10^{-2} \times \sin 90°(-\hat{k}) = 3\text{Nm}(-\hat{k})$$

Torque due to mass m

$$\tau_3 = \text{mg} \times 120 \times 10^{-2} \times \sin 90°(-\hat{k})$$

$$= 12\text{m Nm}(-\hat{k})$$

So $\vec{\tau_1} + \vec{\tau_2} + \vec{\tau_3} = 0 \; (\because \vec{\tau}_{net} = 0)$

$4 - 3 - 12\text{m} = 0 \Rightarrow 12\text{m} = 1$

$$\therefore \text{m} = \frac{1}{12} \text{kg}$$

20. **(b)** Given :

Force, $\vec{F} = 3\hat{j}$ N

Position vector, $\vec{r} = 2\hat{k}$ m

Torque, $\vec{\tau} = \vec{r} \times \vec{F} = 2\hat{k} \times 3\hat{j} = 6(\hat{k} \times \hat{j}) = 6(-\hat{i})$

$\Rightarrow \vec{\tau} = -6\hat{i}$ Nm

21. **(a)** According to the Work-energy theorem,

$$W = \frac{1}{2} I\left(\omega_f^2 - \omega_i^2\right)$$

Given that,

$\theta = 2\pi$ revolution/minute

$\theta = 2\pi \times 2\pi = 4\pi^2$ rad

$$\omega_i = 3 \times \frac{2\pi}{60} \text{ rad/s}$$

$\omega_f = 0$ rad/s

Putting the values of ω_f and ω_i we get

$$\Rightarrow -\tau\theta = \frac{1}{2} \times \frac{1}{2} mr^2\left(0^2 - \omega_i^2\right)$$

$$\Rightarrow -\tau = \frac{\frac{1}{2} \times \frac{1}{2} \times 2 \times \left(4 \times 10^{-2}\right)\left(-3 \times \frac{2\pi}{60}\right)^2}{4\pi^2}$$

$\Rightarrow \tau = 2 \times 10^{-6}$ N-m

22. **(b)** Given, mass of cylinder m = 3kg

R = 40 cm = 0.4 m

F = 30 N ; $\alpha = ?$

As we know, torque $\tau = I\alpha$

$F \times R = MR^2\alpha$

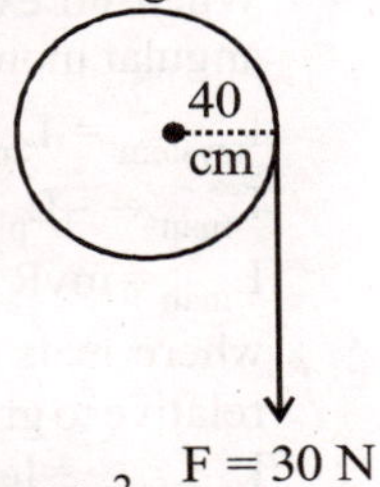

$$\alpha = \frac{F \times R}{MR^2}$$

$$\alpha = \frac{30 \times (0.4)}{3 \times (0.4)^2} \quad \text{or, } \alpha = 25 \text{rad/s}^2$$

23. **(a)** **Given:** Radius of disc, R = 50 cm

Angular acceleration, $\alpha = 2.0$ rads^{-2}; time t = 2s

Particle at periphery (assume) will have both radial (one) and tangential acceleration

$a_t = R\alpha = 0.5 \times 2 = 1$ m/s^2

Angular speed

$\omega = \omega_0 + \alpha t$

$\omega = 0 + 2 \times 2 = 4$ rad/sec

Centripetal acceleration

$a_c = \omega^2 R = (4)^2 \times 0.5 = 16 \times 0.5 = 8$m/s^2

Linear acceleration, $a_t = \alpha.R = 1$m/sec^2

Net acceleration,

$$a_{total} = \sqrt{a_t^2 + a_c^2} = \sqrt{1^2 + 8^2} \approx 8 \text{ m/s}^2$$

> **NOTES**
>
> The acceleration of the particle moving in a circle has two components; tangential acceleration (at) which is along the tangent and radial acceleration (ar) which 15 towards the centre of circle. As the two components are mutually perpendicular therefore, net acceleration of the particle will be
>
> $$a = \sqrt{a_t^2 + a_r^2}$$

24. **(b)** Applying angular momentum conservation

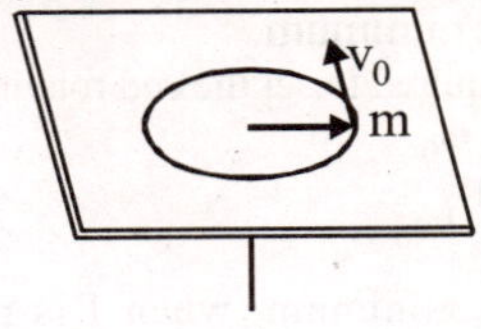

$$mv_0R_0 = (m)(v)\left(\frac{R_0}{2}\right)$$

$\therefore \quad v' = 2v_0$

Therefore, final KE $= \frac{1}{2} m(2v_0)^2 = 2mv_0^2$

25. **(c)** By torque balancing about B

$N_A(d) = W(d - x)$

$$N_A = \frac{W(d-x)}{d}$$

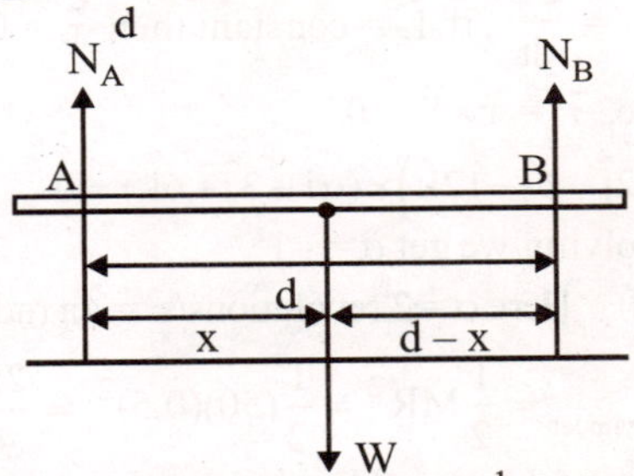

26. **(d)** **Given :** Speed v = 54 kmh^{-1} = 15 ms^{-1}

Moment of inertia, I = 3 kgm^2

Time t = 15s

$$\omega_i = \frac{v}{r} = \frac{15}{0.45} = \frac{100}{3} \qquad [\text{Given } \omega_f = 0]$$

$\omega_f = \omega_i + \alpha t$

$$0 = \frac{100}{3} + (-\alpha)(15) \quad \Rightarrow \quad \alpha = \frac{100}{45}$$

Average torque transmitted by brakes to the wheel,

$$\tau = (I)(\alpha) = 3 \times \frac{100}{45} = 6.66 \text{ kgm}^2\text{s}^{-2}$$

27. **(c)** From figure,

Total moment of inertia of the rod,

$I = m_1x^2 + m_2(L - x)^2$

$I = m_1x^2 + m_2[L^2 + x^2 - 2Lx]$

$I = m_1x^2 + m_2L^2 + m_2x^2 - 2m_2Lx$

As I is minimum, so, $\frac{dI}{dx} = 0$ ... (i)

$$\Rightarrow \frac{dI}{dx} = 2m_1x + 2m_2x - 2m_2L$$

Using equation (i)

$\Rightarrow \quad 2m_1x + 2m_2x - 2m_2L = 0$

$$\Rightarrow \quad x = \frac{m_2L}{(m_1 + m_2)}$$

When I is minimum, the work done by rotating a rod will be minimum.

NOTES

Work required to set the rod rotating with angular velocity ω_0.

$$K.E. = \frac{1}{2}I\omega_0^2$$

Work is minimum, when I is minimum. I is minimum about the centre of mass.

$\therefore\ m_1x_1 = m_2(L - x)$

$\Rightarrow m_1x = m_2L - M_2x$

$$\therefore\ x = \frac{m_2L}{m_1 + m_2}$$

28. (d) From Newton's second law for rotational motion,

$$\vec{\tau} = \frac{d\vec{L}}{dt}, \text{ if } \vec{L} = \text{constant then } \vec{\tau} = 0$$

So, $\vec{\tau} = \vec{r} \times \vec{F} = 0$

$$\left(2\hat{i} - 6\hat{j} - 12\hat{k}\right) \times (\alpha\hat{i} + 3\hat{j} + 6\hat{k}) = 0$$

Solving we get $\alpha = -1$

29. (d) Here $\alpha = 2$ revolutions/s$^2 = 4\pi$ rad/s^2 (given)

$$I_{cylinder} = \frac{1}{2}MR^2 = \frac{1}{2}(50)(0.5)^2 = \frac{25}{4} \text{ Kg-m}^2$$

As $\tau = I\alpha$ so $TR = I\alpha$

$$\Rightarrow T = \frac{I\alpha}{R} = \frac{\left(\frac{25}{4}\right)(4\pi)}{(0.5)} N = 50\pi N = 157 N$$

30. (d)

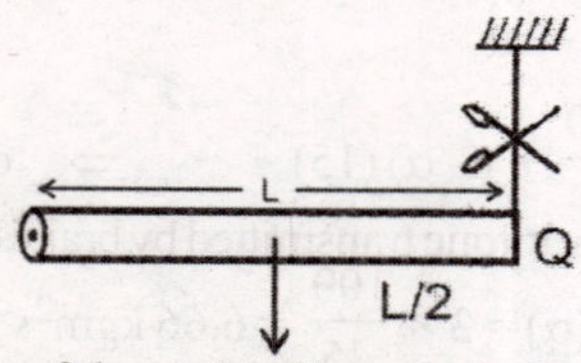

Weight of the rod will produce the torque

$$\tau = mg\frac{L}{2} = I\alpha = \frac{mL^2}{3}\alpha \quad \left[\because I_{rod} = \frac{ML^2}{3}\right]$$

Hence, angular acceleration, $\alpha = \frac{3g}{2L}$

31. (c) Given: $m_1 = 2$ kg $\quad m_2 = 4$ kg

$r_1 = 0.2$ m $\quad r_2 = 0.1$ m

$\omega_1 = 50$ rad s^{-1} $\quad \omega_2 = 200$ rad s^{-1}

As, angular momentum,

$I_1\omega_1 + I_2\omega_2 = (I_1 + I_2)\omega_1$

$$\therefore\ \omega_f = \frac{I_1\omega_1 + I_2\omega_2}{I_1 + I_2} = \frac{\frac{1}{2}m_1r_1^2\omega_1 + \frac{1}{2}m_2r_2^2\omega_2}{\frac{1}{2}m_1r_1^2 + \frac{1}{2}m_2r_2^2}$$

By putting the value of m_1, m_2, r_1, r_2 and solving we get $= 100$ rad s^{-1}

32. (a) $\because \vec{L} = \vec{r} \times \vec{p}$

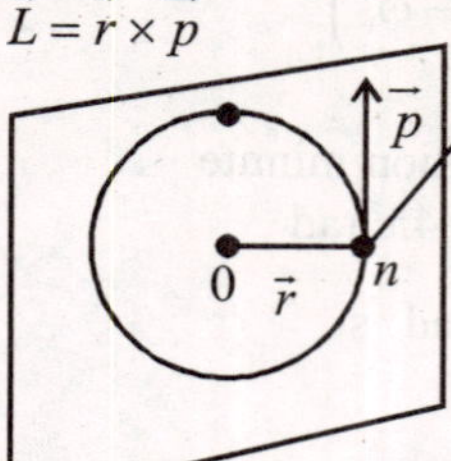

By right hand screw, rule, the direction of $\vec{L}$ is $\perp$ to the plane containing $\vec{r}$ and $\vec{p}$.

33. (a) $F_1x + F_2x = F_3x$

$F_3 = F_1 + F_2$

34. (c) According to the conservation of angular momentum,

When no external torque acts on system, the angular momentum does not change.

$L_{system} = L_{man} + L_{platform} = \text{constant}$

$L_{man} = -L_{platform}$

$L_{man} = mvR$

where m is mass of man, v is speed of man relative to ground, R is the radius of platform,

$L_{platform} = I\omega$

where I is moment of inertia, ω is angular velocity.

$$\Rightarrow \omega = \frac{L_{platform}}{I} = -\frac{L_{man}}{I} = \frac{-mvR}{I}$$

Speed of man relative to platform is

$$v_{platform} = v - \omega R = v + \frac{mvR}{I}R$$

$$= v\left[1 + \frac{mR^2}{I}\right]$$

Time taken by man to complete one revolution is,

$$t = \frac{2\pi R}{v_{platform}} = \frac{2\pi r}{v\left[1 + \frac{mR^2}{I}\right]} = 2 \times \frac{2}{1(1+1)}\pi$$

$\Rightarrow\ t = 2\pi$ second

35. (a) $\theta(t) = 2t^3 - 6t^2$

$$\Rightarrow \frac{d\theta}{dt} = 6t^2 - 12t \Rightarrow \alpha = \frac{d^2\theta}{dt^2} = 12t - 12 = 0$$

$\therefore\ t = 1$ sec.

NOTES When angular acceleration (α) is zero then torque on the wheel becomes zero.

36. (d) By conservation of angular momentum,

$I_t \omega_i = (I_t + I_b)\,\omega_f$

where ω_f is the final angular velocity of disks

$$\therefore \quad \omega_f = \left(\frac{I_t}{I_t + I_b}\right)\omega_i$$

Loss in K.E., ΔK = Initial K.E. – Final K.E.

$$= \frac{1}{2}I_t\omega_i^2 - \frac{1}{2}(I_t + I_b)\omega_f^2$$

$$= \frac{1}{2}I_t\omega_i^2 - \frac{1}{2}(I_t + I_b)\frac{I_t^2}{(I_t + I_b)^2}\omega_i^2$$

$$= \frac{1}{2}\omega_i^2\frac{I_t}{I_t + I_b}(I_t + I_b - I_t) = \frac{1}{2}\omega_i^2\frac{I_t I_b}{I_t + I_b}$$

37. (a) In absence of external torque, L = Iω = constant

$I_1\omega_1 = I_2\omega_2, I_1 = MR^2, I_2 = MR^2 + 2mR^2$

(Moment of inertia of a thin circular ring about an axis vertical to its plane = MR^2)

$$\therefore \quad \omega_2 = \frac{I_1}{I_2}\omega = \frac{M}{M + 2m}\omega.$$

38. (b) $\vec{\tau} = \vec{r} \times \vec{F} \Rightarrow \vec{r}\,.\,\vec{\tau} = 0 \qquad \vec{F}\,.\,\vec{\tau} = 0$

Since, $\vec{\tau}$ is perpendicular to the plane of $\vec{r}$ and $\vec{F}$, hence the dot product of $\vec{\tau}$ with $\vec{r}$ and $\vec{F}$ is zero.

NOTES Radial component of force makes no contribution to the torque. Only transverse component of force contributes to the torque.

39. (c)

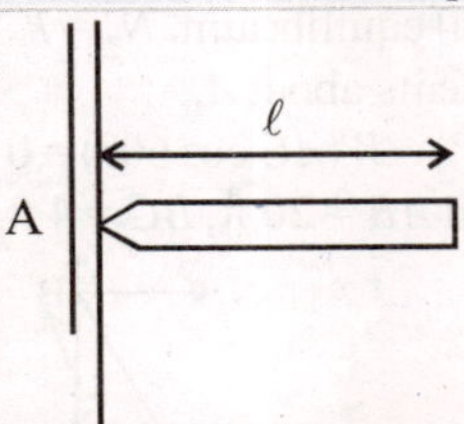

Weight of the rod will produce torque,

$$\tau = mg \times \frac{\ell}{2}$$

Also, $\tau = I\alpha$

where, I is the moment of inertia = $\frac{m\ell^2}{3}$

and α is the angular acceleration

$$\therefore \frac{m\ell^2}{3}\alpha = mg \times \frac{\ell}{2} \Rightarrow \alpha = \frac{3g}{2\ell}$$

40. (a) Angular momentum = Linear momentum × distance of line of action of linear momentum about the origin.

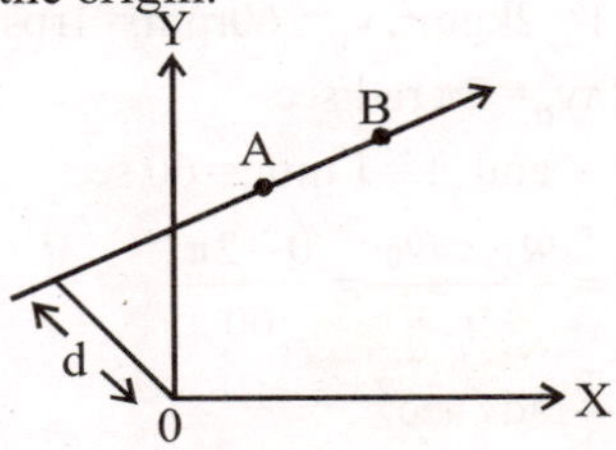

$L_A = p_A \times d, L_B = p_B \times d$

As linear momenta are p_A and p_B equal, therefore, $L_A = L_B$.

41. (d)

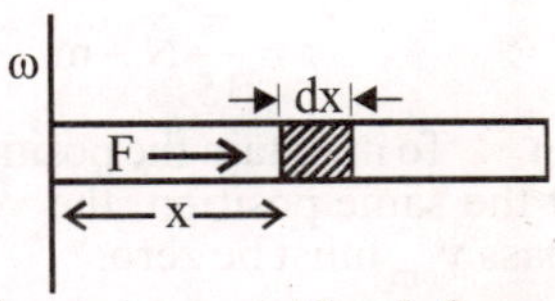

Rotation will create centrifugal force due to which pressure will be created at the other end. Now, this centrifugal force will depend on distance of the liquid from the axis as centrifugal force = $m\omega^2 r$.

Let us consider a small part dx of the liquid column which is at a distance of x.

Mass of this element will be,

$$dm = \frac{M}{L}dx$$

So, force exerted on this element will be,

$\Rightarrow df = dm.\omega^2 x$

As the mass of liquid extends from x = 0 to x = L, the net force on the wall experienced by the mass of the liquid is,

$$\Rightarrow \int_0^f df = \frac{M}{L}\int_0^L \omega^2.x.dx$$

$$\Rightarrow \int_0^f df = \frac{M}{L}\omega^2\int_0^L x.dx$$

$$\Rightarrow f = \frac{M}{L}\omega^2\left[\frac{L^2}{2}\right] = \frac{M\omega^2 L}{2}$$

42. (d) The relation between K.E. of rotation and angular momentum,

$$K=\frac{L^2}{2I} \Rightarrow L^2=2KI \Rightarrow L=\sqrt{2KI}$$

$$\frac{L_1}{L_2}=\sqrt{\left(\frac{K_1}{K_2}\right)\left(\frac{I_1}{I_2}\right)}=\sqrt{\left(\frac{K}{K}\right)\left(\frac{I}{2I}\right)}=\frac{1}{\sqrt{2}}$$

$L_1 : L_2 = 1 : \sqrt{2}$

43. **(d)** $I = 2\text{kgm}^2, \nu_0 = 60\text{rpm} = 1\text{rps}$

$\omega_0 = 2\pi\nu_0 = 2\pi$ rad/sec

$\omega_f = 0$ and $t = 1$ min $= 60$ sec

So, $\alpha = \frac{\omega_f - \omega_0}{t} = \frac{0-2\pi}{60}$

$\alpha = \frac{-\pi}{30}$ rad / sec^2

and Torque, $\tau = I.\alpha$

$$\tau = 2.\left(\frac{-\pi}{30}\right) = \frac{-\pi}{15}$$

$$\tau = \frac{\pi}{15} \text{N}-\text{m}$$

44. **(c)** To maintain the position of centre of mass at the same position, the velocity of centre of mass ν_{cm} must be zero.

That means, $\nu_{cm} = \frac{m_1\nu_1 + m_2\nu_2}{(m_1+m_2)} = 0$

$$\Rightarrow \frac{m_1\frac{dr_1}{dt} + m_2\frac{dr_2}{dt}}{(m_1+m_2)} = 0 \quad \left[\because \nu_1 = \frac{dr_1}{dt}, \; \nu_2 = \frac{dr_2}{dt}\right]$$

$\Rightarrow m_1 dr_1 + m_2 dr_2 = 0$

where, ν_1 and ν_2 are the velocities of the particles and dr_1 and dr_2 are the change in dosplacement of particles. Then, $dr_2 = -\left(\frac{m_1}{m_2}\right)dr_1$

> **NOTES**
> As the centre of mass of a body is a point where the whole mass of the body is concentrated, therefore, sum of moments of masses of all the particles of the body about the centre of mass is zero, i.e., $\sum_{i-1}^{i=n} m_i\vec{r_i} = 0$
>
> For two particle system
>
> $m_1\vec{r_1} + m_2\vec{r_2} = 0$
>
> $\Rightarrow \vec{r_1} = \frac{m_1\vec{r_1}}{m_2}$

45. **(d)** Angular momentum will be conserved

$$I_1\omega = I_1\omega' + I_2\omega' \Rightarrow \omega' = \frac{I_1\omega}{I_1+I_2}$$

46. **(c)** Applying conservation law of angular momentum, $I_1\omega_1 = I_2\omega_2$

$I_2 = (Mr^2) + 4(m)(r^2) = (M+4m)r^2$

(Taking $\omega_1 = \omega$ and $\omega_2 = \omega_1$)

$\Rightarrow Mr^2\omega = (M+4m)r^2\omega_1$

$$\Rightarrow \omega_1 = \frac{M\omega}{M+4m}$$

47. **(b)** If external torque is zero, angular momentum remains conserved.

$L = I\omega =$ constant

> **NOTES**
> External torque is zero because the weight of child acts downward

48. **(b)** As net torque applied is zero.

Hence, $\tau = \frac{dL}{dt}$

$\frac{dL}{dt} = 0$, $L =$ constant.

L (angular momentum) remains conserved.

49. **(d)** $\tau = 1000\,N\text{-}m, I = 200\,\text{kg-}m^2$

$\tau = I.\alpha$ and $\alpha = \left(\frac{\omega_f - \omega_0}{t}\right)$

$\Rightarrow \alpha = \frac{1000}{200} = 5$ rad/sec^2

$\omega_1 = \omega_0 + \alpha t = 0 + 3 \times 5 = 15$ rad/s

50. **(d)** AB is the ladder, let F be the horizontal force and W is the weight of man. Let N_1 and N_2 be normal reactions of ground and wall, respectively. Then for vertical equilibrium

$W = N_1$ .....(1)

For horizontal equilibrium, $N_2 = F$ .....(2)

Taking moments about A,

$N_2(AB \sin 60°) - W(AC \cos 60°) = 0$ ......(3)

Using (2) and $AB = 20$ ft, $BC = 4$ ft, we get

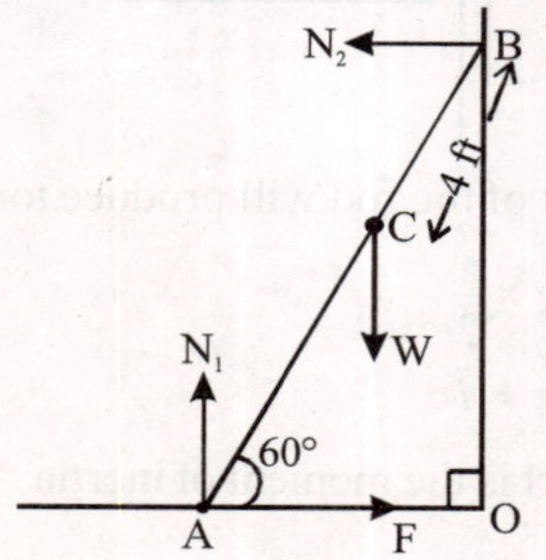

$$F\left(20\times\frac{\sqrt{3}}{2}\right)-W\left(16\times\frac{1}{2}\right)=0$$

$$\Rightarrow F=\frac{8W\times 2}{20\sqrt{3}}=\frac{4W}{5\sqrt{3}}=\frac{150\times 4}{5\sqrt{3}} \text{ pound}$$

$$=40\sqrt{3}=40\times 1.73=69.2 \text{ pound}$$

51. (c) A couple is formed of two equal and opposite forces at some separation; so net force is zero. Hence ,a couple does not produce translatory motion; but it causes change in rotational motion.

52. (a) Let body contain m_1, m_2, m_3 ....m_n masses at distance r_1, r_2, r_3 .... r_n from axis OA.
Angular momentum of body

$$=m_1v_1r_1+m_2v_2r_2 .....+m_nv_nr_n$$
$$=m_1(\omega r_1)r_1+m_2(\omega r_2)r_2......+m_n(\omega r_n)r_n$$
$$=(m_1r_1^2)\omega+(m_2r_2^2)\omega.......+(m_nr_n^2)\omega$$
$$=\left(\sum_{i=1}^{n} m_ir_i^2\right)\omega=I\omega$$

53. (a) Angular momentum $\vec{L}$ is defined as $\vec{L}=\vec{r}\times m(\vec{v})$
So, $\vec{L}$ is an axial vector.

54. (a)

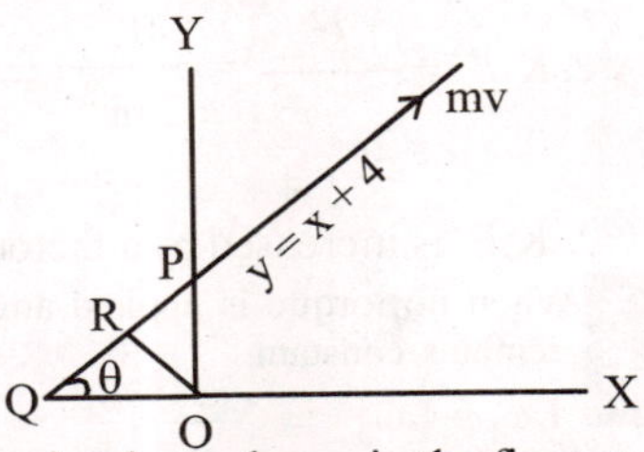

$y=x+4$ line has been shown in the figure when $x=0, y=4$. So, $OP=4$.
The slope of the line can be obtained by comparing with the equation of the straight line
$y=mx+c$
$m=\tan\theta=1$
$\Rightarrow\theta=45°$
$\angle OQP=\angle OPQ=45°$
If we draw a line perpendicular to this line, length of the perpendicular $=OR$

In ΔOPR, $\frac{OR}{OP}=\sin 45°$

$$\Rightarrow OR=OP\sin 45°$$
$$=4\times\frac{1}{\sqrt{2}}=\frac{4}{\sqrt{2}}=2\sqrt{2}$$

Angular momentum of particle going along this line $=r\times mv=2\sqrt{2}\times 5\times 3\sqrt{2}=60$ units

55. (b) Angular momentum about the point of contact with the surface includes the angular momentum about the centre. Due to friction linear momentum is conserved.

56. (b) As per question, 90° arc is removed from circular ring,

$$\therefore \text{Mass removed}=\frac{M}{4}$$

$$\therefore \text{Mass remaining } M-\frac{M}{4}=\frac{3}{4}M$$

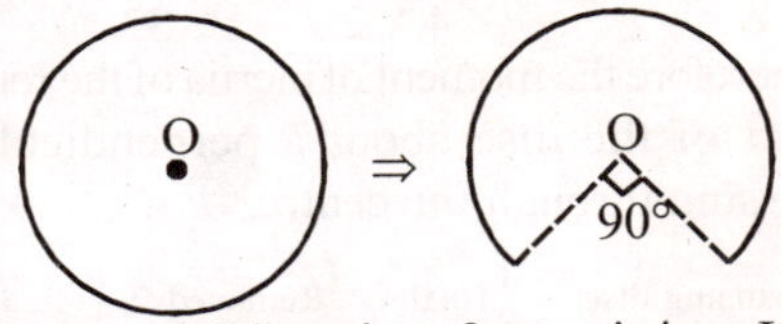

Moment of inertia of remaining $I_{remaining}$

$$=\int dmr^2$$
$$=R^2\int dm \quad (\because r=R)$$
$$\therefore I_{remaining}=\frac{3MR^2}{4}=KMR^2 \therefore K=\frac{3}{4}M$$

NOTES Moment of inertia of a continuous distribution of mass, treating the element of mass dm at position r as particle then dI= dmr² so $I=\int r^2dm$

57. (c) Angular momentum, remains conserved until the torque acting on sphere remains zero.
$\tau_{ex}=0$

So, $\frac{dL}{dt}=0$

i.e., angular momentum L = constant.

58. (a) Here, $I\omega_1+I\omega_2=2I\omega$

$$\Rightarrow \omega=\frac{\omega_1+\omega_2}{2}$$

$$(K.E.)_i=\frac{1}{2}I\omega_1^2+\frac{1}{2}I\omega_2^2$$

$$(K.E.)_f=\frac{1}{2}\times 2I\omega^2=I\left(\frac{\omega_1+\omega_2}{2}\right)^2$$

$$\text{Loss in K.E.}=(K.E)_f-(K.E)_i=\frac{1}{4}I(\omega_1-\omega_2)^2$$

59. **(b)** Moment of inertia of complete disc about point 'O'.

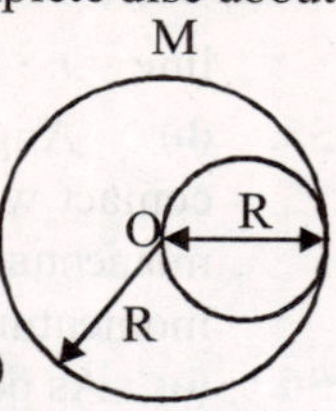

$I_{\text{Total disc}} = \dfrac{MR^2}{2}$

Mass of removed disc

$M_{\text{Removed}} = \dfrac{M}{4}$ (Mass $\propto$ area)

Moment of inertia of removed disc about point 'O'.

I_{Removed} (about same perpendicular axis)

$= I_{cm} + mx^2$

$= \dfrac{M}{4}\dfrac{(R/2)^2}{2} + \dfrac{M}{4}\left(\dfrac{R}{2}\right)^2 = \dfrac{3MR^2}{32}$

Therefore the moment of inertia of the remaining part of the disc about a perpendicular axis passing through the centre,

$I_{\text{Remaing disc}} = I_{\text{Total}} - I_{\text{Removed}}$

$= \dfrac{MR^2}{2} - \dfrac{3}{32}MR^2 = \dfrac{13}{32}MR^2$

> **NOTES** Moment of inertia of a part of a rigid body (Symmetrically cut from the whose mass) is the same as that of the whole body when whole mass is replaced by mass of that part.

60. **(c)** Moment of inertia of shell 1 along diameter

$I_{\text{diameter}} = \dfrac{2}{3}Mr^2$

Moment of inertia of shell 2 = M.I of shell 3

$= I_{\text{tangential}} = \dfrac{2}{3}Mr^2 + Mr^2 = \dfrac{5}{3}Mr^2$

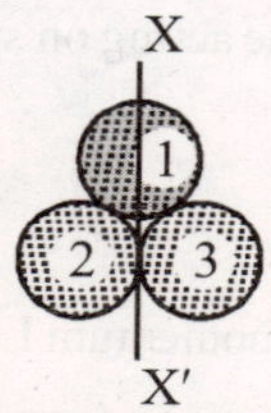

So, I of the system along XX'

$= I_{\text{diameter}} + (I_{\text{tangential}}) \times 2$

or, $I_{\text{total}} = \dfrac{2}{3}Mr^2 + \left(\dfrac{5}{3}Mr^2\right) \times 2$

$= \dfrac{12}{3}Mr^2 = 4Mr^2$

61. **(a)** $\because I = MK^2 \therefore K = \sqrt{\dfrac{I}{M}}$

$I_{\text{ring}} = MR^2$ and $I_{\text{disc}} = \dfrac{1}{2}MR^2$

$$\frac{K_1}{K_2} = \sqrt{\frac{I_1}{I_2}} = \sqrt{\frac{MR^2}{\left(\dfrac{MR^2}{2}\right)}} = \sqrt{2} : 1$$

62. **(a)** According to parallel axis theorem of the moment of Inertia

$I = I_{cm} + md^2$

d is maximum for point B so I_{max} about B.

63. **(b)** By theorem of parallel axes,

$I = I_{cm} + Md^2$

$I = I_0 + M(L/2)^2 = I_0 + ML^2/4$

64. **(c)** $\text{K.E.} = \dfrac{L^2}{2I}$

The angular momentum L remains conserved about the centre.

That is, L = constant.

$I = mr^2$

$\therefore \text{K.E.} = \dfrac{L^2}{2mr^2}$

In 2nd case, $\text{K.E.} = \dfrac{L^2}{2(mr'^2)}$

But $r' = \dfrac{r}{2}$

$\therefore K.E' = \dfrac{L^2}{2m.\dfrac{r^2}{4}} = \dfrac{4L^2}{2mr^2} \Rightarrow \text{K.E.}' = 4\,\text{K.E.}$

$\therefore$ K.E. is increased by a factor of 4.

> **NOTES** When no torque is applied angular momentum remains constant.
>
> $I_1\omega_1 = I_2\omega_2 \Rightarrow \dfrac{\omega_2}{\omega_1} = \dfrac{I_1}{I_2}$
>
> $\therefore \dfrac{E_2}{E_1} = \dfrac{\frac{1}{2}I_2\omega_2^2}{\frac{1}{2}I_1\omega_1^2} = \dfrac{I_2}{I_1}\left(\dfrac{I_1}{I_2}\right)^2 = \dfrac{I_2}{I_2}$
>
> So, if moment of inertia (mr^2) decreases K.E. of rotation increases and vice versa.

65. **(d)** Moment of inertia of a thin rod of length l about an axis passing through centre and perpendicular to the rod $= \dfrac{1}{12}Ml^2$.

Thus moment of inertia of the frame.

$\dfrac{ml^2}{12} + \dfrac{ml^2}{4} = \dfrac{4ml^2}{12} = \dfrac{ml^2}{3}$

Total M.I. $= 4 \times \dfrac{ml^2}{3}$

66. (b) Mass of each part = M/2
Length of each part = L/2

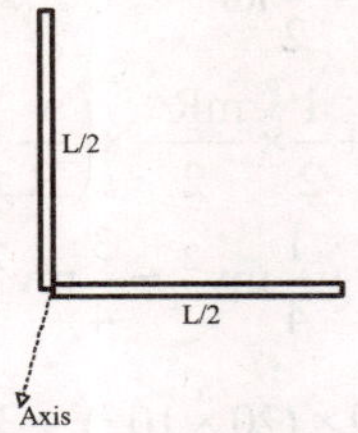

Total M.I. = Sum of M.I.s of both parts

$$= \left(\frac{M}{2}\right)\left(\frac{L}{2}\right)^2 \times \frac{1}{3} + \left(\frac{M}{2}\right)\left(\frac{L}{2}\right)^2 \times \frac{1}{3}$$

$$= 2 \times \frac{M}{2} \times \frac{L^2}{4} \times \frac{1}{3} = \frac{ML^2}{12}$$

67. (c) M.I. of a uniform circular disc of radius 'R' and mass 'M' about an axis passing though C.M. and normal to the disc is

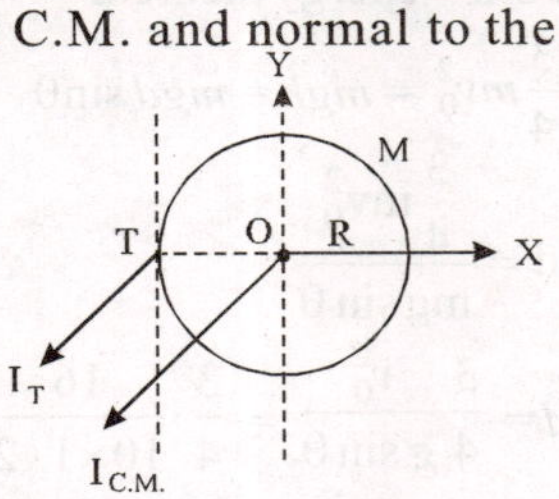

$$I_{C.M.} = \frac{1}{2} MR^2$$

From parallel axis theorem,

$$I_T = I_{C.M.} + MR^2 = \frac{1}{2}MR^2 + MR^2 = \frac{3}{2}MR^2$$

68. (d)

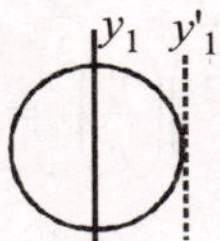

Circular disc [1]

$$I_{y_1} = \frac{MR^2}{4} \quad \therefore I'_{y_1} = \frac{MR^2}{4} + MR^2 = \frac{5}{4}MR^2$$

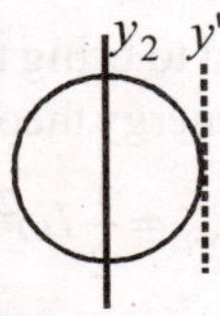

Circular ring [2]

$$I_{y_2} = \frac{MR^2}{2} \quad \therefore I'_{y_2} = \frac{MR^2}{2} + MR^2 = \frac{3}{2}MR^2$$

$$I'_{y_1} = MK_1^2, \; I'_{y_2} = MK_2^2$$

$$\therefore \frac{K_1^2}{K_2^2} = \frac{I'_{y_1}}{I'_{y_2}} \Rightarrow K_1 : K_2 = \sqrt{5} : \sqrt{6}$$

69. (d) $I_{AX} = m(AB)^2 + m(OC)^2$
$= m\ell^2 + m(\ell \cos 60°)^2$
$= m\ell^2 + m\ell^2/4 = 5/4\, m\ell^2$

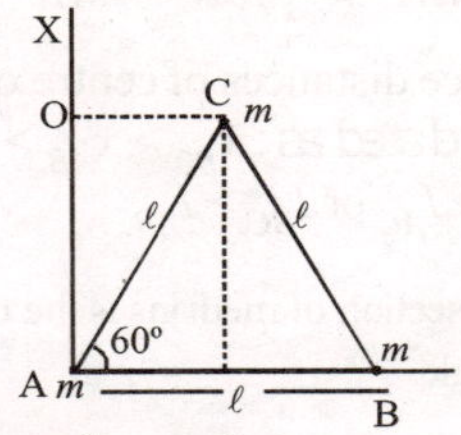

70. (d) As we know that,

$$\text{Rotational energy} = \frac{1}{2}I(\omega)^2 = \frac{1}{2}(mK^2)\omega^2 \text{ And,}$$

$$\text{Linear kinetic energy} = \frac{1}{2}m\omega^2 R^2$$

$\therefore$ Required fraction,

$$= \frac{\frac{1}{2}(mK^2)\omega^2}{\frac{1}{2}(mK^2)\omega^2 + \frac{1}{2}m\omega^2 R^2} = \left(\frac{K^2}{K^2 + R^2}\right)$$

71. (b) Moment of inertia $= \int r^2 dm$.

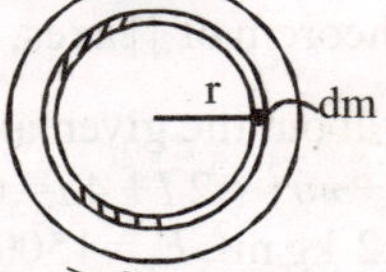

$\therefore$ Since, $\rho_{iron} > \rho_{aluminium}$

So, whole of aluminium is kept in the core and the iron at the outer rim of the disc.

72. (b)

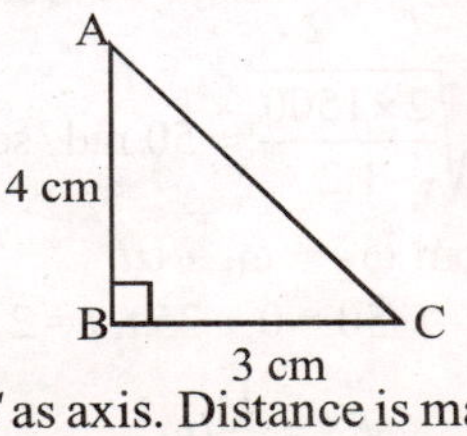

If we choose *BC* as axis. Distance is maximum. Hence, Moment of Inertia is maximum.

> **NOTES**
> Moment of Inertia depend upon mass and distribution of masses as $I = \Sigma mr^2$.
> Further, as the distance of masses is more, more is the moment of Inertia.
> $\therefore I_2 > I_1, I_2 > I_3$

73. (c) Moment of inertia of disc about its diameter is $I_d = \frac{1}{4}MR^2$

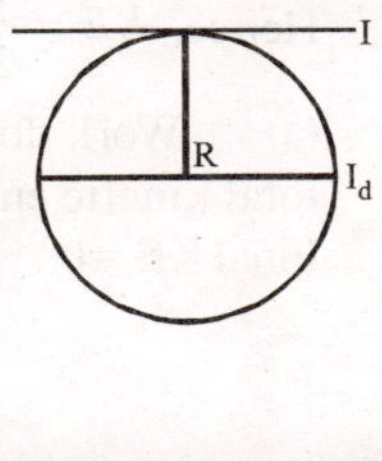

MI of disc about a tangent passing through rim and in the plane of disc is

$$I = I_d + MR^2 = \frac{1}{4}MR^2 + MR^2 = \frac{5}{4}MR^2$$

74. **(b)** Since distances of centre of mass from the sides are related as : $x_{BC} > x_{AB} > x_{AC}$, therefore, $I_{BC} > I_{AB} > I_{AC}$ or $I_{BC} > I_{AB}$.

NOTES: The intersection of medians is the centre of mass of the triangle.

75. **(d)** The M.I. is minimum about EG because mass distribution is at minimum distance from EG.

NOTES: Among so many parallel axis, moment of inertia of the body is the minimum about on axis which is passing through the centre of mass of the body.

76. **(c)** M.I of uniform circular disc about its diameter = I

According to theorem of perpendicular axes,

M.I. of disc about its axis $= \frac{1}{2}mr^2 = 2I$

Applying theorem of || axes, $\left(\because I = \frac{1}{4}mr^2\right)$

M.I of disc about the given axis

$$= 2I + mr^2 = 2I + 4I = 6I$$

77. **(b)** $I = 1.2 \text{ kg m}^2, E_r = 1500 \text{ J}$, $\alpha = 25 \text{ rad/sec}^2, \omega_1 = 0, t = ?$

As $E_r = \frac{1}{2}I\omega^2, \Rightarrow \omega = \sqrt{\frac{2E_r}{I}}$

$$= \sqrt{\frac{2 \times 1500}{1.2}} = 50 \text{ rad/sec}$$

From $\omega_2 = \omega_1 + \alpha t$

$$50 = 0 + 25t, t = 2 \text{ seconds}$$

78. **(c)** $E_r = \frac{1}{2}I\omega^2$

$$I = \frac{2E_r}{\omega^2} = \frac{2 \times 360}{30 \times 30} = 0.8 \text{ kg m}^2$$

79. **(a)** Kinetic energy $= \frac{1}{2}I\omega^2$

and for ring $I = mr^2$

Hence, $KE = \frac{1}{2}mr^2\omega^2$

80. **(a)** Work done to stop the disc = change in total kinetic energy of disc

Final KE = 0

Initial KE = Translational K.E. + Rotational K.E.

$$= \frac{1}{2}mv^2 + \frac{1}{2}I\omega^2$$

$$= \frac{1}{2}mv^2 + \frac{1}{2} \times \frac{mR^2}{2} \times \left(\frac{v}{R}\right)^2$$

$$= \frac{1}{2}mv^2 + \frac{1}{4}mv^2 = \frac{3}{4}mv^2$$

$$= \frac{3}{4} \times 100 \times (20 \times 10^{-2})^2 = 3J$$

$|\Delta KE| = 3J$

81. **(a)** Kinetic energy at the bottom

$$= \frac{1}{2}mv_0^2 + \frac{1}{2}I\omega^2 = \frac{3}{4}mv_0^2$$

As $I_{\text{solid cylinder}} = \frac{1}{2}MR^2$ and $V = R\omega$

From work – energy theorem

$$\frac{3}{4}mv_0^2 = mgh = mgd\sin\theta$$

$$d = \frac{\frac{3}{4}mv_0^2}{mg\sin\theta}$$

$$\Rightarrow \quad d = \frac{3}{4}\frac{v_0^2}{g\sin\theta} = \frac{3}{4} \cdot \frac{16}{10 \times 1/2} = 2.4 \text{ m}$$

82. **(b)** In rolling motion, rotational kinetic energy.

$$K_t = \frac{1}{2}mv^2$$

And, $K_t + K_r = \frac{1}{2}mv^2 + \frac{1}{2}I\omega^2$

$$= \frac{1}{2}mv^2 + \frac{1}{2}\left(\frac{2}{5}mr^2\right)\left(\frac{v}{r}\right)^2 = \frac{7}{10}mv^2$$

$$\therefore \quad \frac{K_t}{K_t + K_r} = \frac{\frac{1}{2}mv^2}{\frac{7}{10}mv^2} = \frac{5}{7}$$

83. **(a)** Work done required to bring them rest

$\Delta W = \Delta KE$ (work-energy theorem)

$$\Delta W = \frac{1}{2}I\omega^2 \left(\Delta kE_{rot} = \frac{1}{2}I\omega^2\right)$$

or, $\Delta W \propto I$ (for same ω)

$I_{\text{solid sphere}} = \frac{2}{5}MR^2$, $I_{\text{Disk}} = \frac{1}{2}MR^2$

$I_{\text{Ring}} = MR^2 \quad \therefore \; W_C > W_B > W_A$

84. **(b)** Time of descent $\propto \frac{K^2}{R^2}$

Order of value of $\frac{K^2}{R^2}$

for disc; $\frac{K^2}{R^2} = \frac{1}{2} = 0.5$

for sphere; $\frac{K^2}{R^2} = \frac{2}{5} = 0.4$

(sphere) < (disc)

$\therefore$ Sphere reaches first

NOTES The value $\frac{K^2}{R^2}$ is a measure of moment of inertia. Its. value is fixed for a particular shape of the body and independent of mass and readius of the body.

85. **(a)** For solid sphere rolling without slipping on inclined plane, acceleration

$$a_1 = \frac{g \sin\theta}{\left(1 + \frac{K^2}{R^2}\right)}$$

For solid sphere slipping on inclined plane without rolling, acceleration

$$a_2 = g \sin\theta$$

Therefore required ratio $= \frac{a_1}{a_2}$

$$= \frac{1}{\left(1 + \frac{K^2}{R^2}\right)} = \frac{1}{\left(1 + \frac{2}{5}\right)} = \frac{5}{7}$$

86. **(c)**

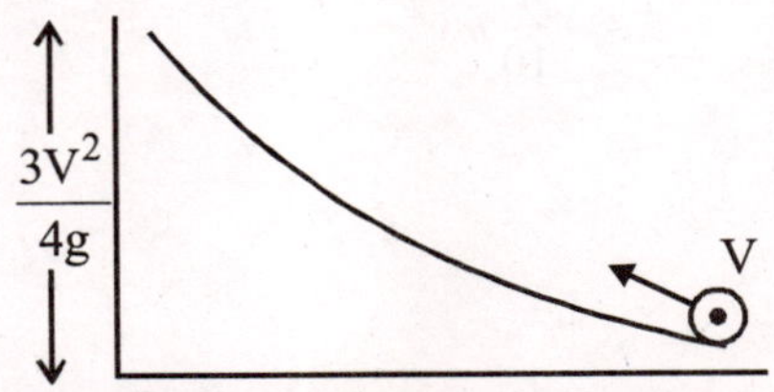

From law of conservation of mechanical energy,

$$\Rightarrow \frac{1}{2}I\omega^2 + 0 + \frac{1}{2}mv^2 = mg \times \frac{3v^2}{4g}$$

$$\Rightarrow \frac{1}{2}I\omega^2 = \frac{3}{4}mv^2 - \frac{1}{2}mv^2$$

$$\Rightarrow \frac{1}{2}I\omega^2 = \frac{mv^2}{2}\left(\frac{3}{2} - 1\right)$$

or, $\frac{1}{2}I\left(\frac{V^2}{R^2}\right) = \frac{mv^2}{4}$ or, $I = \frac{1}{2}mR^2$

Hence, object is a disc.

87. **(d)** Net work done by frictional force when drum rolls down without slipping is zero.

$$W_{net} = 0$$

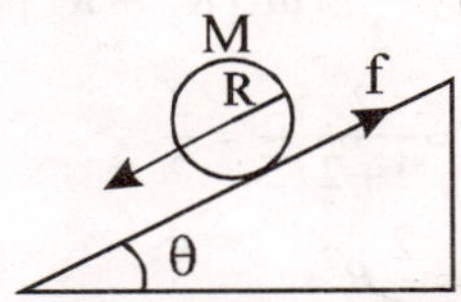

$$W_{trans.} + W_{rot.} = 0;\ \Delta K_{trans.} + \Delta K_{rot.} = 0$$

$$\Delta K_{trans} = -\Delta K_{rot}.$$

i.e., converts translation energy to rotational energy.

88. **(b)** $K.E. = \frac{1}{2}I\omega^2 + \frac{1}{2}mv^2$

$$K.E. = \frac{1}{2}\left(\frac{1}{2}mr^2\right)\omega^2 + \frac{1}{2}mv^2$$

$$= \frac{1}{4}mv^2 + \frac{1}{2}mv^2 = \frac{3}{4}mv^2$$

Now, gain in K.E. = Loss in P.E.

$$\frac{3}{4}mv^2 = mgh \Rightarrow v = \sqrt{\left(\frac{4}{3}\right)gh}$$

89. **(b)**

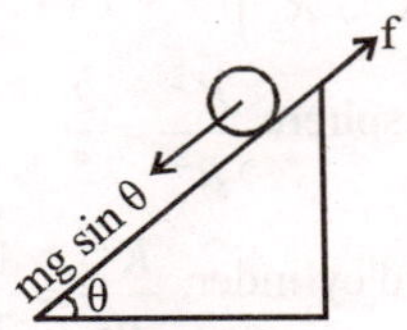

Torque, $I\alpha = f.R.$

Using Newton's IInd law, $mg \sin\theta - f = ma$

$\because$ pure rolling is there, $a = R\alpha$

$$\Rightarrow mg\sin\theta - \frac{I\alpha}{R} = ma$$

$$\Rightarrow mg\sin\theta - \frac{Ia}{R^2} = ma \quad \left(\because \alpha = \frac{a}{R}\right)$$

or, acceleration, $a = \frac{mg\sin\theta}{(I/R^2 + m)}$

Using, $s = ut + \frac{1}{2}at^2$

or, $s = \frac{1}{2}at^2 \Rightarrow t \propto \frac{1}{\sqrt{a}}$

t minimum means a should be more. This is possible when I is minimum which is the case for solid cylinder.

Therefore, solid cylinder will reach the bottom first.

90. (b) $\frac{K_r}{E} = \frac{\frac{1}{2}MK^2\omega^2}{\frac{1}{2}M\omega^2[K^2+R^2]} = \frac{K^2}{K^2+R^2}$

$= \frac{2/5}{1+2/5} = \frac{2}{7}$

Here, $K^2 = \frac{2}{5}R^2$

> **NOTES** When a body is executing only linear motion, its K.E. is given by $E_t = \frac{1}{2}mv^2$ when a body is rolling without slipping, its centre of mas has linear motion too.
> Therefore, its total kinetic energy
> $E = E_t + E_r$
> $= \frac{1}{2}mv^2 + \frac{1}{2}I\omega^2$
> $= \frac{1}{2}mv^2 + \frac{1}{2}m(K^2)\frac{v^2}{R^2}$
> $= \frac{1}{2}mv^2\left(1+\frac{K^2}{R^2}\right)$

91. (c) $a = \frac{g\sin\theta}{(1+K^2/R^2)} = \frac{g\sin 30°}{1+1} = \frac{g}{4}$

92. (a) For solid sphere, $\frac{K^2}{R^2} = \frac{2}{5}$

For disc and solid cylinder, $\frac{K^2}{R^2} = \frac{1}{2}$

As $\frac{K^2}{R^2}$ for solid sphere is smallest, it takes minimum time to reach the bottom of the incline

> **NOTES** When bodies of different shapes are allowed to roll down an incline, then velocity at the bottom of incline, acceleration and time taken in reaching the bottom depends on $\frac{K^2}{R^2}$. If the body has less moment of inertia then $\frac{K^2}{R^2}$ will be less, then it will take less time in reaching the bottom.

93. (a) P.E. = total K.E.

$mgh = \frac{7}{10}mv^2$

$\therefore v = \sqrt{\frac{10gh}{7}}$

94. (d) $E = E_t + E_r = \frac{1}{2}mv^2 + \frac{1}{2}I\omega^2$

$= \frac{1}{2}mv^2 + \frac{1}{2}\times\left(\frac{2}{5}mr^2\right)\omega^2$

$= \frac{1}{2}mv^2 + \frac{1}{5}mv^2 = \frac{7}{10}mv^2$

$\therefore \frac{E_t}{E} = \frac{\frac{1}{2}mv^2}{\frac{7}{10}mv^2} = \frac{5}{7}$

7 Gravitation

Trend Analysis with Important Topics & Sub-Topics

		2021		2020		2019		2018		2017		2016	
Topic Name	**Sub-Topic**	QNS.	LOD	QNS.	LOD	QNS.	LOD	QNS.	LOD	QNS.	LOD	QNS.	LOD
Kepler's Laws of Planetary Motion	Kepler's Law of Planetary Motion							1	E				
Newton's Universal Law of Gravitation Acceleration due to Gravity	Variation of g with height and depth			1	A								
	Gravitational Force, Value of g on Earth's Surface									1	E		
Gravitational Field, Potential and Energy	Gravitational Potential Energy					1	A						
	Gravitational Potential											1	D
Motion of Satellites, Escape Speed and Orbital Velocity	Escape Speed	2	E, A									1	A

LOD - Level of Difficulty | E - Easy | A - Average | D - Difficult | Qns - No. of Questions

Topic 1 Kepler's Laws of Planetary Motion

1. The kinetic energies of a planet in an elliptical orbit about the Sun, at positions A, B and C are K_A, K_B and K_C, respectively. AC is the major axis and SB is perpendicular to AC at the position of the Sun S as shown in the figure. Then *[2018]*

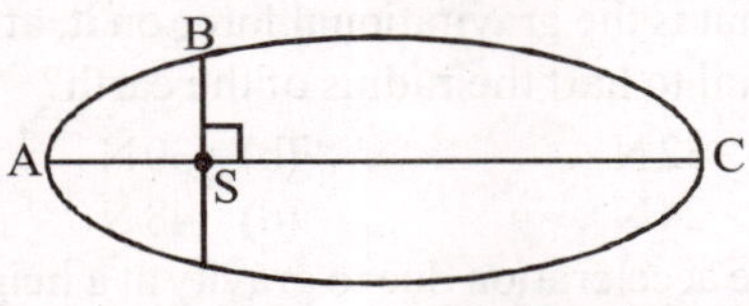

(a) $K_A < K_B < K_C$ (b) $K_A > K_B > K_C$
(c) $K_B > K_A > K_C$ (d) $K_B < K_A < K_C$

2. Kepler's third law states that square of period of revolution (T) of a planet around the sun, is proportional to third power of average distance r between sun and planet *i.e.*, $T^2 = Kr^3$ here K is constant. If the masses of sun and planet are M and m respectively then as per Newton's law of gravitation force of attraction between them is $F = \frac{GMm}{r^2}$, here G is gravitational constant. The relation between G and K is described as *[2015]*

(a) $GMK = 4\pi^2$ (b) $K = G$
(c) $K = \frac{1}{G}$ (d) $GK = 4\pi^2$

3. The figure shows elliptical orbit of a planet m about the sun S. The shaded area SCD is twice the shaded area SAB. If t_1 is the time for the planet to move from C to D and t_2 is the time to move from A to B then : *[2009]*

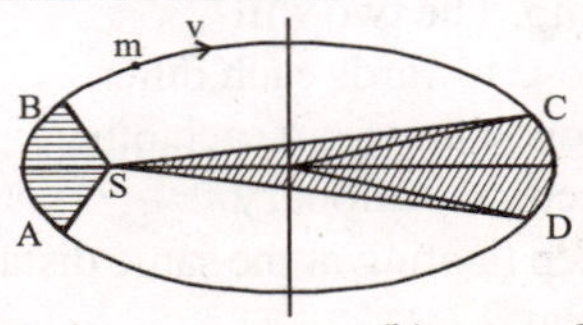

(a) $t_1 = 4t_2$ (b) $t_1 = 2t_2$
(c) $t_1 = t_2$ (d) $t_1 > t_2$

4. The period of revolution of planet A around the Sun is 8 times that of B. The distance of A from the Sun is how many times greater than that of B from the Sun? *[1997]*
(a) 2 (b) 3 (c) 4 (d) 5

5. The distance of Neptune and Saturn from the sun is nearly 10^{13} and 10^{12} meter respectively. Assuming that they move in circular orbits, their periodic times will be in the ratio *[1994]*
(a) 10 (b) 100
(c) $10\sqrt{10}$ (d) 1000

6. A satellite A of mass m is at a distance of r from the surface of the earth. Another satellite B of mass 2m is at a distance of 2r from the earth's centre. Their time periods are in the ratio of *[1993]*
(a) 1:2 (b) 1:16
(c) 1:32 (d) $1:2\sqrt{2}$

7. The distance of two planets from the sun are 10^{13} and 10^{12} metres respectively. The ratio of time periods of these two planets is *[1988]*
(a) $\frac{1}{\sqrt{10}}$ (b) 100
(c) $10\sqrt{10}$ (d) $\sqrt{10}$

8. The largest and the shortest distance of the earth from the sun are r_1 and r_2. Its distance from the sun when it is at perpendicular to the major-axis of the orbit drawn from the sun is *[1988]*
(a) $\frac{r_1+r_2}{4}$ (b) $\frac{r_1+r_2}{r_1-r_2}$
(c) $\frac{2r_1r_2}{r_1+r_2}$ (d) $\frac{r_1+r_2}{3}$

Topic 2 Newton's Universal Law of Gravitation

9. Two astronauts are floating in gravitation free space after having lost contact with their spaceship. The two will *[2017]*
(a) move towards each other.
(b) move away from each other.
(c) become stationary
(d) keep floating at the same distance between them.

10. Two spherical bodies of mass M and 5 M and radii R and 2 R released in free space with initial separation between their centres equal to 12 R. If they attract each other due to gravitational force only, then the distance covered by the smaller body before collision is *[2015]*
(a) 4.5 R (b) 7.5 R
(c) 1.5 R (d) 2.5 R

11. A spherical planet has a mass M_P and diameter D_P. A particle of mass m falling freely near the surface of this planet will experience an acceleration due to gravity, equal to : **[2012]**
(a) $4GM_P/D_P^2$ (b) GM_Pm/D_P^2
(c) GM_P/D_P^2 (d) $4GM_Pm/D_P^2$

12. Two spheres of masses m and M are situated in air and the gravitational force between them is F. The space around the masses is now filled with a liquid of specific gravity 3. The gravitational force will now be *[2003]*
(a) $\frac{F}{9}$ (b) $3F$
(c) F (d) $\frac{F}{3}$

13. What will be the formula of the mass in terms of g, R and G (R = radius of earth) *[1996]*
(a) $g^2\frac{R}{G}$ (b) $G\frac{R^2}{g}$
(c) $G\frac{R}{g}$ (d) $g\frac{R^2}{G}$

Topic 3 Acceleration due to Gravity

14. A body weighs 72 N on the surface of the earth. What is the gravitational force on it, at a height equal to half the radius of the earth? *[2020]*
(a) 32 N (b) 30 N
(c) 24 N (d) 48 N

15. The acceleration due to gravity at a height 1 km above the earth is the same as at a depth d below the surface of earth. Then *[2017]*
(a) $d = 1$ km (b) $d = \frac{3}{2}$ km
(c) $d = 2$ km (d) $d = \frac{1}{2}$ km

16. The height at which the weight of a body becomes 1/16th, its weight on the surface of earth (radius R), is : *[2012]*
(a) $5R$ (b) $15R$
(c) $3R$ (d) $4R$

17. A roller coaster is designed such that riders experience "weightlesness" as they go round the top of a hill whose radius of curvature is 20 m. The speed of the car at the top of the hill is between: *[2008]*
(a) 14 m/s and 15 m/s (b) 15 m/s and 16 m/s
(c) 16 m/s and 17 m/s (d) 13 m/s and 14 m/s

18. Imagine a new planet having the same density as that of earth but it is 3 times bigger than the earth in size. If the acceleration due to gravity on the surface of earth is g and that on the surface of the new planet is g', then *[2005]*
(a) $g' = g/9$ (b) $g' = 27g$
(c) $g' = 9g$ (d) $g' = 3g$

19. The density of a newly discovered planet is twice that of earth. The acceleration due to gravity at the surface of the planet is equal to that at the surface of the earth. If the radius of the earth is R, the radius of the planet would be
(a) ½R (b) 2R *[2004]*
(c) 4R (d) 1/4R

20. The acceleration due to gravity on the planet A is 9 times the acceleration due to gravity on planet B. A man jumps to a height of 2m on the surface of A. What is the height of jump by the same person on the planet B? *[2003]*
(a) $\frac{2}{3}$m (b) $\frac{2}{9}$m
(c) 18m (d) 6m

21. Assuming earth to be a sphere of uniform density, what is the value of 'g' in a mine 100 km below the earth's surface? (Given, $R = 6400$ km) *[2001]*
(a) $9.65\ m/s^2$ (b) $7.65\ m/s^2$
(c) $5.06\ m/s^2$ (d) $3.10\ m/s^2$

22. A body weighs 72 N on the surface of the earth. What is the gravitational force on it due to earth at a height equal to half the radius of the earth from the surface? *[2000]*
(a) 32 N (b) 28 N
(c) 16 N (d) 72 N

23. In a rocket a seconds pendulum is mounted. Its period of oscillation decreases when the rocket *[1991]*
(a) comes down with uniform acceleration
(b) moves round the earth in a geostationary orbit
(c) moves up with a uniform velocity
(d) moves up with uniform acceleration

Topic 4 Gravitational Field, Potential and Energy

24. The work done to raise a mass m from the surface of the earth to a height h, which is equal to the radius of the earth, is : *[2019]*
(a) mgR (b) 2 mgR
(c) $\frac{1}{2}$mgR (d) $\frac{3}{2}$mgR

25. Assuming that the gravitational potential energy of an object at infinity is zero, the change in potential energy (final – initial) of an object of mass m, when taken to a height h from the surface of earth (of radius R), is given by, *[NEET Odisha 2019]*
(a) $\frac{GMm}{R+h}$ (b) $-\frac{GMm}{R+h}$
(c) $\frac{GMmh}{R(R+h)}$ (d) mgh

26. At what height from the surface of earth the gravitational potential and the value of g are $-5.4 \times 10^7\ J\ kg^{-1}$ and $6.0\ ms^{-2}$ respectively? Take the radius of earth as 6400 km : *[2016]*
(a) 2600 km (b) 1600 km
(c) 1400 km (d) 2000 km

27. Dependence of intensity of gravitational field (E) of earth with distance (r) from centre of earth is correctly represented by: *[2014]*

(a) 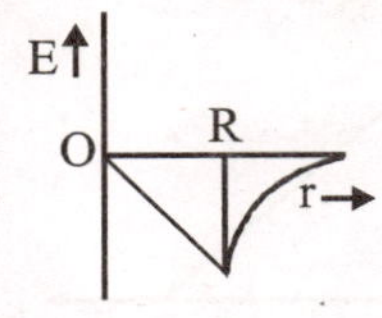

(b) 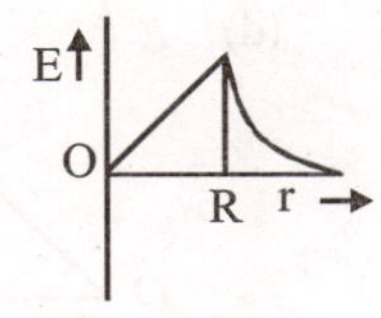

(c) 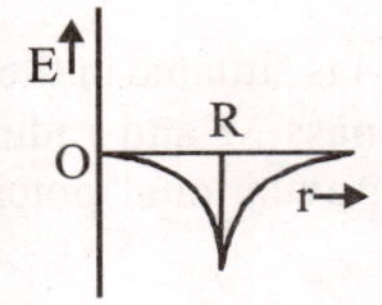

(d) 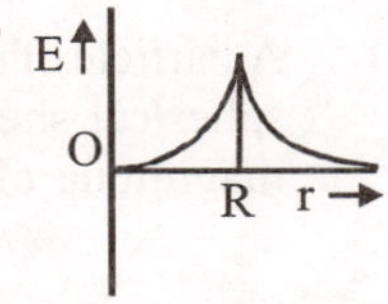

28. A body of mass '*m*' is taken from the earth's surface to the height equal to twice the radius (R) of the earth. The change in potential energy of body will be *[2013]*

(a) $\frac{2}{3}$ mgR (b) 3 mgR

(c) $\frac{1}{3}$ mgR (d) mg2R

29. Infinite number of bodies, each of mass 2 kg are situated on *x*-axis at distances 1m, 2m, 4m, 8m, ... respectively, from the origin. The resulting gravitational potential due to this system at the origin will be *[2013]*

(a) $-\frac{8}{3}$G (b) $-\frac{4}{3}$G

(c) – 4 G (d) –G

30. Which one of the following plots represents the variation of gravitational field on a particle with distance *r* due to a thin spherical shell of radius *R*? (*r* is measured from the centre of the spherical shell) *[2012M]*

(a)
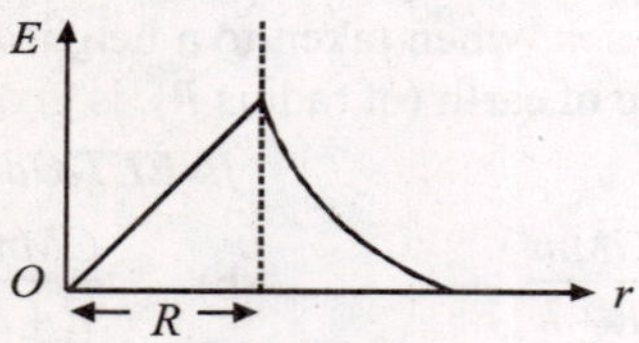

(b)
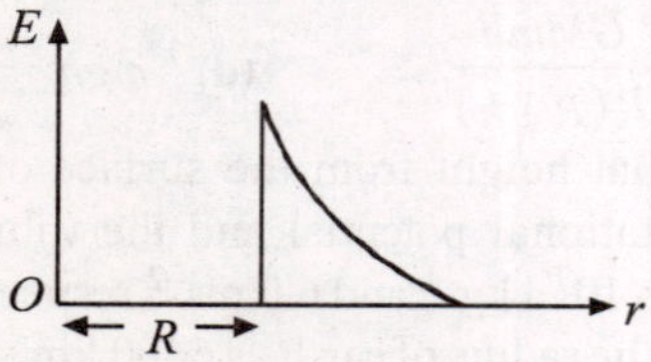

(c)
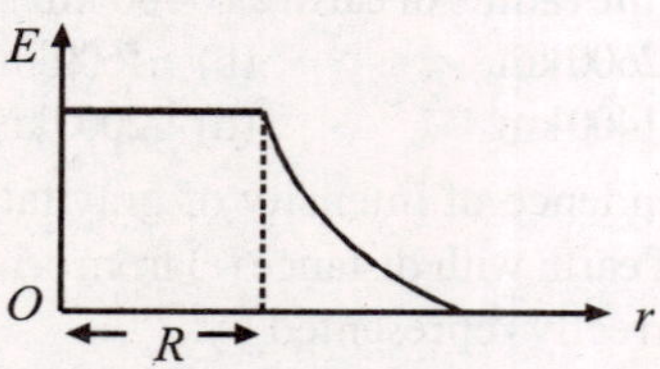

(d)
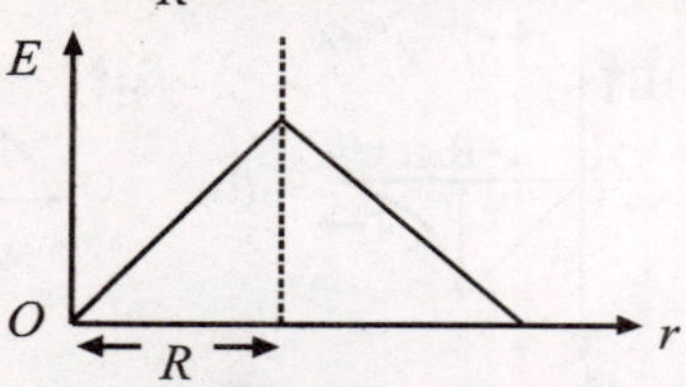

31. A particle of mass M is situated at the centre of spherical shell of mass M and radius a. The magnitude of the gravitational potential at a point situated at a/2 distance from the centre, will be *[2011M]*

(a) $\frac{2GM}{a}$ (b) $\frac{3GM}{a}$

(c) $\frac{4GM}{a}$ (d) $\frac{GM}{a}$

32. A particle of mass *M* is situated at the centre of a spherical shell of same mass and radius *a*. The gravitational potential at a point situated at $\frac{a}{2}$ distance from the centre, will be: *[2010]*

(a) $-\frac{3GM}{a}$ (b) $-\frac{2GM}{a}$

(c) $-\frac{GM}{a}$ (d) $-\frac{4GM}{a}$

33. The Earth is assumed to be a sphere of radius *R*. A platform is arranged at a height *R* from the surface of the Earth. The velocity of a body from this platform is *f v*, where *v* is its velocity from the surface of the Earth. The value of *f* is *[2006]*

(a) $\frac{1}{\sqrt{2}}$ (b) $\frac{1}{3}$

(c) $\frac{1}{2}$ (d) $\sqrt{2}$

34. Assuming the radius of the earth as *R*, the change in gravitational potential energy of a body of mass m, when it is taken from the earth's surface to a height 3*R* above its surface, is *[2002]*

(a) 3 *mg R* (b) $\frac{3}{4}mg\,R$

(c) 1 *mg R* (d) $\frac{3}{2}mg\,R$

35. The potential energy of a satellite, having mass m and rotating at a height of 6.4×10^6 m from the earth surface, is *[2001]*

(a) $-mgR_e$ (b) $-0.67\,mgR_e$

(c) $-0.5\,mgR_e$ (d) $-0.33\,mgR_e$

36. With what velocity should a particle be projected so that its height becomes equal to radius of earth? *[2001]*

(a) $\left(\frac{GM}{R}\right)^{1/2}$ (b) $\left(\frac{8GM}{R}\right)^{1/2}$

(c) $\left(\frac{2GM}{R}\right)^{1/2}$ (d) $\left(\frac{4GM}{R}\right)^{1/2}$

Topic 5 Motion of Satellites, Escape Speed and Orbital Velocity

37. The escape velocity from the Earth's surface is v. The escape velocity from the surface of another planet having a radius, four times that of Earth and same mass density is [2021]
(a) $4v$ (b) v
(c) $2v$ (d) $3v$

38. A particle of mass 'm' is projected with a velocity $v = kV_e$ ($k < 1$) from the surface of the earth. (V_e = escape velocity)
The maximum height above the surface reached by the particle is [2021]
(a) $\frac{Rk^2}{1-k^2}$ (b) $R\left(\frac{k}{1-k}\right)^2$
(c) $R\left(\frac{k}{1+k}\right)^2$ (d) $\frac{R^2k}{1+k}$

39. The time period of a geostationary satellite is 24 h, at a height $6R_E$ (R_E is radius of earth) from surface of earth. The time period of another satellite whose height is 2.5 R_E from surface will be, [NEET Odisha 2019]
(a) $\frac{12}{2.5}h$ (b) $6\sqrt{2}\,h$
(c) $12\sqrt{2}\,h$ (d) $\frac{24}{2.5}h$

40. The ratio of escape velocity at earth (v_e) to the escape velocity at a planet (v_p) whose radius and mean density are twice as that of earth is : [2016]
(a) 1 : 2 (b) $1 : 2\sqrt{2}$
(c) 1 : 4 (d) $1 : \sqrt{2}$

41. A satellite S is moving in an elliptical orbit around the earth. The mass of the satellite is very small compared to the mass of the earth. Then, [2015 RS]
(a) the total mechanical energy of S varies periodically with time.
(b) the linear momentum of S remains constant in magnitude.
(c) the acceleration of S is always directed towards the centre of the earth.
(d) the angular momentum of S about the centre of the earth changes in direction, but its magnitude remains constant.

42. A remote - sensing satellite of earth revolves in a circular orbit at a height of 0.25×10^6 m above the surface of earth. If earth's radius is 6.38×10^6 m and g = 9.8 ms^{-2}, then the orbital speed of the satellite is: [2015 RS]
(a) 8.56 km s^{-1} (b) 9.13 km s^{-1}
(c) 6.67 km s^{-1} (d) 7.76 km s^{-1}

43. A black hole is an object whose gravitational field is so strong that even light cannot escape from it. To what approximate radius would earth (mass = 5.98×10^{24} kg) have to be compressed to be a black hole? [2014]
(a) 10^{-9} m (b) 10^{-6} m
(c) 10^{-2} m (d) 100m

44. The radius of a planet is twice the radius of earth. Both have almost equal average mass-densities. If V_P and V_E are escape velocities of the planet and the earth, respectively, then [NEET Kar. 2013]
(a) $V_E = 1.5V_P$ (b) $V_P = 1.5V_E$
(c) $V_P = 2V_E$ (d) $V_E = 3V_P$

45. A particle of mass 'm' is kept at rest at a height $3R$ from the surface of earth, where 'R' is radius of earth and 'M' is mass of earth. The minimum speed with which it should be projected, so that it does not return back, is (g is acceleration due to gravity on the surface of earth) [NEET Kar. 2013]
(a) $\left(\frac{GM}{R}\right)^{\frac{1}{2}}$ (b) $\left(\frac{GM}{2R}\right)^{\frac{1}{2}}$
(c) $\left(\frac{gR}{4}\right)^{\frac{1}{2}}$ (d) $\left(\frac{2g}{4}\right)^{\frac{1}{2}}$

46. A geostationary satellite is orbiting the earth at a height of $5R$ above that surface of the earth, R being the radius of the earth. The time period of another satellite in hours at a height of $2R$ from the surface of the earth is : [2012]
(a) 5 (b) 10
(c) $6\sqrt{2}$ (d) $\frac{6}{\sqrt{2}}$

47. If v_e is escape velocity and v_0 is orbital velocity of a satellite for orbit close to the earth's surface, then these are related by : [2012M]
(a) $v_0 = \sqrt{2}v_e$ (b) $v_0 = v_e$
(c) $v_e = \sqrt{2v_0}$ (d) $v_e = \sqrt{2}v_0$

48. A planet moving along an elliptical orbit is closest to the sun at a distance r_1 and farthest away at a distance of r_2. If v_1 and v_2 are the linear velocities at these points respectively, then the ratio $\frac{v_1}{v_2}$ is **[2011]**

(a) $(r_1/r_2)^2$ (b) r_2/r_1
(c) $(r_2/r_1)^2$ (d) r_1/r_2

49. A particle of mass m is thrown upwards from the surface of the earth, with a velocity u. The mass and the radius of the earth are, respectively, M and R. G is gravitational constant and g is acceleration due to gravity on the surface of the earth. The minimum value of u so that the particle does not return back to earth, is **[2011M]**

(a) $\sqrt{\frac{2GM}{R}}$ (b) $\sqrt{\frac{2GM}{R^2}}$
(c) $\sqrt{2gR^2}$ (d) $\sqrt{\frac{2GM}{R^2}}$

50. The radii of circular orbits of two satellites A and B of the earth, are 4R and R, respectively. If the speed of satellite A is 3 V, then the speed of satellite B will be: **[2010]**

(a) 3 V/4 (b) 6 V
(c) 12 V (d) 3 V/2

51. Two satellites of earth, S_1 and S_2 are moving in the same orbit. The mass of S_1 is four times the mass of S_2. Which one of the following statements is true? **[2007]**

(a) The potential energies of earth satellites in the two cases are equal.
(b) S_1 and S_2 are moving with the same speed.
(c) The kinetic energies of the two satellites are equal.
(d) The time period of S_1 is four times that of S_2.

52. For a satellite moving in an orbit around the earth, the ratio of kinetic energy to potential energy is **[2005]**

(a) $\frac{1}{2}$ (b) $\frac{1}{\sqrt{2}}$
(c) 2 (d) $\sqrt{2}$

53. The escape velocity on the surface of earth is 11.2 km/s. What would be the escape velocity on the surface of another planet of the same mass but 1/4 times the radius of the earth?

(a) 22.4 km/s (b) 44.8 km/s **[2000]**
(c) 5.6 km/s (d) 11.2 km/s

54. The escape velocity of a sphere of mass m is given by (G = Universal gravitational constant; M = Mass of the earth and R_e = Radius of the earth) **[1999]**

(a) $\sqrt{\frac{GM}{R_e}}$ (b) $\sqrt{\frac{2GM}{R_e}}$
(c) $\sqrt{\frac{2GMm}{R_e}}$ (d) $\sqrt{\frac{2GM + R_e}{R_e}}$

55. The escape velocity of a body on the surface of the earth is 11.2 km/s. If the earth's mass increases to twice its present value and the radius of the earth becomes half, the escape velocity would become **[1997]**

(a) 44.8 km/s
(b) 22.4 km/s
(c) 11.2 km/s (remains unchanged)
(d) 5.6 km/s

56. A ball is dropped from a satellite revolving around the earth at a height of 120 km. The ball will **[1996]**

(a) continue to move with same speed along a straight line tangentially to the satellite at that time
(b) continue to move with the same speed along the original orbit of satellite
(c) fall down to earth gradually
(d) go far away in space

57. The escape velocity from the surface of the earth is v_e. The escape velocity from the surface of a planet whose mass and radius are three times those of the earth, will be **[1995]**

(a) v_e (b) $3v_e$
(c) $9v_e$ (d) $1/3v_e$

58. A satellite in force free space sweeps stationary interplanetary dust at a rate dM/dt = αv where M is the mass and v is the velocity of the satellite and α is a constant. What is the deceleration of the satellite? **[1994]**

(a) $-\alpha v^2$ (b) $-\alpha v^2/2M$
(c) $-\alpha v^2/2M$ (d) $-2\alpha v^2/M$

59. The escape velocity from earth is 11.2 km/s. If a body is to be projected in a direction making an angle 45° to the vertical, then the escape velocity is *[1993]*
(a) 11.2×2 km/s (b) 11.2 km/s
(c) $11.2/\sqrt{2}$ km/s (d) $11.2\sqrt{2}$ km/s

60. The mean radius of earth is R, its angular speed on its own axis is ω and the acceleration due to gravity at earth's surface is g. What will be the radius of the orbit of a geostationary satellite? *[1992]*
(a) $(R^2g/\omega^2)^{1/3}$ (b) $(Rg/\omega^2)^{1/3}$
(c) $(R^2\omega^2/g)^{1/3}$ (d) $(R^2g/\omega)^{1/3}$

61. A satellite of mass m is orbiting around the earth in a circular orbit with a velocity v. What will be its total energy? *[1991]*
(a) $(3/4)mv^2$ (b) $(1/2)mv^2$
(c) mv^2 (d) $-(1/2)mv^2$

62. A planet is moving in an elliptical orbit around the sun. If T, V, E and L stand respectively for its kinetic energy, gravitational potential energy, total energy and magnitude of angular momentum about the centre of force, which of the following is correct? *[1990]*
(a) T is conserved
(b) V is always positive
(c) E is always negative
(d) L is conserved but direction of vector L changes continuously

63. If the gravitational force between two objects were proportional to $1/R$ (and not as $1/R^2$) where R is separation between them, then a particle in circular orbit under such a force would have its orbital speed v proportional to *[1989]*
(a) $1/R^2$ (b) R^0
(c) R^1 (d) $1/R$

64. For a satellite escape velocity is 11 km/s. If the satellite is launched at an angle of 60° with the vertical, then escape velocity will be *[1989]*
(a) 11 km/s (b) $11\sqrt{3}$ km/s
(c) $\frac{11}{\sqrt{3}}$ km/s (d) 33 km/s

ANSWER KEY

1	(b)	9	(a)	17	(a)	25	(c)	33	(a)	41	(c)	49	(a)	56	(b)	63	(b)
2	(a)	10	(b)	18	(d)	26	(a)	34	(b)	42	(d)	50	(b)	57	(a)	64	(a)
3	(b)	11	(a)	19	(a)	27	(b)	35	(c)	43	(c)	51	(b)	58	(c)		
4	(c)	12	(c)	20	(c)	28	(a)	36	(a)	44	(c)	52	(a)	59	(b)		
5	(c)	13	(d)	21	(a)	29	(c)	37	(a)	45	(b)	53	(a)	60	(a)		
6	(d)	14	(a)	22	(a)	30	(b)	38	(a)	46	(c)	54	(b)	61	(d)		
7	(c)	15	(c)	23	(d)	31	(b)	39	(b)	47	(d)	55	(b)	62	(c)		
8	(c)	16	(c)	24	(c)	32	(a)	40	(b)	48	(b)						

Hints & Solutions

1. **(b)** Speed of the planet will be maximum when its distance from the sun is minimum as mvr = constant.

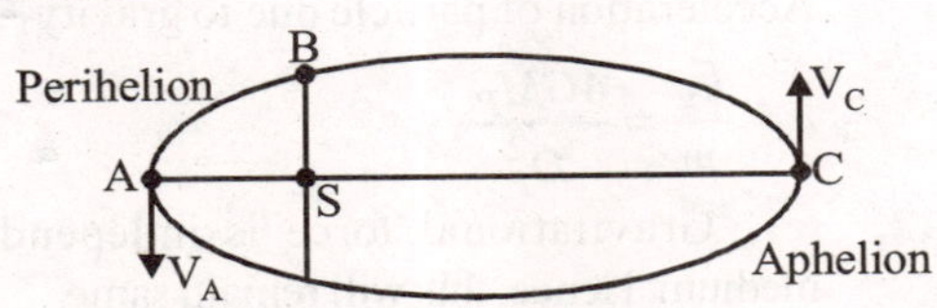

Point A is perihelion and C is aphelion.
Clearly, $V_A > V_B > V_C$
So, $K_A > K_B > K_C$

2. **(a)** As we know, orbital speed, $v_{orb} = \sqrt{\frac{GM}{r}}$

Time period $T = \frac{2\pi r}{v_{orb}} = \frac{2\pi r}{\sqrt{GM}}\sqrt{r}$

Squarring both sides,

$$T^2 = \left(\frac{2\pi r\sqrt{r}}{\sqrt{GM}}\right)^2 = \frac{4\pi^2}{GM}.r^3$$

$$\Rightarrow \frac{T^2}{r^3} = \frac{4\pi^2}{GM} = K$$

$$\Rightarrow GMK = 4\pi^2$$

3. **(b)** According to Kepler's law, the areal velocity of a planet around the sun always remains constant.

SCD : $A_1 - t_1$ (areal velocity constant)

SAB : $A_2 - t_2$

$$\frac{A_1}{t_1} = \frac{A_2}{t_2};$$

$$t_1 = t_2 . \frac{A_1}{A_2} \qquad \text{(given } A_1 = 2A_2)$$

$$= t_2 . \frac{2A_2}{A_2}$$

$$\therefore \quad t_1 = 2t_2$$

NOTES

The area covered by radius vector in dt seconds = $\frac{1}{2}r^2 d\theta$

$$\text{Area velocity} = \frac{1}{2}r^2\frac{d\theta}{dt} = \frac{1}{2}r^2\omega \qquad \left(\because \omega\frac{d\theta}{dt}\right)$$

$$= \frac{1}{2}rv \qquad (\because v = \omega r)$$

It follows that speed of the planet is maximum when it is closest to the sun and is minimum when the planet is farthest from the sun.

4. **(c)** Let T_A and T_B be time period of A and B about sun.

$$T_A = 8T_B$$

$$\frac{T_A}{T_B} = 8 \qquad ...(1)$$

According to Kepler's Law, $T^2 \propto r^3$

$$\left(\frac{T_A}{T_B}\right)^2 = \frac{(r_A)^3}{(r_B)^3} \Rightarrow \left(\frac{r_A}{r_B}\right)^3 = 64 \Rightarrow \frac{r_A}{r_B} = 4$$

5. **(c)** $T^2 \propto R^3$ (According to Kepler's law)

$T_1^2 \propto (10^{13})^3$ and $T_2^2 \propto (10^{12})^3$

$$\therefore \quad \frac{T_1^2}{T_2^2} = (10)^3 \text{ or } \frac{T_1}{T_2} = 10\sqrt{10}$$

6. **(d)** Also, $T^2 \propto r^3$.

NOTES

Time period does not depend on the mass.

7. **(c)** $T^2 \propto R^3$ (Kepler's law)

$$\frac{T_1^2}{T_2^2} = \left(\frac{10^{13}}{10^{12}}\right)^3 \Rightarrow \frac{T_1}{T_2} = 10\sqrt{10}$$

8. **(c)** Applying the properties of ellipse, we have

$$\frac{2}{R} = \frac{1}{r_1} + \frac{1}{r_2} = \frac{r_1 + r_2}{r_1 r_2}$$

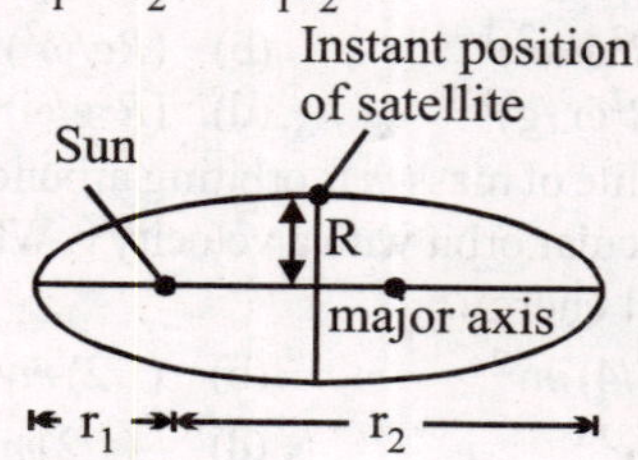

$$R = \frac{2r_1 r_2}{r_1 + r_2}$$

9. **(a)** Both the astronauts are in the condition of weightlessness. Gravitational force between them pulls towards each other. Hence Astronauts move towards each other under mutual gravitional force.

10. **(b)** Let the distance moved by spherical body of mass M is x_1 and by spherical body of mass 5m is x_2

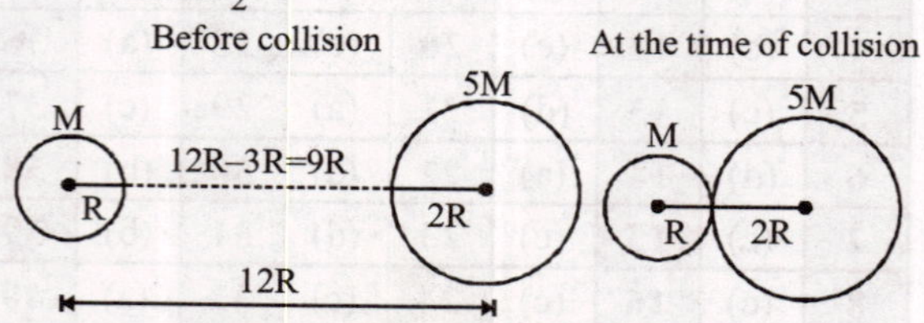

As their C.M. will remain stationary

So, $(M)(x_1) = (5M)(x_2)$ or, $x_1 = 5x_2$

$x_1 + x_2 = 9R$

So, $x_1 = 7.5\,R$

11. **(a)** Gravitational attraction force on particle B,

$$F_g = \frac{GM_P m}{(D_P/2)^2}$$

Acceleration of particle due to gravity

$$a = \frac{F_g}{m} = \frac{4GM_P}{D_P^2}$$

12. **(c)** Gravitational force is independent of medium, Hence, this will remain same.

NOTES

Since, $f = G\dfrac{M_1M_2}{d^2}$ therefore the affecting factor are:

(i) Mass of the two bodies *i.e.*, $f \alpha M_1M_2$ When mass increases gravitational force increase.

(ii) The distance between thier centers; *i.e.*, $f \alpha \dfrac{1}{d^2}$ when the distance increases gravitational force decreases.

13. **(d)** We know that $mg = \dfrac{GMm}{R^2}$

$$\therefore g = \frac{GM}{R^2} \Rightarrow M = \frac{gR^2}{G}$$

14. **(a)** Weight of a body on the surface of the earth,

$W_S = mgs = 72\text{ N}$

Acceleration due to gravity, g varies with height,

$h = \dfrac{R}{2}$ (given)

$$W_h = \frac{mg}{\left(1+\frac{h}{R}\right)^2} = \frac{72}{\left(1+\frac{R/2}{R}\right)^2} = \frac{72}{(3/2)^2}$$

$$= \frac{4}{9}\times 72 = 32\text{ N}$$

15. **(c)**

Above earth surface	Below earth surface
$g_h = g\left(1-\dfrac{2h}{R_e}\right)$	$g_d = g\left(1-\dfrac{d}{R_e}\right)$

According to question, $g_h = g_d$

$$g\left(1-\frac{2h}{R_e}\right) = g\left(1-\frac{d}{R_e}\right)$$

Clearly,

$d = 2h = 2\text{ km}$

NOTES

Graph of g, R and d:

(i) g at surface of earth is maximum

(ii) g at centre of earth is zero

(iii) g is maximum at poles and minimum at the equator.

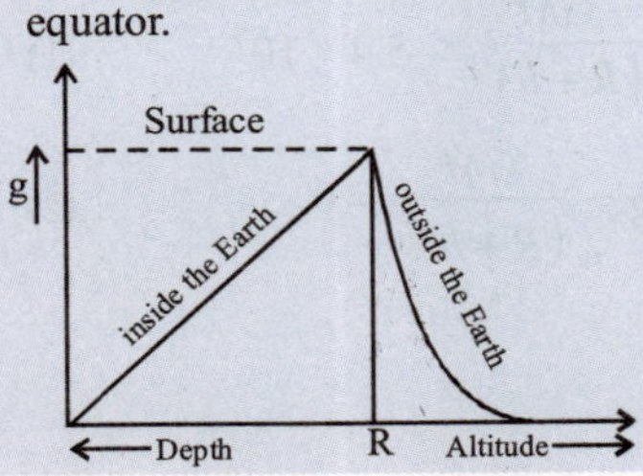

16. **(c)** $\dfrac{GM_m}{(R+h)^2} = \dfrac{1}{16}\dfrac{GM_m}{R^2}$

$$\Rightarrow \frac{1}{(R+h)^2} = \frac{1}{16R^2}$$

$$\Rightarrow \frac{R}{(R+h)} = \frac{1}{4}$$

$$\Rightarrow \frac{R+h}{R} = 4$$

$$\Rightarrow h = 3R$$

17. **(a)** For the riders to experience weightlessness at the top of the hill, the weight of the rider must be balanced by the centripetal force.

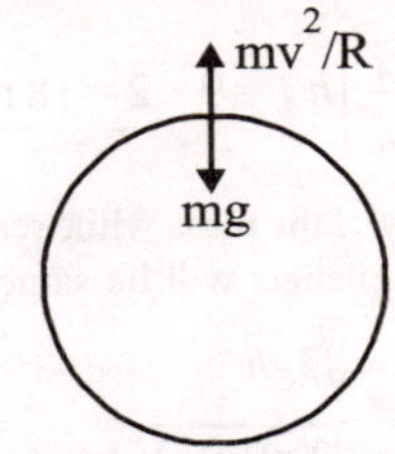

i.e., $mg = \dfrac{mv^2}{R}$

$$\Rightarrow v = \sqrt{gR} = \sqrt{10\times 20} = 14.1\text{ ms}^{-1}$$

Hence, the speed of the car should be between 14 ms^{-1} and 15 ms^{-1}.

18. **(d)** We know that

$$g = \frac{GM}{R^2} = \frac{G\left(\frac{4}{3}\pi R^3\right)\rho}{R^2} = \frac{4}{3}\pi GR\rho$$

g depends on radius of the planet

$$\frac{g'}{g} = \frac{R'}{R} = \frac{3R}{R} = 3 \text{ [As G \& } \rho \text{ are constants]}$$

$$\therefore g' = 3g$$

NOTES

If the radius of a planet decreases by n%, keeping its mass unchanged, the acceleration due to gravity on its surface increases by $2n$%

$$\frac{\Delta g}{g} = -\frac{2\Delta R}{R}$$

$= -2(-n)$

$= 2n\%$

19. **(a)** $g = \dfrac{GM}{R^2}$. Also, $M = \rho \times \dfrac{4}{3}\pi R^3$

$\therefore \quad g = G\frac{4}{3}\rho\pi R.$

At the surface of planet, $g_p = \frac{4}{3}G(2\rho)\pi R'$,

At the surface of the earth $g_e = \frac{4}{3}G\rho\pi R$

$g_e = g_p \Rightarrow \rho R = 2\rho R' \Rightarrow R' = R/2$

20. **(c)** Applying conservation of total mechanical energy principle

$$\frac{1}{2}mv^2 = mg_Ah_A = mg_Bh_B$$

$$\Rightarrow g_Ah_A = g_Bh_B$$

$$\Rightarrow h_B = \left(\frac{g_A}{g_B}\right)h_A = 9 \times 2 = 18 \text{ m}$$

NOTES

The velocity of the mass while reaching the surface of both the planets will be same.

$$\Rightarrow \sqrt{2g'h'} = \sqrt{2gh}$$

$$\Rightarrow \sqrt{2gh'} = \sqrt{2 \times 9g \times 2}$$

$$\Rightarrow 2h' = 36 \Rightarrow h' = 18\text{m}$$

21. **(a)** We know that effective gravity g' at depth below earth surface is given by

$$g' = g\left(1 - \frac{d}{R}\right)$$

Here, $d = 100$ km, $R = 6400$ km,

$$\therefore \quad g' = 9.8\left(1 - \frac{100}{6400}\right) = 9.65 \text{ m/s}^2$$

22. **(a)** $mg = 72$ N (body weight on the surface)

$$g = \frac{GM}{R^2}$$

At a height $H = \frac{R}{2}$,

$$g' = \frac{GM}{\left(R + \frac{R}{2}\right)^2}$$

$$= \frac{4\ GM}{9\ R^2}$$

Body weight at height $H = \frac{R}{2}$,

$$mg' = m \times \frac{4}{9}\left(\frac{GM}{R^2}\right)$$

$$= m \times \frac{4}{9} \times g = \frac{4}{9}mg$$

$$= \frac{4}{9} \times 72 = 32 \text{ N}$$

23. **(d)** $T = 2\pi\sqrt{l/g}$. When the rocket accelerates upwards g increases to $(g + a)$.

24. **(c)** Mass to be raised $= m$

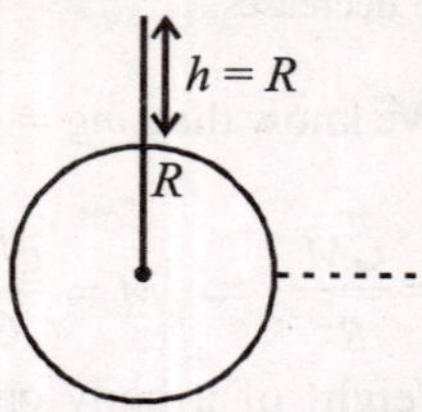

Potential energy at the surface of the earth

$$U_{\text{surface}} = \frac{-GMm}{R}$$

Potential energy at a height from the surface of the earth $h = \text{R}$

$$U_{\text{height}} = \frac{-GMm}{2R}$$

According to work-energy theorem, work done = change in PE

$$\therefore \quad \text{W} = \text{U}_{\text{height}} - \text{U}_{\text{surface}}$$

$$\Rightarrow \frac{-GMm}{2R} - \left(-\frac{GMm}{R}\right)$$

$$= \frac{GMm}{2R} = \frac{gR^2m}{2R} = \frac{mgR}{2} \quad (\because \text{GM} = \text{gR}^2)$$

25. **(c)** As $U = -\frac{GMm}{r}$

$$\therefore \quad (\text{P.E})_\text{A} = -\frac{GMm}{R}; (\text{P.E})_\text{B} = -\frac{GMm}{R+h}$$

$$\therefore \quad \Delta U = (\text{P.E})_\text{B} - (\text{P.E})_\text{A}$$

$$= -\frac{GMm}{R+h} + \frac{GMm}{R} = \frac{GMmh}{(R)(R+h)}$$

26. **(a)** As we know, gravitational potential (V) and acceleration due to gravity (g) with height

$$\text{V} = \frac{-GM}{(R+h)} = -5.4 \times 10^7 \quad \ldots(1)$$

$$\text{and } g = \frac{GM}{(R+h)^2} = 6 \quad \ldots(2)$$

Dividing (1) by (2)

$$\Rightarrow \frac{\frac{-GM}{(R+h)}}{\frac{GM}{(R+h)^2}} = \frac{-5.4\times10^7}{6}$$

$$\Rightarrow \frac{5.4\times10^7}{(R+h)} = 6$$

$\Rightarrow$ $(R+h) = 9000$ km so, $h = 2600$ km

27. **(b)** First when $(r < R)$ $E \propto r$ and then when $r \geq R$ $E \propto \frac{1}{r^2}$. Hence graph (b) correctly dipicts.

> **NOTES** The field strength of uniform solid sphere within it decreases linearly with 'r' and becomes zero as we reach at the centre of the sphere.

28. **(a)** Initial P. E., $U_i = \frac{-GMm}{R}$,

Final P.E., $U_f = \frac{-GMm}{3R}$ [$\because$ R' = R + 2R = 3R]

$\therefore$ Change in potential energy,

$$\Delta U = \frac{-GMm}{3R} + \frac{GMm}{R}$$

$$= \frac{GMm}{R}\left(1-\frac{1}{3}\right) = \frac{2}{3}\frac{GMm}{R} = \frac{2}{3}mgR$$

$$\left(\because \frac{GMm}{R} = mgR\right)$$

> **NOTES** $\Delta U = \frac{mgh}{1+\frac{h}{R}}$
>
> By placing the value of h = 2R we get
>
> $\Delta U = \frac{2}{3}$ mgR.
>
> Work done against gravitational forces in taking a body of mass m from the surface of earth to a height h is the change in potential energy of the body and is given by
>
> $$\Delta u = \frac{GMmh}{R(R+h)}$$
>
> (a) if $h << R$, then $W = \frac{GMmh}{R(R+h)}$
>
> (b) if $h = R$, then $W = \frac{GMm}{2R} = \frac{1}{2}mgh$

29. **(c)**

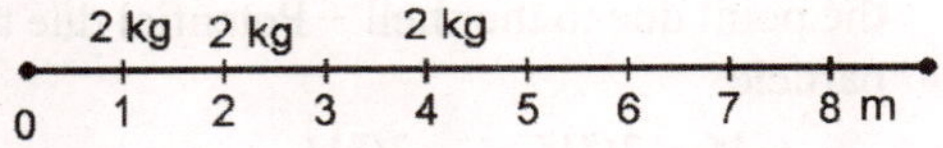

Gravitational potential $V = \frac{-Gm}{r}$

$$V_0 = -\frac{G\times2}{1} - \frac{G\times2}{2} - \frac{G\times2}{4} - \frac{G\times2}{8} --- \infty$$

$$V_0 = -2G\left[1+\frac{1}{2}+\frac{1}{4}+\frac{1}{8}+....\infty\right]$$

$$V_0 = -2G\times\frac{1}{\left(1-\frac{1}{2}\right)} = -2G\times\frac{1}{\frac{1}{2}} = -4G$$

30. **(b)** The Gravitational field due to a thin spherical shell of radius R at distance r.

$E = \frac{GM}{r^2}$ (If $r > R$)

For $r = R$ *i.e.*, on the surface of the shell

$E = \frac{GM}{R^2}$

For $r < R$ *i.e.*, inside the shell

$E = 0$

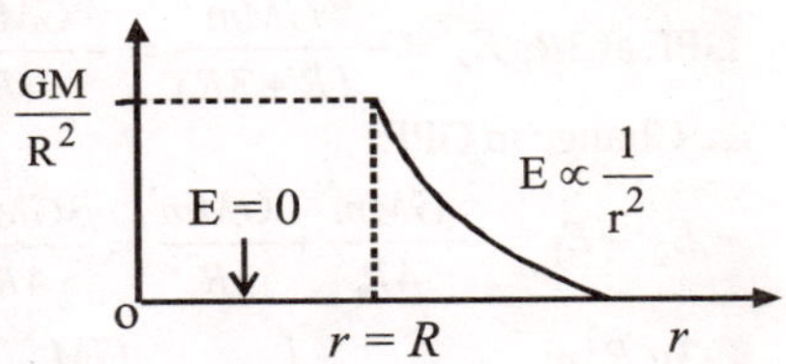

31. **(b)** $V_P = V_{\text{spherical shell}} + V_{\text{particle}}$

$$= \frac{GM}{a} + \frac{GM}{a/2} = \frac{3GM}{a}$$

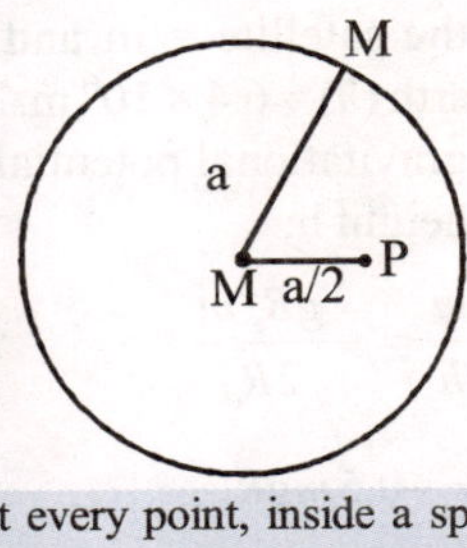

> **NOTES** At every point, inside a spherical shell of radius R, the gravitational potential is same as that on that surface of the shell, *i.e.*,
>
> $$V = -\frac{GM}{R}$$

32. (a) Potential at the given point = Potential at the point due to the shell + Potential due to the particle

$$= -\frac{GM}{a} - \frac{2GM}{a} = -\frac{3GM}{a}$$

33. (a) Potential energy at height $R = -\frac{GMm}{2R}$

If m be the mass of a body which is thrown with velocity v_e so that it goes out of gravitational field from distance R, then

$$\frac{1}{2}mv_e^2 = \frac{GM}{2R}.m \Rightarrow v_e = \sqrt{\frac{GM}{R}}$$

or, $v_e = \sqrt{gR}$

Now, $v = \sqrt{2gR}$, So, $v = \sqrt{2}\,v_e$

or, $v_e = \frac{v}{\sqrt{2}}$

Comparing it with given equation, $f = \frac{1}{\sqrt{2}}$.

34. (b) Gravitational potential energy (GPE) on the surface of earth,

$$E_1 = -\frac{GMm}{R}$$

GPE at $3R$, $E_2 = -\frac{GMm}{(R+3R)} = -\frac{GMm}{4R}$

$\therefore$ Change in GPE

$$= E_2 - E_1 = -\frac{GMm}{4R} + \frac{GMm}{R} = \frac{3GMm}{4R}$$

$$= \frac{3gR^2m}{4R} \qquad \left(\because g = \frac{GM}{R^2}\right)$$

$$= \frac{3}{4}mgR$$

35. (c) Mass of the satellite = m and height of satellite from earth $(h) = 6.4 \times 10^6$ m.

We know that gravitational potential energy of the satellite at height h,

$$GPE = -\frac{GM_e m}{R_e + h} = -\frac{gR_e^2 m}{2R_e}$$

$$= -\frac{gR_e m}{2} = -0.5\,mgR_e$$

(where, $GM_e = gR_e^2$ and $h = R_e$)

36. (a) From conservation of energy

$$\frac{1}{2}mu^2 - \frac{GMm}{R} = \frac{1}{2}m \times (0)^2 - \frac{GMm}{(R+R)}$$

$$\Rightarrow u^2 = \frac{2GM}{R} - \frac{2GM}{2R} = \frac{GM}{R}$$

$$\Rightarrow u = \sqrt{\frac{GM}{R}}$$

NOTES

Use $v^2 = \frac{2gh}{\left(1+\frac{h}{R}\right)}$ where $h = R$

$v = \sqrt{gR} = \sqrt{\frac{GM}{R}}$

37. (a) Escape velocity from the earth's surface

$$v_e = \sqrt{\frac{2GM}{R}} = \sqrt{\frac{2G\rho\frac{4}{3}\pi R^3}{R}} \qquad \left[\because \rho = \frac{M}{V}\right]$$

or, $v_e = \sqrt{\frac{8G\rho\pi}{3}R^2}$ or, $v_e \propto R$ ($\because$ same density)

$\therefore \frac{v}{v_1} = \frac{R}{4R}$ or, $v_1 = 4v$

NOTES

Escape velocity is double if the radius of the earth is double keeping the density unchanged. Escape velocity depends on the mass and size of the planet.

38. (a) From law of conservation of mechanical energy,

$$\frac{1}{2}mv^2 - \frac{GmM}{R} = -\frac{GmM}{(R+h)}$$

$$\Rightarrow \frac{v^2}{2} = \frac{GM}{R} - \frac{(GM)}{(R+h)} = \frac{h}{R(R+h)}GM$$

$$\Rightarrow \frac{1}{2}k^2V_e^2 = \frac{GMh}{R(R+h)}$$

We know, escape velocity $V_e = \sqrt{\frac{2GM}{R}}$

$$\Rightarrow \frac{1}{2}k^2\left(\frac{2GM}{R}\right) = \frac{(GM)}{R(R+h)} \text{ or, } k^2 = \frac{h}{(R+h)}$$

$Rk^2 + hk^2 = h \Rightarrow Rk^2 = h(1-k^2)$

$$\therefore h = \frac{Rk^2}{(1-k^2)}$$

39. (b) $T^2 \propto r^3$

$T^2 \propto (R_E + h)^3$

$$\frac{T_1^2}{T_2^2} = \frac{(R_E + 6R_E)^3}{(R_E + 2.5R_E)^3}$$

$$\frac{T_1^2}{T_2^2} = \frac{7^3}{\left(\frac{7}{2}\right)^3}$$

$T_2 = 6\sqrt{2}\,h$

40. (b) As we know, escape velocity,

$$v_e = \sqrt{\frac{2GM}{R}} = \sqrt{\frac{2G}{R}\cdot\left(\frac{4}{3}\pi R^3 \rho\right)}$$

$v_e \propto R\sqrt{\rho}$

$$\therefore \quad \frac{V_e}{V_p} = \frac{R_e}{R_p}\sqrt{\frac{\rho_e}{\rho_p}}$$

$$\Rightarrow \quad \frac{V_e}{V_p} = \frac{R_e}{2R_e}\sqrt{\frac{\rho_e}{2\rho_e}}$$

$\therefore$ Ratio $\frac{V_e}{V_p} = 1 : 2\sqrt{2}$

41. (c) The gravitational force on the satellite will be aiming towards the centre of the earth so acceleration of the satellite will also be aiming towards the centre of the earth.

42. (d) **Given:** Height of the satellite from the earth's surface $h = 0.25 \times 10^6$m

Radius of the earth $R = 6.38 \times 10^6$m

Acceleration due to gravity $g = 9.8$ m/s^2

Orbital velocity, $v_0 = ?$

$$v_0 = \sqrt{\frac{GM}{(R+h)}} = \sqrt{\frac{GM}{R^2}\cdot\frac{R^2}{(R+h)}}$$

$$= \sqrt{\frac{9.8 \times 6.38 \times 6.38}{6.63 \times 10^6}}$$

$= 7.76$ km/s $\quad \left[\because \frac{GM}{R^2} = g\right]$

43. (c) From question,

Escape velocity

$$= \sqrt{\frac{2GM}{R}} = c = \text{speed of light}$$

$$\Rightarrow \quad R = \frac{2GM}{c^2}$$

$$= \frac{2 \times 6.6 \times 10^{-11} \times 5.98 \times 10^{24}}{(3 \times 10^8)^2}\,\text{m}$$

$= 10^{-2}$ m

44. (c) Escape velocity,

$$V_e = \sqrt{\frac{2GM_E}{R_E}} = \sqrt{\frac{2G}{R_E}\left(\frac{4}{3}\pi R_E^3 \rho_E\right)} = R\sqrt{\frac{8}{3}\pi G\rho_E}$$

$$\Rightarrow V_e \propto R \Rightarrow \frac{V_P}{V_E} = \frac{R_P}{R_E} = 2$$

$\Rightarrow V_P = 2V_E.$

45. (b) As we know, the minimum speed with which a body is projected so that it does not return back is called escape speed.

$$V_e = \sqrt{\frac{2GM}{r}} = \sqrt{\frac{2GM}{R+h}} = \sqrt{\frac{2GM}{4R}}$$

$$= \left(\frac{GM}{2R}\right)^{\frac{1}{2}} \quad (\because h = 3R)$$

46. (c) According to Kelpner's law of period

$T^2 \propto R^3$

$$\frac{T_1^2}{T_2^2} = \frac{R_1^3}{R_2^3} = \frac{(6R)^3}{(3R)^3} = 8$$

$$\frac{24 \times 24}{T_2^2} = 8$$

$$T_2^2 = \frac{24 \times 24}{8} = 72 = 36 \times 2$$

$T_2 = 6\sqrt{2}$

47. (d) $v_e = \sqrt{\frac{2GM}{R}} \Rightarrow v_0 = \sqrt{\frac{GM}{R}}$

$v_e = \sqrt{2}v_0$

The orbital velocity of a satellite at a height h above the surface of earth,

$$v_0 = \sqrt{\frac{GM}{(R+h)}} = \sqrt{\frac{gR^2}{(R+h)}} \quad (\because GM = gR^2)$$

Here, M = mass of earth,

R = radius of earth,

g = acceleration due to If the satellite is very close to the surrface of earth gravity earth on surface of then $h = 0$

$$\therefore \quad v_0 = \sqrt{\frac{gR^2}{R}} = \sqrt{gR} = \sqrt{\frac{GM}{R}}$$

48. (b) Angular momentum is conserved

$\therefore L_1 = L_2$

$\Rightarrow mr_1v_1 = mr_2v_2$

$\Rightarrow r_1v_1 = r_2v_2$

$\Rightarrow \frac{v_1}{v_2} = \frac{r_2}{r_1}$

49. (a) The velocity u should be equal to the escape velocity. That is,

$u = \sqrt{2gR}$

But $g = \frac{GM}{R^2}$

$\therefore u = \sqrt{2 \cdot \frac{GM}{R^2} \cdot R} \Rightarrow \sqrt{\frac{2GM}{R}}$

50. (b) Orbital velocity of a satellite in a circular orbit of radius a is given by

$v = \sqrt{\frac{GM}{a}}$

$\Rightarrow v \propto \sqrt{\frac{1}{a}}$

$\Rightarrow \frac{v_2}{v_1} = \sqrt{\frac{a_1}{a_2}}$

$\therefore v_2 = v_1\sqrt{\frac{4R}{R}} = 2v_1 = 6V$

51. (b) Since orbital velocity of satellite is $v = \sqrt{\frac{GM}{r}}$, it does not depend upon the mass of the satellite.

Therefore, both satellites will move with same speed.

52. (a) K.E. of satellite moving in an orbit around the earth is

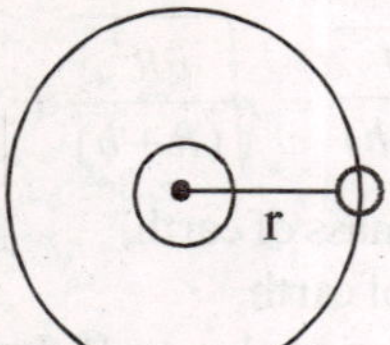

$$K = \frac{1}{2}mv^2 = \frac{1}{2}m\left(\sqrt{\frac{GM}{r}}\right)^2 = \frac{GMm}{2r}$$

P.E. of satellite and earth system is

$$U = -\frac{GMm}{r} \Rightarrow \frac{K}{U} = \frac{\frac{GMm}{2r}}{-\frac{GMm}{r}} = -\frac{1}{2}$$

53. (a) $v_{\text{earth}} = \sqrt{\frac{2GM_e}{R_e}}$

$v_{\text{planet}} = \sqrt{\frac{2GM_p}{R_p}} = \sqrt{\frac{2GM_e}{R_e/4}} = \sqrt{\frac{8GM_e}{R_e}}$

$\frac{v_{\text{planet}}}{v_{\text{earth}}} = \sqrt{\frac{8GM_e}{R_e}} \times \sqrt{\frac{R_e}{2GM_e}} = 2$

$\therefore v_{\text{planet}} = 2 \times v_{\text{earth}} = 2 \times 11.2 = 22.4$ km/s

54. (b) Escape velocity is the minimum velocity with which a body is projected to escape from earth's gravitational field

$\frac{1}{2}mv_e^2 = \frac{GMm}{R_e} \Rightarrow v_e = \sqrt{\frac{2GM}{R_e}}$

55. (b) Escape velcocity

$v_e = \sqrt{\frac{2GM_e}{R_e}}, \ v'_e = \sqrt{\frac{2GM'_e}{R'_e}}$

$\therefore \frac{v'_e}{v_e} = \sqrt{\frac{M'_e}{M_e} \times \frac{R_e}{R'_e}}$

Given $M'_e = 2M_e$ and $R'_e = \frac{R_e}{2}$

$\therefore \frac{v'_e}{v_e} = \sqrt{\frac{2M_e}{M_e} \times \frac{R_e}{R_e/2}} = \sqrt{4} = 2$

$v'_e = 2v_e = 2 \times 11.2 = 22.4$ km/s

56. (b) The orbital speed of satellite is independent of mass of satellite, so the ball will behave as a satellite and will continue to move with the same speed in the original orbit.

57. (a) Escape velocity on surface of earth

$v_e = \sqrt{\frac{2GM_e}{R_e}} \propto \sqrt{\frac{M_e}{R_e}}$.

$\therefore \frac{v_e}{v_P} = \sqrt{\frac{M_e}{R_e}} \times \sqrt{\frac{R_p}{M_p}}$

$= \sqrt{\frac{M_e}{R_e}} \times \sqrt{\frac{3R_e}{3M_e}} = \frac{1}{1} = 1$

or, $v_P = v_e$.

58. (c) $F = \left(\frac{dM}{dt}\right) v = \alpha v^2 \qquad \left(\because \frac{dM}{dt} = \alpha v\right)$

$\therefore$ Retardation $= \frac{-F}{M} = -\frac{\alpha v^2}{M}$

59. (b) Escape velocity does not depend on the angle of projection.

NOTES

The value of escape velocity is given by

$V_e = \sqrt{\frac{2GM}{R}}$

Here, M = mass of earth, G = gravitational constant
R = radius of earth

$V_e = \sqrt{\frac{2GM}{R^2}}$

$= \sqrt{\frac{2 \times G}{R^2} \times \frac{4}{3}\pi R^3 \rho} \qquad \left(\because M = \frac{4}{3}\pi R^3 \rho\right)$

$V_e = R\sqrt{\frac{8\pi G\rho}{3}}$

Thus, value of escape velocity depends upon the mass M, density ρ and radius R of earth/planet from which the body is projected and independent of angle of projection.

60. (a) $T = \frac{2\pi r}{v_0} = \frac{2\pi r}{(gR^2/r)^{1/2}} = \frac{2\pi r^{3/2}}{\sqrt{gR^2}} = \frac{2\pi}{\omega}$

Hence, $r^{3/2} = \frac{\sqrt{gR^2}}{\omega}$ or $r^3 = \frac{gR^2}{\omega^2}$

or, $r = (gR^2/\omega^2)^{1/3}$

61. (d) Total energy $= -\text{K E} = \frac{\text{PE}}{2}$

$\text{K.E} = \frac{-1}{2}mv^2$

62. (c) In a circular or elliptical orbital motion, torque is always acting parallel to displacement or velocity. So, angular momentum is conserved. In attractive field, potential energy is negative. Kinetic energy changes as velocity increase when distance is less. So, option (c) is correct.

63. (b) $F = \frac{k}{R} = \frac{Mv^2}{R}$. Hence $v \propto R^0$

64. (a) Since, escape velocity $\left(v_e = \sqrt{2gR_e}\right)$ is independent of angle of projection, so it will not change.

8 Mechanical Properties of Solids

Trend Analysis with Important Topics & Sub-Topics

Topic Name	Sub-Topic	2021 QNS.	2021 LOD	2020 QNS.	2020 LOD	2019 QNS.	2019 LOD	2018 QNS.	2018 LOD	2017 QNS.	2017 LOD	2016 QNS.	2016 LOD
Hooke's Law & Young's Modulus of Elasticity	Hooke's Law & Young's Modulus and Stress-strain Curve			1	E	1	A	1	A				
Bulk and Rigidity Modulus & Work Done in Stretching a Wire	Bulk Modulus									1	A		

LOD - Level of Difficulty | E - Easy | A - Average | D - Difficult | Qns - No. of Questions

Topic 1 Hooke's Law & Young's Modulus of Elasticity

1. A wire of length L, area of cross section A is hanging from a fixed support. The length of the wire changes to L_1 when mass M is suspended from its free end. The expression for Young's modulus is : *[2020]*

(a) $\frac{Mg(L_1-L)}{AL}$ (b) $\frac{MgL}{AL_1}$

(c) $\frac{MgL}{A(L_1-L)}$ (d) $\frac{MgL_1}{AL}$

2. The stress-strain curves are drawn for two different materials *X* and *Y*. It is observed that the ultimate strength point and the fracture point are close to each other for material *X* but are far apart for material *Y*.
We can say that materials *X* and *Y* are likely to be (respectively), *[NEET Odisha 2019]*

(a) Plastic and ductile
(b) Ductile and brittle
(c) Brittle and ductile
(d) Brittle and plastic

3. Two wires are made of the same material and have the same volume. The first wire has cross-sectional area A and the second wire has cross-sectional area 3A. If the length of the first wire is increased by $\Delta\ell$ on applying a force F, how much force is needed to stretch the second wire by the same amount? *[2018]*

(a) 9F (b) 6F
(c) F (d) 4F

4. The Young's modulus of steel is twice that of brass. Two wires of same length and of same area of cross section, one of steel and another of brass are suspended from the same roof. If we want the lower ends of the wires to be at the same level, then the weights added to the steel and brass wires must be in the ratio of : *[2015 RS]*

(a) 2 : 1 (b) 4 : 1
(c) 1 : 1 (d) 1 : 2

5. Copper of fixed volume 'V; is drawn into wire of length '*l*'. When this wire is subjected to a constant force 'F', the extension produced in the wi re is 'Δl'. Which of the following graphs is a straight line? *[2014]*

(a) Δl versus $\frac{1}{l}$ (b) Δl versus l^2

(c) Δl versus $\frac{1}{l^2}$ (d) Δl versus l

6. The following four wires are made of the same material. Which of these will have the largest extension when the same tension is applied ? *[2013]*

(a) Length = 100 cm, diameter = 1 mm
(b) Length = 200 cm, diameter = 2 mm
(c) Length = 300 cm, diameter = 3 mm
(d) Length = 50 cm, diameter = 0.5 mm

7. If the ratio of diameters, lengths and Young's modulus of steel and copper wires shown in the figure are p, q and s respectively, then the corresponding ratio of increase in their lengths would be *[NEET Kar. 2013]*

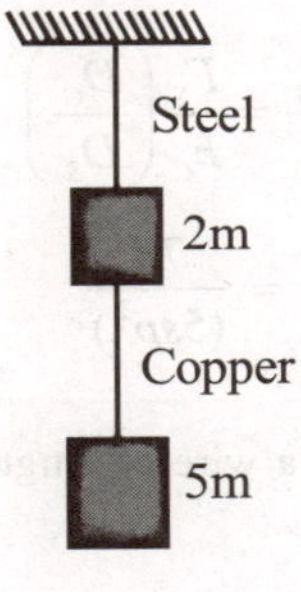

(a) $\frac{7q}{(5sp)}$
(b) $\frac{5q}{(7sp^2)}$
(c) $\frac{7q}{(5sp^2)}$
(d) $\frac{2q}{(5sp)}$

8. Two wires A and B are of the same material. Their lengths are in the ratio 1 : 2 and the diameter are in the ratio 2 : 1. If they are pulled by the same force, then increase in length will be in the ratio *[1988]*

(a) 2 : 1 (b) 1 : 4
(c) 1 : 8 (d) 8 : 1

Topic 2 Bulk and Rigidity Modulus & Work Done in Stretching a Wire

9. When a block of mass M is suspended by a long wire of length L, the length of the wire becomes (L + l). The elastic potential energy stored in the extended wire is : *[2019]*

(a) Mgl (b) MgL
(c) $\frac{1}{2}$Mgl (d) $\frac{1}{2}$MgL

10. The bulk modulus of a spherical object is 'B'. If it is subjected to uniform pressure 'p', the fractional decrease in radius is *[2017]*

(a) $\frac{B}{3p}$ (b) $\frac{3p}{B}$
(c) $\frac{p}{3B}$ (d) $\frac{p}{B}$

11. When an elastic material with Young's modulus Y is subjected to stretching stress S, elastic energy stored per unit volume of the material is *[1989]*

(a) YS / 2 (b) $S^2Y/2$
(c) $S^2/2Y$ (d) S / 2Y

ANSWER KEY

1	(c)	3	(a)	5	(b)	7	(c)	9	(c)	11	(c)						
2	(c)	4	(a)	6	(d)	8	(c)	10	(c)								

Hints & Solutions

1. **(c)** Stress $= \frac{\text{force}}{\text{cross-section area}} = \frac{Mg}{A}$

Strain $= \frac{\text{change in length}}{\text{original length}} = \frac{\Delta L}{L} = \frac{L_1 - L}{L}$

Young's modulus, $Y = \frac{\text{stress}}{\text{strain}} = \frac{MgL}{A(L_1 - L)}$

2. **(c)** Fracture point and ultimate strength point is close for material X, hence X is brittle in nature and both points are far apart for material Y hence it is ductile.

3. **(a)** Wire 1:

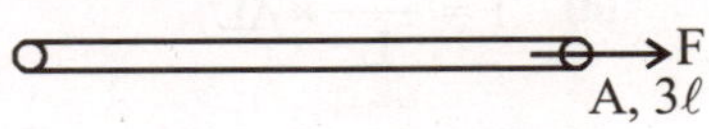

$$\Delta \ell = \left(\frac{F}{AY}\right) 3\ell \qquad ...(i)$$

Wire 2: 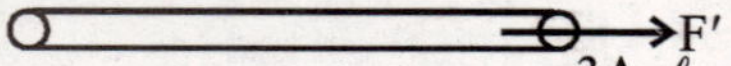

$$\Delta\ell = \left(\frac{F'}{3AY}\right)\ell \quad \text{...(ii)}$$

From equation (i) & (ii),

$$\Delta\ell = \left(\frac{F}{AY}\right)3\ell = \left(\frac{F'}{3AY}\right)\ell \text{ or, } F' = 9F$$

4. **(a)** Young's modulus $Y = \frac{W}{A}.\frac{l}{\Delta l}$

$$\frac{W_1}{Y_1} = \frac{W_2}{Y_2}$$

[$\because$ A, l, Δl same for both brass and steel]

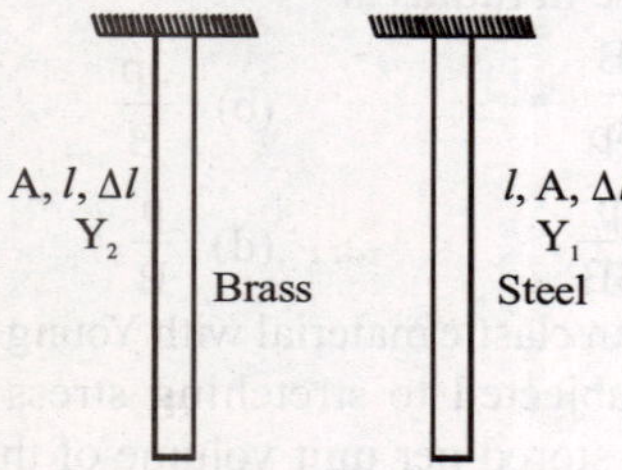

$$\frac{W_1}{W_2} = \frac{Y_1}{Y_2} = 2 \quad [Y_{steel}/Y_{brass} = 2 \text{ given}]$$

5. **(b)** As $Y = \dfrac{\frac{F}{A}}{\frac{\Delta l}{l}} \Rightarrow \Delta l = \dfrac{Fl}{AY}$

But $V = Al$ so $A = \dfrac{V}{l}$

Therefore $\Delta l = \dfrac{Fl^2}{VY} \propto l^2$

Hence graph of Δl versus l^2 will give a straight line.

6. **(d)** $F = \dfrac{YA}{L} \times \Delta L$

So, extension, $\Delta L \propto \dfrac{L}{A} \propto \dfrac{L}{D^2}$

[$\because$ F and Y are constant]

$$\Delta L_1 \propto \frac{100}{1^2} \propto 100 \text{ and } \Delta L_2 \propto \frac{200}{2^2} \propto 50$$

$$\Delta L_3 \propto \frac{300}{3^2} \propto \frac{100}{3} \text{ and } \Delta L_4 \propto \frac{50}{\frac{1}{4}} \propto 200$$

The ratio of $\dfrac{L}{D^2}$ is maximum for case (d).

Hence, option (d) is correct.

7. **(c)** From formula,

Increase in length $\Delta L = \dfrac{FL}{AY} = \dfrac{4FL}{\pi D^2 Y}$

$$\frac{\Delta L_S}{\Delta L_C} = \frac{F_S}{F_C}\left(\frac{D_C}{D_S}\right)^2 \frac{Y_C}{Y_S}\frac{L_S}{L_C} = \frac{7}{5} \times \left(\frac{1}{p}\right)^2 \left(\frac{1}{s}\right) q$$

$$= \frac{7q}{(5sp^2)}$$

> **NOTES**
>
> If a wire of length l is suspended from a rigid support and A is the area of cross–section of the wire, then mass of the wire, m = A$l\rho$. Then extension in the wire due to its own weight can be find as follows.
>
> Young Modulus, $Y = \dfrac{mg}{A} \times \dfrac{(l/2)}{\Delta l}$
>
> Here, L = l/2 because the weight of wire acts at the mid point of wire.
>
> $\therefore$ Extension in the wire due to its own weight
>
> $$\Delta l = \frac{l^2 \rho g}{2Y}$$

8. **(c)** We know that Young's modulus

$$Y = \frac{F}{\pi r^2} \times \frac{L}{\ell}$$

Since Y, F are same for both the wires, we have,

$$\ell \propto \frac{L}{r^2}$$

$$\text{or, } \frac{\ell_1}{\ell_2} = \frac{r_2^2 \times L_1}{r_1^2 \times L_2} = \frac{(D_2/2)^2 \times L_1}{(D_1/2)^2 \times L_2}$$

$$\text{or,}\ \frac{\ell_1}{\ell_2} = \frac{D_2^2 \times L_1}{D_1^2 \times L_2} = \frac{D_2^2}{(2D_2)^2} \times \frac{L_2}{2L_2} = \frac{1}{8}$$

So, $\ell_1 : \ell_2 = 1:8$

9. **(c)** Here,

$Kx_0 = Mg$

where K = force constant

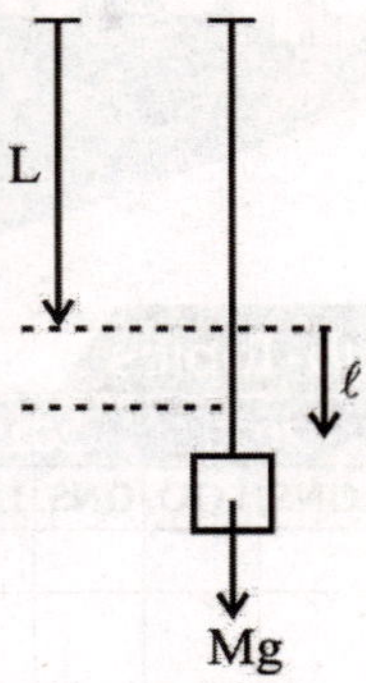

$$\Delta E = \frac{1}{2}Kx_0^2$$

$$= \frac{1}{2}\frac{Mg}{x_0} \times x_0^2$$

$$= \frac{1}{2}Mgx_0$$

Stored elastic potential energy in extended wire, $= \frac{1}{2}Mg\ell$ [here $x_0 = \ell$]

10. **(c)** Bulk modulus is given by

$$B = \frac{p}{\left(\frac{\Delta V}{V}\right)} \quad \text{or} \quad \frac{\Delta V}{V} = \frac{p}{B}$$

$3\frac{\Delta R}{R} = \frac{p}{B}$ (here, $\frac{\Delta R}{R}$ = fractional decreases in radius)

$$\Rightarrow \frac{\Delta R}{R} = \frac{p}{3B}$$

> **NOTES**
>
> If the object of mass m and radius r is sorrounded by a liquid in a cylindrical container
>
> $$\frac{\Delta R}{R} = \frac{1}{3}\frac{\Delta V}{V} \quad \text{...(i)}$$
>
> Bulk modulus, $K = -V\frac{\Delta P}{\Delta V}$
>
> $$\therefore \frac{\Delta V}{V} = \frac{\Delta P}{K} = \frac{mg}{AK} \quad \left(\because \Delta P = \frac{mg}{A}\right)$$
>
> $$\therefore \frac{\Delta R}{R} = \frac{1}{3}\frac{mg}{AK} \quad \text{(Here A = area of object)}$$

11. **(c)** Energy stored per unit volume

$$= \frac{1}{2} \times \text{stress} \times \text{strain}$$

$$= \frac{1}{2} \times \text{stress} \times (\text{stress/Young's modulus})$$

$$= \frac{1}{2} \times (\text{stress})^2/(\text{Young's modulus}) = \frac{S^2}{2Y}$$

9 Mechanical Properties of Fluids

Trend Analysis with Important Topics & Sub-Topics

Topic Name	Sub-Topic	2021 QNS.	2021 LOD	2020 QNS.	2020 LOD	2019 QNS.	2019 LOD	2018 QNS.	2018 LOD	2017 QNS.	2017 LOD	2016 QNS.	2016 LOD
Pressure, Density, Pascal's Law & Archimedes' Principle	Fluid Pressure									1	A		
	Pressure Density & Archimedes' Principle	1	A									1	A
Fluid Flow, Reyonld's Number & Bernoulli's Principle													
Viscosity & Terminal Velocity	Stoke's Formula & Terminal Velocity							1	A				
	Velocity of Efflux					1	A						
Surface Tension, Surface Energy & Capillarity	Pressure Inside Soap Bubble			1	A	1	A						
	Capillary Action			1	A								

LOD - Level of Difficulty | E - Easy | A - Average | D - Difficult | Qns - No. of Questions

Topic 1 Pressure, Density Pascal's Law & Archimedes' Principle

1. The velocity of a small ball of mass M and density d, when dropped in a container filled with glycerine becomes constant after some time. If the density of glycerine is $\frac{d}{2}$, then the viscous force acting on the ball will be

   (a) $2\,Mg$ (b) $\frac{Mg}{2}$ *[2021]*

   (c) Mg (d) $\frac{3}{2}Mg$

2. In a u-tube as shown in the fig. water and oil are in the left side and right side of the tube respectively. The heights from the bottom for water and oil columns are 15 cm and 20 cm respectively. The density of the oil is [take π water $= 1000 kg/m^3$] *[NEET Odisha 2019]*

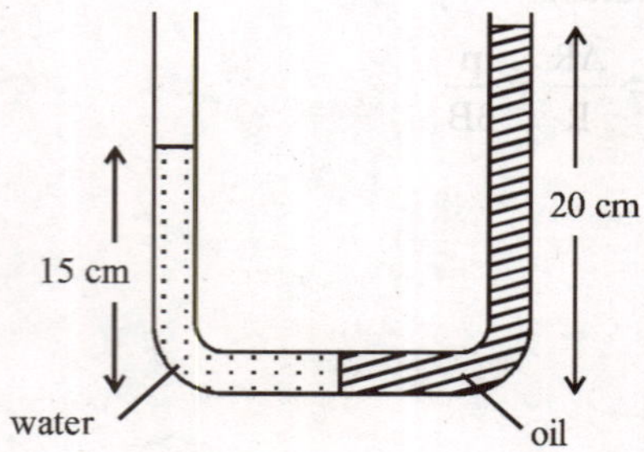

   (a) $1333\,kg/m^3$ (b) $1200\,kg/m^3$

   (c) $750\,kg/m^3$ (d) $1000\,kg/m^3$

3. A U tube with both ends open to the atmosphere, is partially filled with water. Oil, which is immiscible with water, is poured into one side until it stands at a distance of 10 mm above the water level on the other side. Meanwhile the water rises by 65 mm from its original level (see diagram). The density of the oil is *[2017]*

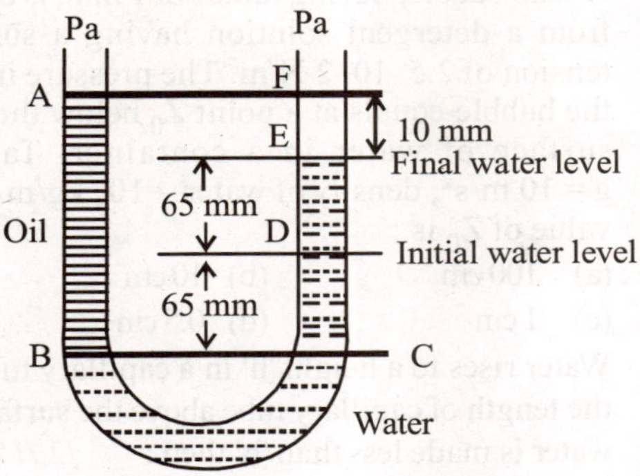

(a) $425\ kg\ m^{-3}$ (b) $800\ kg\ m^{-3}$
(c) $928\ kg\ m^{-3}$ (d) $650\ kg\ m^{-3}$

4. Two non-mixing liquids of densities ρ and nρ (n > 1) are put in a container. The height of each liquid is h. A solid cylinder of length L and density d is put in this container. The cylinder floats with its axis vertical and length pL(p < 1) in the denser liquid. The density d is equal to : *[2016]*
(a) $\{1+(n+1)p\}\rho$ (b) $\{2+(n+1)p\}\rho$
(c) $\{2+(n-1)p\}\rho$ (d) $\{1+(n-1)p\}\rho$

5. The approximate depth of an ocean is 2700 m. The compressibility of water is $45.4 \times 10^{-11}\ Pa^{-1}$ and density of water is $10^3\ kg/m^3$. What fractional compression of water will be obtained at the bottom of the ocean ? *[2015]*
(a) 1.0×10^{-2} (b) 1.2×10^{-2}
(c) 1.4×10^{-2} (d) 0.8×10^{-2}

6. In rising from the bottom of a lake, to the top, the temperature of an air bubble remains unchanged, but its diameter gets doubled. If *h* is the barometric height (expressed in m of mercury of relative density ρ) at the surface of the lake, the depth of the lake is *[1994]*
(a) 8 ρh m (b) 7ρh m
(c) 9 ρh m (d) 12 ρh m

7. The compressibility of water is 4×10^{-5} per unit atmospheric pressure. The decrease in volume of 100 cm^3 of water under a pressure of 100 atmosphere will be *[1990]*
(a) $0.4\ cm^3$ (b) $4 \times 10^{-5}\ cm^3$
(c) $0.025\ cm^3$ (d) $0.004\ cm^3$

Topic 2 Fluid Flow, Reyonld's Number & Bernoulli's Principle

8. A small hole of area of cross-section $2\ mm^2$ is present near the bottom of a fully filled open tank of height 2 m. Taking $g = 10\ m/s^2$, the rate of flow of water through the open hole would be nearly: *[2019]*
(a) $12.6\ 10^{-6}\ m^3/s$ (b) $8.9\ 10^{-6}\ m^3/s$
(c) $2.23\ 10^{-6}\ m^3/s$ (d) $6.4\ 10^{-6}\ m^3/s$

9. A wind with speed 40 m/s blows parallel to the roof of a house. The area of the roof is 250 m^2. Assuming that the pressure inside the house is atmosphere pressure, the force exterted by the wind on the roof and the direction of the force will be ($\rho_{air} = 1.2\ kg/m^3$) *[2015]*
(a) 4.8×10^5 N, upwards
(b) 2.4×10^5 N, upwards
(c) 2.4×10^5 N, downwards
(d) 4.8×10^5 N, downwards

10. The cylindrical tube of a spray pump has radius, R, one end of which has n fine holes, each of radius r. If the speed of the liquid in the tube is V, the speed of the ejection of the liquid through the holes is : *[2015 RS]*
(a) $\frac{VR^2}{nr^2}$ (b) $\frac{VR^2}{n^3r^2}$
(c) $\frac{V^2R}{nr}$ (d) $\frac{VR^2}{n^2r^2}$

11. A fluid is in streamline flow across a horizontal pipe of variable area of cross section. For this which of the following statements is correct? *[NEET Kar. 2013]*
(a) The velocity is minimum at the narrowest part of the pipe and the pressure is minimum at the widest part of the pipe
(b) The velocity is maximum at the narrowest part of the pipe and pressure is maximum at the widest part of the pipe
(c) Velocity and pressure both are maximum at the narrowest part of the pipe
(d) Velocity and pressure both are maximum at the widest part of the pipe

Topic 3 Viscosity & Terminal Velocity

12. Two small spherical metal balls, having equal masses, are made from materials of densities ρ_1 and ρ_2 ($\rho_1 = 8\rho_2$) and have radii of 1 mm and 2 mm, respectively, they are made to fall vertically (from rest) in a viscous medium whose coefficient of viscosity equals η and whose density is $0.1\rho_2$. The ratio of their terminal velocities would be, *[NEET Odisha 2019]*
(a) $\frac{79}{36}$ (b) $\frac{79}{72}$ (c) $\frac{19}{36}$ (d) $\frac{39}{72}$

13. A small sphere of radius 'r' falls from rest in a viscous liquid. As a result, heat is produced due to viscous force. The rate of production of heat when the sphere attains its terminal velocity, is proportional to *[2018]*
(a) r^3 (b) r^2
(c) r^4 (d) r^5

14. The terminal velocity v_r of a small steel ball of radius r falling under gravity through a column of a viscous liquid of coefficient of viscosity η depends on mass of the ball m, acceleration due to gravity g, coefficient of viscosity η and radius r. Which of the following relations is dimensionally correct?
(a) $v_r \propto \frac{mgr}{\eta}$ (b) $v_r \propto mg\eta r$ *[1992]*
(c) $v_r \propto \frac{mg}{r\eta}$ (d) $v_r \propto \frac{\eta mg}{r}$

Topic 4 Surface Tension, Surface Energy & Capillarity

15. A capillary tube of radius r is immersed in water and water rises in it to a height h. The mass of the water in the capillary is 5 g. Another capillary tube of radius 2r is immersed in water. The mass of water that will rise in this tube is : *[2020]*
(a) 5.0 g (b) 10.0 g
(c) 20.0 g (d) 2.5 g

16. A soap bubble, having radius of 1 mm, is blown from a detergent solution having a surface tension of 2.5 10–2 N/m. The pressure inside the bubble equals at a point Z_0 below the free surface of water in a container. Taking $g = 10 \text{ m/s}^2$, density of water = 10_3 kg/m$_3$, the value of Z_0 is : *[2019]*
(a) 100 cm (b) 10 cm
(c) 1 cm (d) 0.5 cm

17. Water rises to a height 'h' in a capillary tube. If the length of capillary tube above the surface of water is made less than 'h' then : *[2015 RS]*
(a) water rises upto the top of capillary tube and stays there without overflowing
(b) water rises upto a point a little below the top and stays there
(c) water does not rise at all.
(d) Water rises upto the tip of capillary tube and then starts overflowing like fountain.

18. A certain number of spherical drops of a liquid of radius 'r' coalesce to form a single drop of radius 'R' and volume 'V'. If 'T' is the surface tension of the liquid, then : *[2014]*
(a) energy = $4VT\left(\frac{1}{r}-\frac{1}{R}\right)$ is released
(b) energy = $3VT\left(\frac{1}{r}+\frac{1}{R}\right)$ is absorbed
(c) energy = $3VT\left(\frac{1}{r}-\frac{1}{R}\right)$ is released
(d) energy is neither released nor absorbed

19. The wetability of a surface by a liquid depends primarily on *[2013]*
(a) surface tension
(b) density
(c) angle of contact between the surface and the liquid
(d) viscosity

20. The angle of contact between pure water and pure glass, is *[1996]*
(a) 0º (b) 45º
(c) 90º (d) 135º

ANSWER KEY

1	(b)	4	(d)	7	(a)	10	(a)	13	(d)	16	(c)	19	(c)						
2	(c)	5	(b)	8	(a)	11	(b)	14	(c)	17	(a)	20	(a)						
3	(c)	6	(b)	9	(b)	12	(a)	15	(b)	18	(c)								

Hints & Solutions

1. **(b)** Let F_v = viscous force and F_B = Bouyant force acting on the ball.

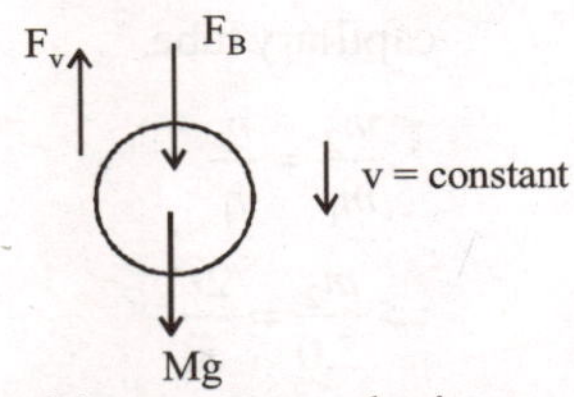

When body moves with constant velocity

$Mg = F_B + F_v$ [a = 0]

$F_v = Mg - F_B$

$= dVg - \frac{d}{2}.Vg$ (M = dVg) V = volume of ball.

or $F_v = \frac{d}{2}Vg = \frac{M}{2}g$ [$\because$ M = dV]

> **NOTES** Due to motion of a body through the fluid, the fluid in contact with the body is dragged with it causes relative motion in fluid layers hear the body and hence viscous force starts.

2. **(c)** Pressure at A = Pressure at B.

$P_a + 0.15 \times 10^3 \times g = P_a + 0.20 \times d_0\, g$

$d_0 = \frac{0.15 \times 10^3}{0.20} = 0.75 \times 10^3 = 750\ kg/m^3$

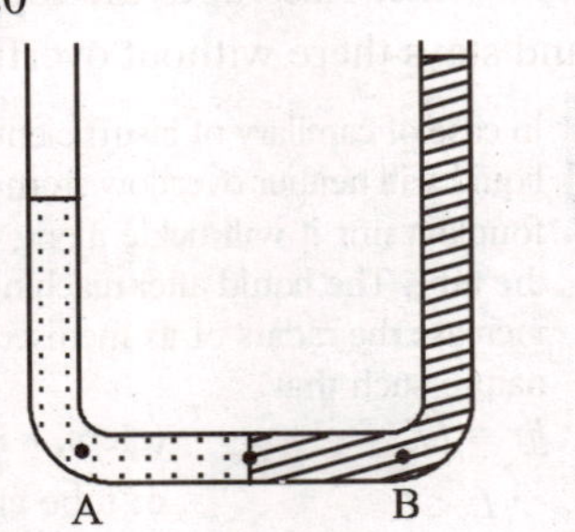

3. **(c)** Here, $h_{oil} \times \rho_{oil} \times g = h_{water} \times \rho_{water} \times g$

$\rho_0 g \times 140 \times 10^{-3} = \rho_w g \times 130 \times 10^{-3}$

$\rho_{oil} = \frac{130}{140} \times 10^3 \approx 928 kg/m^3$ [$\because \rho_w = 1\ kgm^{-3}$]

4. **(d)** As we know,

Pressure P = Vdg

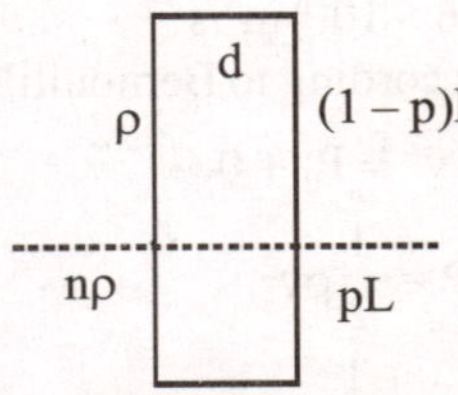

Here, $LAdg = (pL)A(n\rho)g + (1-p)LA\rho g$

$\Rightarrow d = (1-p)\rho + pn\rho = [1 + (n-1)p]\rho$

> **NOTES** When a body of volume *V* floats in a liquid then Weight of the body = up thrust
> $\Rightarrow V\rho g = V_{in}\sigma g$
> Here, V = volume of body
> ρ = density of body
> σ = density of liquid
> V_{in} = volume of the body immersed in liquid

5. **(b)** Compressibility of water,

$K = 45.4 \times 10^{-11}\ Pa^{-1}$

density of water $P = 10^3\ kg/m^3$

depth of ocean, h = 2700 m

We have to find $\frac{\Delta V}{V} = ?$

As we know, compressibility,

$K = \frac{1}{B} = \frac{(\Delta V/V)}{P}$ [$P \doteq \rho gh$]

So, $(\Delta V/V) = K\rho gh$

$= 45.4 \times 10^{-11} \times 10^3 \times 10 \times 2700 = 1.2258 \times 10^{-2}$

6. **(b)** $(h\rho g + H \times 1 \times g)\frac{4}{3}\pi r^3 = h\rho g \times \frac{4}{3}\pi(2r)^3$

This gives $H = 7h\rho$

7. **(a)** $K = \frac{1}{B} = \frac{\Delta V/V}{P}$. Here, P = 100 atm,

$K = 4 \times 10^{-5}$ and $V = 100\ cm^3$.

Hence, $\Delta V = 0.4\ cm^3$

8. **(a)** Volumetric flow rate of liquid

$Q = au = a\sqrt{2gh}$

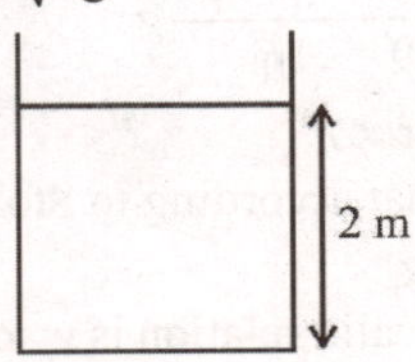

$= 2\times10^{-6}\,m^2 \times \sqrt{2\times10\times2}\ m/s$

$= 2\times2\times3.14\times10^{-6}\,m^3/s$

$= 12.56\times10^{-6}\,m^3/s$

$= 12.6\times10^{-6}\,m^3/s$

9. **(b)** According to Bernoulli's theorem,

$P + \frac{1}{2}\rho v^2 = P_0 + 0$

So, $\Delta P = \frac{1}{2}\rho v^2$

$F = \Delta P A = \frac{1}{2}\rho v^2 A$

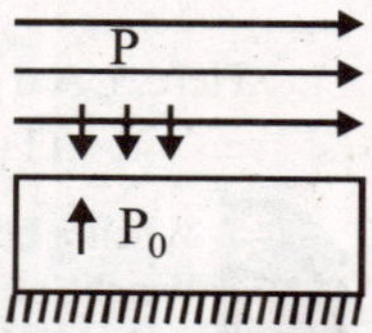

$= \frac{1}{2}\times1.2\times40\times40\times250$

$= 2.4\times10^5\,N$ (upwards)

10. **(a)** Inflow rate of volume of the liquid = Outflow rate of volume of the liquid

$\pi R^2 V = n\pi r^2(v) \Rightarrow v = \frac{\pi R^2 V}{n\pi r^2} = \frac{VR^2}{nr^2}$

11. **(b)** According to Bernoulli's theorem,

$P + \frac{1}{2}\rho v^2$ = constant and Av = constant

If A is minimum, v is maximum, P is minimum.

12. **(a)** $V_{T_1} = \frac{2\times(1)^2}{9\eta}(\rho_1 - 0.1\rho_2)g$

$V_{T_1} = \frac{2\times1}{9\eta}(8\rho_2 - 0.1\rho_2)g$

$V_{T_2} = \frac{2\times(2)^2}{9\eta}(\rho_2 - 0.1\rho_2)g$

$\therefore \frac{V_{T_1}}{V_{T_2}} = \frac{7.9}{4(0.9)} = \frac{79}{36}$

13. **(d)** Power = rate of production of heat = F.V

$= 6\pi\eta r V_T \cdot V_T = 6\pi\eta r V_T^2$

($\because F = 6\pi\eta V_T r$ stoke's formula)

$V_T \propto r^2$

$\because \quad V_T = \frac{2}{9}\frac{r^2(\rho-\sigma)}{\eta}g$

$\therefore \quad$ Power $\propto r^5$

14. **(c)** Note that according to Stoke's law

$6\pi\eta r v_r = mg$

Hence, the valid relation is $v_r \propto mg/r\eta$

15. **(b)** Force of surface tension balances the weight of water in capillary tube.

$F_S = 2\pi r T\cos\theta = mg$

Here, T and θ are constant.

So, $m \propto r$

Let m_1 and m_2 be the mass of water in two capillary tube.

$\therefore \frac{m_2}{m_1} = \frac{r_2}{r_1}$

$\Rightarrow \frac{m_2}{5.0} = \frac{2r}{r} \qquad (\because r_2 = 2r)$

$\Rightarrow m_2 = 10.0\,g$

16. **(c)** Pressure inside the soap bubble

$= P_0 + \frac{4T}{R}$

And pressure at a point below the surface of water $= \rho g Z_0 + P_0$

[where P_0 = atmospheric pressure]

By equating these pressure we get,

$P_0 + \frac{4T}{R} = P_0 + \rho g Z_0$

$Z_0 = \frac{4T}{R\times\rho g}$

$Z_0 = \frac{4\times2.5\times10^{-2}}{10^{-3}\times1000\times10}\,m$

$Z_0 = 1\,cm$

17. **(a)** Water rises upto the top of capillary tube and stays there without overflowing.

> **NOTES**
>
> In case of capillary of insufficient length i.e. $L \angle h$ the liquid will neither overflow from the upper end like a fountain nor it will tickle along the vertical sides of the tube. The liquid after reaching the upper end will increase the radius of its meniscus without changing nature such that
>
> $hr = Lr'$ (Here, r is the intial radius of tube and r' is the new radius)
>
> $\because L < h$
>
> $\because r' > r$

18. **(c)** Volume same $\Rightarrow n\left(\frac{4}{3}\pi r^3\right) = \frac{4}{3}\pi R^3$

$\Rightarrow n = \frac{R^3}{r^3}$

Change in energy $= T\Delta A = T[4\pi R^2 - n4\pi r^2]$

$$= 4\pi T\left[R^2 - \frac{R^3}{r^3}r^2\right]$$

$$= 3\left(\frac{4}{3}\pi R^3\right)T\left[\frac{1}{R} - \frac{1}{r}\right]$$

$$= 3VT\left[\frac{1}{R} - \frac{1}{r}\right](R > r)$$

$$= 3VT\left[\frac{1}{r} - \frac{1}{R}\right] \text{ is released}$$

NOTES

Energy released when n drops of radius r coalesec to form a body drop of radius R,

Energy released = $4\pi R^3 T\left[\frac{1}{r} \frac{-1}{R}\right]$

If this energy get converted into kinetic energy of big drop, then

$$\frac{1}{2}mv^2 = 4\pi R^3 T\left[\frac{1}{r} \frac{-1}{R}\right]$$

$$\Rightarrow \frac{1}{2}\left[\frac{4}{3}\pi R^3 d\right]V^2 = 4\pi R^3 T\left[\frac{1}{r} \frac{-1}{R}\right]$$

$$\Rightarrow v2 = \frac{6T}{d}\left[\frac{1}{r} \frac{-1}{R}\right]$$

$$\Rightarrow \text{Velocity of big drop } V = \sqrt{\frac{6T}{d}\left(\frac{1}{r} \frac{-1}{R}\right)}$$

(Where, d = density of big liquid drop)

19. (c) Wetability of a surface by a liquid primarily depends on angle of contact between the surface and liquid.

If angle of contact is acute liquids wet the solid and vice-versa.

20. (a) We know that angle of contact is the angle between the tangent to liquid surface at the point of contact and solid surface inside the liquid. In case of pure water and pure glass, the angle of contact is zero.

10 Thermal Properties of Matter

Trend Analysis with Important Topics & Sub-Topics

Topic Name	Sub-Topic	2021 QNS.	2021 LOD	2020 QNS	2020 LOD	2019 QNS.	2019 LOD	2018 QNS.	2018 LOD	2017 QNS.	2017 LOD	2016 QNS.	2016 LOD
Thermometry & Thermal Expansion	Linear expansion of solids					1	A					1	A
Calorimetry & Heat Transfer	Heat Supplied through solids									1	A		
	Wein's & Stefan's law							1	A	1	A	1	A
	Principle of Calorimetry											1	A
Newton's Law of Cooling	Newton's Law of Cooling	1	A										

LOD - Level of Difficulty | E - Easy | A - Average | D - Difficult | Qns - No. of Questions

Topic 1 Thermometry & Thermal Expansion

1. A copper rod of 88 cm and an aluminium rod of unknown length have their increase in length independent of increase in temperature. The length of aluminium rod is : ($\alpha_{Cu} = 1.7 \times 10^{-5}\,K^{-1}$ and $\alpha_{Al} = 2.2 \times 10^{-5} K^{-1}$) *[2019]*
   (a) 6.8 cm (b) 113.9 cm
   (c) 88 cm (d) 68 cm

2. Coefficient of linear expansion of brass and steel rods are α_1 and α_2. Lengths of brass and steel rods are ℓ_1 and ℓ_2 respectively. If $(\ell_2 - \ell_1)$ is maintained same at all temperatures, which one of the following relations holds good? *[2016, 1999, 1995]*
   (a) $\alpha_1 \ell_2{}_2 = \alpha_2 \ell_1{}_2$ (b) $\alpha_1 \ell_2 = \alpha_2 \ell_1$
   (c) $\alpha_1 \ell_2 = \alpha_2 \ell_1$ (d) $\alpha_1 \ell_1 = \alpha_2 \ell_2$

3. The value of coefficient of volume expansion of glycerine is $5 \times 10^{-4}\,K^{-1}$. The fractional change in the density of glycerine for a rise of 40°C in its temperature, is: *[2015 RS]*
   (a) 0.020 (b) 0.025
   (c) 0.010 (d) 0.015

4. The density of water at 20°C is 998 kg/m^3 and at 40°C 992 kg/m^3. The coefficient of volume expansion of water is *[NEET Kar. 2013]*
   (a) 10^{-4}/°C (b) 3×10^{-4}/°C
   (c) 2×10^{-4}/°C (d) 6×10^{-4}/°C

5. On a new scale of temperature (which is linear) and called the W scale, the freezing and boiling points of water are 39° W and 239° W respectively. What will be the temperature on the new scale, corresponding to a temperature of 39° C on the Celsius scale? *[2008]*
   (a) 78° W (b) 117° W
   (c) 200° W (d) 139° W

6. If the cold junction of a thermo-couple is kept at 0°C and the hot junction is kept at T°C then the relation between neutral temperature (T_n) and temperature of inversion (T_i) is *[2007]*
   (a) $T_n = 2T_i$ (b) $T_n = T_i - T$
   (c) $T_n = T_i + T$ (d) $T_n = T_i/2$

7. The temperature of inversion of a thermo-couple is 620°C and the neutral temperature is 300°C. What is the temperature of cold junction? *[2005]*
   (a) 320°C (b) 20°C
   (c) –20°C (d) 40°C

8. Mercury thermometer can be used to measure temperature upto *[1992]*

(a) 260°C (b) 100°C
(c) 357°C (d) 500°C

9. A centigrade and a Fahrenheit thermometer are dipped in boiling water. The water temperature is lowered until the Fahrenheit thermometer registers 140°. What is the fall in temperature as registered by the Centigrade thermometer? *[1990]*

(a) 80° (b) 60°
(c) 40° (d) 30°

Topic 2 Calorimetry & Heat Transfer

10. A deep rectangular pond of surface area A, containing water (density = ρ), specific heat (capacity = s), is located in a region where the outside air temperature is at a steady value of –26°C. The thickness of the frozen ice layer in this pond, at a certain instant is x.
Taking the thermal conductivity of ice as K, and its specific latent heat of fusion as L, the rate of increase of the thickness of ice layer, at this instant, would be given by *[NEET Odisha 2019]*

(a) $26K/\rho x(L+4s)$ (b) $26K/\rho x(L-4s)$
(c) $26K/(\rho x^2 L)$ (d) $26K/(\rho x L)$

11. The power radiated by a black body is P and it radiates maximum energy at wavelength, λ_0. If the temperature of the black body is now changed so that it radiates maximum energy at wavelength $\frac{3}{4}\lambda_0$, the power radiated by it becomes nP. The value of n is *[2018]*

(a) $\frac{3}{4}$ (b) $\frac{4}{3}$
(c) $\frac{81}{256}$ (d) $\frac{256}{81}$

12. Two rods A and B of different materials are welded together as shown in figure. Their thermal conductivities are K_1 and K_2. The thermal conductivity of the composite rod will be : *[2017]*

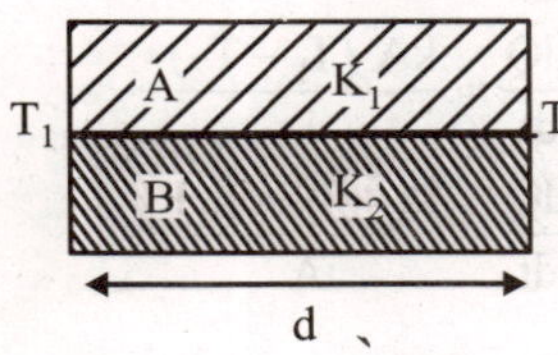

(a) $\frac{3(K_1+K_2)}{2}$ (b) K_1+K_2
(c) $2(K_1+K_2)$ (d) $\frac{K_1+K_2}{2}$

13. A spherical black body with a radius of 12 cm radiates 450 watt power at 500 K. If the radius were halved and the temperature doubled, the power radiated in watt would be : *[2017]*

(a) 450 (b) 1000
(c) 1800 (d) 225

14. A black body is at a temperature of 5760 K. The energy of radiation emitted by the body at wavelength 250 nm is U_1, at wavelength 500 nm is U_2 and that at 1000 nm is U_3. Wien's constant, $b = 2.88 \times 10^6$ nmK. Which of the following is correct? *[2016]*

(a) $U_1 = 0$ (b) $U_3 = 0$
(c) $U_1 > U_2$ (d) $U_2 > U_1$

15. A piece of ice falls from a height h so that it melts completely. Only one-quarter of the heat produced is absorbed by the ice and all energy of ice gets converted into heat during its fall. The value of h is : *[2016]*
[Latent heat of ice is 3.4×10^5 J/kg and g = 10 N/kg]

(a) 34 km (b) 544 km
(c) 136 km (d) 68 km

16. On observing light from three different stars P, Q and R, it was found that intensity of violet colour is maximum in the spectrum of P, the intensity of green colour is maximum in the spectrum of R and the intensity of red colour is maximum in the spectrum of Q. If T_P, T_Q and T_R are the respective absolute temperature of P, Q and R, then it can be concluded from the above observations that *[2015]*

(a) $T_P > T_R > T_Q$ (b) $T_P < T_R < T_Q$
(c) $T_P < T_Q < T_R$ (d) $T_P > T_Q > T_R$

17. The two ends of a metal rod are maintained at temperatures 100°C and 110°C. The rate of heat flow in the rod is found to be 4.0 J/s. If the ends are maintained at temperatures 200°C and 210°C, the rate of heat flow will be *[2015]*

(a) 16.8 J/s (b) 8.0 J/s
(c) 4.0 J/s (d) 44.0 J/s

18. Steam at 100°C is passed into 20 g of water at 10°C. When water acquires a temperature of 80°C, the mass of water present will be: *[2014]*
[Take specific heat of water = 1 cal g^{-1} °C^{-1} and latent heat of steam = 540 cal g^{-1}]
(a) 24 g (b) 31.5 g
(c) 42.5 g (d) 22.5 g

19. A piece of iron is heated in a flame. It first becomes dull red then becomes reddish yellow and finally turns to white hot. The correct explanation for the above observation is possible by using *[2013]*
(a) Wien's displacement law
(b) Kirchoff's law
(c) Newton's law of cooling
(d) Stefan's law

20. Two metal rods 1 and 2 of same lengths have same temperature difference between their ends. Their thermal conductivities are K_1 and K_2 and cross sectional areas A_1 and A_2, respectively. If the rate of heat conduction in rod 1 is four times that in rod 2, then *[NEET Kar. 2013]*
(a) $K_1A_1 = K_2A_2$ (b) $K_1A_1 = 4K_2A_2$
(c) $K_1A_1 = 2K_2A_2$ (d) $4K_1A_1 = K_2A_2$

21. If the radius of a star is R and it acts as a black body, what would be the temperature of the star, in which the rate of energy production is Q? *[2012]*
(a) $Q/4\pi R^2\sigma$ (b) $(Q/4\pi R^2\sigma)^{-1/2}$
(c) $(4\pi R^2 Q/\sigma)^{1/4}$ (d) $(Q/4\pi R^2\sigma)^{1/4}$
(σ stands for Stefan'S constant)

22. Liquid oxygen at 50 K is heated to 300 K at constant pressure of 1 atm. The rate of heating is constant. Which one of the following graphs represents the variation of temperature with time? *[2012]*

(a)
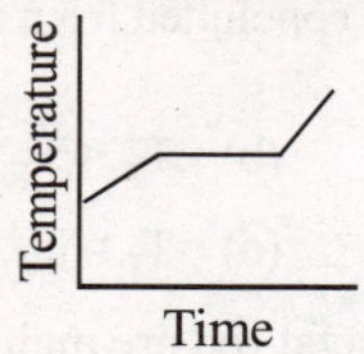

(b)
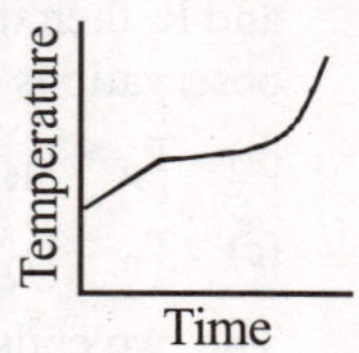

(c)
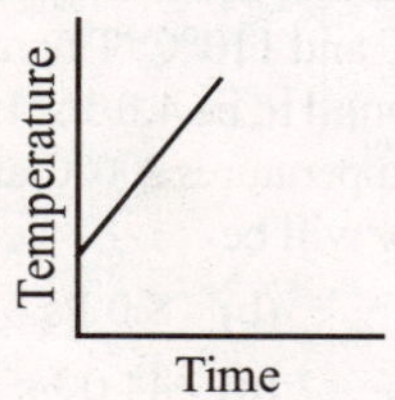

(d)
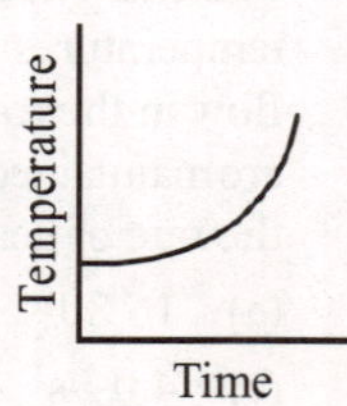

23. A slab of stone of area 0.36 m^2 and thickness 0.1 m is exposed on the lower surface to steam at 100°C. A block of ice at 0°C rests on the upper surface of the slab. In one hour 4.8 kg of ice is melted. The thermal conductivity of slab is : (Given latent heat of fusion of ice = 3.36 × 10^5 J kg^{-1}.) : *[2012M]*
(a) 1.24 J/m/s/°C (b) 1.29 J/m/s/°C
(c) 2.05 J/m/s/°C (d) 1.02 J/m/s/°C

24. A cylindrical metallic rod in thermal contact with two reservoirs of heat at its two ends conducts an amount of heat Q in time t. The metallic rod is melted and the material is formed into a rod of half the radius of the original rod. What is the amount of heat conducted by the new rod, when placed in thermal contact with the two reservoirs in time t? *[2010]*
(a) $\dfrac{Q}{4}$ (b) $\dfrac{Q}{16}$
(c) $2Q$ (d) $\dfrac{Q}{2}$

25. The total radiant energy per unit area, normal to the direction of incidence, received at a distance R from the centre of a star of radius r, whose outer surface radiates as a black body at a temperature T K is given by: *[2010]*
(a) $\dfrac{\sigma r^2 T^4}{R^2}$ (b) $\dfrac{\sigma r^2 T^4}{4\pi r^2}$
(c) $\dfrac{\sigma r^4 T^4}{r^4}$ (d) $\dfrac{4\pi\sigma r^2 T^4}{R^2}$
(where σ is Stefan's constant)

26. The two ends of a rod of length L and a uniform cross-sectional area A are kept at two temperatures T_1 and T_2 ($T_1 > T_2$). The rate of heat transfer, $\dfrac{dQ}{dt}$ through the rod in a steady state is given by: *[2009]*
(a) $\dfrac{dQ}{dt} = \dfrac{k(T_1 - T_2)}{LA}$
(b) $\dfrac{dQ}{dt} = kLA\,(T_1 - T_2)$
(c) $\dfrac{dQ}{dt} = \dfrac{kA\,(T_1 - T_2)}{L}$
(d) $\dfrac{dQ}{dt} = \dfrac{kL\,(T_1 - T_2)}{A}$

27. A black body at 227°C radiates heat at the rate of 7 cals/cm^{2}s. At a temperature of 727°C, the rate of heat radiated in the same units will be: *[2009]*

(a) 50 (b) 112
(c) 80 (d) 60

28. An electric kettle takes 4A current at 220 V. How much time will it take to boil 1 kg of water from temperature 20° C? The temperature of boiling water is 100° C. *[2008]*

(a) 6.3 min (b) 8.4 min
(c) 12.6 min (d) 4.2 min

29. A black body is at 727° C. It emits energy at a rate which is proportional to *[2007]*

(a) $(1000)^4$ (b) $(1000)^2$
(c) $727)^4$ (d) $(727)^2$

30. Assuming the sun to have a spherical outer surface of radius r, radiating like a black body at temperature t°C, the power received by a unit surface, (normal to the incident rays) at a distance R from the centre of the sun is *[2007]*

(a) $\dfrac{r^2\sigma(t+273)^4}{4\pi R^2}$ (b) $\dfrac{16\pi^2 r^2\sigma t^4}{R^2}$

(c) $\dfrac{r^2\sigma(t+273)^4}{R^2}$ (d) $\dfrac{4\pi r^2\sigma t^4}{R^2}$

where σ is the Stefan's constant.

31. A black body at 1227°C emits radiations with maximum intensity at a wavelength of 5000Å. If the temperature of the body is increased by 1000°C, the maximum intensity will be observed at *[2006]*

(a) 5000Å (b) 6000Å
(c) 3000Å (d) 4000Å

32. Which of the following circular rods (given radius r and length l), each made of the same material and whose ends are maintained at the same temperature will conduct most heat? *[2005]*

(a) $r = r_0;\ l = l_0$ (b) $r = 2r_0;\ l = l_0$
(c) $r = r_0;\ l = 2l_0$ (d) $r = 2r_0;\ l = 2l_0$

33. If λ_m denotes the wavelength at which the radiative emission from a black body at a temperature T K is maximum, then *[2004]*

(a) $\lambda_m \propto T^{-1}$
(b) $\lambda_m \propto T^4$
(c) λ_m is independent of T
(d) $\lambda_m \propto T$

34. Consider a compound slab consisting of two different materials having equal thicknesses and thermal conductivities K and $2K$, respectively. The equivalent thermal conductivity of the slab is *[2003]*

(a) $\dfrac{4}{3}K$ (b) $\dfrac{2}{3}K$
(c) $\sqrt{3}\,K$ (d) $3\,K$

35. Wien's law is concerned with *[2002]*

(a) relation between emissivity and absorptivity of a radiating surface
(b) total radiation, emitted by a hot surface
(c) an expression for spectral distribution of energy of a radiation from any source
(d) a relation between the temperature of a black body and the wavelength at which there is maximum radiant energy per unit wavelength

36. Radiation from which of the following sources, approximates black body radiation best? *[2002]*

(a) A tungsten lamp
(b) Sodium flame
(c) Hot lamp black
(d) A hole in a cavity, maintained at constant temperature

37. Two rods of thermal conductivities K_1 and K_2, cross-sections A_1 and A_2 and specific heats S_1 and S_2 are of equal lengths. The temperatures of two ends of each rod are T_1 and T_2. The rate of flow of heat at the steady state will be equal if *[2002]*

(a) $\dfrac{K_1}{A_1S_1} = \dfrac{K_2}{A_2S_2}$ (b) $K_1A_1 = K_2A_2$
(c) $K_1S_1 = K_2S_2$ (d) $A_1S_1 = A_2S_2$

38. A black body has maximum wavelength λ_m at temperature 2000 K. Its corresponding wavelength at temperature 3000 K will be *[2001]*

(a) $\dfrac{3}{2}\lambda_m$ (b) $\dfrac{2}{3}\lambda_m$

(c) $\dfrac{4}{9}\lambda_m$ (d) $\dfrac{9}{4}\lambda_m$

39. A cylindrical rod having temperature T_1 and T_2 at its end. The rate of flow of heat is Q_1 cal/sec. If all the linear dimensions are doubled keeping temperature constant, then the rate of flow of heat Q_2 will be *[2001]*

(a) $4Q_1$ (b) $2Q_1$
(c) $Q_1/4$ (d) $Q_1/2$

40. The presence of gravitational field is required for the heat transfer by *[2000]*
(a) conduction
(b) stirring of liquids
(c) natural convection
(d) radiation

41. If 1 g of steam is mixed with 1 g of ice, then the resultant temperature of the mixture is *[1999]*
(a) 270°C (b) 230°C
(c) 100°C (d) 50°C

42. The radiant energy from the sun, incident normally at the surface of earth is 20 k cal/m^2 min. What would have been the radiant energy, incident normally on the earth, if the sun had a temperature, twice of the present one? *[1998]*
(a) 160 k cal/m^2 min (b) 40 k cal/m^2 min
(c) 320k cal/m^2 min (d) 80 k cal/m^2 min

43. A black body is at temperature of 500 K. It emits energy at rate which is proportional to *[1997]*
(a) $(500)^4$ (b) $(500)^3$
(c) $(500)^2$ (d) 500

44. If the temperature of the sun is doubled, the rate of energy received on earth will be increased by a factor of *[1993]*
(a) 2 (b) 4
(c) 8 (d) 16

45. Thermal capacity of 40 g of aluminium ($s = 0.2$ cal /g K) is *[1990]*
(a) 168 joule/°C (b) 672 joule/°C
(c) 840 joule/°C (d) 33.6 joule/°C

46. 10 gm of ice cubes at 0°C are released in a tumbler (water equivalent 55 g) at 40°C. Assuming that negligible heat is taken from the surroundings, the temperature of water in the tumbler becomes nearly (L = 80 cal/g) *[1988]*
(a) 31°C (b) 22°C
(c) 19°C (d) 15°C

Topic 3 Newton's Law of Cooling

47. A cup of coffee cools from 90°C to 80°C in t minutes, when the room temperature is 20°C. The time taken by a similar cup of coffee to cool from 80°C to 60°C at a room temperature same at 20°C is *[2021]*
(a) $\frac{5}{13}t$ (b) $\frac{13}{10}t$
(c) $\frac{13}{5}t$ (d) $\frac{10}{13}t$

48. An object kept in a large room having air temperature of 25°C takes 12 minutes to cool from 80°C to 70°C.
The time taken to cool for the same object from 70°C to 60°C would be nearly. *[NEET Odisha 2019]*
(a) 15 min (b) 10 min
(c) 12 min (d) 20 min

49. Certain quantity of water cools from 70°C to 60°C in the first 5 minutes and to 54°C in the next 5 minutes. The temperature of the surroundings is: *[2014]*
(a) 45°C (b) 20°C
(c) 42°C (d) 10°C

50. A beaker full of hot water is kept in a room. If it cools from 80°C to 75°C in t_1 minutes, from 75° C to 70°C in t_2 minutes and from 70°C to 65°C in t_3 minutes, then *[1995]*
(a) $t_1 = t_2 = t_3$ (b) $t_1 < t_2 = t_3$
(c) $t_1 < t_2 < t_3$ (d) $t_1 > t_2 > t_3$

51. A body cools from 50.0°C to 48°C in 5s. How long will it take to cool from 40.0°C to 39°C? Assume the temperature of surroundings to be 30.0°C and Newton's law of cooling to be valid. *[1994]*
(a) 2.5 s (b) 10 s

ANSWER KEY

1	(d)	7	(c)	13	(c)	19	(a)	25	(a)	31	(c)	37	(b)	43	(a)	49	(a)
2	(d)	8	(c)	14	(d)	20	(b)	26	(c)	32	(b)	38	(b)	44	(d)	50	(c)
3	(a)	9	(c)	15	(c)	21	(d)	27	(b)	33	(a)	39	(b)	45	(d)	51	(b)
4	(b)	10	(d)	16	(a)	22	(a)	28	(a)	34	(a)	40	(c)	46	(b)		
5	(b)	11	(d)	17	(c)	23	(a)	29	(a)	35	(d)	41	(c)	47	(c)		
6	(d)	12	(d)	18	(d)	24	(b)	30	(c)	36	(d)	42	(c)	48	(a)		

Hints & Solutions

1. **(d)** $l'_{Cu} = l_{Cu}(1 + \alpha_{Cu}\Delta T)$

$\Delta l_{Cu} = l_{Cu}\alpha_{Cu}\Delta T$

$l'_{Al} = l_{Al}(1 + \alpha_{Al}\Delta T)$

$\Delta l_{Al} = l_{Al}\alpha_{Al}\Delta T$

Since, change in length is independent of temperature

$\therefore \alpha_{Cu}l_{Cu} = \alpha_{Al}l_{Al}$

$\Rightarrow 1.7 \times 10^{-5} \times 88\text{ cm} = 2.2 \times 10^{-5} \times l_{Al}$

$\Rightarrow l_{Al} = \dfrac{1.7 \times 88}{2.2} = 68\text{ cm}$

2. **(d)** From question, $(\ell_2 - \ell_1)$ is maintained same at all temperatures hence change in length for both rods should be same

i.e., $\Delta\ell_1 = \Delta\ell_2$

As we know, coefficient of linear expansion,

$\alpha = \dfrac{\Delta\ell}{\ell_0\Delta T}$

$\ell_1\alpha_1\Delta T = \ell_2\alpha_2\Delta T$

$\ell_1\alpha_1 = \ell_2\alpha_2$

3. **(a)** From question,

Rise in temperature $\Delta t = 40°C$

Coefficient of volume expansion

$\gamma = 5 \times 10^{-4} K^{-1}$

Fractional change in the density $\dfrac{\Delta\rho}{\rho_0} = \dfrac{\rho_0 - \rho_T}{\rho_0}$

$\rho_T = \rho_0(1 - \gamma\Delta t)$

$\Rightarrow \dfrac{\Delta\rho}{\rho_0} = \dfrac{\rho_0 - \rho_T}{\rho_0} = \gamma\Delta T = (5 \times 10^{-4})(40) = 0.020$

NOTES

Most substances (solid & liquid) expand when they are heated, so volume of a given mass of a substance increases on heating so density decreases as

$\rho \propto \dfrac{1}{V} \Rightarrow \dfrac{\rho'}{\rho} = \dfrac{V}{V'} = \dfrac{V}{V + \Delta V}$

$= \dfrac{V}{V + \gamma V A\theta} = \dfrac{1}{1 + \gamma\Delta\theta}$

4. **(b)** Given

$T_1 = 20°C$, $T_2 = 40°C$

$\rho_{20} = 998\text{ kg/m}^3$, $\rho_{40} = 992\text{ kg/m}^3$

As $\rho_{T2} = \dfrac{\rho_{T1}}{1 + \gamma\Delta T} = \dfrac{\rho_{T1}}{1 + \gamma(T_2 - T_1)}$

$\therefore\ 992 = \dfrac{998}{1 + \gamma(40 - 20)} \Rightarrow 992 = \dfrac{998}{1 + 20\gamma}$

$\Rightarrow 20\gamma = \dfrac{998}{992} - 1 = \dfrac{6}{992}$

$\Rightarrow \gamma = \dfrac{6}{992} \times \dfrac{1}{20} = 3 \times 10^{-4}/°C$

5. **(b)** For different temperature scales, we have

$\dfrac{x - \text{L.F.P}}{\text{U.F.P.} - \text{L.F.P.}} = \text{constant}$

Where L.F.P $\Rightarrow$ Lower Fixed point

U.H.F. $\Rightarrow$ Upper fixed point

where x is the measurement at that scale. Here, if C and W be the measurements on Celsius and W scale then,

$\dfrac{C - 0}{100 - 0} = \dfrac{W - 39}{239 - 39}$ $(\because C = 39°C)$

$\Rightarrow W = \dfrac{39 \times 200}{100} + 39 = 78 + 39 = 117°\text{ W}$

6. **(d)** Since $T_n = \dfrac{T_i + T_c}{2}$ = Neutral temperature

$T_n = \dfrac{T_i + 0}{2} = \dfrac{T_i}{2}$

$[T_c = 0°C$ = temperature of cold junction$]$

7. **(c)** $\theta_n = \dfrac{\theta_c + \theta_i}{2}$

$\therefore \theta_c = 2\theta_n - \theta_i = 2(300) - 620 = -20°C$

8. **(c)** Mercury thermometer is based on the principle of change of volume with rise of temperature and can measure temperatures ranging from –30°C to 357°C.

9. **(c)** Using $\dfrac{F - 32}{180} = \dfrac{C}{100}$

$\Rightarrow \dfrac{140 - 32}{180} = \dfrac{C}{100}$

$\Rightarrow C = 60$

Temperature of boiling water = 100°C

We get, fall in temperature = 100 – 60 = 40°C

NOTES: Centigrade (c) Fahrenheit (F) and Kelvin (K) scales of temperature are related as

$$\frac{C-0}{100}=\frac{F-32}{212-32}$$

$$=\frac{K-273.15}{373.15-273.15}$$

10. (d) We assume that at any instant thickness of ice is x. And time taken to form additional thickness (dx) is dt.

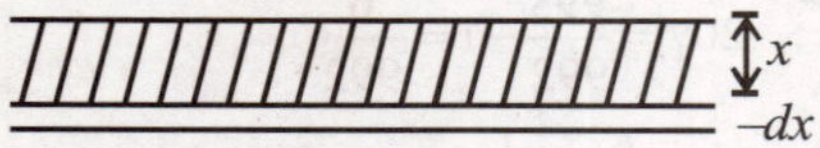

$$mL=\frac{KA[26-0]dt}{x}$$

$$\Rightarrow (Adx)\rho L=\frac{KA(26)\,dt}{x}$$

$$\text{So, } \frac{dx}{dt}=\frac{26K}{x\rho L}$$

11. (d) From Wien's law

$\lambda_{max}T=\text{constant}$

i.e., $\lambda_{max_1}T_1=\lambda_{max_2}T_2$

$$\Rightarrow \lambda_0 T=\frac{3\lambda_0}{4}T'$$

$$\Rightarrow T'=\frac{4}{3}T$$

Power radiated $P\propto T^4$

$$\text{So, } \frac{P_2}{P_1}=n=\left(\frac{T'}{T}\right)^4=\left(\frac{4}{3}\right)^4=\frac{256}{81}$$

12. (d) Equivalent thermal conductivity of the composite rod in parallel combination will be,

$$K=\frac{K_1A(T_1-T_2)}{d}+\frac{K_2A(T_1-T_2)}{d}$$

$$\frac{K_{EQ}2A(T_1-T_2)}{d}=\frac{A(T_1-T_2)}{d}[K_1+K_2]$$

Hence equivalent thermal conductivities for two rods of equal area is given by

$$K_{EQ}=\frac{k_1+k_2}{2}$$

13. (c) Given $r_1=12$ cm, $r_2=6$ cm

$T_1=500$ K and $T_2=2\times 500=1000$ K

$P_1=450$ watt

Rate of power loss $P\propto r^2T^4$

$$\frac{P_1}{P_2}=\frac{r_1^2T_1^4}{r_2^2T_2^4}$$

$$P_2=P_1\frac{r_2^2T_2^4}{r_1^2T_1^4}$$

Solving we get, $P_2=1800$ watt

14. (d) According to Wein's displacement law, maximum amount of emitted radiation

corresponding to $\lambda_m=\frac{b}{T}$

$$\lambda_m=\frac{2.88\times10^6\text{ nmK}}{5760\text{K}}=500\text{ nm}$$

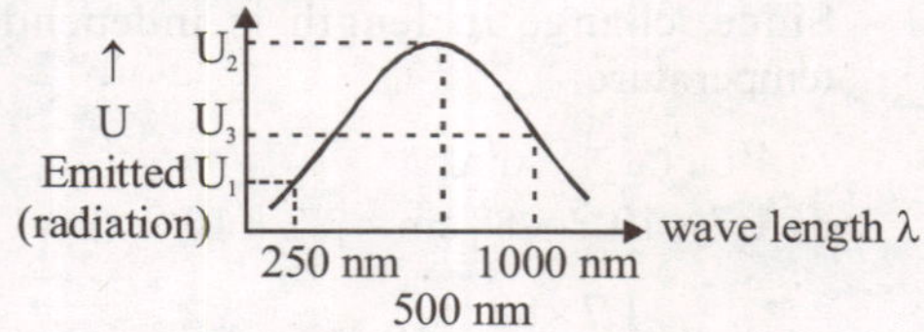

From the graph $U_1<U_2>U_3$

NOTES: At a given temperature, intensity of heat radiation increases with wavelength, reaches a maximum at a particular wavelength and with further increase in wavelength it decreases.

15. (c) According to question only one-quarter of the heat produced by falling piece of ice is absorbed in the melting of ice.

$$\text{i.e., } \frac{mgh}{4}=mL$$

$$\Rightarrow h=\frac{4L}{g}=\frac{4\times3.4\times10^5}{10}=136\text{ km}.$$

16. (a) From Wein's displacement law

$\lambda_m\times T=\text{constant}$

P – max. intensity is at violet

$\Rightarrow \lambda_m$ is minimum $\Rightarrow$ temp maximum

R – max. intensity is at green

$\Rightarrow \lambda_m$ is moderate $\Rightarrow$ temp moderate

Q – max. intensity is at red $\Rightarrow \lambda_m$ is maximum $\Rightarrow$ temp. minimum *i.e.,* $T_p>T_R>T_Q$

17. (c) As the temperature difference $\Delta T=10°C$ as well as the thermal resistance is same for both the cases, so thermal current or rate of heat flow will also be same for both the cases.

18. (d) According to the principle of calorimetry.

Heat lost = Heat gained

$mL_v+ms_w\Delta\theta=m_ws_w\Delta\theta$

$\Rightarrow\ m\times540+m\times1\times(100-80)$

$\Rightarrow\ 20\times1\times(80-10)$

$\Rightarrow\ m=2.5$ g

Therefore total mass of water at 80°C

$=(20+2.5)\text{ g}=22.5\text{ g}$

19. **(a)** Wein's displacement law

According to this law

$$\lambda_{max} \propto \frac{1}{T}$$

or, $\lambda_{max} \times T = \text{constant}$

So, as the temperature increases λ decreases.

20. **(b)** $Q_1 = 4Q_2$ (Given)

$$\Rightarrow \frac{K_1A_1\Delta t}{L} = 4\frac{K_2A_2\Delta t}{L} \Rightarrow K_1A_1 = 4K_2A_2.$$

21. **(d)** Stefan's law for black body radiation

$Q = \sigma e A T^4$

$$T = \left[\frac{Q}{\sigma(4\pi R^2)}\right]^{1/4}$$

Here $e = 1$

$A = 4\pi R^2$

22. **(a)** Initially liquid oxygen will gain the temperature up to its boiling temperature then it change its state to gas. After this again its temperature will increase, so corresponding graph will be graph (a).

23. **(a)** According to condition,

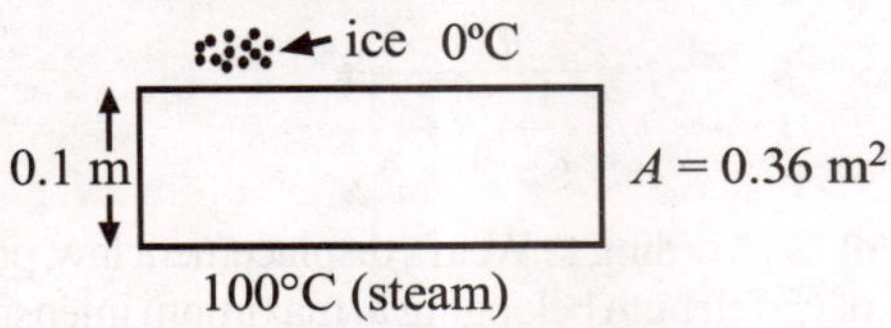

Rate of heat given by steam = Rate of heat taken by ice

where K = Thermal conductivity of the slab

m = Mass of the ice

L = Latent heat of melting/fusion

A = Area of the slab

$$\frac{dQ}{dt} = \frac{KA(100-0)}{1} = m\frac{dL}{dt}$$

$$\frac{K \times 100 \times 0.36}{0.1} = \frac{4.8 \times 3.36 \times 10^5}{60 \times 60}$$

$K = 1.24$ J/m/s/°C

NOTES

In electrical conduction

$$I = \frac{dq}{dt} = \frac{V_1 - V_2}{R} = \frac{\sigma A}{\ell}(V_1 - V_2)$$

In thermal conduction,

$$H = \frac{d\theta}{dt} = \frac{\theta_1 - \theta_2}{R} = \frac{kA}{\ell}(\theta_1 - \theta_2)$$

K = Thermal conductivity of conductor.

24. **(b)** The rate of heat flow is given by

$$\frac{Q}{t} = K.A.\frac{\Delta T}{\ell}$$

Area of Original rod $A = \pi R^2$;

Area of new rod $A' = \dfrac{\pi R^2}{4}$

$$A' = \frac{A}{4}$$

Volume of original rod will be equal to the volume of new rod.

So, $A'\ell' = A\ell$

or,

$$\therefore \pi R^2 \ell = \pi\left(\frac{R}{2}\right)^2 \ell'$$

$$\Rightarrow \frac{\ell'}{\ell} = \frac{R^2}{\left(\frac{R^2}{4}\right)} = 4$$

$$\therefore \frac{Q'}{Q} = \frac{A'}{A}\frac{\ell}{\ell'} = \frac{1}{4}.\frac{1}{4} = \frac{1}{16}$$

$$\therefore Q' = \frac{Q}{16}$$

25. **(a)** $E = \dfrac{S}{S_0}\sigma T^4 = \dfrac{4\pi r^2}{4\pi R^2}\sigma T^4$

$$= \sigma\frac{r^2}{R^2}T^4$$

26. **(c)** $\dfrac{dQ}{dt} = \dfrac{kA(T_1 - T_2)}{L}$

[($T_1 - T_2$) is the temperature difference]

NOTES

Rate of heat flow is called heat current,

$H = \dfrac{dQ}{dt}$ and $H = \dfrac{(T_1 - T_2)}{R}$

and thermal resistance, $R = \dfrac{L}{kA}$

27. **(b)** According to Stefan's law $E = \sigma T^4$,

$T_1 = 500$ K

$T_2 = 1000$ K

$$\frac{E_2}{E_1} = \left(\frac{T_2}{T_1}\right)^4 = \left(\frac{1000}{500}\right)^4 = 16$$

$\therefore \quad E_2 = 16 \times 7 = 112$ cal / cm²s

28. **(a)** Heat required to raise the temperature of 1kg water from 20°C to 100°C is given by

$Q = ms\Delta\theta = (1\times 4200\times 80)$ J

Power of kettle (P) = VI = (220 × 4)W

$$\therefore \text{Time taken} = \frac{Q}{P} = \frac{1\times 4200\times 80}{220\times 4}$$

$$= 381.81 \text{ sec} = 6.36 \text{ min}$$

29. **(a)** According to Stefan's law,

$E \propto T^4$

$E \propto (t+273)^4 K$ [727°C = (727 + 273)K]

$E \propto (727+273)^4 K$

$E \propto (1000)^4 K$

30. **(c)** Power radiated by the sun at t°C

$= \sigma(t+273)^4 4\pi r^2$

Power received by a unit surface

$$= \frac{\sigma(t+273)^4 4\pi r^2}{4\pi R^2} = \frac{r^2\sigma(t+273)^4}{R^2}$$

31. **(c)** Applying Wein's displacement law,

$\lambda_m T$ = constant

5000 Å × (1227 + 273) = (2227 + 273) × λ_m

$$\lambda_m = \frac{5000\times 1500}{1000} = 3000\text{Å}$$

32. **(b)** From given option

(i) $r = 2r_0, l = 2l_0$

$$\therefore R = \frac{2\ell_0}{K\pi(2r_0)^2} = \frac{\ell_0}{2K\pi r_0^2}$$

(ii) $r = 2r_0, l = l_0$

$$\therefore R = \frac{\ell_0}{K\pi(2r_0)^2} = \frac{\ell_0}{4K\pi r_0^2}$$

(iii) $r = r_0, l = 2l_0$

$$\therefore R = \frac{2\ell_0}{K\pi r_0^2} = \frac{2\ell_0}{K\pi r_0^2}$$

(iv) $r = r_0, l = l_0$

$$\therefore R = \frac{\ell_0}{K\pi r_0^2} = \frac{\ell_0}{K\pi r_0^2}$$

It is clear that for option (b) resistance is minimum, hence, heat flow will be maximum.

(i) Rate of heat flow is directly proportional to area

(ii) inversely proportional to length.

∴ Heat flow will be maximum when r is maximum and ℓ is minimum.

NOTES

We know that $Q = \dfrac{T_H - T_L}{R}$

Also, Thermal resistance $R = \dfrac{\ell}{KA} = \dfrac{\ell}{K\pi r^2}$

Heat flow will be maximum when thermal resistance is minimum.

33. **(a)** From Wein's displacement law

$\lambda_m T$ = constant

$\Rightarrow \lambda_m \propto T^{-1}$

34. **(a)** In series, equivalent thermal conductivity

$$K_{eq} = \frac{2K_1K_2}{K_1+K_2}$$

$$\text{or, } K_{eq} = \frac{2\times K\times 2K}{K+2K} = \frac{4}{3}K$$

NOTES

In electricity equivalent resistance in series grouping $R_s = R_1 + R_2 + R_3 \ldots + R_n$

In heat-thermal resistance in series $R_s = R_1 + R_2 + R_3 \ldots + R_n$

$$\frac{\ell_1 + \ell_2 + \ldots \ell_n}{Ks} = \frac{\ell_1}{K_1A} + \frac{\ell_2}{K_2A} + \ldots + \frac{\ell_n}{K_nA}$$

For n no. of rods of equal length

$$K_s = \frac{n}{\frac{1}{K_1} + \frac{1}{K_2} \ldots + \frac{1}{K_n}}$$

35. **(d)** According to Wein's displacement law, product of wavelength belonging to maximum intensity and temperature is constant *i.e.*, $\lambda_m T$ = constant.

36. **(d)**

37. **(b)** Rate of heat flow for one rod

$$= \frac{K_1A_1(T_1 - T_2)}{d} \quad [d \to \text{Length}]$$

Rate of heat flow for other rod

$$= \frac{K_2A_2(T_1 - T_2)}{d}$$

In steady state, $\dfrac{K_1A_1(T_1 - T_2)}{d}$

$$= \frac{K_2A_2(T_1 - T_2)}{d} \Rightarrow K_1A_1 = K_2A_2$$

38. **(b)** According to Wein's displacement law,

$\lambda_m T = 2.88\times 10^{-3}$

When T = 2000 K,

$\lambda_m(2000) = 2.88\times 10^{-3}$ ....(1)

When T = 3000 K,

$\lambda'_m(3000) = 2.88\times 10^{-3}$ ....(2)

Dividing (1) by (2),

$$\frac{2}{3}\frac{\lambda_m}{\lambda'_m}=1 \Rightarrow \frac{\lambda_m}{\lambda'_m}=\frac{3}{2} \Rightarrow \lambda'_m=\frac{2}{3}\lambda_m$$

39. **(b)** $Q=\frac{KA(\theta_1-\theta_2)t}{l}$

Rate of heat flow,

$$H=\frac{Q}{t}=\frac{KA(\theta_1-\theta_2)}{l} \text{ i.e., } H \propto \frac{A}{l}$$

Dimensions of area $A = [L^2]$, dimensions of distance $l = [L]$

$$\therefore\ H \propto L \Rightarrow \frac{H_2}{H_1}=\frac{L_2}{L_1}=2 \Rightarrow H_2=2H_1$$

40. **(c)** Natural convection

NOTES: In convection, the temperature gradient exists in the vertical direction and not in the horizontal direction. So, up and down movement of particles takes place which depends on the weight or gravity.

41. **(c)** Heat required by ice at 0°C to reach a temperature of 100°C = $mL + mc\Delta\theta$

$= 1 \times 80 + 1 \times 1 \times (100 - 0) = 180$ cal

Heat available with 1 g steam to condense into 1 g of water at 100°C = 536 cal.

Obviously the whole steam will not be condensed and ice will attain a temperature of 100°C; so the temperature of mixture = 100°C.

42. **(c)** According to Stefan's law, $E \propto T^4$

$$\frac{E_1}{E_2}=\frac{T_1^4}{T_2^4} \text{ or } \frac{20}{E_2}=\frac{T^4}{2^4T^4}$$

$\Rightarrow E_2 = 320$ kcal/m^2. min.

43. **(a)** According to Stefan's Law $E=\sigma eAT^4$

$E \propto T^4$; so, $E \propto (500)^4$

44. **(d)** Amount of energy radiated $\propto T^4$

45. **(d)** Thermal capacity = ms = 40 × 0.2 = 8 cal/°C

= 4.2 × 8 J = 33.6 joules/°C

46. **(b)** Let the final temperature be T

Heat gained by ice = mL + m × s × $(T - 0)$

= 10 × 80 + 10 × 1 × T

Heat lost by water = 55 × 1× (40 – T)

By using law of calorimetery,

800 + 10 T = 55 × (40 – T)

⇒ T = 21.54°C = 22°C

47. **(c)** From Newton's law of cooling

$$-K\left(\frac{T_1+T_2}{2}-T_s\right)=\frac{T_1-T_2}{\Delta t}$$

T_1 and T_2 are initial and final temperature and T_s = surrounding temperature.

$$\Rightarrow -K\left[\frac{(90+80)}{2}-20\right]=\frac{90-80}{t}$$

$$\Rightarrow -K(65)=\frac{10}{t} \Rightarrow K=\frac{-2}{13t}$$

When cool from 80°C to 60°C

$$-K\left(\frac{80+60}{2}-20\right)=\frac{(80-60)}{t_1}$$

$$\Rightarrow -K(50)=\frac{20}{t_1} \Rightarrow \frac{2}{13t}(50)=\frac{20}{t_1}$$

$$\therefore t_1=\frac{13t}{5}$$

48. **(a)** According to Newton's law of cooling

$$\frac{(T_1-T_2)}{t}=K\left(\frac{T_1+T_2}{2}-T_0\right)$$

$$\frac{(80-70)}{12}=K(75-25) \quad ...(i)$$

$$\frac{(70-60)}{t'}=K(65-25) \quad ...(ii)$$

$\Rightarrow t' = 12 \times = 15$ min

49. **(a)** Let the temperature of surroundings be θ_0

By Newton's law of cooling

$$\frac{\theta_1-\theta_2}{t}=k\left[\frac{\theta_1+\theta_2}{2}-\theta_0\right]$$

$$\Rightarrow \frac{70-60}{5}=k\left[\frac{70+60}{2}-\theta_0\right]$$

$$\Rightarrow 2=k[65-\theta_0] \quad ...(i)$$

Similarly, $$\frac{60-54}{5}=k\left[\frac{60+54}{2}-\theta_0\right]$$

$$\Rightarrow \frac{6}{5}=k[57-\theta_0] \quad ...(ii)$$

By dividing (i) by (ii) we have

$$\frac{10}{6}=\frac{65-\theta_0}{57-\theta_0} \Rightarrow \theta_0=45^{\circ}$$

50. **(c)** Let θ_0 be the temperature of the surrounding. Then

$$\frac{80-75}{t_1}=k\left(\frac{80+75}{2}-\theta_0\right)$$

or, $\frac{5}{t_1}=k(77.5-\theta_0)$

or, $t_1=\frac{5}{k(77.5-\theta_0)}$ ...(1)

Similarly, $t_2=\frac{5}{k(72.5-\theta_0)}$ ...(2)

and $t_3=\frac{5}{k(67.5-\theta_0)}$ ...(3)

From (1), (2) & (3), it is obvious that

$t_1<t_2<t_3$

51. **(b)** Rate of cooling $\propto$ temperature difference between system and surrounding.

As the temperature difference is halved, so the rate of cooling will also be halved.

So time taken will be doubled

$t=2\times 5$ sec. = 10 sec.

11 Thermodynamics

Trend Analysis with Important Topics & Sub-Topics

Topic Name	Sub-Topic	2021 QNS.	2021 LOD	2020 QNS	2020 LOD	2019 QNS.	2019 LOD	2018 QNS.	2018 LOD	2017 QNS.	2017 LOD	2016 QNS.	2016 LOD
First Law of Thermodynamics	Ist Law of Thermodynamics							1	E				
Specific Heat Capacity & Thermodynamic Processes	Thermodynamic Processes			1	A	1	E	1	A	1	E	1	E
Carnot Engine, Refrigerator & Second Law of Thermodynamics	Efficiency of Heat Engine							1	E	1	E		
	COP of Refrigerator											1	A

LOD - Level of Difficulty | E - Easy | A - Average | D - Difficult | Qns - No. of Questions

Topic 1 First Law of Thermodynamics

1. 1 g of water, of volume 1 cm3 at 100°C, is converted into steam at same temperature under normal atmospheric pressure (» 1×10^5 Pa). The volume of steam formed equals 1671 cm^3. If the specific latent heat of vaporisation of water is 2256 J/g, the change in internal energy is, *[NEET Odisha 2019]*
   (a) 2256 J (b) 2423 J
   (c) 2089 J (d) 167 J

2. A sample of 0.1 g of water at 100°C and normal pressure (1.013×10^5 Nm^{-2}) requires 54 cal of heat energy to convert to steam at 100°C. If the volume of the steam produced is 167.1 cc, the change in internal energy of the sample, is *[2019]*
   (a) 104.3 J (b) 208.7 J
   (c) 84.5 J (d) 42.2 J

3. A system is taken from state *a* to state *c* by two paths *adc* and *abc* as shown in the figure. The internal energy at *a* is $U_a = 10$ J. Along the path *adc* the amount of heat absorbed $\delta Q_1 = 50$ J and the work done $\delta W_1 = 20$ J whereas along the path abc the heat absorbed $\delta Q_2 = 36$ J. The amount of work done along the path *abc* is *[NEET Kar. 2013]*

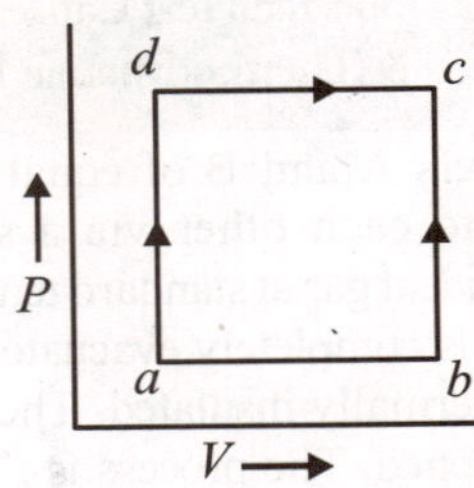

   (a) 6 J (b) 10 J
   (c) 12 J (d) 36 J

4. An ideal gas goes from state *A* to state *B* via three different processes as indicated in the *P-V* diagram : *[2012M]*

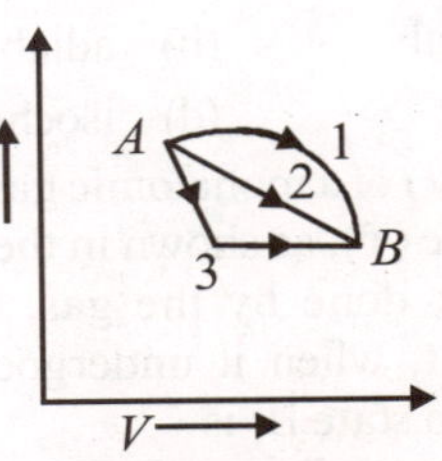

If Q_1, Q_2, Q_3 indicate the heat a absorbed by the

gas along the three processes and ΔU_1, ΔU_2, ΔU_3 indicate the change in internal energy along the three processes respectively, then

(a) $Q_1 > Q_2 > Q_3$ and $\Delta U_1 = \Delta U_2 = \Delta U_3$
(b) $Q_3 > Q_2 > Q_1$ and $\Delta U_1 = \Delta U_2 = \Delta U_3$
(c) $Q_1 = Q_2 = Q_3$ and $\Delta U_1 > \Delta U_2 > \Delta U_3$
(d) $Q_3 > Q_2 > Q_1$ and $\Delta U_1 > \Delta U_2 > \Delta U_3$

5. The internal energy change in a system that has absorbed 2 kcals of heat and done 500 J of work is: *[2009]*

(a) 6400 J (b) 5400 J
(c) 7900 J (d) 8900 J

6. 110 joules of heat is added to a gaseous system whose internal energy is 40J. Then the amount of external work done is *[1993]*

(a) 150 J (b) 70 J
(c) 110 J (d) 40 J

7. Which of the following is not thermodynamical function ? *[1993]*

(a) Enthalpy (b) Work done
(c) Gibb's energy (d) Internal energy

8. First law of thermodynamics is consequence of conservation of *[1988]*

(a) work (b) energy
(c) heat (d) all of these

Topic 2 Specific Heat Capacity & Thermodynamic Processes

9. Two cylinders A and B of equal capacity are connected to each other via a stop cock. A contains an ideal gas at standard temperature and pressure. B is completely evacuated. The entire system is thermally insulated. The stop cock is suddenly opened. The process is : *[2020]*

(a) adiabatic (b) isochoric
(c) isobaric (d) isothermal

10. In which of the following processes, heat is neither absorbed nor released by a system ? *[2019]*

(a) isothermal (b) adiabatic
(c) isobaric (d) isochoric

11. The volume (V) of a monatomic gas varies with its temperature (T), as shown in the graph. The ratio of work done by the gas, to the heat absorbed by it, when it undergoes a change from state A to state B, is *[2018]*

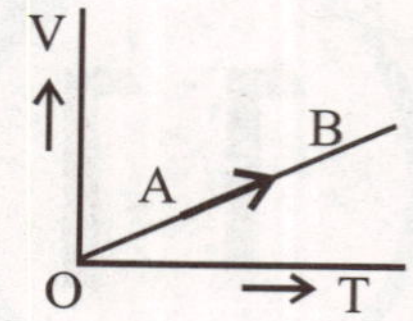

(a) $\frac{2}{5}$ (b) $\frac{2}{3}$
(c) $\frac{2}{7}$ (d) $\frac{1}{3}$

12. Thermodynamic processes are indicated in the following diagram : *[2017]*

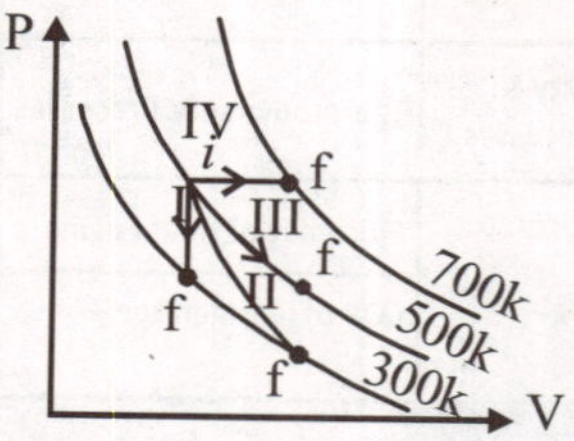

Match the following

	Column-1		Column-2
P.	Process I	A.	Adiabatic
Q.	Process II	B.	Isobaric
R.	Process III	C.	Isochoric
S.	Process IV	D.	Isothermal

(a) P → C, Q → A, R → D, S → B
(b) P → C, Q → D, R → B, S → A
(c) P → D, Q → B, R → A, S → C
(d) P → A, Q → C, R → D, S → B

13. A gas is compressed isothermally to half its initial volume. The same gas is compressed separately through an adiabatic process until its volume is again reduced to half. Then : *[2016]*

(a) Compressing the gas isothermally will require more work to be done.
(b) Compressing the gas through adiabatic process will require more work to be done.
(c) Compressing the gas isothermally or adiabatically will require the same amount of work.
(d) Which of the case (whether compression through isothermal or through adiabatic process) requires more work will depend upon the atomicity of the gas.

14. Figure below shows two paths that may be taken by a gas to go from a state A to a state C.

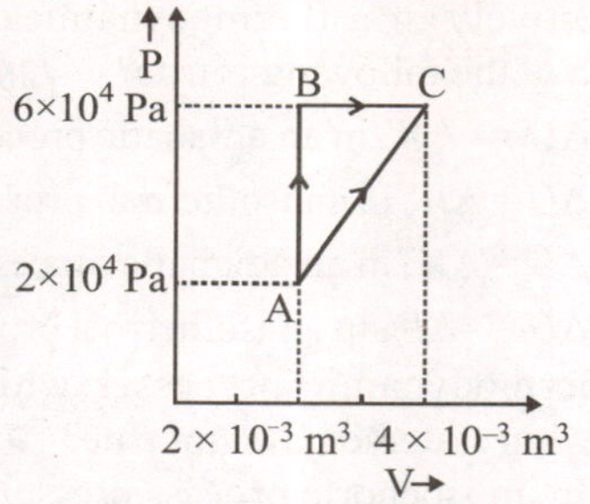

In process AB, 400 J of heat is added to the system and in process BC, 100 J of heat is added to the system. The heat absorbed by the system in the process AC will be *[2015]*

(a) 500 J (b) 460 J
(c) 300 J (d) 380 J

15. An ideal gas is compressed to half its initial volume by means of several processes. Which of the process results in the maximum work done on the gas? *[2015 RS]*

(a) Isobaric (b) Isochoric
(c) Isothermal (d) Adiabatic

16. A monoatomic gas at a pressure P, having a volume V expands isothermally to a volume 2V and then adiabatically to a volume 16V. The final pressure of the gas is : (take $\gamma = \frac{5}{3}$) *[2014]*

(a) 64P (b) 32P
(c) $\frac{P}{64}$ (d) 16P

17. A thermodynamic system undergoes cyclic process ABCDA as shown in fig. The work done by the system in the cycle is : *[2014]*

(a) P_0V_0
(b) $2P_0V_0$
(c) $\frac{P_0V_0}{2}$
(d) Zero

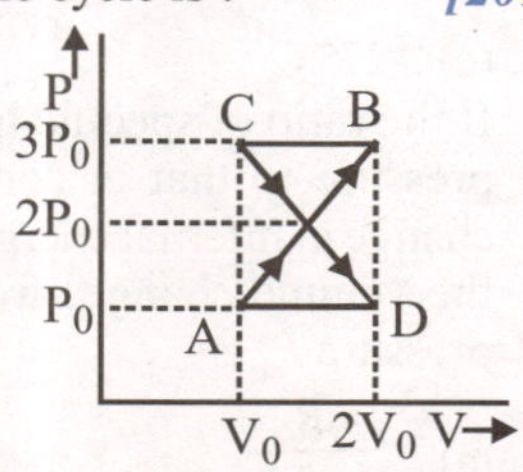

18. A gas is taken through the cycle $A \to B \to C \to A$, as shown in figure. What is the net work done by the gas? *[2013]*

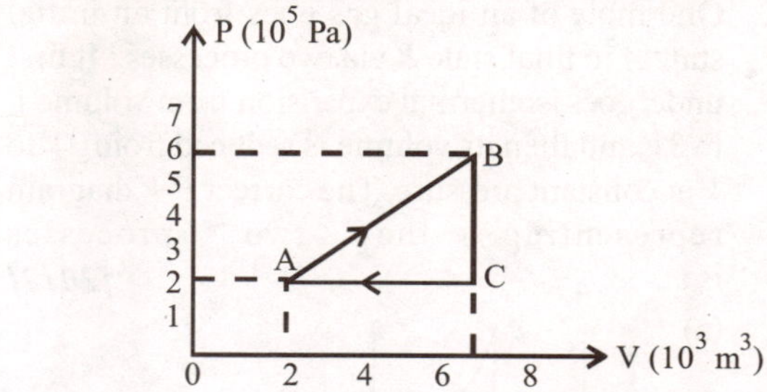

(a) 1000 J (b) zero
(c) −2000 J (d) 2000 J

19. During an adiabatic process, the pressure of a gas is found to be proportional to the cube of its temperature. The ratio of $\frac{C_p}{C_v}$ for the gas is *[2013]*

(a) 2 (b) $\frac{5}{3}$
(c) $\frac{3}{2}$ (d) $\frac{4}{3}$

20. Which of the following relations does not give the equation of an adiabatic process, where terms have their usual meaning? *[NEET Kar. 2013]*

(a) $P^{\gamma}T^{1-\gamma}$ = constant
(b) $P^{1-\gamma}\,T^{\gamma}$ = constant
(c) PV^{γ} = constant
(d) $TV^{\gamma-1}$ = constant

21. A thermodynamic system is taken through the cycle *ABCD* as shown in figure. Heat rejected by the gas during the cyclic process is : *[2012]*

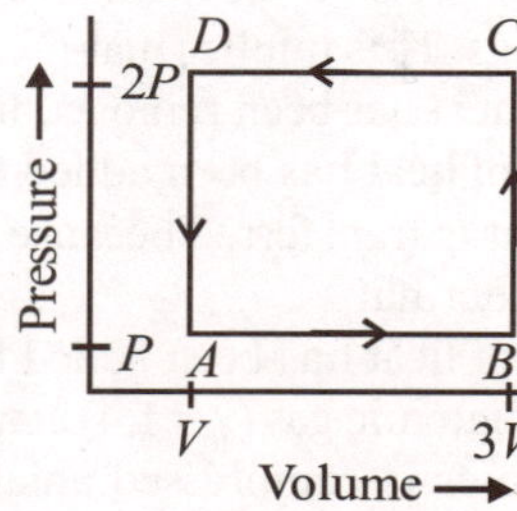

(a) 2PV (b) 4PV
(c) $\frac{1}{2}$PV (d) PV

22. One mole of an ideal gas goes from an initial state A to final state B via two processes : It first undergoes isothermal expansion from volume V to $3V$ and then its volume is reduced from $3V$ to V at constant pressure. The correct P-V diagram representing the two processes is : *[2012]*

(a)
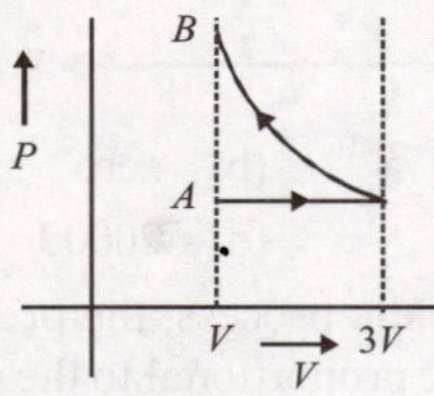

(b)
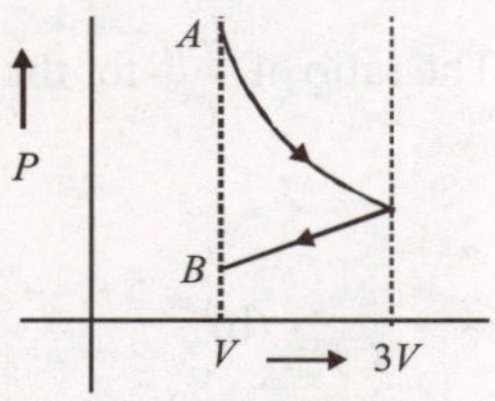

(c)
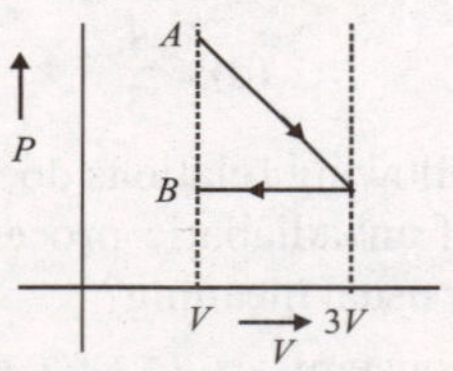

(d)
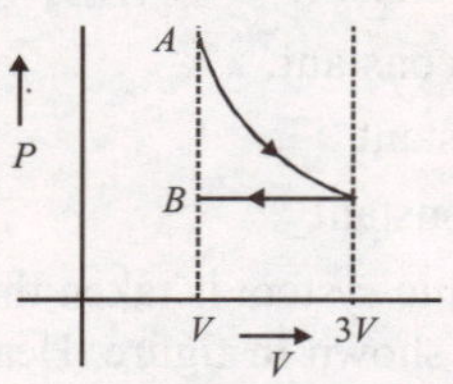

23. During an isothermal expansion, a confined ideal gas does –150 J of work against its surroundings. This implies that *[2011]*
(a) 150 J heat has been removed from the gas
(b) 300 J of heat has been added to the gas
(c) no heat is transferred because the process is isothermal
(d) 150 J of heat has been added to the gas

24. A mass of diatomic gas ($\gamma = 1.4$) at a pressure of 2 atmospheres is compressed adiabatically so that its temperature rises from 27°C to 927°C. The pressure of the gas in final state is *[2011M]*
(a) 28 atm (b) 68.7 atm
(c) 256 atm (d) 8 atm

25. If ΔU and ΔW represent the increase in internal energy and work done by the system respectively in a thermodynamical process, which of the following is true? *[2010, 1998]*
(a) $\Delta U = -\Delta W$, in an adiabatic process
(b) $\Delta U = \Delta W$, in an isothermal process
(c) $\Delta U = \Delta W$, in an adiabatic process
(d) $\Delta U = -\Delta W$, in an isothermal process

26. In thermodynamic processes which of the following statements is not true? *[2009]*
(a) In an isochoric process pressure remains constant
(b) In an isothermal process the temperature remains constant
(c) In an adiabatic process PV^{γ} = constant
(d) In an adiabatic process the system is insulated from the surroundings

27. If Q, E and W denote respectively the heat added, change in internal energy and the work done in a closed cyclic process, then: *[2008]*
(a) W = 0 (b) Q = W = 0
(c) E = 0 (d) Q = 0

28. One mole of an ideal gas at an initial temperature of T K does 6R joules of work adiabatically. If the ratio of specific heats of this gas at constant pressure and at constant volume is 5/3, the final temperature of gas will be *[2004]*
(a) $(T-4)$ K (b) $(T+2.4)$ K
(c) $(T-2.4)$ K (d) $(T+4)$ K

29. An ideal gas at 27ºC is compressed adiabatically to $\frac{8}{27}$ of its original volume. The rise in temperature is $\left(\gamma = \frac{5}{3}\right)$ *[1999]*
(a) 475ºC (b) 402ºC
(c) 275ºC (d) 175ºC

30. If the ratio of specific heat of a gas at constant pressure to that at constant volume is γ, the change in internal energy of a mass of gas, when the volume changes from V to $2V$ at constant pressure P, is *[1998]*
(a) $\frac{R}{(\gamma-1)}$ (b) PV
(c) $\frac{PV}{(\gamma-1)}$ (d) $\frac{\gamma PV}{(\gamma-1)}$

31. A sample of gas expands from volume V_1 to V_2. The amount of work done by the gas is greatest, when the expansion is *[1997]*
(a) adiabatic (b) isobaric
(c) isothermal (d) equal in all cases

32. An ideal gas undergoing adiabatic change has the following pressure-temperature relationship *[1996]*
(a) $P^{\gamma-1}T^{\gamma}$ = constant
(b) $P^{\gamma}T^{\gamma-1}$ = constant
(c) $P^{\gamma}T^{1-\gamma}$ = constant
(d) $P^{1-\gamma}T^{\gamma}$ = constant

33. A diatomic gas initially at 18ºC is compressed adiabatically to one eighth of its original volume. The temperature after compression will be
(a) 18ºC (b) 668.4ºK *[1996]*
(c) 395.4ºC (d) 144ºC

34. An ideal gas *A* and a real gas *B* have their volumes increased from *V* to 2*V* under isothermal conditions. The increase in internal energy *[1993]*
(a) will be same in both *A* and *B*
(b) will be zero in both the gases
(c) of *B* will be more than that of *A*
(d) of *A* will be more than that of *B*

35. A thermodynamic system is taken from state A to B along *ACB* and is brought back to A along *BDA* as shown in the *PV* diagram. The net work done during the complete cycle is given by the area *[1992]*

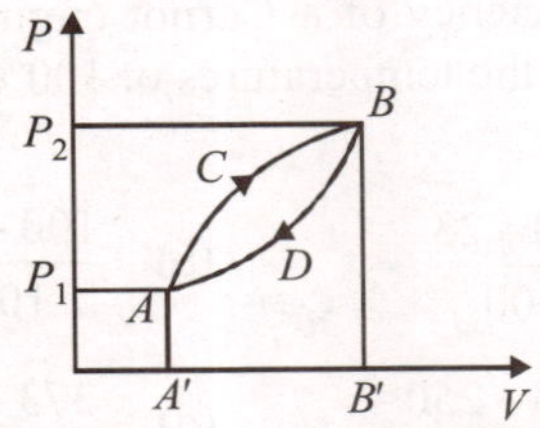

(a) $P_1ACBP_2P_1$ (b) $ACBB'A'A$
(c) $ACBDA$ (d) $ADBB'A'A$

36. A thermodynamic process is shown in the figure. The pressures and volumes corresponding to some points in the figure are

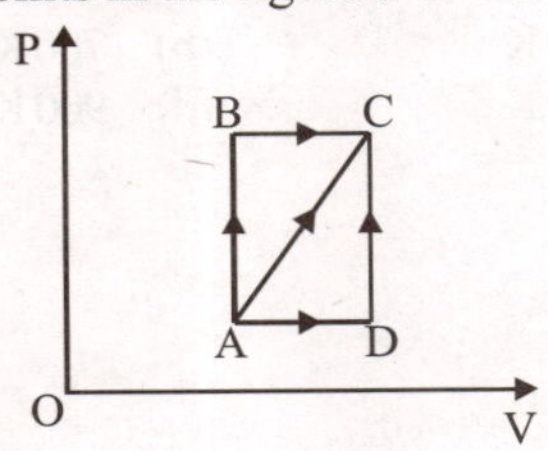

$P_A = 3 \times 10^4\ Pa$
$V_A = 2 \times 10^{-3}\ m^3$
$P_B = 8 \times 10^4\ Pa$
$V_D = 5 \times 10^{-3}\ m^3$.
In process *AB*, 600 J of heat is added to the system and in process BC, 200 J of heat is added to the system. The change in internal energy of the system in process *AC* would be *[1991]*
(a) 560 J (b) 800 J
(c) 600 J (d) 640 J

37. At 27° C a gas is compressed suddenly such that its pressure becomes (1/8) of original pressure. Final temperature will be (γ = 5/3)
(a) 420 K (b) 300K *[1989]*
(c) –142°C (d) 327ºC

Topic 3 Carnot Engine, Refrigerator & Second Law of Thermodynamics

38. The efficiency of an ideal heat engine working between the freezing point and boiling point of water, is *[2018]*
(a) 26.8% (b) 20%
(c) 12.5% (d) 6.25%

39. A carnot engine having an efficiency of $\frac{1}{10}$ as heat engine, is used as a refrigerator. If the work done on the system is 10 J, the amount of energy absorbed from the reservoir at lower temperature is : *[2017, 2015]*
(a) 90 J (b) 99 J
(c) 100 J (d) 1 J

40. A refrigerator works between 4°C and 30°C. It is required to remove 600 calories of heat every second in order to keep the temperature of the refrigerated space constant. The power required is:
(Take 1 cal = 4.2 joules) *[2016]*
(a) 2.365 W (b) 23.65 W
(c) 236.5 W (d) 2365 W

41. The coefficient of performance of a refrigerator is 5. If the inside temperature of freezer is –20°C, then the temperature of the surroundings to which it rejects heat is *[2015 RS]*
(a) 41°C (b) 11°C
(c) 21°C (d) 31°C

42. Two Carnot engines A and B are operated in series. The engine A receives heat from the source at temperature T_1 and rejects the heat to the sink at temperature T. The second engine B receives the heat at temperature T and rejects to its sink at temperature T_2. For what value of T the efficiencies of the two engines are equal? *[NEET Kar. 2013]*

(a) $\frac{T_1+T_2}{2}$ (b) $\frac{T_1-T_2}{2}$

(c) T_1T_2 (d) $\sqrt{T_1T_2}$

43. When 1 kg of ice at 0°C melts to water at 0°C, the resulting change in its entropy, taking latent heat of ice to be 80 cal/°C, is *[2011]*

(a) 273 cal/K (b) 8 × 104 cal/K

(c) 80 cal/K (d) 293 cal/K

44. An engine has an efficiency of 1/6. When the temperature of sink is reduced by 62°C, its efficiency is doubled. Temperature of the source is *[2007]*

(a) 37°C (b) 62°C

(c) 99°C (d) 124°C

45. A Carnot engine whose sink is at 300 K has an efficiency of 40%. By how much should the temperature of source be increased so as to increase, its efficiency by 50% of original efficiency ? *[2006]*

(a) 325 K (b) 250 K

(c) 380 K (d) 275 K

46. Which of the following processes is reversible? *[2005]*

(a) Transfer of heat by conduction

(b) Transfer of heat by radiation

(c) Isothermal compression

(d) Electrical heating of a nichrome wire

47. An ideal gas heat engine operates in Carnot cycle between 227°C and 127°C. It absorbs 6×10^4 cals of heat at higher temperature. Amount of heat converted to work is *[2005]*

(a) 4.8×10^4 cals (b) 6×10^4 cals

(c) 2.4×10^4 cals (d) 1.2×10^4 cals

48. A Carnot engine whose efficiency is 50% has an exhaust temperature of 500 K. If the efficiency is to be 60% with the same intake temperature, the exhaust temperature must be (in K) *[2002]*

(a) 800 (b) 200

(c) 400 (d) 600

49. An ideal gas heat engine operates in a Carnot cycle between 227°C and 127°C. It absorbs 6 kcal at the higher temperature. The amount of heat (in kcal) converted into work is equal to *[2002]*

(a) 1.2 (b) 4.8

(c) 3.5 (d) 1.6

50. The temperature of source and sink of a heat engine are 127°C and 27°C respectively. An inventor claims its efficiency to be 26%, then: *[2001]*

(a) it is impossible

(b) it is possible with high probability

(c) it is possible with low probability

(d) data are insufficient.

51. A reversible engine converts one-sixth of the heat input into work. When the temperature of the sink is reduced by 62°C, the efficiency of the engine is doubled. The temperatures of the source and sink are *[2000]*

(a) 99°C, 37°C (b) 80°C, 37°C

(c) 95°C, 37°C (d) 90°C, 37°C

52. The efficiency of a Carnot engine operating between the temperatures of 100°C and –23°C will be *[1997]*

(a) $\frac{100+23}{100}$ (b) $\frac{100-23}{100}$

(c) $\frac{373+250}{373}$ (d) $\frac{373-250}{373}$

53. An ideal carnot engine, whose efficiency is 40% receives heat at 500 K. If its efficiency is 50%, then the intake temperature for the same exhaust temperature is *[1995]*

(a) 600 K (b) 700 K

(c) 800 K (d) 900 K

ANSWER KEY

1	(c)	7	(b)	13	(b)	19	(c)	25	(a)	31	(b)	37	(c)	43	(d)	49	(a)
2	(b)	8	(b)	14	(b)	20	(a)	26	(a)	32	(d)	38	(a)	44	(c)	50	(a)
3	(a)	9	(a)	15	(d)	21	(a)	27	(c)	33	(b)	39	(a)	45	(b)	51	(a)
4	(a)	10	(b)	16	(c)	22	(d)	28	(a)	34	(b)	40	(c)	46	(c)	52	(d)
5	(c)	11	(a)	17	(d)	23	(d)	29	(b)	35	(c)	41	(d)	47	(d)	53	(a)
6	(b)	12	(a)	18	(a)	24	(c)	30	(c)	36	(a)	42	(d)	48	(c)		

Hints Solutions

1. **(c)** $\Delta Q = 2256 \times 1 = 2256$ J
$\Delta W = P[V_{steam} - V_{water}]$ $[\because dw = pdv]$
$= 1 \times 105[1671 - 1] \times 106 = 167$ J
As $\Delta Q = \Delta U + \Delta W$
$2256 = \Delta U + 167$
$\Delta U = 2089$ J

2. **(b)** Using first law of thermodynamics equation,
$\Delta Q = \Delta U + \Delta W$
$\Rightarrow$ 54×4.18
$= \Delta U + 1.013 \times 10^5(167.1 \times 10^{-6} - 0)$
$(\because \Delta W = P\Delta V)$
$\Rightarrow$ $\Delta U = 208.7$ J

3. **(a)** From first law of thermodynamics
$Q_{adc} = \Delta U_{adc} + W_{adc}$
$50\text{ J} = \Delta U_{adc} + 20\text{ J}$
$\Delta U_{adc} = 30\text{ J}$
Again, $Q_{abc} = \Delta U_{abc} + W_{abc}$
$W_{abc} = Q_{abc} - \Delta U_{abc}$
$= Q_{abc} - \Delta U_{adc}$
$= 36\text{ J} - 30\text{ J}$
$= 6\text{ J}$

4. **(a)** Initial and final condition is same for all process
$\Delta U_1 = \Delta U_2 = \Delta U_3$
from first law of thermodynamics
$\Delta Q = \Delta U + \Delta W$
Work done
$\Delta W_1 > \Delta W_2 > \Delta W_3$ (Area of P.V. graph)
So $\Delta Q_1 > \Delta Q_2 > \Delta Q_3$

5. **(c)** According to first law of thermodynamics
$Q = \Delta U + W$
$\Delta U = Q - W$
$= 2 \times 4.2 \times 1000 - 500 = 8400 - 500$
$= 7900$ J

6. **(b)** $\Delta Q = \Delta U + \Delta W$
$\Rightarrow \Delta W = \Delta Q - \Delta U = 110 - 40 = 70$ J

7. **(b)** Work done is not a thermodynamical function.

> **NOTES**
> The fundamental thermodynamic equations follow from five primary thermodynamic definitions and describe internal energy, enthalpy, Helmholtz energy and Gibb's energy.

8. **(b)** The first law of thermodynamics is just a conservation of energy.

9. **(a)** Entire system is thermally insulated. So, no heat exchange with surrounding will take place. Hence, process will be adiabatic.

10. **(b)** In adiabatic process, there is no exchange of heat, with the surroundings.

11. **(a)** Gas is monatomic, so $C_p = \frac{5}{2}R$
Given process is isobaric
$\therefore$ $dQ = n\,C_p\,dT$
$\Rightarrow$ $dQ = n\left(\frac{5}{2}R\right)dT$
$dW = P\,dV = n\,RdT$
$\therefore$ Required ratio $= \frac{dW}{dQ} = \frac{nRdT}{n\left(\frac{5}{2}R\right)dT} = \frac{2}{5}$

12. **(a)** Process I volume is constant hence, it is isochoric
In process IV, pressure is constant hence, it is isobaric

13. **(b)** W_{ext} = negative of area with volume-axis
W(adiabatic) > W(isothermal)

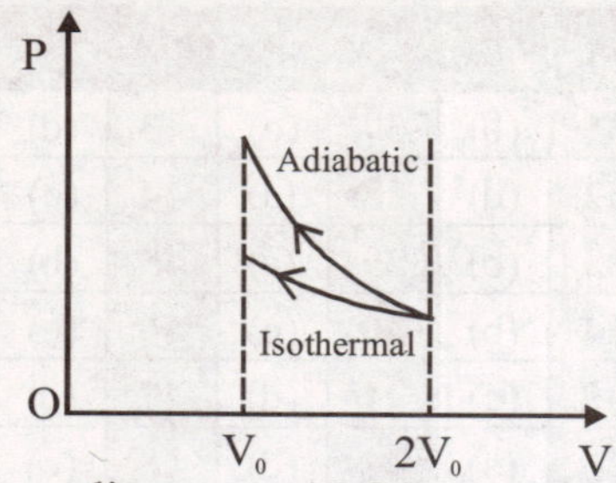

14. **(b)** In cyclic process ABCA

$Q_{cycle} = W_{cycle}$

$Q_{AB} + Q_{BC} + Q_{CA}$ = ar. of ΔABC

$+ 400 + 100 + Q_{C\to A} = \frac{1}{2}(2 \times 10^{-3})(4 \times 10^{4})$

$\Rightarrow \quad Q_{C\to A} = -460\text{ J} \quad \Rightarrow \quad Q_{A\to C} = +460\text{ J}$

15. **(d)**

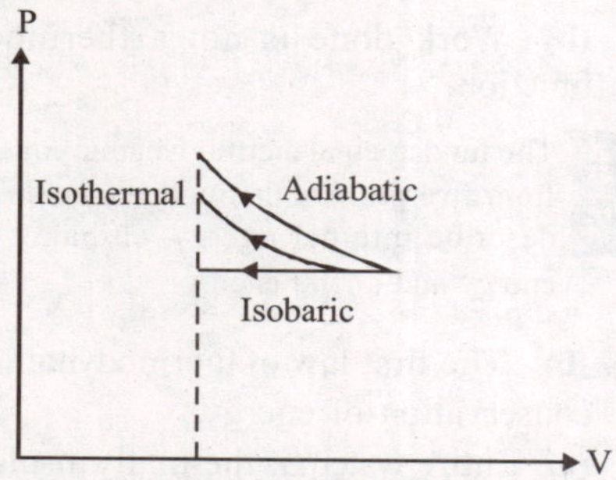

Since area under the curve is maximum for adiabatic process so, work done (W = PdV) on the gas will be maximum for adiabatic process.

16. **(c)** For isothermal process $P_1V_1 = P_2V_2$

$\Rightarrow \quad PV = P_2(2V) \Rightarrow P_2 = \frac{P}{2}$

For adiabatic process

$P_2V_2^{\gamma} = P_3V_3^{\gamma}$

$\Rightarrow \quad \left(\frac{P}{2}\right)(2v)^{\gamma} = P_3 16v)^{\gamma}$

$\Rightarrow \quad P_3 = \frac{3}{2}\left(\frac{1}{8}\right)^{5/3} = \frac{P}{64}$

17. **(d)** Work done by the system in the cycle = Area under P-V curve and V-axis

$= \frac{1}{2}(2P_0 - P_0)(2V_0 - V_0) +$

$\left[-\left(\frac{1}{2}\right)(3P_0 - 2P_0)(2V_0 - V_0)\right]$

$= \frac{P_0V_0}{2} - \frac{P_0V_0}{2} = 0$

18. **(a)** W_{net} = Area of triangle ABC

$= \frac{1}{2}\, AC \times BC$

$= \frac{1}{2} \times 5 \times 10^{-3} \times 4 \times 10^{5} = 1000\text{ J}$

19. **(c)** According to question $P \propto T^3$

But as we know for an adiabatic process the pressure $P \propto T^{\frac{\gamma}{\gamma-1}}$.

So, $\frac{\gamma}{\gamma-1} = 3 \Rightarrow \gamma = \frac{3}{2}$ or, $\frac{C_p}{C_v} = \frac{3}{2}$

20. **(a)** Adiabatic equations of state are

PV^{γ} = constant $\qquad TV^{\gamma-1}$ = constant

$P^{1-\gamma}T^{\gamma}$ = constant.

NOTES: When a thermodynamic system undergoes a change in such a way that to exchange of heat takes place between system and surroundings, then the process is adiabatic process. In this process *P*, *V* and *T* change but $\Delta Q = 0$.

21. **(a)** ∵ Internal energy is the state function.

∴ In cyclie process; $\Delta U = 0$

According to 1st law of thermodynamics

$\Delta Q = \Delta U + W$

So heat absorbed

$\Delta Q = W$ = Area under the curve

$= -(2V)(P) = -2PV$

So heat rejected = 2PV

22. **(d)** 1st process is isothermal expansion which is only correct shown in option (d)

2nd process is isobaric compression which is correctly shown in option (d)

23. **(d)** According to question work done by gas is given –150J so that according to it answer will be (d).

24. **(c)** $T_1 = 273 + 27 = 300\text{K}$

$T_2 = 273 + 927 = 1200\text{K}$

For adiabatic process,

$P^{1-\gamma}T^{\gamma}$ = constant

$\Rightarrow P_1^{1-\gamma}T_1^{\gamma} = P_2^{1-\gamma}T_2^{\gamma}$

$\Rightarrow \left(\frac{P_2}{P_1}\right)^{1-\gamma} = \left(\frac{T_1}{T_2}\right)^{\gamma} \Rightarrow \left(\frac{P_1}{P_2}\right)^{1-\gamma} = \left(\frac{T_2}{T_1}\right)^{\gamma}$

$\Rightarrow \left(\frac{P_1}{P_2}\right)^{1-1.4} = \left(\frac{1200}{300}\right)^{1.4} \Rightarrow \left(\frac{P_1}{P_2}\right)^{-0.4} = (4)^{1.4}$

$\Rightarrow \left(\frac{P_2}{P_1}\right)^{0.4} = 4^{1.4}$

$$\Rightarrow P_2 = P_1\, 4^{\left(\frac{1.4}{0.4}\right)} = P_1\, 4^{\left(\frac{7}{2}\right)}$$

$$\Rightarrow P_1(2^7) = 2 \times 128 = 256 \text{ atm}$$

25. (a) By first law of thermodynamics,

$\Delta Q = \Delta U + \Delta W$

In adiabatic process, $\Delta Q = 0$

$\therefore \quad \Delta U = -\Delta W$

In isothermal process, $\Delta U = 0$

$\therefore \quad \Delta Q = \Delta W$

26. (a) In an isochoric process volume remains constant whereas pressure remains constant in isobaric process.

27. (c) In a cyclic process, the initial state coincides with the final state. Hence, the change in internal energy is zero, as it depends only on the initial and final states.

NOTES

In case of cyclic process, $u_f = u_i \;\therefore\; \Delta u = u_f - u_i = 0$

And also $\Delta u \propto \Delta T \;\therefore\; \Delta T = 0$

28. (a) $T_1 = T,\ W = 6R$ joules, $\gamma = \frac{5}{3}$

$$W = \frac{P_1V_1 - P_2V_2}{\gamma - 1} = \frac{nRT_1 - nRT_2}{\gamma - 1}$$

$$= \frac{nR(T_1 - T_2)}{\gamma - 1}$$

$$n = 1,\ T_1 = T \Rightarrow \frac{R(T - T_2)}{5/3 - 1} = 6R$$

$$\Rightarrow T_2 = (T - 4)\text{K}$$

29. (b) T = 27°C = 300 K

$$\gamma = \frac{5}{3};\quad V_2 = \frac{8}{27}V_1;\quad \frac{V_1}{V_2} = \frac{27}{8}$$

From adiabatic process we know that

$$T_1V_1^{\gamma-1} = T_2V_2^{\gamma-1}$$

$$\frac{T_2}{T_1} = \left(\frac{V_1}{V_2}\right)^{\gamma-1} = \left(\frac{27}{8}\right)^{\frac{5}{3}-1}$$

$$\frac{T_2}{T_1} = \frac{9}{4} \Rightarrow T_2 = \frac{9}{4} \times T_1 = \frac{9}{4} \times 300 = 675\,K$$

$$T_2 = 675 - 273\text{°C} = 402\text{°C}$$

30. (c) Change in internal energy is equal to work done in adiabatic system

$\Delta W = -\Delta U$ (Expansion in the system)

$$= -\frac{1}{\gamma - 1}(P_1V_1 - P_2V_2)$$

$$\Delta U = \frac{1}{1-\gamma}(P_2V_2 - P_1V_1)$$

Here, $V_1 = V, V_2 = 2V$

$$\therefore \quad \Delta U = \frac{1}{1-\gamma}[P \times 2V - PV] = \frac{PV}{1-\gamma}$$

$$\Rightarrow \Delta U = -\frac{PV}{\gamma - 1}$$

31. (b) In thermodynamics for same change in volume, the work done is maximum in isobaric process because in $P-V$ graph, area enclosed by curve and volume axis is maximum in isobaric process.

32. (d) We know that in adiabatic process,

$PV^\gamma = \text{constant}$ ....(1)

From ideal gas equation, we know that

$PV = nRT$

$$V = \frac{nRT}{P} \quad \text{....(2)}$$

Putting the value from equation (2) in equation (1),

$$P\left(\frac{nRT}{P}\right)^\gamma = \text{constant}$$

$$P^{(1-\gamma)}\, T^\gamma = \text{constant}$$

33. (b) Initial temperature $(T_1) = 18\text{°C} = 291$ K

Let Initial volume $(V_1) = V$

Final volume $(V_2) = \frac{V}{8}$

According to adiabatic process,

$TV^{\gamma-1} = \text{constant}$

According to question, $T_1V_1^{\gamma-1} = T_2V_2^{\gamma-1}$

$$\Rightarrow T_2 = 293\left(\frac{V_1}{V_2}\right)^{\gamma-1}$$

$$\Rightarrow T_2 = 293(8)^{\frac{7}{5}-1} = 293 \times 2.297 = 668.4\text{K}$$

$$\left[\text{For diatomic gas } \gamma = \frac{C_p}{C_v} = \frac{7}{5}\right]$$

34. (b) Under isothermal conditions, there is no change in internal energy.

35. (c) Work done = Area under curve $ACBDA$

36. (a) BC is isobaric process,

$\therefore \quad W = P_B \times (V_D - V_A) = 240$ J

$\Delta Q = 600 + 200 = 800$ J

Using $\Delta Q = \Delta U + \Delta W$

$\Rightarrow \Delta U = \Delta Q - \Delta W = 800 - 240 = 560$ J

NOTES: Since *AB* is an isochoric process i.e., $\Delta V = 0$ so, no work is done.

37. **(c)** $T_1^{\gamma} P_1^{1-\gamma} = T_2^{\gamma} P_2^{1-\gamma}$

$$\Rightarrow \left(\frac{T_2}{T_1}\right)^{\gamma} = \left(\frac{P_1}{P_2}\right)^{1-\gamma}$$

$$\Rightarrow T_2 = T_1 . \left(\frac{P_1}{P_2}\right)^{\frac{1-\gamma}{\gamma}} = 300 \times (8)^{-2/5} = 142°\text{C}$$

38. **(a)** Efficiency of ideal heat engine, $\eta = \left(1 - \frac{T_2}{T_1}\right)$

Sink temperature, $T_2 = 100°\text{C} = 100 + 273 = 373$ K

Source temperature, $T_1 = 0°\text{C} = 0 + 273 = 273$ K

Percentage efficiency, $\%\eta = \left(1 - \frac{T_2}{T_1}\right) \times 100$

$$= \left(1 - \frac{273}{373}\right) \times 100 = \left(\frac{100}{373}\right) \times 100 = 26.8\%$$

39. **(a)** Given, efficiency of engine, $\eta = \frac{1}{10}$

work done on system. W = 10J

Coefficient of performance of refrigerator

$$\beta = \frac{Q_2}{W} = \frac{1-\eta}{\eta} = \frac{1 - \frac{1}{10}}{\frac{1}{10}} = \frac{\frac{9}{10}}{\frac{1}{10}} = 9$$

Energy absorbed from reservoir

$Q_2 = \beta W \qquad Q_2 = 9 \times 10 = 90$ J

NOTES: A refrigerator works along the reverse direction of a heat engine.

40. **(c)** **Given:** $T_2 = 4°\text{C} = 4 + 273 = 277$ K

$T_1 = 30°\text{C} = 30 + 273 = 303$ K

$$\beta = \frac{Q_2}{W} = \frac{T_2}{T_1 - T_2}$$

$$\therefore \quad \beta = \frac{600 \times 4.2}{W} = \frac{277}{303 - 277}$$

$\Rightarrow$ W = 236.5 joule

$$\text{Power P} = \frac{W}{t} = \frac{236.5 \text{ joule}}{1 \text{ sec}} = 236.5 \text{ watt.}$$

41. **(d)** Coefficient of performance,

$$\text{Cop} = \frac{T_2}{T_1 - T_2}$$

$$5 = \frac{273 - 20}{T_1 - (273 - 20)} = \frac{253}{T_1 - 253}$$

$5T_1 - (5 \times 253) = 253$

$5T_1 = 253 + (5 \times 253) = 1518$

$$\therefore T_1 = \frac{1518}{5} = 303.6$$

or, $T_1 = 303.6 - 273 = 30.6 \cong 31°\text{C}$

42. **(d)** Efficiency of engine *A*, $\eta_1 = 1 - \frac{T}{T_1}$,

Efficiency of engine *B*, $\eta_2 = 1 - \frac{T_2}{T}$

Here, $\eta_1 = \eta_2$

$$\therefore \frac{T}{T_1} = \frac{T_2}{T} \Rightarrow T = \sqrt{T_1 T_2}$$

43. **(d)** Change in entropy is given by

$$\Delta S = \frac{dQ}{T} \text{ or } \Delta S = \frac{\Delta Q}{T} = \frac{mL_f}{273}$$

$$\Delta S = \frac{1000 \times 80}{273} = 293 \text{ cal/K.}$$

44. **(c)** Since efficiency of engine is $\eta = 1 - \frac{T_2}{T_1}$

According to problem,

$$\frac{1}{6} = 1 - \frac{T_2}{T_1} \qquad \text{.......... (1)}$$

When the temperature of the sink is reduced by 62°C, its efficiency is doubled

$$2\left(\frac{1}{6}\right) = 1 - \frac{T_2 - 62}{T_1} \qquad \text{.......... (2)}$$

Solving (1) and (2) $\qquad T_2 = 372$ K

$T_1 = 99°\text{C}$ = Temperature of source.

45. **(b)** We know that efficiency of Carnot Engine

$$\eta = \frac{T_1 - T_2}{T_1}$$

where, T_1 is temp. of source and T_2 is temp. of sink

$$\therefore \quad 0.40 = \frac{T_1 - 300}{T_1} \Rightarrow T_1 - 300 = 0.40 T_1$$

$$0.6 T_1 = 300 \Rightarrow T_1 = \frac{300}{.6} = \frac{3000}{6} = 500K$$

Now efficiency to be increased by 50%

$\therefore \quad 0.60 = \frac{T_1 - 300}{T_1} \Rightarrow T_1 - 300 = 0.6T_1$

$0.4T_1 = 300 \Rightarrow T_1 = \frac{300}{.4} = \frac{300 \times 10}{4} = 750$

Increase in temp $= 750 - 500 = 250\,K$

46. (c) Isothermal process occur very slowly so it is quasi-static and hence it is reversible.

NOTES

The conditions for reversibility
- Must take place infinitely slowly
- The temperature of the system must not differ appreciably
- There must be complete absence of dissipative forces such as friction, viscosity, electric resistance, etc.

47. (d) We know that

efficiency of carnot engine

$= 1 - \frac{T_2}{T_1} = 1 - \frac{400}{500} = \frac{1}{5}$

$[\because T_1 = (273 + 227)\text{K} = 500\,\text{K}$

and $T_2 = (273 + 127)\text{K} = 400\,\text{K}]$

$\text{Efficiency of Heat engine} = \frac{\text{Work output}}{\text{Heat input}}$

or, $\frac{1}{5} = \frac{\text{work output}}{6 \times 10^4}$

$\Rightarrow$ work output $= 1.2 \times 10^4$ cal

48. (c) $\eta = 1 - \frac{T_2}{T_1}$ or $\frac{50}{100} = 1 - \frac{500}{T_1}$

$\Rightarrow T_1 = 1000\,K$

Also, $\frac{60}{100} = 1 - \frac{T_2}{1000} \Rightarrow T_2 = 400\,K$

49. (a) $\text{Efficiency} = \frac{T_1 - T_2}{T_1}$

$T_1 = 227 + 273 = 500\,\text{K}$

$T_2 = 127 + 273 = 400\,\text{K}$

$\eta = \frac{500 - 400}{500} = \frac{1}{5}$

Hence, output work

$= (\eta) \times \text{Heat input} = \frac{1}{5} \times 6 = 1.2\ \text{kcal}$

NOTES

Efficiency of heat engine: $n = \frac{\text{Work done}(W)}{\text{Heat in put}(Q_1)}$

$= 1 - \frac{Q_2}{Q_1} = 1 - \frac{T_2}{T_1}$

50. (a) $\eta = 1 - \frac{300}{400} = \frac{100}{400} = \frac{1}{4}$ $\eta = \frac{1}{4} \times 100 = 25\%$

Hence, it is not possible to have efficiency more than 25%.

51. (a) Initially the efficiency of the engine was $\frac{1}{6}$ which increases to $\frac{1}{3}$ when the sink temperature reduces by 62° C.

$\eta = \frac{1}{6} = 1 - \frac{T_2}{T_1}$, when T_2 = sink temperature

T_1 = source temperature $\Rightarrow T_2 = \frac{5}{6}T_1$

Secondly,

$\frac{1}{3} = 1 - \frac{T_2 - 62}{T_1} = 1 - \frac{T_2}{T_1} + \frac{62}{T_1} = 1 - \frac{5}{6} + \frac{62}{T_1}$

or, $T_1 = 62 \times 6 = 372\text{K} = 372 - 273 = 99°\text{C}$

& $T_2 = \frac{5}{6} \times 372 = 310\,\text{K} = 310 - 273 = 37°\,\text{C}$

52. (d) $\eta = 1 - \frac{T_1}{T_2}$

$T_1 = -23°\text{C} = 250\,\text{K},\ T_2 = 100°\text{C} = 373\text{K}$

$\eta = 1 - \frac{250}{373} = \frac{373 - 250}{373}$

53. (a) Efficiency of carnot engine (η_1) = 40% = 0.4; Initial intake temperature (T_1) = 500K and new efficiency (η_2) = 50% = 0.5.

Efficiency $(\eta) = 1 - \frac{T_2}{T_1}$ or $\frac{T_2}{T_1} = 1 - \eta$.

Therefore in first case, $\frac{T_2}{500} = 1 - 0.4 = 0.6$.

$\Rightarrow T_2 = 0.6 \times 500 = 300\text{K}$

And in second case, $\frac{300}{T_1} = 1 - 0.5 = 0.5$

$\Rightarrow T_1 = \frac{300}{0.5} = 600\,K$

12 Kinetic Theory

Trend Analysis with Important Topics & Sub-Topics

Topic Name	Sub-Topic	2021 QNS.	2021 LOD	2020 QNS.	2020 LOD	2019 QNS.	2019 LOD	2018 QNS.	2018 LOD	2017 QNS.	2017 LOD	2016 QNS.	2016 LOD
Kinetic Theory of an Ideal Gas & Gas Laws	Ideal Gas Equation			1	A								
Speeds of Gas, Pressure & Kinetic Energy	Pressure, Kinetic Energy					1	E						
	RMS Speed Formula	1	A					1	D			1	A
Degree of Freedom, Specific Heat Capacity & Mean Free Path	Law of Equipartition of Energy, Degree of Freedom									1	A		
	Mean free path formula			1	E								

LOD - Level of Difficulty   E - Easy   A - Average   D - Difficult   Qns - No. of Questions

Topic 1 Kinetic Theory of an Ideal Gas & Gas Laws

1. A cylinder contains hydrogen gas at pressure of 249 kPa and temperature 27°C. *[2020]*
Its density is : ($R = 8.3\ J\ mol^{-1}\ K^{-1}$)
(a) $0.2\ kg/m^3$ (b) $0.1\ kg/m^3$
(c) $0.02\ kg/m^3$ (d) $0.5\ kg/m^3$

2. Two vessels separately contain two ideal gases A and B at the same temperature. The pressure of A being twice that of B. Under such conditions, the density of A is found to be 1.5 times the density of B. The ratio of molecular weight of A and B is : *[2015 RS]*
(a) $\frac{3}{4}$ (b) 2
(c) $\frac{1}{2}$ (d) $\frac{2}{3}$

3. In the given (V – T) diagram, what is the relation between pressure P_1 and P_2 ? *[2013]*
(a) $P_2 > P_1$
(b) $P_2 < P_1$
(c) Cannot be predicted
(d) $P_2 = P_1$

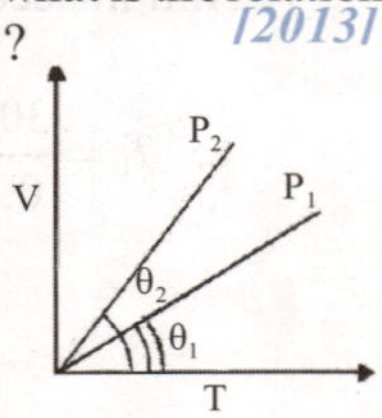

4. At 10° C the value of the density of a fixed mass of an ideal gas divided by its pressure is x. At 110°C this ratio is: *[2008]*
(a) x (b) $\frac{383}{283}x$
(c) $\frac{10}{110}x$ (d) $\frac{283}{383}x$

5. When volume of system is increased twice and temperature is decreased half of its initial temperature, then pressure becomes
(a) 2 times (b) 4 times
(c) $\frac{1}{2}$ times (d) $\frac{1}{4}$ times

6. The equation of state for 5 g of oxygen at a pressure P and temperature T, when occupying a volume V, will be *[2004]*
(a) PV = (5/16) RT
(b) PV = (5/32) RT
(c) PV = 5 RT
(d) PV = (5/2) RT
(where R is the gas constant).

7. A gas at 27°C temperature and 30 atmospheric pressure is allowed to expand to the atmospheric pressure. If the volume becomes 10 times its initial volume, then the final temperature becomes
(a) 100°C (b) 173°C [2001]
(c) 273°C (d) –173°C

8. The equation of state, corresponding to 8g of O_2 is [1994]
(a) $PV = 8RT$ (b) $PV = RT/4$
(c) $PV = RT$ (d) $PV = RT/2$

9. According to kinetic theory of gases, at absolute zero temperature [1990]
(a) water freezes
(b) liquid helium freezes
(c) molecular motion stops
(d) liquid hydrogen freezes

10. At constant volume, temperature is increased then [1989]
(a) collision on walls will be less
(b) number of collisions per unit time will increase
(c) collisions will be in straight lines
(d) collisions will not change.

Topic 2 Speeds of Gas, Pressure & Kinetic Energy

11. Match Column - I and Column - II and choose the correct match from the given choices. [2021]

Column - I	Column - II
(A) Root mean square speed of gas molecules	(i) $\frac{1}{3}nm\bar{v}^2$
(B) Pressure exerted by ideal gas	(ii) $\sqrt{\frac{3RT}{M}}$
(C) Average kinetic energy of a molecule	(iii) $\frac{5}{2}RT$
(D) Total internal energy of 1 mole of a diatomic gas	(iv) $\frac{3}{2}k_BT$

(a) A - (iii), B - (ii), C - (i), D - (iv)
(b) A - (iii), B - (i), C - (iv), D - (ii)
(c) A - (ii), B - (iii), C - (iv), D - (i)
(d) A - (ii), B - (i), C - (iv), D - (iii)

12. Increase in temperature of a gas filled in a container would lead to: [2019]
(a) increase in its mass
(b) increase in its kinetic energy
(c) decrease in its pressure
(d) decrease in intermolecular distance

13. At what temperature will the rms speed of oxygen molecules become just sufficient for escaping from the Earth's atmosphere?
(Given :
Mass of oxygen molecule (m) = 2.76×10^{-26} kg
Boltzmann's constant $k_B = 1.38 \times 10^{-23}$ JK^{-1}) [2018]
(a) 2.508×10^4 K (b) 8.360×10^4 K
(c) 1.254×10^4 K (d) 5.016×10^4 K

14. The molecules of a given mass of a gas have r.m.s. velocity of 200 ms^{-1} at 27°C and 1.0×10^5 Nm^{-2} pressure. When the temperature and pressure of the gas are respectively, 127°C and 0.05×10^5 Nm^{-2}, the r.m.s. velocity of its molecules in ms^{-1} is : [2016]
(a) $100\sqrt{2}$ (b) $\frac{400}{\sqrt{3}}$
(c) $\frac{100\sqrt{2}}{3}$ (d) $\frac{100}{3}$

15. In a vessel, the gas is at a pressure P. If the mass of all the molecules is halved and their speed is doubled, then the resultant pressure will be [NEET Kar. 2013]
(a) $4P$ (b) $2P$ (c) P (d) $P/2$

16. At 0 K, which of the following properties of a gas will be zero? [1996]
(a) Kinetic energy (b) Potential energy
(c) Density (d) Mass

17. The temperature of a gas is raised from 27°C to 927°C. The root mean square speed is [1994]
(a) $\sqrt{(927/27)}$ times the earlier value
(b) remain the same
(c) gets halved
(d) get doubled

18. If C_s be the velocity of sound in air and C be the r.m.s velocity, then [1994]
(a) $C_s < C$ (b) $C_s = C$
(c) $C_s = C(\gamma/3)^{1/2}$ (d) none of these

19. Relation between pressure (P) and energy (E) of a gas is [1991]
(a) $P = \frac{2}{3}E$ (b) $P = \frac{1}{3}E$
(c) $P = \frac{1}{2}E$ (d) $P = 3E$

20. Three containers of the same volume contain three different gases. The masses of the molecules are m_1, m_2 and m_3 and the number of molecules in their respective containers are N_1, N_2 and N_3. The gas pressure in the containers are P_1, P_2 and P_3 respectively. All the gases are now mixed and put in one of these containers. The pressure P of the mixture will be *[1991]*

(a) $P<(P_1+P_2+P_3)$

(b) $P=\frac{P_1+P_2+P_3}{3}$

(c) $P=P_1+P_2+P_3$

(d) $P>(P_1+P_2+P_3)$

21. N molecules each of mass m of a gas A and 2N molecules each of mass 2m of gas B are contained in the same vessel which is maintained at temperature T. The mean square velocity of molecules of B type is v^2 and the mean square rectangular component of the velocity of A type is denoted by ω^2. Then ω^2/v^2 *[1991]*

(a) 2 (b) 1

(c) 1/3 (d) 2/3

22. Two containers A and B are partly filled with water and closed. The volume of A is twice that of B and it contains half the amount of water in B. If both are at the same temperature, the water vapour in the containers will have pressure in the ratio of *[1988]*

(a) 1 : 2 (b) 1 : 1

(c) 2 : 1 (d) 4 : 1

Topic 3 Degree of Freedom, Specific Heat Capacity & Mean Free Path

23. The mean free path for a gas, with molecular diameter d and number density n can be expressed as : *[2020]*

(a) $\frac{1}{\sqrt{2}n\pi d^2}$ (b) $\frac{1}{\sqrt{2}n^2\pi d^2}$

(c) $\frac{1}{\sqrt{2}n^2\pi^2 d^2}$ (d) $\frac{1}{\sqrt{2}n\pi d}$

24. The value of $\gamma=\left(\frac{C_p}{C_v}\right)$, for hydrogen, helium and another ideal diatomic gas X (whose molecules are not rigid but have an additional vibrational mode), are respectively equal to, *[NEET Odisha 2019]*

(a) $\frac{7}{5},\frac{5}{3},\frac{7}{5}$ (b) $\frac{7}{5},\frac{5}{3},\frac{9}{7}$

(c) $\frac{5}{3},\frac{7}{5},\frac{9}{7}$ (d) $\frac{5}{3},\frac{7}{5},\frac{7}{5}$

25. A gas mixture consists of 2 moles of O_2 and 4 moles of Ar at temperature T. Neglecting all vibrational modes, the total internal energy of the system is :- *[2017]*

(a) 15 RT (b) 9 RT

(c) 11 RT (d) 4 RT

26. One mole of an ideal diatomic gas undergoes a transition from A to B along a path AB as shown in the figure.

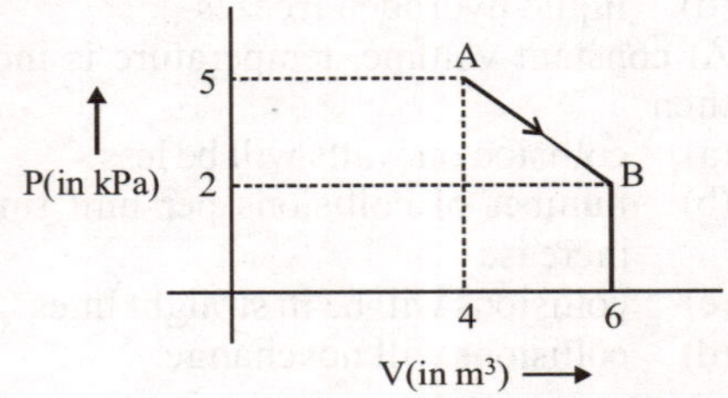

The change in internal energy of the gas during the transition is: *[2015]*

(a) – 20 kJ (b) 20 J

(c) –12 kJ (d) 20 kJ

27. The ratio of the specific heats $\frac{C_p}{C_v}=\gamma$ in terms of degrees of freedom (n) is given by *[2015]*

(a) $\left(1+\frac{n}{3}\right)$ (b) $\left(1+\frac{2}{n}\right)$

(c) $\left(1+\frac{n}{2}\right)$ (d) $\left(1+\frac{1}{n}\right)$

28. 4.0 g of a gas occupies 22.4 litres at NTP. The specific heat capacity of the gas at constant volume is $5.0 JK^{-1}$. If the speed of sound in this gas at NTP is $952 ms^{-1}$, then the heat capacity at constant pressure is (Take gas constant R = $8.3 JK^{-1} mol^{-1}$) *[2015 RS]*

(a) $7.5 JK^{-1} mol^{-1}$ (b) $7.0 JK^{-1} mol^{-1}$

(c) $8.5 JK^{-1} mol^{-1}$ (d) $8.0 JK^{-1} mol^{-1}$

29. The mean free path of molecules of a gas, (radius 'r') is inversely proportional to : *[2014]*

(a) r^3 (b) r^2

(c) r (d) $\sqrt{r}$

30. The amount of heat energy required to raise the temperature of 1 g of Helium at NTP, from T_1K to T_2K is [2013]

(a) $\frac{3}{2}N_a k_B(T_2 - T_1)$ (b) $\frac{3}{4}N_a k_B(T_2 - T_1)$

(c) $\frac{3}{4}N_a k_B \frac{T_2}{T_1}$ (d) $\frac{3}{8}N_a k_B(T_2 - T_1)$

31. The molar specific heats of an ideal gas at constant pressure and volume are denoted by C_p and C_v, respectively. If $\gamma = \frac{C_p}{C_v}$ and R is the universal gas constant, then C_v is equal to [2013]

(a) $\frac{R}{(\gamma - 1)}$ (b) $\frac{(\gamma - 1)}{R}$

(c) γR (d) $\frac{1+\gamma}{1-\gamma}$

32. The molar specific heat at constant pressure of an ideal gas is (7/2) R. The ratio of specific heat at constant pressure to that at constant volume is [2006]

(a) 8/7 (b) 5/7

(c) 9/7 (d) 7/5

33. If γ be the ratio of specific heats of a perfect gas, the number of degrees of freedom of a molecule of the gas is [2000]

(a) $\frac{25}{2}(\gamma - 1)$ (b) $\frac{3\gamma - 1}{2\gamma - 1}$

(c) $\frac{2}{\gamma - 1}$ (d) $\frac{9}{2}(\gamma - 1)$

34. The degree of freedom of a molecule of a triatomic gas is [1999]

(a) 2 (b) 4

(c) 6 (d) 8

35. The number of translational degrees of freedom for a diatomic gas is [1993]

(a) 2 (b) 3

(c) 5 (d) 6

36. If for a gas, $\frac{R}{C_v} = 0.67$, the gas is made up of molecules which are [1992]

(a) diatomic

(b) mixture of diatomic and polyatomic molecules

(c) monoatomic

(d) polyatomic

37. For hydrogen gas, $C_p - C_v = a$ and for oxygen gas, $C_p - C_v = b$, so the relation between a and b is given by [1991]

(a) $a = 16\,b$ (b) $16\,b = a$

(c) $a = 4\,b$ (d) $a = b$

38. For a certain gas the ratio of specific heats is given to be $\gamma = 1.5$. For this gas [1990]

(a) $C_V = 3R/J$ (b) $C_P = 3R/J$

(c) $C_P = 5R/J$ (d) $C_V = 5R/J$

39. One mole of an ideal monoatomic gas requires 207 J heat to raise the temperature by 10 K when heated at constant pressure. If the same gas is heated at constant volume to raise the temperature by the same 10 K, the heat required is

[Given the gas constant $R = 8.3$ J/ mol. K] [1990]

(a) 198.7 J (b) 29 J

(c) 215.3 J (d) 124 J

40. A polyatomic gas with n degrees of freedom has a mean energy per molecule given by [1989]

(a) $\frac{nkT}{N}$ (b) $\frac{nkT}{2N}$

(c) $\frac{nkT}{2}$ (d) $\frac{3kT}{2}$

ANSWER KEY

1	(a)	6	(b)	11	(d)	16	(a)	21	(d)	26	(a)	31	(a)	36	(c)		
2	(a)	7	(d)	12	(b)	17	(d)	22	(b)	27	(b)	32	(d)	37	(d)		
3	(b)	8	(b)	13	(b)	18	(c)	23	(a)	28	(d)	33	(c)	38	(b)		
4	(d)	9	(c)	14	(b)	19	(a)	24	(b)	29	(b)	34	(c)	39	(d)		
5	(d)	10	(b)	15	(b)	20	(c)	25	(c)	30	(d)	35	(b)	40	(c)		

Hints & Solutions

1. (a) Given : Pressure $P = 249\ k\ Pa$
$= 249 \times 10^3\ N/m^2$
Mass of hydrogen, $M = 2g = 2 \times 10^{-3}$ kg
Temperature, $T = 27 + 273 = 300$ K
Using, ideal gas equation, $PV = nRT$

$$PM = \rho RT \Rightarrow \rho = \frac{PM}{RT} \quad \left(\because \rho = \frac{M}{V}\right)$$

$$\therefore \rho = \frac{(249 \times 10^3)(2 \times 10^{-3})}{8.3 \times 300} = 0.2\ kg/m^3.$$

2. (a) From $PV = nRT$

$$P_A = \frac{\rho_A M_A}{RT} \text{ and } P_B = \frac{\rho_B M_B}{RT}$$

From question,

$$\frac{P_A}{P_B} = \frac{\rho_A}{\rho_B}\frac{M_A}{M_B} = 2\frac{M_A}{M_B} = \frac{3}{2}$$

So, $\dfrac{M_A}{M_B} = \dfrac{3}{4}$

3. (b) $P_1 > P_2$

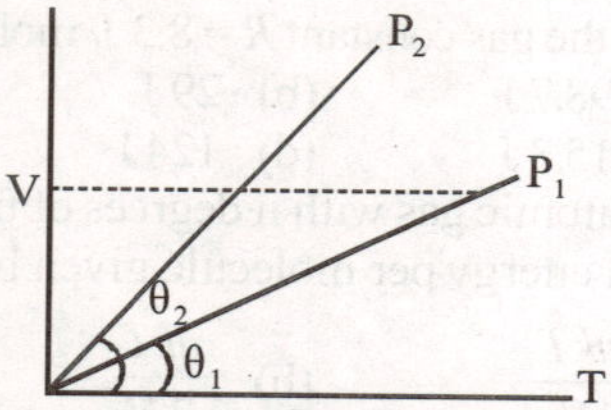

As V = constant $\Rightarrow P \propto T$
Hence from V–T graph $P_1 > P_2$

NOTES

From the Gay - Lussac's law, volume remain constant, the pressure of a given mass of a gas is directly proportional to its absolute temperature.

$\therefore P \propto T$

$$\frac{P}{T} = \text{constant} \Rightarrow \frac{P_1}{T_1} = \frac{P_2}{T_2}$$

4. (d) Let the mass of the gas be m.
At a fixed temperature and pressure, volume is fixed.

Density of the gas, $r = \dfrac{m}{V}$

Now $\dfrac{r}{P} = \dfrac{m}{PV} = \dfrac{m}{nRT}$

$\Rightarrow \dfrac{m}{nRT} = x$ (By question)

$\Rightarrow xT = \text{constant} \Rightarrow x_1T_1 = x_2T_2$

$$\Rightarrow x_2 \Rightarrow \frac{x_1T_1}{T_2} = \frac{283}{383}x \quad \left[\begin{array}{l}\text{Given} \\ T_1 = 283K \\ T_2 = 383K\end{array}\right]$$

5. (d)

6. (b) $n = \dfrac{5}{32}$ $\therefore PV = \dfrac{5}{32}RT$ [$\because PV = nRT$]

7. (d) Given : Initial temperature of gas
$(T_1) = 27°C = 300$ K
Initial pressure $(P_1) = 30$ atm
Initial volume $(V_1) = V$
Final pressure $(P_2) = 1$ atm
Final volume $(V_2) = 10\ V$.
We know from the general gas equation that

$$\frac{P_1V_1}{T_1} = \frac{P_2V_2}{T_2} \text{ or, } \frac{30 \times V}{300} = \frac{1 \times 10V}{T_2}$$

or, $T_2 = 100\ K = -173°C$.

8. (b) 8g of oxygen is equivalent to (1/4) mole.

So, $PV = \dfrac{1}{4}RT$

9. (c)

10. (b) As the temperature increases, the average velocity increases. So, the collisions are faster.

11. (d) Root mean square RMS speed of gas molecule

$$V_{RMS} = \sqrt{\frac{3\ RT}{M}}$$

Pressure exerted by ideal gas $P = \dfrac{1}{3}nm\overline{v}^2$

Average kinetic energy of a molecule $= \dfrac{3}{2}k_BT$

Total internal energy of a gas is $(U) = \dfrac{1}{2}nfRT$

Here, for one mole n = 1 and
f = 5 for diatomic

$\therefore U = \frac{5}{2}RT$

12. (b) $U = \frac{F}{2}nRT$

As $U \propto T$

$\therefore$ Increase in temperature would lead to the increase in kinetic energy of gas.

13. (b) Let at temperature T rms speed of oxygen molecules become just sufficient for escaping from the Earth's atmosphere

$V_{escape} = 11200$ m/s

Also, $V_{rms} = V_{escape} = \sqrt{\frac{3k_BT}{m_{O_2}}} = 11200$ m/s

Putting value of K_B and m_{O_2} we get,

$T = 8.360 \times 10^4$ K

14. (b) Here $v_1 = 200$ m/s;
temperature $T_1 = 27°C = 27 + 273 = 300$ k
temperature $T_2 = 127°C = 127 + 273 = 400$ k, $v_2 = ?$

R.M.S. Velocity, $v \propto \sqrt{T}$

$\Rightarrow \frac{v_2}{200} = \sqrt{\frac{400}{300}}$

$\Rightarrow v_2 = \frac{200 \times 2}{\sqrt{3}}$ m/s $\Rightarrow v_2 = \frac{400}{\sqrt{3}}$ m/s

15. (b) $\because P = \frac{1}{3}\frac{mN}{V}v_{rms}^2$

where m = mass of each molecules
N = total number of molecules
V = volume of the gas

When mass is halved and speed is doubled then

Resultant pressure, $P' = \frac{1}{3} \times \left(\frac{m}{2}\right) \times \frac{N}{V}(2v_{rms})^2$
$= 2P$

16. (a) At 0 K, molecular motion stops. Hence, kinetic energy of molecules becomes zero.

17. (d) $v_{rms} \propto \sqrt{T}$
As temperature increases from 300 K to 1200 K that is four times, so, v_{rms} will be doubled.

18. (c) Velocity of sound $(C_S) = \sqrt{\frac{\gamma P}{\rho}}$

R.M.S. velocity of gas molecules (C) = $\sqrt{\frac{3P}{\rho}}$

$\frac{C_s}{C} = \sqrt{\frac{\gamma P}{\rho} \times \frac{\rho}{3P}} = \sqrt{\frac{\gamma}{3}} \quad \Rightarrow C_s = C \times \sqrt{\frac{\gamma}{3}}$

19. (a) $P = \frac{1}{3}Nmc^2 = \frac{2}{3} \times \left(\frac{1}{2}Nm\right)c^2 = \frac{2}{3}K.E$

20. (c) According to Dalton's law of partial pressures, we have $P = P_1 + P_2 + P_3$

> **NOTES**
> Dalton's law will be valid only if the gases contained are non reacting.
> For n non-reacting gases, $P = P_1 + P_2 + P_3 + ... P_n$
> Here P = Pressure exerted by mixture
> $P_1, P_2, P_3 ..... P_n$ = Partial pressure of component gas.

21. (d) The mean square velocity of A type molecules $= \omega^2 + \omega^2 + \omega^2 = 3\omega^2$

Therefore, $\frac{1}{2}m(3\omega^2) = \frac{1}{2}(2m)v^2$

This gives $\omega^2 / v^2 = 2/3$

> **NOTES**
> Mean kinetic energy of the two types of molecules should be equal.

22. (b) Vapour pressure does not depend on the amount of substance. It depends on the temperature alone.

23. (a) Mean free path for a gas $\lambda_m = \frac{1}{\sqrt{2}n\pi d^2}$

Here, d = diameter of a gas molecule and, n = molecular density.

24. (b) For Hydrogen f = 5
For Helium f = 3
For gas X
f = 7.

$\gamma = 1 + \frac{2}{f} = 1 + \frac{2}{7} = \frac{9}{7}$

So, $\left(\frac{7}{5}, \frac{5}{3}, \frac{9}{7}\right)$

25. (c) Internal energy of the system is given by

$U = \frac{f}{2}nRT$

Degree of freedom

$F_{diatomic} = 5$

$f_{monoatomic} = 3$
and, number of moles
$n(O_2) = 2$
$n(Ar) = 4$

$$U_{total} = \frac{5}{2}(2)RT + \frac{3}{2}(4)RT = 11RT$$

26. **(a)** Change in internal energy from $A \to B$

$$\Delta U = \frac{f}{2} nR\Delta T = \frac{f}{2} nR(T_f - T_i)$$

$$= \frac{5}{2}\{P_f V_f - P_i V_i\}$$

$$= \frac{5}{2}\{2 \times 10^3 \times 6 - 5 \times 10^3 \times 4\}$$

$$= \frac{5}{2}\{12 - 20\} \times 10^3 \text{ J} = 5 \times (-4) \times 10^3 \text{ J}$$

$\Delta U = -20$ KJ

NOTES (As gas is diatomic $\therefore f = 5$)

27. **(b)** Let 'n' be the degree of freedom

$$\gamma = \frac{C_p}{C_v} = \frac{\left(\frac{n}{2}+1\right)R}{\left(\frac{n}{2}\right)R} = \left(1 + \frac{2}{n}\right)$$

28. **(d)** Molar mass of the gas = 4g/mol
Speed of sound

$$v = \sqrt{\frac{\gamma RT}{m}} \Rightarrow 952 = \sqrt{\frac{\gamma \times 3.3 \times 273}{4 \times 10^{-3}}}$$

$$\Rightarrow \quad \gamma = 1.6 = \frac{16}{10} = \frac{8}{5}$$

Also, $\gamma = \frac{C_P}{C_V} = \frac{8}{5}$

So, $C_P = \frac{8 \times 5}{5} = 8 \text{JK}^{-1}\text{mol}^{-1}$

[$C_V = 5.0 \text{ JK}^{-1}$ given]

29. **(b)** Mean free path $\lambda_m = \frac{1}{\sqrt{2}\pi d^2 n}$

where d = diameter of molecule and d = 2r

$$\therefore \quad \lambda_m \propto \frac{1}{r^2}$$

30. **(d)** From first law of thermodynamics,

$$\Delta Q = \Delta U + \Delta W = \frac{3}{2} \cdot \frac{1}{4} R(T_2 - T_1) + 0$$

$$= \frac{3}{8} N_a K_B (T_2 - T_1) \; [\because K = \frac{R}{N}]$$

31. **(a)** $C_p - C_v = R \Rightarrow C_p = C_v + R$

$$\because \gamma = \frac{C_p}{C_v} = \frac{C_v + R}{C_v} = \frac{C_v}{C_v} + \frac{R}{C_v}$$

$$\Rightarrow \gamma = 1 + \frac{R}{C_v} \Rightarrow \frac{R}{C_v} = \gamma - 1$$

$$\Rightarrow C_v = \frac{R}{\gamma - 1}$$

32. **(d)** $C_P = \frac{7}{2}R$; $C_V = C_P - R = \frac{7}{2}R - R = \frac{5}{2}R$

$$\frac{C_P}{C_V} = \frac{7/2\,R}{5/2\,R} = \frac{7}{5}$$

33. **(c)** We know that ratio of specific heats,

$$\gamma = 1 + \frac{2}{n} \; or \; n = \frac{2}{\gamma - 1}$$

[where n = Degree of freedom]

34. **(c)** No. of degree of freedom = $3K - N$
where K is no. of atom and N is the number of relations between atoms. For triatomic gas,

$K = 3, N = {}^3C_2 = 3$

No. of degree of freedom = 3 (3) – 3 = 6

35. **(b)** Number of translational degrees of freedom are same for all types of gases.

36. **(c)** Since $\frac{R}{C_V} = 0.67 \Rightarrow C_V = \frac{3}{2}R$, hence gas is monoatomic.

NOTES For a gas having f degree of freedom specific heat at constant volume, $C_v = \frac{1}{2}fR$ specific heat at constant pressure,

$$C_p = \left(\frac{f}{2} + 1\right)R$$

where, R = universal gas constant

37. **(d)** Both are diatomic gases and $C_p - C_v = R$ for all gases.

38. (b) $\gamma = \frac{C_P}{C_V} = \frac{15}{10} = \frac{3}{2} \Rightarrow C_V = \frac{2}{3}C_P$

$C_P - C_V = \frac{R}{J} \Rightarrow C_P - \frac{2}{3}C_P = \frac{R}{J}$

$\Rightarrow \frac{C_P}{3} = \frac{R}{J} \Rightarrow C_P = \frac{3R}{J}$

39. (d) $C_p = \frac{5}{2}R$ and $C_v = \frac{3}{2}R$

We know that $Q_v = nC_v\Delta T$ and $Q_p = nC_p\Delta T$

$\Rightarrow \frac{Q_v}{Q_p} = \frac{3}{5}$.

Given $Q_p = 207\text{ J} \Rightarrow Q_v \cong 124\text{ J}$

40. (c) According to law of equipartition of energy, the energy per degree of freedom is $\frac{1}{2}kT$.

For a polyatomic gas with n degrees of freedom, the mean energy per molecule $= \frac{1}{2}nkT$

13 Oscillations

Trend Analysis with Important Topics & Sub-Topics

Topic Name	Sub-Topic	2021 QNS.	2021 LOD	2020 QNS.	2020 LOD	2019 QNS.	2019 LOD	2018 QNS.	2018 LOD	2017 QNS.	2017 LOD	2016 QNS.	2016 LOD
Displacement, Phase, Velocity & Acceleration of SHM	Phase difference between displacement and acceleration, y projection of radius vector			1	A	1	E						
Energy in Simple Harmonic Motion	Average velocity, amplitude in one vibration					2	A						
Time Period, Frequency, Simple Pendulum & Spring Pendulum	Time period, $T = 2\pi/\omega$ Potential energy frequency	2	A					1	E	1	E		
	Spring constant, spring connected in series and parallel									1	D		
Damped SHM, Forced Ocillations & Resonance													

LOD - Level of Difficulty | E - Easy | A - Average | D - Difficult | Qns - No. of Questions

Topic 1 Displacement, Phase, Velocity & Acceleration of SHM

1. The phase difference between displacement and acceleration of a particle in a simple harmonic motion is : *[2020]*

   (a) $\frac{3\pi}{2}$ rad (b) $\frac{\pi}{2}$ rad

   (c) zero (d) π rad

2. The radius of circle, the period of revolution, initial position and sense of revolution are indicated in the fig. *[2019]*

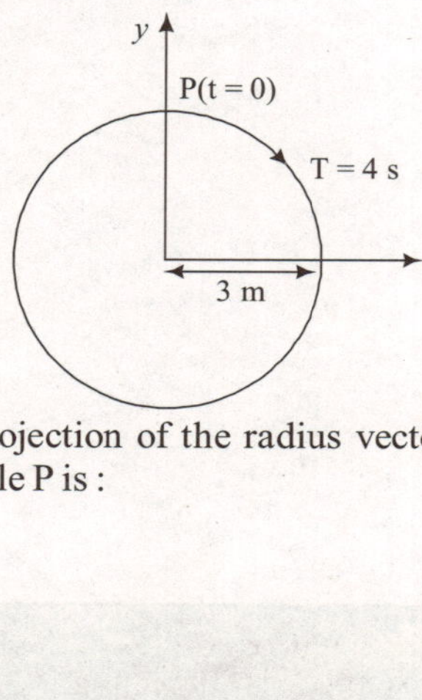

   y - projection of the radius vector of rotating particle P is :

   (a) $y(t) = -3\cos 2\pi t$, where y in m

   (b) $y(t) = 4\sin\left(\frac{\pi t}{2}\right)$, where y in m

   (c) $y(t) = 3\left(\frac{3\pi t}{2}\right)$, cos where y in m

   (d) $y(t) = 3\cos\left(\frac{\pi t}{2}\right)$, where y in m

3. Average velocity of a particle executing SHM in one complete vibration is: *[2019]*

   (a) $\frac{A\omega}{2}$ (b) $A\omega$

   (c) $\frac{A\omega^2}{2}$ (d) zero

4. The displacement of a particle executing simple harmonic motion is given by
   $y = A_0 + A\sin\omega t + B\cos\omega t$. *[2019]*
   Then the amplitude of its oscillation is given by:

   (a) $A_0 + \sqrt{A^2 + B^2}$ (b) $\sqrt{A^2 + B^2}$

   (c) $\sqrt{A_0^2 + (A + B)^2}$ (d) A+B

5. The distance covered by a particle undergoing SHM in one time period is (amplitude = A), *[NEET Odisha 2019]*

(a) 4A (b) Zero
(c) A (d) 2A

6. When two displacements represented by $y_1 = a\sin(\omega t)$ and $y_2 = b\cos(\omega t)$ are superimposed the motion is: *[2015]*

(a) simple harmonic with amplitude $\frac{a}{b}$
(b) simple harmonic with amplitude $\sqrt{a^2 + b^2}$
(c) simple harmonic with amplitude $\frac{(a+b)}{2}$
(d) not a simple harmonic

7. The oscillation of a body on a smooth horizontal surface is represented by the equation,

$X = A\cos(\omega t)$

where X = displacement at time t

ω = frequency of oscillation

Which one of the following graphs shows correctly the variation of 'a' with 't'? *[2014]*

(a) 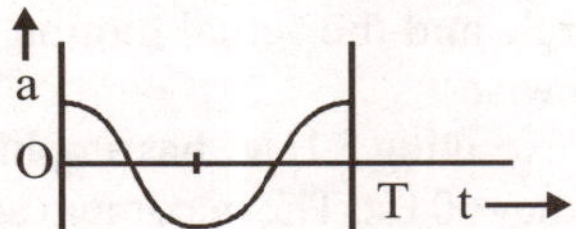

(b) 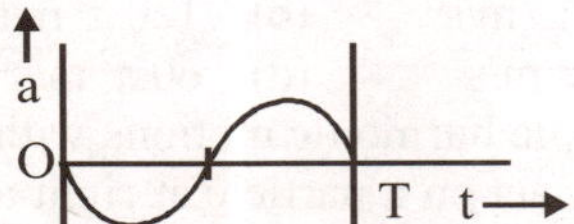

(c) 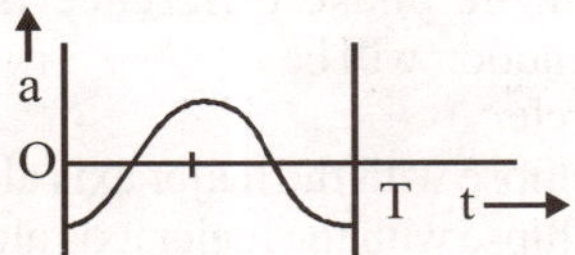

(d) 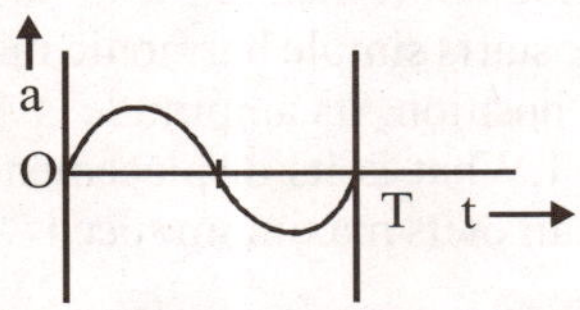

8. A particle of mass *m* oscillates along *x*-axis according to equation $x = a\sin\omega t$. The nature of the graph between momentum and displacement of the particle is *[NEET Kar. 2013]*

(a) straight line passing through origin
(b) circle
(c) hyperbola
(d) ellipse

9. The equation of a simple harmonic wave is given by

$$y = 3\sin\frac{\pi}{2}(50t - x)$$

Where *x* and *y* are in meters and *t* is in seconds. The ratio of maximum particle velocity to the wave velocity is *[2012M]*

(a) 2π (b) $\frac{3}{2}\pi$
(c) 3π (d) $\frac{2}{3}\pi$

10. A particle of mass m is released from rest and follows a parabolic path as shown. Assuming that the displacement of the mass from the origin is small, which graph correctly depicts the position of the particle as a function of time *[2011]*

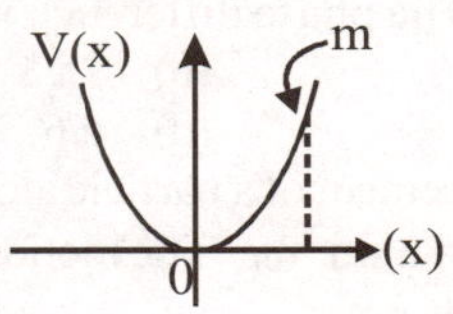

(a) 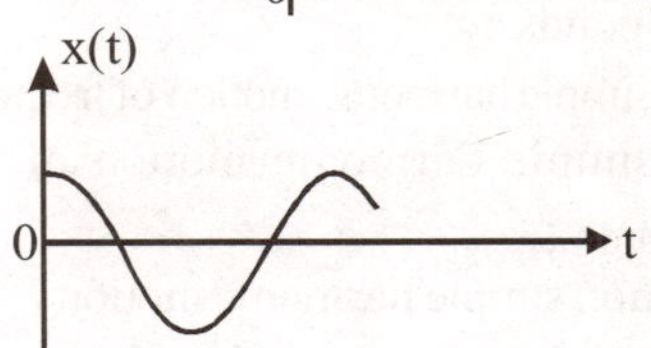

(b) 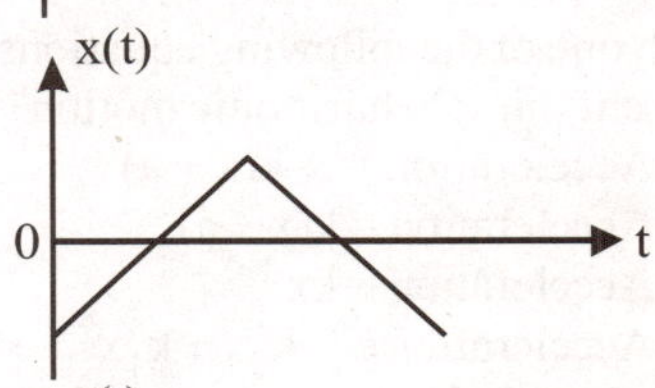

(c) 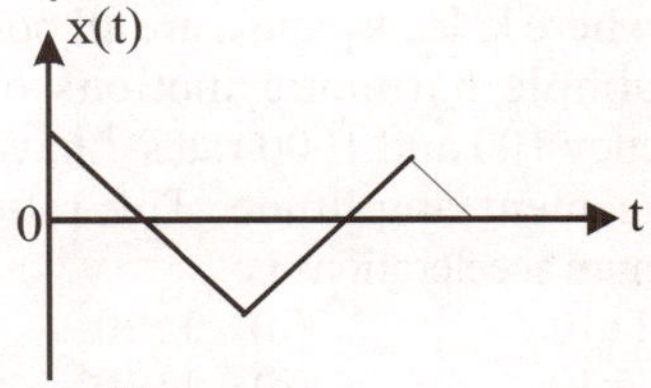

(d) 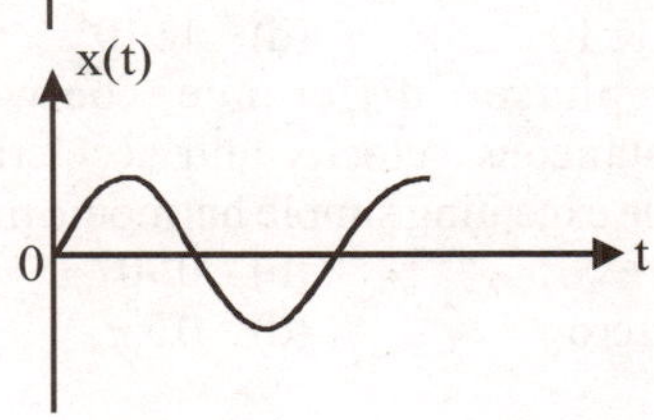

11. Out of the following functions, representing motion of a particle, which represents SHM? [2011]

(A) $y = \sin \omega t - \cos \omega t$

(B) $y = \sin^3 \omega t$

(C) $y = 5\cos\left(\frac{3\pi}{4} - 3\omega t\right)$

(D) $y = 1 + \omega t + \omega^2 t^2$

(a) Only (A)
(b) Only (D) does not represent SHM
(c) Only (A) and (C)
(d) Only (A) and (B)

12. Two particles are oscillating along two close parallel straight lines side by side, with the same frequency and amplitudes. They pass each other, moving in opposite directions when their displacement is half of the amplitude. The mean positions of the two particles lie on a straight line perpendicular to the paths of the two particles. The phase difference is [2011M]

(a) 0 (b) $2\pi/3$
(c) π (d) $\pi/6$

13. The displacement of a particle along the x-axis is given by $x = a \sin^2 \omega t$. The motion of the particle corresponds to: [2010]

(a) simple harmonic motion of frequency ω / π
(b) simple harmonic motion of frequency $3\omega / 2\pi$
(c) non simple harmonic motion
(d) simple harmonic motion of frequency $\omega / 2\pi$

14. Which one of the following equations of motion represents simple harmonic motion? [2009]

(a) Acceleration $= -k(x + a)$
(b) Acceleration $= k(x + a)$
(c) Acceleration $= kx$
(d) Acceleration $= -k_0 x + k_1 x^2$
where k, k_0, k_1 and a are all postive.

15. Two simple harmonic motions of angular frequency 100 and 1000 rad s^{-1} have the same displacement amplitude. The ratio of their maximum accelerations is: [2008]

(a) $1:10$ (b) $1:10^2$
(c) $1:10^3$ (d) $1:10^4$

16. The phase difference between the instantaneous velocity and acceleration of a particle executing simple harmonic motion is [2007]

(a) π (b) $0.707\,\pi$
(c) zero (d) $0.5\,\pi$

17. A particle executing simple harmonic motion of amplitude 5 cm has maximum speed of 31.4 cm/s. Its oscillation per second [2005]

(a) 4 (b) 3
(c) 2 (d) 1

18. Which one of the following statements is true for the speed v and the acceleration a of a particle executing simple harmonic motion ? [2004]

(a) When *v* is maximum, *a* is zero
(b) When *v* is maximum, *a* is maximum
(c) Value of a is zero, whatever may be the value of *v*
(d) When *v* is zero, a is zero

19. Two simple harmonic motions act on a particle. These harmonic motions are $x = A\cos(\omega t + \delta)$, $y = A\cos(\omega t + \alpha)$ when $\delta = \alpha + \frac{\pi}{2}$, the resulting motion is [2000]

(a) a circle and the actual motion is clockwise
(b) an ellipse and the actual motion is counterclockwise
(c) an elllipse and the actual motion is clockwise
(d) a circle and the actual motion is counter clockwise

20. A particle executing S.H.M. has amplitude 0.01m and frequency 60 Hz. The maximum acceleration of the particle is [1999]

(a) $144\,\pi^2\,m/s^2$ (b) $120\,\pi^2\,m/s^2$
(c) $80\,\pi^2\,m/s^2$ (d) $60\,\pi^2\,m/s^2$

21. Two simple harmonic motions with the same frequency act on a particle at right angles i.e., along x and y axis. If the two amplitudes are equal and the phase difference is $\pi/2$, the resultant motion will be [1997]

(a) a circle
(b) an ellipse with the major axis along y-axis
(c) an ellipse with the major axis along x-axis
(d) a straight line inclined at 45° to the x-axis

22. A particle starts simple harmonic motion from the mean position. Its amplitude is A and time period is T. What is its displacement when its speed is half of its maximum speed [1996]

(a) $\frac{\sqrt{2}}{3}A$ (b) $\frac{\sqrt{3}}{2}A$

(c) $\frac{2}{\sqrt{3}}A$ (d) $\frac{A}{\sqrt{2}}$

23. Which one of the following is a simple harmonic motion? *[1994]*
(a) Ball bouncing between two rigid vertical walls
(b) Particle moving in a circle with uniform speed
(c) Wave moving through a string fixed at both ends
(d) Earth spinning about its own axis.

24. A particle is subjected to two mutually perpendicular simple harmonic motions such that its x and y coordinates are given by

$$x = 2\sin\omega t; \quad y = 2\sin\left(\omega t + \frac{\pi}{4}\right)$$

The path of the particle will be *[1994]*
(a) a straight line (b) a circle
(c) an ellipse (d) a parabola

25. The composition of two simple harmonic motions of equal periods at right angle to each other and with a phase difference of π results in the displacement of the particle along *[1990]*
(a) circle (b) figures of eight
(c) straight line (d) ellipse

26. A particle moving along the X-axis, executes simple harmonic motion then the force acting on it is given by *[1988]*
(a) $-A\,kx$ (b) $A\cos(kx)$
(c) $A\exp(-kx)$ (d) Akx
where, A and k are positive constants.

Topic 2 Energy in Simple Harmonic Motion

27. The particle executing simple harmonic motion has a kinetic energy $K_0\cos^2\omega t$. The maximum values of the potential energy and the total energy are respectively *[2007]*
(a) $K_0/2$ and K_0 (b) K_0 and $2K_0$
(c) K_0 and K_0 (d) 0 and $2K_0$.

28. The potential energy of a long spring when stretched by 2 cm is U. If the spring is stretched by 8 cm, the potential energy stored in it is *[2006]*
(a) 8 U (b) 16U
(c) U/4 (d) 4 U

29. A particle of mass m oscillates with simple harmonic motion between points x_1 and x_2, the equilibrium position being O. Its potential energy is plotted. It will be as given below in the graph. *[2003]*

(a) 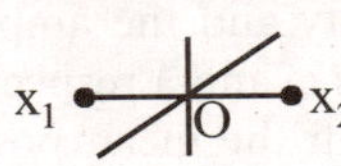

(b) 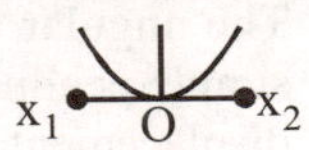

(c) 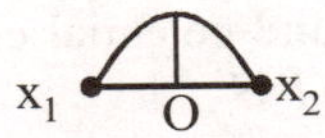

(d) 

30. The potential energy of a simple harmonic oscillator when the particle is half way to its end point is *[2003]*
(a) $\frac{1}{2}E$ (b) $\frac{2}{3}E$
(c) $\frac{1}{8}E$ (d) $\frac{1}{4}E$
where E is the total energy

31. A particle is executing a simple harmonic motion of amplitude 'a'. Its potential energy is maximum when the displacement from the position of the maximum kinetic energy is *[2002]*
(a) 0 (b) $\pm a$
(c) $\pm a/2$ (d) $-a/2$

32. There is a body having mass m and performing S.H.M. with amplitude a. There is a restoring force $F = -kx$. The total energy of body depends upon *[2001]*
(a) k, x (b) k, a
(c) k, a, x (d) k, a, v

33. In a simple harmonic motion, when the displacement is one-half the amplitude, what fraction of the total energy is kinetic? *[1996]*
(a) 0 (b) $\frac{1}{4}$
(c) $\frac{1}{2}$ (d) $\frac{3}{4}$

34. A linear harmonic oscillator of force constant 2×10^6 N/m and amplitude 0.01 m has a total mechanical energy of 160 J. Its *[1996]*
(a) potential energy is 160 J
(b) potential energy is 100 J
(c) potential energy is zero
(d) potential energy is 120 J

35. A body executes S.H.M with an amplitude A. At what displacement from the mean position is the potential energy of the body is one fourth of its total energy? *[1993]*
(a) A/4 (b) A/2
(c) 3A/4
(d) Some other fraction of A.

36. The angular velocity and the amplitude of a simple pendulum is ω and a respectively. At a displacement x from the mean position if its kinetic energy is T and potential energy is V, then the ratio of T to V is [1991]

(a) $\frac{(a^2 - x^2\omega^2)}{x^2\omega^2}$ (b) $\frac{x^2\omega^2}{(a^2 - x^2\omega^2)}$

(c) $\frac{(a^2 - x^2)}{x^2}$ (d) $\frac{x^2}{(a^2 - x^2)}$

Topic 3 Time Period, Frequency, Simple Pendulum & Spring Pendulum

37. A body is executing simple harmonic motion with frequency 'n', the frequency of its potential energy is [2021]
(a) $4n$ (b) n
(c) $2n$ (d) $3n$

38. A spring is stretched by 5 cm by a force 10 N. The time period of the oscillations when a mass of 2 kg is suspended by it is [2021]
(a) 0.628 s (b) 0.0628 s
(c) 6.28 s (d) 3.14 s

39. A truck is stationary and has a bob suspended by a light string, in a frame attached to the truck. The truck suddenly moves to the right with an acceleration of a. The pendulum will tilt [NEET Odisha 2019]
(a) to the left and angle of inclination of the pendulum with the vertical is $\tan^{-1}\left(\frac{g}{a}\right)$
(b) to the left and angle of inclination of the pendulum with the vertical is $\sin^{-1}\left(\frac{g}{a}\right)$
(c) to the left and angle of inclination of the pendulum with the vertical is $\tan^{-1}\left(\frac{a}{g}\right)$
(d) to the left and angle of inclination of the pendulum with the vertical is $\sin^{-1}\left(\frac{a}{g}\right)$

40. A mass falls from a height 'h' and its time of fall 't' is recorded in terms of time period T of a simple pendulum. On the surface of earth it is found that $t = 2T$. The entire set up is taken on the surface of another planet whose mass is half of that of earth and radius the same. Same experiment is repeated and corresponding times noted as t' and T'.
Then we can say [NEET Odisha 2019]
(a) $t' = 2T'$ (b) $t' = T'$
(c) $t' > 2T'$ (d) $t' < 2T'$

41. A pendulum is hung from the roof of a sufficiently high building and is moving freely to and fro like a simple harmonic oscillator. The acceleration of the bob of the pendulum is 20 m/s^2 at a distance of 5 m from the mean position. The time period of oscillation is [2018]
(a) 2π s (b) π s
(c) 1 s (d) 2 s

42. A particle executes linear simple harmonic motion with an amplitude of 3 cm. When the particle is at 2 cm from the mean position, the magnitude of its velocity is equal to that of its acceleration. Then its time period in seconds is [2017]

(a) $\frac{\sqrt{5}}{2\pi}$ (b) $\frac{4\pi}{\sqrt{5}}$

(c) $\frac{2\pi}{\sqrt{3}}$ (d) $\frac{\sqrt{5}}{\pi}$

43. A spring of force constant k is cut into lengths of ratio 1 : 2 : 3. They are connected in series and the new force constant is k'. Then they are connected in parallel and force constant is k''. Then k' : k'' is [2017]
(a) 1 : 9 (b) 1 : 11
(c) 1 : 14 (d) 1 : 6

44. A particle is executing SHM along a straight line. Its velocities at distances x_1 and x_2 from the mean position are V_1 and V_2, respectively. Its time period is [2015]

(a) $2\pi\sqrt{\frac{x_2^2 - x_1^2}{V_1^2 - V_2^2}}$ (b) $2\pi\sqrt{\frac{V_1^2 + V_2^2}{x_1^2 + x_2^2}}$

(c) $2\pi\sqrt{\frac{V_1^2 - V_2^2}{x_1^2 - x_2^2}}$ (d) $2\pi\sqrt{\frac{x_1^2 - x_2^2}{V_1^2 - V_2^2}}$

45. A particle is executing a simple harmonic motion. Its maximum acceleration is a and maximum velocity is b. [2015 RS]
Then its time period of vibration will be :

(a) $\frac{\alpha}{\beta}$ (b) $\frac{\beta^2}{\alpha}$

(c) $\frac{2\pi\beta}{\alpha}$ (d) $\frac{\beta^2}{\alpha^2}$

46. The period of oscillation of a mass M suspended from a spring of negligible mass is T. If along with it another mass M is also suspended, the period of oscillation will now be [2010]
(a) T (b) $T/\sqrt{2}$
(c) 2T (d) $\sqrt{2}$T

47. A simple pendulum performs simple harmonic motion about x = 0 with an amplitude a and time period T. The speed of the pendulum at $x = \frac{a}{2}$ will be: [2009]
(a) $\frac{\pi a}{T}$ (b) $\frac{3\pi^2 a}{T}$
(c) $\frac{\pi a\sqrt{3}}{T}$ (d) $\frac{\pi a\sqrt{3}}{2T}$

48. A block of mass M is attached to the lower end of a vertical spring. The spring is hung from a ceiling and has force constant value k. The mass is released from rest with the spring initially unstretched. The maximum extension produced in the length of the spring will be: [2009]
(a) 2 Mg/k (b) 4 Mg/k
(c) Mg/2k (d) Mg/k

49. A point performs simple harmonic oscillation of period T and the equation of motion is given by x = a sin (ωt + π/6). After the elapse of what fraction of the time period the velocity of the point will be equal to half of its maximum velocity? [2008]
(a) T/8 (b) T/6
(c) T/3 (d) T/12

50. A mass of 2.0 kg is put on a flat pan attached to a vertical spring fixed on the ground as shown in the figure. The mass of the spring and the pan is negligible. When pressed slightly and released the mass executes a simple harmonic motion. The spring constant is 200 N/m. What should be the minimum amplitude of the motion so that the mass gets detached from the pan (take $g = 10$ m/s^2)? [2007]

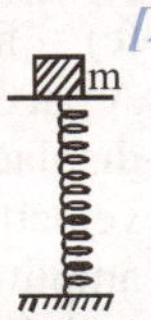

(a) 10.0 cm
(b) any value less than 12.0 cm
(c) 4.0 cm
(d) 8.0 cm

51. A particle executes simple harmonic oscillation with an amplitude a. The period of oscillation is *T*. The minimum time taken by the particle to travel half of the amplitude from the equilibrium position is [2007]
(a) *T*/8 (b) *T*/12
(c) *T*/2 (d) *T*/4

52. Two springs of spring constants k_1 and k_2 are joined in series. The effective spring constant of the combination is given by [2004]
(a) $k_1k_2/(k_1 + k_2)$ (b) k_1k_2
(c) $(k_1 + k_2)/2$ (d) $k_1 + k_2$

53. The time period of a mass suspended from a spring is *T*. If the spring is cut into four equal parts and the same mass is suspended from one of the parts, then the new time period will be [2003]
(a) 2*T* (b) $\frac{T}{4}$
(c) 2 (d) $\frac{T}{2}$

54. A body of mass M, executes vertical SHM with periods t_1 and t_2, when separately attached to spring A and spring B respectively. The period of SHM, when the body executes SHM, as shown in the figure is t_0. Then [2002]

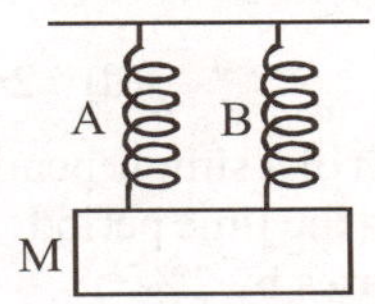

(a) $t_0^{-1} = t_1^{-1} + t_2^{-1}$ (b) $t_0 = t_1 + t_2$
(c) $t_0^2 = t_1^2 + t_2^2$ (d) $t_0^{-2} = t_1^{-2} + t_2^{-2}$

55. The amplitude of a pendulum executing simple harmonic motion falls to 1/3 the original value after 100 oscillations. The amplitude falls to *S* times the original value after 200 oscillations, where *S* is [2002]
(a) 1/9 (b) 1/2
(c) 2/3 (d) 1/6

56. A simple pendulum has a metal bob, which is negatively charged. If it is allowed to oscillate above a positively charged metallic plate, then its time period will [2001]
(a) increase (b) decrease
(c) become zero (d) remain the same

57. Masses M_A and M_B hanging from the ends of strings of lengths L_A and L_B are executing simple harmonic motions. If their frequencies are $f_A = 2f_B$, then [2000]
(a) $L_A = 2L_B$ and $M_A = M_B/2$
(b) $L_A = 4L_B$ regardless of masses
(c) $L_A = L_B/4$ regardless of masses
(d) $L_A = 2L_B$ and $M_A = 2M_B$

58. The time period of a simple pendulum is 2 seconds. If its length is increased by 4-times, then its period becomes *[1999]*
(a) 16 s (b) 12 s
(c) 8 s (d) 4 s

59. Two simple pendulums of length 5m and 20m respectively are given small linear displacement in one direction at the same time. They will again be in the phase when the pendulum of shorter length has completed .... oscillations *[1998]*
(a) 5 (b) 1
(c) 2 (d) 3

60. A mass m is vertically suspended from a spring of negligible mass; the system oscillates with a frequency n. What will be the frequency of the system, if a mass 4 m is suspended from the same spring? *[1998]*
(a) $\frac{n}{4}$ (b) $4n$
(c) $\frac{n}{2}$ (d) $2n$

61. If the length of a simple pendulum is increased by 2%, then the time period *[1997]*
(a) increases by 2%
(b) decreases by 2%
(c) increases by 1%
(d) decreases by 1%

62. A hollow sphere is filled with water. It is hung by a long thread. As the water flows out of a hole at the bottom, the period of oscillation will *[1997]*
(a) first increase and then decrease
(b) first decrease and then increase
(c) go on increasing
(d) go on decreasing

63. If a simple harmonic oscillator has got a displacement of 0.02 m and acceleration equal to 2.0 m/s^2 at any time, the angular frequency of the oscillator is equal to *[1992]*
(a) 10 rad/s (b) 0.1 rad/s
(c) 100 rad/s (d) 1 rad/s

64. A simple harmonic oscillator has an amplitude A and time period T. The time required by it to travel from $x = A$ to $x = A/2$ is *[1992]*
(a) T/6 (b) T/4
(c) T/3 (d) T/2

65. A body is executing S.H.M. When the displacements from the mean position are 4cm and 5 cm, the corresponding velocities of the body are 10 cm per sec and 8 cm per sec. Then the time period of the body is *[1991]*
(a) 2π sec (b) $\pi/2$ sec
(c) π sec (d) $(3\pi/2)$sec

66. A simple pendulum is suspended from the roof of a trolley which moves in a horizontal direction with an acceleration a, then the time period is given by $T = 2\pi\sqrt{(l/g)}$, where g is equal to *[1991]*
(a) g (b) $g - a$
(c) $g + a$ (d) $\sqrt{(g^2 + a^2)}$

67. A mass m is suspended from a two coupled springs, connected in series. The force constant for springs are k_1 and k_2. The time period of the suspended mass will be *[1990]*
(a) $T = 2\pi\sqrt{\frac{m}{k_1 - k_2}}$ (b) $T = 2\pi\sqrt{\frac{mk_1k_2}{k_1 + k_2}}$
(c) $T = 2\pi\sqrt{\frac{m}{k_1 + k_2}}$ (d) $T = 2\pi\sqrt{\frac{m(k_1 + k_2)}{k_1k_2}}$

Topic 4 Damped SHM, Forced Oscillations & Resonance

68. The damping force on an oscillator is directly proportional to the velocity. The unit of the constant of proportionality is: **[2012]**
(a) kgms^{-1} (b) kgms^{-2}
(c) kgs^{-1} (d) kgs

69. In case of a forced vibration, the resonance wave becomes very sharp when the *[2003]*
(a) quality factor is small
(b) damping force is small
(c) restoring force is small
(d) applied periodic force is small

70. Resonance is an example of *[1999]*
(a) tuning fork (b) forced vibration
(c) free vibration (d) damped vibration

71. A particle, with restoring force proportional to displacement and resistve force proportional to velocity is subjected to a force $F \sin \omega_0$. If the amplitude of the particle is maximum for $\omega = \omega_1$ and the energy of the particle is maximum for $\omega = \omega_2$, then *[1989]*
(a) $\omega_1 = \omega_0$ and $\omega_2 \neq \omega_0$
(b) $\omega_1 = \omega_0$ and $\omega_2 = \omega_0$
(c) $\omega_1 \neq \omega_0$ and $\omega_2 = \omega_0$
(d) $\omega_1 \neq \omega_0$ and $\omega_2 \neq \omega_0$

ANSWER KEY

1	(d)	9	(b)	17	(d)	25	(c)	33	(d)	41	(b)	49	(d)	57	(c)	65	(c)
2	(d)	10	(a)	18	(a)	26	(a)	34	(b)	42	(b)	50	(a)	58	(d)	66	(d)
3	(d)	11	(c)	19	(d)	27	(c)	35	(b)	43	(b)	51	(b)	59	(c)	67	(d)
4	(b)	12	(b)	20	(a)	28	(b)	36	(c)	44	(a)	52	(a)	60	(c)	68	(c)
5	(a)	13	(a)	21	(a)	29	(b)	37	(c)	45	(c)	53	(d)	61	(c)	69	(b)
6	(b)	14	(a)	22	(b)	30	(d)	38	(a)	46	(d)	54	(d)	62	(a)	70	(b)
7	(c)	15	(b)	23	(c)	31	(b)	39	(c)	47	(c)	55	(a)	63	(a)	71	(c)
8	(d)	16	(d)	24	(c)	32	(b)	40	(a)	48	(d)	56	(b)	64	(a)		

Hints & Solutions

1. **(d)** Displacement equation of a SHM

$$y = A\sin(\omega t + \phi)$$

$\therefore$ Velocity, $v = \dfrac{dy}{dt} = A\omega\cos(\omega t + \phi)$

Acceleration, $a = \dfrac{dv}{dt}$

or, $a = -A\omega^2 \sin(\omega t + \phi)$

$\therefore a = A\omega^2 \sin(\omega t + \phi + \pi)$

Hence, phase difference between displacement and acceleration is π.

2. **(d)** At t = 0, y = 3, which is maximum displacement so equation will be cosine function.

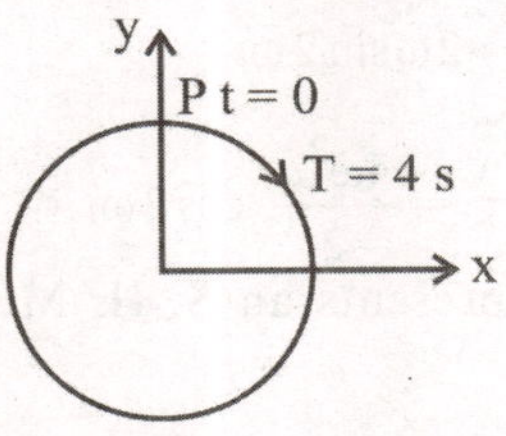

$$\omega = \frac{2\pi}{T} = \frac{2\pi}{4} = \frac{\pi}{2} \text{ rad/s} \quad (\because T = 4 \text{ s})$$

$$y = a\cos\omega t \Rightarrow y = 3\cos\frac{\pi}{2}t$$

3. **(d)** Displacement of the particle in one complete vibration is zero, so, average velocity in one complete vibration will be

$$= \frac{\text{Displacement}}{\text{Time interval}} = \frac{y_f - y_i}{T} = 0$$

4. **(b)**

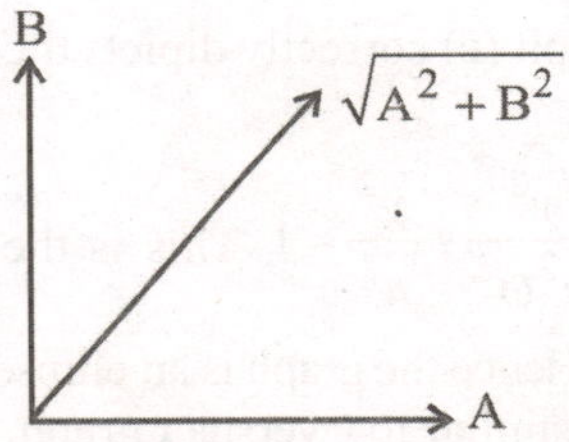

Given equation

$y = A_0 + A\sin\omega t + B\sin\omega t$

Now assume $(y - A_0) = \gamma$

$y - A_0 = A\sin\omega t + B\sin\omega t$

$\gamma = A\sin\omega t + B\cos\omega t$

$= \sqrt{A^2 + B^2}\,\sin(\omega t + \phi)$

which is S.H.M.

where cos

and $\sin\phi = \dfrac{B}{\sqrt{A^2 + B^2}}$

so, resultant amplitude

$= \sqrt{A^2 + B^2}$

5. **(a)** In one time period total distance travelled by the particle is 4A.

6. **(b)** The two displacements equations are

$y_1 = a\sin(\omega t)$

and $y_2 = b\cos(\omega t) = b\sin\left(\omega t + \dfrac{\pi}{2}\right)$

$y_{eq} = y_1 + y_2$

$= a\sin\omega t + b\cos\omega t$

$$= a\sin\omega t + b\sin\left(\omega t + \frac{\pi}{2}\right)$$

Since the frequencies for both SHMs are same, resultant motion will be SHM.

Now amplitude, $A_{eq} = \sqrt{a^2 + b^2 + 2ab\cos\frac{\pi}{2}}$

$$\Rightarrow A_{eq} = \sqrt{a^2 + b^2}$$

7. (c) Displacement, x = A cos (ωt) (given)

Velocity, $v = \frac{dx}{dt} = -A\omega\sin(\omega t)$

Acceleration, $a = \frac{dv}{dt} = -A\omega^2\cos\omega t$

Hence graph (c) correctly dipicts the variation of a with t.

8. (d) As $\frac{v^2}{a^2\omega^2} + \frac{y^2}{a^2} = 1$ This is the equation of ellipse. Hence the graph is an ellipse. *P* versus *x* graph is similar to *v* versus *x* graph.

9. (b) $y = 3\sin\frac{\pi}{2}(50t - x)$

$y = 3\sin\left(25\pi t - \frac{\pi}{2}x\right)$ on comparing with the standard wave equation

$y = a\sin(\omega t - kx)$

Wave velocity $v = \frac{\omega}{k} = \frac{25\pi}{\pi/2} = 50$ m/sec.

The velocity of particle

$$v_p = \frac{\partial y}{\partial t} = 75\pi\cos\left(25\pi t - \frac{\pi}{2}x\right)$$

$$v_{p\,max} = 75\pi$$

then $\frac{v_{p\,max}}{v} = \frac{75\pi}{50} = \frac{3\pi}{2}$

10. (a) The given velocity-position graph depicts that the motion of the particle is SHM.
In SHM, at t = 0, v = 0 and $x = x_{max}$
So, option (a) is correct.

11. (c) Only functions given in (A) & (C) represent SHM.

NOTES $y = A\sin\omega t + B\cos\omega t \Rightarrow y = a\sin(\omega t + \phi)$. This is also a equation of S.H.M where $\phi = \sqrt{A^2 + B^2}$

$A = a\cos\phi$, $B = a\sin\phi$ and $\theta = \tan^{-1}\frac{B}{A}$

12. (b) Equation of SHM is given by
$x = A\sin(\omega t + \delta)$
$(\omega t + \delta)$ is called phase.

When $x = \frac{A}{2}$, then $\sin(\omega t + \delta) = \frac{1}{2}$

$$\Rightarrow \omega t + \delta = \frac{\pi}{6}$$

or $\phi_1 = \frac{\pi}{6}$

For second particle,

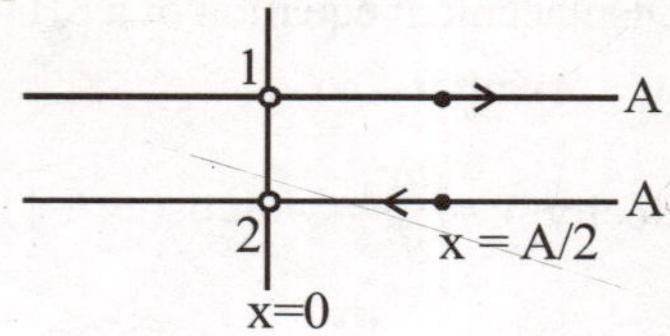

$$\phi_2 = \pi - \frac{\pi}{6} = \frac{5\pi}{6}$$

$$\therefore \phi = \phi_2 - \phi_1$$

$$= \frac{4\pi}{6} = \frac{2\pi}{3}$$

13. (a) $x = a\sin^2\omega t = \frac{a}{2}(1 - \cos 2\omega t)$

$$\frac{dx}{dt} = \frac{a}{2}\,2\omega\sin 2\omega t$$

$$\Rightarrow \frac{d^2x}{dt^2} = \frac{4\omega^2 a}{2}\cdot\cos 2\omega t$$

This represents an S. H. M. of frequency $= \frac{\omega}{\pi}$

14. (a) acceleration, a = – kX, where, X = x + a.
In simple harmonic motion acceleration is directly proportional to the displacement from the mean position. Also the acceleration is in the opposite direction of displacement.

15. (b) Maximum acceleration of a particle in the simple harmonic motion is directly proportional to the square of angular frequency *i.e.*,

i.e., $a \propto \omega^2$

$\therefore \frac{a_1}{a_2} = \frac{\omega_1^2}{\omega_2^2} = \frac{(100)^2}{(1000)^2} = \frac{1}{100}$

$\Rightarrow a_1 : a_2 = 1 : 10^2.$

16. (d) Let $y = A \sin \omega t$

$v_{inst} = \frac{dy}{dt} = A\omega \cos \omega t = A\omega \sin(\omega t + \pi/2)$

Acceleration $= -A\omega^2 \sin \omega t$
$= A\omega^2 \sin(\pi + \omega t)$

$\therefore \phi = \frac{\pi}{2} = 0.5\pi$

17. (d) In S.H.M., $v_{max} = A\,\omega = A(2\pi f)$

$f = \frac{v_{max}}{2\pi A} = \frac{31.4}{2(3.14)\times 5} = 1$ oscillation per sec.

18. (a) When v is maximum, a is zero

$v = \omega\sqrt{A^2 - x^2}$; $a = \omega^2 x$

v_{max} at $x = 0$, but at $x = 0$, $a = 0$

19. (d) $x = A\cos(\omega t + \delta)$

$y = A\cos(\omega t + \alpha)$ ....(1)

When $\delta = \alpha + \frac{\pi}{2}$

$x = A\cos\left(\frac{\pi}{2} + \omega t + \alpha\right)$

$x = -A\sin(\omega t + \alpha)$ ....(2)

Squaring (1) and (2) and then adding
$x^2 + y^2 = A^2 [\cos^2(\omega t + \alpha) + \sin^2(\omega t + \alpha)]$
or $x^2 + y^2 = A^2$, which is the equation of a circle.
The present motion is anticlockwise.

20. (a) Amplitude (A) = 0.01 m, Frequency = 60 Hz
Maximum acceleration

$= A\omega^2 = 0.01 \times (2\pi n)^2$

$= 0.01 \times 4\pi^2 \times 60 \times 60 = 144\pi^2$ m/sec^2

21. (a) Equation of two simple harmonic motions

$y = A\sin(\omega t + \phi)$ ....(1)

$x = A\sin\left(\omega t + \phi + \frac{\pi}{2}\right)$

$\Rightarrow x = A\cos(\omega t + \phi)$ ....(2)

On squaring and adding equations (1) and (2),
$x^2 + y^2 = A^2$
This is an equation of a circle. Hence, resulting motion will be a circular motion.

22. (b) $v_{max} = A\omega$ when $v = \frac{v_{max}}{2} = \frac{A\omega}{2}$

$v = \omega\sqrt{A^2 - y^2}$

$\Rightarrow \frac{A\omega}{2} = \omega\sqrt{A^2 - y^2} \Rightarrow y = \pm\frac{\sqrt{3}}{2}A$

> **NOTES** The displacement at which the speed is n times the maximum speed is given by $y = a\sqrt{1 - n^2}$

23. (c) A wave moving through a string fixed at both ends, is a transverse wave formed as a result of simple harmonic motion of particles of the string.

24. (c) As phase difference $= \frac{\pi}{4}$, the resultant path of particle is an ellipse.

25. (c) $x = a\sin\omega t$
and $y = b\sin(\omega t + \pi) = -b\sin\omega t$

or, $\frac{x}{a} = -\frac{y}{b}$ or $y = \frac{b}{a}x$

It is an equation of a straight line.

> **NOTES** If two mutually perpendicular S.H.M's of same frequency be $x = 9$, $\sin\omega t$ and $y = a_2 \sin(\omega t + \phi)$ then general equation of Lissajous figure is
>
> $\frac{x^2}{a_1^2} + \frac{y_2}{a_2^2} - \frac{2xy}{a_1 a_2}\cos\phi = \sin^2\phi$
>
> For $\phi = \pi \frac{x^2}{a_1^2} + \frac{y^2}{a_2^2} - \frac{2xy}{a_1 a_2}\cos(\pi) = \sin 2(\pi)$
>
> $\Rightarrow \frac{x}{a_1} + \frac{y}{a_2} = 0 \quad \frac{x}{a} + \frac{y}{a^2} = 0 \Rightarrow y = \frac{-a_2}{a_1}x$

26. (a) For simple harmonic motion, $F = -Kx$.
Here, $K = Ak$.

27. (c) We have, $U + K = E$
where, U = potential energy, K = Kinetic energy, E = Total energy.
Also, we know that, in S.H.M., when potential energy is maximum, K.E. is zero and vice-versa.

$\therefore U_{max} + 0 = E \Rightarrow U_{max} = E$

Further,

$K.E. = \frac{1}{2}m\omega^2 a^2 \cos^2 \omega t$

But by question, $K.E. = K_0 \cos^2 \omega t$

$$\therefore K_0 = \frac{1}{2} m\omega^2 a^2$$

Hence, total energy, $E = \frac{1}{2} m\omega^2 a^2 = K_0$

$$\therefore U_{max} = K_0 \;\&\; E = K_0.$$

28. **(b)** The potential energy of a spring $= \frac{1}{2} kx^2$

$$U = \frac{1}{2} k.(2)^2 = 4 \times \frac{1}{2} k$$

For x = 8 cm,

$$\text{Energy stored} = \frac{1}{2} k.(8)^2 = 64 \times \frac{1}{2} k$$

$$= 64 \times \frac{U}{4} = 16U$$

29. **(b)** Potential energy $\propto$ (Amplitude)^2
$\propto$ (Displacement)^2

P.E. is maximum at maximum distance.
P.E. is zero at equilibrium point.
P.E. curve is parabolic.

30. **(d)** $P.E. = \frac{1}{2} kx^2 = E$

At half way

$$P.E. = \frac{1}{2} k\left(\frac{x}{2}\right)^2 = \frac{\frac{1}{2} kx^2}{4} = \frac{E}{4}$$

31. **(b)** P.E. of particle executing S.H.M. $= \frac{1}{2} m\omega^2 x^2$

At $x = a$, P.E. is maximum i.e. $= \frac{1}{2} m\omega^2 a^2$

$$K.E. = \frac{1}{2} m\omega^2 \;(a^2 - x^2)$$

At $x = 0$, K.E. is maximum. Hence, displacement from position of maximum Kinetic energy $= \pm a$.

32. **(b)** Total Energy of body performing simple harmonic motion $= \frac{1}{2} m\omega^2 a^2 = \frac{1}{2} ka^2$

where $k = m\omega^2$

NOTES

$$T.E. = K.E. + P.E. = \frac{1}{2} m(a^2 - x^2)\omega^2 + \frac{1}{2} m\omega^2 x^2$$

Hence energy depends upon amplitude and k (spring constant).

33. **(d)** Total energy of particle executing S.H.M. of amplitude (A).

$$E = \frac{1}{2} m\omega^2 A^2$$

K.E. of the particle

$$= \frac{1}{2} m\omega^2 \left(A^2 - \frac{A^2}{4}\right) \quad \left(\text{when } x = \frac{A}{2}\right)$$

$$= \frac{1}{2} m\omega^2 \times \frac{3}{4} A^2 = \frac{1}{2} \times \frac{3}{4} m\omega^2 A^2$$

$$\text{Clearly, } \frac{\text{KE}}{\text{Total Energy}} = \frac{3}{4}$$

34. **(b)** Force constant $k = 2 \times 10^6$ N/m

Amplitude $(x) = 0.01$ m

$$\text{Potential Energy} = \frac{1}{2} kx^2$$

$$= \frac{1}{2} \times (2 \times 10^6) \times (0.01)^2 = 100 \text{ J}$$

35. **(b)** $P.E = \frac{1}{2} m\omega^2 x^2.$

$$\text{Total energy } E = \frac{1}{2} m\omega^2 A^2$$

$$\text{P.E.} = \frac{1}{4} E \Rightarrow \frac{1}{2} m\omega^2 x^2 = \frac{1}{8} m\omega^2 A^2$$

$$\therefore x = \frac{1}{2} A.$$

36. **(c)** P.E., $V = \frac{1}{2} m\omega^2 x^2$

and K.E., $T = \frac{1}{2} m\omega^2 (a^2 - x^2)$

$$\therefore \frac{T}{V} = \frac{a^2 - x^2}{x^2}$$

37. **(c)** Displacement of particle executing SHM $y = A\sin(\omega t + \phi)$ ...(i)

$$\text{Potential energy } U = \frac{1}{2} ky^2$$

$$\text{or, } U = \frac{1}{2} kA^2 \sin^2(\omega t + \phi) \quad \text{...(ii)}$$

From eq. I and II,
Time period of $y = A\sin(\omega t + \phi)$

$$T_1 = \frac{2\pi}{\omega} \;\therefore\; \text{frequency } n_1 = \frac{\omega}{2\pi}$$

and time period of $y^2 = A^2\sin^2(\omega t + \phi)$

$$T_2 = \frac{\pi}{\omega} \quad \therefore \text{ frequency } n_2 = \frac{\omega}{\pi}$$

$\therefore n_2 = 2n_1 = 2n$

NOTES Potential energy varies periodically with double the frequency of SHM. It is maximum and equal to total energy at extreme position and minimum at mean position.

38. **(a)** Spring constant

$$k = \frac{F}{x} = \frac{10}{5\times10^{-2}} = 200\,\text{N/m}$$

And for spring-mass system undergoing SHM

$$T = 2\pi\sqrt{\frac{m}{k}} = 2\pi\sqrt{\frac{2}{200}} = \frac{2\pi}{10} = 0.628\text{s}$$

$(\because m = 2\text{kg given})$

39. **(c)** $T\sin\theta = ma$

$T\cos\theta = mg$

$$\tan\theta = \frac{a}{g}$$

$$\theta = \tan^{-1}\left(\frac{a}{g}\right)$$

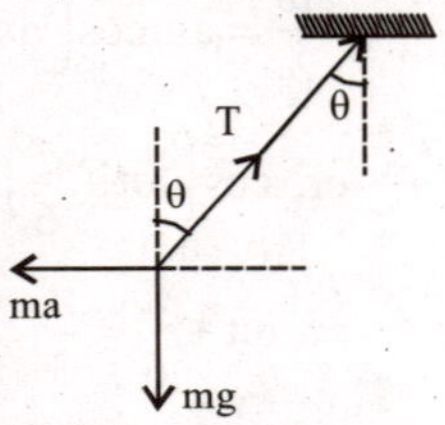

40. **(a)** At surface of earth time taken in falling h distance.

$$t = \sqrt{\frac{2h}{g}}$$

and $T = 2\pi\sqrt{\frac{l}{g}}$

Given $t = 2T$

$$\frac{t}{T} = 2$$

For surface of other planet

$$g' = \frac{g}{2}$$

Time taken in falling h distance

$$t' = \sqrt{\frac{2h}{g'}} = \sqrt{2}t$$

and $T' = 2\pi\sqrt{\frac{l}{g'}} = \sqrt{2}T$

Here $\frac{t'}{T'} = \frac{\sqrt{2}t}{\sqrt{2}T} = 2$

$\therefore \quad t' = 2T$

41. **(b)** From question, acceleration, $a = 20\text{ m/s}^2$, and displacement, $y = 5\text{m}$

$|a| = \omega^2 y$

$\Rightarrow \quad 20 = \omega^2(5)$

$\Rightarrow \quad \omega = 2\text{ rad/s}$

Time period of pendulum,

$$T = \frac{2\pi}{\omega} = \frac{2\pi}{2} = \pi\,\text{s}$$

42. **(b)** Given, Amplitude $A = 3$ cm

When particle is at $x = 2$ cm

According to question, magnitude of velocity = acceleration

$$\omega\sqrt{A^2 - x^2} = x\omega^2$$

$$\sqrt{(3)^2 - (2)^2} = 2\left(\frac{2\pi}{T}\right)$$

$$\sqrt{5} = \frac{4\pi}{T} \Rightarrow T = \frac{4\pi}{\sqrt{5}}$$

43. **(b)** Let ℓ be the complete length of the spring. Length when cut in ratio, 1 : 2 : 3 are

$\frac{\ell}{6}, \frac{\ell}{3}$ and $\frac{\ell}{2}$

Spring constant (k) $\propto \frac{1}{\text{length}(\ell)}$

Spring constant for given segments

$k_1 = 6k, k_2 = 3k$ and $k_3 = 2k$

When they are connected **in series**

$$\frac{1}{k'} = \frac{1}{6k} + \frac{1}{3k} + \frac{1}{2k}$$

$$\Rightarrow \quad \frac{1}{k'} = \frac{6}{6k}$$

$\therefore$ Force constant $k' = k$

And when they are connected **in parallel**

$k'' = 6k + 3k + 2k$

$\Rightarrow k'' = 11k$

Then the ratios

$\frac{k'}{k''} = \frac{1}{11}$ i.e., $k' : k'' = 1 : 11$

NOTES If a spring of force constant K is divided into n equal parts then spring constant of each part will become nk and if these n parts connected in parallel then $k_{eff} = n^2 k$.

44. (a) As we know, for particle undergoing SHM,

$$V = \omega\sqrt{A^2 - X^2}$$

$$V_1^2 = \omega^2 (A^2 - x_1^2)$$

$$V_2^2 = \omega^2 (A^2 - x_2^2)$$

Substracting we get,

$$\frac{V_1^2}{\omega^2} + x_1^2 = \frac{V_2^2}{\omega^2} + x_2^2$$

$$\Rightarrow \frac{V_1^2 - V_2^2}{\omega^2} = x_2^2 - x_1^2$$

$$\Rightarrow \omega = \sqrt{\frac{V_1^2 - V_2^2}{x_2^2 - x_1^2}}$$

$$\Rightarrow T = 2\pi\sqrt{\frac{x_2^2 - x_1^2}{V_1^2 - V_2^2}}$$

45. (c) As, we know, in SHM

Maximum acceleration of the particle, $\alpha = A\omega^2$

Maximum velocity, $\beta = A\omega$

$$\Rightarrow \omega = \frac{\alpha}{\beta}$$

$$\Rightarrow T = \frac{2\pi}{\omega} = \frac{2\pi\beta}{\alpha} \quad \left[\because \omega = \frac{2\pi}{T}\right]$$

46. (d) $T = 2\pi\sqrt{\frac{M}{K}}$

$$\therefore \frac{T_1}{T_2} = \sqrt{\frac{M_1}{M_2}}$$

$$\therefore T_2 = T_1\sqrt{\frac{M_2}{M_1}} = T_1\sqrt{\frac{2M}{M}}$$

$$T_2 = T_1\sqrt{2} = \sqrt{2}\,T \text{ (where } T_1 = T)$$

> **NOTES**
>
> If the spring has a mass M and mass m is suspended from it, effective mass is given by $m_{eff} = m + \frac{M}{3}$
>
> If the two mass m_1 and m_2 are connected by a spring and made to oscillate on horizontal surface, reduced mass $\frac{1}{m_r} = \frac{1}{m_1} + \frac{1}{m_2}$
>
> $\therefore$ Time period $T = 2\pi\sqrt{\frac{m_r}{k}}$

47. (c) Speed $v = \omega\sqrt{a^2 - x^2}$, $x = \frac{a}{2}$

$$\therefore \quad v = \omega\sqrt{a^2 - \frac{a^2}{4}} = \omega\sqrt{\frac{3a^2}{4}} = \frac{2\pi}{T}\frac{a\sqrt{3}}{2} = \frac{\pi a\sqrt{3}}{T}$$

48. (d) Restoring force, $f' = -kx$

where x is the extension produced in the spring.

Weight of the mass acting downward = Mg.

In equilibrium

$$kx = Mg \quad \text{or} \quad x = \frac{Mg}{k}$$

49. (d) We have $x = a\sin\left(\omega t + \frac{\pi}{6}\right)$

$$\therefore \text{Velocity, } v = \frac{dx}{dt} = a\omega\cos\left(\omega t + \frac{\pi}{6}\right)$$

Maximum velocity = $a\omega$

According to question,

$$\frac{a\omega}{2} = a\omega\cos\left(\omega t + \frac{\pi}{6}\right)$$

$$\text{or, } \cos\left(\omega t + \frac{\pi}{6}\right) = \frac{1}{2} = \cos 60° \text{ or } \cos\frac{\pi}{3}$$

$$\Rightarrow \omega t + \frac{\pi}{6} = \frac{\pi}{3}$$

$$\omega t = \frac{\pi}{3} - \frac{\pi}{6} \text{ or, } \omega t = \frac{\pi}{6}$$

$$\text{or, } \frac{2\pi}{T}.t = \frac{\pi}{6} \Rightarrow t = \frac{T}{12}$$

50. (a) Mass gets detached at the upper extreme position when pan returns to its mean position.

At that point, $R = mg - m\omega^2 a = 0$

i.e. $g = \omega^2 a$

$$\Rightarrow a = g/\omega^2 = mg/k$$

$$\Rightarrow a = \frac{2 \times 10}{200} \quad \left[As = \omega^2 = \frac{k}{m}\right]$$

$$\Rightarrow a = 1/10\,m = 10\text{cm}$$

51. (b) Displacement from the mean position

$$y = a\sin\left(\frac{2\pi}{T}\right)t$$

According to problem $y = a/2$

$$a/2 = a\sin\left(\frac{2\pi}{T}\right)t$$

$$\Rightarrow \frac{\pi}{6} = \left(\frac{2\pi}{T}\right)t \Rightarrow t = T/12$$

This is the minimum time taken by the particle to travel half of the amplitude from the equilibrium position.

52. (a) $\frac{1}{k_{eq}} = \frac{1}{k_1} + \frac{1}{k_2} \Rightarrow \frac{1}{k_{eq}} = \frac{k_1 + k_2}{k_1 k_2}$

$\Rightarrow k_{eq} = \frac{k_1 k_2}{k_1 + k_2}$

53. (d) $T = 2\pi\sqrt{\frac{m}{k}}$

When a spring is cut into n parts

Spring constant for each part = nk

Here, $n = 4$

$T_1 = 2\pi\sqrt{\frac{m}{4k}} = \frac{T}{2}$

54. (d) The time period of spring mass system,

$T = 2\pi\sqrt{\frac{m}{K}}$

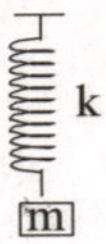

$\therefore \quad t_1 = 2\pi\sqrt{\frac{m}{k_1}}$ .....(i)

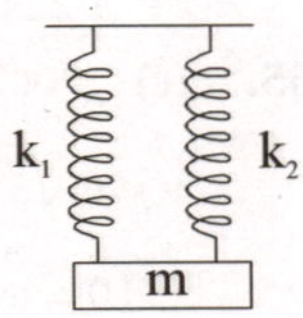

$t_2 = 2\pi\sqrt{\frac{m}{k_2}}$ .....(ii)

When springs are connected in parallel then

$K_{eff} = k_1 + k_2$

So, $t_0 = 2\pi\sqrt{\frac{m}{K_{eff}}} \Rightarrow 2\pi\sqrt{\frac{m}{(k_1 + k_2)}}$ .... (iii)

from (i), $\frac{1}{t_1^2} = \frac{1}{4\pi^2} \times \frac{k_1}{m}$

from (ii), $\frac{1}{t_2^2} = \frac{1}{4\pi^2} \times \frac{k_2}{m}$

from (iii), $\frac{1}{t_0^2} = \frac{1}{4\pi^2} \times \frac{(k_1 + k_2)}{m}$

from above expressions,

$\frac{1}{t_0^2} = \frac{1}{t_1^2} + \frac{1}{t_2^2}$

NOTES

$t = 2\pi\sqrt{\frac{m}{k}}$

$\Rightarrow \quad k = \text{Const.}t^{-2}$

Here the springs are joined in parallel. So

$k_0 = k_1 + k_2$

where k_0 is resultant force constant

$\therefore \quad \text{Const.}t_0^{-2} = \text{Const.}t_1^{-2} + \text{Const.}t_2^{-2}$

or, $t_0^{-2} = t_1^{-2} + t_2^{-2}$

55. (a) In harmonic oscillator, amplitude falls exponentially.

After 100 oscillations amplitude falls to $\frac{1}{3}$ times.

$\therefore$ After next 100 oscillations i.e., after 200 oscillations amplitude falls to $\left(\frac{1}{3}\right)^2 = \frac{1}{9}$ times.

56. (b) We know that the time period (T)

$= 2\pi\sqrt{\frac{l}{g'}} \propto \frac{1}{\sqrt{g'}} \qquad g' = g_{effective}$

Since the negatively charged bob is attracted by the positively charged plate, therefore acceleration due to gravity will increase and time period will decrease.

57. (c) $f_A = \frac{1}{2\pi}\sqrt{\frac{g}{L_A}}$

and $f_B = \frac{f_A}{2} = \frac{1}{2\pi}\sqrt{\frac{g}{L_B}}$

$\therefore \quad \frac{f_A}{f_{A/2}} = \frac{1}{2\pi}\sqrt{\frac{g}{L_A}} \times 2\pi\sqrt{\frac{L_B}{g}}$

$\Rightarrow 2 = \sqrt{\frac{L_B}{L_A}} \Rightarrow 4 = \frac{L_B}{L_A}$,

regardless of mass.

58. (d) $T = 2\pi\sqrt{\frac{\ell}{g}} \qquad T \propto \sqrt{\ell}$

If ℓ is increased by 4 times, time period will increase by two times.

59. (c) Let the pendulums be in phase after t sec of start. Within this time, if the bigger pendulum executes n oscillations, the smaller one will have executed $(n + 1)$ oscillations.

Now, the time of n oscillation = $2\pi\sqrt{\frac{20}{g}} \times n$

& the time of $(n+1)$ oscillation

$$= 2\pi\sqrt{\frac{5}{g}} \times (n+1)$$

To be in phase

$$2\pi\sqrt{\frac{20}{g}} \times n = 2\pi\sqrt{\frac{5}{g}} \times (n+1)$$

or, $2n = n+1$

or, $n = 1$

Hence, the no. of oscillations executed by shorter pendulum $= n+1 = 1+1 = 2$

60. (c) $n = \frac{1}{2\pi}\sqrt{\frac{k}{m}}$

$$n' = \frac{1}{2\pi}\sqrt{\frac{k}{4m}} = \frac{1}{2} \times \frac{1}{2\pi}\sqrt{\frac{k}{m}}$$

On putting the value of n we get $n' = \frac{n}{2}$

61. (c) We know that $T = 2\pi\sqrt{\frac{l}{g}}$

$$\frac{\Delta T}{T} \times 100 = \frac{1}{2}\frac{\Delta l}{l} \times 100$$

If length is increased by 2%., time period increases by 1%.

NOTES

(i) If g is constant and length varies by n %. Then % change in time period $\frac{\Delta T}{T} \times 100 = \frac{n}{2}\%$

(ii) If l is constant and g varies by n %. Then % change in time period $\frac{\Delta T}{T} \times 100 = \frac{n}{2}\%$

Valid only for percentage less than 10%.

62. (a) Time period of simple pendulum

$$T = 2\pi\sqrt{\left(\frac{l}{g}\right)} \propto \sqrt{l} \text{ where } l \text{ is effective length.}$$

[i.e distance between centre of suspension and centre of gravity of bob]

Initially, centre of gravity is at the centre of sphere. When water leaks the centre of gravity goes down until it is half filled; then it begins to go up and finally it again goes at the centre. That is effective length first increases and then decreases. As $T \propto \sqrt{l}$, so time period first increases and then decreases.

63. (a) $\omega^2 = \frac{\text{acceleration}}{\text{displacement}} = \frac{2.0}{0.02}$

$\omega^2 = 100$ or $\omega = 10$ rad/s

64. (a) For S.H.M., $x = A \sin\left(\frac{2\pi}{T}t\right)$

When $x = A$, $A = A \sin\left(\frac{2\pi}{T}.t\right)$

$$\therefore \sin\left(\frac{2\pi}{T}.t\right) = 1$$

$$\Rightarrow \sin\left(\frac{2\pi}{T}.t\right) = \sin\left(\frac{\pi}{2}\right) \Rightarrow t = (T/4)$$

When $x = \frac{A}{2}, \frac{A}{2} = A \sin\left(\frac{2\pi}{T}.t\right)$

or, $\sin\frac{\pi}{6} = \sin\left(\frac{2\pi}{T}t\right)$ or $t = (T/12)$

Now, time taken to travel from $x = A$ to $x = A/2$ is $(T/4 - T/12) = T/6$

65. (c) For S.H.M., Velocity,

$v = \omega\sqrt{a^2 - x^2}$ at displacement x.

$\Rightarrow 10 = \omega\sqrt{a^2 - 16}$ .....(1)

and $8 = \omega\sqrt{a^2 - 25}$ .....(2)

Dividing, $\frac{5^2}{4^2} = \frac{a^2 - 16}{a^2 - 25} = \frac{25}{16}$

or, $16a^2 - 256 = 25a^2 - 625 \Rightarrow 9a^2 = 369$

$$\therefore a^2 = \frac{369}{9}$$

Putting this value in equation (2) mentioned above,

$$10 = \omega\sqrt{\frac{369}{9} - 16} \Rightarrow 10 = \omega\sqrt{\frac{225}{9}}$$

or, $\omega = \frac{10 \times 3}{15} = 2$ radian / sec.

Time period $= \frac{2\pi}{\omega} = \frac{2\pi}{2} = \pi$ sec.

66. (d) The effective value of acceleration due to gravity is $\sqrt{(a^2 + g^2)}$

NOTES

If a simple pendulum is suspended from a trolley that is moving with constant speed v around a circle of radius r $T = 2\pi\sqrt{\dfrac{l}{\sqrt{g^2 + \left(\dfrac{v^2}{r}\right)^2}}}$

67. **(d)** The effective spring constant of two springs in series; $K = \dfrac{k_1k_2}{k_1 + k_2}$.

Time period,

$$T = 2\pi\sqrt{\frac{m}{K}} = 2\pi\sqrt{\frac{m(k_1 + k_2)}{k_1k_2}}$$

68. **(c)** $F \propto v \Rightarrow \quad F = kv$

$$k = \frac{F}{v} \Rightarrow [k] = \frac{[\text{kgms}^{-2}]}{[\text{ms}^{-1}]} = \text{kgs}^{-1}$$

69. **(b)** The resonance wave becomes very sharp when damping force is small.

70. **(b)** We know that if frequency of an external forced oscillation is equal to the natural frequency of the body, then amplitude of the forced oscillation of the body becomes very large. This phenomenon is known as resonant vibration. Therefore, resonance is an example of forced vibration.

71. **(c)** At maximum energy of the particle, velocity resonance takes place, which occurs when frequency of external periodic force is equal to natural frequency of undamped vibrations, i.e. $\omega_2 = \omega_0$.

Further, amplitude resonance takes place at a frequency of external force which is less than the frequency of undamped natural vibrations, *i.e.*, $\omega_1 \neq \omega_0$.

14 Waves

Trend Analysis with Important Topics & Sub-Topics

Topic Name	Sub-Topic	2021 QNS.	2021 LOD	2020 QNS.	2020 LOD	2019 QNS.	2019 LOD	2018 QNS.	2018 LOD	2017 QNS.	2017 LOD	2016 QNS.	2016 LOD
Basic of Mechanical Waves, Progressive & Stationary Waves													
Vibration of String & Organ Pipe	Resonance Produced at Air							1	A				
	Open and Closed Organ Pipe							1	D	1	A	1	E
Beats, Interference & Superposition of Waves	Beat Frequency			1	A								
Musical Sound & Doppler's Effect	Doppler Effect Formula									1	E	1	A

LOD - Level of Difficulty E - Easy A - Average D - Difficult Qns - No. of Questions

Topic 1 Basic of Mechanical Waves, Progressive & Stationary Waves

1. A wave travelling in the +ve x-direction having displacement along y-direction as 1m, wavelength 2π m and frequency $\frac{1}{\pi}$ Hz is represented by *[2013]*
   (a) $y = \sin(2\pi x - 2\pi t)$
   (b) $y = \sin(10\pi x - 20\pi t)$
   (c) $y = \sin(2\pi x + 2\pi t)$
   (d) $y = \sin(x - 2t)$

2. Two waves are represented by the equations $y_1 = a \sin(\omega t + kx + 0.57)$ m and $y_2 = a \cos(\omega t + kx)$ m, where x is in meter and t in sec. The phase difference between them is *[2011]*
   (a) 1.0 radian (b) 1.25 radian
   (c) 1.57 radian (d) 0.57 radian

3. Sound waves travel at 350 m/s through a warm air and at 3500 m/s through brass. The wavelength of a 700 Hz acoustic wave as it enters brass from warm air *[2011]*
   (a) decreases by a factor 10
   (b) increases by a factor 20
   (c) increases by a factor 10
   (d) decreases by a factor 20

4. A transverse wave is represented by $y = A \sin(\omega t - kx)$. For what value of the wavelength is the wave velocity equal to the maximum particle velocity? *[2010]*
   (a) $\frac{\pi A}{2}$ (b) πA
   (c) $2\pi A$ (d) A

5. A wave in a string has an amplitude of 2 cm. The wave travels in the + ve direction of x axis with a speed of 128 m/sec and it is noted that 5 complete waves fit in 4 m length of the string. The equation describing the wave is *[2009]*
   (a) $y = (0.02)\text{ m} \sin(15.7x - 2010t)$
   (b) $y = (0.02)\text{ m} \sin(15.7x + 2010t)$
   (c) $y = (0.02)\text{ m} \sin(7.85x - 1005t)$
   (d) $y = (0.02)\text{ m} \sin(7.85x + 1005t)$

6. The wave described by $y = 0.25 \sin(10\pi x - 2\pi t)$, where x and y are in meters and t in seconds, is a wave travelling along the: [2008]
(a) –ve x direction with frequency 1 Hz.
(b) +ve x direction with frequency π Hz and wavelength $\lambda = 0.2$ m.
(c) +ve x direction with frequency 1 Hz and wavelength $\lambda = 0.2$ m
(d) –ve x direction with amplitude 0.25 m and wavelength $\lambda = 0.2$ m

7. Which one of the following statements is true ? [2006]
(a) The sound waves in air are longitudinal while the light waves are transverse
(b) Both light and sound waves in air are longitudinal
(c) Both light and sound waves can travel in vacuum
(d) Both light and sound waves in air are transverse

8. A transverse wave propagating along x-axis is represented by $y(x,t) = 8.0\sin(0.5\pi x - 4\pi t - \frac{\pi}{4})$ where x is in metres and t is in seconds. The speed of the wave is [2006]
(a) 0.5π m/s (b) $\frac{\pi}{4}$ m/s
(c) 8 m/s (d) 4π m/s

9. The time of reverberation of a room A is one second. What will be the time (in seconds) of reverberation of a room, having all the dimensions double of those of room A? [2006]
(a) 4 (b) $\frac{1}{2}$
(c) 1 (d) 2

10. A point source emits sound equally in all directions in a non–absorbing medium. Two points P and Q are at distances of 2 m and 3 m respectively from the source. The ratio of the intensities of the waves at P and Q is [2005]
(a) 3 : 2 (b) 2 : 3
(c) 9 : 4 (d) 4 : 9

11. The phase difference between two waves, represented by
$y_1 = 10^{-6} \sin\{100\,t + (x/50) + 0.5\}$ m
$y_2 = 10^{-6} \cos\{100\,t + (x/50)\}$ m
where x is expressed in metres and t is expressed in seconds, is approximately [2004]
(a) 1.5 radians (b) 1.07 radians
(c) 2.07 radians (d) 0.5 radians

12. The equation for a transverse wave travelling along the positive x-axis with amplitude 0.2 m, velocity $v = 360\ \text{ms}^{-1}$ and wavelength $\lambda = 60$ m can be written as [2002]
(a) $y = 0.2\sin\left[2\pi\left(6t - \frac{x}{60}\right)\right]$
(b) $y = 0.2\sin\left[\pi\left(6t + \frac{x}{60}\right)\right]$
(c) $y = 0.2\sin\left[\pi\left(6t - \frac{x}{60}\right)\right]$
(d) $y = 0.2\sin\left[2\pi\left(6t + \frac{x}{60}\right)\right]$

13. The equation of a wave is represented by: $y = 10^{-4}\sin\left[100t - \frac{x}{10}\right]$. The velocity of the wave will be [2001]
(a) 100 m/s (b) 250 m/s
(c) 750 m/s (d) 1000 m/s

14. A transverse wave is represented by the equation $y = y_0 \sin\frac{2\pi}{\lambda}(vt - x)$
For what value of λ is the maximum particle velocity equal to two times the wave velocity? [1998]
(a) $\lambda = 2\pi y_0$ (b) $\lambda = \frac{\pi y_0}{3}$
(c) $\lambda = \frac{\pi y_0}{2}$ (d) $\lambda = \pi y_0$

15. A standing wave having 3 nodes and 2 antinodes is formed between two atoms having a distance 1.21 Å between them. The wavelength of the standing wave is [1998]
(a) 1.21 Å (b) 2.42 Å
(c) 6.05 Å (d) 3.63 Å

16. In a sinusoidal wave, the time required for a particular point to move from maximum displacement to zero displacement is 0.170 sec. The frequency of the wave is [1998]
(a) 1.47 Hz (b) 0.36 Hz
(c) 0.73 Hz (d) 2.94 Hz

17. The speed of a wave in a medium is 960 m/s. If 3600 waves are passing through a point in the medium in 1 min., then the wavelength of the wave is *[1997]*
(a) 8m (b) 12m
(c) 16m (d) 20m

18. The equation of a travelling wave is
$y = 60 \cos(180t - 6x)$
where y is in microns, t in second and x in metres. The ratio of maximum particle velocity to velocity of wave propagation is *[1997]*
(a) 3.6 (b) 3.6×10^{-4}
(c) 3.6×10^{-6} (d) 3.6×10^{-11}

19. Two sound waves having a phase difference of 60° have path difference of *[1996]*
(a) 2λ (b) $\frac{\lambda}{2}$
(c) $\frac{\lambda}{3}$ (d) $\frac{\lambda}{6}$

20. What is the effect of humidity on sound waves when humidity increases? *[1996]*
(a) speed of sound waves is more
(b) speed of sound waves is less
(c) speed of sound waves remains same
(d) speed of sound waves becomes zero

21. The equation of a sound wave is given as: $y = 0.0015 \sin(62.4x + 316t)$. The wavelength of this wave is *[1996]*
(a) 0.4 unit (b) 0.3 unit
(c) 0.2 unit (d) 0.1 unit

22. From a wave equation: $y = 0.5\sin\frac{2\pi}{3.2}(64t - x)$, the frequency of the wave is *[1995]*
(a) 5 Hz (b) 15 Hz
(c) 20 Hz (d) 25 Hz

23. Two waves are approaching each other with a velocity of 20 m/s and frequency n. The distance between two consecutive nodes is *[1995]*
(a) $\frac{20}{n}$ (b) $\frac{10}{n}$
(c) $\frac{5}{n}$ (d) $\frac{n}{10}$

24. The speed of a wave in a medium is 760 m/s. If 3600 waves are passing through a point in the medium in 2 min, then their wavelength is *[1995]*
(a) 13.8 m (b) 25.3 m
(c) 41.5 m (d) 57.2 m

25. A hospital uses an ultrasonic scanner to locate tumours in a tissue. The operating frequency of the scanner is 4.2 MHz. The speed of sound in a tissue is 1.7 km/s. The wavelength of sound in tissue is close to *[1995]*
(a) 4×10^{-4} m (b) 8×10^{-4} m
(c) 4×10^{-3} m (d) 8×10^{-3} m

26. A standing wave is represented by $y = A\sin(100t)\cos(0.01x)$, where y and A are in millimetre, t in seconds and x is in metre. Velocity of wave is *[1994]*
(a) 10^4 m/s
(b) 1 m/s
(c) 10^{-4} m/s
(d) not derivable from above data

27. Which of the following equations represent a wave? *[1994]*
(a) $y = A \sin \omega t$
(b) $y = A\cos kx$
(c) $y = A \sin(at - bx + c)$
(d) $y = A(\omega t - kx)$

28. A wave of frequency 100 Hz is sent along a string towards a fixed end. When this wave travels back after reflection, a node is formed at a distance of 10 cm from the fixed end of the string. The speeds of incident (and reflected) waves are *[1994]*
(a) 5 m/s (b) 10 m/s
(c) 20 m/s (d) 40 m/s

29. The temperature at which the speed of sound becomes double as was at 27°C is *[1993]*
(a) 273°C (b) 0°C
(c) 927°C (d) 1027°C

30. With the propagation of a longitudinal wave through a material medium, the quantities transmitted in the propagation direction are *[1992]*
(a) Energy, momentum and mass
(b) Energy
(c) Energy and mass
(d) Energy and linear momentum

31. The frequency of sinusoidal wave $y = 0.40 \cos[2000t + 0.80]$ would be *[1992]*
(a) 1000π Hz (b) 2000 Hz
(c) 20 Hz (d) $\frac{1000}{\pi}$ Hz

32. Velocity of sound waves in air is 330 m/s. For a particular sound wave in air, a path difference of 40 cm is equivalent to phase difference of 1.6 π. The frequency of this wave is *[1990]*
(a) 165 Hz (b) 150 Hz
(c) 660 Hz (d) 330 Hz

33. The transverse wave represented by the equation
$y = 4\sin\left(\frac{\pi}{6}\right)\sin(3x - 15t)$ has *[1990]*
(a) amplitude = 4
(b) wavelength $= 4\frac{\pi}{3}$
(c) speed of propagation = 5
(d) period $= \frac{\pi}{15}$

34. When sound waves travel from air to water, which of the following remains constant? *[1989]*
(a) Velocity (b) Wavelength
(c) Frequency (d) All of the above

35. Equation of a progressive wave is given by
$y = 4\sin\left[\pi\left(\frac{t}{5} - \frac{x}{9}\right) + \frac{\pi}{6}\right]$
Then which of the following is correct? *[1988]*
(a) $v = 5$ cm (b) $\lambda = 18$ cm
(c) $a = 0.04$ cm (d) f = 50 Hz

36. The velocity of sound in any gas depends upon *[1988]*
(a) wavelength of sound only
(b) density and elasticity of gas
(c) intensity of sound waves only
(d) amplitude and frequency of sound

Topic 2 Vibration of String & Organ Pipe

37. In a guitar, two strings A and B made of same material are slightly out of tune and produce beats of frequency 6 Hz. When tension in B is slightly decreased, the beat frequency increases to 7 Hz. If the frequency of A is 530 Hz, the original frequency of B will be *[2020]*
(a) 524 Hz (b) 536 Hz
(c) 537 Hz (d) 523 Hz

38. A tuning fork with frequency 800 Hz produces resonance in a resonance column tube with upper end open and lower end closed by water surface. Successive resonance are observed at lengths 9.75 cm, 31.25 cm and 52.75 cm. The speed of sound in air is, ***[NEET Odisha 2019]***
(a) 172 m/s (b) 500 m/s
(c) 156 m/s (d) 344 m/s

39. A tuning fork is used to produce resonance in a glass tube. The length of the air column in this tube can be adjusted by a variable piston. At room temperature of 27°C two successive resonances are produced at 20 cm and 73 cm of column length. If the frequency of the tuning fork is 320 Hz, the velocity of sound in air at 27°C is *[2018]*
(a) 330 m/s (b) 339 m/s
(c) 300 m/s (d) 350 m/s

40. The fundamental frequency in an open organ pipe is equal to the third harmonic of a closed organ pipe. If the length of the closed organ pipe is 20 cm, the length of the open organ pipe is *[2018]*
(a) 13.2 cm (b) 8 cm
(c) 16 cm (d) 12.5 cm

41. The two nearest harmonics of a tube closed at one end and open at other end are 220 Hz and 260 Hz. What is the fundamental frequency of the system? *[2017]*
(a) 20 Hz (b) 30 Hz
(c) 40 Hz (d) 10 Hz

42. An air column, closed at one end and open at the other, resonates with a tuning fork when the smallest length of the column is 50 cm. The next larger length of the column resonating with the same tuning fork is : *[2016]*
(a) 66.7 cm (b) 100 cm
(c) 150 cm (d) 200 cm

43. A uniform rope of length L and mass m_1 hangs vertically from a rigid support. A block of mass m_2 is attached to the free end of the rope. A transverse pulse of wavelength λ_1 is produced at the lower end of the rope. The wavelength of the pulse when it reaches the top of the rope is λ_2 the ratio λ_2/λ_1 is *[2016]*
(a) $\sqrt{\frac{m_1}{m_2}}$ (b) $\sqrt{\frac{m_1+m_2}{m_2}}$
(c) $\sqrt{\frac{m_2}{m_1}}$ (d) $\sqrt{\frac{m_1+m_2}{m_1}}$

44. The fundamental frequency of a closed organ pipe of length 20 cm is equal to the second overtone of an organ pipe open at both the ends. The length of organ pipe open at both the ends is *[2015]*
(a) 100 cm (b) 120 cm
(c) 140 cm (d) 80 cm

45. A string is stretched between two fixed points separated by 75.0 cm. It is observed to have resonant frequencies of 420 Hz and 315 Hz. There are no other resonant frequencies between these two. The lowest resonant frequency for this string is : *[2015 RS]*
(a) 205 Hz (b) 10.5 Hz
(c) 105 Hz (d) 155 Hz

46. If n_1, n_2 and n_3 are the fundamental frequencies of three segments into which a string is divided, then the original fundamental frequency n of the string is given by : *[2014, 2012, 2000]*
(a) $\frac{1}{n}=\frac{1}{n_1}+\frac{1}{n_2}+\frac{1}{n_3}$
(b) $\frac{1}{\sqrt{n}}=\frac{1}{\sqrt{n_1}}+\frac{1}{\sqrt{n_2}}+\frac{1}{\sqrt{n_3}}$
(c) $\sqrt{n}=\sqrt{n_1}+\sqrt{n_2}+\sqrt{n_3}$
(d) $n=n_1+n_2+n_3$

47. The number of possible natural oscillation of air column in a pipe closed at one end of length 85 cm whose frequencies lie below 1250 Hz are : (velocity of sound = 340 ms^{-1}) *[2014]*
(a) 4 (b) 5
(c) 7 (d) 6

48. If we study the vibration of a pipe open at both ends, then which of the following statements is not true ? *[2013]*
(a) Odd harmonics of the fundamental frequency will be generated
(b) All harmonics of the fundamental frequency will be generated
(c) Pressure change will be maximum at both ends
(d) Antinode will be at open end

49. The length of the wire between two ends of a sonometer is 100 cm. What should be the positions of two bridges below the wire so that the three segments of the wire have their fundamental frequencies in the ratio of 1 : 3 : 5? *[NEET Kar. 2013]*
(a) $\frac{1500}{23}$ cm, $\frac{2000}{23}$ cm
(b) $\frac{1500}{23}$ cm, $\frac{500}{23}$ cm
(c) $\frac{1500}{23}$ cm, $\frac{300}{23}$ cm
(d) $\frac{300}{23}$ cm, $\frac{1500}{23}$ cm

50. A string of 7 m length has a mass of 0.035 kg. If tension in the string is 60.5 N, then speed of a wave on the string is *[2001, 1989]*
(a) 77 m/s (b) 102 m/s
(c) 110 m/s (d) 165 m/s

51. An organ pipe P_1 closed at one end vibrating in its first overtone and another pipe P_2, open at both ends vibrating in its third overtone are in resonance with a given tuning fork. The ratio of lengths of P_1 and P_2 respectively are given by *[1997]*
(a) 1 : 2 (b) 1 : 3
(c) 3 : 8 (d) 3 : 4

52. A cylindrical resonance tube open at both ends, has a fundamental frequency, '*f*' in air. If half of the length is dipped vertically in water, the fundamental frequency of the air column will be *[1997]*
(a) $2f$ (b) $3f/2$
(c) f (d) $f/2$

53. A stretched string resonates with tuning fork frequency 512 Hz when length of the string is 0.5 m. The length of the string required to vibrate resonantly with a tuning fork of frequency 256 Hz would be *[1993]*
(a) 0.25 m (b) 0.5 m
(c) 1 m (d) 2 m

54. A closed organ pipe (closed at one end) is excited to support the third overtone. It is found that air in the pipe has *[1991]*
(a) three nodes and three antinodes
(b) three nodes and four antinodes
(c) four nodes and three antinodes
(d) four nodes and four antinodes

Topic 3 Beats, Interference & Superposition of Waves

55. A source of unknown frequency gives 4 beats/s, when sounded with a source of known frequency 250 Hz. The second harmonic of the source of unknown frequency gives five beats per second, when sounded with a source of frequency 513 Hz. The unknown frequency is *[2013]*
(a) 246 Hz (b) 240 Hz
(c) 260 Hz (d) 254 Hz

56. Two sources *P* and *Q* produce notes of frequency 660 Hz each. A listener moves from *P* to *Q* with a speed of 1 ms^{-1}. If the speed of sound is 330 m/s, then the number of beats heard by the listener per second will be *[NEET Kar. 2013]*
(a) zero (b) 4
(c) 8 (d) 2

57. Two sources of sound placed close to each other are emitting progressive waves given by $y_1 = 4 \sin 600\pi t$ and $y_2 = 5 \sin 608\pi t$. An observer located near these two sources of sound will hear : *[2012]*
(a) 4 beats per second with intensity ratio 25 : 16 between waxing and waning.
(b) 8 beats per second with intensity ratio 25 : 16 between waxing and waning
(c) 8 beats per second with intensity ratio 81 : 1 between waxing and waning
(d) 4 beats per second with intensity ratio 81 : 1 between waxing and waning

58. Two identical piano wires kept under the same tension T have a fundamental frequency of 600 Hz. The fractional increase in the tension of one of the wires which will lead to occurrence of 6 beats/s when both the wires oscillate together would be *[2011M]*
(a) 0.02 (b) 0.03
(c) 0.04 (d) 0.01

59. A tuning fork of freqqency 512 Hz makes 4 beats per second with the vibrating string of a piano. The beat frequency decreases to 2 beats per sec when the tension in the piano string is slightly increased. The frequency of the piano string before increasing the tension was *[2010]*
(a) 510 Hz (b) 514 Hz
(c) 516 Hz (d) 508 Hz

60. Each of the two strings of length 51.6 cm and 49.1 cm are tensioned separately by 20 N force. Mass per unit length of both the strings is same and equal to 1 g/m. When both the strings vibrate simultaneously the number of beats is *[2009]*
(a) 7 (b) 8
(c) 3 (d) 5

61. Two sound waves with wavelengths 5.0 m and 5.5m respectively, each propagate in a gas with velocity 330 m/s. We expect the following number of beats per second *[2006]*
(a) 0 (b) 1
(c) 6 (d) 12

62. Two vibrating tuning forks produce progressive waves given by $y_1 = 4 \sin 500\pi t$ and $y_2 = 2 \sin 506\pi t$. Number of beats produced per minute is *[2005]*
(a) 360 (b) 180
(c) 60 (d) 3

63. Two sound sources emitting sound each of wavelength λ are fixed at a given distance apart. A listener moves with a velocity u along the line joining the two sources. The number of beats heard by him per second is *[2000]*
(a) $\frac{u}{2\lambda}$ (b) $\frac{2u}{\lambda}$
(c) $\frac{u}{\lambda}$ (d) $\frac{u}{3\lambda}$

64. Two waves of lengths 50 cm and 51 cm produce 12 beats per sec. The velocity of sound is *[1999]*
(a) 306 m/s (b) 331 m/s
(c) 340 m/s (d) 360 m/s

65. Two waves of the same frequency and intensity superimpose each other in opposite phases. After the superposition, the intensity and frequency of waves will *[1996]*
(a) increase (b) decrease
(c) remain constant (d) become zero

66. A source of sound gives 5 beats per second, when sounded with another source of frequency 100/sec. The second harmonic of the source, together with a source of frequency 205/sec gives 5 beats per second. What is the frequency of the source? *[1995]*
(a) 95 sec^{-1} (b) 100 sec^{-1}
(c) 105 sec^{-1} (d) 205 sec^{-1}

67. For production of beats the two sources must have [1992]
(a) different frequencies and same amplitude
(b) different frequencies
(c) different frequencies, same amplitude and same phase
(d) different frequencies and same phase

68. A wave has S.H.M whose period is 4 seconds while another wave which also possess SHM has its period 3 seconds. If both are combined, then the resultant wave will have the period equal to [1993]
(a) 4 seconds (b) 5 seconds
(c) 12 seconds (d) 3 seconds

69. If the amplitude of sound is doubled and the frequency is reduced to one fourth, the intensity of sound at the same point will be [1989]
(a) increasing by a factor of 2
(b) decreasing by a factor of 2
(c) decreasing by a factor of 4
(d) unchanged

Topic 4 Musical Sound & Doppler's Effect

70. Two cars moving in opposite directions approach each other with speed of 22 m/s and 16.5 m/s respectively. The driver of the first car blows a horn having a frequency 400 Hz. The frequency heard by the driver of the second car is [velocity of sound 340 m/s] : [2017]
(a) 361 Hz (b) 411 Hz
(c) 448 Hz (d) 350 Hz

71. A siren emitting a sound of frequency 800 Hz moves away from an observer towards a cliff at a speed of $15 ms^{-1}$. Then, the frequency of sound that the observer hears in the echo reflected from the cliff is : [2016]
(Take velocity of sound in air = $330 ms^{-1}$)
(a) 765 Hz (b) 800 Hz
(c) 838 Hz (b) 885 Hz

72. A source of sound S emitting waves of frequency 100 Hz and an observor O are located at some distance from each other. The source is moving with a speed of $19.4 ms^{-1}$ at an angle of 60° with the source observer line as shown in the figure. The observor is at rest. The apparent frequency observed by the observer is (velocity of sound in air $330 ms^{-1}$) [2015 RS]

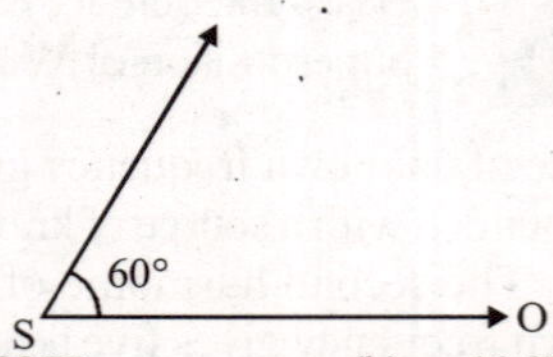

(a) 103 Hz (b) 106 Hz
(c) 97 Hz (d) 100 Hz

73. A speeding motorcyclist sees trafic jam ahead of him. He slows down to 36 km/hour. He finds that traffic has eased and a car moving ahead of him at 18 km/hour is honking at a frequency of 1392 Hz. If the speeds of sound is 343 m/s, the frequency of the honk as heard by him will be : [2014]
(a) 1332 Hz (b) 1372 Hz
(c) 1412 Hz (d) 1464 Hz

74. A train moving at a speed of $220 ms^{-1}$ towards a stationary object, emits a sound of frequency 1000 Hz. Some of the sound reaching the object gets reflected back to the train as echo. The frequency of the echo as detected by the driver of the train is : [2012M]
(speed of sound in air is $330 ms^{-1}$)
(a) 3500 Hz (b) 4000 Hz
(c) 5000 Hz (d) 3000 Hz

75. The driver of a car travelling with speed 30 m/sec towards a hill sounds a horn of frequency 600 Hz. If the velocity of sound in air is 330 m/s, the frequency of reflected sound as heard by driver is [2009]
(a) 555.5 Hz (b) 720 Hz
(c) 500 Hz (d) 550 Hz

76. A car is moving towards a high cliff. The car driver sounds a horn of frequency f. The reflected sound heard by the driver has frequency $2f$. If v be the velocity of sound, then the velocity of the car, in the same velocity units, will be [2004]
(a) $v/2$ (b) $v/\sqrt{2}$
(c) $v/3$ (d) $v/4$

77. An observer moves towards a stationary source of sound with a speed 1/5th of the speed of sound. The wavelength and frequency of the sound emitted are λ and f respectively. The apparent frequency and wavelength recorded by the observer are respectively. [2003]

(a) 0.8f, 0.8λ (b) 1.2*f*, 1.2λ
(c) 1.2*f*, λ (d) *f*, 1.2λ

78. A whistle of frequency 385 Hz rotates in a horizontal circle of radius 50 cm at an angular speed of 20 radians s^{-1}. The lowest frequency heard by a listener a long distance away at rest with respect to the centre of the circle, given velocity of sound equal to 340 ms^{-1}, is *[2002]*
(a) 396 Hz (b) 363 Hz
(c) 374 Hz (d) 385 Hz

79. A source and an observer move away from each other, with a velocity of 10m/s with respect to ground. If the observer finds the frequency of sound coming from the source as 1950 Hz, then original frequency of source is (velocity of sound in air = 340 m/s) *[2001]*
(a) 1950 Hz (b) 2068 Hz
(c) 2132 Hz (d) 2486 Hz

80. A vehicle, with a horn of frequency n is moving with a velocity of 30 m/s in a direction perpendicular to the straight line joining the observer and the vehicle. The observer perceives the sound to have a frequency $n + n_1$. Then (if the sound velocity in air is 300 m/s) *[1998]*
(a) $n_1 = 10n$ (b) $n_1 = 0$
(c) $n_1 = 0.1n$ (d) $n_1 = -0.1n$

81. A star, which is emitting radiation at a wavelength of 5000 Å, is approaching the earth with a velocity of 1.50×10^6 m/s. The change in wavelength of the radiation as received on the earth is *[1996]*
(a) 0.25 Å (b) 2.5 Å
(c) 25 Å (d) 250 Å

82. Two trains move towards each other with the same speed. The speed of sound is 340 m/s. If the height of the tone of the whistle of one of them heard on the other changes 9/8 times, then the speed of each train should be *[1991]*
(a) 20 m/s (b) 2 m/s
(c) 200 m/s (d) 2000 m/s

ANSWER KEY

1	(d)	11	(b)	21	(d)	31	(d)	41	(a)	51	(c)	61	(c)	71	(c)	81	(c)
2	(a)	12	(a)	22	(c)	32	(c)	42	(c)	52	(c)	62	(b)	72	(a)	82	(a)
3	(c)	13	(d)	23	(b)	33	(c)	43	(b)	53	(d)	63	(b)	73	(c)		
4	(c)	14	(d)	24	(b)	34	(c)	44	(b)	54	(d)	64	(a)	74	(c)		
5	(c)	15	(a)	25	(a)	35	(b)	45	(c)	55	(d)	65	(c)	75	(b)		
6	(c)	16	(a)	26	(a)	36	(b)	46	(a)	56	(b)	66	(c)	76	(c)		
7	(a)	17	(c)	27	(c)	37	(a)	47	(d)	57	(d)	67	(b)	77	(c)		
8	(c)	18	(b)	28	(c)	38	(d)	48	(c)	58	(a)	68	(c)	78	(c)		
9	(d)	19	(d)	29	(c)	39	(b)	49	(a)	59	(d)	69	(c)	79	(b)		
10	(c)	20	(a)	30	(b)	40	(a)	50	(c)	60	(a)	70	(c)	80	(b)		

Hints & Solutions

1. **(d)** The standard equation of a wave travelling along x-axis (+ve direction) is given by, $Y = A \sin(kx - \omega t)$ where,
angular frequency, $\omega = 2\pi f$

$$\Rightarrow \frac{2\pi}{\pi} = 2 \qquad [\because f = \frac{1}{\pi}]$$

angular wave number, $k = \frac{2\pi}{\lambda} \Rightarrow \frac{2\pi}{2\pi} = 1$
$[\because \lambda = 2\pi]$

$\therefore Y = 1 \sin(x - 2t)$ [∵ Amplitude, A = 1 m]

NOTES: If the sign between t and x terms is negative the wave is propagating along +(ve) X-axis and if the sign is positive then wave moves in –(ve) X-axis direction.

2. **(a)** Here, $y_1 = a\sin(\omega t + kx + 0.57)$
so, $\phi_1 = 0.57$
and $y_2 = a\cos(\omega t + kx)$

$$= a \sin\left[\frac{\pi}{2} + (\omega t + kx)\right]$$

so, $\phi_2 = \pi/2$

Phase difference, $\Delta\phi = \phi_2 - \phi_1$

$$= \frac{\pi}{2} - 0.57 = \frac{3.14}{2} - 0.57 = 1.57 - 0.57$$

$$= 1 \text{ radian}$$

3. (c) We have, $v = n\lambda$

$\Rightarrow v \propto \lambda$ (as n remains constant)

Thus, as v increases 10 times, λ also increases 10 times.

4. (c) $y = A \sin(\omega t - kx)$

Particle velocity,

$$v_p = \frac{dy}{dt} = A\,\omega\cos(\omega t - kx)$$

$\therefore \quad v_{p\,max} = A\omega$

wave velocity $= \frac{\omega}{k}$

$\therefore \quad A\omega = \frac{\omega}{k}$

i.e., $A = \frac{1}{k}$ But $k = \frac{2\pi}{\lambda}$

$\therefore \quad \lambda = 2\pi A$

5. (c) $A = 2\text{cm}, \frac{\omega}{k} = 128\,\text{ms}^{-1}, 5\lambda = 4, \lambda = \frac{4}{5}\text{m}$

$y = A \sin(kx - \omega t)$,

$$k = \frac{2\pi}{\lambda} \Rightarrow \frac{2\pi \times 5}{4} = \frac{31.4}{4} = 7.85$$

$y = 0.02$ m $\sin(7.85x - 1005\,t)$

$\omega = 128 \times 7.85 = 1005$

6. (c) $y = 0.25 \sin(10\pi x - 2\pi t)$

Comparing this equation with the standard wave equation

$y = a \sin(kx - \omega t)$

We get, $k = 10\pi$

$$\Rightarrow \frac{2\pi}{\lambda} = 10\pi \Rightarrow \lambda = 0.2 \text{ m}$$

And $\omega = 2\pi$ or, $2\pi v = 2\pi \Rightarrow v = 1\text{Hz}$.

The sign inside the bracket is negative, hence the wave travels in +ve x-direction.

7. (a) Sound waves in air are longitudinal and the light waves are transverse.

NOTES

Sound waves are longitudinal waves and require medium for their propagation.

Light waves *i.e.*, electromagnetic waves are transverse in nature and do not require medium for their propagation.

8. (c) Speed of a wave represented by the equation

$y(x, t) = A \sin(kx - \omega t + \phi)$ is $v = \frac{\omega}{k}$

By comparison, $\omega = 4\pi$; $k = 0.5\pi$

$$v = \frac{\omega}{k} = \frac{4\pi}{0.5\pi} = 8 \text{ m/sec}$$

9. (d) The time gap between the initial direct note and the reflected note upto the minimum andibility level is called reverberation time.

Sabine has shown that standard reverberation time for an auditorium is given by the formula

$$T_R = K\frac{V}{\alpha S}$$

Where, V is volume of the auditorium, S is the surface area. So, $T_R = \frac{K.V}{\alpha\, S} = 1$ (given)

$$T_R = \frac{K.l^3}{6\alpha l^2}$$

(Assuming auditorium to be cubic in shape)

$$= \frac{K}{6\alpha}l$$

So, $T_R \,\alpha\, l$

If dimension is doubled, reverberation time t will be doubled. So, New $T_R = 2$ sec.

10. (c) Intensity = Energy/sec/unit area

Area $\propto r^2 \Rightarrow I \propto 1/r^2$

$$\Rightarrow \frac{I_1}{I_2} = \frac{r_2^2}{r_1^2} = \frac{3^2}{2^2} = \frac{9}{4}$$

11. (b) $y_1 = 10^{-6} \sin(100\,t + x/50 + 0.5)$m

$= 10^{-6} \cos(100\,\text{t} + x/50 - \pi/2 + 0.5)$m

$y_2 = 10^{-6} \cos(100\,t + x/50)$m

$\therefore \quad \phi = \pi/2 - 0.5 = 1.07$ rad

12. (a) As we know that, velocity $v = n\lambda$

$$\Rightarrow \text{frequency, } n = \frac{v}{\lambda} = \frac{360}{60} = 6$$

Amplitude, $a = 0.2$

For a wave travelling along positive x-axis

$y = a\sin(\omega t - kx)$

$= a\sin\left(2\pi nt - \frac{2\pi x}{\lambda}\right)$

$= a\sin 2\pi\left(nt - \frac{x}{\lambda}\right) = 0.2\sin 2\pi\left(6t - \frac{x}{60}\right)$

13. (d) $y = 10^{-4}\sin\left[100t - \frac{x}{10}\right]$

frequency, $n = \left(\frac{100}{2\pi}\right)\text{sec}^{-1}$, $\lambda = 2\pi 10$ m

velocity $v = \lambda n = \left(\frac{100}{2\pi}\right)(2\pi\times 10) = 1000$ m/s

14. (d) $y = y_0 \sin\frac{2\pi}{\lambda}(vt - x)$

Particle velocity,

$\frac{dy}{dt} = y_0 \times \frac{2\pi}{\lambda} v \cos\frac{2\pi}{\lambda}(vt - x)$

Maximum particle velocity $= y_0 \times \frac{2\pi v}{\lambda}$

Wave velocity $= v$ [given]

So, $y_0 \times \frac{2\pi v}{\lambda} = 2v$

$\lambda = \pi \cdot y_0$

15. (a) Let ℓ be length of string

$\ell = \left(\frac{\lambda}{2}\right)2 \Rightarrow \lambda = \ell$

Hence, the wave length of standing wave

$\Rightarrow \lambda = \ell = 1.21$Å

16. (a) Time taken to move from maximum to zero displacement $= \frac{T}{4}$

$\therefore$ Time period $T = 4 \times 0.170$ second

$\therefore$ Frequency, $n = \frac{1}{T} = \frac{1}{4\times 0.170} = 1.47$ Hz

17. (c) Given speed of wave $(v) = 960$m/s

Frequency of wave $(f) = 3600/$ min

$= \frac{3600}{60}$ rev/sec $= 60$ rev per sec.

Wavelength of waves $(\lambda) = \frac{v}{f} = \frac{960}{60} = 16$m.

18. (b) $y = 60\cos(180t - 6x)$ ....(1)

$\omega = 180, k = 6 \Rightarrow \frac{2\pi}{\lambda} = 6$

Wave velocity,

$v = \frac{\omega}{k} = \frac{2\pi}{T} \times \frac{\lambda}{2\pi} = \frac{180}{6} = 30$ m/s

Differentiating (1) w.r.t. t,

Particle velocity,

$v_p = \frac{dy}{dt} = -60 \times 180\sin(180t - 6x)$

$v_{p\,max} = 60 \times 180$ μm/s
$= 10800$ μm/s $= 0.0108$ m/s

$\frac{v_{p\,max}}{v} = \frac{0.0108}{30} = 3.6 \times 10^{-4}$

NOTES

Wave velocity (v) is the distance travelled by the disturbance in one second. It only depends on the properties of the medium and is independent of time and position.

$V = n\lambda = \frac{\lambda}{T} = \frac{\omega\lambda}{2\pi} = \frac{\omega}{k}$

19. (d) Phase difference $= 60° = \frac{\pi}{3}$

Path difference $= \frac{\lambda}{2\pi}$ (phase diff.)

$= \frac{\lambda}{2\pi} \times \frac{\pi}{3} = \frac{\lambda}{6}$

20. (a) Velocity of sound $= \sqrt{\frac{\gamma RT}{M}}$

When water vapour are present in air, average molecular weight of air decreases and hence velocity increases.

21. (d) $y = 0.0015\sin(62.4x + 316t)$

On comparing with $y = A\sin(\omega t + kx)$

$\omega = 316, k = 62.4$

$\Rightarrow k = \frac{2\pi}{\lambda} = 62.4 \Rightarrow \lambda = 0.1$ unit

22. (c) Given, $y = 0.5\sin\frac{2\pi}{3.2}(64t - x)$

Standard equation of the wave is :

$y = a\sin\frac{2\pi}{\lambda}(vt - x)$

Comparing the given equation with the standard equation, we get $v = 64$ and $\lambda = 3.2$. Therefore,

frequency $= \frac{64}{3.2} = 20$ Hz.

23. (b) Distance between two successive nodes

$\Rightarrow \frac{\lambda}{2} = \frac{v}{2n} \Rightarrow \frac{20}{2n} = \frac{10}{n}$ $\left(\text{where } \lambda = \frac{v}{n}\right)$

24. (b) Speed of the wave (v) = 760 m/s;
Number of waves = 3600
and time taken (t) = 2 min = 120 sec.
Frequecny of waves

$$(n) = \frac{3600}{120} = 30/\text{sec}.$$

$$\therefore \text{wavelength of waves } (\lambda) = \frac{v}{n} = \frac{760}{30} = 25.3\text{m}$$

25. (a) Frequency (n) = 4.2 MHz = 4.2×10^6 Hz and speed of sound (v) = 1.7 km/s = 1.7×10^3 m/s. Wave length of sound in tissue,

$$\lambda = \frac{v}{n} = \frac{1.7\times10^3}{4.2\times10^6} = 4\times10^{-4} m$$

26. (a) The wave equation is $y = A\sin(\omega t)\cos(kx)$;
$c = \omega/k = 100/0.01 = 10^4$ m/s.

27. (c) $y = A \sin(at - bx + c)$ represents a wave, where a may correspond to ω and b may correspond to K.

28. (c) As fixed end is a node, therefore, distance between two consecutive nodes

$$= \frac{\lambda}{2} = 10 \text{ cm.}$$

λ = 20 cm = 0.2 m
As $v = n\lambda$ $\therefore$ $v = 100 \times 0.2 = 20$ m/s

29. (c) We have $c \propto \sqrt{T}$

$$\therefore \frac{c_1}{c_2} = \sqrt{\frac{T_1}{T_2}}$$

$$\Rightarrow \frac{c_1}{2c_2} = \sqrt{\frac{(27+273)k}{T_2}}$$

$\Rightarrow T_2 = 1200\ k = 927°$C

30. (b) With the propagation of a longitudinal wave, energy alone is propagated.

NOTES Energy is propagated along with the wave motion without any net transport of the matter.

31. (d) Given $y = 0.40 \cos[2000t + 0.80]$
Comparing with the standard equation
$y = A \cos(2\pi\nu t + \phi)$
We get angular frequency, $2\pi\nu = 2000$

$$\nu \Rightarrow \frac{2000}{2\pi} = \frac{1000}{\pi}\text{Hz}$$

32. (c) From $\Delta x = \dfrac{\lambda}{2\pi}\Delta\phi$,

$$\lambda = 2\pi\frac{\Delta x}{\Delta\phi} = \frac{2\pi(0.4)}{1.6\pi} = 0.5\text{m}$$

$$f = \frac{v}{\lambda} = \frac{330}{0.5} = 660 \text{ Hz}$$

33. (c) Given $y = 4\sin\left(\frac{\pi}{6}\right)\sin(3x - 15t)$
Compare the given equation with standard form

$$y = A\sin\left[\frac{2\pi x}{\lambda} - \frac{2\pi t}{T}\right]$$

where, $\dfrac{2\pi}{\lambda} = 3, \lambda = \dfrac{2\pi}{3}$ and $\dfrac{2\pi}{T} = 15$

$$T = \frac{2\pi}{15}$$

Speed of propagation,

$$v = \frac{\lambda}{T} = \frac{2\pi/3}{2\pi/15} = 5$$

34. (c)

35. (b) The standard equation of a progressive wave is

$$y = a\sin\left[2\pi\left(\frac{t}{T} - \frac{x}{\lambda}\right) + \phi\right]$$

The given equation can be written as

$$y = 4\sin\left[2\pi\left(\frac{t}{10} - \frac{x}{18}\right) + \frac{\pi}{6}\right]$$

$\therefore$ $a = 4$ cm, $T = 10$ s, $\lambda = 18$ cm and $\phi = \pi/6$
Hence, (b) is correct.

36. (b) Velocity of sound in any gas depends upon density and elasticity of gas.

37. (a) Frequency of string, $f = \dfrac{1}{2l}\sqrt{\dfrac{T}{m}}$

Frequency $\propto \sqrt{\text{Tension}}$
Difference of f_A and f_B is 6 Hz.
If tension decreases, f_B decreases and becomes f'_B.
Now, difference of f_A and f'_B = 7 Hz (increases)
So, $f_A > f_B$
$f_A - f_B = 6$ Hz
$\Rightarrow f_A = 530$ Hz $\Rightarrow f_B = 524$ Hz (original)

38. (d) $l_1 = 9.75$ cm
$l_2 = 31.25$ cm
$l_3 = 52.75$ cm
e = end correction

$$\frac{\lambda}{4} + e = 9.75 \text{ cm} \quad ...(i)$$

$$\frac{3\lambda}{4} + e = 31.25 \text{ cm} \quad ...(ii)$$

$$\frac{3\lambda}{4} - \frac{\lambda}{4} = 31.25 - 9.75$$

$$\frac{\lambda}{2} = 21.5$$

$$\lambda = 43 \text{ cm}$$

$$v = \lambda \times f$$

$$v = 34400 \times 10^{-2}$$

$$\therefore \quad v = 344 \text{ ms}^{-1}$$

39. **(b)** Two successive resonance are produced at 20 cm and 73 cm of column length

$$\therefore \quad \frac{\lambda}{2} = (73 - 20) \times 10^{-2} \text{ m}$$

$$\Rightarrow \quad \lambda = 2 \times (73 - 20) \times 10^{-2}$$

Velocity of sound, $v = n\lambda$

$= 2 \times 320\,[73 - 20] \times 10^{-2}$

$= 339.2 \text{ ms}^{-1}$

40. **(a)** For closed organ pipe, third harmonic

$$n = \frac{(2N-1)V}{4\ell} = \frac{3V}{4\ell} \quad (\because N = 2)$$

For open organ pipe, fundamental frequenty

$$n = \frac{NV}{2\ell} = \frac{V}{2\ell'} \quad (\because N = 1)$$

According to question, $\frac{3V}{4\ell} = \frac{V}{2\ell'}$

$$\Rightarrow \quad \ell' = \frac{4\ell}{3\times 2} = \frac{2\ell}{3} = \frac{2\times 20}{3} = 13.33 \text{ cm}$$

41. **(a)** Difference in two successive frequencies of closed pipe,

$$\frac{2v}{4l} \Rightarrow 260 - 220 = 40 \text{ Hz}$$

$$\text{or } \frac{2v}{4l} = 40 \text{ Hz}$$

$$\Rightarrow \quad \frac{v}{4\ell} = 20\text{Hz}$$

Which is the fundamental frequency of system of closed organ pipe.

42. **(c)** For a closed organ pipe first minimum resonating length

$$L_1 = \frac{\lambda}{4} = 50 \text{ cm}$$

$\therefore$ Next or second resonating length, L_2

$$\Rightarrow \quad \frac{3\lambda}{4} = 150 \text{ cm}$$

If an open pipe is half submerged in water, it will become a closed organ pipe of length half that of a open pipe. It's fundamental frequency will become

$$n' = \frac{V}{4\left(\frac{\ell}{2}\right)} = \frac{V}{2\ell} \text{ equal to that of open pipe.}$$

43. **(b)** From figure, tension $T_1 = m_2 g$

$T_2 = (m_1 + m_2)g$

As we know

Velocity $\propto \sqrt{T}$

So, $\lambda \propto \sqrt{T}$

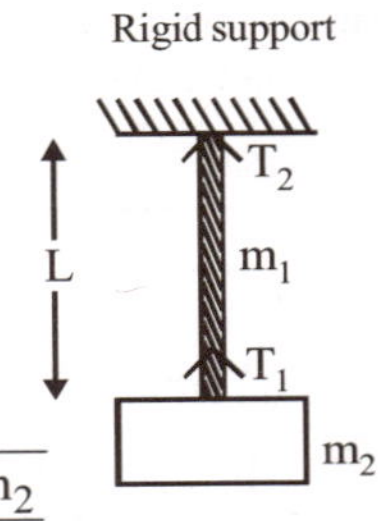

$$\Rightarrow \quad \frac{\lambda_1}{\lambda_2} = \frac{\sqrt{T_1}}{\sqrt{T_2}}$$

$$\Rightarrow \quad \frac{\lambda_2}{\lambda_1} = \sqrt{\frac{m_1 + m_2}{m_2}}$$

44. **(b)** Fundamental frequency of closed organ pipe,

$$v_c = \frac{v}{4l_c}$$

Fundamental frequency of open organ pipe

$$v_0 = \frac{v}{2l_0}$$

Second overtone frequency of open organ pipe

$$= \frac{3v}{2l_0}$$

From question, $\frac{v}{4l_c} = \frac{3v}{2l_0}$

$$\Rightarrow \quad l_0 = 6l_c = 6 \times 20 = 120 \text{ cm}$$

45. **(c)** Resonant frequencies, for a string fixed at both ends, will be

$$f_n = \frac{nv}{2L} \quad \text{where, } n = 1, 2, 3.....$$

The difference between two consecutive resonant frequency is,

$$A_{fn} = f_{n+1} - f_n \Rightarrow \frac{(n+1)v}{2L} - \frac{nv}{2L} = \frac{v}{2L}$$

which is also the lowest resonant frequency (n = 1).

Thus, for the given string the lowest resonant frequency will be

$= 420 \text{ Hz} - 315 \text{ Hz} = 105 \text{ Hz}$

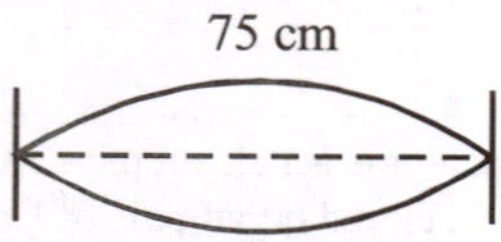

The difference between any two successive frequencies will be 'n'
According to question, n = 420 – 315 = 105 Hz
So the lowest frequency of the string is 105 Hz.

> **NOTES** In a stretched string all multiples of frequencies can be obtained *i.e.*, if fundamental frequency is n then higher frequencies will be 2n, 3n, 4n ...

46. (a) 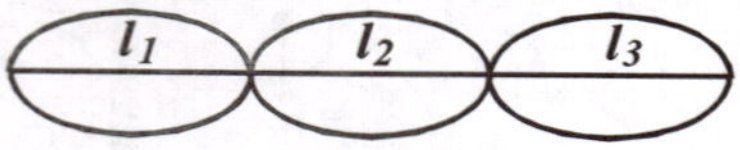

$$n = \frac{1}{2l}\sqrt{\frac{T}{m}}$$

or, $n \propto \frac{1}{l}$ or nl = constant, K

$\therefore n_1 l_1 = K,$

$n_2 l_2 = K, n_3 l_3 = K$

Also, $l = l_1 + l_2 + l_3$

or, $\frac{K}{n} = \frac{K}{n_1} + \frac{K}{n_2} + \frac{K}{n_3}$

or, $\frac{1}{n} = \frac{1}{n_1} + \frac{1}{n_2} + \frac{1}{n_3}$

47. (d) In case of closed organ pipe frequency,

$$f_n = (2n + 1)\frac{v}{4l}$$

for n = 0, f_0 = 100 Hz
n = 1, f_1 = 300 Hz
n = 2, f_2 = 500 Hz
n = 3, f_3 = 700 Hz
n = 4, f_4 = 900 Hz
n = 5, f_5 = 1100 Hz
n = 6, f_6 = 1300 Hz

Hence possible natural oscillation whose frequencies < 1250 Hz = 6(n = 0, 1, 2, 3, 4, 5)

48. (c) Pressure change will be minimum at both ends. In fact, pressure variation is maximum at $\ell/2$ because the displacement node is pressure antinode.

49. (a) From formula, $f = \frac{1}{x}\sqrt{\frac{T}{m}}$

$\Rightarrow f \propto \frac{1}{l}$

$\therefore l_1 : l_2 : l_3 = \frac{1}{f_1} : \frac{1}{f_2} : \frac{1}{f_3}$

$= f_2 f_3 : f_1 f_3 : f_1 f_2$

[Given: $f_1 : f_2 : f_3 = 1 : 3 : 5$]

= 15 : 5 : 3

Therefore the positions of two bridges below the wire are

$\frac{15\times100}{15+5+3}$ cm and $\frac{15\times100+5\times100}{15+5+3}$ cm

i.e., $\frac{1500}{23}$ cm, $\frac{2000}{23}$ cm

50. (c) Given : Length (l) = 7 m
Mass (M) = 0.035 kg and tension (T) = 60.5 N.
Therefore, mass of string per unit length (m) =

$\frac{0.035}{7} = 0.005$ kg/m

speed of wave, $= \sqrt{\frac{T}{m}}$

$\Rightarrow \sqrt{\frac{60.5}{0.005}} = 110$ m/s

51. (c) We know that the length of pipe closed at one end for first overtone $(l_1) = \frac{3\lambda}{4}$
and length of the open pipe for third overtone

$(l_2) = \frac{4\lambda}{2} = 2\lambda$.

Therefore, the ratio of lengths $\frac{l_1}{l_2} = \frac{3\lambda/4}{2\lambda} = \frac{3}{8}$
or $l_1 : l_2 = 3 : 8$.

> **NOTES** In an open pipe all harmonics are present whereas in a closed organ pipe only alternate harmonics of frequencies are present. Hence musical sound produced by an open organ pipe is sweeter than that produced by a closed organ pipe.

52. (c) Fundamental frequency of open pipe,

$f = \frac{v}{2l}$

When half of tube is filled with water, then the length of air column becomes half $\left(l' = \frac{l}{2}\right)$ and the pipe becomes closed.
So, new fundamental frequency

$$f' = \frac{v}{4l'} = \frac{v}{4\left(\frac{l}{2}\right)} = \frac{v}{2l}$$

Clearly $f' = f$.

53. (d) $f = \frac{1}{2l}\left[\frac{T}{\mu}\right]^{\frac{1}{2}}$. When f is halved, the length is doubled.

54. (d) Third overtone has a frequency 7 n, which means $L = \frac{7\lambda}{4}$ = three full loops + one half loop, which would make four nodes and four antinodes.

55. (d) When sounded with a source of known frequency fundamental frequency
= 250 ± 4 Hz = 254 Hz or 246 Hz
2nd harmonic if unknown frequency (suppose) 254 Hz = 2 × 254 = 508 Hz
As it gives 5 beats
∴ 508 + 5 = 513 Hz
Hence, unknown frequency is 254 Hz

56. (b) $\frac{\Delta f}{f} = \frac{v}{c}$

$$\Rightarrow \frac{(\text{Beats})/2}{f} = \frac{v}{c}$$

$$\Rightarrow \text{Beats} = \frac{2fv}{c} = 4$$

57. (d) $2\pi f_1 = 600\pi$

$f_1 = 300$ ... (1)

$2\pi f_2 = 608\pi$

$f_2 = 304$ ... (2)

$|f_1 - f_2| = 4$ beats

$$\frac{I_{max}}{I_{min}} = \frac{(A_1 + A_2)^2}{(A_1 + A_2)^2} = \frac{(5+4)^2}{(5-4)^2} = \frac{81}{1}$$

where A_1, A_2 are amplitudes of given two sound wave.

58. (a) For fundamental mode, $f = \frac{1}{2\ell}\sqrt{\frac{T}{\mu}}$

Taking logarithm on both sides, we get

$$\Rightarrow \log f = \log\left(\frac{1}{2\ell}\right) + \log\left(\sqrt{\frac{T}{\mu}}\right)$$

$$\Rightarrow \log\left(\frac{1}{2\ell}\right) + \frac{1}{2}\log\left(\frac{T}{\mu}\right)$$

or $\log f = \log\left(\frac{1}{2\ell}\right) + \frac{1}{2}[\log T - \log \mu]$

Differentiating both sides, we get

$\frac{df}{f} = \frac{1}{2}\frac{dT}{T}$ (as ℓ and μ are constants)

$$\Rightarrow \frac{dT}{T} = 2 \times \frac{df}{f}$$

Here df = 6

f = 600 Hz

$$\therefore \frac{dT}{T} = \frac{2 \times 6}{600} = 0.02$$

59. (d) The frequency of the piano string = 512 ± 4 = 516 or 508. When the tension is increased, beat frequency decreases to 2, it means that frequency of the string is 508 as frequency of string increases with tension.

60. (a) The frequency of vibration of a string is given by, $f = \frac{1}{2l}\sqrt{\frac{T}{m}}$ where m is mass per unit length.

$$f_1 = \frac{1}{2l_1}\sqrt{\frac{T}{m}},\ f_2 = \frac{1}{2l_2}\sqrt{\frac{T}{m}},$$

$$f_2 - f_1 = \frac{1}{2}\sqrt{\frac{T}{m}}\frac{(l_1 - l_2)}{l_1 l_2}$$

$$\sqrt{\frac{T}{m}} = \sqrt{\frac{20}{10^{-3}}} = \sqrt{2} \times 10^2 = 1.414 \times 100$$

$= 141.4$

$$\frac{l_1 - l_2}{l_1 l_2} = \frac{(51.6 - 49.1) \times 10^2}{51.6 \times 49.1}$$

$$= \frac{2.5 \times 10^2}{50 \times 50} = \frac{1}{10}$$

$$\therefore \quad f_2 - f_1 = \frac{1}{2} \times 141.4 \times \frac{1}{10} = 7 \text{ beats}$$

61. (c) Frequencies of sound waves are $\frac{330}{5}$ and $\frac{330}{5.5}$

i.e., 66 Hz and 60 Hz

Frequency of beat = 66 – 60 = 6 per second

NOTES Beat frequency is the number of beats produced per second.

$n = n_1 \sim n_2$

62. **(b)** Equation of progressive wave is given by $y = A\sin 2\pi f t$

Given $y_1 = 4\sin 500\,\pi t$ and $y_2 = 2\sin 506\pi t$.

Comparing the given equations with equation of progressive wave, we get

$2f_1 = 500 \Rightarrow f_1 = 250$

$2f_2 = 506 \Rightarrow f_2 = 253$

Beats $= f_2 - f_1 = 253 - 250 = 3$ beats/sec

$= 3 \times 60 = 180$ beats/minute.

63. **(b)** Frequency received by listener from the rear source,

$$n' = \frac{v-u}{v} \times n = \frac{v-u}{v} \times \frac{v}{\lambda} = \frac{v-u}{\lambda}$$

Frequency received by listener from the front source,

$$n'' = \frac{v+u}{v} \times \frac{v}{\lambda} = \frac{v+u}{\lambda}$$

No. of beats = n" – n'

$$= \frac{v+u}{\lambda} - \frac{v-u}{\lambda} = \frac{v+u-v+u}{\lambda} = \frac{2u}{\lambda}$$

64. **(a)** Given : Wavelength of first wave (λ_1)

= 50 cm = 0.5 m

Wavelength of second wave (λ_2)

= 51 cm = 0.51m

frequency of beats per sec (n) = 12.

We know that the frequency of beats,

$$n = 12 = \frac{v}{\lambda_1} - \frac{v}{\lambda_2}$$

$$\Rightarrow 12 = v\left[\frac{1}{0.5} - \frac{1}{0.51}\right]$$

$$= v[2 - 1.9608] = v \times 0.0392$$

$$\text{or, } v = \frac{12}{0.0392} = 306 \text{ m/s}$$

[where, v = velocity of sound]

65. **(c)** After the superposition, frequency and intensity of waves will remain constant.

66. **(c)** Frequency of first source with 5 beats/ sec = 100 Hz and frequency of second source with 5 beats/sec = 205 Hz. The frequency of the first source = 100 ± 5 = 105 or 95 Hz. Therefore, frequency of second harmonic of source = 210 Hz or 190 Hz. As the second harmonic gives 5 beats/ second with the sound of frequency 205 Hz, therefore, frequency of second harmonic source should be 210 Hz or frequency of source = 105 Hz.

67. **(b)** For production of beats different frequencies are essential. The different amplitudes affect the minimum and maximum amplitude of the beats and different phases affect the time of occurrence of minimum and maximum.

68. **(c)** Beats are produced. Frequency of beats will be $\frac{1}{3} - \frac{1}{4} = \frac{1}{12}$. Hence time period = 12 s.

NOTES Beat period (T) is the time interval between two successive beats.

$$T = \frac{1}{\text{Beat frequency}} = \frac{1}{n_1 - n_2}$$

69. **(c)** Intensity is proportional to (amplitude)^2 and also intensity α (frequency)^2. Therefore, intensity becomes $\frac{2^2}{4^2} = \frac{1}{4}$th

70. **(c)** As we known from Doppler's Effect

$$f_{apprent} = f_0\left[\frac{v+v_0}{v-v_s}\right] = 400\left[\frac{340+16.5}{340-22}\right]$$

$f_{apprent} = 448$ Hz

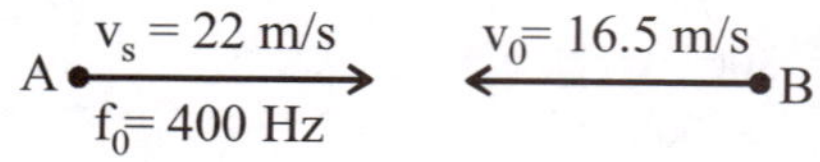

71. **(c)** According to Doppler's effect in sound

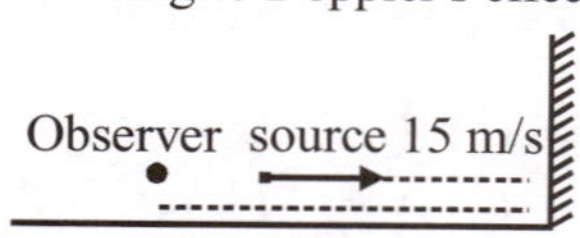

Apparent frequency,

$$n' = \frac{v}{v - v_s} n_0$$

$$= \frac{330}{330-15}(800) = \frac{330 \times 800}{315} = 838 \text{ Hz}$$

The frequency of sound observer hears in the echo reflected from the cliff is 838 Hz.

72. **(a)** Here, original frequency of sound, $f_0 = 100$ Hz

Speed of source $V_s = 19.4\cos 60° = 9.7$

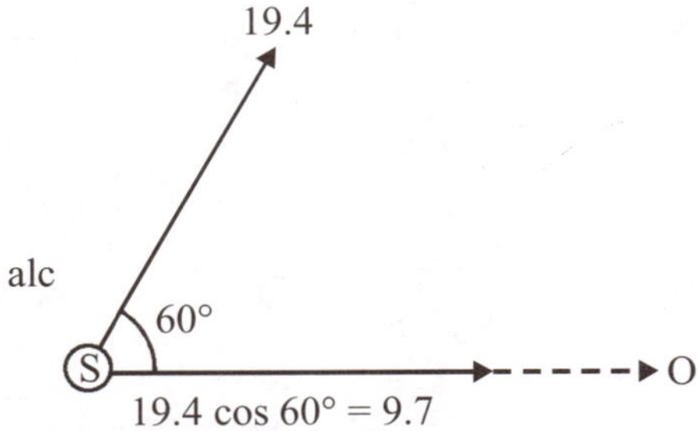

From Doppler's formula

$$f' = f_0\left(\frac{V - V_0}{V - V_s}\right)$$

$$f' = 100\left(\frac{V - 0}{V - (+9.7)}\right)$$

$$f' = 100\frac{V}{V\left(1 - \frac{9.7}{V}\right)} = \frac{100}{\left(1 - \frac{9.7}{330}\right)}$$

$= 103\text{Hz}$

Apparent frequency ff = 103 Hz

73. **(c)** According to Doppler's effect Apparent frequency,

$$n' = n\left(\frac{v + v_0}{v + v_s}\right) = 1392\left(\frac{343 + 10}{343 + 5}\right).$$

$= 1412$ Hz

74. **(c)** Frequency of the echo detected by the driver of the train is (According to Doppler effect in sound)

$$f' = \left(\frac{v + u}{v - u}\right)f$$

where f = original frequency of source of sound f' = Apparent frequency of source because of the relative motion between source and observer.

$$f' = \left(\frac{330 + 220}{330 - 220}\right)1000 = 5000 \text{ Hz}$$

NOTES: Velocity of sound direction is always taken from source to observer. All the velocities in the direction of v are positive and in the opposite direction of v are taken negative.

75. **(b)** From the given condition

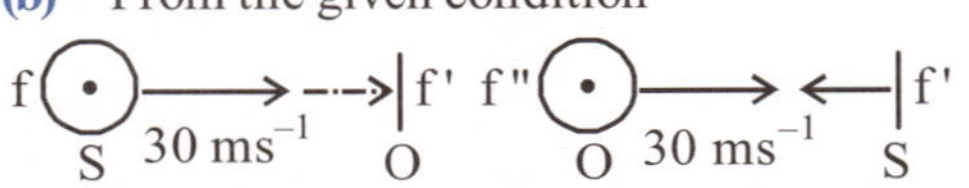

f' is the apparent frequency received by an observer at the hill. f″ is the frequency of the reflected sound as heard by driver.

$$f' = \frac{v}{v - 30}f \text{ and}$$

$$f'' = \frac{v + 30}{v}f' \Rightarrow \frac{v + 30}{v - 30}f = \frac{360}{300} \times 600$$

$= 720$ Hz

76. **(c)** Let f' be the frequency of sound heard by cliff.

$$\therefore f = \frac{vf}{v - v_c} \quad ....(1)$$

Now, for the reflected wave, cliff acts as a source,

$$\therefore 2f = \frac{f'(v + v_c)}{v} \quad ....(2)$$

$$2f = \frac{(v + v_c)f}{v - v_c}$$

$$\Rightarrow 2v - 2v_c = v + v_c \text{ or } \frac{v}{3} = v_c$$

77. **(c)** Wavelength remains constant (unchanged) in this case.

$$f_{apparent} = \left(\frac{u + u/5}{u}\right)f = \frac{6}{5}f = 1.2f$$

78. **(c)** Velocity of source, $v_s = r\omega$ $= 0.50 \times 20 = 10 \text{ ms}^{-1}$

$$n' = \left(\frac{v}{v + v_s}\right)n = \frac{340 \times 385}{340 + 10} = 374 \text{ Hz}$$

79. **(b)** According to Doppler's effect

$$n' = \left(\frac{v - v_0}{v - v_s}\right)n = \left(\frac{340 - 10}{340 + 10}\right)n = \frac{330}{350} \times 1950$$

= `068 Hz

80. **(b)** As the source is not moving towards or away from the observer in a straight line, so the Doppler's effect will not be observed by the observer.

NOTES: If source and observer both are relatively at rest and if speed of sound is increased then frequency heard by observer will not change.

81. **(c)** Given : Wavelength (λ) = 5000 Å velocity of star (v) = 1.5×10^6 m/s.

We know that wavelength of the approaching star $(\lambda') = \lambda\left(\frac{c - v}{c}\right)$

or, $\frac{\lambda'}{\lambda} = \frac{c-v}{c} = 1 - \frac{v}{c}$

or, $\frac{v}{c} = 1 - \frac{\lambda'}{\lambda} = \frac{\lambda - \lambda'}{\lambda} = \frac{\Delta\lambda}{\lambda}$

Therefore, $\Delta\lambda = \lambda \times \frac{v}{c} = 5000 \times \frac{1.5 \times 10^6}{3 \times 10^8} = 25\,\text{Å}$

[where $\Delta\lambda$ = Change in the wavelength]

82. **(a)** Here, $\nu' = \frac{9}{8}\nu$

Source and observer are moving in opposite direction, therefore, apparent frequency

$$\nu' = \nu \times \frac{(v+u)}{(v-u)}$$

$$\frac{9}{8}\nu = \nu \times \left(\frac{340+u}{340-u}\right)$$

$\Rightarrow 9 \times 340 - 9u = 8 \times 340 + 8u$

$\Rightarrow 17u = 340 \times 1 \Rightarrow u = \frac{340}{17} = 20$ m/sec.

15 Electric Charges and Fields

Trend Analysis with Important Topics & Sub-Topics

		2021		2020		2019		2018		2017		2016	
Topic Name	**Sub-Topic**	**QNS.**	**LOD**	**QNS.**	**LOD**	**QNS.**	**LOD**	**QNS.**	**LOD**	**QNS.**	**LOD**	**QNS.**	**LOD**
Charges & Coulomb's Law	Charge density Coulomb's Law	1	A			1	A					1	D
	Electrostatic Force							1	D	1	A		
Electric Field, Electric Field Lines & Dipole	Electric Field			1	A	1	A						
	Dipole moment	1	A										
Electric Flux & Gauss's Law	Gauss's Law					1	E						

LOD - Level of Difficulty | E - Easy | A - Average | D - Difficult | Qns - No. of Questions

Topic 1 Charges & Coulomb's Law

1. Two charged spherical conductors of radius R_1 and R_2 are connected by a wire. Then the ratio of surface charge densities of the spheres (σ_1/σ_2) is *[2021]*

   (a) $\frac{R_1^2}{R_2^2}$ (b) $\frac{R_1}{R_2}$

   (c) $\frac{R_2}{R_1}$ (d) $\sqrt{\left(\frac{R_1}{R_2}\right)}$

2. Two point charges A and B, having charges +Q and –Q respectively, are placed at certain distance apart and force acting between them is F. If 25% charge of A is transferred to B, then force between the charges becomes : *[2019]*

   (a) F (b) $\frac{9F}{16}$

   (c) $\frac{16F}{9}$ (d) $\frac{4F}{3}$

3. Two metal spheres, one of radius *R* and the other of radius 2*R* respectively have the same surface charge density σ. They are brought in contact and separated. What will be the new surface charge densities on them? *[NEET Odisha 2019]*

   (a) $\sigma_1 = \frac{5}{3}\sigma, \sigma_2 = \frac{5}{6}\sigma$ (b) $\sigma_1 = \frac{5}{6}\sigma, \sigma_2 = \frac{5}{2}\sigma$

   (c) $\sigma_1 = \frac{5}{2}\sigma, \sigma_2 = \frac{5}{6}\sigma$ (d) $\sigma_1 = \frac{5}{2}\sigma, \sigma_2 = \frac{5}{3}\sigma$

4. Suppose the charge of a proton and an electron differ slightly. One of them is – e, the other is (e + Δe). If the net of electrostatic force and gravitational force between two hydrogen atoms placed at a distance d (much greater than atomic size) apart is zero, then Δe is of the order of [Given mass of hydrogen $m_h = 1.67 \times 10^{-27}$ kg]

   (a) 10^{-23}C (b) 10^{-37} C *[2017]*

   (c) 10^{-47} C (d) 10^{-20} C

5. Two identical charged spheres suspended from a common point by two massless strings of lengths *l*, are initially at a distance d (d << *l*) apart because of their mutual repulsion. The charges begin to leak from both the spheres at a constant rate. As a result, the spheres approach each other with a velocity v. Then v varies as a function of the distance x between the spheres, as : *[2016]*

   (a) $v \propto x^{\frac{1}{2}}$ (b) $v \propto x$

   (c) $v \propto x^{-\frac{1}{2}}$ (d) $v \propto x^{-1}$

6. Two pith balls carrying equal charges are suspended from a common point by strings of equal length. The equilibrium separation between them is r. Now the strings are rigidly clamped at half the height. The equilibrium separation between the balls now become *[2013]*

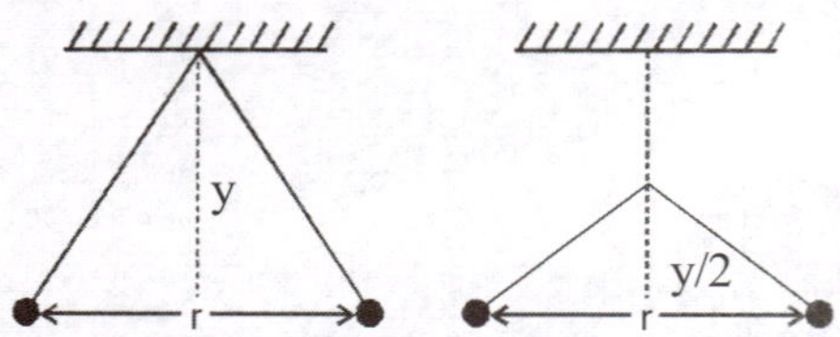

(a) $\left(\frac{r}{\sqrt[3]{2}}\right)$ (b) $\left(\frac{2r}{\sqrt{3}}\right)$

(c) $\left(\frac{2r}{3}\right)$ (d) $\left(\frac{r}{\sqrt{2}}\right)^2$

7. A charge 'q' is placed at the centre of the line joining two equal charges 'Q'. The system of the three charges will be in equilibrium if 'q' is equal to *[NEET Kar. 2013, 1995]*

(a) $Q/2$ (b) $-Q/4$

(c) $Q/4$ (d) $-Q/2$

8. Two metallic spheres of radii 1 cm and 3 cm are given charges of -1×10^{-2} C and 5×10^{-2} C, respectively. If these are connected by a conducting wire, the final charge on the bigger sphere is : *[2012M]*

(a) 2×10^{-2} C (b) 3×10^{-2} C

(c) 4×10^{-2} C (d) 1×10^{-2} C

9. Two positive ions, each carrying a charge q, are separated by a distance d. If F is the force of repulsion between the ions, the number of electrons missing from each ion will be (e being the charge of an electron) *[2010]*

(a) $\frac{4\pi\varepsilon_0 Fd^2}{e^2}$ (b) $\sqrt{\frac{4\pi\varepsilon_0 Fe^2}{d^2}}$

(c) $\sqrt{\frac{4\pi\varepsilon_0 Fd^2}{e^2}}$ (d) $\frac{4\pi\varepsilon_0 Fd^2}{q^2}$

10. An electron is moving round the nucleus of a hydrogen atom in a circular orbit of radius r. The Coulomb force $\vec{F}$ between the two is *[2003]*

(a) $K\frac{e^2}{r^3}\vec{r}$ (b) $K\frac{e^2}{r^2}\hat{r}$

(c) $-K\frac{e^2}{r^3}\hat{r}$ (d) $-K\frac{e^2}{r^3}\vec{r}$

$\left(\text{where } K=\frac{1}{4\pi\varepsilon_0}\right)$

11. When air is replaced by a dielectric medium of force constant K, the maximum force of attraction between two charges, separated by a distance *[1999]*

(a) decreases K-times
(b) increases K-times
(c) remains unchanged
(d) becomes $\frac{1}{K^2}$ times

12. Point charges $+4q$, $-q$ and $+4q$ are kept on the X-axis at points $x=0$, $x=a$ and $x=2a$ respectively. Then *[1988]*

(a) only $-q$ is in stable equilibrium
(b) none of the charges is in equilibrium
(c) all the charges are in unstable equilibrium
(d) all the charges are in stable equilibrium.

Topic 2 Electric Field, Electric Field Lines & Dipole

13. Polar molecules are the molecules *[2021]*

(a) having a permanent electric dipole moment
(b) having zero dipole moment
(c) acquire a dipole moment only in the presence of electric field due to displacement of charges
(d) acquire a dipole moment only when magnetic field is absent

14. A spherical conductor of radius 10 cm has a charge of 3.2×10^{-7} C distributed uniformly. What is the magnitude of electric field at a point 15 cm from the centre of the sphere? *[2020]*

$\left(\frac{1}{4\pi\epsilon_0}=9\times10^9\,\text{Nm}^2/\text{C}^2\right)$

(a) 1.28×10^5 N/C (b) 1.28×10^6 N/C

(c) 1.28×10^7 N/C (d) 1.28×10^4 N/C

15. Two parallel infinite line charges with linear charge densities $+\lambda$ C/m and $-\lambda$ C/m are placed at a distance of 2R in free space. What is the electric field mid-way between the two line

charges? *[2019]*

(a) zero (b) $\frac{2\lambda}{\pi \in_0 R}$ N/C

(c) $\frac{\lambda}{\pi \in_0 R}$ N/C (d) $\frac{\lambda}{2\pi\varepsilon_0 R}$ N/C

16. A hollow metal sphere of radius R is uniformly charged. The electric field due to the sphere at a distance r from the centre : *[2019]*
(a) increases as r increases for r < R and for r > R
(b) zero as r increases for r < R, decreases as r increases for r > R
(c) zero as r increases for r < R, increases as r increases for r > R
(d) decreases as r increases for r < R and for r > R

17. An electron falls from rest through a vertical distance h in a uniform and vertically upward directed electric field E. The direction of electric field is now reversed, keeping its magnitude the same. A proton is allowed to fall from rest in it through the same vertical distance h. The time of fall of the electron, in comparison to the time of fall of the proton is *[2018]*
(a) smaller (b) 5 times greater
(c) equal (d) 10 times greater

18. A toy car with charge q moves on a frictionless horizontal plane surface under the influence of a uniform electric field $\vec{E}$. Due to the force q $\vec{E}$, its velocity increases from 0 to 6 m/s in one second duration. At that instant the direction of the field is reversed. The car continues to move for two more seconds under the influence of this field. The average velocity and the average speed of the toy car between 0 to 3 seconds are respectively *[2018]*
(a) 2 m/s, 4 m/s (b) 1 m/s, 3 m/s
(c) 1.5 m/s, 3 m/s (d) 1 m/s, 3.5 m/s

19. An electric dipole of dipole moment p is aligned parallel to a uniform electric field E. The energy required to rotate the dipole by 90° is *[NEET Kar. 2013]*
(a) pE^2 (b) p^2E
(c) pE (d) infinity

20. An electric dipole of moment 'p' is placed in an electric field of intensity 'E'. The dipole acquires a position such that the axis of the dipole makes an angle θ with the direction of the field. Assuming that the potential energy of the dipole to be zero when = 90°, the torque and the potential energy of the dipole will respectively be : *[2012]*
(a) $pE\sin\theta, -pE\cos\theta$
(b) $pE\sin\theta, -2pE\cos\theta$
(c) $pE\sin\theta, 2pE\cos\theta$
(d) $pE\cos\theta, -pE\cos\theta$

21. The electric potential V at any point (x, y, z), all in meters in space is given by $V = 4x^2$ volt. The electric field at the point (1, 0, 2) in volt/meter is *[2011M]*
(a) 8 along positive X-axis
(b) 16 along negative X-axis
(c) 16 along positive X-axis
(d) 8 along negative X-axis

22. The electric field at a distance $\frac{3R}{2}$ from the centre of a charged conducting spherical shell of radius R is E. The electric field at a distance $\frac{R}{2}$ from the centre of the sphere is *[2010]*
(a) $\frac{E}{2}$ (b) zero
(c) E (d) $\frac{E}{2}$

23. The mean free path of electrons in a metal is 4×10^{-8} m. The electric field which can give on an average 2 eV energy to an electron in the metal will be in units of V/m *[2009]*
(a) 5×10^{-11} (b) 8×10^{-11}
(c) 5×10^{7} (d) 8×10^{7}

24. A thin conducting ring of radius R is given a charge +Q. The electric field at the centre O of the ring due to the charge on the part AKB of the ring is E. The electric field at the centre due to the charge on the part ACDB of the ring is *[2008]*

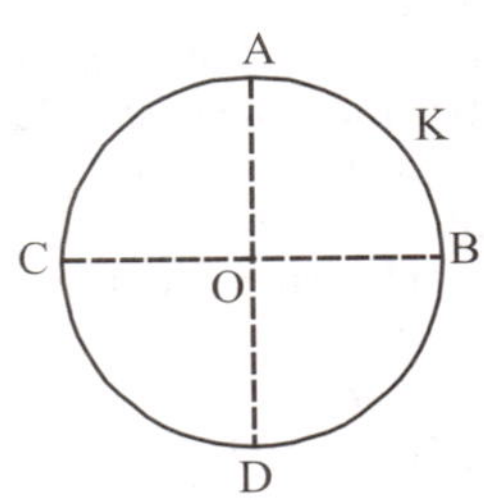

(a) E along KO (b) E along OK
(c) E along KO (d) 3 E along OK

25. Three point charges $+q$, $-q$ and $+q$ are placed at points $(x=0, y=a, z=0)$, $(x=0, y=0, z=0)$ and $(x=a, y=0, z=0)$ respectively. The magnitude and direction of the electric dipole moment vector of this charge assembly are [2007]

(a) $\sqrt{2}qa$ along the line joining points $(x=0, y=0, z=0)$ and $(x=a, y=a, z=0)$
(b) qa along the line joining points $(x=0, y=0, z=0)$ and $(x=a, y=a, z=0)$
(c) $\sqrt{2}qa$ along +ve x direction
(d) $\sqrt{2}qa$ along +ve y direction

26. The electric intensity due to a dipole of length 10 cm and having a charge of 500 μC, at a point on the axis at a distance 20 cm from one of the charges in air, is [2001]

(a) 6.25×10^7 N/C (b) 9.28×10^7 N/C
(c) 13.1×10^{11} N/C (d) 20.5×10^7 N/C

27. If a dipole of dipole moment $\vec{p}$ is placed in a uniform electric field $\vec{E}$, then torque acting on it is given by [2001]

(a) $\vec{\tau} = \vec{p}.\vec{E}$ (b) $\vec{\tau} = \vec{p} \times \vec{E}$
(c) $\vec{\tau} = \vec{p} + \vec{E}$ (d) $\vec{\tau} = \vec{p} - \vec{E}$

28. A semi-circular arc of radius 'a' is charged uniformly and the charge per unit length is λ. The electric field at the centre of this arc is [2000]

(a) $\frac{\lambda}{2\pi\varepsilon_0 a}$ (b) $\frac{\lambda}{2\pi\varepsilon_0 a^2}$
(c) $\frac{\lambda}{4\pi^2\varepsilon_0 a}$ (d) $\frac{\lambda^2}{2\pi\varepsilon_0 a}$

29. A hollow insulated conduction sphere is given a positive charge of 10 μC. What will be the electric field at the centre of the sphere if its radius is 2 metres? [1998]

(a) zero (b) 5 μCm^{-2}
(c) 20 μCm^{-2} (d) 8 μCm^{-2}

30. A point Q lies on the perpendicular bisector of an electrical dipole of dipole moment p. If the distance of Q from the dipole is r (much larger than the size of the dipole), then the electric field at Q is proportional to [1998]

(a) p^{-1} and r^{-2} (b) p and r^{-2}
(c) p^2 and r^{-3} (d) p and r^{-3}

31. From a point charge, there is a fixed point A. At A, there is an electric field of 500 V/m and potential difference of 3000 V. Distance between point charge and A will be [1997]

(a) 6m (b) 12m
(c) 16m (d) 24m

32. Intensity of an electric field (E) depends on distance r, due to a dipole, is related as [1996]

(a) $E \propto \frac{1}{r}$ (b) $E \propto \frac{1}{r^2}$
(c) $E \propto \frac{1}{r^3}$ (d) $E \propto \frac{1}{r^4}$

33. The formation of a dipole is due to two equal and dissimilar point charges placed at a [1996]

(a) short distance (b) long distance
(c) above each other (d) none of these

34. An electric dipole, consisting of two opposite charges of 2×10^{-6} C each separated by a distance 3 cm is placed in an electric field of 2×10^5 N/C. Torque acting on the dipole is [1995]

(a) 12×10^{-1} Nm (b) 12×10^{-2} Nm
(c) 12×10^{-3} Nm (d) 12×10^{-4} Nm

35. There is an electric field E in x-direction. If the work done on moving a charge of 0.2 C through a distance of 2 m along a line making an angle 60° with x-axis is 4J, then what is the value of E?

(a) 3 N/C (b) 4 N/C [1995]
(c) 5 N/C (d) 20 N/C

Topic 3 Electric Flux & Gauss's Law

36. A sphere encloses an electric dipole with charges $\pm 3 \times 10^{-6}$ C. What is the total electric flux across the sphere? [NEET Odisha 2019]

(a) 6×10^{-6} Nm^2/C (b) -3×10^{-6} Nm^2/C
(c) Zero (d) 3×10^{-6} Nm^2/C

37. The electric field in a certain region is acting radially outward and is given by E = Aa. A charge contained in a sphere of radius 'a' centred at the origin of the field, will be given by [2015]

(a) $A\,\varepsilon_0\,a^2$ (b) $4\,\pi\varepsilon_0\,Aa^3$
(c) $\varepsilon_0\,Aa^3$ (d) $4\,\pi\varepsilon_0\,Aa^2$

38. What is the flux through a cube of side 'a' if a point charge of q is at one of its corner : [2012]

(a) $\frac{2q}{\varepsilon_0}$ (b) $\frac{q}{8\varepsilon_0}$
(c) $\frac{q}{\varepsilon_0}$ (d) $\frac{q}{2\varepsilon_0}$

39. A charge Q is enclosed by a Gaussian spherical surface of radius R. If the radius is doubled, then the outward electric flux will [2011]
(a) increase four times
(b) be reduced to half
(c) remain the same
(d) be doubled

40. A square surface of side L meter in the plane of the paper is placed in a uniform electric field E (volt/m) acting along the same plane at an angle θ with the horizontal side of the square as shown in Figure. The electric flux linked to the surface, in units of volt. m, is [2010]

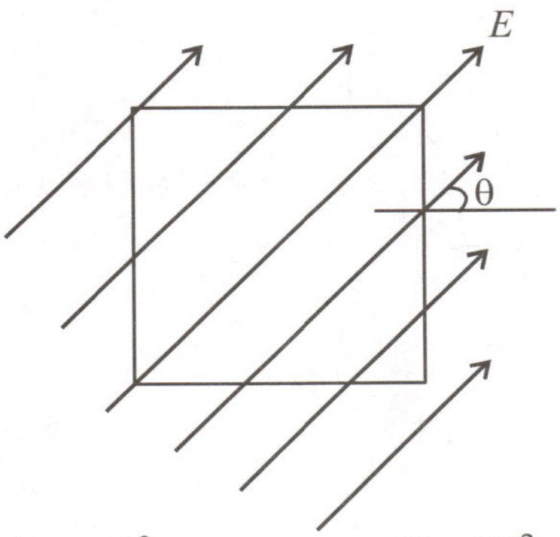

(a) EL^2 (b) $EL^2 \cos\theta$
(c) $EL^2 \sin\theta$ (d) zero

41. A hollow cylinder has a charge q coulomb within it. If ϕ is the electric flux in units of voltmeter associated with the curved surface B, the flux linked with the plane surface A in units of voltmeter will be [2007]

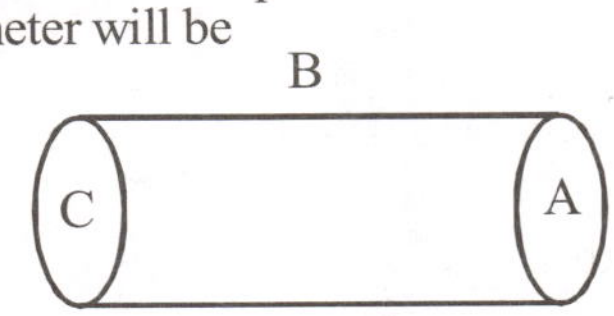

(a) $\frac{q}{2\varepsilon_0}$ (b) $\frac{\phi}{3}$
(c) $\frac{q}{\varepsilon_0} - \phi$ (d) $\frac{1}{2}\left(\frac{q}{\varepsilon_0} - \phi\right)$

42. A square surface of side L metres is in the plane of the paper. A uniform electric field $\vec{E}$ (volt / m), also in the plane of the paper, is limited only to the lower half of the square surface (see figure). The electric flux in SI units associated with the surface is [2006]

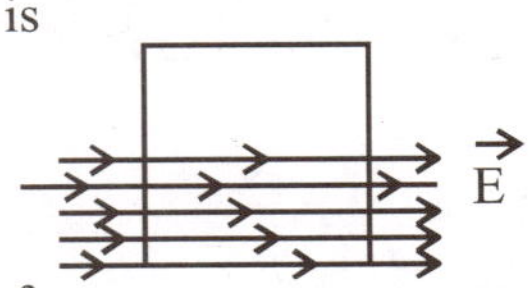

(a) $EL^2/2$ (b) zero
(c) EL^2 (d) $EL^2/(2\varepsilon_0)$

43. A charge q is located at the centre of a cube. The electric flux through any face is [2003]
(a) $\frac{q}{6(4\pi\varepsilon_0)}$ (b) $\frac{2\pi q}{6(4\pi\varepsilon_0)}$
(c) $\frac{4\pi q}{6(4\pi\varepsilon_0)}$ (d) $\frac{\pi q}{6(4\pi\varepsilon_0)}$

44. A charge $Q\,\mu C$ is placed at the centre of a cube, the flux coming out from any surface will be [2001]
(a) $\frac{Q}{6\varepsilon_0} \times 10^{-6}$ (b) $\frac{Q}{6\varepsilon_0} \times 10^{-3}$
(c) $\frac{Q}{24\varepsilon_0}$ (d) $\frac{Q}{8\varepsilon_0}$

45. A charge Q is placed at the corner of a cube. The electric flux through all the six faces of the cube is [2000]
(a) $Q/3\varepsilon_0$ (b) $Q/6\varepsilon_0$
(c) $Q/8\varepsilon_0$ (d) Q/ε_0

46. A point charge $+q$ is placed at mid point of a cube of side 'L'. The electric flux emerging from the cube is [1996]
(a) $\frac{q}{\varepsilon_0}$ (b) $\frac{6qL^2}{\varepsilon_0}$
q

ANSWER KEY

1	(c)	6	(a)	11	(a)	16	(b)	21	(d)	26	(a)	31	(a)	36	(c)	41	(d)	46	(a)
2	(b)	7	(b)	12	(c)	17	(a)	22	(b)	27	(b)	32	(c)	37	(b)	42	(b)		
3	(a)	8	(b)	13	(a)	18	(b)	23	(c)	28	(a)	33	(a)	38	(b)	43	(c)		
4	(b)	9	(c)	14	(a)	19	(c)	24	(b)	29	(a)	34	(c)	39	(c)	44	(a)		
5	(c)	10	(d)	15	(c)	20	(a)	25	(a)	30	(d)	35	(d)	40	(d)	45	(c)		

Hints & Solutions

1. **(c)** Charge density $\sigma = \dfrac{\text{charge}}{\text{area}}$

When two conductors are connected by a conducting wire, then they have same potential i.e., $V_1 = V_2$

$$\therefore \frac{1}{4\pi\varepsilon_0}\frac{Q_1}{R_1} = \frac{1}{4\pi\varepsilon_0}\frac{Q_2}{R_2}$$

$$\Rightarrow \frac{1}{4\pi\varepsilon_0}\frac{Q_1}{R_1}\times\frac{R_1}{R_1} = \frac{1}{4\pi\varepsilon_0}\frac{Q_2}{R_2}\times\frac{R_2}{R_2}$$

$$\Rightarrow \frac{Q_1R_1}{4\pi R_1^2\varepsilon_0} = \frac{Q_2R_2}{4\pi R_2^2\varepsilon_0}$$

$$\therefore \frac{Q_1}{4\pi R_1^2}\Big/\frac{Q_2}{4\pi R_2^2} = \frac{R_2}{R_1}$$

or, $\sigma_1/\sigma_2 = R_2/R_1$

2. **(b)** We know that,

$F = \dfrac{kQ^2}{r^2}$

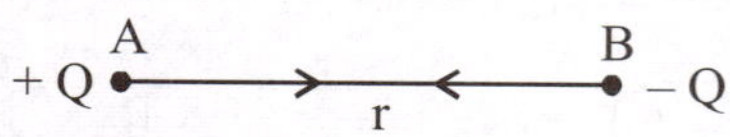

If 25% of charges of A transfer to B then

$$q_A = Q - \frac{Q}{4} = \frac{3Q}{4} \text{ and } q_B = -Q + \frac{Q}{4} = \frac{-3Q}{4}$$

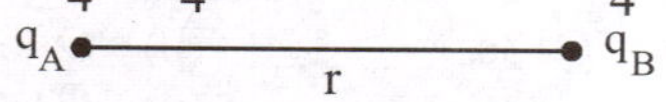

$$F_1 = \frac{kq_Aq_B}{r^2}$$

$$\Rightarrow F_1 = \frac{k\left(\frac{3Q}{4}\right)^2}{r^2} = \frac{9}{16}\frac{kQ}{r^2} = \frac{9F}{16}$$

3. **(a)** $Q_1 = \sigma 4\pi {R_1}^2 = \sigma 4\pi R^2$

$Q_2 = \sigma 4\pi(2R)^2 = \sigma 16\pi R^2$

After redistribution

$$\frac{Q_1'}{Q_2'} = \frac{R}{2R} \Rightarrow Q_2' = 2Q_1' \quad \text{...(i)}$$

$$Q_1' + Q_2' = 20\sigma\pi R^2 \quad \text{...(ii)}$$

From eq. (i) and (ii)

$$Q_1' = \frac{20}{3}\sigma\pi R^2 \Rightarrow \sigma_1' = \frac{5}{3}\sigma$$

$$Q_2' = \frac{40}{3}\sigma\pi R^2 \Rightarrow \sigma_2' = \frac{5}{6}\sigma$$

4. **(b)** Net charge on one H atom $= {}^-e + (e + \Delta e) = \Delta e$

According to question, the net electrostatic force (F_E) = gravitational force (F_G)

$F_E = F_G$

$$\text{or} \quad \frac{1}{4\pi\varepsilon_0}\frac{\Delta e^2}{d^2} = \frac{Gm^2}{d^2}$$

$$\Rightarrow \Delta e = m\sqrt{\frac{G}{k}} \quad \left(\frac{1}{4\pi\varepsilon_0} = k = 9\times10^9\right)$$

$$= 1.67\times10^{-27}\sqrt{\frac{6.67\times10^{-11}}{9\times10^9}}$$

$\Delta e \approx 1.436 \times 10^{-37}$ C

5. **(c)** From figure $T\cos\theta = mg$

$$T\sin\theta = \frac{kq^2}{x^2}$$

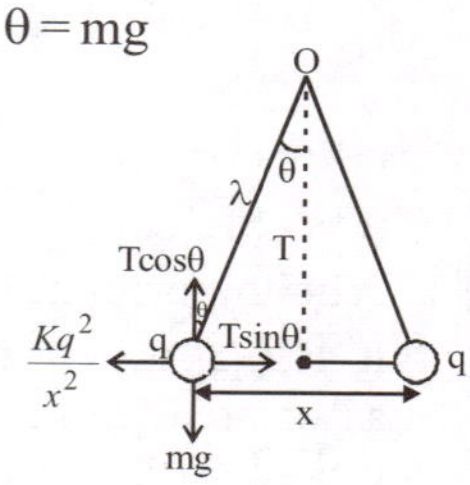

$$\text{so,} \quad \tan\theta = \frac{F_e}{mg} \simeq \theta$$

$$\tan\theta = \frac{kq^2}{x^2mg}$$

As θ is small $\tan\theta \approx \sin\theta = \dfrac{x}{2l}$

$$\therefore \quad \frac{kq^2}{x^2mg} = \frac{x}{2\ell}$$

or $\quad x^3 \propto q^2 \quad \text{...(1)}$

or $\quad x^{3/2} \propto q \quad \text{...(2)}$

Differentiating eq. (1) w.r.t. time

$$3x^2\frac{dx}{dt} \propto 2q\frac{dq}{dt} \text{ but } \frac{dq}{dt} \text{ is constant}$$

so $x^2(v) \propto q$

Replace q from eq. (2)

$x^2(v) \propto x^{3/2}$ or $v \propto x^{-1/2}$

6. **(a)**

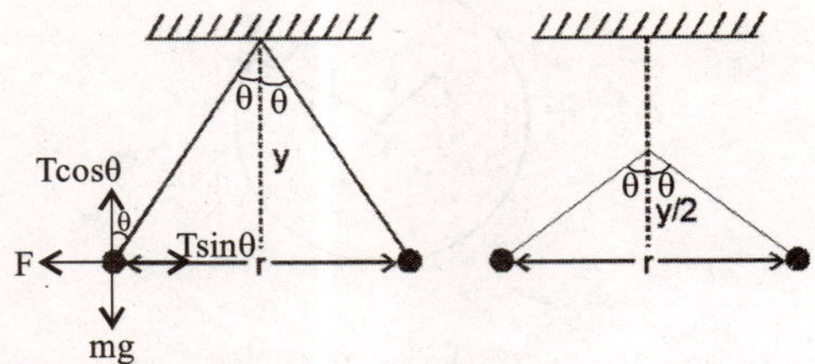

$T\cos\theta = mg$
$T\sin\theta = F$

$$\therefore \quad \tan\theta = \frac{F}{mg}$$

From figure, $\tan\theta = \frac{F_e}{mg}$

$$\Rightarrow \frac{r/2}{y} = \frac{\frac{kq^2}{r^2}}{mg}$$

$[\because F = \frac{kq^2}{r^2}$ from coulomb's law$]$

$\Rightarrow r^3 \propto y$

When string is clamped at half the height

$r'^3 \propto \frac{y}{2}$

Dividing the above two equations, we have

$$\frac{r'}{r} = \frac{1}{2^{1/3}} \Rightarrow r' = \frac{r}{\sqrt[3]{2}}$$

7. **(b)** The system of three charges will be in equilibrium.

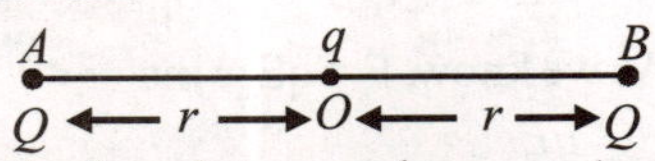

For this, force between charge at A and B + force between charge at point O and either at A or B is zero.

i.e., $\frac{KQ^2}{(2r)^2} + \frac{KQq}{(r)^2} = 0$

By solving we get,

$$q = -\frac{Q}{4}.$$

8. **(b)** At equilibrium potential of both sphere becomes same if charge of sphere one x and other sphere $(Q-x)$ then where $Q = 4 \times 10^{-2}\,C$

$V_1 = V_2$

$$\frac{kx}{1\,\text{cm}} = \frac{k(Q-x)}{3\,\text{cm}}$$

$3x = Q - x \Rightarrow 4x = Q$

$$x = \frac{Q}{4} = \frac{4\times10^{-2}}{4}C = 1\times10^{-2}$$

$Q' = Q - x = 3\times10^{-2}C$

NOTES

Charge given to a conductor always resides on its outer surface. If surface is uniform the charge distributes uniformly on the surface and for irregular surface the distribution of charge *i.e.*, charge density is not uniform. It is maximum where the radius of curvature is maximum and vice-versa

i.e., $\sigma \propto \frac{1}{R}$.

9. **(c)** Let n be the number of electrons missing.

$$F = \frac{1}{4\pi\varepsilon_0}\cdot\frac{q^2}{d^2}$$

$$\Rightarrow \quad q = \sqrt{4\pi\varepsilon_0 d^2 F} = ne$$

$$\therefore \quad n = \sqrt{\frac{4\pi\varepsilon_0 F d^2}{e^2}}$$

NOTES

If the charge of an electron ($= 1.6\times10^{-19}$ C) is taken as elementary unit *i.e.*, quanta of charge, the charge on any body will be some integral multiple of e

i.e., $Q = \pm ne$ where $n = 1, 2, 3....$

10. **(d)** Charges ($-e$) on electron and (e) on proton exert a force of attraction given by

$$\text{Force} = (K)\frac{(-e)(e)}{r^2}\hat{r} = \frac{-Ke^2}{r^3}\vec{r} \quad \left(\because \hat{r} = \frac{\vec{r}}{|r|}\right)$$

NOTES

Magnitude of Coulomb force is given by $\frac{1}{4\pi\varepsilon_0}\frac{q_2q_2}{r^2}$,

but in vector form $\vec{F} = \frac{1}{4\pi\varepsilon_0}\frac{q_1q_2}{r^3}\vec{r}$

11. **(a)** In air, $F_{air} = \frac{1}{4\pi\varepsilon_0}\frac{q_1q_2}{r^2}$

In medium, $F_m = \frac{1}{4\pi\varepsilon_0}\frac{q_1q_2}{Kr^2}$

$$\therefore \frac{F_m}{F_{air}} = \frac{1}{K} \Rightarrow F_m = \frac{F_{air}}{K} \text{ (decreases K-times)}$$

12. (c) Net force on each of the charge due to the other charges is zero. However, disturbance in any direction other than along the line on which the charges lie, will not make the charges return.

13. (a) In polar molecules, like $HCl \rightarrow H^{\delta+} Cl^{\delta-}$ the centre of positive charges does not coincide with the centre of negative charges. Hence, permanent electric dipole exists.

14. (a) If the charge on a spherical conductor of radius *R* is *Q*, then electric field at distance *r* from centre is

$E = 0$ (if $r \angle R$)

$$E = \frac{1}{4\pi \epsilon_0} \frac{Q}{r^2} \quad \text{(if } r \geq R\text{)}$$

Electric field at a distance 15 cm from the centre of sphere will be

$$E = \frac{9 \times 10^9 \times 3.2 \times 10^{-7}}{225 \times 10^{-4}}$$

$$= 0.128 \times 10^6 = 1.28 \times 10^5 \text{ N/C}$$

15. (c) Electric field due to line charge (1)

$$\vec{E}_1 = \frac{\lambda}{2\pi\varepsilon_0 R} \hat{i} \text{N/C}$$

Electric field due to line charge (2)

$$\vec{E}_2 = \frac{\lambda}{2\pi\varepsilon_0 R} \hat{i} \text{N/C}$$

$$\vec{E}_{net} = \vec{E}_1 + \vec{E}_2$$

$$= \frac{\lambda}{2\pi\varepsilon_0 R} \hat{i} + \frac{\lambda}{2\pi\varepsilon_0 R} \hat{i}$$

$$= \frac{\lambda}{\pi\varepsilon_0 R} \hat{i} \text{N/C}$$

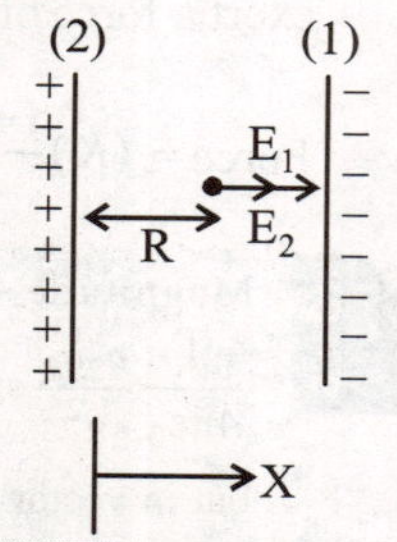

NOTES

Both $\vec{E}_1$ and $\vec{E}_2$ are in the same direction.

Electric field at a point due to *+ve* charge (*+q*) acts away from the charge and due to negative charge (*–q*) it acts towards the charge.

16. (b) Charge Q will be distributed over the surface of hollow metal sphere.

(i) For r < R (inside)

[At a point inside the hollow sphere]

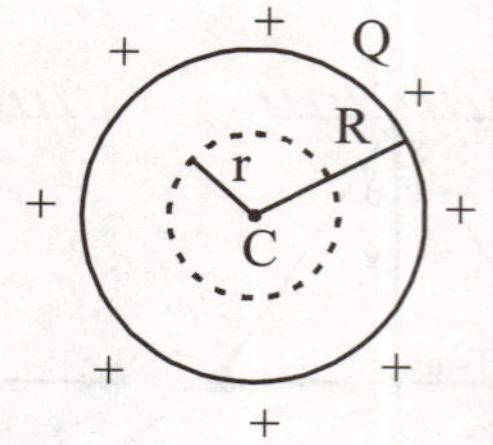

By Gauss's law, $\oint \vec{E}_{in} \cdot \vec{dS} = \frac{q_{en}}{\varepsilon_0} = 0$

As enclosed charge is = 0

So, $E_{in} = 0$ the electric field inside the hollow sphere is always zero.

(ii) For r > R (outside)

[At a point outside hollow sphere]

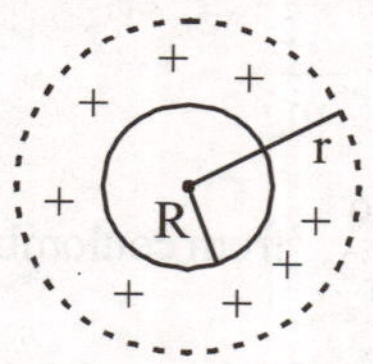

By Gauss's law, $\oint \vec{E}_0 \cdot \vec{dS} = \frac{q_{en}}{\varepsilon_0}$ $(Q\ q_{en} = Q)$

$$\therefore E_0 4\pi r^2 = \frac{Q}{\varepsilon_0}$$

$$\therefore E_0 \propto \frac{1}{r^2}$$

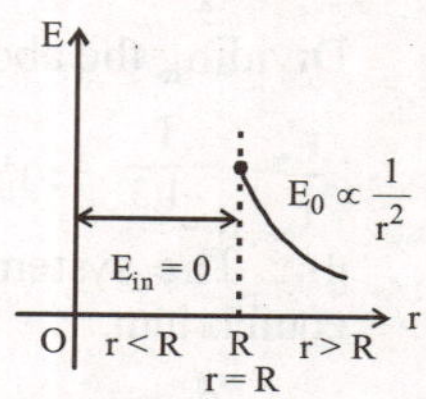

17. (a) As we know, F = qE = ma

$$\Rightarrow a = \frac{qE}{m}$$

$$h = \frac{1}{2}\frac{qE}{m}t^2$$

$$\therefore \quad t = \sqrt{\frac{2hm}{qE}}$$

i.e., time $t \propto \sqrt{m}$ as 'q' is same for electron and proton.

Since, electron has smaller mass so it will take smaller time.

18. (b)

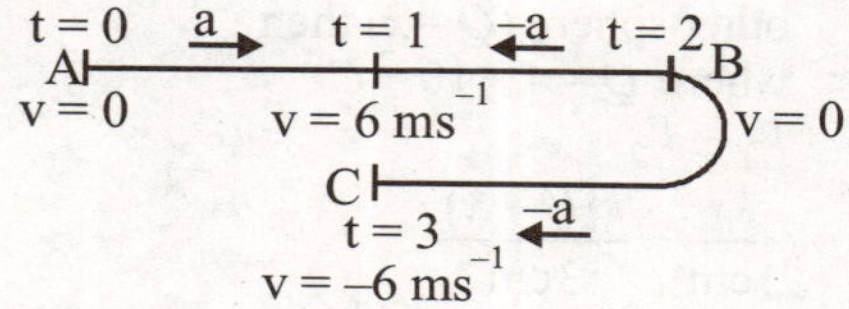

Acceleration, $a = \frac{v - u}{t} = \frac{6 - 0}{1}$

$= 6 \text{ ms}^{-2}$

For t = 0 to t = 1 s,

$$S_1 = \frac{1}{2} \times 6(1)^2 = 3 \text{ m} \quad ...(i)$$

For t = 1 s = to t = 2 s,

$$S_2 = 6.1 - \frac{1}{2} \times 6(1)^2 = 3\text{m} \quad ...(ii)$$

For t = 2 s to t = 3 s,

$$S_3 = 0 - \frac{1}{2} \times 6(1)^2 = -3 \text{ m} \quad (iii)$$

Total displacement $S = S_1 + S_2 + S_3 = 3$ m

$$\text{Average velocity} = \frac{3}{3} = 1 \text{ ms}^{-1}$$

Total distance travelled = 9 m

$$\text{Average speed} = \frac{9}{3} = 3 \text{ ms}^{-1}$$

19. **(c)** When electric dipole is aligned parallel $\theta = 0°$ and the dipole is rotated by 90° *i.e.*, $\theta = 90°$.

Energy required to rotate the dipole

$W = U_f - U_i = (-pE \cos 90°) - (-pE \cos 0°)$
$= pE.$

20. **(a)** The torque on the dipole is given as

$\tau = pE \sin \theta$

The potential energy of the dipole in the electric field is given as

$U = -\ pE \cos \theta$

21. **(d)** $\vec{E} = -\left[\frac{dV}{dx}\hat{i} + \frac{dV}{dy}\hat{j} + \frac{dV}{dz}\hat{k}\right]$

Given, $V = 4x^2$

$$\therefore \quad E = -i\frac{d(4x^2)}{dx}$$

$$= -8x\,\hat{i} \text{ volt/meter}$$

$$\therefore \vec{E}_{(1,0,2)} = -8\hat{i}\ \text{V/m}$$

22. **(b)** Electric field at a point inside a charged conducting spherical shell is zero.

23. **(c)** Energy = 2eV

$\Rightarrow \ eV_o = 2eV \quad \Rightarrow V_o = 2$ volt

$$E = \frac{V_o}{d} = \frac{2}{4 \times 10^{-8}}$$

$$= 0.5 \times 10^8 = 5 \times 10^7 \text{ Vm}^{-1}$$

24. **(b)** By the symmetry of the figure, the electric fields at O due to the portions AC and BD are equal in magnitude and opposite in direction. So, they cancel each other.

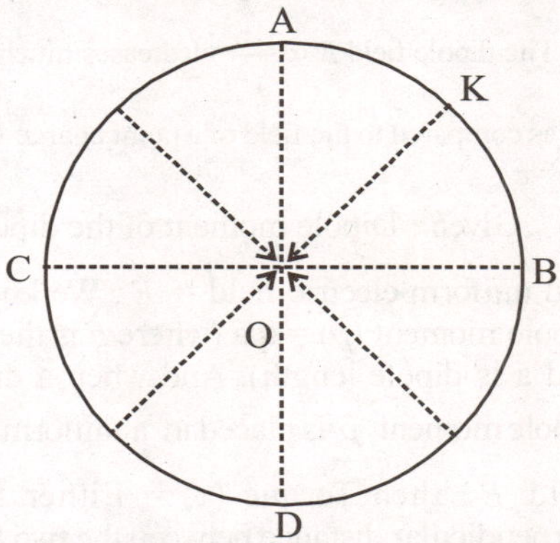

Similarly, the field at O due to CD and AKB are equal in magnitude but opposite in direction. Therefore, the electric field at the centre due to the charge on the part ACDB is E along OK.

25. **(a)** Three point charges $+q$, $-2q$ and $+q$ are placed at points B $(x = 0, y = a, z = 0)$, O $(x = 0, y = 0, z = 0)$ and $A(x = a, y = 0, z = 0)$

The system consists of two dipole moment vectors due to ($+q$ and $-q$) and again due to ($+q$ and $-q$) charges having equal magnitudes qa units – one along $\overrightarrow{OA}$ and other along $\overrightarrow{OB}$. Hence, net dipole moment,

$p_{\text{net}} = \sqrt{(qa)^2 + (qa)^2} = \sqrt{2}qa$ along $\overrightarrow{OP}$ at an angle 45° with positive X-axis.

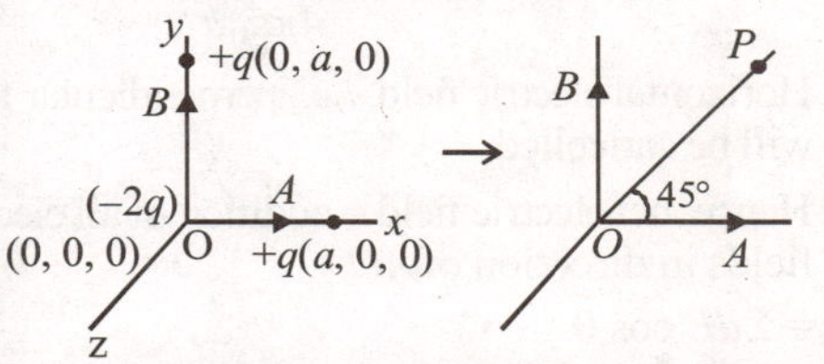

26. **(a)** Given : Length of the dipole $(2l) = 10$cm $= 0.1$m or $l = 0.05$ m

Charge on the dipole (q) = 500 μC = 500 × 10^{-6} C and distance of the point on the axis from the mid-point of the dipole $(r) = 20 + 5 = 25$ cm = 0.25 m. We know that the electric field intensity due to dipole on the given point (E) =

$$\frac{1}{4\pi\varepsilon_0} \times \frac{2(q.2l)r}{(r^2 - l^2)^2}$$

$$= 9 \times 10^9 \times \frac{2(500 \times 10^{-6} \times 0.1) \times 0.25}{[(0.25)^2 - (0.05)^2]^2}$$

$$= \frac{225 \times 10^3}{3.6 \times 10^{-3}} = 6.25 \times 10^7 \ N/C$$

> NOTES: The dipole field $E \propto \frac{1}{r^3}$ decreases much rapidly as compared to the field of a point charge $E \propto \frac{1}{r^2}$

27. **(b)** Given : Dipole moment of the dipole = $\vec{p}$ and uniform electric field = $\vec{E}$. We know that dipole moment (p) = q.a (where q is the charge and a is dipole length). And when a dipole of dipole moment $\vec{p}$ is placed in a uniform electric field $\vec{E}$, then Torque (τ) = Either force × Perpendicular distance between the two forces = $qaE\sin\theta$ or $\tau = pE\sin\theta$ or $\vec{\tau} = \vec{p} \times \vec{E}$ (vector form)

28. **(a)** λ = linear charge density;

Charge on elementary portion $dx = \lambda\, dx$.

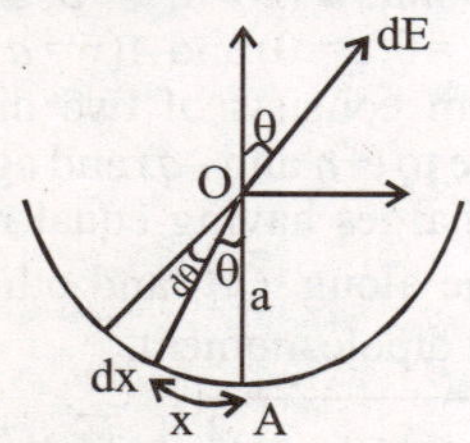

Electric field at O, $dE = \frac{\lambda dx}{4\pi\varepsilon_0 a^2}$

Horizontal electric field, *i.e.*, perpendicular to AO, will be cancelled.

Hence, net electric field = addition of all electrical fields in direction of AO

$= \Sigma dE \cos\theta$

$$\Rightarrow E = \int \frac{\lambda dx}{4\pi\varepsilon_0 a^2} \cos\theta$$

Also, $d\theta = \frac{dx}{a}$ or $dx = a d\theta$

$$E = \int_{-\pi/2}^{\pi/2} \frac{\lambda \cos\theta d\theta}{4\pi\varepsilon_0 a} = \frac{\lambda}{4\pi\varepsilon_0 a}[\sin\theta]_{-\pi/2}^{\pi/2}$$

$$= \frac{\lambda}{4\pi\varepsilon_0 a}[1-(-1)] = \frac{\lambda}{2\pi\varepsilon_0 a}$$

29. **(a)** Charge resides on the outer surface of a conducting hollow sphere of radius R. We consider a spherical surface of radius $r <$ R.

By Gauss's theorem,

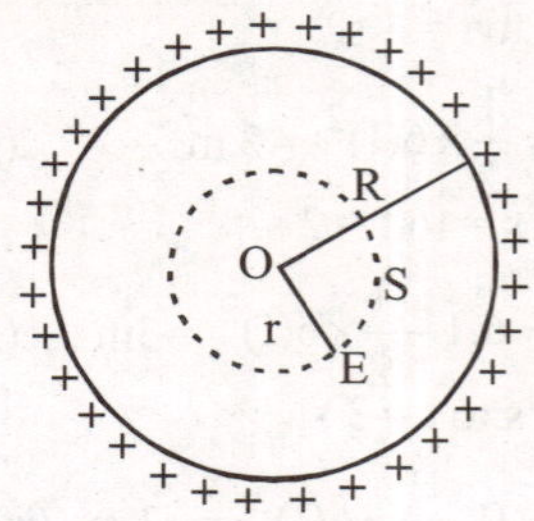

$$\oint \vec{E}.\vec{ds} = \frac{1}{\varepsilon_0} \times \text{charge enclosed}$$

or, $E.4\pi r^2 = \frac{1}{\varepsilon_0} \times 0 \Rightarrow E = 0$

i.e. electric field inside a hollow sphere is zero.

> NOTES: Electric field inside the charged conducting sphere or shell, $E_{in} = 0$
> And for uniformly charged non-conducting sphere,
> $E_{in} = \frac{1}{4\pi\varepsilon_0}\frac{Q_r}{R^3}$

30. **(d)** Electric field at equatorial point

$$Eeq = \frac{1p}{4\pi E_o (r^2 + l^2)^{3/2}}$$

$$E = \frac{p}{4\pi\varepsilon_0 . r^3} \quad \text{[For } r >> l\text{]}$$

Apparently, $E \propto p$ and $E \propto \frac{1}{r^3} \propto r^{-3}$.

31. **(a)** Given : Electric field (E) = 500 V/m and potential difference (V) = 3000 V.

We know that electric field

(E) = 500 = $\frac{V}{d}$ or $d = \frac{3000}{500} = 6$ m

[where d = Distance between point charge and A]

32. **(c)** Intensity of electric field due to a Dipole

$$\text{E} = \frac{p}{4\pi\varepsilon_0 r^3}\sqrt{3\cos^2\theta + 1} \Rightarrow E \propto \frac{1}{r^3}$$

33. **(a)** Dipole is formed when two equal and

34. (c) Charges (q) = 2×10^{-6} C,
Distance
(d) = 3 cm = 3×10^{-2} m and
electric field (E)
= 2×10^5 N/C.
Torque (τ) = pE = qdE
T = $(2 \times 10^{-6}) \times (3 \times 10^{-2}) \times (2 \times 10^5)$
= 12×10^{-3} N–m.

35. (d) Charge (q) = 0.2 C;
Distance (d) = 2m; Angle θ = 60° and work done (W) = 4J.
Work done in moving the charge (W)
= $F.d \cos\theta = qEd\cos\theta$

$$\text{or, } E = \frac{W}{qd\cos\theta} = \frac{4}{0.2 \times 2 \times \cos 60^\circ} = \frac{4}{0.4 \times 0.5}$$
= 20 N/C.

36. (c) $\phi_{Total} = \frac{q_{enclosed}}{\varepsilon_0}$

$$\phi_{Total} = \frac{0}{\varepsilon_0} = 0$$

37. (b) Net flux emitted from a spherical surface of radius a according to Gauss's theorem

$$\phi_{net} = \frac{q_{in}}{\varepsilon_0}$$

$$\text{or, } (Aa)(4\pi a^2) = \frac{q_{in}}{\varepsilon_0}$$

So, $q_{in} = 4\pi\varepsilon_0 A a^3$

NOTES
Electric flux is equal to the product of an area element A and the perpendicular component of electric field E integrated over a surface.
$\phi = E.A \cos\theta$

38. (b) Eight identical cubes are required to arrange so that this charge is at centre of the cube formed so flux.

$$\phi = \frac{q}{8\varepsilon_0}$$

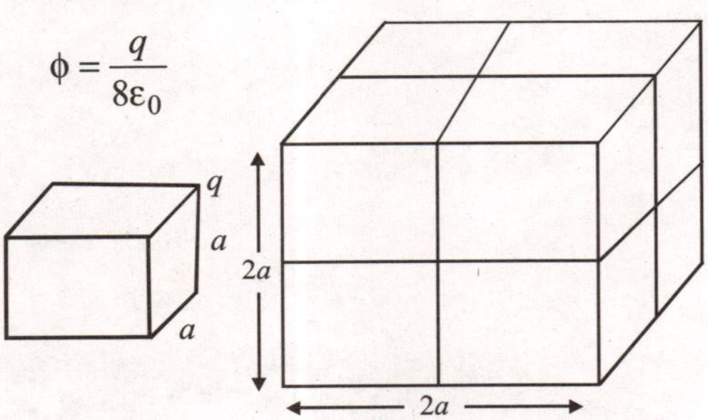

39. (c) By Gauss's theorem,

$$\phi = \frac{Q_{in}}{\varepsilon_0}$$

Thus, the net flux depends only on the charge enclosed by the surface. Hence, there will be no effect on the net flux if the radius of the surface is doubled.

40. (d) Electric flux, $\phi = EA\cos\theta$, where θ = angle between E and normal to the surface.
Here $\theta = \frac{\pi}{2}$
$\Rightarrow \phi = 0$

41. (d) Since $\phi_{total} = \phi_A + \phi_B + \phi_C = \frac{q}{\varepsilon_0}$,
where q is the total charge.
As shown in the figure, flux associated with the curved surface B is $\phi = \phi_B$
Let us assume flux linked with the plane surfaces A and C be
$\phi_A = \phi_C = \phi'$
Therefore,

$$\frac{q}{\varepsilon_0} = 2\phi' + \phi_B = 2\phi' + \phi$$

$$\Rightarrow \phi' = \frac{1}{2}\left(\frac{q}{\varepsilon_0} - \phi\right)$$

42. (b) Flux = $\vec{E}.\vec{A}$.
$\vec{E}$ is electric field vector & $\vec{A}$ is area vector.
Here, angle between $\vec{E}$ & $\vec{A}$ is 90°.
So, $\vec{E}.\vec{A} = 0$; Flux = 0

43. (c) Cube has 6 faces. Flux through any face is given by

$$\phi = \frac{q}{6\varepsilon_0} = \frac{q4\pi}{6(4\pi\varepsilon_0)}$$

NOTES
If charge (Q) is kept at the centre of cube

$$\phi_{total} = \frac{Q}{\varepsilon_0}, \phi_{face} = \frac{Q}{6\varepsilon_0}$$

$$\phi_{corner} = \frac{Q}{8\varepsilon_0}, \quad \phi_{edge} = \frac{Q}{12\varepsilon_0}$$

44. (a) Total flux out of all six faces as per Gauss's theorem should be $\frac{Q \times 10^{-6}}{\varepsilon_0}$

Therefore, flux coming out of each face

$$= \frac{Q}{6 \epsilon_0} \times 10^{-6}\ C$$

45. (c) According to Gauss's theorem, electric flux through a closed surface $= \frac{Q}{\varepsilon_0}$ where q is charge enclosed by the surface.

The charge is kept at a corner of a cube so it can be visualised as being at the centre of 8 cubes.

$\therefore$ The flux through each cube will be $\frac{1}{8}$ of the flux $\frac{\theta}{E_o}$

$\therefore$ Flux through all six faces $= \frac{\theta}{8E_o}$

46. (a) By Gauss theorem

$$\text{Total electric flux} = \frac{\text{Total charge inside cube}}{\varepsilon_0}$$

$$\Rightarrow \phi = \frac{q}{\varepsilon_0}$$

16 Electrostatic Potential and Capacitance

Trend Analysis with Important Topics & Sub-Topics

Topic Name	Sub-Topic	2021 QNS.	2021 LOD	2020 QNS.	2020 LOD	2019 QNS.	2019 LOD	2018 QNS.	2018 LOD	2017 QNS.	2017 LOD	2016 QNS.	2016 LOD
Electrostatic Potential & Equipotential Surfaces	Electrostatic pot. & field	1	A	1	E								
	Electrostatic pot. and dipole			1	E								
	Equipotential Surfaces									1	E		
Electric Potential Energy & Work Done in Carrying a Charge		1	A										
Capacitors, Capacitance, Grouping of Capacitors & Energy Stored in a Capacitor	Capacitance of Capacitor			1	A								
	Electrostatic Force between Isolated Parallel Plates							1	A				
	Grouping of Capacitors	1	E										
	Capacitance of a capacitor & Energy stored in a capacitor	1	A							1	A	1	A

LOD - Level of Difficulty | E - Easy | A - Average | D - Difficult | Qns - No. of Questions

Topic 1 Electrostatic Potential & Equipotential Surfaces

1. Twenty seven drops of same size are charged at 200 V each. They combine to form a bigger drop. Calculate the potential of the bigger drop. *[2021]*

(a) 1980 V (b) 660 V
(c) 1320 V (d) 1520 V

2. In a certain region of space with volume 0.2 m^3, the electric potential is found to be 5 V throughout. The magnitude of electric field in this region is : *[2020]*

(a) 0.5 N/C (b) 1 N/C
(c) 5 N/C (d) zero

3. A short electric dipole has a dipole moment of 16×10^{-9} C m. The electric potential due to the dipole at a point at a distance of 0.6 m from the centre of the dipole, situated on a line making an angle of 60° with the dipole axis is : *[2020]*

$$\left(\frac{1}{4\pi \epsilon_0} = 9\times10^{9}\ \text{N m}^2/\text{C}^2\right)$$

(a) 200 V (b) 400 V
(c) zero (d) 50 V

4. The diagrams below show regions of equipotentials. *[2017]*

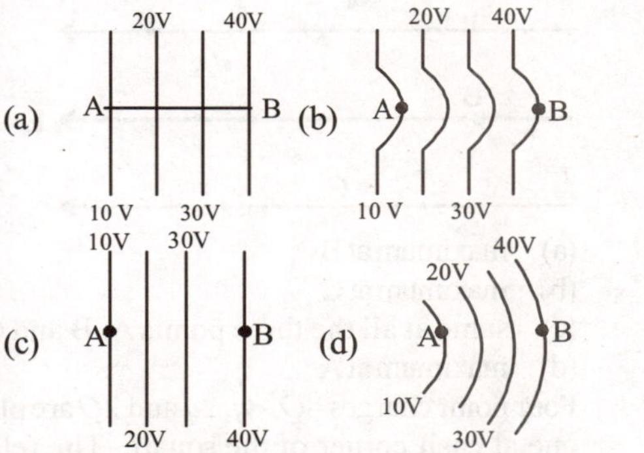

A positive charge is moved from A to B in each diagram.

(a) In all the four cases the work done is the same
(b) Minimum work is required to move q in figure (a)
(c) Maximum work is required to move q in figure (b)
(d) Maximum work is required to move q in figure (c)

5. If potential (in volts) in a region is expressed as V(x, y, z) = 6 xy – y + 2yz, the electric field (in N/C) at point (1, 1, 0) is : *[2015 RS]*

(a) $-(6\hat{i}+5\hat{j}+2\hat{k})$ (b) $-(2\hat{i}+3\hat{j}+\hat{k})$

(c) $-(6\hat{i}+9\hat{j}+\hat{k})$ (d) $-(3\hat{i}+5\hat{j}+3\hat{k})$

6. In a region, the potential is represented by $V(x, y, z) = 6x - 8xy - 8y + 6yz$, where V is in volts and x, y, z are in metres. The electric force experienced by a charge of 2 coulomb situated at point (1, 1, 1) is : *[2014]*

(a) $6\sqrt{5}$ N (b) 30 N

(c) 24 N (d) $4\sqrt{35}$ N

7. A conducting sphere of radius R is given a charge Q. The electric potential and the electric field at the centre of the sphere respectively are: *[2014]*

(a) Zero and $\dfrac{Q}{4\pi\varepsilon_0 R^2}$

(b) $\dfrac{Q}{4\pi\varepsilon_0 R}$ and Zero

(c) $\dfrac{Q}{4\pi\varepsilon_0 R}$ and $\dfrac{Q}{4\pi\varepsilon_0 R^2}$

(d) Both are zero

8. A, B and C are three points in a uniform electric field. The electric potential is *[2013]*

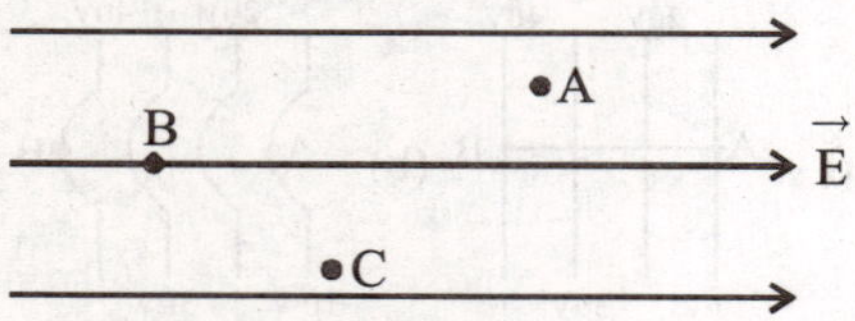

(a) maximum at B

(b) maximum at C

(c) same at all the three points A, B and C

(d) maximum at A

9. Four point charges $-Q, -q, 2q$ and $2Q$ are placed, one at each corner of the square. The relation between Q and q for which the potential at the centre of the square is zero is : *[2012]*

(a) $Q = -q$ (b) $Q = -\dfrac{1}{q}$

(c) $Q = q$ (d) $Q = \dfrac{1}{q}$

10. Four electric charges +q, +q, –q and – q are placed at the corners of a square of side 2L (see figure). The electric potential at point A, midway between the two charges +q and +q, is *[2011]*

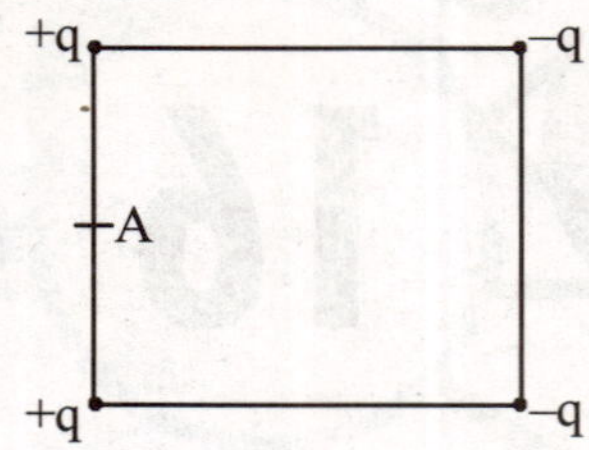

(a) $\dfrac{1}{4\pi\varepsilon_0}\dfrac{2q}{L}(1+\sqrt{5})$

(b) $\dfrac{1}{4\pi\varepsilon_0}\dfrac{2q}{L}\left(1+\dfrac{1}{\sqrt{5}}\right)$

(c) $\dfrac{1}{4\pi\varepsilon_0}\dfrac{2q}{L}\left(1-\dfrac{1}{\sqrt{5}}\right)$

(d) zero

11. Three concentric spherical shells have radii a, b and c (a < b < c) and have surface charge densities σ, – σ and σ respectively. If V_A, V_B and V_C denotes the potentials of the three shells, then for c = a + b, we have *[2009]*

(a) $V_C = V_B \neq V_A$ (b) $V_C \neq V_B \neq V_A$

(c) $V_C = V_B = V_A$ (d) $V_C = V_A \neq V_B$

12. The electric potential at a point (x, y, z) is given by $V = -x^2y - xz^3 + 4$. The electric field $\vec{E}$ at that point is *[2009]*

(a) $\vec{E} = \hat{i}\,2xy + \hat{j}(x^2+y^2) + \hat{k}(3xz - y^2)$

(b) $\vec{E} = \hat{i}z^3 + \hat{j}\,xyz + \hat{k}z^2$

(c) $\vec{E} = \hat{i}(2xy - z^3) + \hat{j}\,xy^2 + \hat{k}\,3z^2x$

(d) $\vec{E} = \hat{i}(2xy + z^3) + \hat{j}x^2 + \hat{k}\,3xz^2$

13. The electric potential at a point in free space due to a charge Q coulomb is $Q \times 10^{11}$ volts. The electric field at that point is *[2008]*

(a) $4\pi\varepsilon_0 Q \times 10^{22}$ volt/m

(b) $12\pi\varepsilon_0 Q \times 10^{20}$ volt/m

(c) $4\pi\varepsilon_0 Q \times 10^{20}$ volt/m

(d) $12\pi\varepsilon_0 Q \times 10^{22}$ volt/m

14. A solid spherical conductor is given a charge. The electrostatic potential of the conductor is *[2002]*

(a) constant throughout the conductor

(b) largest at the centre

(c) largest on the surface

(d) largest somewhere between the centre and the surface

15. A hollow metal sphere of radius 10 cm is charged such that the potential on its surface is 80 V. The potential at the centre of the sphere is [1994]

(a) zero (b) 80 V
(c) 800 V (d) 8 V

Topic 2 Electric Potential Energy & Work Done in Carrying a Charge

16. A dipole is placed in an electric field as shown. In which direction will it move? [2021]

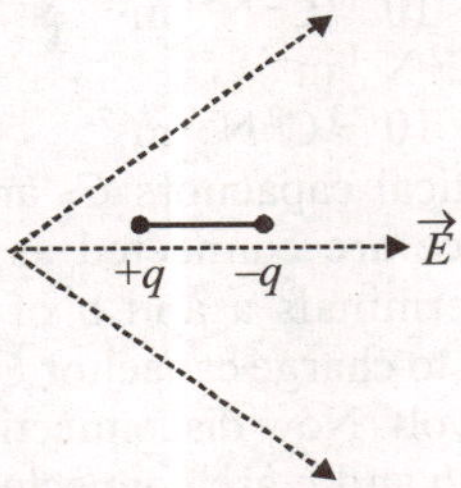

(a) Towards the right as its potential energy will increase.
(b) Towards the left as its potential energy will increase.
(c) Towards the right as its potential energy will decrease.
(d) Towards the left as its potential energy will decrease.

17. The potential energy of particle in a force field is $U = \frac{A}{r^2} - \frac{B}{r}$, where A and B are positive constants and r is the distance of particle from the centre of the field. For stable equilibrium, the distance of the particle is : [2012]

(a) $B/2A$ (b) $2A/B$
(c) A/B (d) B/A

18. Three charges, each +q, are placed at the corners of an isosceles triangle ABC of sides BC, AC and AB. D and E are the mid points of BC and CA. The work done in taking a charge Q from D to E is [2011M]

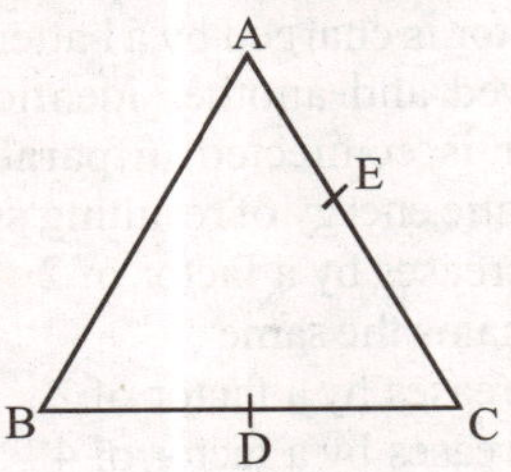

(a) $\frac{3qQ}{8\pi \epsilon_0 a}$ (b) $\frac{qQ}{4\pi \epsilon_0 a}$
(c) zero (d) $\frac{3qQ}{4\pi \epsilon_0 a}$

19. Charges $+q$ and $-q$ are placed at points A and B respectively which are a distance $2L$ apart, C is the midpoint between A and B. The work done in moving a charge $+Q$ along the semicircle CRD is [2007]

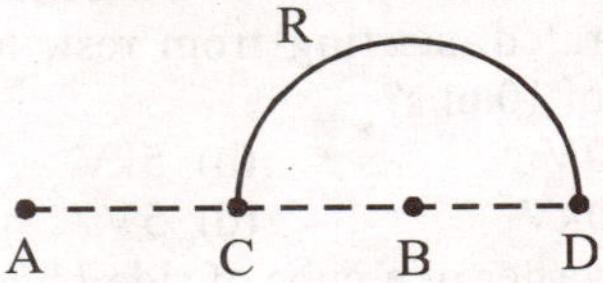

(a) $\frac{qQ}{2\pi\varepsilon_0 L}$ (b) $\frac{qQ}{6\pi\varepsilon_0 L}$
(c) $-\frac{qQ}{6\pi\varepsilon_0 L}$ (d) $\frac{qQ}{4\pi\varepsilon_0 L}$

20. As per the diagram, a point charge $+q$ is placed at the origin O. Work done in taking another point charge $-Q$ from the point A [coordinates $(0, a)$] to another point B [coordinates $(a, 0)$] along the straight path AB is: [2005]

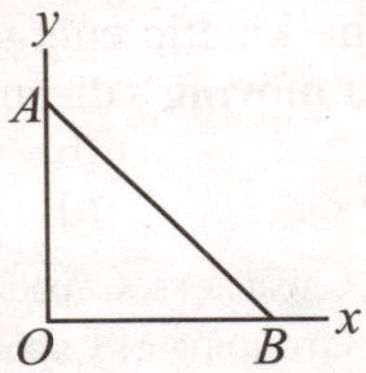

(a) zero (b) $\left(\frac{-qQ}{4\pi\varepsilon_0}\frac{1}{a^2}\right)\sqrt{2}a$
(c) $\left(\frac{qQ}{4\pi\varepsilon_0}\frac{1}{a^2}\right)\cdot\frac{a}{\sqrt{2}}$ (d) $\left(\frac{qQ}{4\pi\varepsilon_0}\frac{1}{a^2}\right)\cdot\sqrt{2}a$

21. Two charges q_1 and q_2 are placed 30 cm apart, as shown in the figure. A third charge q_3 is moved along the arc of a circle of radius 40 cm from C to D. The change in the potential energy of the system is $\frac{q_3}{4\pi \epsilon_0}k$, where k is [2005]

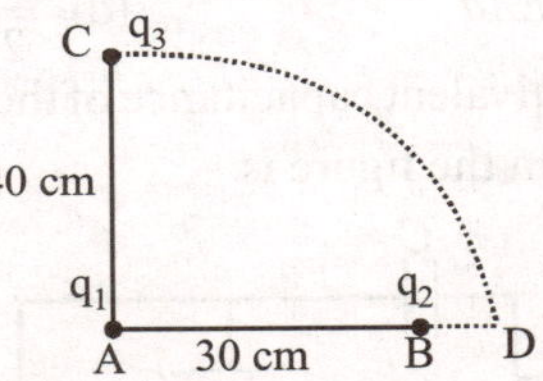

(a) $8q_1$ (b) $6q_1$
(c) $8q_2$ (d) $6q_2$

22. An electric dipole has the magnitude of its charge as q and its dipole moment is p. It is placed in uniform electric field E. If its dipole moment is along the direction of the field, the force on it and its potential energy are respectively *[2004]*
(a) $q.E$ and max. (b) 2 q.E and min.
(c) $q.E$ and p.E (d) zero and min.

23. A bullet of mass 2 g is having a charge of 2μC. Through what potential difference must it be accelerated, starting from rest, to acquire a speed of 10 m/s? *[2004]*
(a) 50 V (b) 5 kV
(c) 50 kV (d) 5 V

24. Each corner of a cube of side l has a negative charge, $-q$. The electrostatic potential energy of a charge q at the centre of the cube is *[2002]*
(a) $-\frac{4q^2}{\sqrt{2}\pi\varepsilon_0 l}$ (b) $\frac{\sqrt{3}q^2}{4\pi\varepsilon_0 l}$
(c) $\frac{4q^2}{\sqrt{2}\pi\varepsilon_0 l}$ (d) $-\frac{4q^2}{\sqrt{3}\pi\varepsilon_0 l}$

25. A particle of mass m and charge q is placed at rest in a uniform electric field E and then released. The kinetic energy attained by the particle after moving a distance y is *[1998]*
(a) qEy^2 (b) $qE^2 y$
(c) qEy (d) $q^2 Ey$

Topic 3 Capacitors, Capacitance, Grouping of Capacitors & Energy Stored in a Capacitor

26. A parallel plate capacitor has a uniform electric field '$\vec{E}$' in the space between the plates. If the distance between the plates is 'd' and the area of each plate is 'A', the energy stored in the capacitor is (ε_0 = permittivity of free space) *[2021]*
(a) $\frac{E^2 Ad}{\varepsilon_0}$ (b) $\frac{1}{2}\varepsilon_0 E^2$
(b) $\varepsilon_0 EAd$ (d) $\frac{1}{2}\varepsilon_0 E^2 Ad$

27. The equivalent capacitance of the combination shown in the figure is *[2021]*

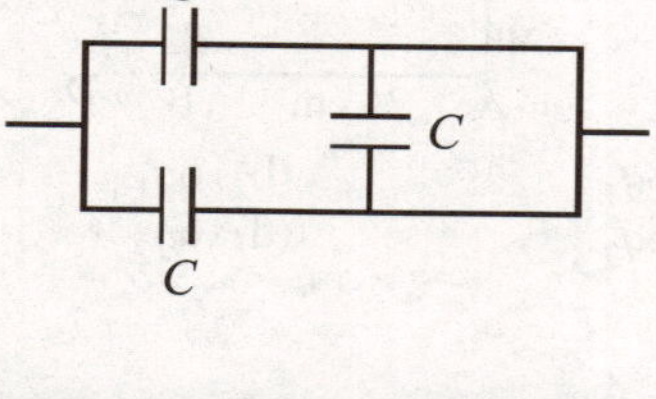

(a) $\frac{3C}{2}$ (b) $3C$
(c) $2C$ (D) $\frac{C}{2}$

28. The capacitance of a parallel plate capacitor with air as medium is 6 μF. With the introduction of a dielectric medium, the capacitance becomes 30 μF. The permittivity of the medium is : *[2020]*
($\epsilon_0 = 8.85 \times 10^{-12} C^2 N^{-1} m^{-2}$)
(a) $1.77 \times 10^{-12} C^2 N^{-1} m^{-2}$
(b) $0.44 \times 10^{-10} C^2 N^{-1} m^{-2}$
(c) $5.00 C^2 N^{-1} m^{-2}$
(d) $0.44 \times 10^{-13} C^2 N^{-1} m^{-2}$

29. Two identical capacitors C_1 and C_2 of equal capacitance are connected as shown in the circuit. Terminals a and b of the key k are connected to charge capacitor C_1 using battery of emf V volt. Now disconnecting a and b the terminals b and c are connected. Due to this, what will be the percentage loss of energy? *[NEET Odisha 2019]*

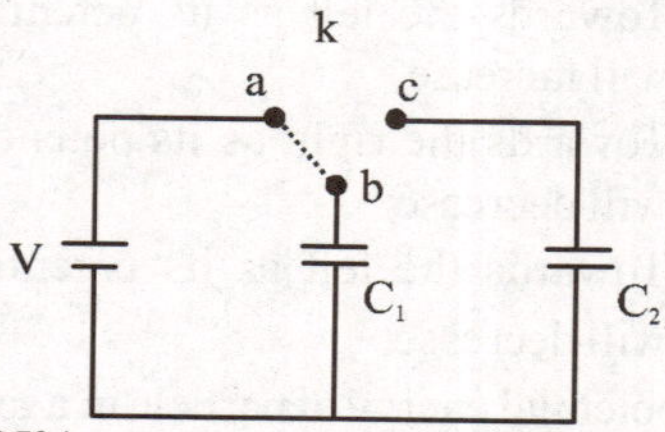

(a) 25% (b) 75%
(c) 0% (d) 50%

30. The electrostatic force between the metal plates of an isolated parallel plate capacitor C having a charge Q and area A, is *[2018]*
(a) independent of the distance between the plates
(b) linearly proportional to the distance between the plates
(c) inversely proportional to the distance between the plates
(d) proportional to the square root of the distance between the plates

31. A capacitor is charged by a battery. The battery is removed and another identical uncharged capacitor is connected in parallel. The total electrostatic energy of resulting system : *[2017]*
(a) decreases by a factor of 2
(b) remains the same
(c) increases by a factor of 2
(d) increases by a factor of 4

32. 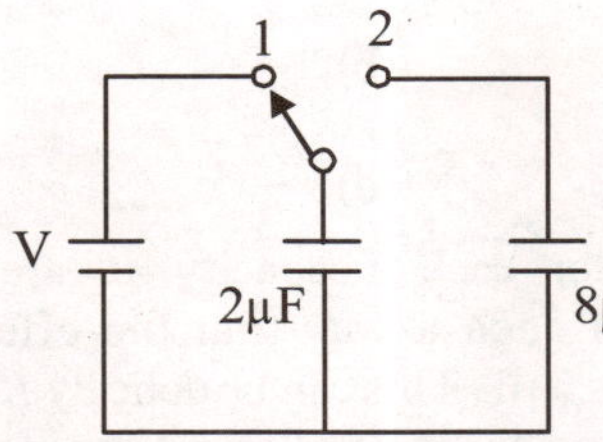

A capacitor of $2\mu F$ is charged as shown in the diagram. When the switch S is turned to position 2, the percentage of its stored energy dissipated is : *[2016]*

(a) 0% (b) 20%
(c) 75% (d) 80%

33. A parallel plate air capacitor of capacitance C is connected to a cell of emf V and then disconnected from it. A dielectric slab of dielectric constant K, which can just fill the air gap of the capacitor, is now inserted in it. Which of the following is incorrect ? *[2015]*

(a) The energy stored in the capacitor decreases K times.
(b) The chance in energy stored is $\frac{1}{2}CV^2\left(\frac{1}{K}-1\right)$
(c) The charge on the capacitor is not conserved.
(d) The potential difference between the plates decreases K times.

34. A parallel plate air capacitor has capacity 'C' distance of separation between plates is 'd' and potential difference 'V' is applied between the plates. Force of attraction between the plates of the parallel plate air capacitor is : *[2015 RS]*

(a) $\frac{CV^2}{2d}$ (b) $\frac{CV^2}{d}$
(c) $\frac{C^2V^2}{2d^2}$ (d) $\frac{C^2V^2}{2d^2}$

35. Two thin dielectric slabs of dielectric constants K_1 and K_2 $(K_1 < K_2)$ are inserted between plates of a parallel plate capacitor, as shown in the figure. The variation of electric field 'E' between the plates with distance 'd' as measured from plate P is correctly shown by : *[2014]*

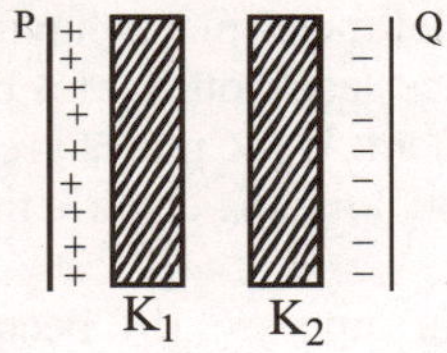

(a) 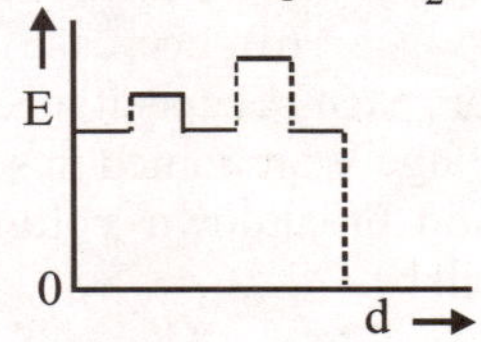

(b) 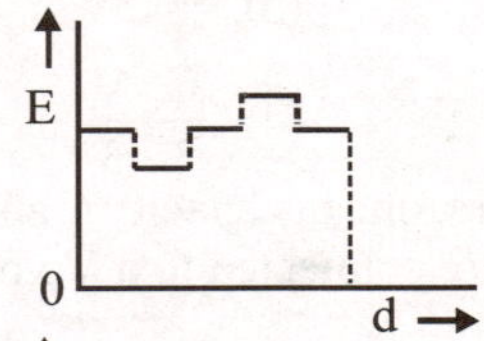

(c) 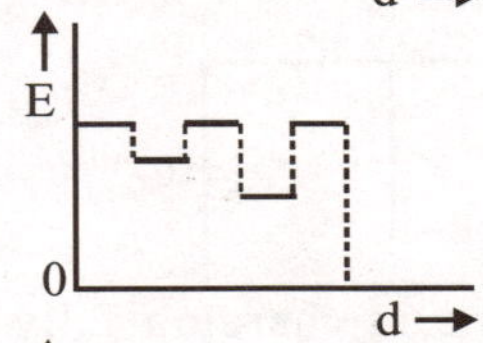

(d) 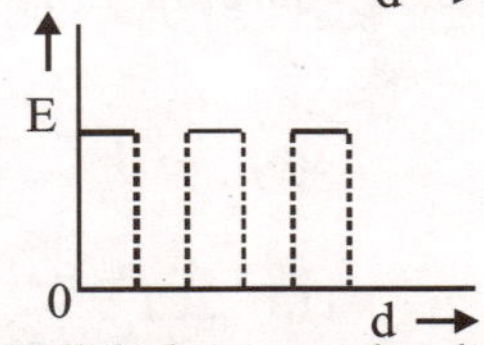

36. A parallel plate capacitor has a uniform electric field E in the space between the plates. If the distance between the plates is d and area of each plate is A, the energy stored in the capacitor is : *[2012M, 2011, 2008]*

(a) $\frac{1}{2}\varepsilon_0 E^2$ (b) E^2Ad/ε_0
(c) $\frac{1}{2}\varepsilon_0 E^2 Ad$ (d) $\varepsilon_0 EAd$

37. A series combination of n_1 capacitors, each of value C_1, is charged by a source of potential difference 4 V. When another parallel combination of n_2 capacitors, each of value C_2, is charged by a source of potential difference V, it has the same (total) energy stored in it, as the first combination has. The value of C_2, in terms of C_1, is then *[2010]*

(a) $\frac{2C_1}{n_1 n_2}$ (b) $16\frac{n_2}{n_1}C_1$
(c) $2\frac{n_2}{n_1}C_1$ (d) $\frac{16C_1}{n_1 n_2}$

38. Two parallel metal plates having charges + Q and –Q face each other at a certain distance between them. If the plates are now dipped in kerosene oil tank, the electric field between the plates will *[2010]*

(a) remain same (b) become zero
(c) increases (d) decrease

39. Three capacitors each of capacitance C and of breakdown voltage V are joined in series. The capacitance and breakdown voltage of the combination will be *[2009]*

(a) $3C, \frac{V}{3}$ (b) $\frac{C}{3}, 3V$
(c) $3C, 3V$ (d) $\frac{C}{3}, \frac{V}{3}$

40. Two condensers, one of capacity C and other of capacity C/2 are connected to a V-volt battery, as shown.

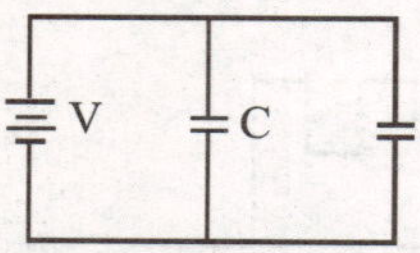

The work done in charging fully both the condensers is *[2007]*

(a) $\frac{1}{4}CV^2$ (b) $\frac{3}{4}CV^2$
(c) $\frac{1}{2}CV^2$ (d) $2CV^2$.

41. A parallel plate air capacitor is charged to a potential difference of V volts. After disconnecting the charging battery the distance between the plates of the capacitor is increased using an insulating handle. As a result the potential difference between the plates *[2006]*

(a) does not change (b) becomes zero
(c) increases (d) decreases

42. A network of four capacitors of capacity equal to $C_1 = C$, $C_2 = 2C$, $C_3 = 3C$ and $C_4 = 4C$ are connected to a battery as shown in the figure. The ratio of the charges on C_2 and C_4 is: *[2005]*

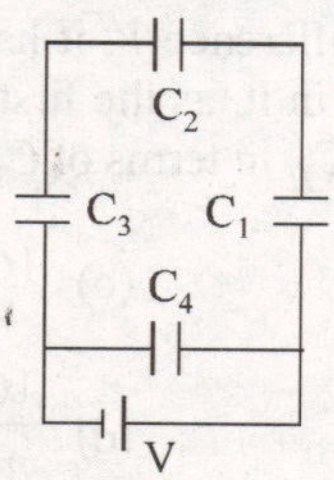

(a) $\frac{4}{7}$ (b) $\frac{3}{22}$
(c) $\frac{7}{4}$ (d) $\frac{22}{3}$

43. Three capacitors each of capacity 4μF are to be connected in such a way that the effective capacitance is 6 μF. This can be done by *[2003]*

(a) connecting two in parallel and one in series
(b) connecting all of them in series
(c) connecting them in parallel
(d) connecting two in series and one in parallel

44. A capacitor C_1 is charged to a potential difference V. The charging battery is then removed and the capacitor is connected to an uncharged capacitor C_2. The potential difference across the combination is *[2002]*

(a) $\frac{VC_1}{(C_1+C_2)}$ (b) $V\left(1+\frac{C_2}{C_1}\right)$
(c) $V\left(1+\frac{C_1}{C_2}\right)$ (d) $\frac{VC_2}{(C_1+C_2)}$

45. In a parallel plate capacitor, the distance between the plates is d and potential difference across the plates is V. Energy stored per unit volume between the plates of capacitor is *[2001]*

(a) $\frac{Q^2}{2V^2}$ (b) $\frac{1}{2}\varepsilon_0\frac{V^2}{d^2}$
(c) $\frac{1}{2}\frac{V^2}{\varepsilon_0 d^2}$ (d) $\frac{1}{2}\varepsilon_0\frac{V^2}{d}$

46. The capacity of a parallel plate condenser is 10 μF when the distance between its plates is 8 cm. If the distance between the plates is reduced to 4 cm then the capacity of this parallel plate condenser will be *[2001]*

(a) 5 μF (b) 10 μF
(c) 20 μF (d) 40 μF

47. Energy stored in a capacitor is *[2001]*

(a) $\frac{1}{2}QV$ (b) QV (c) $\frac{1}{QV}$ (d) $\frac{2}{QV}$

48. A capacitor is charged to store an energy U. The charging battery is disconnected. An identical capacitor is now connected to the first capacitor in parallel. The energy in each of the capacitor is *[2000]*

(a) $U/2$ (b) $3U/2$
(c) U (d) $U/4$

49. A parallel plate condenser with oil between the plates (dielectric constant of oil $K = 2$) has a capacitance C. If the oil is removed, then capacitance of the capacitor becomes

(a) $\sqrt{2}C$ (b) $2C$ [1999, 97]

(c) $\frac{C}{\sqrt{2}}$ (d) $\frac{C}{2}$

50. What is the effective capacitance between points X and Y? [1999]

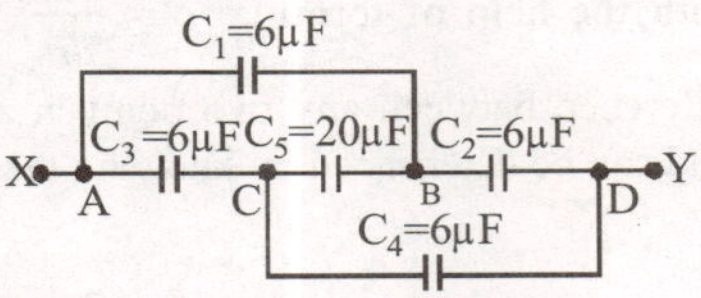

(a) 24 μF (b) 18 μF

(c) 12 μF (d) 6 μF

51. If the potential of a capacitor having capacity 6 μF is increased from 10 V to 20 V, then increase in its energy will be [1995]

(a) 4×10^{-4} J (b) 4×10^{-4} J

(c) 9×10^{-4} J (d) 12×10^{-6} J

52. The four capacitors, each of 25μ F are connected as shown in fig. The dc voltmeter reads 200 V. The charge on each plate of capacitor is [1994]

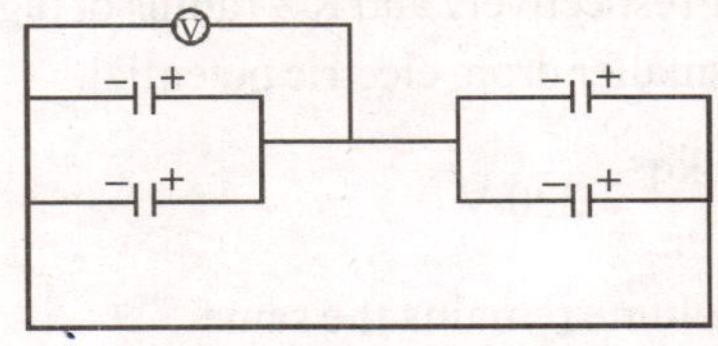

(a) $\pm 2 \times 10^{-3}$ C (b) $\pm 5 \times 10^{-3}$ C

(c) $\pm 2 \times 10^{-2}$ C (d) $\pm 5 \times 10^{-2}$ C

53. A 4μF conductor is charged to 400 volts and then its plates are joined through a resistance of 1 kΩ. The heat produced in the resistance is

(a) 0.16 J (b) 1.28 J [1989]

(c) 0.64 J (d) 0.32 J

ANSWER KEY

1	(a)	7	(b)	13	(a)	19	(c)	25	(c)	31	(a)	37	(d)	43	(d)	49	(d)
2	(d)	8	(a)	14	(a)	20	(a)	26	(d)	32	(d)	38	(d)	44	(a)	50	(d)
3	(a)	9	(a)	15	(b)	21	(c)	27	(c)	33	(c)	39	(b)	45	(b)	51	(c)
4	(a)	10	(c)	16	(c)	22	(d)	28	(b)	34	(a)	40	(b)	46	(c)	52	(b)
5	(a)	11	(d)	17	(b)	23	(c)	29	(d)	35	(c)	41	(c)	47	(a)	53	(d)
6	(d)	12	(d)	18	(c)	24	(d)	30	(a)	36	(c)	42	(b)	48	(d)		

Hints & Solutions

1. **(a)** Let charge and radius of smaller drop is q and r respectively and R = radius of bigger drop

For smaller drop, electric potential,

$$V = \frac{kq}{r} = 220\ V$$

As volume remains the same

$$\left(\frac{4}{3}\pi r^3\right) \times 27 = \frac{4}{3}\pi R^3 \Rightarrow R = \sqrt[3]{27}r = 3r$$

And from conservation of charge,

Q = 27q

∴ Potential of bigger drop,

$$V_{\text{big drop}} = \frac{kQ}{R} = \frac{k(27q)}{3r} = 9\left(\frac{kq}{r}\right)$$

$= 9 \times 220 = 1980$ V

2. **(d)** Since, electric potential is constant throughout the volume, hence electric field, $E = -\frac{dV}{dr} = 0.$

3. **(a)** Given,

Dipole moment of short electric dipole,

$p = 16 \times 10^{-9}$ C m.

Distance from centre of dipole, $r = 0.6$ m

Electric potential, $V = \frac{kp\cos\theta}{r^2}$

$$\Rightarrow V = \frac{9 \times 10^9 \times 16 \times 10^{-9} \times \cos 60}{0.36} = 200\ V$$

4. **(a)** As the regions are of equipotentials, so

Work done W = qΔV

ΔV is same in all the cases hence work - done will also be same in all the cases.

5. **(a)** Potential in a region is given by,

$V = 6xy - y + 2yz$

As we know the relation between electric potential and electric field is

$$\vec{E} = -\left(\frac{\partial V}{\partial x}\hat{i} + \frac{\partial V}{\partial y}\hat{j} + \frac{\partial V}{\partial z}\hat{k}\right)$$

where $V = 6xy - y + 2yz$

$$\vec{E} = -\left[(6y\hat{i} + (6x - 1 + 2z)\hat{j} + (2y)\hat{k}\right]$$

$$\vec{E}_{(1,1,0)} = -(6\hat{i} + 5\hat{j} + 2\hat{k})$$

NOTES

With the help of formula $E = \frac{-dv}{dr}$, potential difference between any two point in an electric field can be determined by knowing the boundary conditions $dV = -\int_{r_1}^{r_2} \vec{E}\,\vec{dr} = -\int_{r_1}^{r_2} E.dr\cos\theta$

6. **(d)** $\vec{E} = -\frac{\partial V}{\partial x}\hat{i} - \frac{\partial V}{\partial y}\hat{j} - \frac{\partial V}{\partial z}\hat{k}$

where $V(x, y, z) = 6x - 8xy - 8y + 6yz$

$= -[(6 - 8y)\hat{i} + (-8x - 8 + 6z)\hat{j} + (6y)\hat{k}]$

At (1, 1, 1), $\vec{E} = 2\hat{i} + 10\hat{j} - 6\hat{k}$

$\Rightarrow \quad (\vec{E}) = \sqrt{(2)^2 + (10)^2 + (-6)^2} = \sqrt{140} = 2\sqrt{35}$

$\therefore \quad F = q\vec{E} = 2 \times 2\sqrt{35} = 4\sqrt{35}$

7. **(b)** Due to conducting sphere

At centre, electric field E = 0

but electric potential, $V = \frac{Q}{4\pi \epsilon_0 R}$

8. **(a)** Potential at B, V_B is maximum

$$V_B > V_C > V_A$$

As in the direction of electric field potential decreases.

9. **(a)** Let the side length of square be '*a*' then potential at centre *O* is

$$AC = BD = \sqrt{a^2 + a^2} = a\sqrt{2}$$

$$OA = OB = OC = OD = \frac{a\sqrt{2}}{2} = \frac{a}{\sqrt{2}}$$

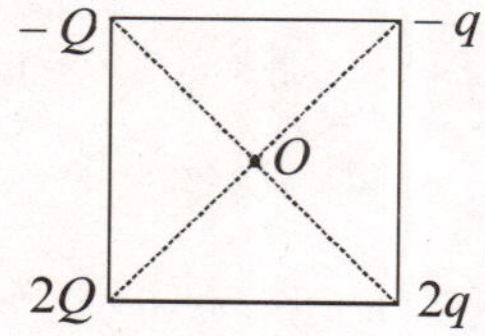

$$V = \frac{k(-Q)}{\left(\frac{a}{\sqrt{2}}\right)} + \frac{k(-q)}{\frac{a}{\sqrt{2}}} + \frac{k(2q)}{\frac{a}{\sqrt{2}}} + \frac{k(2Q)}{\frac{a}{\sqrt{2}}} = 0$$

(Given)

$$= -Q - q + 2q + 2Q = 0 \Rightarrow Q + q = 0$$

$$\Rightarrow Q = -q$$

10. **(c)** Distance of point A from the two +q charges = L.

Distance of point A from the two –q charges $= \sqrt{L^2 + (2L)^2} = \sqrt{5}L$.

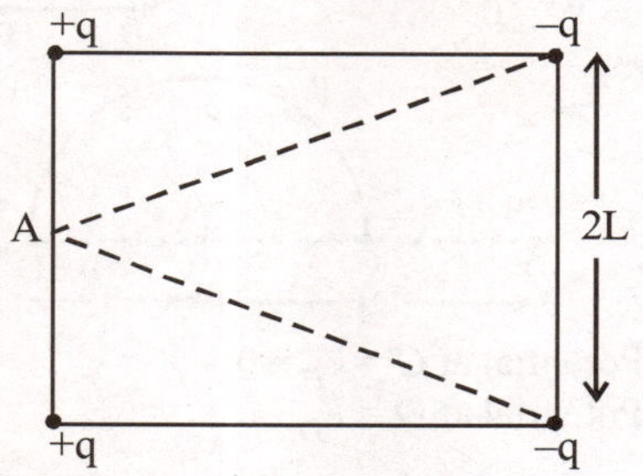

$$\therefore V_A = \left(\frac{Kq}{L} \times 2\right) - \left(\frac{Kq}{\sqrt{5}L} \times 2\right)$$

$$= \frac{2Kq}{L}\left[1 - \frac{1}{\sqrt{5}}\right]$$

$$= \frac{1}{4\pi \epsilon_0} . \frac{2q}{L}\left(1 - \frac{1}{\sqrt{5}}\right)$$

11. **(d)** c = a + b.

$$V_A = \frac{1}{4\pi E_o}\left[\frac{qA}{a} + \frac{qB}{b} + \frac{qC}{a}\right]$$

$$= \frac{1}{4\pi E_o}\left[\frac{4\pi a^2 \sigma}{a} - \frac{4\pi b^2 \sigma}{b} + \frac{4\pi c^2 \sigma}{c}\right]$$

$$= \frac{4\pi}{4\pi E_o}\left[\frac{a^2 \sigma}{a} - \frac{b^2 \sigma}{b} + \frac{c^2 \sigma}{c}\right]$$

$$\Rightarrow V_A = \frac{\sigma a}{\varepsilon_0} - \frac{\sigma b}{\varepsilon_0} + \frac{\sigma c}{\varepsilon_0} = \frac{\sigma}{\varepsilon_0}[c - (b - a)]$$

Similarly $V_B = \frac{1}{4\pi E_o}\left[\frac{qA}{a} + \frac{qB}{b} + \frac{qC}{c}\right]$

$$\Rightarrow V_B = \frac{1}{4\pi E_o}\left[\frac{4\pi\sigma a^2}{b} - \frac{4\pi\sigma b^2}{b} + \frac{4\pi\sigma c^2}{c}\right]$$

$$V_B = \frac{-\sigma b}{\varepsilon_0} + \frac{1}{4\pi\varepsilon_0} . \frac{\sigma \times 4\pi a^2}{b} + \frac{\sigma c}{\varepsilon_0}$$

$$= \frac{\sigma}{\varepsilon_0}\left[c - \frac{(b^2 - a^2)}{b}\right]$$

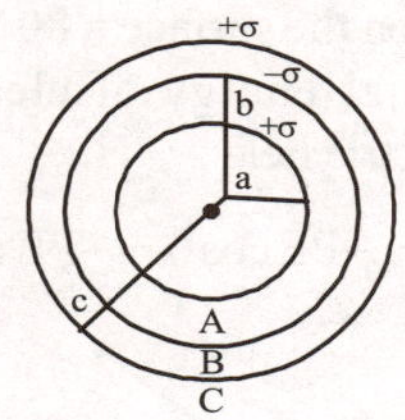

$$V_C = \frac{4\pi}{4\pi E_o}\left[\frac{a^2\sigma}{c} - \frac{b^2\sigma}{c} + \frac{\sigma c^2}{c}\right]$$

$$V_C = \frac{\sigma c}{\varepsilon_0} - \frac{1}{4\pi\varepsilon_0} . \frac{\sigma \times 4\pi b^2}{c} + \frac{1}{4\pi\varepsilon_0} . \frac{\sigma \times 4\pi a^2}{c}$$

$$= \frac{\sigma}{\varepsilon_0}\left[c - \frac{(b^2 - a^2)}{c}\right]$$

$$= \frac{\sigma}{\varepsilon_0}[c - (b - a)]$$

$$V_A = V_C \neq V_B$$

12. **(d)** The electric field at a point is equal to negative of potential gradient at that point.

$$\vec{E} = -\frac{\partial V}{\partial r} = \left[-\frac{\partial V}{\partial x}\hat{i} - \frac{\partial V}{\partial y}\hat{j} - \frac{\partial V}{\partial z}\hat{k}\right]$$

$$= [(2xy + z^3)\,\hat{i} + \hat{j}\,x^2 + \hat{k}\,3xz^2]$$

13. **(a)** Given that, $V = Q \times 10^{11}$ volts

Electric potential at point is given by

$$V = \frac{1}{4\pi \epsilon_0} . \frac{Q}{r} \text{ or, } Q \times 10^{11} = \frac{1}{4\pi \epsilon_0} . \frac{Q}{r}$$

$$\Rightarrow r = \frac{1}{4\pi \epsilon_0} . 10^{-11} \text{m.}$$

$$\text{As } |E| = \frac{|V|}{r} = \frac{Q \times 10^{11}}{\frac{1}{4\pi \epsilon_0} . 10^{-11}}$$

$$= 4\pi \epsilon_0 Q \times 10^{22} \text{ volt m}^{-1}.$$

14. (a) Electric potential is constant (equal to $\frac{kq}{R}$, where $k=\frac{1}{4\pi\epsilon_0}$) within or on the surface of conductor.

15. (b) Potential at the centre of the sphere = potential on the surface = 80 V.

16. (c) Potential energy of electric dipole in external electric field

$$U=-\vec{P}.\vec{E}=-PE\cos\theta=-PE\cos 180°$$

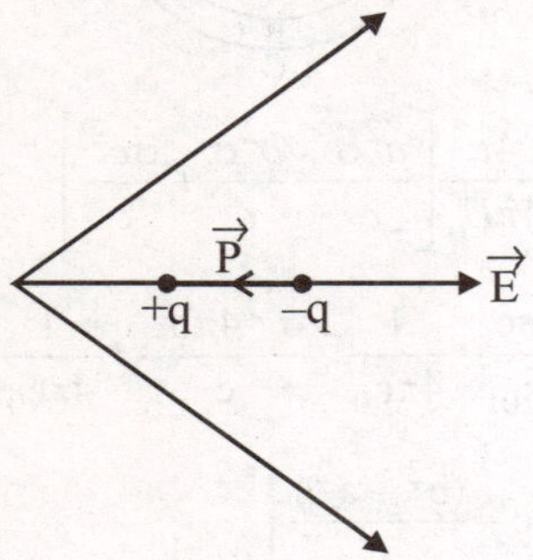

since angle between $\vec{E}$ and $\vec{P}$ is 180°

$\therefore$ U = +PE

On moving towards right electric field strength decrease $\therefore$ U decreases.

Net force on electric dipole is towards right and net torque acting on it is zero so move towards right.

NOTES

Potential energy of a dipole U = –PE cos θ, θ = 0° ⇒ stable equilibrium and U = –PE i.e., minimum value.

For θ = 90°, U = 0 and dipole is not in equilibrium.

For θ = 180° U = PE i.e., maximum value and dipole is in unstable equilibrium.

17. (b) for equilibrium

$$\frac{dU}{dr}=0 \quad\Rightarrow\quad \frac{-2A}{r^3}+\frac{B}{r^2}=0 \quad r=\frac{2A}{B}$$

for stable equilibrium

$\frac{d^2U}{dr^2}$ should be positive for the value of r.

Here $\frac{d^2U}{dr^2}=\frac{6A}{r^4}-\frac{2B}{r^3}$ is +ve value for

$$r=\frac{2A}{B}$$

NOTES

After displacing a charged particle from its equilibrium position, if it return back then it is said to be in stable equilibrium. If U is the potential energy then in case of stable equilibrium $\frac{d_2U}{dx^2}$ is positive *i.e.*, U is minimum.

18. (c) AC = BC

$V_D=V_E$

We have,

$W=Q(V_D-V_E)$

$\Rightarrow W=0$

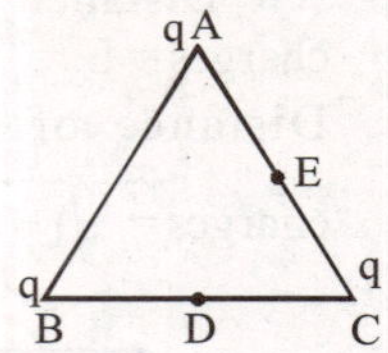

19. (c)

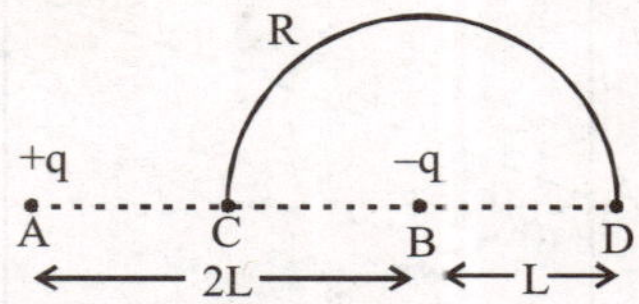

Potential at $C=V_C=0$

Potential at $D=V_D$

$$=k\left(\frac{-q}{L}\right)+\frac{kq}{3L}=-\frac{2}{3}\frac{kq}{L}$$

Potential difference

$$V_D-V_C=\frac{-2}{3}\frac{kq}{L}=\frac{1}{4\pi\epsilon_0}\left(-\frac{2}{3}.\frac{q}{L}\right)$$

$\Rightarrow$ Work done $=Q(V_D-V_C)$

$$=-\frac{2}{3}\times\frac{1}{4\pi\varepsilon_0}\frac{qQ}{L}=\frac{-qQ}{6\pi\varepsilon_0 L}$$

20. (a) We know that potential energy of two charge system is given by

$$U=\frac{1}{4\pi\epsilon_0}\frac{q_1q_2}{r}$$

According to question,

$$U_A=\frac{1}{4\pi\epsilon_0}\frac{(+q)(-Q)}{a}=-\frac{1}{4\pi\varepsilon_0}\frac{Qq}{a}$$

and $U_B=\frac{1}{4\pi\epsilon_0}\frac{(+q)(-Q)}{a}=-\frac{1}{4\pi\varepsilon_0}\frac{Qq}{a}$

$\Delta U=U_B-U_A=0$

We know that for conservative force,

$W=-\Delta U=0$

21. (c) We know that potential energy of discrete system of charges is given by

$$U=\frac{1}{4\pi\epsilon_0}\left(\frac{q_1q_2}{r_{12}}+\frac{q_2q_3}{r_{23}}+\frac{q_3q_1}{r_{31}}\right)$$

Here, $r_{12} = AB = 0.3m, r_{31} = AC = 0.4m$
According to question,

$$U_{initial} = \frac{1}{4\pi \epsilon_0}\left(\frac{q_1q_2}{0.3} + \frac{q_2q_3}{0.5} + \frac{q_3q_1}{0.4}\right)$$

$$r_{23} = \sqrt{(0.3)^2 + (0.4)^2} = 0.5m$$

When the charge moved to point D

$r_{12} = 0.3$ m
$r_{23} = 0.1$ m
$r_{31} = 0.4$ m

$$U_{final} = \frac{1}{4\pi \epsilon_0}\left(\frac{q_1q_2}{0.3} + \frac{q_2q_3}{0.1} + \frac{q_3q_1}{0.4}\right)$$

$$U_{final} - U_{initial} = \frac{1}{4\pi \epsilon_0}\left(\frac{q_1q_2}{0.1} - \frac{q_2q_3}{0.5}\right)$$

$$= \frac{1}{4\pi \epsilon_0}\left[10q_2q_3 - 2q_2q_3\right] = \frac{q_3}{4\pi \epsilon_0}(8q_2)$$

NOTES When we use the formula of potential energy of system of charges, then we consider that potential energy of the system at infinity is zero.

22. **(d)** When the dipole is in the direction of field then net force is $qE + (-qE) = 0$

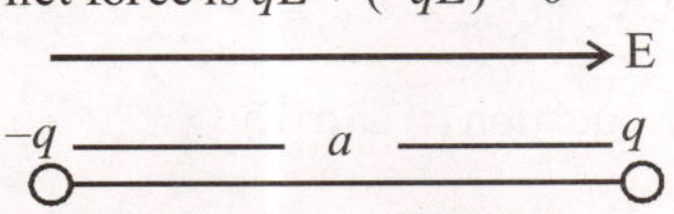

and its potential energy is minimum
$= -$ P.E. $= -qaE$

23. **(c)** $qV = \frac{1}{2}mv^2$

$$\Rightarrow 2 \times 10^{-6} \times V = \frac{1}{2} \times \frac{2}{1000} \times 10 \times 10$$

$\therefore$ V = 50 kV

24. **(d)** Length of body diagonal = $\sqrt{3}\ell$

$\therefore$ Distance of centre of cube from each corner

$$r = \frac{\sqrt{3}}{2}\ell$$

P.E. at centre
= 8 × Potential Energy of the charge (+q) due to charge (−q) at one corner

$$= 8 \times \frac{Kq \times (-q)}{r}$$

$$= 8 \times \frac{1}{4\pi\varepsilon_0 \sqrt{3}\, l} \times 2 \times q \times (-q) = \frac{-4q^2}{\sqrt{3}\pi\varepsilon_0 l}$$

25. **(c)** K.E. = Force × distance $= qE.y$

26. **(d)** Energy density i.e. $\frac{\text{energy}}{\text{volume}}$ with electric field

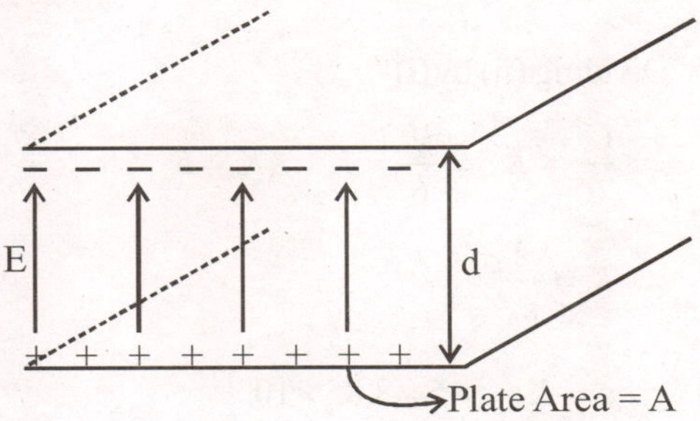

$$u = \frac{dU}{dV} = \frac{1}{2}\varepsilon_0 E^2$$

$$\Rightarrow dU = \frac{1}{2}\varepsilon_0 E^2 dV$$

$\therefore$ Energy stored in the capacitor

$$U = \int dU = \int \frac{1}{2}\varepsilon_0 E^2 dV$$

$$= \frac{1}{2}\varepsilon_0 E^2 \int dV = \frac{1}{2}\varepsilon_0 E^2 V$$

$$\therefore U = \frac{1}{2}\varepsilon_0 E^2 Ad \qquad [\because V = Ad]$$

27. **(c)** The circuit is as shown is figure,

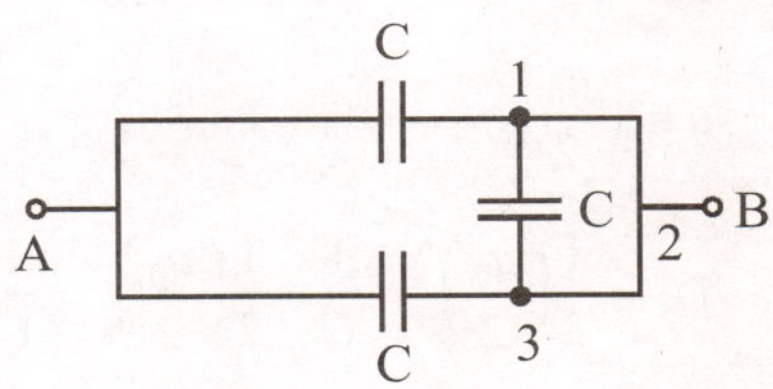

Points 1, 2, 3 are connected by conducting wire so at these points same potential.
So the capacitor is short circuited. So it does not store any charge.
Therefore equivalent capacitance $C_{AB} = C + C = 2C$

28. **(b)** Capacitance of a parallel plate capacitor with air is

$$C = \frac{\varepsilon_0 A}{d} \qquad \ldots(i)$$

Here, A = area of plates of capacitor,
d = distance between the plates

Capacitance of a same parallel plate capacitor with introduction of dielectric medium of dielectric constant K is

$$C' = \frac{K\varepsilon_0 A}{d} \qquad \ldots(ii)$$

Dividing (ii) by (i)

$$\Rightarrow \frac{C'}{C} = K \Rightarrow \frac{30}{6} = K \Rightarrow K = 5$$

$$\Rightarrow K = \frac{\varepsilon}{\varepsilon_0}$$

$$\Rightarrow \varepsilon = K\varepsilon_0 = 5 \times 8.85 \times 10^{-12}$$

$$= 0.44 \times 10^{-10}\ C^2N^{-1}m^{-2}$$

29. (d)

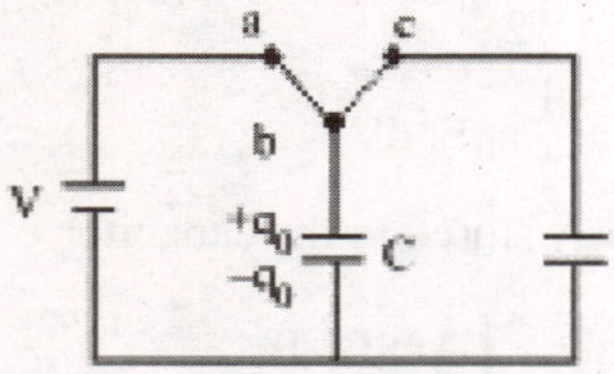

$$U_i = \frac{1}{2}CV^2$$

On switching key at point c

$$\frac{q_0 - q}{C} = \frac{q}{C}$$

$$2q = q_0$$

$$q = \left(\frac{q_0}{2}\right)$$

$$U_f = \frac{1}{2}\left(\frac{q_0}{2}\right)^2 \times \frac{1}{C} + \frac{1}{2}\left(\frac{q_0}{2}\right)^2 \times \frac{1}{C}$$

$$U_f = \frac{1}{4}CV^2$$

loss = 50%

30. (a) Electrostatic force between the metal plates

$$F_{plate} = \frac{Q^2}{2A\varepsilon_0}$$

For isolated capacitor Q = constant

Clearly, F is independent of the distance between plates.

31. (a) When battery is replaced by another uncharged capacitor

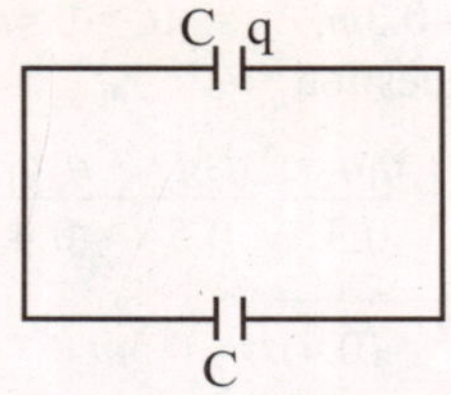

As uncharged capacitor is connected parallel

So, $C' = 2C$

and $V_c = \dfrac{q_1 + q_2}{C_1 + C_2}$

$$V_c = \frac{q + 0}{C + C} = \frac{CV}{C + C} \qquad [\because q = cv]$$

$$\Rightarrow \quad V_c = \frac{V}{2}$$

Initial Energy of system, $U_i = \dfrac{1}{2}CV^2$ ... (i)

Final energy of system, $U_f = \dfrac{1}{2}(2C)\left(\dfrac{V}{2}\right)^2$

$$= \frac{1}{2}CV^2\left(\frac{1}{2}\right) \qquad \ldots(ii)$$

From equation (i) and (ii)

$$U_f = \frac{1}{2}U_i.$$

i.e., Total electrostatic energy of resulting system decreases by a factor of 2

32. (d) **When S and 1 are connected**

The 2μF capacitor gets charged. The potential difference across its plates will be V.

The potential energy stored in 2 μF capacitor

$$U_i = \frac{1}{2}CV^2 = \frac{1}{2} \times 2 \times V^2 = V^2$$

When S and 2 are connected

The 8μF capacitor also gets charged. During this charging process current flows in the wire and some amount of energy is dissipated as heat. The energy loss is

$$\Delta U = \frac{1}{2}\frac{C_1 C_2}{C_1 + C_2}(V_1 - V_2)^2$$

Here, $C_1 = 2\mu F$, $C_2 = 8\ \mu F$, $V_1 = V$, $V_2 = 0$

$$\therefore\ \Delta U = \frac{1}{2} \times \frac{2 \times 8}{2 + 8}(V - 0)^2 = \frac{4}{5}V^2$$

The percentage of the energy dissipated

$$= \frac{\Delta U}{U_i} \times 100 = \frac{\frac{4}{5}V^2}{V^2} \times 100 = 80\%$$

33. **(c)** Capacitance of the capacitor, $C = \frac{Q}{V}$

After inserting the dielectric, new capacitance $C' = K.C$

New potential difference

$$V' = \frac{V}{K}$$

$$U_i = \frac{1}{2}CV^2 = \frac{Q^2}{2C} \quad (\because Q = CV)$$

$$U_f = \frac{Q^2}{2C} = \frac{Q^2}{2KC} = \frac{C^2V^2}{2KC} = \left(\frac{U_i}{K}\right)$$

$$\Delta U = U_f - U_i = \frac{1}{2}CV^2\left\{\frac{1}{K} - 1\right\}$$

As the capacitor is isolated, so charge will remain conserved and p.d. between two plates of the capacitor

$$V' = \frac{Q}{KC} = \frac{V}{K}$$

NOTES: When the battery remains connected, the capacity of capacitor and charge stored is changed. When the battery disconnected, charge remains same but potential and capacitance are changed.

34. **(a)** Force of attraction between the plates,

$$F = qE$$

$$= q \times \frac{\sigma}{2\epsilon_0}$$

$$= q\frac{q}{2A\epsilon_0} \quad \left[\because E = \frac{6}{2E_o}\right]$$

$$= \frac{q^2}{2\left(\frac{\epsilon_0 A}{d}\right) \times d} = \frac{C^2V^2}{2Cd} = \frac{CV^2}{2d}$$

Here, $C = \frac{\epsilon_0 A}{d}$, $q = CV$, $A = \text{area}$

NOTES: To find the force of attraction between the plates, we use the concept that work done in displacing the plates against the force is equal to the increase in energy of the capacitor.

35. **(c)** Electric field, $E \propto \frac{1}{K}$

As $K_1 < K_2$ so $E_1 > E_2$

Hence graph (c) correctly dipicts the variation of electric field E with distance d.

36. **(c)** The energy stored by a capacitor

$$U = \frac{1}{2}CV^2 \quad \text{...(i)}$$

V is the p.d. between two plates of the capacitor potential difference $V = E.d.$

The capacitance of the parallel plate capacitor

$$C = \frac{A\varepsilon_0}{d}$$

Substituting the value of C in equation (i)

$$U = \frac{1}{2}\frac{A\varepsilon_0}{d}(Ed)^2 = \frac{1}{2}A\varepsilon_0 E^2 d$$

37. **(d)** In series, $C_{eff} = \frac{C_1}{n_1}$

$\therefore$ Energy stored,

$$E_S = \frac{1}{2}C_{eff}V_S^2 = \frac{1}{2}\frac{C_1}{n_1}16V^2$$

$$= 8V^2\frac{C_1}{n_1}$$

In parallel, $C_{eff} = n_2 C_2$

$\therefore$ Energy stored, $E_p = \frac{1}{2}n_2C_2V^2$

According to question $E_s = E_p$

$$\therefore \quad \frac{8V^2C_1}{n_1} = \frac{1}{2}n_2C_2V^2$$

$$\Rightarrow \quad C_2 = \frac{16C_1}{n_1 n_2}$$

38. **(d)** Electric field between two parallel plates placed in vacuum

$$E = \frac{\sigma}{\varepsilon_o} = \frac{Q}{A\varepsilon_o}$$

In a medium of dielectric constant K

$$E' = \frac{\sigma^-}{E_o K}$$

For kerosene $K > 1$, $\therefore E' < E$

39. **(b)** In series combination of capacitors

$$V_{eff} = V + V + V = 3V$$

$$\frac{1}{C_{eff}} = \frac{1}{C} + \frac{1}{C} + \frac{1}{C}$$

$$\Rightarrow \quad C_{eff} = \frac{C}{3}$$

Thus, the capacitance and breakdown voltage of the combination will be $\frac{C}{3}$ and 3V.

40. **(b)** Work done = Change in energy

$$= \frac{1}{2}\left(C + \frac{C}{2}\right)V^2 = \frac{1}{2}\left(\frac{3C}{2}\right)V^2 = \frac{3CV^2}{4}$$

41. **(c)** If we increase the distance between the plates its capacity decreases resulting in higher potential as we know $Q = CV$. Since Q is constant (battery has been disconnected), on decreasing C, V will increase.

42. **(b)**

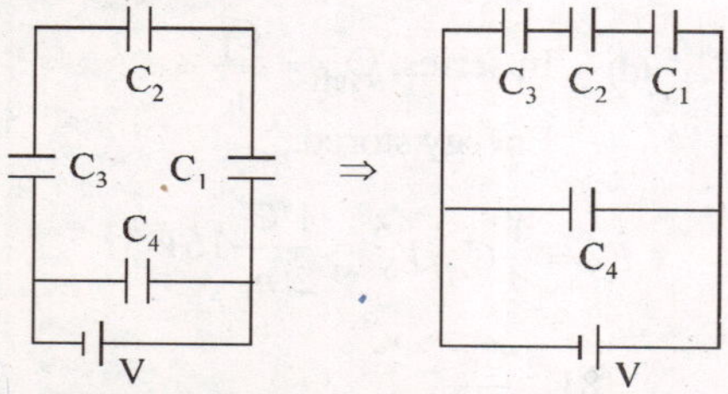

Equivalent capacitance for three capacitors (C_1, C_2 & C_3) in series is given by

$$\frac{1}{C_{eq.}} = \frac{1}{C_1} + \frac{1}{C_2} + \frac{1}{C_3} = \frac{C_2C_3 + C_3C_1 + C_1C_2}{C_1C_2C_3}$$

$$\Rightarrow \quad C_{eq.} = \frac{C_1C_2C_3}{C_1C_2 + C_2C_3 + C_3C_1}$$

$$\Rightarrow \quad C_{eq.} = \frac{C(2C)(3C)}{C(2C) + (2C)(3C) + (3C)C} = \frac{6}{11}C$$

$\Rightarrow$ Charge on capacitors (C_1, C_2 & C_3) in series

$$= C_{eq}V = \frac{6C}{11}V$$

Charge on capacitor $C_4 = C_4V = 4CV$

$$\frac{\text{Charge on } C_2}{\text{Charge on } C_4} = \frac{\frac{6C}{11}V}{4CV} = \frac{6}{11} \times \frac{1}{4} = \frac{3}{22}$$

43. **(d)** For series, $C' = \frac{C_1 \times C_2}{C_1 + C_2} = \frac{4 \times 4}{4 + 4} = 2\mu F$

For parallel, $C_{eq} = C' + C_3 = 2 + 4 = 6\mu F$

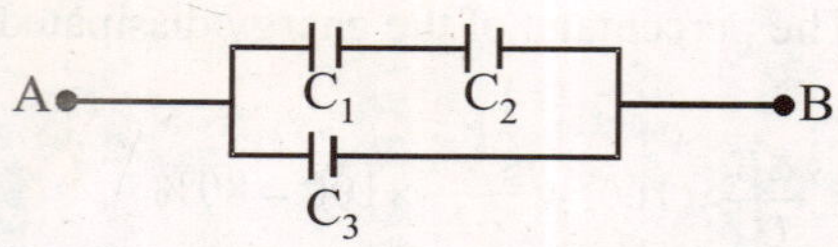

NOTES If C_P is the effective capacity when n identical capacitors are connected in parallel and C_s is their effective capacity when connected in series then

$$\frac{C_P}{C_s} = n^2$$

44. **(a)** Charge $Q = C_1V$

Total capacity of combination (parallel)

$$C = C_1 + C_2 \quad \text{P.D.} = \frac{Q}{C} = \frac{C_1V}{C_1 + C_2}$$

45. **(b)** Energy stored per unit volume

$$= \frac{1}{2}\varepsilon_0E^2 = \frac{1}{2}\varepsilon_0\left(\frac{V}{d}\right)^2 = \frac{1}{2}\varepsilon_0\frac{V^2}{d^2} \left(\because E = \frac{V}{d}\right)$$

46. **(c)** $C = 10\,\mu F \qquad d = 8$ cm

$C' = ? \qquad d' = 4$ cm

$$C = \frac{A \in_0}{d} \quad \Rightarrow \quad C \propto \frac{1}{d}$$

If d is halved then C will be doubled.

Hence, $C' = 2C = 2 \times 10\,\mu F = 20\,\mu F$

47. **(a)** Energy stored in capacitor

$$= \frac{1}{2}CV^2 = \frac{1}{2}\frac{Q}{V}.V^2 \quad (Q = CV) = \frac{1}{2}QV$$

48. **(d)** In first case, when capacitor C connected with battery.

$$\text{Energy stored, } u_1 = \frac{1}{2}CV^2 = \frac{1}{2}\frac{Q^2}{C}$$

When battery is disconnected and another capacitor of same capacity is connected in parallel to first capacitor, then

$C_{eq} = C' = 2C$

$$\text{Voltage across each capacitor, } V' = \frac{Q}{2C}$$

$$\therefore \quad \text{Energy stored, } V_2 = \frac{1}{2}CV^2$$

$$= \frac{1}{2}C\left(\frac{Q}{2C}\right)^2 = \frac{1}{4}.\frac{1}{2}\frac{Q^2}{C} = \frac{U_1}{4}$$

49. **(d)** When oil is placed between space of plates

$$C=\frac{2A\varepsilon_0}{d}\ ...(1)\ \left[\because C=\frac{KA\varepsilon_0}{d},\ \text{if } K=2\right]$$

When oil is removed $C'=\frac{A\varepsilon_0}{d}$ ...(2)

On comparing both equations, we get $C' = C/2$

> **NOTES**
> When dielectric is partially filled between the plates of a parallel plate capacitor then its capacitance increases but potential difference decreases. To maintain the capacitance and potential difference of capacitor as before separatoin betweeen the plates has to be increased by d. In such case
>
> $k=\frac{t}{t-d}$ (Here, t = seperation between plates.)

50. **(d)** From given circuit,

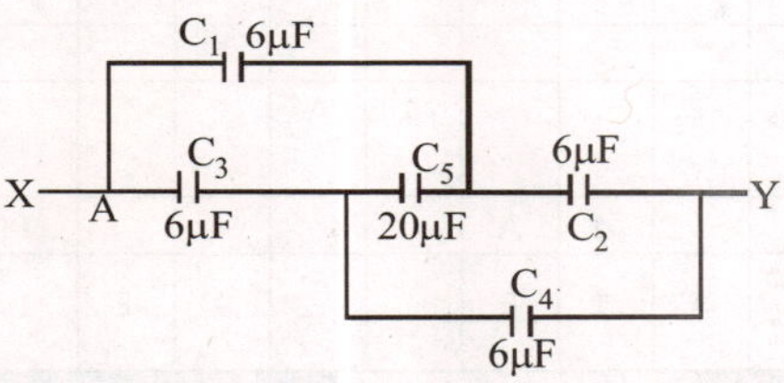

Equivalent circuit

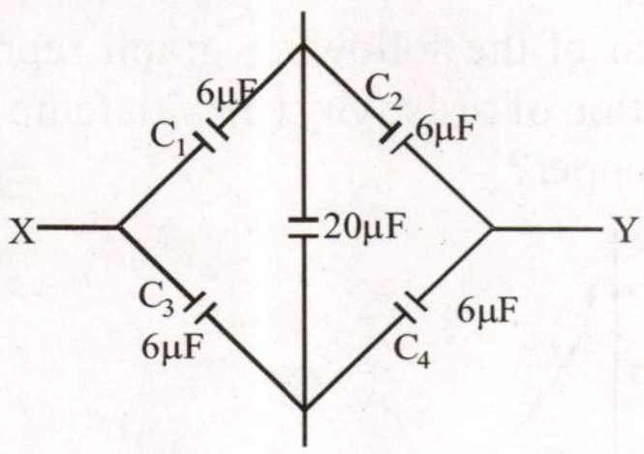

Here, $\frac{C_1}{C_3}=\frac{C_2}{C_4}$

Hence, no charge will flow through 20μF

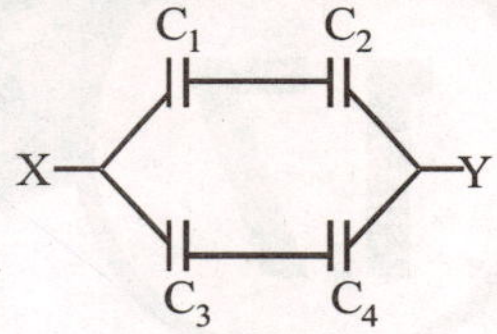

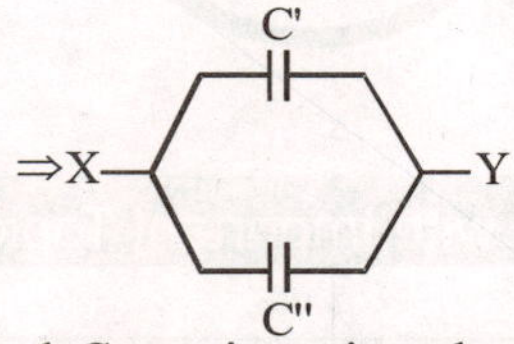

C_1 and C_2 are in series, also C_3 and C_4 are in series.
Hence, $C' = 3\ \mu F$, $C'' = 3\ \mu F$
C' and C'' are in parallel.
Hence net capacitance = $C' + C'' = 3 + 3 = 6\ \mu F$

51. **(c)** Capacitance of capacitor (C) = 6 μF = $6\times 10^{-6}\ F$; Initial potential (V_1) = 10 V and final potential (V_2) = 20 V.
The increase in energy (ΔU)

$$=\frac{1}{2}C(V_2^2-V_1^2)$$

$$=\frac{1}{2}\times(6\times10^{-6})\times[(20)^2-(10)^2]$$

$$=(3\times10^{-6})\times300=9\times10^{-4}\ \text{J}.$$

52. **(b)** Charge on each plate of each capacitor

$$Q=\pm CV=\pm 25\times10^{-6}\times200$$

$$=\pm5\times10^{-3}C$$

53. **(d)** The energy stored in the capacitor

$$=\frac{1}{2}CV^2=\frac{1}{2}\times4\times10^{-6}\times(400)^2=0.32\ \text{J}$$

This energy will be converted into heat in the resistor.

17 Current Electricity

Trend Analysis with Important Topics & Sub-Topics

Topic Name	Sub-Topic	2021		2020		2019		2018		2017		2016	
		QNS.	LOD	QNS.	LOD	QNS.	LOD	QNS.	LOD	QNS.	LOD	QNS.	LOD
Electric Current, Drift of Electrons, Ohm's Law, Resistance & Resistivity	Resistivity vs temperature			1	A								
	Colour code of resistor			1	E			1	E				
	Mobility & Drift velocity	1	E	1	E								
	Resistance & Temp. coeff. of resistance			1	E					1	E		
Combination of Resistances		2	E, A										
Kirchhoff's Laws, Cells, Thermo emf & Electrolysis	Kirchhoff's rule & Grouping of resistances							1	A				
Heating Effects of Current	Heating effect of current					1	A					1	A
Wheatstone Bridge & Different Measuring Instruments	Meter bridge & other measuring instrument	1	E	1	A	1	E			1	E	1	A

LOD - Level of Difficulty | E - Easy | A - Average | D - Difficult | Qns - No. of Questions

Topic 1: Electric Current, Drift of Electrons, Ohm's Law, Resistance & Resistivity

1. **Column-I** gives certain physical terms associated with flow of current through a metallic conductor.
Column-II gives some mathematical relations involving electrical quantities. Match **Column-I** and **Column-II** with appropriate relations. *[2021]*

Column-I	Column-II
(A) Drift velocity	(i) $\frac{m}{ne^2\rho}$
(B) Electrical resistivity	(ii) nev_d
(C) Relaxation period	(iii) $\frac{eE}{m}\tau$
(D) Current density	(iv) $\frac{E}{J}$

(a) A - (iii), B - (ii), C - (iv), D - (i)
(b) A - (iii), B - (iv), C - (i), D - (ii)
(c) A - (iii), B - (iv), C - (ii), D - (i)
(d) A - (iii), B - (i), C - (iv), D - (ii)

2. Which of the following graph represents the variation of resistivity (ρ) with temperature (T) for copper? *[2020]*

(a)
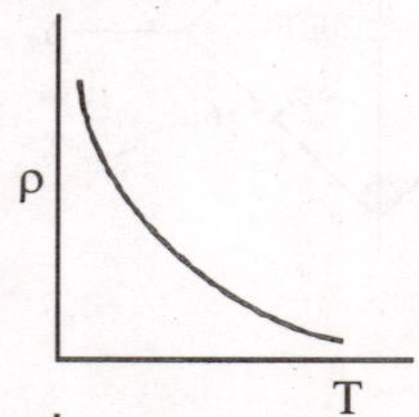

(b)
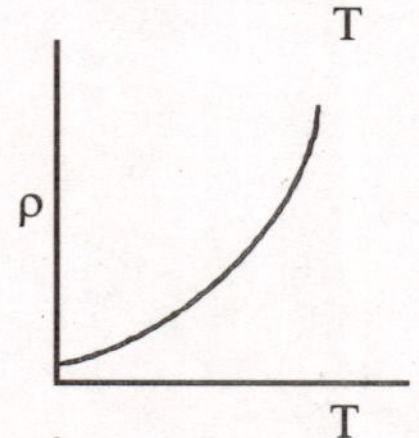

(c)
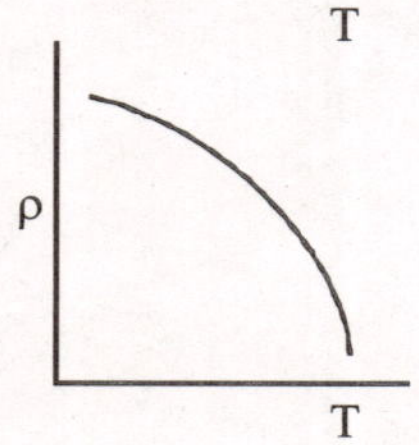

(d) 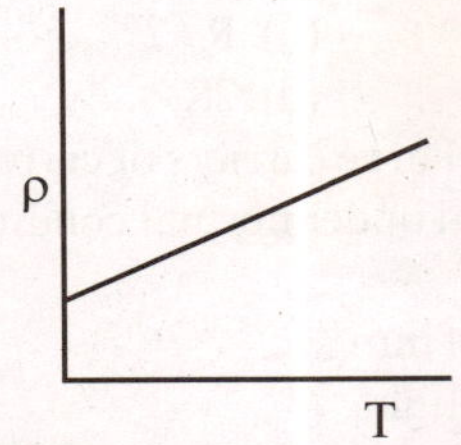

3. The solids which have the negative temperature coefficient of resistance are: *[2020]*
(a) insulators only
(b) semiconductors only
(c) insulators and semiconductors
(d) metals

4. The color code of a resistance is given below

The values of resistance and tolerance, respectively, are *[2020]*
(a) 47 kΩ, 10% (b) 4.7 kΩ, 5%
(c) 470 Ω, 5% (d) 470 kΩ, 5%

5. A charged particle having drift velocity of 7.5×10^{-4} m s^{-1} in an electric field of 3×10^{-10} Vm^{-1}, has a mobility in $m^2 V^{-1} s^{-1}$ of : *[2020]*
(a) 2.5×10^6 (b) 2.5×10^{-6}
(c) 2.25×10^{-15} (d) 2.25×10^{15}

6. A carbon resistor of (47 ± 4.7) kΩ is to be marked with rings of different colours for its identification. The colour code sequence will be *[2018]*
(a) Violet – Yellow – Orange – Silver
(b) Yellow – Violet – Orange – Silver
(c) Green – Orange – Violet – Gold
(d) Yellow – Green – Violet – Gold

7. The resistance of a wire is 'R' ohm. If it is melted and stretched to 'n' times its original length, its new resistance will be :- *[2017]*
(a) $\frac{R}{n}$ (b) n^2R
(c) $\frac{R}{n^2}$ (d) nR

8. Across a metallic conductor of non-uniform cross section a constant potential difference is applied. The quantity which remains constant along the conductor is : *[2015]*
(a) current (b) drift velocity
(c) electric field (d) current density

9. A wire of resistance 4 Ω is stretched to twice its original length. The resistance of stretched wire would be *[2013]*
(a) 4 Ω (b) 8 Ω (c) 16 Ω (d) 2 Ω

10. Two rods are joined end to end, as shown. Both have a cross-sectional area of 0.01 cm^2. Each is 1 meter long. One rod is of copper with a resistivity of 1.7×10^{-6} ohm-centimeter, the other is of iron with a resistivity of 10^{-5} ohm-centimeter.
How much voltage is required to produce a current of 1 ampere in the rods? *[NEET Kar. 2013]*

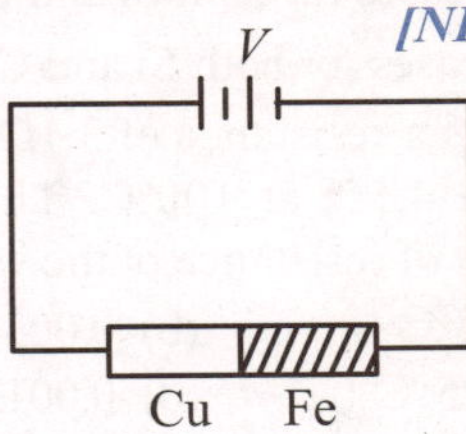

(a) 0.117 V (b) 0.00145 V
(c) 0.0145 V (d) 1.7×10^{-6} V

11. A wire of a certain material is stretched slowly by ten per cent. Its new resistance and specific resistance become respectively: *[2008]*
(a) 1.2 times, 1.3 times
(b) 1.21 times, same
(c) both remain the same
(d) 1.1 times, 1.1 times

12. The electric resistance of a certain wire of iron is *R*. If its length and radius are both doubled, then
(a) the resistance and the specific resistance, will both remain unchanged *[2004]*
(b) the resistance will be doubled and the specific resistance will be halved
(c) the resistance will be halved and the specific resistance will remain unchanged
(d) the resistance will be halved and the specific resistance will be doubled

13. A 6 volt battery is connected to the terminals of the three metre long wire of uniform thickness and resistance of 100 ohm. The difference of potential between two points on the wire separated by a distance of 50 cm will be *[2004]*

(a) 1.5 volt (b) 3 volt
(c) 3 volt (d) 1 volt

14. The resistivity (specific resistance) of a copper wire *[2002]*
(a) increases with increase in its temperature
(b) decreases with increase in its cross-section
(c) increases with increase in its length
(d) increases with increase in its cross-section

15. Si and Cu are cooled to a temperature of 300 K, then resistivity? *[2001]*
(a) For Si increases and for Cu decreases
(b) For Cu increases and for Si decreases
(c) Decreases for both Si and Cu
(d) Increases for both Si and Cu

16. A wire has a resistance of 3.1Ω at 30°C and a resistance 4.5Ω at 100°C. The temperature coefficient of resistance of the wire *[2001]*
(a) $0.008\,°C^{-1}$ (b) $0.0034\,°C^{-1}$
(c) $0.0025\,°C^{-1}$ (d) $0.0012\,°C^{-1}$

17. The resistance of a discharge tube is *[1999]*
(a) zero (b) ohmic
(c) non-ohmic (d) infinity

18. There are three copper wires of length and cross sectional area (L, A), $(2L, \frac{1}{2}A)$, $(\frac{1}{2}L, 2A)$. In which case is the resistance minimum? *[1997]*
(a) It is the same in all three cases
(b) Wire of cross-sectional area $2A$
(c) Wire of cross-sectional area A
(d) Wire of cross-sectional area $\frac{1}{2}A$

19. If the resistance of a conductor is 5 Ω at 50°C and 7Ω at 100°C, then the mean temperature coefficient of resistance (of the material) is *[1996]*
(a) 0.001/°C (b) 0.004/°C
(c) 0.006/°C (d) 0.008/°C

20. If a negligibly small current is passed through a wire of length 15 m and of resistance 5Ω having uniform cross-section of 6×10^{-7} m^2, then coefficient of resistivity of material, is *[1996]*
(a) 1×10^{-7} Ω-m (b) 2×10^{-7} Ω-m
(c) 3×10^{-7} Ω-m (d) 4×10^{-7} Ω-m

21. If a wire of resistance R is melted and recasted to half of its length, then the new resistance of the wire will be *[1995]*
(a) R/4 (b) R/2
(c) R (d) 2R

22. The velocity of charge carriers of current (about 1 amp) in a metal under normal conditions is of the order of *[1991]*
(a) a fraction of mm/sec
(b) velocity of light
(c) several thousand metres/second
(d) a few hundred metres per second

23. The masses of the three wires of copper are in the ratio of 1 : 3 : 5 and their lengths are in the ratio of 5 : 3 : 1. The ratio of their electrical resistance is *[1988]*
(a) 1:3:5 (b) 5:3:1
(c) 1:25:125 (d) 125:15:1

Topic 2 Combination of Resistances

24. The effective resistance of a parallel connection that consists of four wires of equal length, equal area of cross-section and same material is 0.25 Ω. What will be the effective resistance if they are connected in series? *[2021]*
(a) 4 Ω (b) 0.25 Ω
(c) 0.5 Ω (d) 1 Ω

25. Three resistors having resistances r_1, r_2 and r_3 are connected as shown in the given circuit.

The ratio $\frac{i_3}{i_1}$ of currents in terms of resistances used in the circuit is *[2021]*

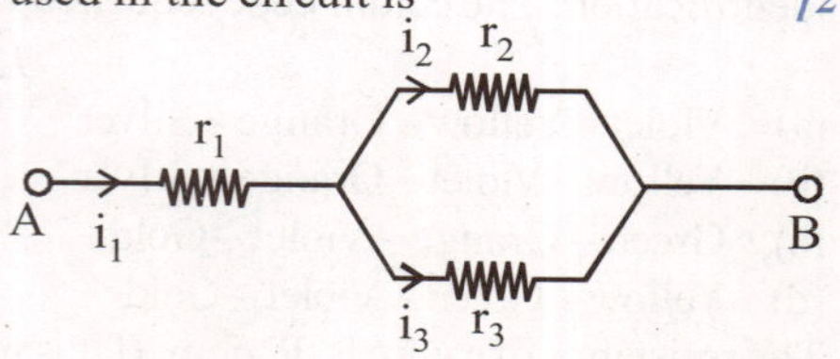

(a) $\frac{r_2}{r_1+r_3}$ (b) $\frac{r_1}{r_2+r_3}$
(c) $\frac{r_2}{r_2+r_3}$ (d) $\frac{r_1}{r_1+r_2}$

26. A set of 'n' equal resistors, of value 'R' each, are connected in series to a battery of emf 'E' and internal resistance 'R'. The current drawn is I. Now, the 'n' resistors are connected in parallel to the same battery. Then the current drawn from battery becomes 10 I. The value of'n' is *[2018]*

(a) 10 (b) 11
(c) 9 (d) 20

27. A, B and C are voltmeters of resistance R, 1.5 R and 3R respectively as shown in the figure. When some potential difference is applied between X and Y, the voltmeter readings are V_A, V_B and V_C respectively. Then *[2015]*

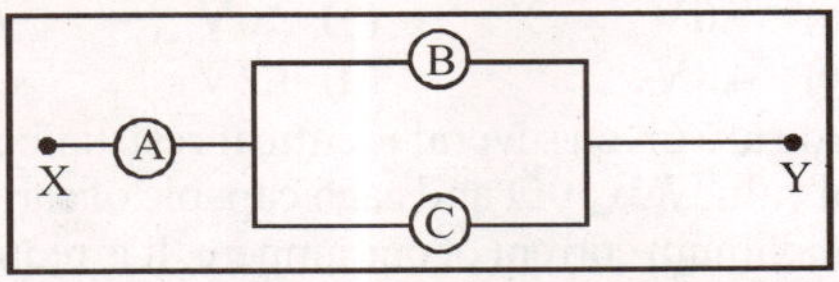

(a) $V_A \neq V_B = V_C$ (b) $V_A = V_B \neq V_C$
(c) $V_A \neq V_B \neq V_C$ (d) $V_A = V_B = V_C$

28. Two metal wires of identical dimension are connected in series. If σ_1 and σ_2 are the conductivities of the metal wires respectively, the effective conductivity of the combination is : *[2015 RS]*

(a) $\frac{\sigma_1 + \sigma_2}{2\sigma_1\sigma_2}$ (b) $\frac{\sigma_1 + \sigma_2}{\sigma_1\sigma_2}$

(c) $\frac{\sigma_1\sigma_2}{\sigma_1 + \sigma_2}$ (d) $\frac{2\sigma_1\sigma_2}{\sigma_1 + \sigma_2}$

29. A 12 cm wire is given a shape of a right angled triangle *ABC* having sides 3 cm, 4 cm and 5 cm as shown in the figure. The resistance between two ends (*AB, BC, CA*) of the respective sides are measured one by one by a multi-meter. The resistances will be in the ratio of *[NEET Kar. 2013]*

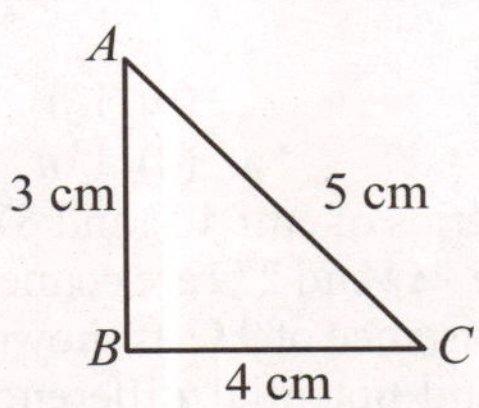

(a) 3 : 4 : 5 (b) 9 : 16 : 25
(c) 27 : 32 : 35 (d) 21 : 24 : 25

30. A ring is made of a wire having a resistance $R_0 = 12\ \Omega$. Find the points *A* and *B* as shown in the figure, at which a current carrying conductor should be connected so that the resistance *R* of the sub-circuit between these points is equal to $\frac{8}{3}\Omega$. *[2012]*

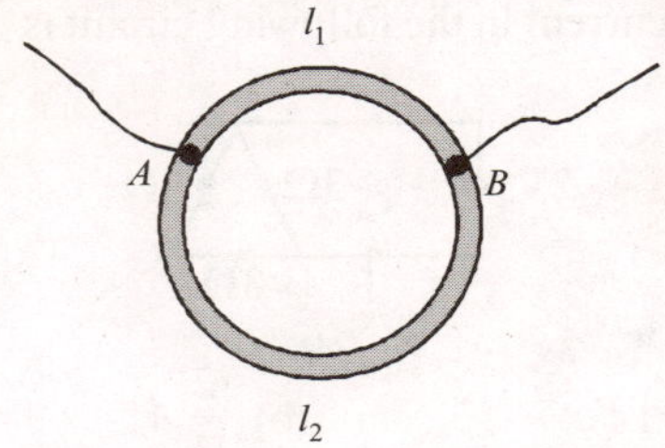

(a) $\frac{l_1}{l_2} = \frac{5}{8}$ (b) $\frac{l_1}{l_2} = \frac{1}{3}$

(c) $\frac{l_1}{l_2} = \frac{3}{8}$ (d) $\frac{l_1}{l_2} = \frac{1}{2}$

31. A wire of resistance 12 ohms per meter is bent to form a complete circle of radius 10 cm. The resistance between its two diametrically opposite points, A and B as shown in the figure, is *[2009]*

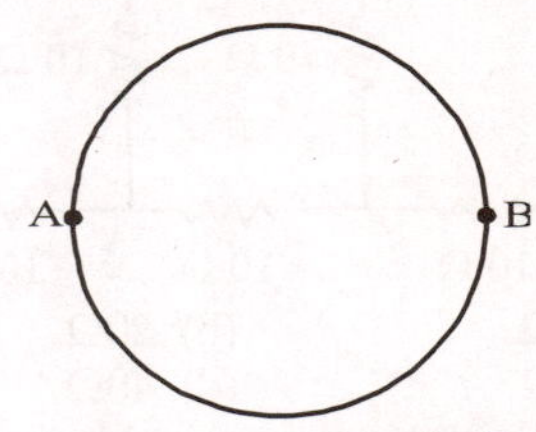

(a) 3Ω (b) $6\pi\Omega$
(c) 6Ω (d) $0.6\pi\Omega$

32. When a wire of uniform cross–section *a*, length *l* and resistance *R* is bent into a complete circle, resistance between any two of diametrically opposite points will be *[2005]*

(a) $\frac{R}{4}$ (b) $4R$

(c) $\frac{R}{8}$ (d) $\frac{R}{2}$

33. Resistances n, each of r ohm, when connected in parallel give an equivalent resistance of *R* ohm. If these resistances were connected in series, the combination would have a resistance in ohms, equal to *[2004]*

(a) nR (b) n^2R
(c) R/n^2 (d) R/n

34. The current (I) in the given circuit is *[1999]*

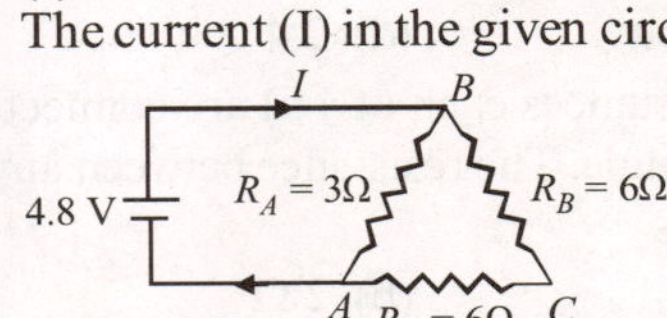

(a) 1.6 A (b) 2.0 A
(c) 0.32 A (d) 3.2 A

35. The current in the following circuit is *[1997]*

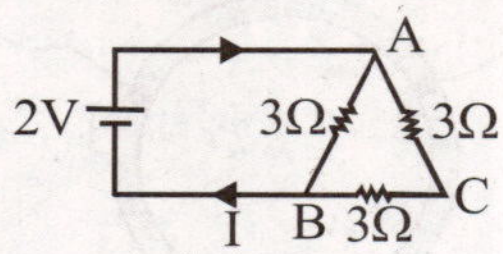

(a) 1A (b) $\frac{2}{3}A$

(c) $\frac{2}{9}A$ (d) $\frac{1}{8}A$

36. What will be the equivalent resistance of circuit shown in figure between two points A and D *[1996]*

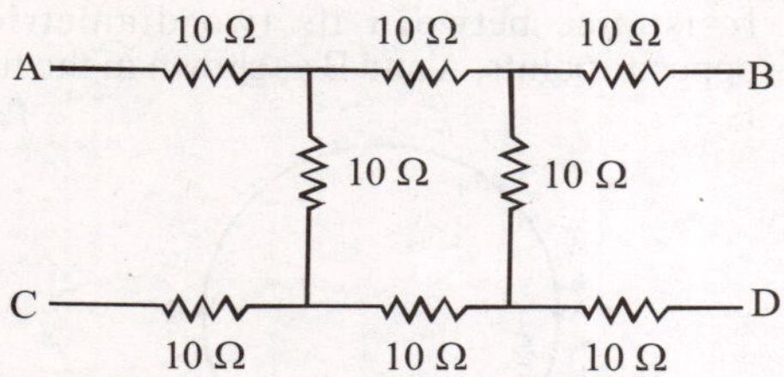

(a) 10Ω (b) 20Ω
(c) 30Ω (d) 40Ω

37. Two wires of the same metal have same length, but their cross-sections are in the ratio 3:1. They are joined in series. The resistance of thicker wire is 10Ω. The total resistance of the combination will be *[1995]*

(a) 10 Ω (b) 20 Ω
(c) 40 Ω (d) 100 Ω

38. In the circuit shown in Fig, the current in 4 Ω resistance is 1.2 *A*. What is the potential difference between *B* and *C*. *[1994]*

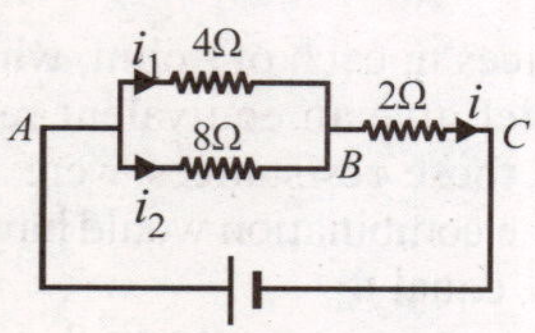

(a) 3.6 volt (b) 6.3 volt
(c) 1.8 volt (d) 2.4 volt

39. Three resistances each of 4 Ω are connected to form a triangle. The resistance between any two terminals is *[1993]*

(a) 12 Ω (b) 2 Ω
(c) 6Ω (d) 8/3Ω

40. Current through 3 Ω resistor is

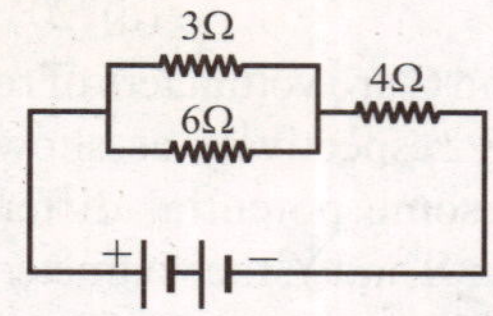

0.8 amp., then potential drop through 4Ω resistor is

(a) 9.6V (b) 2.6V *[1993]*
(c) 4.8V (d) 1.2V

41. You are given several identical resistances each of value $R = 10Ω$ and each capable of carrying a maximum current of one ampere. It is required to make a suitable combination of these resistances of 5Ω which can carry a current of 4 ampere. The minimum number of resistances of the type *R* that will be required for this job is *[1990]*

(a) 4 (b) 10
(c) 8 (d) 20

42. In the network shown in the Fig, each resistance is 1Ω. The effective resistance between *A* and *B* is *[1990]*

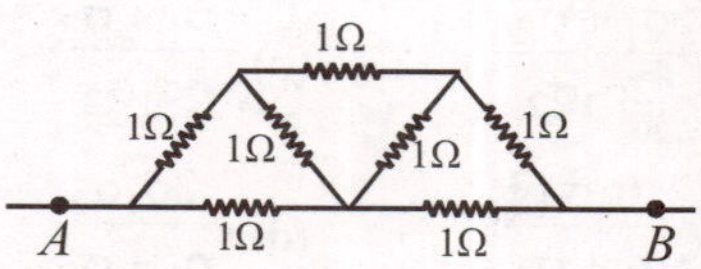

(a) $\frac{4}{3}Ω$ (b) $\frac{3}{2}Ω$

(c) 7 Ω (d) $\frac{8}{7}Ω$

43. *n* equal resistors are first connected in series and then connected in parallel. What is the ratio of the maximum to the minimum resistance ? *[1989]*

(a) n (b) $1/n^2$
(c) n^2 (d) $1/n$

44. Two batteries of emf 4 V and 8V with internal resistance 1 Ω and 2 Ω are connected in a circuit with a resistance of 9 Ω as shown in figure. The current and potential difference between the points *P* and *Q* are *[1988]*

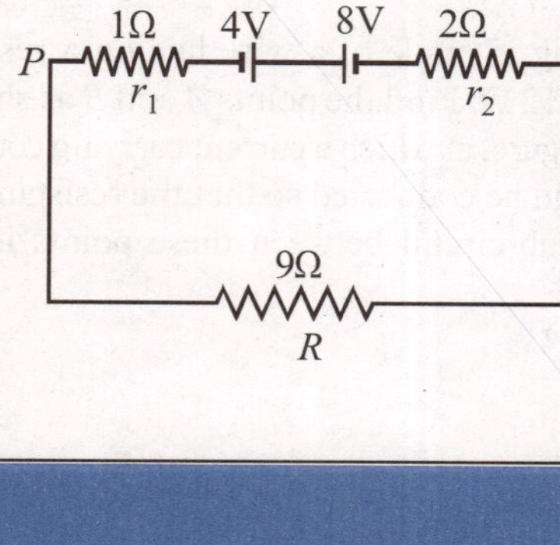

(a) $\frac{1}{3}A$ and 3 V (b) $\frac{1}{6}A$ and 4 V

(c) $\frac{1}{9}A$ and 9 V (d) $\frac{1}{12}A$ and 12V

Topic 3 Kirchhoff's Laws, Cells, Thermo emf & Electrolysis

45. The internal resistance of a 2.1 V cell which gives a current of 0.2 A through a resistance of 10 Ω is [2013]

(a) 0.5 Ω (b) 0.8 Ω
(c) 1.0 Ω (d) 0.2 Ω

46. Cell having an emf ε and internal resistance r is connected across a variable external resistance R. As the resistance R is increased, the plot of potential difference V across R is given by : [2012M]

(a)
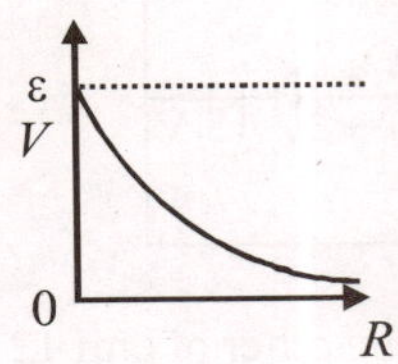

(b)
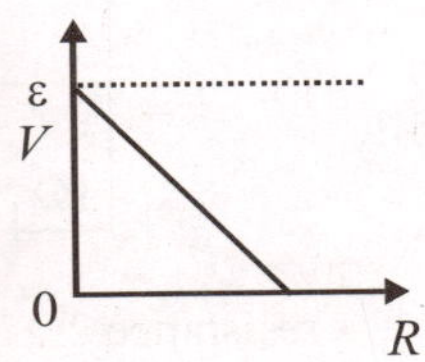

(c)
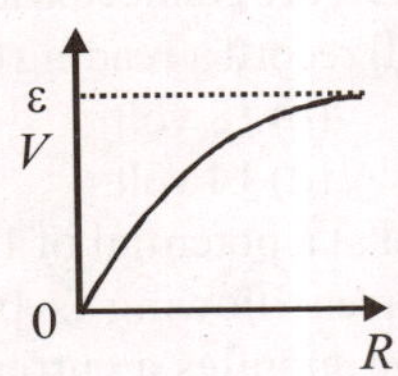

(d)
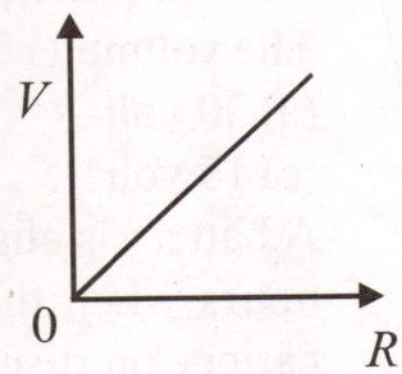

47. A current of 2A flows through a 2Ω resistor when connected across a battery. The same battery supplies a current of 0.5 A when connected across a 9Ω resistor. The internal resistance of the battery is [2011]

(a) 0.5 Ω (b) 1/3 Ω
(c) 1/4 Ω (d) 1 Ω

48. The rate of increase of thermo–e.m.f. with temperature at the neutral temperature of a thermocouple [2011]

(a) is positive
(b) is zero
(c) depends upon the choice of the two materials of the thermocouple
(d) is negative

49. A thermocouple of negligible resistance produces an e.m.f. of 40 μV/°C in the linear range of temperature. A galvanometer of resistance 10 ohm whose sensitivity is 1 μA/div, is employed with the termocouple. The smallest value of temperature difference that can be detected by the system will be [2011M]

(a) 0.5°C (b) 1°C
(c) 0.1°C (d) 0.25°C

50. In the circuit shown in the figure, if potential at point A is taken to be zero, the potential at point B is [2011M]

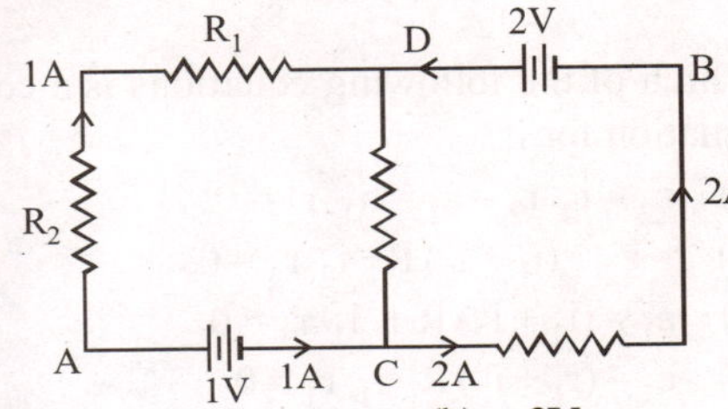

(a) –1V (b) +2V
(c) –2V (d) +1V

51. In producing chlorine by electrolysis 100 kW power at 125 V is being consumed. How much chlorine per minute is liberated? (E.C.E. of chlorine is 0.367×10^{-6} kg/C) [2010]

(a) 1.76×10^{-3} kg (b) 9.67×10^{-3} kg
(c) 17.61×10^{-3} kg (d) 3.67×10^{-3} kg

52. Consider the following two statements:
(A) Kirchhoff's junction law follows from the conservation of charge.
(B) Kirchhoff's loop law follows from the conservation of energy.

Which of the following is correct? [2010]

(a) Both (A) and (B) are wrong
(b) (a) is correct and (B) is wrong
(c) (a) is wrong and (B) is correct
(d) Both (A) and (B) are correct

53. The thermo e.m.f E in volts of a certain thermocouple is found to vary with temperature difference θ in °C between the two junctions according to the relation [2010]

$$E = 30\theta - \frac{\theta^2}{15}$$

The neutral temperature for the thermocouple will be

(a) 30° C (b) 450° C
(c) 400 °C (d) 225° C

54. See the electric circuit shown in the figure.

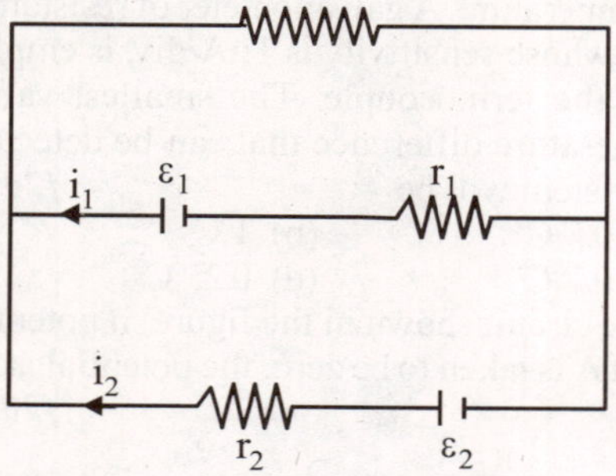

Which of the following equations is a correct equation for it? *[2009]*

(a) $\varepsilon_2 - i_2 r_2 - \varepsilon_1 - i_1 r_1 = 0$
(b) $-\varepsilon_2 - (i_1 + i_2)R + i_2 r_2 = 0$
(c) $\varepsilon_1 - (i_1 + i_2)R + i_1 r_1 = 0$
(d) $\varepsilon_1 - (i_1 + i_2)R - i_1 r_1 = 0$

55. A student measures the terminal potential difference (V) of a cell (of emf E and internal resistance r) as a function of the current (I) flowing through it. The slope and intercept, of the graph between V and I, then, respectively, equal: *[2009]*

(a) – r and E (b) r and – E
(c) – E and r (d) E and – r

56. A cell can be balanced against 110 cm and 100 cm of potentiometer wire, respectively with and without being short circuited through a resistance of 10Ω. Its internal resistance is *[2008]*

(a) 1.0 ohm (b) 0.5 ohm
(c) 2.0 ohm (d) zero

57. A steady current of 1.5 amp flows through a copper voltameter for 10 minutes. If the electrochemical equivalent of copper is 30×10^{-5} g coulomb^{-1}, the mass of copper deposited on the electrode will be *[2007]*

(a) 0.50 g (b) 0.67 g
(c) 0.27 g (d) 0.40 g.

58. In producing chlorine through electrolysis, 100 watt power at 125 V is being consumed. How much chlorine per minute is liberated? E.C.E. of chlorine is 0.367×10^{-6} kg/ coulomb. *[2006]*

(a) 21.3 mg (b) 24.3 mg
(c) 13.6 mg (d) 17.6 mg

59. Kirchhoff's first and second laws for electrical circuits are consequences of *[2006]*

(a) conservation of electric charge and energy respectively
(b) conservation of electric charge
(c) conservation of energy and electric charge respectively
(d) conservation of energy

60. Two cells, having the same e.m.f., are connected in series through an external resistance *R*. Cells have internal resistances r_1 and r_2 $(r_1 > r_2)$ respectively. When the circuit is closed, the potential difference across the first cell is zero. The value of R is *[2006]*

(a) $\frac{r_1 + r_2}{2}$ (b) $\frac{r_1 - r_2}{2}$
(c) $r_1 + r_2$ (d) $r_1 - r_2$

61. Two batteries, one of emf 18 volt and internal

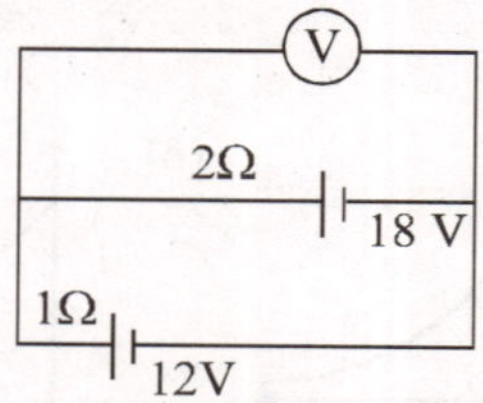

resistance 2Ω and the other of emf 12 volt and internal resistance 1Ω, are connected as shown. The voltmeter V will record a reading of *[2005]*

(a) 30 volt (b) 18 volt
(c) 15 volt (d) 14 volt

62. A battery is charged at a potential of 15V for 8 hours when the current flowing is 10A. The battery on discharge supplies a current of 5A for 15 hours. The mean terminal voltage during discharge is 14V. The "watt-hour" efficiency of the battery is *[2004]*

(a) 87.5% (b) 82.5%
(c) 80% (d) 90%

63. The potential difference between the terminals of a cell in an open circuit is 2.2 V. When a resistor of 5Ω is connected across the terminals of the cell, the potential difference between the terminals of the cell is found to be 1.8 V. The internal resistance of the cell is *[2002]*

(a) $\frac{7}{12}\Omega$ (b) $\frac{10}{9}\Omega$
(c) $\frac{9}{10}\Omega$ (d) $\frac{12}{7}\Omega$

64. In electrolysis, the amount of mass deposited or liberated at an electrode is directly proportional to *[2000]*
(a) square of electric charge
(b) amount of charge
(c) square of current
(d) concentration of electrolyte

65. A car battery has e.m.f. 12 volt and internal resistance 5×10^{-2} ohm. If it draws 60 amp current, the terminal voltage of the battery will be *[2000]*
(a) 15 volt (b) 3 volt
(c) 5 volt (d) 9 volt

66. If nearly 10^5 coulombs liberate 1 gm-equivalent of aluminium, then the amount of aluminium (equivalent weight 9), deposited through electrolysis in 20 minutes by a current of 50 amp. will be *[1998]*
(a) 0.6 gm (b) 0.09 gm
(c) 5.4 gm (d) 10.8 gm

67. Kirchoff's first law, i.e. $\Sigma i = 0$ at a junction, deals with the conservation of *[1992, 1997]*
(a) charge (b) energy
(c) momentum (d) angular momentum

68. Direct current is passed through a copper sulphate solution using platinum electrodes. The elements liberated at the electrodes are *[1993]*
(a) copper at anode and sulphur at cathode
(b) sulphur at anode and copper at cathode
(c) oxygen at anode and copper at cathode
(d) copper at anode and oxygen at cathode

69. Faraday's laws are consequence of conservation of *[1991]*
(a) energy
(b) energy and magnetic field
(c) charge
(d) magnetic field

Topic 4 Heating Effects of Current

70. Six similar bulbs are connected as shown in the figure with a DC source of emf E, and zero internal resistance.
The ratio of power consumption by the bulbs when (i) all are glowing and (ii) in the situation when two from section A and one from section B are glowing, will be:

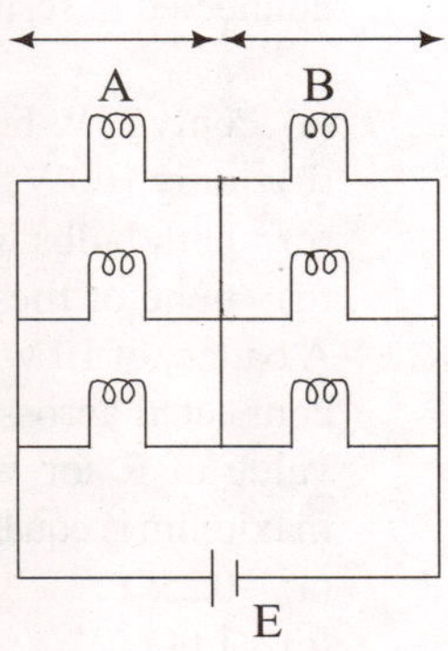

(a) 4 : 9 (b) 9 : 4 *[2019]*
(c) 1 : 2 (d) 2 : 1

71. The charge flowing through a resistance R varies with time t as $Q = at - bt^2$, where a and b are positive constants. The total heat produced in R is: *[2016]*
(a) $\frac{a^3R}{6b}$ (b) $\frac{a^3R}{3b}$
(c) $\frac{a^3R}{2b}$ (d) $\frac{a^3R}{b}$

72. Two cities are 150 km apart. Electric power is sent from one city to another city through copper wires. The fall of potential per km is 8 volt and the average resistance per km is 0.5 Ω. The power loss in the wires is : *[2014]*
(a) 19.2 W (b) 19.2 kW
(c) 19.2 J (d) 12.2 kW

73. Ten identical cells connected in series are needed to heat a wire of length one meter and radius '*r*' by 10ºC in time '*t*'. How many cells will be required to heat the wire of length two meter of the same radius by the same temperature in time '*t*'? *[NEET Kar. 2013]*
(a) 10 (b) 20
(c) 30 (d) 40

74. If voltage across a bulb rated 220 Volt-100 Watt drops by 2.5% of its rated value, the percentage of the rated value by which the power would decrease is : *[2012]*
(a) 20% (b) 2.5%
(c) 5% (d) 10%

75. The power dissipated in the circuit shown in the figure is 30 Watts. The value of *R* is: *[2012M]*

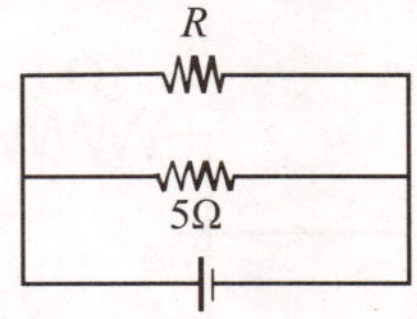

(a) 20 Ω
(b) 15 Ω
(c) 10 Ω
(d) 30 Ω

76. If power dissipated in the 9-Ω resistor in the circuit shown is 36 watt, the potential difference across the 2-Ω resistor is *[2011]*

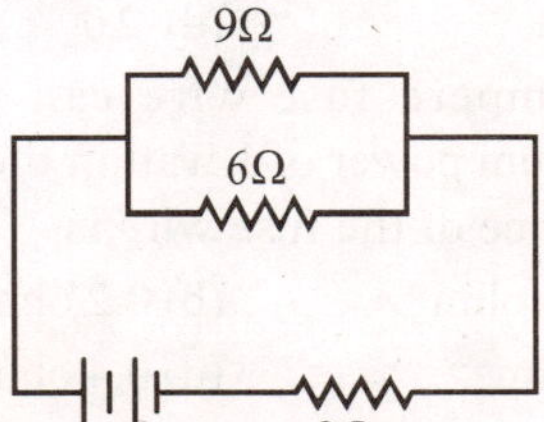

(a) 4 volt (b) 8 volt
(c) 10 volt (d) 2 volt

77. A current of 3 amp flows through the 2Ω resistor shown in the circuit. The power dissipated in the 5-Ω resistor is: [2008]

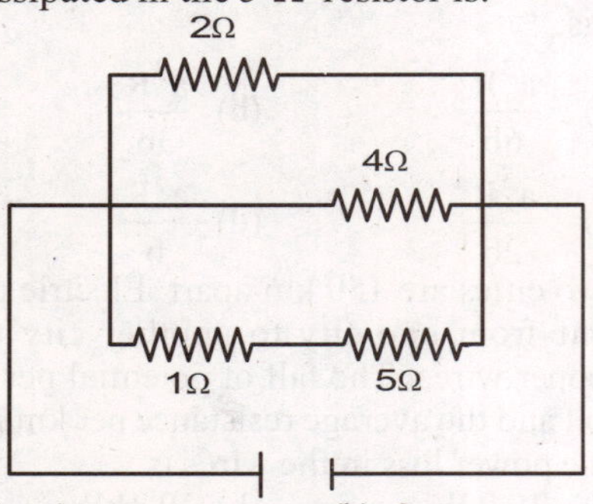

(a) 4 watt (b) 2 watt
(c) 1 watt (d) 5 watt

78. The total power dissipated in watts in the circuit shown here is [2007]

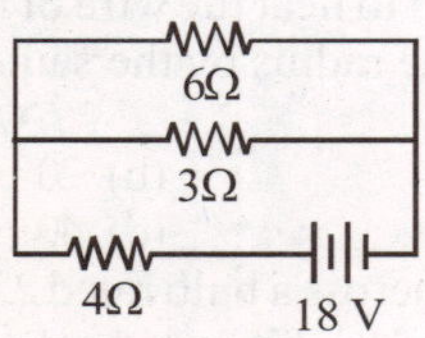

(a) 40 (b) 54
(c) 4 (d) 16

79. Power dissipated across the 8Ω resistor in the circuit shown here is 2 watt. The power dissipated in watt units across the 3Ω resistor is

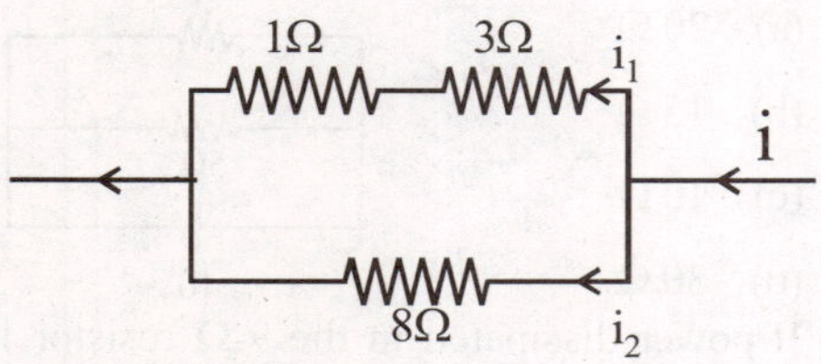

(a) 1.0 (b) 0.5 [2006]
(c) 3.0 (d) 2.0

80. A 5–ampere fuse wire can withstand a maximum power of 1 watt in the circuit. The resistance of the fuse wire is [2005]
(a) 0.04 ohm (b) 0.2 ohm
(c) 5 ohm (d) 0.4 ohm

81. When three identical bulbs of 60 watt, 200 volt rating are connected in series to a 200 volt supply, the power drawn by them will be
(a) 20 watt (b) 60 watt [2004]
(c) 180 watt (d) 10 watt

82. In India electricity is supplied for domestic use at 220 V. It is supplied at 110 V in USA. If the resistance of a 60 W bulb for use in India is R, the resistance of a 60 W bulb for use in USA will be [2004]
(a) R/2 (b) R
(c) 2R (d) R/4

83. An electric kettle has two heating coils. When one of the coils is connected to an a.c. source, the water in the kettle boils in 10 minutes. When the other coil is used, the water boils in 40 minutes. If both the coils are connected in parallel, the time taken by the same quantity of water to boil will be [2003]
(a) 15 min (b) 8 min
(c) 4 min (d) 25 min

84. Two 220 volt, 100 watt bulbs are connected first in series and then in parallel. Each time the combination is connected to a 220 volt a.c. supply line. The power drawn by the combination in each case respectively will be [2003]
(a) 50 watt, 200 watt (b) 50 watt, 100 watt
(c) 100 watt, 50 watt (d) 200 watt, 150 watt

85. Fuse wire is a wire of [2003]
(a) low resistance and high melting point
(b) high resistance and high melting point
(c) high resistance and low melting point
(d) low resistance and low melting point

86. If 25W, 220 V and 100 W, 220 V bulbs are connected in series across a 440 V line, then [2001]
(a) only 25W bulb will fuse
(b) only 100W bulb will fuse
(c) both bulbs will fuse
(d) none of these

87. A battery of 10 V and internal resistance 0.5Ω is connected across a variable resistance R. The value of R for which the power delivered is maximum is equal to [2001, 1992]
(a) 0.25Ω (b) 0.5Ω
(c) 1.0Ω (d) 2.0Ω

88. Two electric bulbs, one of 200 V, 40W and other of 200 V, 100W are connected in a domestic circuit. Then *[2000]*
(a) they have equal resistance
(b) the resistance of 40W bulb is more than 100W bulb
(c) the resistance of 100W bulb is more than 40 W bulb
(d) they have equal current through them

89. Three equal resistors connected across a source of e.m.f. together dissipate 10 watt of power. What will be the power dissipated in watts if the same resistors are connected in parallel across the same source of e.m.f.? *[1998]*
(a) 10 (b) $\frac{10}{3}$
(c) 30 (d) 90

90. A 5°C rise in temperature is observed in a conductor by passing a current. When the current is doubled the rise in temperature will be approximately *[1998]*
(a) 10°C (b) 16°C
(c) 20°C (d) 12°C

91. A (100 W, 200 V) bulb is connected to a 160V power supply. The power consumption would be *[1997]*
(a) 125 W (b) 100 W
(c) 80 W (d) 64 W

92. A heating coil is labelled 100 W, 220 V. The coil is cut in half and the two pieces are joined in parallel to the same source. The energy now liberated per second is *[1995]*
(a) 25 J (b) 50 J
(c) 200 J (d) 400 J

93. A 4 μF capacitor is charged to 400 volts and then its plates are joined through a resistance of 1k Ω. The heat produced in the resistance is *[1994]*
(a) 0.16 J (b) 1.28 J
(b) 0.64 J (d) 0.32 J

94. Two identical batteries each of e.m.f 2V and internal resistance 1Ω are available to produce heat in an external resistance by passing a current through it. The maximum power that can be developed across R using these batteries is *[1990]*
(a) 3.2 W (b) 2.0 W
(c) 1.28 W (d) $\frac{8}{9}$ W

95. Forty electric bulbs are connected in series across a 220 V supply. After one bulb is fused the remaining 39 are connected again in series across the same supply. The illumination will be *[1989]*
(a) more with 40 bulbs than with 39
(b) more with 39 bulbs than with 40
(c) equal in both the cases
(d) in the ratio $40^2 : 39^2$

96. A current of 2 A, passing through a conductor produces 80 J of heat in 10 seconds. The resistance of the conductor in ohm is *[1989]*
(a) 0.5 (b) 2
(c) 4 (d) 20

Topic 5 Wheatstone Bridge & Different Measuring Instruments

97. In a potentiometer circuit a cell of EMF 1.5 *V* gives balance point at 36 cm length of wire. If another cell of *EMF* 2.5 *V* replaces the first cell, then at what length of the wire, the balance point occurs? *[2021]*
(a) 62 cm (b) 60 cm
(c) 21.6 cm (d) 64 cm

98. A resistance wire connected in the left gap of a metre bridge balances a 10 Ω resistance in the right gap at a point which divides the bridge wire in the ratio 3 : 2. If the length of the resistance wire is 1.5 m, then the length of 1 Ω of the resistance wire is : *[2020]*
(a) 1.0×10^{-1} m (b) 1.5×10^{-1} m
(c) 1.5×10^{-2} m (d) 1.0×10^{-2} m

99. In the circuits shown below, the readings of the voltmeters and the ammeters will be : *[2019]*

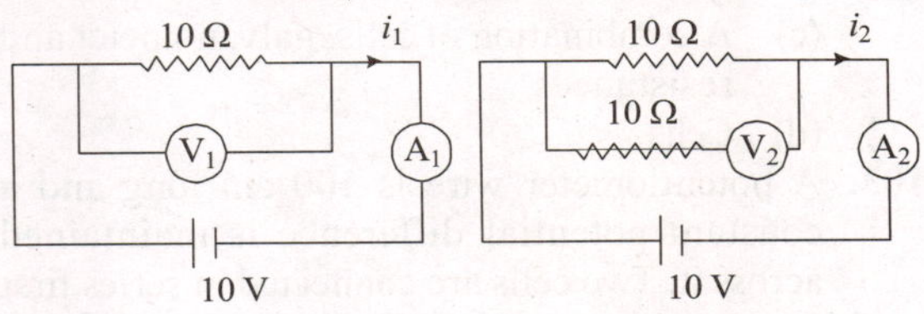

Circuit 1 Circuit 2

(a) $V_2 > V_1$ and $i_1 = i_2$
(b) $V_1 = V_2$ and $i_1 > i_2$
(c) $V_1 = V_2$ and $i_1 = i_2$
(d) $V_2 > V_1$ and $i_1 > i_2$

100. The reading of an ideal voltmeter in the circuit shown is, *[NEET Odisha 2019]*

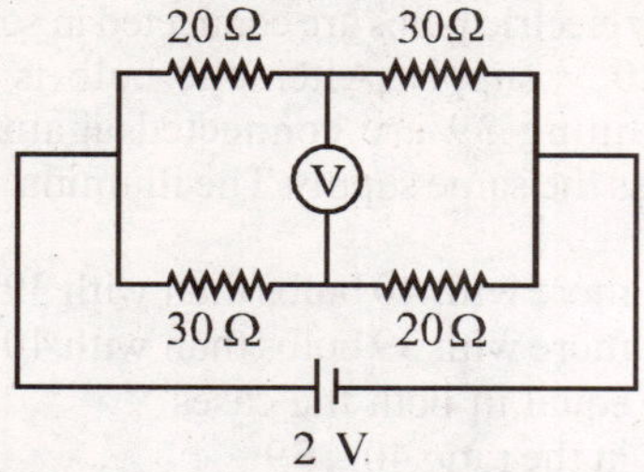

(a) 0.4 V (b) 0.6 V
(c) 0 V (d) 0.5 V

101. The metre bridge shown is in balance position with $\frac{P}{Q}=\frac{l_1}{l_2}$. If we now interchange the position of galvanometer and cell, will the bridge work? If yes, what will be balance condition? *[NEET Odisha 2019]*

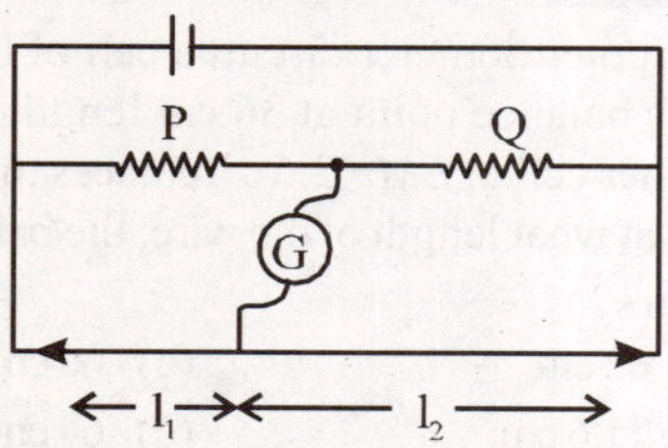

(a) yes, $\frac{P}{Q}=\frac{l_1}{l_2}$ (b) yes, $\frac{P}{Q}=\frac{l_2-l_1}{l_2+l_1}$

(c) no, no null point (d) yes, $\frac{P}{Q}=\frac{l_2}{l_1}$

102. A potentiometer is an accurate and versatile device to make electrical measurements of E.M.F. because the method involves *[2017]*
(a) Potential gradients
(b) A condition of no current flow through the galvanometer
(c) A combination of cells, galvanometer and resistances
(d) Cells

103. A potentiometer wire is 100 cm long and a constant potential difference is maintained across it. Two cells are connected in series first to support one another and then in opposite direction. The balance points are obtained at 50 cm and 10 cm from the positive end of the wire in the two cases. The ratio of emf's is : *[2016]*
(a) 5 : 1 (b) 5 : 4
(c) 3 : 4 (d) 3 : 2

104. A potentiometer wire has length 4 m and resistance 8Ω. The resistance that must be connected in series with the wire and an accumulator of e.m.f. 2V, so as to get a potential gradient 1 mV per cm on the wire is *[2015]*
(a) 40 Ω (b) 44 Ω
(c) 48 Ω (d) 32 Ω

105. A potentiometer wire of length L and a resistance r are connected in series with a battery of e.m.f. E_0 and a resistance r_1. An unknown e.m.f. E is balanced at a length l of the potentiometer wire. The e.m.f. E will be given by : *[2015 RS]*

(a) $\frac{E_0 r}{(r+r_1)}\cdot\frac{l}{L}$ (b) $\frac{E_0 l}{L}$

(c) $\frac{LE_0 r}{(r+r_1)l}$ (d) $\frac{LE_0 r}{l r_1}$

106. The resistances in the two arms of the meter bridge are 5Ω and RΩ, respectively. When the resistance R is shunted with an equal resistance, the new balance point is at $1.6\, l_1$. The resistance 'R' is : *[2014]*

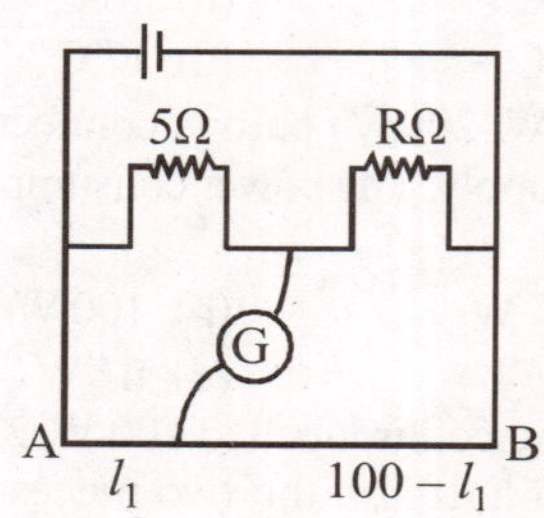

(a) 10Ω (b) 15Ω
(c) 20Ω (d) 25Ω

107. A potentiometer circuit has been set up for finding the internal resistance of a given cell. The main battery used across the potentiometer wire, has an emf of 2.0 V and a negligible internal resistance. The potentiometer wire itself is 4m long, When the resistace R, connected across the given cell, has values of
(i) infinity (ii) 9.5Ω
The balancing lengths', on the potentiometer wire are found to be 3 m and 2.85 m, respectively. The value of internal resistance of the cell is *[2014]*
(a) 0.25Ω (b) 0.95Ω
(c) 0.5Ω (d) 0.75Ω

108. The resistance of the four arms P, Q, R and S in a Wheatstone's bridge are 10 ohm, 30 ohm, 30 ohm and 90 ohm, respectively. The e.m.f. and internal resistance of the cell are 7 volt and 5 ohm respectively. If the galvanometer resistance is 50 ohm, the current drawn from the cell will be *[2013]*

(a) 0.2 A (b) 0.1 A
(c) 2.0 A (d) 1.0 A

109. In the circuit shown the cells *A* and *B* have negligible resistances. For $V_A = 12V$, $R_1 = 500\Omega$ and $R = 100\Omega$ the galvanometer (*G*) shows no deflection. The value of V_B is : *[2012]*

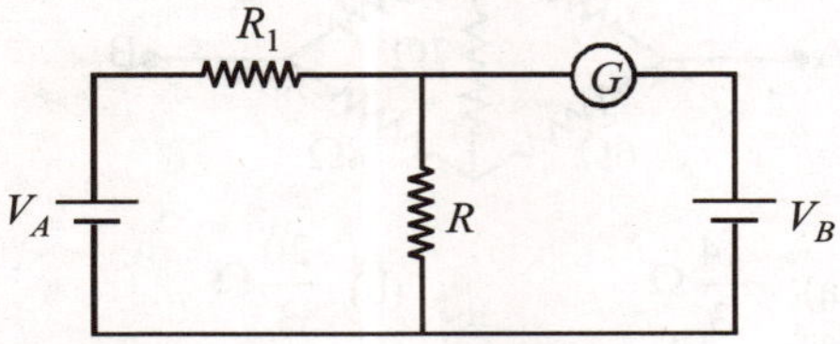

(a) 4 *V* (b) 2 *V*
(c) 12 *V* (d) 6 *V*

110. A potentiometer circuit is set up as shown. The potential gradient, across the potentiometer wire, is *k* volt/cm and the ammeter, present in the circuit, reads 1.0 *A* when two way key is switched off. The balance points, when the key between the terminals (i) 1 and 2 (ii) 1 and 3, is plugged in, are found to be at lengths l_1 cm and l_2 cm respectively. The magnitudes, of the resistors *R* and *X*, in ohms, are then, equal, respectively, to *[2010]*

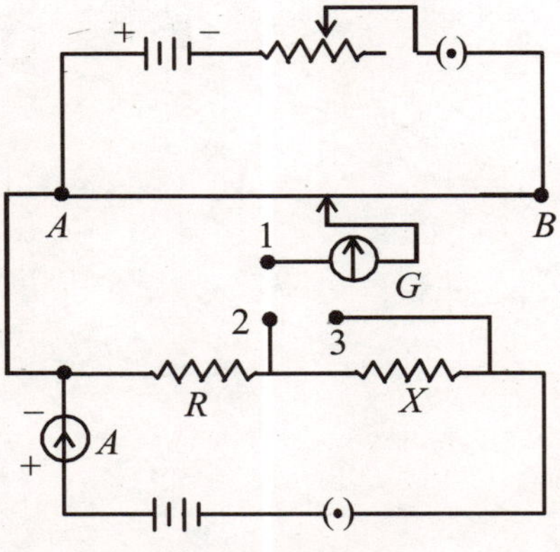

(a) $k\,(l_2 - l_1)$ and $k\,l_2$
(b) $k\,l_1$ and $k\,(l_2 - l_1)$
(c) $k\,(l_2 - l_1)$ and $k\,l_1$
(d) $k\,l_1$ and $k\,l_2$

111. Three resistances *P*, *Q*, *R* each of 2Ω and an unknown resistance S form the four arms of a Wheatstone bridge circuit. When a resistance of 6Ω is connected in parallel to *S* the bridge gets balanced. What is the value of *S*?

(a) 3Ω (b) 6Ω *[2007]*
(c) 1Ω (d) 2Ω

112. In the circuit shown, if a conducting wire is connected between points A and B, the current in this wire will *[2006]*

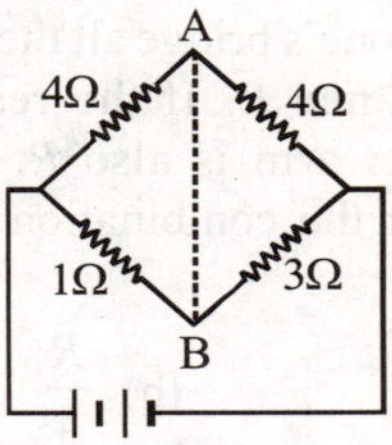

(a) flow in the direction which will be decided by the value of V
(b) be zero
(c) flow from *B* to *A*
(d) flow from *A* to *B*

113. For the network shown in the Fig. the value of the current i is *[2005]*

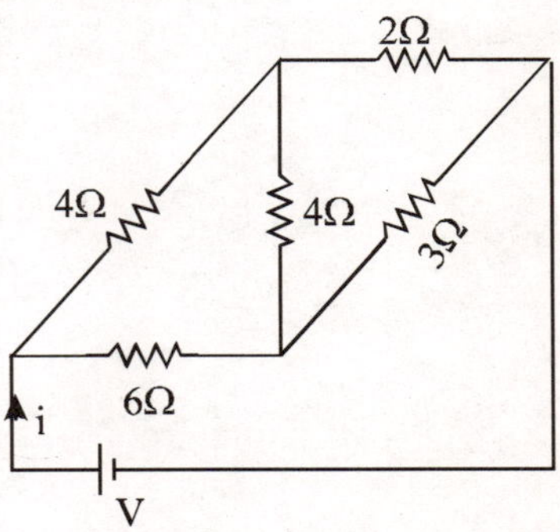

(a) $\frac{9V}{35}$ (b) $\frac{18V}{5}$ (c) $\frac{5V}{9}$ (d) $\frac{5V}{18}$

114. Five equal resistances each of resistance R are connected as shown in the figure. A battery of V volts is connected between *A* and *B*. The current flowing in *AFCEB* will be *[2004]*

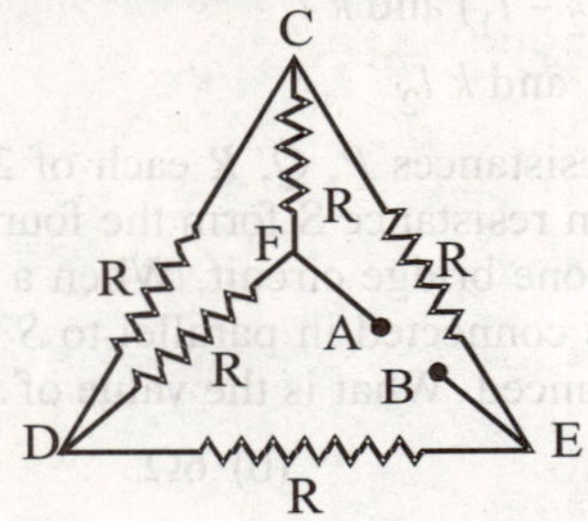

(a) $\frac{2V}{R}$ (b) $\frac{3V}{R}$

(c) $\frac{V}{R}$ (d) $\frac{V}{2R}$

115. In a Wheatstone's bridge all the four arms have equal resistance R. If the resistance of the galvanometer arm is also R, the equivalent resistance of the combination as seen by the battery is *[2003]*

(a) $2R$ (b) $\frac{R}{4}$

(c) $\frac{R}{2}$ (d) R

116. If specific resistance of a potentiometer wire is 10^{-7} Ωm, the current flow through it is 0.1 A and the cross-sectional area of wire is 10^{-6} m^2 then potential gradient will be *[2001]*

(a) 10^{-2} volt/m (b) 10^{-4} volt/m

(c) 10^{-6} volt/m (d) 10^{-8} volt/m

117. Potentiometer measures potential more accurately because *[2000]*

(a) it measures potential in the open circuit

(b) it uses sensitive galvanometer for null deflection

(c) it uses high resistance potentiometer wire

(d) it measures potential in the closed circuit

118. Five resistances have been connected as shown in the figure. The effective resistance between A and B is *[2000]*

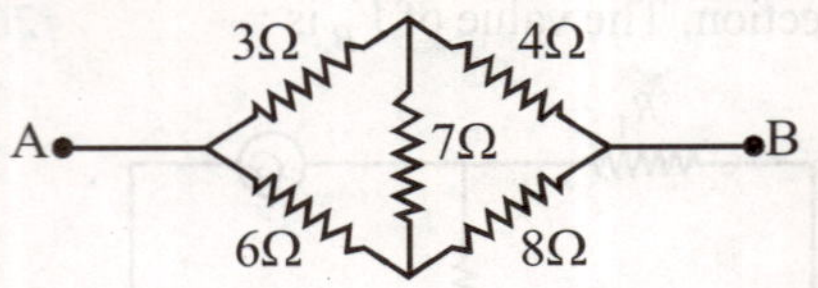

(a) $\frac{14}{3}\Omega$ (b) $\frac{20}{3}\Omega$

(c) 14Ω (d) 21Ω

119. In a metre-bridge, the balancing length from the left end when standard resistance of 1 Ω is in right gap is found to be 20 cm. The value of unknown resistance is *[1999]*

(a) 0.25Ω (b) 0.4Ω

(c) 0.5Ω (d) 4Ω

ANSWER KEY																	
1	(b)	15	(a)	29	(c)	43	(c)	57	(c)	71	(a)	85	(c)	99	(c)	113	(d)
2	(b)	16	(a)	30	(d)	44	(a)	58	(d)	72	(b)	86	(a)	100	(a)	114	(d)
3	(c)	17	(c)	31	(d)	45	(a)	59	(a)	73	(b)	87	(b)	101	(a)	115	(d)
4	(c)	18	(b)	32	(a)	46	(c)	60	(d)	74	(c)	88	(b)	102	(b)	116	(a)
5	(a)	19	(d)	33	(b)	47	(b)	61	(d)	75	(c)	89	(d)	103	(d)	117	(a)
6	(b)	20	(b)	34	(b)	48	(b)	62	(a)	76	(c)	90	(c)	104	(d)	118	(a)
7	(b)	21	(a)	35	(a)	49	(d)	63	(b)	77	(d)	91	(d)	105	(a)	119	(a)
8	(a)	22	(a)	36	(c)	50	(d)	64	(b)	78	(b)	92	(d)	106	(b)		
9	(c)	23	(d)	37	(c)	51	(c)	65	(a)	79	(c)	93	(d)	107	(c)		
10	(a)	24	(a)	38	(a)	52	(d)	66	(c)	80	(a)	94	(b)	108	(a)		
11	(b)	25	(c)	39	(d)	53	(d)	67	(a)	81	(a)	95	(b)	109	(b)		
12	(c)	26	(a)	40	(c)	54	(d)	68	(c)	82	(d)	96	(b)	110	(b)		
13	(d)	27	(d)	41	(c)	55	(a)	69	(a)	83	(b)	97	(b)	111	(a)		
14	(a)	28	(d)	42	(d)	56	(a)	70	(b)	84	(a)	98	(a)	112	(c)		

Hints & Solutions

1. **(b)** Drift velocity, $v_d = \frac{I}{neA} = \frac{eE\tau}{m}$

Electrical resistivity, $\rho = \frac{1}{\sigma} = \frac{E}{J}$ $\left(\because \sigma = \frac{J}{E}\right)$

Relaxation period, $\tau = \frac{m}{ne^2\rho}$

Current density, $J = \frac{I}{A} = nev_d$

NOTES

Relation between drift velocity, v_d, electrical resistivity(ρ) relaxation period (τ) and current density(J)

$$v_d = \frac{i}{neA} = \frac{J}{ne} = \frac{\sigma E}{ne} = \frac{E}{\rho ne} = \frac{v}{\rho l ne};$$

$$R\frac{l}{A} = \frac{m}{ne^2\tau} \cdot \frac{l}{A}$$

2. **(b)** For metals like copper, at temperature much lower than 0°C, graph deviates considerably from a straight line.

3. **(c)** For metals temperature coefficient of resistance is positive while for insulators and semiconductors, temperature coefficient of resistance is negative.

4. **(c)** According to colour coding -

Yellow	Violet	Brown	Gold
4	7	1	5%

First colour gives first digit, second colour gives the second digit and third colour gives the multiplier and fourth colour gives tolerance.

$\therefore$ Resistance,

$$R = 47 \times 10^1 \pm 5\% = 470 \pm 5\% \Omega$$

5. **(a)** Given,

Drift velocity of charged particle,

$V_d = 7.5 \times 10^{-4}$ m/s

Electric field, $E = 3 \times 10^{-10}$ Vm^{-1}

Mobility, $\mu = \frac{V_d}{E} = \frac{7.5 \times 10^{-4}}{3 \times 10^{-10}} = 2.5 \times 10^6 \ m^2V^{-1}s^{-1}$

6. **(b)** Colour code for carbon resistor

0 Black

1	Brown	Tolerance :	±5% Gold
2	Red		±10% Silver
3	Orange		±20% No colour
4	Yellow		
5	Green		
6	Blue		
7	Violet		
8	Grey		
9	White		

$(47 \pm 4.7)\,k\Omega = 47 \times 10^3 \pm 10\%$

$\therefore$ Yellow - Violet - Orange - Silver

7. **(b)** We know that, $R = \frac{\rho \ell}{A}$

or $R = \frac{\rho \ell^2}{\text{Volume}} \Rightarrow R \propto \ell^2$

According to question $\ell_2 = n\ell_1$

$$\frac{R_2}{R_1} = \frac{n^2 l_1^2}{l_1^2}$$

or, $\frac{R_2}{R_1} = n^2$

$\Rightarrow R_2 = n^2 R_1$

8. **(a)** Here, metallic conductor can be considered as the combination of various conductors connected in series. And in series combination current remains same.

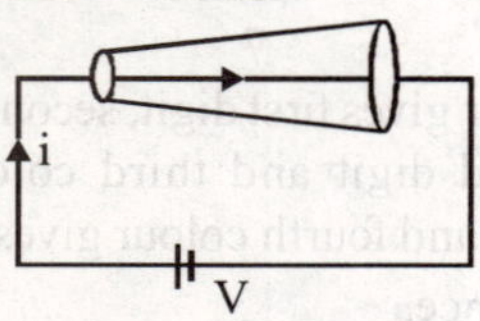

9. **(c)** Resistance $R = \frac{\rho \ell}{A} = 4\,\Omega$

$\because \ell' = 2\,\ell$

On stretching, volume of wire remains constant

$\therefore \ell A' = \ell' A'$

$\therefore A' = \frac{A}{2}$

$\therefore R' \; \rho \frac{2\ell}{\frac{A}{2}} = 4R = 4 \times 4\,\Omega = 16\,\Omega \Rightarrow \ell A = 2\ell A'$

Therefore the resistance of new wire becomes $16\,\Omega$

10. **(a)** Copper rod and iron rod are joined in series.

$\therefore R = R_{Cu} + R_{Fe} = (\rho_1 + \rho_2)\frac{\ell}{A} \quad \left(\because R = \rho \frac{\ell}{A}\right)$

From ohm's law $V = RI$

$= (1.7 \times 10^{-6} \times 10^{-2} + 10^{-5} \times 10^{-2}) \div 0.01 \times 10^{-4}$ volt

$= 0.117$ volt $(\because I = 1A)$

11. **(b)** Resistance of a wire is given by $R = \rho \frac{l}{a}$

If the length is increased by 10% then new

length $l' = l + \frac{1}{10} = \frac{11}{10} l$

In that case, area of cross-section of wire would decrease by 10%

$\therefore$ New area of cross-section

$A' = A - \frac{A}{10} = \frac{9}{10} A$

$$\therefore R' = \rho \frac{\ell'}{A'} = \rho \frac{\frac{11}{10} l}{\frac{9}{10} A}$$

$R' = \frac{11}{9} \rho \frac{l}{R} \qquad R' = 1.21R$

The new resistance increases by 1.21 times. The specific resistance (resistivity) remains unchanged as it depends on the nature of the material of the wire.

NOTES After stretching if length of a conductor increases by $x\%$ then resistance will increase by $2x\%$ (Valid only if $x < 10\%$)

12. **(c)** $R = \frac{\rho \ell_1}{A_1}$, now $\ell_2 = 2\ell_1$

$A_2 = \pi (r_2)^2 = \pi (2r_1)^2 = 4\pi r_1^2 = 4A_1$

$\therefore \quad R_2 = \frac{\rho(2\ell_1)}{4A_1} = \frac{\rho \ell_1}{2A_1} = \frac{R}{2}$

$\therefore$ Resistance is halved, but specific resistance remains the same.

13. **(d)** $R \propto \ell$

For 300 cm, $R = 100\,\Omega$

For 50 cm, $R' = \frac{100}{300} \times 50 = \frac{50}{3}\,\Omega$

$\therefore IR = 6$

$\Rightarrow IR' = \frac{6}{R} \times R' = \frac{6}{100} \times \frac{50}{3} = 1$ volt.

14. (a) Resistivity of copper wire increases with increase in temprature as $\rho_t = \rho_0(1+\alpha t)$

> **NOTES** Copper being a metal has positive coefficient of resistivity.

15. (a) For metals $P_t = P_o(1 + \propto .t)$
So resistivity of copper decreases with decrease in temperature. whereas of semiconductor or *si* increases with decreases in temperature.

16. (a) $R_1 = 3.1\,\Omega$ at $t = 30°C$
$R_2 = 4.5\,\Omega$ at $t = 100°C$
We have, $R = R_0(1+\alpha t)$
$\therefore R_1 = R_0[1+\alpha(30)]$
$R_2 = R_0[1+\alpha(100)]$
$$\Rightarrow \frac{R_1}{R_2} = \frac{1+30\alpha}{1+100\alpha}$$
$$\Rightarrow \frac{3.1}{4.5} = \frac{1+30\alpha}{1+100\alpha} \Rightarrow \alpha = 0.008°C^{-1}$$

17. (c) In discharge tube the current is due to flow of positive ions and electrons. Moreover, secondary emission of electrons is also possible. So V-I curve is non-linear; hence resistance is non-ohmic.

18. (b) $R = \rho\frac{l}{A}$
$$R_1 = \rho\frac{L}{A} \quad ...(1)$$
$$R_2 = \rho\frac{2L}{A}\times 2 \quad ...(2)$$
$$R_3 = \rho\frac{L}{2.2A} = \frac{\rho L}{4A} \quad ...(3)$$
$\Rightarrow R_3 < R_1 < R_2$

19. (d) Resistance of a material at two different temperatures, T_1 and T_2 such that $T_2 > T$, are related as
$R_{T_2} = R_{T_1}[1+\alpha(T_2 - T_1)]$
Using this relation for the present case, we get
$R_{100} = R_{50}[1+\alpha(100-50)]$
$\Rightarrow 7 = 5[1+\alpha(50)]$
$$\Rightarrow \alpha = \frac{(7-5)}{250} = 0.008/°C$$

20. (b) Given : Length of wire $(l) = 15$m
Area (A) $= 6\times10^{-7}\ m^2$
Resistance (R) $= 5\Omega$.
We know that resistance of the wire material
$$R = \rho\frac{l}{A}$$
$$\Rightarrow 5 = \rho\times\frac{15}{6\times10^{-7}} = 2.5\times10^7\rho$$
$$\Rightarrow \rho = \frac{5}{2.5\times10^7} = 2\times10^{-7}\ \Omega - m$$
[where ρ = coefficient of resistivity]

21. (a) Initial resistance $(R_1) = R$; Initial length is ℓ_1 and final length $(\ell_2) = 0.5\ \ell$. Volume of a wire $= \ell.A$. Since the volume of the wire remains the same after recasting, therefore $\ell_1.A_1 = \ell_2 A_2$
or $\frac{\ell_1}{\ell_2} = \frac{A_2}{A_1}$ or $\frac{\ell}{0.5\ell} = \frac{A_2}{A_1}$ or $\frac{A_2}{A_1} = 2$.
We also know that resistance of a wire (R)
$R = \rho\times\frac{\ell}{A}$.; $R \propto \frac{\ell}{A}$
$$\therefore \frac{R_1}{R_2} = \frac{\ell_1}{\ell_2}\times\frac{A_2}{A_1} = \frac{\ell}{0.5\ell}\times 2 = 4$$
or, $R_2 = \frac{R_1}{4} = \frac{R}{4}$.

> **NOTES** When wires are drawn from same volume but with different area of cross-section, then
> $$R \propto \frac{1}{(\text{Area of cross-section})^2}$$

22. (a)

23. (d) $R = \frac{\rho l}{\pi r^2}$. But $m = \pi r^2 ld$ $\therefore \pi r^2 = \frac{m}{ld}$
$$\therefore R = \frac{\rho l^2 d}{m},\ R_1 = \frac{\rho l_1^2 d}{m_1},\ R_2 = \frac{\rho l_2^2 d}{m_2}$$
$$R_3 = \frac{\rho l_3^2 d}{m_3}$$
$$R_1 : R_2 : R_3 = \frac{l_1^2}{m_1} : \frac{l_2^2}{m_2} : \frac{l_3^2}{m_3}$$
$$R_1 : R_2 : R_3 = \frac{25}{1} : \frac{9}{3} : \frac{1}{5} = 125 : 15 : 1$$

24. **(a)** Resistance of each wire in this case will be same as l & A are same and made of same material.

$\therefore R_1 = R_2 = R_3 = R_4 = R$

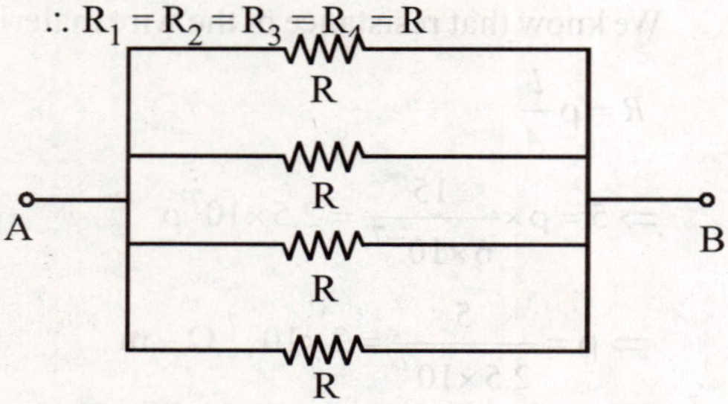

In parallel, $\frac{1}{R_P} = \frac{1}{R_1} + \frac{1}{R_2} + \frac{1}{R_3} + \frac{1}{R_4}$

$\therefore R_P = \frac{R}{4} \Rightarrow 0.25 = \frac{R}{4}$

$\therefore R = 1\,\Omega$

When arranged in series then equivalent resistance.

$R_S = R + R + R + R = 4R$

$\therefore R_S = 4 \times 1 = 4\Omega$

25. **(c)** Resistance r_2 and r_3 are in parallel so potential difference across $r_2 = r_3, = V$

$\therefore i_2 r_2 = i_3 r_3 \Rightarrow i_2 = \frac{i_3 r_3}{r_2}$ ... (i)

Current $i_1 = i_2 + i_3 \Rightarrow i_1 = \left(\frac{r_3}{r_2} + 1\right) i_3$

$\therefore \frac{i_3}{i_1} = \frac{r_2}{r_2 + r_3}$

26. **(a)** In series grouping equivalent resistance

$R_{series} = nR$

In parallel grouping equivalent resistance

$R_{parallel} = \frac{R}{n}$

$I = \frac{E}{nR + R}$ ...(i)

$10\,I = \frac{E}{\frac{R}{n} + R}$ ...(ii)

Dividing eq. (ii) by (i),

$$10 = \frac{(n+1)R}{\left(\frac{1}{n} + 1\right)R}$$

Solving we get, n = 10

27. **(d)** Effective resistance of B and C

$$= \frac{R_B \cdot R_C}{R_B + R_C} = \frac{1.5R \times 3R}{1.5R + 3R} = \frac{4.5R^2}{4.5R} = R$$

i.e., equal to resistance of voltmeter A.

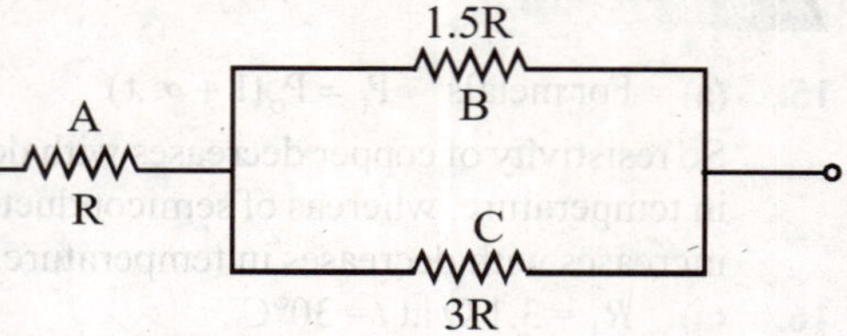

In parallel potential difference is same so, $V_B = V_C$ and in series current is same

So, $V_A = V_B = V_C$

28. **(d)** In figure, two metal wires of identical dimension are connected in series

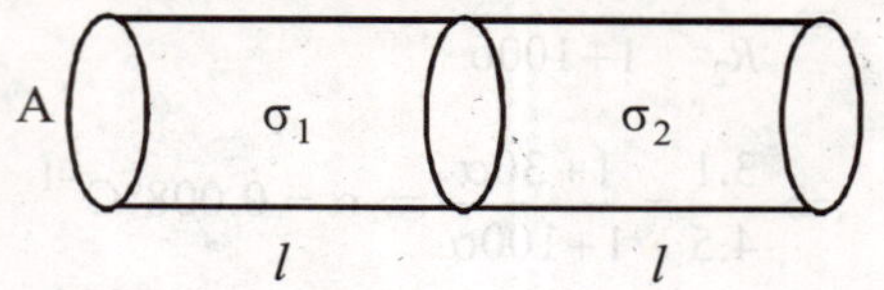

$$R_{eq} = \frac{l}{\sigma_1 A} + \frac{l}{\sigma_2 A} = \frac{l_{eq}}{\sigma_{eq} A_{eq}}$$

$$\frac{2l}{\sigma_{eq} A} = \frac{l}{A}\left(\frac{\sigma_1 + \sigma_2}{\sigma_1 \sigma_2}\right)$$

$$\therefore \sigma_{eq} = \frac{2\sigma_1 \sigma_2}{\sigma_1 + \sigma_2}$$

29. **(c)** Resistance is directly proportional to length,

$$\frac{1}{R_{AB}} = \frac{1}{3} + \frac{1}{4+5} = \frac{(4+5)+3}{(3)(4+5)}$$

$$R_{AB} = \frac{3 \times (4+5)}{3 + (4+5)} = \frac{27}{12}$$

Similarly,

$$R_{BC} = \frac{4 \times (3+5)}{4 + (3+5)} = \frac{32}{12}$$

$$R_{AC} = \frac{5 \times (3+4)}{5 + (3+4)} = \frac{35}{12}$$

$\therefore R_{AB} : R_{BC} : R_{AC} = 27 : 32 : 35$

30. **(d)** Let x is the resistance per unit length then

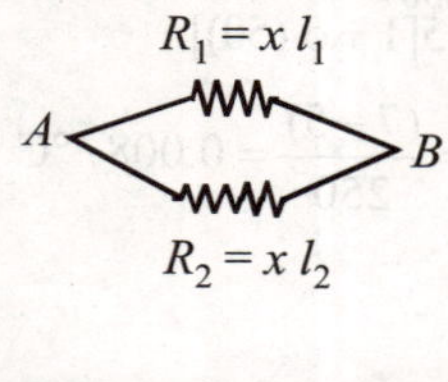

equivalent resistance $R = \dfrac{R_1R_2}{R_1+R_2} = \dfrac{(xl_1)(xl_2)}{xl_1+xl_2}$

$$\Rightarrow \quad \frac{8}{3} = x\frac{l_1l_2}{l_1+l_2}$$

$$\frac{8}{3} = x\frac{l_1}{\frac{l_1}{l_2}+1} \quad \text{...(i)}$$

also $R_0 = xl_1 + x\,l_2$

$$12 = x(l_1+l_2)$$

$$12 = xl_2\left(\frac{l_1}{l_2}+1\right) \quad \text{...(ii)}$$

$$\frac{(i)}{(ii)} \Rightarrow \frac{\frac{8}{3}}{\frac{12}{1}} = \frac{\dfrac{xl_1}{\left(\frac{l_1}{l_2}+1\right)}}{xl_2\left(\frac{l_1}{l_2}+1\right)} = \frac{l_1}{l_2\left(\frac{l_1}{l_2}+1\right)^2}$$

$$\left(\frac{l_1}{l_2}+1\right)^2 \times \frac{8}{36} = \frac{l_1}{l_2}$$

$$(y^2+1+2y)\times\frac{8}{36} = y \;(\text{where } y = \frac{l_1}{l_2})$$

$$8y^2+8+16y = 36y$$

$$\Rightarrow \quad 8y^2-20y+8=0$$

$$\Rightarrow \quad 2y^2-5y+2=0$$

$$\Rightarrow \quad 2y^2-4y-y+2=0$$

$$\Rightarrow \quad 2y(y-2)-1(y-2)=0$$

$$\Rightarrow \quad (2y-1)(y-2)=0$$

$$\Rightarrow \quad y = \frac{l_1}{l_2} = \frac{1}{2} \text{ or } 2$$

31. (d)

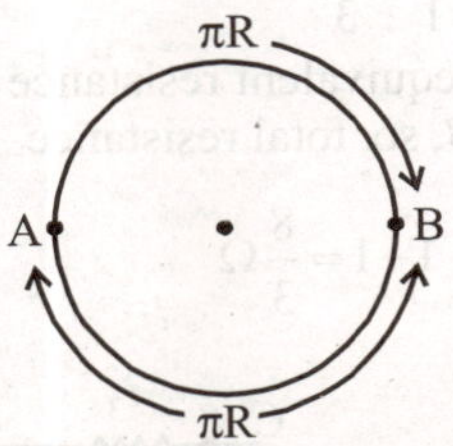

Length of wire $= 2\pi r = 2\pi(0.1) = 0.2\pi$m

Resistances of complete circle/wire $= 12\times 0.2\pi = 2.4\pi\Omega$

$\therefore$ Resistance of each semi-circle $= 1.2\pi\,\Omega$

Hence, equivalent resistance,

$$R_{AB} = \frac{1.2\pi}{2} = 0.6\pi\,\Omega$$

32. (a)

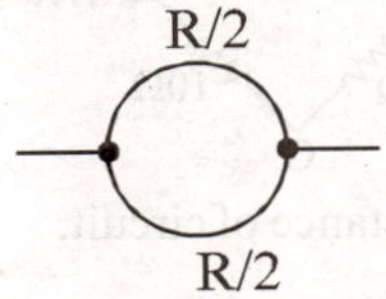

$$R_{eq} = \frac{R_1R_2}{R_1+R_2} = \frac{\left(\frac{R}{2}.\frac{R}{2}\right)}{\frac{R}{2}+\frac{R}{2}} \quad \therefore\; R_{eq} = \frac{R}{4}$$

NOTES: If a wire of resistance R is bent in the form of circle, the effective resistance between the ends of diameter is $R/4$

33. (b) $R = \dfrac{r}{n} \Rightarrow r = nR$

When connected in series, $R_{eq} = nr$

$= n(nR) = n^2R$

34. (b) In circuit, R_B and R_C are in series, so, $R_s = 6+6 = 12\,\Omega$. This $12\,\Omega$ resistance is in parallel with $R_A = 3\,\Omega$,

So, equivalent resistance of circuit

$$R = \frac{3\times12}{3+12} = \frac{36}{15} = \frac{12}{5}\Omega$$

$\therefore$ Current in circuit, $I = \dfrac{V}{R}$

$$= \frac{4.8}{\left(\frac{12}{5}\right)} = \frac{4.8\times5}{12} = 2\text{A}$$

35. (a) Resistance of ACB, $R' = 3\Omega + 3\Omega = 6\Omega$.

For net resistance between A and B; $R' = 6\Omega$ and 3Ω are in parallel.

$$R_{AB} = \frac{3\times6}{3+6} = \frac{18}{9} = 2\Omega$$

Current in circuit $(I) = \dfrac{E}{R_{AB}} = \dfrac{2}{2} = 1A$

36. (c)

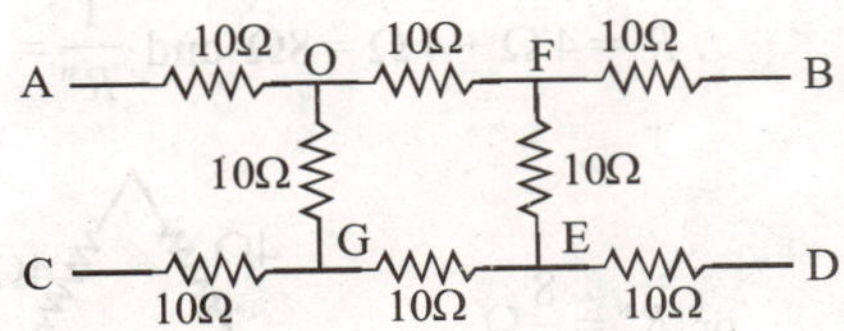

Equivalent Circuit

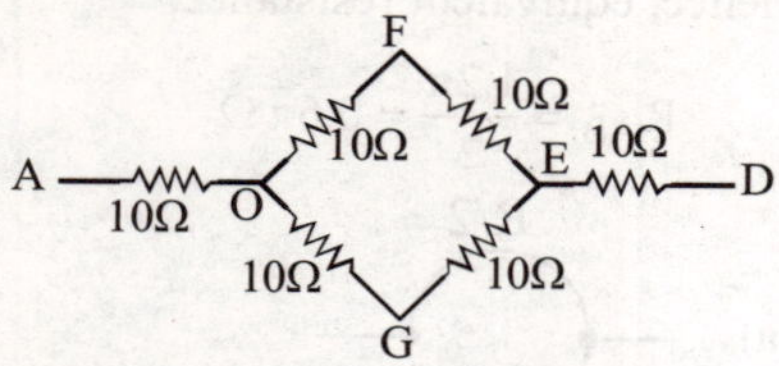

Equivalent Resistance of circuit,

$$= 10\Omega + \frac{20\times 20}{20+20} + 10 = 10+10+10 = 30\Omega$$

37. **(c)** Length of each wire = ℓ; Area of thick wire (A_1) = 3A; Area of thin wire (A_2) = A and resistance of thick wire (R_1) = 10 Ω. Resistance

$$(R) = \rho\frac{\ell}{A} \propto \frac{1}{A} \text{ (if } \ell \text{ is constant)}$$

$$\therefore \frac{R_1}{R_2} = \frac{A_2}{A_1} = \frac{A}{3A} = \frac{1}{3}$$

or, $R_2 = 3R_1 = 3\times 10 = 30\,\Omega$

The equivalent resistance of these two resistors in series

$= R_1 + R_2 = 30 + 10 = 40\Omega.$

38. **(a)** The potential difference across 4Ω resistance is given by

$V = 4\times i_1 = 4\times 1.2 = 4.8$ volt

So, the potential across 8Ω resistance is also 4.8 volt.

Current $i_2 = \frac{V}{8} = \frac{4.8}{8} = 0.6$ amp

Current in 2Ω resistance $i = i_1 + i_2$

$\therefore i = 1.2 + 0.6 = 1.8$ amp

Potential difference across 2Ω resistance

$V_{BC} = 1.8\times 2 = 3.6$volts

39. **(d)** The two resistances are connected in series and the resultant is connected in parallel with the third resistance.

$$\therefore R' = 4\,\Omega + 4\,\Omega = 8\Omega \text{ and } \frac{1}{R''} = \frac{1}{8} + \frac{1}{4} = \frac{3}{8}$$

or $R'' = \frac{8}{3}\Omega$

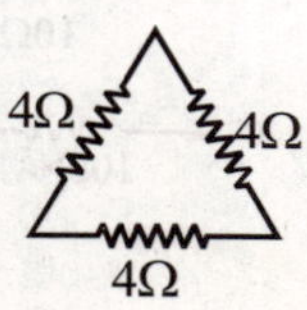

NOTES

If three identical resistors each of resistance R are connected in the form of a triangle, the equivalent resistance between the ends of a side is equal to $\left(\frac{2R}{3}\right)$.

40. **(c)** Voltage across 3 Ω resistance = 3 × 0.8 = 2.4V This voltage is the same across 6 Ω resistance. Hence current through this resistance

$$i = \frac{V}{R} = \frac{2.4}{6} = 0.4 \text{ amp}$$

Total current in the circuit

= 0.8 + 0.4 = 1.2 amp

Voltage across 4 Ω resistance

= 4 × 1.2 = 4.8 volts

41. **(c)** To carry a current of 4 ampere, we need four paths, each carrying a current of one ampere. Let r be the resistance of each path. These are connected in parallel. Hence, their equivalent resistance will be r/4. According to the given problem $\frac{r}{4} = 5$ or $r = 20\Omega$.

For this propose two resistances should be connected. There are four such combinations. Hence, the total number of resistance = 4 × 2 = 8.

42. **(d)** At A current is distributed and at B currents are collected. Between A and B, the distribution is symmetrical. It has been shown in the figure. It appears that current in AO and OB remains same. At O, current i_4 returns back without any change. If we detach O from AB there will not be any change in distribution.

Now, CO & OD will be in series hence its total resistance = 2Ω

It is in parallel with CD, so, equivalent resistance

$$= \frac{2\times 1}{2+1} = \frac{2}{3}\Omega$$

This equivalent resistance is in series with AC & DB, so, total resistance

$$= \frac{2}{3} + 1 + 1 = \frac{8}{3}\Omega$$

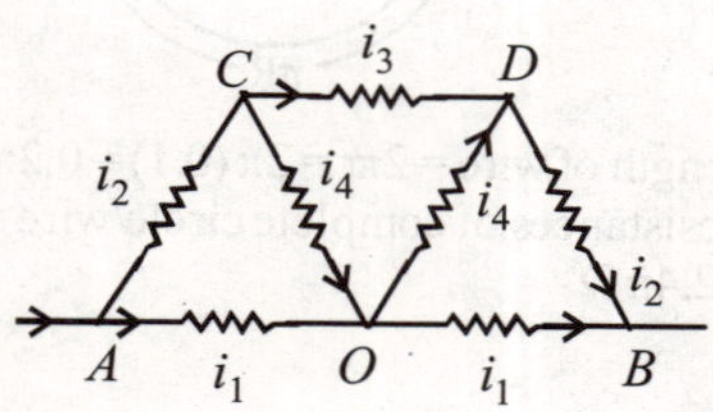

Now $\frac{8}{3}\Omega$ is parallel to AB, that is, 2Ω, so total resistance

$$= \frac{8/3 \times 2}{8/3+2} = \frac{16/3}{14/3} = \frac{16}{14} = \frac{8}{7}\Omega$$

NOTES

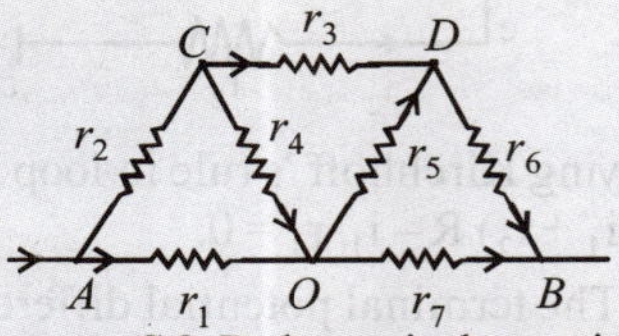

Between C & D, the equivalent resistance is given by

$$1/r = \frac{1}{r_3} + \frac{1}{(r_4 + r_5)} = 1 + \frac{1}{2} = \frac{3}{2}$$

Equivalent resistance along

$$ACDB = 1 + \frac{2}{3} + 1 = \frac{8}{3}$$

∴ Effective resistance between A and B is

$$\frac{1}{R} = \frac{3}{8} + \frac{1}{2} = \frac{7}{8} \text{ or } R = \frac{8}{7}\Omega$$

43. **(c)** In series, $R_s = nR$

In parallel, $\frac{1}{R_p} = \frac{1}{R} + \frac{1}{R} + ...n$ terms

$\therefore R_s/R_p = n^2/1 = n^2$

44. **(a)** $I = \frac{8-4}{1+2+9} = \frac{4}{12} = \frac{1}{3}A$;

$$V_P - V_Q = 4 - \frac{1}{3} \times 3 = 3 \text{ volt}$$

45. **(a)** Given : emf $\varepsilon = 2.1$ V
$I = 0.2$ A, $R = 10\Omega$
Internal resistance r = ?
From formula.
$\varepsilon - Ir = V = IR$
$2.1 - 0.2r = 0.2 \times 10$
$2.1 - 0.2r = 2$ or $0.2r = 0.1$

$$\Rightarrow r = \frac{0.1}{0.2} = 0.5\,\Omega$$

NOTES

$$i = \frac{\varepsilon}{r+R} \Rightarrow 0.2 = \frac{2.1}{r+10}$$

$$\Rightarrow 2.1 = 0.2r + 2 \Rightarrow r = \frac{1}{2} = 0.5\,\Omega$$

46. **(c)** The current through the resistance R

$$I = \left(\frac{\varepsilon}{R+r}\right)$$

The potential difference across R

$$V = IR = \left(\frac{\varepsilon}{R+r}\right)R$$

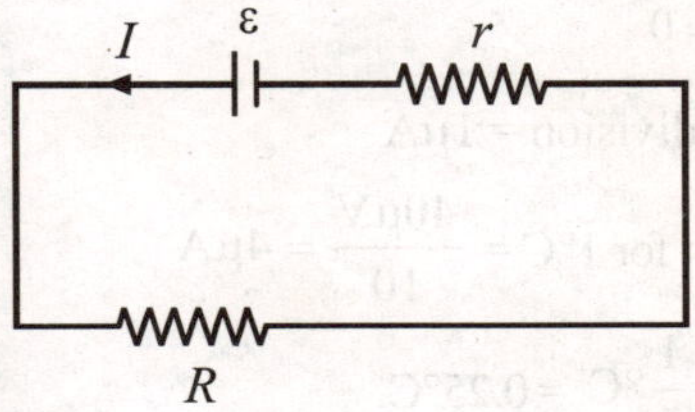

$$V = \frac{\varepsilon}{\left(1+\frac{r}{R}\right)}$$

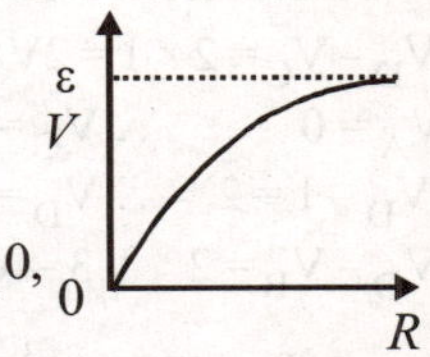

when $R = 0$, $V = 0$,
$R = \infty$, $V = \varepsilon$
Thus V increases as R increases upto certain limit, but it does not increase further.

47. **(b)** Let the internal resistance of the battery be r. Then the current flowing through the circuit is given by

$$i = \frac{E}{R+r}$$

In first case,

$$2 = \frac{E}{2+r} \quad ...(1)$$

In second case,

$$0.5 = \frac{E}{9+r} \quad ...(2)$$

From (1) & (2),
$4 + 2r = 4.5 + 0.5r$

$$\Rightarrow 1.5r = 0.5 \Rightarrow r = \frac{1}{3}\Omega.$$

NOTES

If I_1 be the current in a circuit with an external resistance R_1 and I_2 be the current in the circuit with external resistance R, then internal resistance of cell can be find using

$$r = \frac{I_2R_2 - I_1R_1}{I_1 - I_2}$$

48. (b) We have, $e = at + bt^2$

$$\Rightarrow \frac{de}{dt} = a + 2bt$$

At neutral temperature,

$$t = -\frac{a}{2b}$$

$$\therefore \frac{de}{dt} = 0$$

49. (d) 1 division = $1\mu A$

$$\text{Current for } 1°C = \frac{40\mu V}{10} = 4\mu A$$

$$1\mu A = \frac{1}{4}°C = 0.25°C.$$

50. (d) Current from D to C = 1A

$\therefore V_D - V_C = 2 \times 1 = 2V$

$V_A = 0 \qquad \therefore V_C = 1V, \ \therefore V_D - V_C = 2$

$\Rightarrow V_D - 1 = 2 \qquad \therefore V_D = 3V$

$\therefore V_D - V_B = 2 \quad \therefore 3 - V_B = 2 \ \therefore V_B = 1V$

51. (c) $I = \frac{P}{V} = \frac{100 \times 10^3}{125} A = \frac{10^5}{60} A$

$E.C.E. = 0.367 \times 10^{-6}\ kg\ C^{-1}$

Charge per minute = $(I \times 60)\ C$

$$= \frac{10^5 \times 60}{125} C = \frac{6 \times 10^6}{125} C$$

$$\therefore \text{Mass liberated, } = \frac{6 \times 10^6}{125} \times 0.367 \times 10^{-6}$$

$$= \frac{6 \times 1000 \times 0.367 \times 10^{-3}}{125}$$

$= 17.616 \times 10^{-3}\ kg$

52. (d) Junction law follows from conservation of charge and loop law is the conservation of energy

53. (d) $E = 30\theta - \frac{\theta^2}{15}$

For neutral temperature, $\frac{dE}{d\theta} = 0$

$$0 = 30 - \frac{2}{15}\theta$$

$\therefore \theta = 15 \times 15$

$= 225°\ C$

Hence, neutral temperature is 225°C.

54. (d)

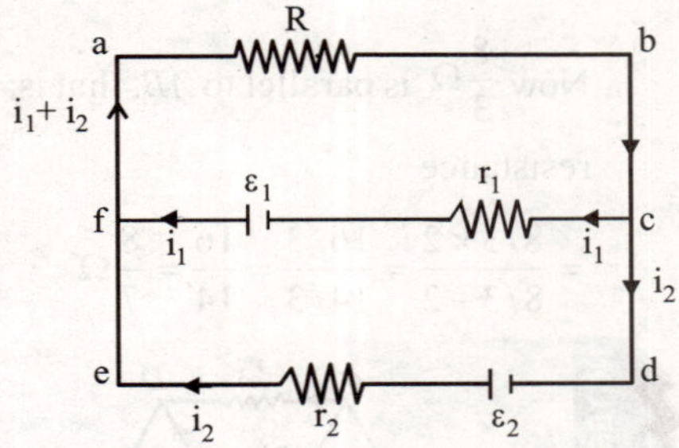

Applying Kirchhoff 's rule in loop **abcfa**

$\varepsilon_1 - (i_1 + i_2) R - i_1 r_1 = 0.$

55. (a) The terminal potential difference of a cell is given by $V + Ir = E$

$V = V_A - V_B$

or $\quad V = E - Ir$

$\Rightarrow \frac{dV}{dI} = -r, \quad$ Also for, $i = 0$ then $V = E$

$\therefore \quad$ slope = – r, intercept = E

56. (a) Here $E > \frac{ER}{R + r}$, hence the lengths 110 cm and 100 cm are interchanged.

Without being short-circuited through R, only the battery E is balanced.

$$E = \frac{V}{L} \times l_1 = \frac{V}{L} \times 110 .....(i)$$

When R is connected across E, $Ri = \frac{V}{L} \times l_2$

or, $R\left(\frac{E}{R + r}\right) = \frac{V}{L} \times 100 .....(ii)$

Dividing (i) by (ii), we get

$$\frac{R + r}{R} = \frac{110}{100}$$

or, $100R + 100r = 110R$

or, $10R = 100r$

$$\therefore r = \frac{10R}{100} = \frac{10 \times 10}{100} (\because R = 10\Omega)$$

$\Rightarrow r = 1\Omega.$

57. (c) We have, $m = ZIt$

where, Z is the electrochemical equivalent of copper.

$\Rightarrow m = 30 \times 10^{-5} \times 1.5 \times 10 \times 60$

$= 0.27$ gm.

58. (d) Power = $V \times I$

$$I = \frac{\text{Power}}{V} = \frac{100}{125}$$

E.C.E. of chlorine is 0.367×10^{-6} kg/coulomb

$$\text{Charge passing in one minute} = \frac{100}{125} \times 60$$
$$= 48 \text{ coulomb}$$

$$\text{Chlorine precipitated} = 0.367 \times 10^{-6} \times 48$$
$$= 17.6 \times 10^{-6} \text{ kg}$$
$$= 17.6 \text{ mg}$$

59. (a) Kirchhoff ' s first law deals with conservation of electrical charge and the second law deals with conservation of electrical energy.

60. (d) Current in the circuit

$$= \frac{E+E}{r_1+r_2+R} = \frac{2E}{r_1+r_2+R}$$

P.D. across first cell = $E - ir_1$

$$= E - \frac{2E \times r_1}{(r_1+r_2)+R}$$

$$\text{Now, } E = \frac{2Er_1}{(r_1+r_2)+R} = 0$$

$$\Rightarrow E = \frac{2Er_1}{r_1+r_2+R} \Rightarrow 2r_1 = r_1 + r_2 + R$$

$$R = r_1 - r_2$$

NOTES

In series grouping of cells their emf's are additive or subtractive while their internal resistances are always additive. If dissimilar plates of cells are connected together their emf's are added to each other while if their similar plates are connected together their emf's are subtractive.

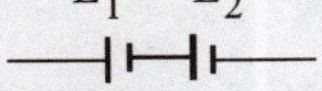

$E_{eq} = E_1 + E_2$

$r_{eq} = r_1 + r_2$

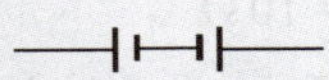

$E_{eq} = E_1 - E_2 \ (E_1 > E_2)$

$r_{eq} = r_1 + r_2$

61. (d) $$V = \frac{\frac{E_1}{r_1} + \frac{E_2}{r_2}}{\frac{1}{r_1} + \frac{1}{r_2}} = \frac{\frac{18}{2} + \frac{12}{1}}{\frac{1}{2} + \frac{1}{1}} = 14\text{V}$$

(Since the cells are in parallel).

62. (a) Efficiency is given by $\eta = \frac{\text{output}}{\text{input}}$

$$= \frac{5 \times 15 \times 14}{10 \times 8 \times 15} = 0.875 \text{ or } 87.5\,\%$$

63. (b) $E = V + ir$

$$2.2 = 1.8 + \frac{1.8}{5} \times r$$

$$\Rightarrow \quad r = \frac{10}{9}\,\Omega$$

64. (b) By Faraday's Ist Law

Amount deposited (m) = Zit = Zq

$m \propto q$

Amount deposited is directly proportional to charge.

65. (a) $E = V - Ir$

$12 = V - 60 \times 5 \times 10^{-2}$

$12 = V - 3$

$\Rightarrow V = 15$ volt

66. (c) $m = Zit,\ 9 = Z \times 10^5,\ Z = 9 \times 10^{-5}$

Again,

$m = Zit = 9 \times 10^{-5} \times 50 \times 20 \times 60 = 5.4$ gm

67. (a) We know from the Kirchhoff 's first law that the algebraic sum of the current meeting at any junction in the circuit is zero (i.e. $\Sigma i = 0$) or the total charge remains constant. Therefore, Kirchhoff's first law at a junction deals with the conservation of charge.

68. (c) In the electrolysis of $CuSO_4$, oxygen is liberated at anode and copper is deposited at cathode.

69. (a) Faraday's laws are based on the conversion of electrical energy into mechanical energy; which is in accordance with the law of conservation of energy.

70. (b) When all bulbs are glowing

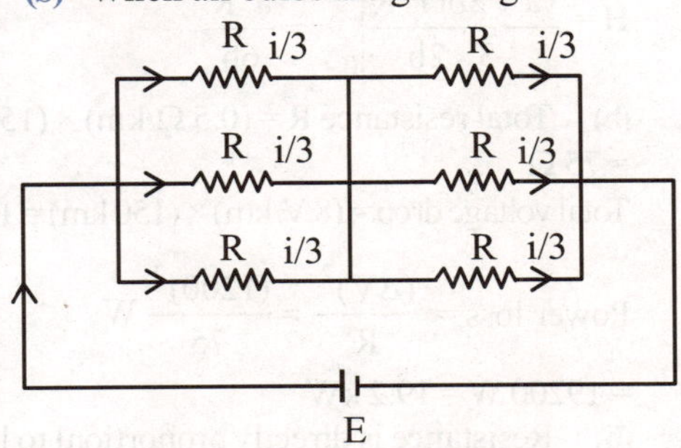

$$R_{eq} = \frac{R}{3} + \frac{R}{3} = \frac{2R}{3}$$

$$\text{Power } (P_i) = \frac{E^2}{R_{eq}} = \frac{3E^2}{2R} \quad \ldots\text{(i)}$$

When two from section A and one from section B are glowing, then

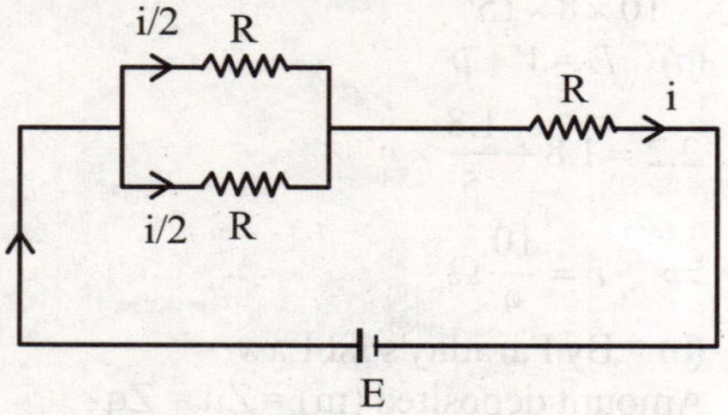

$$R_{eq} = \frac{R}{2} + R = \frac{3R}{2}$$

$$\text{Power } (P_f) = \frac{2E^2}{3R} \quad \ldots(ii)$$

Dividing equation (i) by (ii) we get

$$\frac{P_i}{P_f} = \frac{3E^2 3R}{2R\ 2E^2} = 9:4$$

71. **(a)** **Given:** Charge $Q = at - bt^2$

$\therefore$ Current $i = \frac{\partial Q}{\partial t} = a - 2bt$

$\{\text{for } i = 0 \Rightarrow t = \frac{a}{2b}\}$

From joule's law of heating, heat produced

$dH = i^2 R dt$

$$H = \int_0^{a/2b} (a - 2bt)^2 R dt$$

$$H = \left.\frac{(a-2bt)^3 R}{-3 \times 2b}\right|_0^{\frac{a}{2b}} = \frac{a^3 R}{6b}$$

72. **(b)** Total resistance $R = (0.5\ \Omega/\text{km}) \times (150\ \text{km}) = 75\ \Omega$

Total voltage drop $= (8\ \text{V/km}) \times (150\ \text{km}) = 1200\ \text{V}$

$$\text{Power loss} = \frac{(\Delta V)^2}{R} = \frac{(1200)^2}{75}\ \text{W}$$

$= 19200\ \text{W} = 19.2\ \text{kW}$

73. **(b)** Resistance is directly proportionl to length of the wire. As length is doubled so mass is doubled and resistance is doubled.

We have

$$\frac{(10E)^2}{R} t = mS\Delta T, \text{ Now } \frac{(nE)^2 t}{2R} = (2m)S\Delta T$$

$$\Rightarrow \frac{n^2 E^2 t}{2R} = 2\frac{10^2 E^2 t}{R}$$

$\Rightarrow n = 20$

74. **(c)** Resistance of bulb is constant

$$P = \frac{V^2}{R} \Rightarrow \frac{\Delta p}{p} = \frac{2\Delta V}{V} + \frac{\Delta R}{R}$$

$$\frac{\Delta p}{p} = 2 \times 2.5 + 0 = 5\%$$

75. **(c)** The power dissipated in the circuit.

$$P = \frac{V^2}{R_{eq}} \quad \ldots(i)$$

$V = 10$ volt

$$\frac{1}{R_{eq}} = \frac{1}{R} + \frac{1}{5} = \frac{5+R}{5R}$$

$$R_{eq} = \left(\frac{5R}{5+R}\right)$$

$P = 30$ W

Substituting the values in equation (i)

$$30 = \frac{(10)^2}{\left(\frac{5R}{5+R}\right)}$$

$$\frac{15R}{5+R} = 10$$

$15R = 50 + 10R$

$5R = 50$

$R = 10\ \Omega$

NOTES

For a given voltage V, if resistance is changed from R to $\left(\frac{R}{n}\right)$. Power consumed changes from P to nP.

$$P = \frac{V^2}{R}$$

when $R' = \frac{R}{n}$

then $P' = \frac{V^2}{R/n} = \frac{nV^2}{R} = nP$

76. **(c)** We have, $P = \frac{V^2}{R}$

$$\Rightarrow 36 = \frac{V^2}{9}$$

$\Rightarrow V = 18V$

Current passing through the 9Ω resistor is

$$i_1 = \frac{V}{R} = \frac{18}{9} = 2A$$

The 9Ω and 6Ω resistors are in parallel, therefore

$$i_1 = \frac{6}{9+6} \times i$$

where i is the current delivered by the battery.

$$\therefore i = \frac{2 \times 15}{6} = 5A$$

Thus, potential difference across 2Ω resistor is

$V = iR$

$= 5 \times 2$

$= 10V$

77. **(d)** Clearly, 2Ω, 4Ω and (1+5)Ω resistors are in parallel. Hence, potential difference is same across each of them.

$\therefore I_1 \times 2 = I_2 \times 4 = I_3 \times 6$

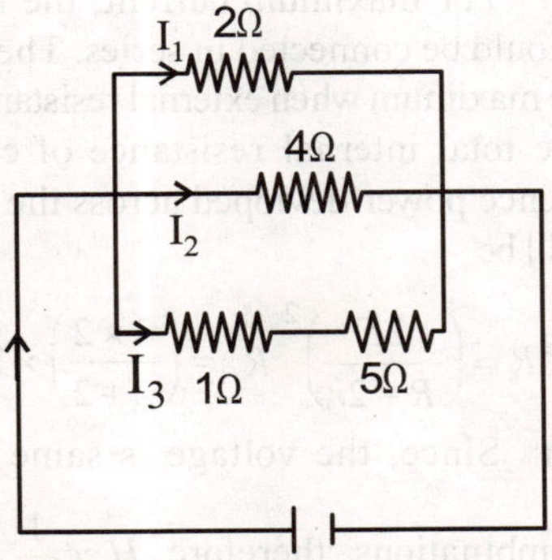

Given $I_1 = 3A$ $\quad \therefore I_1 \times 2 = I_3 \times 6$

Given $I_1 = 3A$.

$\therefore I_1 \times 2 = I_3 \times 6$ provides

$$I_3 = \frac{I_1 \times 2}{6} = \frac{3 \times 2}{6} = 1A.$$

Now, the potential across the 5Ω resistor is

$V = I_3 \times 5 = 1 \times 5 = 5V.$

$\therefore$ the power dissipated in the 5Ω resistor

$$P = \frac{V^2}{R} = \frac{5^2}{5} = 5 \text{watt}.$$

78. **(b)** Power dissipiated = P

$$= \frac{V^2}{R} = \frac{(18)^2}{6} = 54 \text{ W}$$

79. **(c)** Power = $V.I = I^2R$

$$i_2 = \sqrt{\frac{\text{Power}}{R}} = \sqrt{\frac{2}{8}} = \sqrt{\frac{1}{4}} = \frac{1}{2} \text{A}$$

Potential over 8Ω = $Ri_2 = 8 \times \frac{1}{2} = 4V$

This is the potential over parallel branch. So,

$$i_1 = \frac{4}{4} = 1\text{ A}$$

Power of 3Ω = $i_1^2 R = 1 \times 1 \times 3 = 3W$

80. **(a)** $R = \frac{P}{I^2} = \frac{1}{25} = 0.04\Omega$

81. **(a)** $\frac{1}{P_{eq}} = \frac{1}{P_1} + \frac{1}{P_2} + \frac{1}{P_3}$ or $\frac{1}{P_{eq}} = \frac{3}{60}$

$\Rightarrow P_{eq} = 20$ watt.

82. **(d)** $P = \frac{V^2}{R} \Rightarrow R = \frac{V^2}{P} = \frac{(220)^2}{60} = \frac{4(110)^2}{60}$

$$R' = \frac{(110)^2}{60} = \frac{R}{4}$$

83. **(b)** Time $= \frac{10 \times 40}{10 + 40} = \frac{400}{50} = 8 \text{ min}$

84. **(a)** Power $\propto \frac{1}{\text{Resistance}}$

In series combination, resistance doubles. Hence, power will be halved.
In parallel combination, resistance halves. Hence, power will be doubled.

85. **(c)** **Fuse wire :** It is used in a circuit to control the maximum current flowing in circuit. It is a thin wire having **high resistance** and is made up of a material with **low melting point**.

86. **(a)** As for an electric appliance $R = (V_s^2 / W)$, so for same specified voltage V_s

$$\frac{R_{25}}{R_{100}} = \frac{100}{25} = 4$$

i.e, $R_{25} = 4R$ with $R_{100} = R$

Now in series potential divides in proportion to resistance.

So, $V_1 = \frac{R_1}{(R_1 + R_2)} V$

i.e., $V_{25} = \frac{4}{5} \times 440 = 352V$

and $V_2 = \frac{R_2}{(R_1 + R_2)} V$

i.e., $V_{100} = \frac{1}{5} \times 440 = 88V$

From this, it is clear that voltage across 100 W bulb (= 88 V) is lesser than specified (220 V) while across 25 W bulb (= 352 V) is greater than specified (220 V), so, 25 W bulb will fuse.

87. **(b)** Power is maximum when $r = R, R = r = 0.5\Omega$.

> **NOTES**
>
> If a cell of internal resistance r and emf E is connected to an external resistance of radius R, then current given by cell, $I = \frac{E}{R+r}$
>
> Power dissipated in external resistance (load)
>
> $P = VI = (IR)\, I = I^2R$
>
> $\Rightarrow \quad P = \left(\frac{E}{R+r}\right)^2 R$
>
> when $R = r$
>
> Power, $P = \frac{E^2}{4r}$ **(Maximum)**

88. **(b)** $R_1 = \frac{V^2}{P_1} = \frac{200 \times 200}{40} = 1000\,\Omega$

$R_2 = \frac{V^2}{P_2} = \frac{200 \times 200}{100} = 400\,\Omega$

$\therefore R_1$ (for 40W) $> R_2$ (for 100 W)

89. **(d)** In series, Equivalent resistance = $3R$

Power $= \frac{V^2}{3R} \Rightarrow 10 = \frac{V^2}{3R} \Rightarrow V^2 = 30R$

In parallel, $\frac{1}{R'} = \frac{1}{R} + \frac{1}{R} + \frac{1}{R} = \frac{3}{R}$

$\therefore$ Equivalent resistance $R' = \frac{R}{3}$

$\therefore$ Power $= \frac{V^2}{R'} = \frac{30R}{R/3} = 90$ W

> **NOTES**
>
> If voltage V remains same, then power consumed by a n equal resistors in parallel is n^2 times that of power consumed in series.

90. **(c)** Since $H \propto I^2$, doubling the current will produce 4 times heat. Hence, the rise in temperature will also be 4 times i.e., rise in temperature = 4 × 5 = 20ºC.

91. **(d)** Power = 100 W, Voltage = 200 V

Resistance of bulb

$= \frac{V^2}{P} = \frac{200 \times 200}{100} = 400\,\Omega$

When bulb is applied across 160 V,

Current in bulb $= \frac{160}{400} A$

Power consumption $= VI = 160 \times \frac{160}{400} = 64$W

92. **(d)** Power of heating coil = 100 W and voltage (V) = 220 volts. When the heating coil is cut into two equal parts and these parts are joined in parallel, then the resistance of the coil is reduced to one-fourth of the previous value. Therefore energy liberated per second becomes 4 times i.e., 4 × 100 = 400 J.

93. **(d)** The energy stored in the capacitor

$= \frac{1}{2}CV^2 = \frac{1}{2} \times 4 \times 10^{-4} \times 400 \times 400 = 0.32\,J$;

This energy will be converted into heat in the resistor.

94. **(b)** For maximum current, the two batteries should be connected in series. The current will be maximum when external resistance is equal to the total internal resistance of cells i.e. 2Ω. Hence power devloped across the resistacne R will be

$I^2R = \left(\frac{2E}{R+2r}\right)^2 R = \left(\frac{2 \times 2}{2+2}\right) \times 2 = 2W$

95. **(b)** Since, the voltage is same for the two combinations, therefore $H \propto \frac{1}{R}$. Hence, the combination of 39 bulbs will glow more.

96. **(b)** $H = I^2Rt$

or $R = \frac{H}{(I^2t)} = \frac{80}{(2^2 \times 10)} = 2\Omega$

97. **(b)** If ℓ_1 and ℓ_2 are balancing lengths then

$\frac{E_1}{E_2} = \frac{\ell_1}{\ell_2}$

$\therefore \ell_2 = \ell_1\left(\frac{E_2}{E_1}\right) = 36\left(\frac{2.5}{1.5}\right) = 60$cm

98. **(a)** Let R_1 be the resistance of resistance wire.

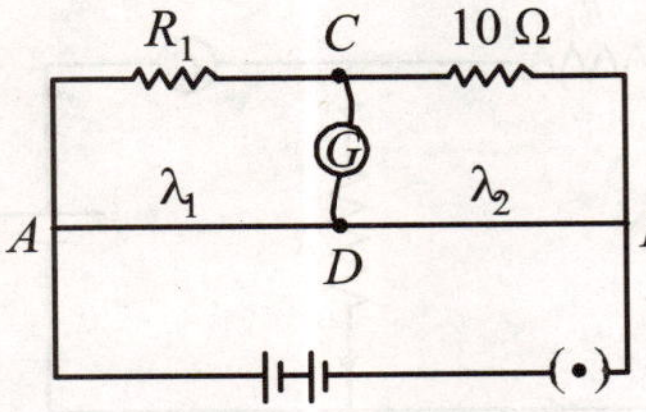

From the balancing condition of metre bridge,

$$\frac{R_1}{10} = \frac{\ell_1}{\ell_2} = \frac{3}{2} \Rightarrow R_1 = \frac{30}{2} = 15\Omega$$

Length of 15Ω resistance wire is 1.5 m.

∴ Length of 1 Ω resistance wire

$$= \frac{1.5}{15} = 0.1\text{ m} = 1.0 \times 10^{-1}\text{ m}$$

99. **(c)** Resistance for ideal voltmeter = ∞

Resistance for ideal ammeter = 0

For I[st] circuit,

$$V_1 = i_1 \times 10 = \frac{10}{10} \times 10 = 10\text{ volt}$$

For IInd circuit,

$$V_2 = i_2 \times 10 = \frac{10}{10} \times 10 = 10\text{ volt}$$

$$V_1 = V_2 \text{ and, } i_1 = i_2 = \frac{10V}{10\Omega} = 1A$$

100. **(a)** Current in first branch $= \frac{2}{50}$ A

Current in second branch $= \frac{2}{50}$ A

ΔV from A to P

$$\Delta V_1 = 2 - \frac{2}{50} \times 20$$

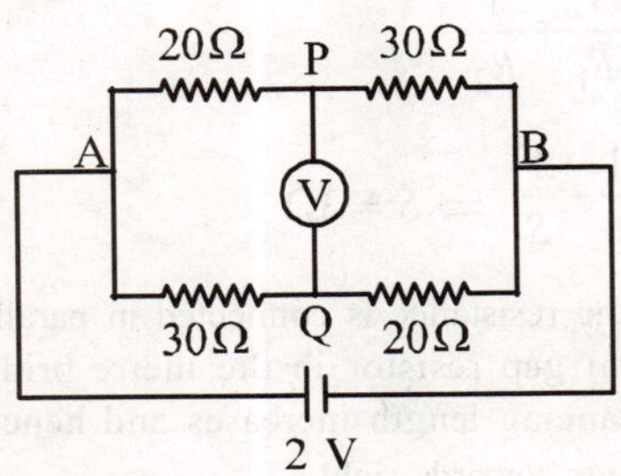

ΔV from A to Q

$$\Delta V_2 = 2 - \frac{2}{50} \times 30$$

⇒ Voltage difference = 0.4 V

101. **(a)** For Balanced Bridge

$$\frac{P}{Q} = \frac{l_1}{l_2}$$

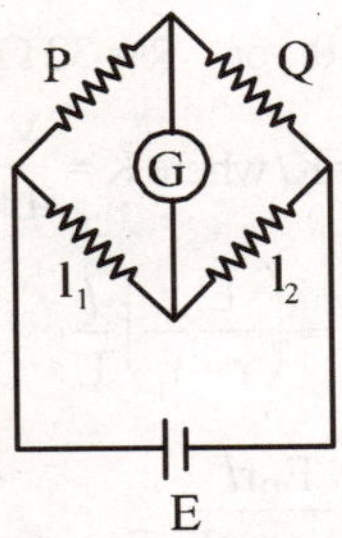

For balanced bridge (finally)

$$\frac{P}{l_1} = \frac{Q}{l_2}$$

$$\frac{P}{Q} = \frac{l_1}{l_2}$$

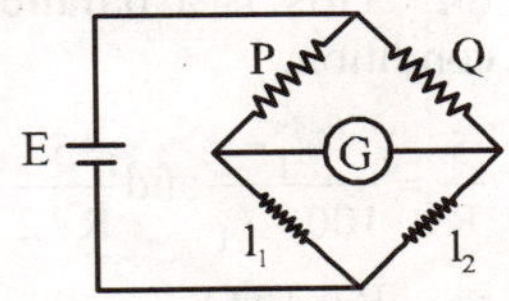

On interchanging galvanometer and battery positions, the balance condition remains unchanged.

102. **(b)** Reading of potentiometer is accurate because during taking reading it does not draw any current from the circuit.

103. **(d)** When two cells are connected in series i.e., $(E_1 + E_2)$ the balance point is at 50 cm. And when two cells are connected in opposite direction i.e., $(E_1 - E_2)$ the balance point is at 10 cm. According to principle of potential

$$\frac{E_1 + E_2}{E_1 - E_2} = \frac{50}{10}$$

$$\Rightarrow \frac{2E_1}{2E_2} = \frac{50+10}{50-10} \Rightarrow \frac{E_1}{E_2} = \frac{3}{2}$$

104. **(d)** Total potential difference across potentiometer wire

$= 10^{-3} \times 400$ volt $= 0.4$ volt

$$\text{potential gradient} = \frac{1\text{mv}}{\text{cm}}$$

$$= 10^{-3}\text{ v/cm} = 10^{-1}\frac{\text{v}}{\text{m}}$$

Let resistance of RΩ connected in series.

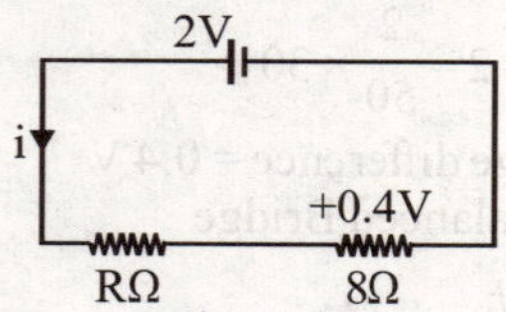

So, $\frac{2}{R+8} = \frac{10^{-1} \times 4}{8} = \frac{1}{20}$

$\Rightarrow \quad R+8=40 \quad \text{or,} \quad R=32\,\Omega$

105. (a) EMF, $E = Kl$ where $K = \frac{V}{L}$ potential gradient

$$K = \frac{V}{L} = \frac{iR}{L} = \left(\frac{E_0 r}{r+r_1}\right)\frac{l}{L}$$

So, $E = Kl = \frac{E_0 r l}{(r+r_1)L}$

106. (b) This is a balanced wheatstone bridge condition,

$$\frac{5}{R} = \frac{\ell_1}{100-\ell_1} \text{ and } \frac{5}{R/2} = \frac{1.6\ell_1}{100-1.6\ell_1}$$

$\Rightarrow \quad R = 15\,\Omega$

107. (c) Internal resistance of the cell,

$$r = \left(\frac{E-V}{V}\right)R = \left(\frac{\ell_1-\ell_2}{\ell_2}\right)R$$

$$= \left(\frac{3-2.85}{2.85}\right)\times(9.5)\,\Omega = 0.5\,\Omega$$

108. (a) Given : V = 7 V

r = 5Ω

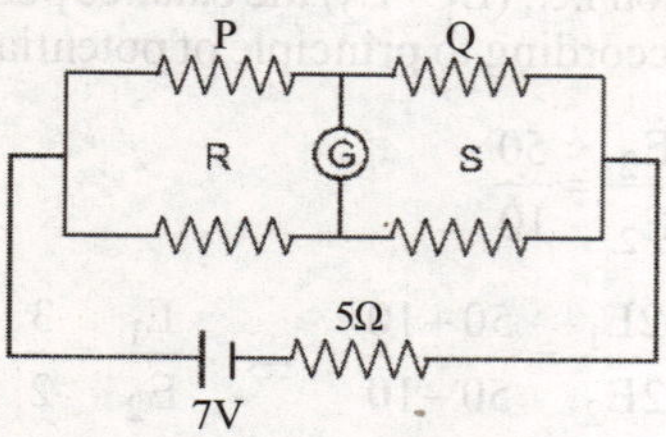

$$R_{eq} = \frac{40\times120}{40+120}\,\Omega$$

$$I = \frac{V}{R} = \frac{7}{5+\frac{40\times120}{40+120}}$$

$$= \frac{7}{5+30} = \frac{1}{5} = 0.2\text{ A.}$$

109. (b)

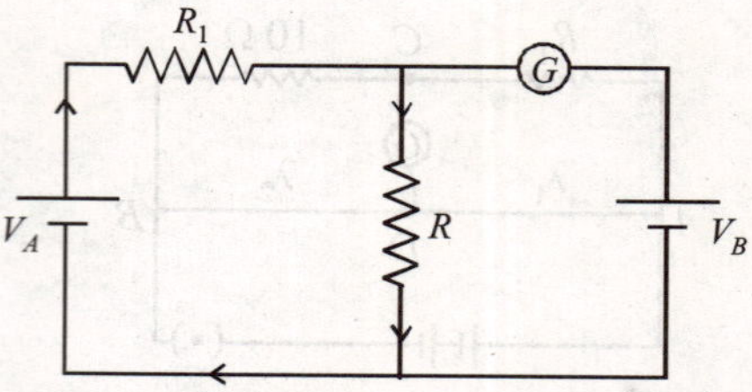

Since deflection in galvanometer is zero so current will flow as shown in the above diagram.

current $I = \frac{V_A}{R_1+R} = \frac{12}{500+100} = \frac{12}{600}$

So $\quad V_B = IR = \frac{12}{600}\times100 = 2V$

110. (b) (i) When key between the terminals 1 and 2 is plugged in,

P.D. across $R = IR = kl_1$

$\Rightarrow R = kl_1$ as $I = 1A$

(ii) When key between terminals 1 and 3 is plugged in,

P.D. across $(X+R) = I(X+R) = kl_2$

$\Rightarrow \quad X + R = kl_2$

$\therefore \quad X = k(l_2 - l_1)$

$\therefore \quad R = kl_1$ and $X = k(l_2 - l_1)$

111. (a) A balanced wheatstone bridge simply requires

$$\frac{P}{Q} = \frac{R}{S} \Rightarrow \frac{2}{2} = \frac{2}{S}$$

Therefore, S should be 2Ω.

A resistance of 6Ω is connected in parallel.

In parallel combination,

$$\frac{1}{R} = \frac{1}{R_1} + \frac{1}{R_2}$$

$$\frac{1}{2} = \frac{1}{6} + \frac{1}{S} \quad \Rightarrow S = 3\Omega$$

NOTES If the resistance is connected in parallel to the right gap resistor in the metre bridge, then balancing length increases and hence jockey moves towards right.

112. (c) Current will flow from B to A

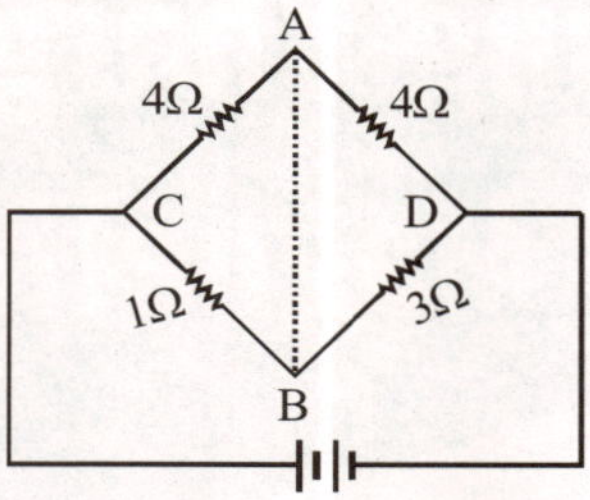

Potential drop over the resistance CA will be more due to higher value of resistance. So potential at A will be less as compared with at B. Hence, current will flow from B to A.

113. (d) It is a balanced Wheatstone bridge. Hence resistance 4Ω can be eliminated.

$$\therefore R_{eq} = \frac{6 \times 9}{6+9} = \frac{18}{5}$$

$$\therefore i = \frac{V}{R_{eq}} = \frac{5V}{18}$$

114. (d) A balanced Whestone's bridge exists between A & B.

$\therefore R_{eq} = R$

Current through circuit = V/R

Current through AFCEB = V/2R

115. (d) Since, Wheatstone's bridge is balanced, then resistance of galvanometer will be uneffective.

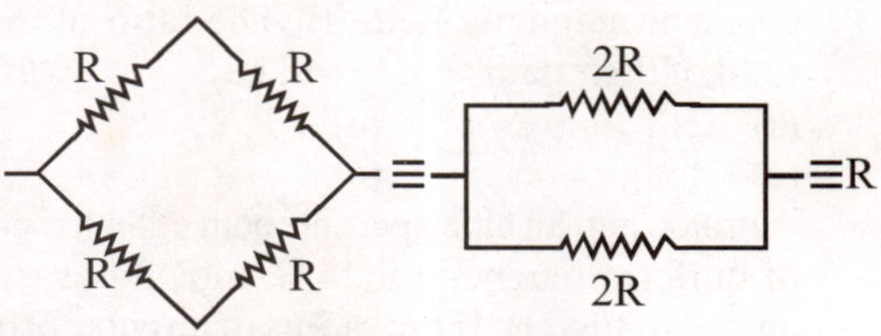

NOTES: Wheat stone bridge is most sensitive if all the arms of bridge have equal resistances.

116. (a) Potential gradient = Potential fall per unit length.

In this case resistance of unit length.

$$R = \frac{\rho l}{A} = \frac{10^{-7} \times 1}{10^{-6}} = 10^{-1}\Omega$$

Potential fall across R is

$$V = I.R = 0.1 \times 10^{-1} = 0.01 \text{ volt/m.}$$

$$= 10^{-2} \text{ volt / m}$$

117. (a) Potentiometer measures potential current more accurately because it measure potential in open circuit and hence error in potential due to internal resistance is removed.

118. (a) It is a balanced wheatstone bridge $\left(\because \frac{3}{4} = \frac{6}{8}\right)$, so the 7Ω resistance is ineffective.

Equivalent resistance of 3Ω and 4Ω = 3 + 4 = 7Ω (series)

Equivalent resistance of 6Ω and 8Ω = 6 + 8 = 14Ω (series)

Equivalent resistance of 7Ω and 14Ω (parallel)

$$= \frac{7 \times 14}{7+14} = \frac{14}{3}\Omega$$

119. (a) Let unknown resistance be X. Then condition of Wheatstone's bridge gives $\frac{X}{R} = \frac{20r}{80r}$, where r is resistance of wire per cm.

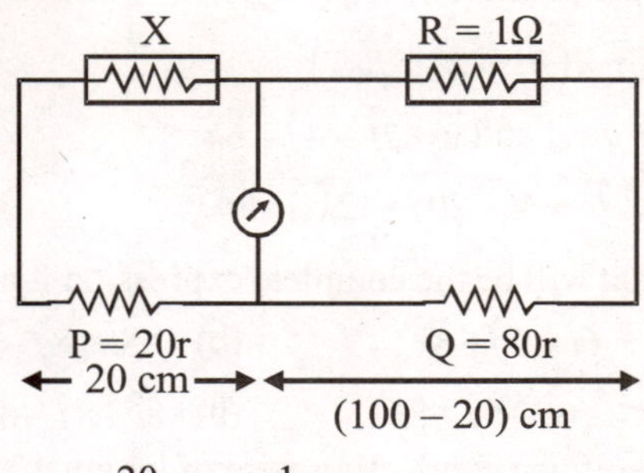

$$\therefore X = \frac{20}{80} \times R = \frac{1}{4} \times 1 = 0.25\Omega$$

18 Moving Charges and Magnetism

Trend Analysis with Important Topics & Sub-Topics

Topic Name	Sub-Topic	2021 QNS.	2021 LOD	2020 QNS.	2020 LOD	2019 QNS.	2019 LOD	2018 QNS.	2018 LOD	2017 QNS.	2017 LOD	2016 QNS.	2016 LOD
Motion of Charged Particle in Magnetic Field & Moment	Radius of Circular path					1	A						
	Lorentz Force Magnetic	2	A										
	Dipole Moment	1	A										
Magnetic Field, Biot-Savart's Law & Ampere's Circuital Law	B_{centre} of Solenoid			1	A								
	B due to Cylindrical Wire Magnetic Force & Field	1	A			1	E					1	A
Force & Torque on a Current Carrying Conductor	Torque on a Current Carrying Loop									1	E		
	Force on a Current Carrying Conductor							1	E	1	E	1	E
Galvanometer and Its Conversion into Ammeter & Voltmeter	Current & Voltage Sensitivity of Galvanometer							1	E				

LOD - Level of Difficulty E - Easy A - Average D - Difficult Qns - No. of Questions

Topic 1 Motion of Charged Particle in Magnetic Field & Moment

1. In the product $\vec{F} = q(\vec{v} \times \vec{B})$
$= q\vec{v} \times (B\hat{i} + B\hat{j} + B_0\hat{k})$
For $q = 1$ and $\vec{v} = 2\hat{i} + 4\hat{j} + 6\hat{k}$
and $\vec{F} = 4\hat{i} - 20\hat{j} + 12\hat{k}$ *[2021]*
What will be the complete expression for $\vec{B}$
(a) $6\hat{i} + 6\hat{j} - 8\hat{k}$ (b) $-8\hat{i} - 8\hat{j} - 6\hat{k}$
(c) $-6\hat{i} - 6\hat{j} - 8\hat{k}$ (d) $8\hat{i} + 8\hat{j} - 6\hat{k}$

2. A uniform conducting wire of length $12a$ and resistance 'R' is wound up as a current carrying coil in the shape of, *[2021]*
(i) an equilateral triangle of side 'a'.
(ii) a square of side 'a'.
The magnetic dipole moments of the coil in each case respectively are
(a) $4\,Ia^2$ and $3\,Ia^2$
(b) $\sqrt{3}\,Ia^2$ and $3\,Ia^2$
(c) $3\,Ia^2$ and Ia^2
(d) $3\,Ia^2$ and $4\,Ia^2$

3. Ionized hydrogen atoms and α-particles with same momenta enters perpendicular to a constant magnetic field, B. The ratio of their radii of their paths $r_H : r_\alpha$ will be : *[2019]*
(a) 2 : 1 (b) 1 : 2
(c) 4 : 1 (d) 1 : 4

4. A proton and an alpha particle both enter a region of uniform magnetic field B, moving at right angles to field B. If the radius of circular orbits for both the particles is equal and the kinetic energy acquired by proton is 1 MeV the energy acquired by the alpha particle will be: *[2015 RS]*
(a) 0.5 MeV (b) 1.5 MeV
(c) 1 MeV (d) 4 MeV

5. An alternating electric field, of frequency v, is applied across the dees (radius = R) of a cyclotron that is being used to accelerate protons (mass = m). The operating magnetic field (B) used in the cyclotron and the kinetic energy (K) of the proton beam, produced by it, are given by : *[2012]*

(a) $B = \frac{m\nu}{e}$ and $K = 2m\pi^2\nu^2R^2$

(b) $B = \frac{2\pi m\nu}{e}$ and $K = m^2\pi\nu R^2$

(c) $B = \frac{2\pi m\nu}{e}$ and $K = 2m\pi^2\nu^2R^2$

(d) $B = \frac{m\nu}{e}$ and $K = m^2\pi\nu R^2$

6. An α-particle moves in a circular path of radius 0.83 cm in the presence of a magnetic field of 0.25 Wb/m^2. The wavelength associated with the particle will be : *[2012]*

(a) 1 Å (b) 0.1 Å
(c) 10 Å (d) 0.01 Å

7. A proton carrying 1 MeV kinetic energy is moving in a circular path of radius R in uniform magnetic field. What should be the energy of an α-particle to describe a circle of same radius in the same field? *[2012M]*

(a) 2 MeV (b) 1 MeV
(c) 0.5 MeV (d) 4 MeV

8. A uniform electric field and uniform magnetic field are acting along the same direction in a certain region. If an electron is projected in the region such that its velocity is pointed along the direction of fields, then the electron *[2011]*

(a) will turn towards right of direction of motion
(b) speed will decrease
(c) speed will increase
(d) will turn towards left direction of motion

9. Two identical bar magnets are fixed with their centres at a distance d apart. A stationary charge Q is placed at P in between the gap of the two magnets at a distance D from the centre O as shown in the Figure *[2010]*

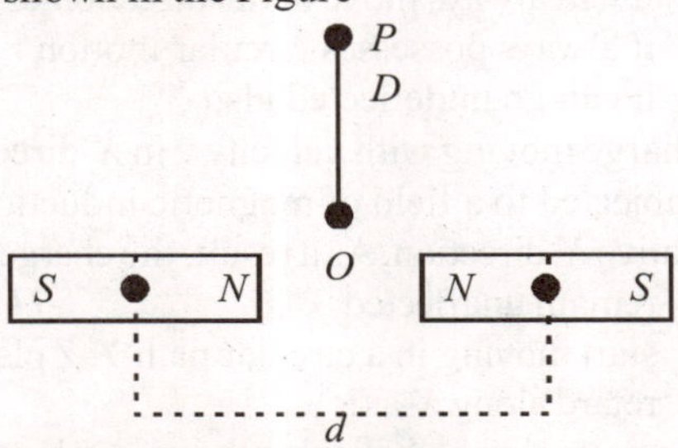

The force on the charge Q is

(a) directed perpendicular to the plane of paper
(b) zero
(c) directed along OP
(d) directed along PO

10. The magnetic force acting on a charged particle of charge $-2\ \mu C$ in a magnetic field of 2T acting in y direction, when the particle velocity is $(2\hat{i} + 3\hat{j}) \times 10^6\ ms^{-1}$, is *[2009]*

(a) 4 N in z direction
(b) 8 N in y direction
(c) 8 N in z direction
(d) 8 N in $-z$ direction

11. Under the influence of a uniform magnetic field, a charged particle moves with constant speed v in a circle of radius R. The time period of rotation of the particle: *[2009]*

(a) depends on R and not on v
(b) is independent of both v and R
(c) depends on both v and R
(d) depends on v and not on R

12. A particle of mass m, charge Q and kinetic energy T enters a transverse uniform magnetic field of induction $\vec{B}$. After 3 seconds, the kinetic energy of the particle will be: *[2008]*

(a) 3T (b) 2T
(c) T (d) 4T

13. A beam of electron passes undeflected through mutually perpendicular electric and magnetic fields. If the electric field is switched off, and the same magnetic field is maintained, the electrons move *[2007]*

(a) in a circular orbit
(b) along a parabolic path
(c) along a straight line
(d) in an elliptical orbit.

14. In a mass spectrometer used for measuring the masses of ions, the ions are initially accelerated by an electric potential V and then made to describe semicircular path of radius R using a magnetic field B. If V and B are kept constant, the ratio $\left(\frac{\text{charge on the ion}}{\text{mass of the ion}}\right)$ will be proportional to *[2007]*

(a) $1/R^2$ (b) R^2
(c) R (d) 1/R

15. A charged paritcle (charge q) is moving in a circle of radius R with uniform speed v. The associated magnetic moment μ is given by *[2007]*

(a) qvR^2 (b) $qvR^2/2$
(c) qvR (d) $qvR/2$

16. Under the influence of a uniform magnetic field a charged particle is moving in a circle of radius

R with constant speed v. The time period of the motion *[2007]*

(a) depends on both R and v
(b) is independent of both R and v
(c) depends on R and not on v
(d) depends on v and not on R

17. An electron moves in a circular orbit with a uniform speed v. It produces a magnetic field B at the centre of the circle. The radius of the circle is proportional to *[2005]*

(a) $\sqrt{\frac{B}{v}}$ (b) $\frac{B}{v}$
(c) $\sqrt{\frac{v}{B}}$ (d) $\frac{v}{B}$

18. A charged particle moves through a magnetic field in a direction perpendicular to it. Then the

(a) velocity remains unchanged *[2003]*
(b) speed of the particle remains unchanged
(c) direction of the particle remains unchanged
(d) acceleration remains unchanged

19. In a certain region of space electric field $\vec{E}$ and magnetic field $\vec{B}$ are perpendicular to each other and an electron enters in region perpendicular to the direction of $\vec{B}$ and $\vec{E}$ both and moves undeflected, then velocity of electron is *[2001]*

(a) $\frac{|\vec{E}|}{|\vec{B}|}$ (b) $\vec{E}\times\vec{B}$
(c) $\frac{|\vec{B}|}{|\vec{E}|}$ (d) $\vec{E}\cdot\vec{B}$

20. A charged particle of charge q and mass m enters perpendicularly in a magnetic field $\vec{B}$. Kinetic energy of the particle is E; then frequency of rotation is *[2001]*

(a) $\frac{qB}{m\pi}$ (b) $\frac{qB}{2\pi m}$
(c) $\frac{qBE}{2\pi m}$ (d) $\frac{qB}{2\pi E}$

21. A proton moving with a velocity 3×10^5 m/s enters a magnetic field of 0.3 tesla at an angle of 30° with the field. The radius of curvature of its path will be (e/m for proton = 10^8 C/kg) *[2000]*

(a) 2 cm (b) 0.5 cm
(c) 0.02 cm (d) 1.25 cm

22. When a proton is accelerated through 1 V, then its kinetic energy will be *[1999]*

(a) 1840 eV (b) 13.6 eV
(c) 1 eV (d) 0.54 eV

23. A positively charged particle moving due east enters a region of uniform magnetic field directed vertically upwards. The particle will

(a) continue to move due east *[1997]*
(b) move in a circular orbit with its speed unchanged
(c) move in a circular orbit with its speed increased
(d) gets deflected vertically upwards.

24. A 10 eV electron is circulating in a plane at right angles to a uniform field at magnetic induction 10^{-4} Wb/m² (= 1.0 gauss). The orbital radius of the electron is *[1996]*

(a) 12 cm (b) 16 cm
(c) 11 cm (d) 18 cm

25. A beam of electrons is moving with constant velocity in a region having simultaneous perpendicular electric and magnetic fields of strength 20 Vm^{-1} and 0.5 T respectively at right angles to the direction of motion of the electrons. Then the velocity of electrons must be *[1996]*

(a) 8 m/s (b) 20 m/s
(c) 40 m/s (d) $\frac{1}{40}$ m/s

26. An electron enters a region where magnetic field (B) and electric field (E) are mutually perpendicular, then *[1994]*

(a) it will always move in the direction of B
(b) it will always move in the direction of E
(c) it always possesses circular motion
(d) it can go undeflected also

27. A charge moving with velocity v in X-direction is subjected to a field of magnetic induction in negative X-direction. As a result, the charge will

(a) remain unaffected *[1993]*
(b) start moving in a circular path Y–Z plane
(c) retard along X-axis
(d) move along a helical path around X-axis

28. A deuteron of kinetic energy 50 keV is describing a circular orbit of radius 0.5 metre in a plane perpendicular to the magnetic field B. The kinetic energy of the proton that describes a circular orbit of radius 0.5 metre in the same plane with the same B is *[1991]*

(a) 25 keV (b) 50 keV
(c) 200 keV (d) 100 keV

29. A uniform magnetic field acts at right angles to the direction of motion of electron. As a result, the electron moves in a circular path of radius 2cm. If the speed of electron is doubled, then the radius of the circular path will be *[1991]*

(a) 2.0 cm (b) 0.5 cm
(c) 4.0 cm (d) 1.0 cm

30. A current carrying coil is subjected to a uniform magnetic field. The coil will orient so that its plane becomes *[1988]*

(a) inclined at 45° to the magnetic field
(b) inclined at any arbitrary angle to the magnetic field
(c) parallel to the magnetic field
(d) perpendicular to the magnetic field

Topic 2 Magnetic Field, Biot-Savart's Law & Ampere's Circuital Law

31. An infinitely long straight conductor carries a current of 5 *A* as shown. An electron is moving with a speed of 10^5 m/s parallel to the conductor. The perpendicular distance between the electron and the conductor is 20 cm at an instant. Calculate the magnitude of the force experienced by the electron at that instant. *[2021]*

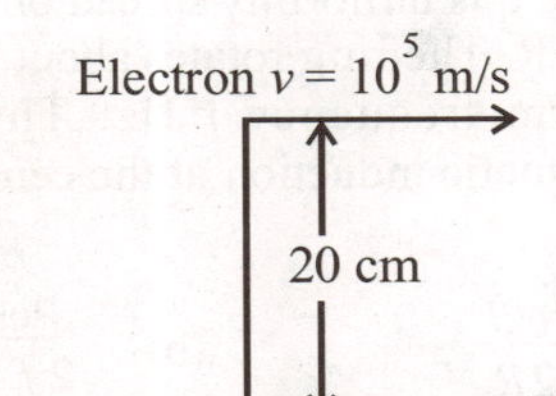

(a) 8×10^{-20} N (b) 4×10^{-20} N
(c) $8\pi \times 10^{-20}$ N (d) $4\pi \times 10^{-20}$ N

32. A thick current carrying cable of radius '*R*' carries current '*I*' uniformly distributed across its cross-section. The variation of magnetic field *B*(*r*) due to the cable with the distance '*r*' from the axis of the cable is represented by *[2021]*

(a) 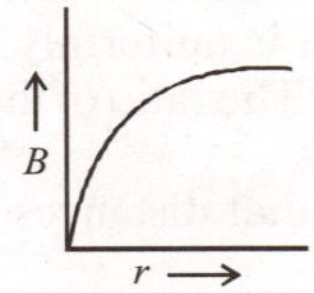

(b) 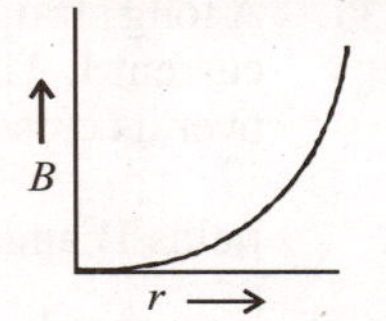

(c) 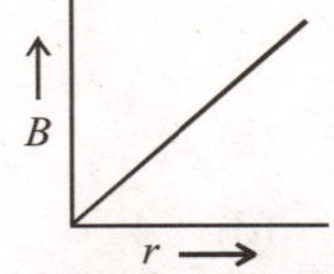

(d) 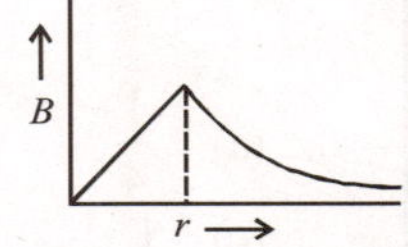

33. A long solenoid of 50 cm length having 100 turns carries a current of 2.5 A. The magnetic field at the centre of the solenoid is : *[2020]*
($\mu_0 = 4\pi \times 10^{-7}$ T m A^{-1})

(a) 3.14×10^{-4} T (b) 6.28×10^{-5} T
(c) 3.14×10^{-5} T (d) 6.28×10^{-4} T

34. A cylindrical conductor of radius R is carrying a constant current. The plot of the magnitude of the magnetic field, B with the distance, *d*, from the centre of the conductor, is correctly represented by the figure : *[2019]*

(a) 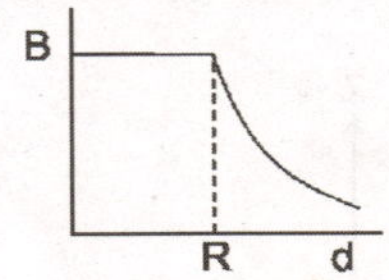

(b) 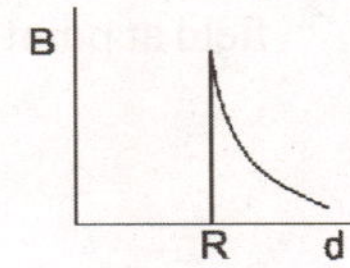

(c) 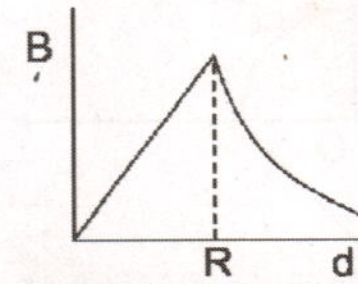

(d) 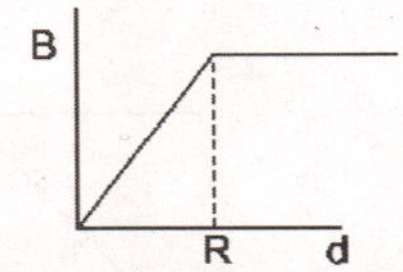

35. Two toroids 1 and 2 have total no. of turns 200 and 100 respectively with average radii 40 cm and 20 cm respectively. If they carry same current *i*, the ratio of the magnetic fields along the two loops is, *[NEET Odisha 2019]*

(a) 1 : 2 (b) 1 : 1
(c) 4 : 1 (d) 2 : 1

36. A straight conductor carrying current *i* splits into two parts as shown in the figure. The radius of the circular loop is *R*. The total magnetic field at the centre *P* of the loop is, *[NEET Odisha 2019]*

(a) $\frac{\mu_0 i}{2R}$, inward
(b) Zero
(c) $3\mu_0 i/32R$, outward
(d) $3\mu_0 i/32R$, inward

37. A long straight wire of radius a carries a steady current I. The current is uniformly distributed over its cross-section. The ratio of the magnetic fields B and B′, at radial distances $\frac{a}{2}$ and 2a respectively, from the axis of the wire is : [2016]

(a) $\frac{1}{4}$ (b) $\frac{1}{2}$

(c) 1 (d) 4

38. An electron moving in a circular orbit of radius r makes n rotations per second. The magnetic field produced at the centre has magnitude: [2015]

(a) Zero (b) $\frac{\mu_0 n^2 e}{r}$

(c) $\frac{\mu_0 ne}{2r}$ (d) $\frac{\mu_0 ne}{2\pi r}$

39. A wire carrying current I has the shape as shown in adjoining figure. Linear parts of the wire are very long and parallel to X-axis while semicircular portion of radius R is lying in Y-Z plane. Magnetic field at point O is : [2015]

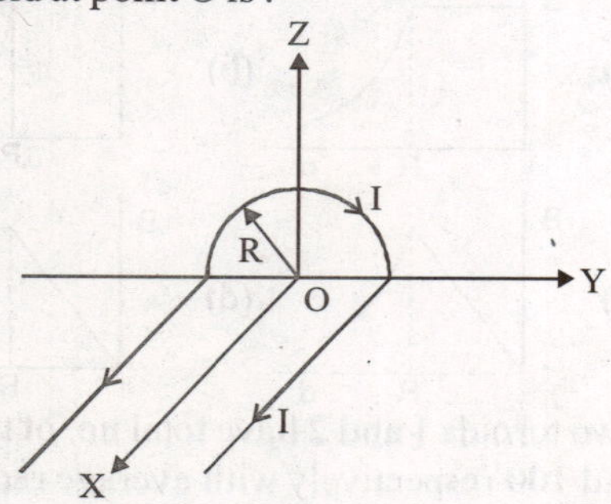

(a) $\vec{B} = -\frac{\mu_0}{4\pi}\frac{I}{R}\left(\mu\hat{i} \times 2\hat{k}\right)$

(b) $\vec{B} = -\frac{\mu_0}{4\pi}\frac{I}{R}\left(\pi\hat{i} + 2\hat{k}\right)$

(c) $\vec{B} = \frac{\mu_0}{4\pi}\frac{I}{R}\left(\pi\hat{i} - 2\hat{k}\right)$

(d) $\vec{B} = \frac{\mu_0}{4\pi}\frac{I}{R}\left(\pi\hat{i} + 2\hat{k}\right)$

40. Two identical long conducting wires AOB and COD are placed at right angle to each other, with one above other such that 'O' is their common point for the two. The wires carry I_1 and I_2 currents respectively. Point 'P' is lying at distance 'd' from 'O' along a direction perpendicular to the plane containing the wires. The magnetic field at the point 'P' will be : [2014]

(a) $\frac{\mu_0}{2\pi d}\left(\frac{I_1}{I_2}\right)$ (b) $\frac{\mu_0}{2\pi d}(I_1 + I_2)$

(c) $\frac{\mu_0}{2\pi d}(I_1^2 - I_2^2)$ (d) $\frac{\mu_0}{2\pi d}(I_1^2 + I_2^2)^{1/2}$

41. When a proton is released from rest in a room, it starts with an initial acceleration a_0 towards west. When it is projected towards north with a speed v_0 it moves with an initial acceleration $3a_0$ towards west. The electric and magnetic fields in the room are respectively [2013]

(a) $\frac{ma_0}{e}$ west, $\frac{2ma_0}{ev_0}$ down

(b) $\frac{ma_0}{e}$ east, $\frac{3ma_0}{ev_0}$ up

(c) $\frac{ma_0}{e}$ east, $\frac{3ma_0}{ev_0}$ down

(d) $\frac{ma_0}{e}$ west, $\frac{2ma_0}{ev_0}$ up

42. Two similar coils of radius R are lying concentrically with their planes at right angles to each other. The currents flowing in them are I and $2I$, respectively. The resultant magnetic field induction at the centre will be: [2012]

(a) $\frac{\sqrt{5}\mu_0 I}{2R}$ (b) $\frac{3\mu_0 I}{2R}$

(c) $\frac{\mu_0 I}{2R}$ (d) $\frac{\mu_0 I}{R}$

43. Charge q is uniformly spread on a thin ring of radius R. The ring rotates about its axis with a uniform frequency f Hz. The magnitude of magnetic induction at the centre of the ring is [2011M, 2010]

(a) $\frac{\mu_0 qf}{2R}$ (b) $\frac{\mu_0 q}{2fR}$

(c) $\frac{\mu_0 q}{2\pi fR}$ (d) $\frac{\mu_0 qf}{2\pi R}$

44. A particle having a mass of 10^{-2} kg carries a charge of 5×10^{-8}C. The particle is given an initial horozontal velocity of 10^5 ms^{-1} in the presence of electric field $\vec{E}$ and magnetic field $\vec{B}$. To keep the particle moving in a horizontal direction, it is necessary that [2010]

(1) $\vec{B}$ should be perpendicular to the direction of velocity and $\vec{E}$ should be along the direction of velocity.

(2) Both $\vec{B}$ and $\vec{E}$ should be along the direction of velocity.

(3) Both $\vec{B}$ and $\vec{E}$ are mutually perpendicular and perpendicular to the direction of velocity.

(4) $\vec{B}$ should be along the direction of velocity and $\vec{E}$ should be perpendicular to the direction of velocity.

Which one of the following pairs of statements is possible?

(a) (2) and (4) (b) (1) and (3)
(c) (3) and (4) (d) (2) and (3)

45. A current loop consists of two identical semicircular parts each of radius R, one lying in the x-y plane and the other in x-z plane. If the current in the loop is i., the resultant magnetic field due to the two semicircular parts at their common centre is *[2010]*

(a) $\dfrac{\mu_0 i}{\sqrt{2}R}$ (b) $\dfrac{\mu_0 i}{2\sqrt{2}R}$

(c) $\dfrac{\mu_0 i}{2R}$ (d) $\dfrac{\mu_0 i}{4R}$

46. Two circular coils 1 and 2 are made from the same wire but the radius of the 1st coil is twice that of the 2nd coil. What potential difference in volts should be applied across them so that the magnetic field at their centres is the same *[2006]*

(a) 4 (b) 6
(c) 2 (d) 3

47. A long solenoid carrying a current produces a magnetic field B along its axis. If the current is double and the number of turns per cm is halved, the new value of the magnetic field is *[2003]*

(a) $4B$ (b) $B/2$
(c) B (d) $2B$

48. A wire carries a current. Maintaining the same current it is bent first to form a circular plane coil of one turn which produces a magnetic field B at the centre of the coil. The same length is now bent more sharply to give a double loop of smaller radius. The magnetic field at the centre of the double loop, caused by the same current is *[2002]*

(a) $4B$ (b) $B/4$
(c) $B/2$ (d) $2B$

49. Two long parallel wires P and Q are both perpendicular to the plane of the paper with distance of 5 m between them. If P and Q carry currents of 2.5 amp and 5 amp respectively in the same direction, then the magnetic field at a point half-way between the wires is *[2000]*

(a) $\dfrac{3\mu_0}{2\pi}$ (b) $\dfrac{\mu_0}{\pi}$

(c) $\dfrac{\sqrt{3}\mu_0}{2\pi}$ (d) $\dfrac{\mu_0}{2\pi}$

50. Magnetic field intensity at the centre of a coil of 50 turns, radius 0.5 m and carrying a current of 2 A is *[1999]*

(a) 0.5×10^{-5} T (b) 1.25×10^{-4} T
(c) 3×10^{-5} T (d) 4×10^{-5} T

51. If a long hollow copper pipe carries a current, then magnetic field is produced *[1999]*

(a) inside the pipe only
(b) outside the pipe only
(c) both inside and outside the pipe
(b) no where

52. A coil of one turn is made of a wire of certain length and then from the same length a coil of two turns is made. If the same current is passed in both the cases, then the ratio of the magnetic inductions at their centres will be *[1998]*

(a) 2 : 1 (b) 1 : 4
(c) 4 : 1 (d) 1 : 2

53. The magnetic field ($d\vec{B}$) due to a small element ($d\ell$) at a distance ($\vec{r}$) and element carrying current i is *[1996]*

(a) $d\vec{B} = \dfrac{\mu_0}{4\pi} i \left(\dfrac{d\vec{\ell} \times \vec{r}}{r}\right)$

(b) $d\vec{B} = \dfrac{\mu_0}{4\pi} i^2 \left(\dfrac{d\vec{\ell} \times \vec{r}}{r^2}\right)$

(c) $d\vec{B} = \dfrac{\mu_0}{4\pi} i^2 \left(\dfrac{d\vec{\ell} \times \vec{r}}{r}\right)$

(d) $d\vec{B} = \dfrac{\mu_0}{4\pi} i \left(\dfrac{d\vec{\ell} \times \vec{r}}{r^3}\right)$

54. Two equal electric currents are flowing perpendicular to each other as shown in the figure. *AB* and *CD* are perpendicular to each other and symmetrically placed with respect to the current flow. Where do we expect the resultant magnetic field to be zero? *[1996]*

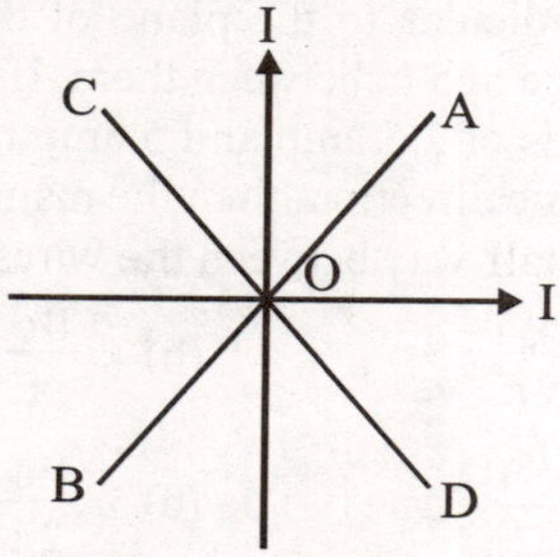

(a) on *AB*
(b) on *CD*
(c) on both *AB* and *CD*
(d) on both *OD* and *BO*

55. A straight wire of diameter 0.5 mm carrying a current of 1 A is replaced by another wire of 1 mm diameter carrying same current. The strength of magnetic field far away is *[1995, 97, 99]*
(a) twice the earlier value
(b) same as the earlier value
(c) one-half of the earlier value
(d) one-quarter of the earlier value

56. At what distance from a long straight wire carrying a current of 12 A will the magnetic field be equal to 3×10^{-5} Wb/m^2? *[1995]*
(a) 8×10^{-2} m (b) 12×10^{-2} m
(c) 18×10^{-2} m (d) 24×10^{-2} m

57. The magnetic field at a distance *r* from a long wire carrying current i is 0.4 tesla. The magnetic field at a distance 2*r* is *[1992]*
(a) 0.2 tesla (b) 0.8 tesla
(c) 0.1 tesla (d) 1.6 tesla

58. The magnetic induction at a point P which is at a distance of 4 cm from a long current carrying wire is 10^{-3} T. The field of induction at a distance 12 cm from the current will be *[1990]*
(a) 3.33×10^{-4} T (b) 1.11×10^{-4} T
(c) 3×10^{-3} T (d) 9×10^{-3} T

59. Energy in a current carrying coil is stored in the form of *[1989]*
(a) electric field
(b) magnetic field
(c) dielectric strength
(d) heat

60. Tesla is the unit of *[1988]*
(a) magnetic flux
(b) magnetic field
(c) magnetic induction
(d) magnetic moment

Topic 3 Force & Torque on a Current Carrying Conductor

61. A metallic rod of mass per unit length 0.5 kg m^{-1} is lying horizontally on a smooth inclined plane which makes an angle of 30° with the horizontal. The rod is not allowed to slide down by flowing a current through it when a magnetic field of induction 0.25 T is acting on it in the vertical direction. The current flowing in the rod to keep it stationary is *[2018]*
(a) 7.14 A (b) 5.98 A
(c) 11.32 A (d) 14.76 A

62. A 250-turn rectangular coil of length 2.1 cm and width 1.25 cm carries a current of 85 μA and subjected to magnetic field of strength 0.85 T. Work done for rotating the coil by 180° against the torque is *[2017]*
(a) 4.55 μJ (b) 2.3 μJ
(c) 1.15 μJ (d) 9.1 μJ

63. An arrangement of three parallel straight wires placed perpendicular to plane of paper carrying same current 'I along the same direction is shown in fig. Magnitude of force per unit length on the middle wire 'B' is given by *[2017]*

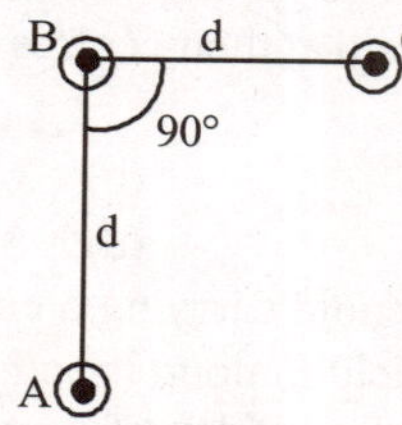

(a) $\frac{2\mu_0 i^2}{\pi d}$ (b) $\frac{\sqrt{2}\mu_0 i^2}{\pi d}$
(c) $\frac{\mu_0 i^2}{\sqrt{2}\pi d}$ (d) $\frac{\mu_0 i^2}{2\pi d}$

64. A square loop ABCD carrying a current i, is placed near and coplanar with a long straight conductor XY carrying a current I, the net force on the loop will be : *[2016]*

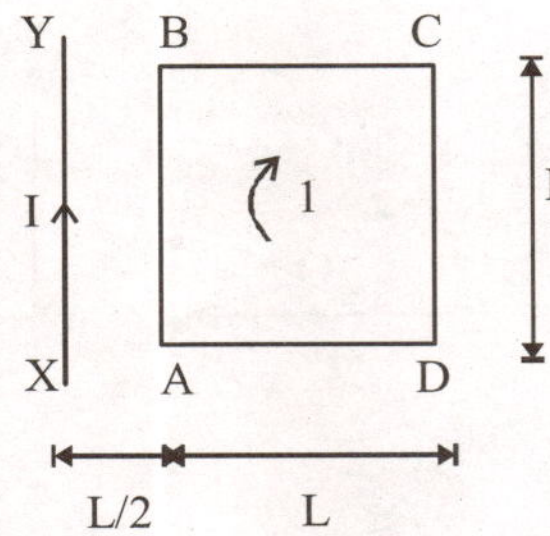

(a) $\frac{2\mu_0 Ii}{3\pi}$ (b) $\frac{\mu_0 Ii}{2\pi}$

(c) $\frac{2\mu_0 IiL}{3\pi}$ (d) $\frac{\mu_0 IiL}{2\pi}$

65. A rectangular coil of length 0.12 m and width 0.1 m having 50 turns of wire is suspended vertically in a uniform magnetic field of strength 0.2 weber/m^2. The coil carries a current of 2A. If the plane of the coil is inclined at an angle of 30° with the direction of the field, the torque required to keep the coil in stable equilibrium will be : *[2015 RS]*

(a) 0.20 Nm (b) 0.24 Nm

(c) 0.12 Nm (d) 0.15 Nm

66. A current loop in a magnetic field *[2013]*

(a) can be in equilibrium in one orientation

(b) can be in equilibrium in two orientations, both the equilibrium states are unstable

(c) can be in equilibrium in two orientations, one stable while the other is unstable

(d) experiences a torque whether the field is uniform or non-uniform in all orientations

67. A long straight wire carries a certain current and produces a magnetic field of 2×10^{-4} $\frac{\text{weber}}{\text{m}^2}$ at a perpendicular distance of 5 cm from the wire. An electron situated at 5 cm from the wire moves with a velocity 10^7 m/s towards the wire along perpendicular to it. The force experienced by the electron will be *[NEET Kar. 2013]*

(charge on electron $= 1.6 \times 10^{-19}$ C)

(a) Zero (b) 3.2 N

(c) 3.2×10^{-16} N (d) 1.6×10^{-16} N

68. A circular coil *ABCD* carrying a current *i* is placed in a uniform magnetic field. If the magnetic force on the segment *AB* is $\vec{F}$, the force on the remaining segment *BCDA* is *[NEET Kar. 2013, 2010]*

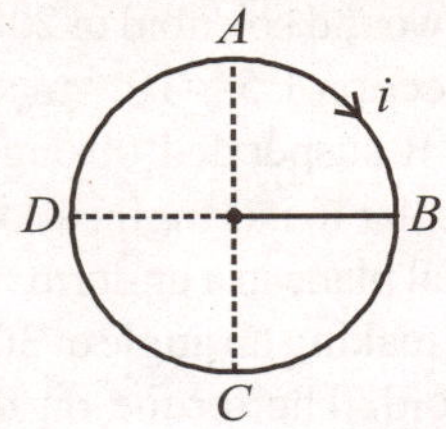

(a) $\vec{F}$ (b) $-\vec{F}$

(c) $3\vec{F}$ (d) $-3\vec{F}$

69. A current carrying loop in the form of a right angle isosceles triangle ABC is placed in a uniform magnetic field acting along AB. If the magnetic force on the arm BC is F, what is the force on the arm AC? *[2011]*

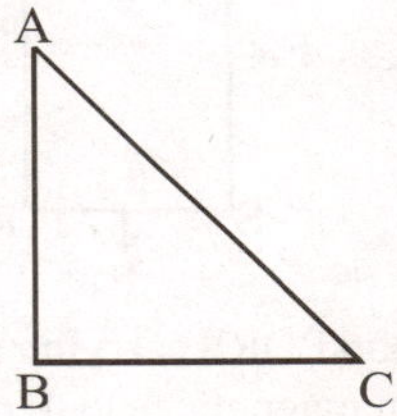

(a) $-\sqrt{2}\,\vec{F}$ (b) $-\vec{F}$

(c) $\vec{F}$ (d) $\sqrt{2}\,\vec{F}$

70. A square loop, carrying a steady current I, is placed in a horizontal plane near a long straight conductor carrying a steady current I_1 at a distance d from the conductor as shown in figure. The loop will experience *[2011M]*

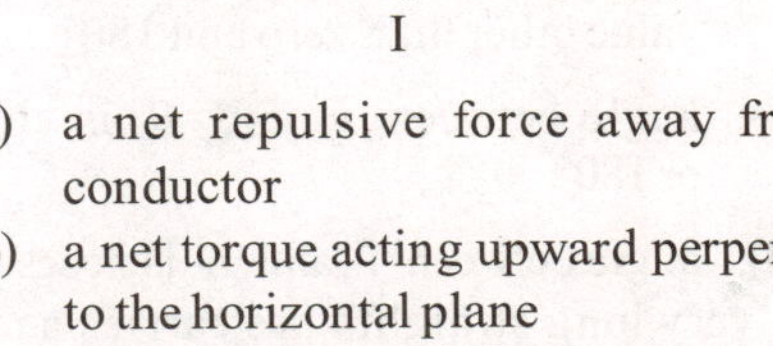

(a) a net repulsive force away from the conductor

(b) a net torque acting upward perpendicular to the horizontal plane

(c) a net torque acting downward normal to the horizontal plane

(d) a net attractive force towards the conductor

71. A closely wound solenoid of 2000 turns and area of cross-section 1.5×10^{-4} m^2 carries a current of 2.0 A. It suspended through its centre and perpendicular to its length, allowing it to turn in a horizontal plane in a uniform magnetic field 5×10^{-2} tesla making an angle of 30° with the axis of the solenoid. The torque on the solenoid will be: *[2010]*

(a) 3×10^{-2} N-m (b) 3×10^{-3} N-m
(c) 1.5×10^{-3} N-m (d) 1.5×10^{-2} N-m

72.

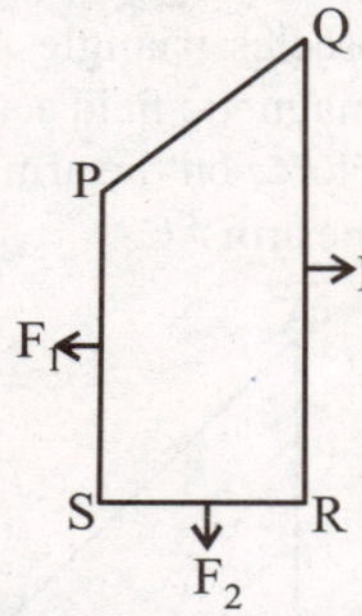

A closed loop PQRS carrying a current is placed in a uniform magnetic field. If the magnetic forces on segments PS, SR, and RQ are F_1, F_2 and F_3 respectively and are in the plane of the paper and along the directions shown, the force on the segment QP is *[2008]*

(a) $F_3 - F_1 - F_2$ (b) $\sqrt{(F_3 - F_1)^2 + F_2^2}$

(c) $\sqrt{(F_3 - F_1)^2 - F_2^2}$ (d) $F_3 - F_1 + F_2$

73. When a charged particle moving with velocity $\vec{v}$ is subjected to a magnetic field of induction $\vec{B}$, the force on it is non-zero. This implies that *[2006]*

(a) angle between $\vec{v}$ and $\vec{B}$ can have any value other than 90°

(b) angle between $\vec{v}$ and $\vec{B}$ can have any value other than zero and 180°

(c) angle between $\vec{v}$ and $\vec{B}$ is either zero or 180°

(d) angle between $\vec{v}$ and $\vec{B}$ is necessarily 90°

74. A very long straight wire carries a current I. At the instant when a charge $+Q$ at point P has velocity $\vec{v}$, as shown, the force on the charge is *[2005]*

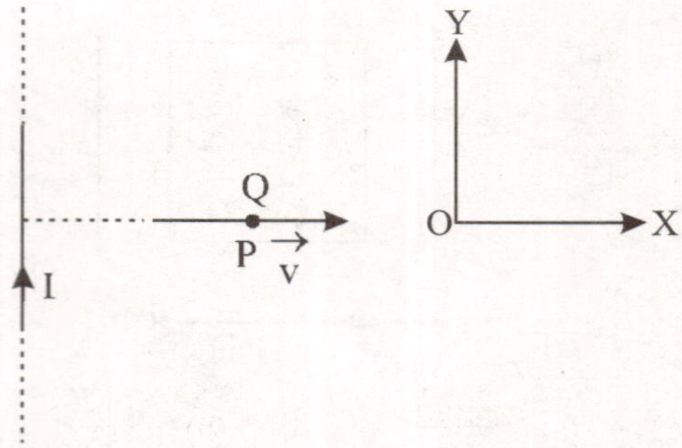

(a) along OY (b) opposite to OY
(c) along OX (d) opposite to OX

75. A particle having charge q moves with a velocity $\vec{v}$ through a region in which both an electric field $\vec{E}$ and a magnetic field $\vec{B}$ are present. The force on the particle is *[2002]*

(a) $q\vec{E} + q(\vec{B} \times \vec{v})$ (b) $q\vec{E}.(\vec{B} \times \vec{v})$

(c) $q\vec{v} + q(\vec{E} \times \vec{B})$ (d) $q\vec{E} + q(\vec{v} \times \vec{B})$

76. Two long parallel wires are at a distance of 1 metre. Both of them carry one ampere of current. The force of attraction per unit length between the two wires is *[1998]*

(a) 2×10^{-7} N/m (b) 2×10^{-8} N/m
(c) 5×10^{-8} N/m (d) 10^{-7} N/m

77. A coil carrying electric current is placed in uniform magnetic field, then *[1993]*

(a) torque is formed
(b) e.m.f is induced
(c) both (a) and (b) are correct
(d) none of the above

78. A straight wire of length 0.5 metre and carrying a current of 1.2 ampere is placed in uniform magnetic field of induction 2 tesla. The magnetic field is perpendicular to the length of the wire. The force on the wire is *[1992]*

(a) 2.4 N (b) 1.2 N
(c) 3.0 N (d) 2.0 N

Topic 4 Galvanometer and Its Conversion into Ammeter & Voltmeter

79. Current sensitivity of a moving coil galvanometer is 5 div/mA and its voltage sensitivity (angular deflection per unit voltage applied) is 20 div/V. The resistance of the galvanometer is *[2018]*

(a) 40 Ω (b) 25 Ω
(c) 500 Ω (d) 250 Ω

80. A circuit contains an ammeter, a battery of 30V and a resistance 40.8Ω all connected in series. If the ammeter has a coil of resistance 480Ω and a shunt of 20Ω, the reading in the ammeter will be: *[2015 RS]*

(a) 0.25 A (b) 2A
(c) 1 A (d) 0.5 A

81. In an ammeter 0.2% of main current passes through the galvanometer. If resistance of galvanometer is G, the resistance of ammeter will be : *[2014]*

(a) $\frac{1}{499}G$ (b) $\frac{499}{500}G$
(c) $\frac{1}{500}G$ (d) $\frac{500}{499}G$

82. A milli voltmeter of 25 milli volt range is to be converted into an ammeter of 25 ampere range. The value (in ohm) of necessary shunt will be : **[2012]**

(a) 0.001 (b) 0.01
(c) 1 (d) 0.05

83. A galvanometer of resistance, G is shunted by a resistance S ohm. To keep the main current in the circuit unchanged, the resistance to be put in series with the galvanometer is *[2011M]*

(a) $\frac{S^2}{(S+G)}$ (b) $\frac{SG}{(S+G)}$
(c) $\frac{G^2}{(S+G)}$ (d) $\frac{G}{(S+G)}$

84. A galvanometer has a coil of resistance 100 ohm and gives a full-scale deflection for 30 *mA* current. It is to work as a voltmeter of 30 volt range, the resistance required to be added will be *[2010]*

(a) 900 Ω (b) 1800 Ω
(c) 500 Ω (d) 1000 Ω

85. A galvanometer having a coil resistance of 60 Ω shows full scale deflection when a current of 1.0 amp passes through it. It can be converted into an ammeter to read currents upto 5.0 amp by *[2009]*

(a) putting in series a resistance of 15Ω
(b) putting in series a resistance of 240Ω
(c) putting in parallel a resistance of 15Ω
(d) putting in parallel a resistance of 240Ω

86. A galvanometer of resistance 50 Ω is connected to battery of 3V along with a resistance of 2950 Ω in series. A full scale deflection of 30 divisions is obtained in the galvanometer. In order to reduce this deflection to 20 divisions, the resistance in series should be *[2008]*

(a) 5050 Ω (b) 5550 Ω
(c) 6050 Ω (d) 4450 Ω

87. The resistance of an ammeter is 13 Ω and its scale is graduated for a current upto 100 amps. After an additional shunt has been connected to this ammeter it becomes possible to measure currents upto 750 amperes by this meter. The value of shunt-resistance is *[2007]*

(a) 2 Ω (b) 0.2 Ω
(c) 2 k Ω (d) 20 Ω

88. A galvanometer acting as a voltmeter will have *[2004, 2002]*

(a) a low resistance in series with its coil.
(b) a high resistance in parallel with its coil
(c) a high resistance in series with its coil
(d) a low resistance in parallel with its coil

89. A galvanometer of 50 ohm resistance has 25 divisions. A current of 4×10^{-4} ampere gives a deflection of one per division. To convert this galvanometer into a voltmeter having a range of 25 volts, it should be connected with a resistance of *[2004, 2002]*

(a) 2450 Ω in series (b) 2500 Ω in series.
(c) 245 Ω in series. (d) 2550 Ω in series.

90. A galvanometer having a resistance of 8 ohms is shunted by a wire of resistance 2 ohms. If the total current is 1 amp, the part of it passing through the shunt will be *[1998]*

(a) 0.25 amp (b) 0.8 amp
(c) 0.2 amp (d) 0.5 amp

91. A galvanometer of resistance 20 Ω gives full scale deflection with a current of 0.004 A. To convert it into an ammeter of range 1 A, the required shunt resistance should be *[1996]*

(a) 0.38Ω (b) 0.21Ω
(c) 0.08Ω (d) 0.05Ω

92. To convert a galvanometer into an ammeter, one needs to connect a *[1992]*

(a) low resistance in parallel
(b) high resistance in parallel
(c) low resistance in series
(d) high resistance in series.

ANSWER KEY																	
1	(c)	12	(c)	23	(b)	34	(c)	45	(b)	56	(a)	67	(c)	78	(b)	89	(a)
2	(b)	13	(a)	24	(c)	35	(b)	46	(None)	57	(a)	68	(b)	79	(d)	90	(b)
3	(a)	14	(a)	25	(c)	36	(b)	47	(c)	58	(a)	69	(b)	80	(d)	91	(c)
4	(c)	15	(d)	26	(d)	37	(c)	48	(a)	59	(b)	70	(d)	81	(c)	92	(a)
5	(c)	16	(b)	27	(a)	38	(c)	49	(d)	60	(b)	71	(d)	82	(a)		
6	(d)	17	(c)	28	(d)	39	(b)	50	(b)	61	(c)	72	(b)	83	(c)		
7	(b)	18	(b)	29	(c)	40	(d)	51	(b)	62	(d)	73	(b)	84	(a)		
8	(b)	19	(a)	30	(d)	41	(a)	52	(b)	63	(c)	74	(a)	85	(c)		
9	(b)	20	(b)	31	(a)	42	(a)	53	(d)	64	(a)	75	(d)	86	(d)		
10	(d)	21	(b)	32	(d)	43	(a)	54	(a)	65	(a)	76	(a)	87	(a)		
11	(b)	22	(c)	33	(d)	44	(d)	55	(b)	66	(c)	77	(a)	88	(c)		

Hints & Solutions

1. (c) Lorentz force, $\vec{F} = q(\vec{v} \times \vec{B})$

$= q\vec{v} \times (B\hat{i} + B\hat{j} + B_0\hat{k})$

$\therefore q = 1,\ \vec{v} = 2\hat{i} + 4\hat{j} + 6\hat{k}$ and

$\vec{F} = 4\hat{i} - 20\hat{j} + 12\hat{k}$ (given)

$\therefore (4\hat{i} - 20\hat{j} + 12\hat{k})$

$= -1 \times [(2\hat{i} + 4\hat{j} + 6\hat{k}) \times (B\hat{i} + B\hat{j} + B_0\hat{k})]$

or $(4\hat{i} - 20\hat{j} + 12\hat{k})$

$= i(4B_0 - 6B) - j(2B_0 - 6B) + \hat{k}(2B - 4B)$

Comparing L.H.S and R.H.S, we get

$4B_0 - 6B = 4 \Rightarrow 2B_0 - 3B = 2$ ...(i)

$-(2B_0 - 6B) = -20 \Rightarrow B_0 - 3B = 10$ ...(ii)

$2B - 4B = 12 \Rightarrow B = -6$ ...(iii)

From eq. (ii) and (iii)

$B = -6$ and $B_0 = -8$

$\therefore$ Complete expression for $\vec{B} = -6\hat{i} - 6\hat{j} - 8\hat{k}$

> NOTES
>
> **NOTE :** In a magnetic field a moving charged particle experiences magnetic force
>
> $\vec{F}_m = q(\vec{v} \times \vec{B})$
>
> $\vec{F}_m = q(\vec{v} \times \vec{B})$ is called Lorentz force.

2. (b) Current in the loop $I = \frac{V}{R}$ equal for both loops.

Magnetic moment of equilateral triangle of side a

$$M_T = \left(\frac{12a}{3a}\right).I.\frac{\sqrt{3}}{4}a^2 = \sqrt{3}Ia^2 \quad [\because M = NIA]$$

Similarly magnetic moment of square of side a

$$= \left(\frac{12a}{4a}\right).Ia^2 \quad [\because M = NIA]$$

$M_S = 3Ia^2$

> NOTES
>
> **NOTE :** A current carrying coil behaves as a bar magnet and its magnetic moment M = NiA, N = no, of turns, i = current through the coil, A = area of the coil. For a given perimeter circular shape has maximum area hence maximum magnetic moment.

3. (a) Radius of the path $= r = \frac{mv}{qB} = \frac{P}{qB}$

For H^+ion, $r_H = \frac{p_H}{eB}$

For α particle

$$r_\alpha = \frac{p_\alpha}{2eB}$$

$$\frac{r_H}{r_\alpha} = \frac{\frac{p}{eB}}{\frac{p}{2eB}} \quad \text{[as given } p_H = p_\alpha = p\text{]}$$

$$\Rightarrow \frac{r_H}{r_\alpha} = \frac{2}{1}$$

4. (c) As we know, $F = qvB = \frac{mv^2}{R}$

$$\therefore \text{R} = \frac{mv}{qB} = \frac{\sqrt{2m(kE)}}{qB}$$

Since R is same so, KE $\propto \frac{q^2}{m}$

Therefore KE of α particle

$$= \frac{q^2}{m} = \frac{(2)^2}{4} = 1 \text{ MeV}$$

5. (c) Time period of cyclotron is

$$T = \frac{1}{\upsilon} = \frac{2\pi m}{eB}; \; B = \frac{2\pi m}{e}\upsilon; \; R = \frac{m\upsilon}{eB} = \frac{p}{eB}$$

$$\Rightarrow p = eBR = e \times \frac{2\pi m\upsilon}{e} R = 2\pi m\upsilon R$$

$$\text{K.E.} = \frac{p^2}{2m} = \frac{(2\pi m\upsilon R)^2}{2m} = 2\pi^2 m\upsilon^2 R^2$$

6. (d) Wavelength

$$\lambda = \frac{h}{p} \Rightarrow \lambda = \frac{h}{mv},$$

h = plank's constant = 6.63×10^{-34} J.S

For circular motion = $F_c = qvB$

$$\Rightarrow \frac{mv^2}{r} = qvB \quad \Rightarrow \quad \frac{mv}{qB} = r$$

$$r = \frac{mv}{qB} \Rightarrow mv = qrB$$

$$\Rightarrow \quad (2e)(0.83 \times 10^{-2})\left(\frac{1}{4}\right)$$

$$\lambda = \frac{6.6 \times 10^{-34} \times 4}{2 \times 1.6 \times 10^{-19} \times 0.83 \times 10^{-2}}$$

$\lambda = 9.93 \times 10^{-34+21} \approx 0.01$ Å

7. (b) According to the principal of circular motion in a magnetic field,

$$F_c = F_m \quad \Rightarrow \quad \frac{mv^2}{R} = qVB$$

$$\Rightarrow \; R = \frac{mv}{qB} = \frac{P}{qB} = \frac{\sqrt{2m.K}}{qB}$$

$$R_\alpha = \frac{\sqrt{2(4m)K'}}{2qB}$$

$$\frac{R}{R_\alpha} = \sqrt{\frac{K}{K'}}$$

but $\quad R = R_\alpha$ (given)

Thus $\quad K = K' = 1$ MeV

8. (b) $\vec{v}$ and $\vec{B}$ are in same direction so that magnetic force on electron becomes zero, only electric force acts. But force on electron due to electric field is opposite to the direction of velocity.

NOTES

(i) If $\vec{v}, \vec{E}$ and $\vec{B}$ are all collinear. In this case the magnetic force acting on the electron is zero and only electric force will act. So, acceleration $\vec{a} = \frac{\vec{F}}{m}$. The electron will undeflected in a straight line path with change in speed.

(ii) When $\vec{v}, \vec{E}$ and $\vec{B}$ are mutually perpendicular.

In this case $\vec{F} = \vec{F}_e + \vec{F}_m = 0$

Here, F_e = electric force

F_m = magnetic force

Here, $\vec{a} = \frac{\vec{F}}{m} = 0$

The particle (electron) will pass with same velocity without any deviation in path.

9. (b) Force on a charged particle is given by $F = qvB$. Here $v = 0$ and also resultant B is zero.

$\therefore$ Force = 0

10. (d) The magnetic force acting on the charged paraticle is given by

$$\vec{F} = q\left(\vec{v} \times \vec{B}\right)$$

$$= (-2 \times 10^{-6})[\{2\hat{i} + 3\hat{j}) \times 10^6\} \times (2\hat{j})]$$

$$= -4(2\hat{k})$$

$$= -8\hat{k}$$

$\therefore$ Force is of 8N along negative z-axis.

11. (b) The time period of the charged particle is given by, $T = \frac{2\pi m}{qB}$

Thus, time period is independent of both v and R.

12. (c) When a charged particle enters a transverse magnetic field it traverse a circular path. Its kinetic energy remains constant.

13. (a) If the electric field is switched off, and the same magnetic field is maintained, the electrons move in a circular orbit and electron will travel a magnetic field perpendicular to its velocity.

14. (a) In mass spectrometer, when ions are accelerated through potential V

$$\frac{1}{2}mv^2 = qV \quad \text{..........(i)}$$

As the magnetic field curves the path of the ions in a semicircular orbit

$$Bqv = \frac{mv^2}{R} \Rightarrow v = \frac{BqR}{m} \quad \text{.......... (ii)}$$

Substituting (ii) in (i)

$$\frac{1}{2}m\left[\frac{BqR}{m}\right]^2 = qV$$

or $\frac{q}{m} = \frac{2V}{B^2R^2}$

Since V and B are constants,

$$\therefore \frac{q}{m} \propto \frac{1}{R^2}$$

15. (d) Magnetic moment, $m = IA$

$$= \frac{qv}{2\pi R}(\pi R^2) = \frac{qvR}{2}$$

$$\left[\because I = \frac{q}{T} \text{ and } T = \frac{2\pi R}{v}\right]$$

16. (b)

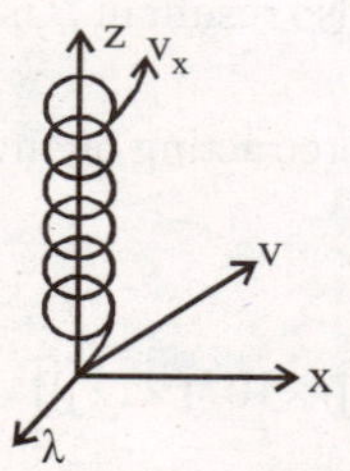

When a test charge q_0 enters a magnetic field $\vec{B}$ directed along z-axis, with a velocity $\vec{v}$ making angles d with the z-axis. The time period of the motion is independent of R and v.

17. (c) Magnetic field produced by moving electron in circular path

$$B = \frac{\mu_o i}{2r}$$

Here, i = current

r = radius of circular path

But $i = \frac{q}{t} = \frac{q}{2\pi r}v \quad \left[\because t = \frac{\text{Distance}}{\text{Velocity}} = \frac{2\pi r}{v}\right]$

$\therefore$ Manetic field at centre, $B = \frac{\mu_o}{2r} \times \frac{qv}{2\pi r}$

$$\Rightarrow B = \frac{\mu_o qv}{4\pi r^2} \Rightarrow r \propto \sqrt{\frac{V}{B}}$$

18. (b) Magnetic force acts perpendicular to the velocity. Hence speed remains constant.

19. (a) Electron moves undeflected if force exerted due to electric field is equal to force due to magnetic field.

$$q|\vec{v}||\vec{B}| = q|\vec{E}| \Rightarrow |\vec{v}| = \frac{|\vec{E}|}{|\vec{B}|}$$

20. (b) For circular path in magnetic field,

$mr\omega^2 = qvB$

$$\Rightarrow \omega^2 = \frac{qvB}{mr} \qquad \text{As } v = r\omega$$

$$\therefore \omega^2 = \frac{q(r\omega)B}{mr} \Rightarrow \omega = \frac{qB}{m}$$

$\therefore$ If ν is frequency of rotation, then

$$\nu = \frac{\omega}{2\pi} \Rightarrow \nu = \frac{qB}{2\pi m}$$

21. (b) $r = \frac{mv\sin\theta}{Be} = \frac{3\times10^5 \sin 30°}{0.3\times10^8}$

$$\frac{3\times10^5\times\frac{1}{2}}{3\times10^7} = 0.5\times10^{-2}\text{ m} = 0.5\text{cm}.$$

NOTES

If the change particle is moving perpendicular to the magnetic field then $qVB = \frac{mv^2}{r}$

$$\Rightarrow r = \frac{mv}{qB}$$

If the charge particle is moving at an, angle (other than 0°, 90°, 180°) to the field, then

$$q(v\sin\theta)B = \frac{m(v\sin\theta)^2}{r} \Rightarrow r = \frac{mv\sin\theta}{qB}$$

22. (c) Potential difference (V) = 1V,

K.E. acquired $= qV$

$= 1.6 \times 10^{-19} \times 1 = 1.6 \times 10^{-19}$ joules $= 1\ eV$

23. (b) In a perpendicular magnetic field, the path of a charged particle is a circle, and the magnetic field does not cause any change in energy.

24. (c) K.E. of electron = 10 eV

$$\Rightarrow \frac{1}{2}mv^2 = 10\ eV$$

$$\Rightarrow \frac{1}{2}(9.1\times10^{-31})v^2 = 10\times1.6\times10^{-19}$$

$$\Rightarrow v^2 = \frac{2\times10\times1.6\times10^{-19}}{9.1\times10^{-31}}$$

$\Rightarrow v^2 = 3.52\times10^{12} \Rightarrow v = 1.88\times10^6$ m

Also, we know that for circular motion

$$\frac{mv^2}{r} = Bev \Rightarrow r = \frac{mv}{Be} = 11\text{ cm}$$

25. (c) The electron moves with constant velocity without deflection. Hence, force due to magnetic field is equal and opposite to force due to electric field.

$$qvB = qE \Rightarrow v = \frac{E}{B} = \frac{20}{0.5} = 40\text{ m/s}$$

26. (d) When the deflection produced by electric field is equal to the deflection produced by magnetic field, then the electron can go undeflected.

27. (a) The force acting on a charged particle in magnetic field is given by

$\vec{F} = q(\vec{v}\times\vec{B})$ or $F = qvB\sin\theta$,
When angle between v and B is 180°,
$F = 0$

Force, $\vec{F} = \vec{q}(\vec{v}\times\vec{B}) = qvB\sin\theta$

When $\theta = 0°, 180°$ then particle will be moving in a direction parallel or antiparallel to the field. In such cases, the trajectory of the particle is a straight line.

28. (d) For a charged particle orbiting in a circular path in a magnetic field

$$\frac{mv^2}{r} = Bqv \Rightarrow v = \frac{Bqr}{m}$$

or, $mv^2 = Bqvr$

Also,

$$E_K = \frac{1}{2}mv^2 = \frac{1}{2}Bqvr = Bq\frac{r}{2}.\frac{Bqr}{m} = \frac{B^2q^2r^2}{2m}$$

For deuteron, $E_1 = \frac{B^2q^2r^2}{2\times2m}$

For proton, $E_2 = \frac{B^2q^2r^2}{2m}$

$$\frac{E_1}{E_2} = \frac{1}{2} \Rightarrow \frac{50keV}{E_2} = \frac{1}{2} \Rightarrow E_2 = 100keV$$

29. (c) $r = \frac{mv}{qB}$ or $r \propto v$

As v is doubled, the radius also becomes double. Hence, radius = 2 × 2 = 4 cm

30. (d) The plane of coil will orient itself so that area vector aligns itself along the magnetic field. So, the plane will orient perpendicular to the magnetic field.

31. (a) Magnetic field due to a current carrying wire at a distance r from the wire.

$$\text{or, } B = \frac{10^{-7}\times2\times5}{20\times10^{-2}} = \frac{1}{2}\times10^{-5}$$

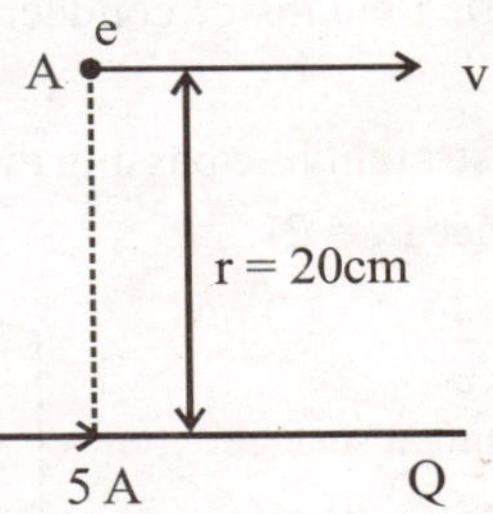

$$B = \frac{\mu_0}{4\pi}\frac{2I}{r} = \frac{4\pi\times10^{-7}\times2\times5}{4\pi\times20\times10^{-2}}$$

$$\text{or, } B = \frac{10^{-7}\times2\times5}{20\times10^{-2}} = \frac{1}{2}\times10^{-5}$$

T upward to the plane of paper
Therefore, force acting on electron
$\vec{F} = q(\vec{v}\times\vec{B})$

$$|\vec{F}| = 1.6\times10^{-19}\times10^5\times\frac{1}{2}\times10^{-5} = 8\times10^{-20}\text{ N}$$

32. (d) From Ampere's circuital law, magnetic field

$B = \frac{\mu_0 I}{2\pi R^2}.r$ If $r < R \Rightarrow B_{inside} \propto r$

$B = \frac{\mu_0 I}{2\pi r}$ If $r \geq R \Rightarrow B_{outside} \propto \frac{1}{r}$

Hence the correct magnetic field B versus distance r graph is as shown below.

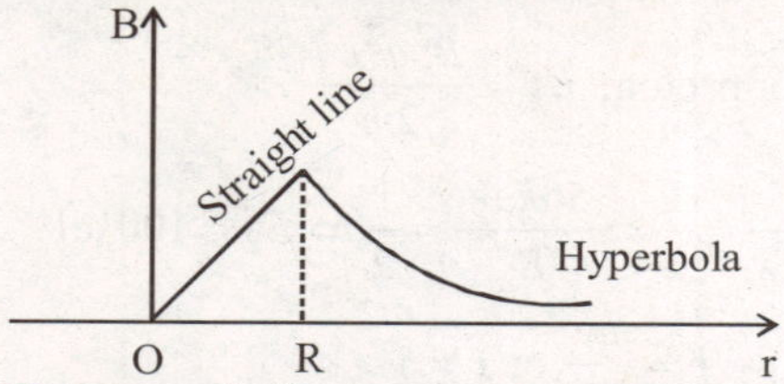

33. (d) Magnetic field at the centre of solenoid,

$$B_{solenoid} = \mu_0 ni$$

Given : No. of turns / length,

$$n = \frac{N}{L} = \frac{100}{50 \times 10^{-2}} = 200 \text{ turns/ m}$$

Current, $I = 2.5$ A

$$\therefore B_{solenoid} = \mu_0 nI = 4\pi \times 10^{-7} \times 200 \times 2.5$$

$$= 6.28 \times 10^{-4} \text{ T}$$

34. (c) Inside ($d < R$)

Magnetic field inside conductor

$B = Kd$ ..... (i)

This is straight line passing through origin

At surface ($d = R$)

$$B = \frac{\mu_0}{2\pi} \frac{1}{d}$$

Maximum at surface

Outside ($d > R$)

$$B = \frac{\mu_0}{2\pi} \frac{1}{d}$$

or $B \propto \frac{1}{d}$ $\therefore$ Hyperbolic

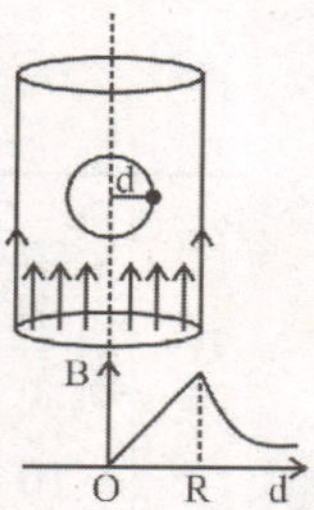

35. (b) $B = \frac{\mu_0 N \cdot I}{2\pi R}$

$$\therefore \frac{B_1}{B_2} = \frac{N_1 R_2}{N_2 R_1} = \frac{200}{100}\frac{20}{40} = 1$$

So, $\frac{B_1}{B_2} = 1$

36. (b) Net magnetic field at point 'P'

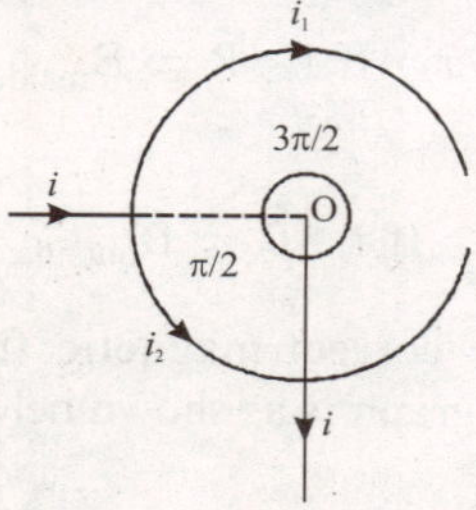

$$B_{net} = \vec{B}_1 + \vec{B}_2$$

Hence, $B_{net} = B_1 - B_2$

$$i_1 = i\left(\frac{\theta}{2\pi}\right) \Rightarrow B_1 = \frac{\mu_0 i_1}{2R}\left(\frac{2\pi - \theta}{2\pi}\right)$$

$$i_2 = i\left(\frac{2\pi - \theta}{2\pi}\right) \Rightarrow B_2 = \frac{\mu_0 i_2}{2R}\left(\frac{\theta}{2\pi}\right)$$

Here $\theta = \frac{\pi}{2}$

$$B_{net} = B_1 - B_2 = 0$$

37. (c) For points inside the wire *i.e.*, ($r \le$ R)

Magnetic field $\text{B} = \frac{\mu_0 Ir}{2\pi R^2}$

For points outside the wire ($r \ge$ R)

Magnetic field, $\text{B}' = \frac{\mu_0 I}{2\pi R}$

$$\therefore \quad \frac{B}{B'} = \frac{\dfrac{\mu_0 I(a/2)}{2\pi a^2}}{\dfrac{\mu_0 I}{2\pi(2a)}} = 1:1$$

38. (c) Radius of circular orbit $= r$

No. of rotations per second $= n$

i.e., $\text{T} = \frac{1}{n}$

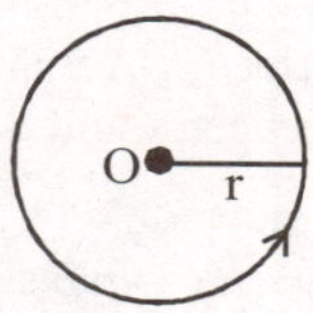

Magnetic field at its centre, B_c =?

As we know, current

$$i = \frac{e}{T} = \frac{e}{(1/n)} = en = \text{equivalent current}$$

Magnetic field at the centre of circular orbit,

$$B_c = \frac{\mu_0 i}{2r} = \frac{\mu_0 ne}{2r}$$

39. (b) Magnetic field due to segment '1'

$$\overrightarrow{\text{B}_1} = \frac{\mu_0 \text{I}}{4\pi \text{R}}[\sin 90° + \sin 0°](-\hat{\text{k}})$$

$$= \frac{-\mu_0 \text{I}}{4\pi \text{R}}(\hat{\text{k}}) = \vec{\text{B}}_3$$

Magnetic field due to segment 2

$$B_2 = \frac{\mu_0 I}{4R}\left(-\hat{i}\right) = \frac{-\mu_0 I}{4\pi R}\left(\pi\hat{i}\right)$$

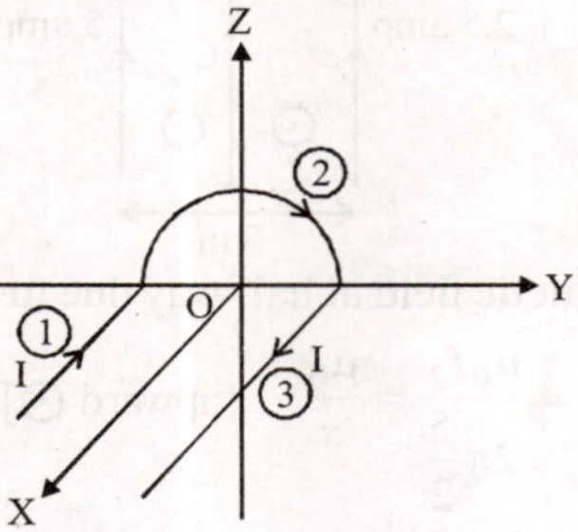

$\therefore$ $\vec{B}$ at centre

$$\vec{B}_c = \vec{B}_1 + \vec{B}_2 + \vec{B}_3 = \frac{-\mu_0 I}{4\pi R}\left(\pi\hat{i} + 2\hat{k}\right)$$

40. **(d)** Net magnetic field, $B = \sqrt{B_1^2 + B_2^2}$

$$= \sqrt{\left(\frac{\mu_0 I_1}{2\pi d}\right)^2 + \left(\frac{\mu_0 I_2}{2\pi d}\right)^2}$$

$$\left(\because B_1 = \frac{\mu_0 I_1}{2\pi d} \text{ and } B_2 = \frac{\mu_0 I_2}{2\pi d}\right)$$

$$= \frac{\mu_0}{2\pi d}\sqrt{I_1^2 + I_2^2}$$

41. **(a)**

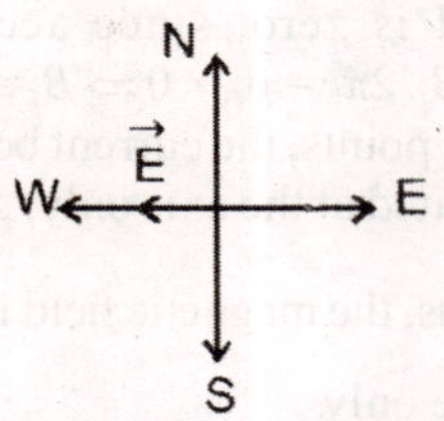

When moves with an acceleration a_0 towards west, electric field

$$E = \frac{F}{q} = \frac{ma_0}{e} \text{ (West)}$$

When moves with an acceleration $3a_0$ towards east, magnetic field

$$B = \frac{2ma_0}{ev_0} \text{ (downward)}$$

42. **(a)**

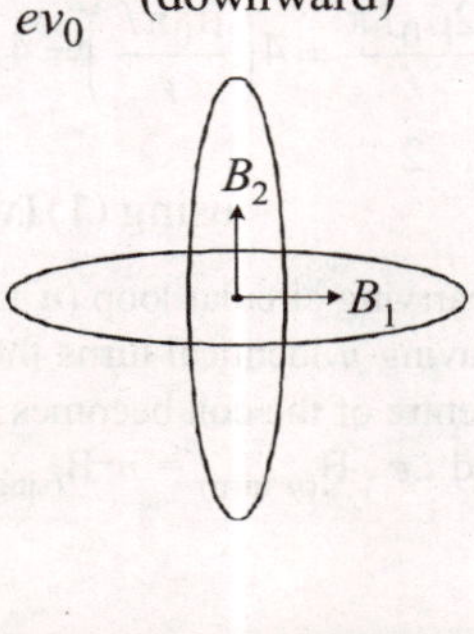

The magnetic field, due the coil, carrying current I Ampere

$$B_1 = \frac{\mu_0 I}{2R}$$

The magnetic field due to the coil, carrying current $2I$ Ampere

$$B_2 = \frac{\mu_0(2I)}{2R}$$

The resultant B

$$B_{net} = \sqrt{B_1^2 + B_2^2 + 2B_1B_2\cos\theta},\ \theta = 90°$$

$$B_{net} = \sqrt{B_1^2 + B_2^2} = \frac{\mu_0(2I)}{2R}\sqrt{1+4}$$

$$= \frac{\sqrt{5}\mu_0 I}{2R}$$

43. **(a)** When the ring rotates about its axis with a uniform frequency f Hz, the current flowing in the ring is

$$I = \frac{q}{T} = qf$$

Magnetic field at the centre of the ring is

$$B = \frac{\mu_0 I}{2R} = \frac{\mu_0 qf}{2R}$$

44. **(d)** Force due to electric field acts along the direction of the electric field but force due to the magnetic field acts along a direction perpendicular to both the velocity of the charged particle and the magnetic field. Hence both statements (2) and (3) are true. In statement (2), magnetic force is zero, so, electric force will keep the particle continue to move in horizontal direction. In statement (3), both electric and magnetic forces will be opposite to each other. If their magnitudes will be equal then the particle will continue horizontal motion.

45. **(b)** Magnetic fields due to the two parts at their common centre are respectively,

$$B_y = \frac{\mu_0 i}{4R} \text{ and } B_z = \frac{\mu_0 i}{4R}$$

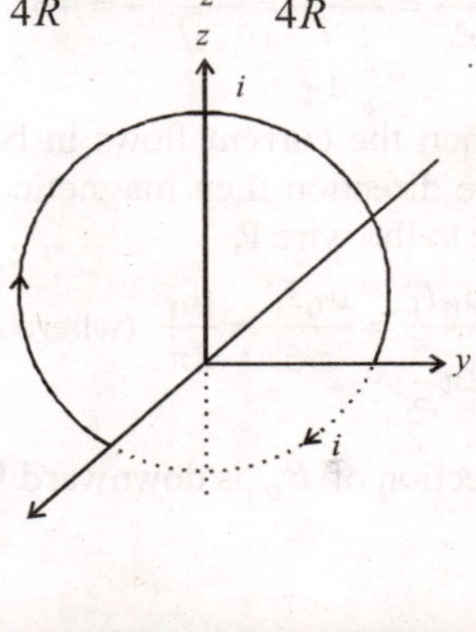

Resultant field $= \sqrt{B_y^2 + B_z^2}$

$$= \sqrt{\left(\frac{\mu_0 i}{4R}\right)^2 + \left(\frac{\mu_0 i}{4R}\right)^{-2}} = \sqrt{2}.\frac{\mu_0 i}{4R} = \frac{\mu_0 i}{2\sqrt{2}R}$$

46. **(None)** If R_1 & R_2 be the radius of the circular wires, $\frac{R_1}{R_2} = \frac{2}{1}$. If same potential is applied on them, current in Ist will be half that in the later. If V potential is applied on them, current in them $= \frac{V}{2R}$ & $\frac{V}{R}$.

Now magnetic field at the centre of circular coil, $= \frac{\mu_0 I}{2r}$

For first wire, field $B_1 = \frac{\mu_0 V}{2R \times 2R}$

For second wire, field $B_2 = \frac{\mu_0 V}{2(R/2) \times R}$

Given $B_1 = B_2$

The given data do not provide any required result. There is a mistake in the framing of the question.

47. **(c)** $B = \mu_0 ni$

$$B_1 = (\mu_0)\left(\frac{n}{2}\right)(2\,i) = \mu_0 ni = B$$

$\Rightarrow B_1 = B$

48. **(a)** Let I be current and l be the length of the wire.

For Ist case : $B = \frac{\mu_0 In}{2r} = \frac{\mu_0 I \times \pi}{l}$

where $2\pi r = l$ and $n = 1$

For IInd Case : $l = 2(2\pi r') \Rightarrow r' = \frac{l}{4\pi}$

$$B' = \frac{\mu_0 nI}{2r'} = \frac{\mu_0 2I}{2\frac{l}{4\pi}} = \frac{4\mu_0 I\pi}{l} = 4B$$

49. **(d)** When the current flows in both wires in the same direction then magnetic field at half way due to the wire P,

$$\vec{B}_p = \frac{\mu_0 I_1}{2\pi\frac{5}{2}} = \frac{\mu_0 I_1}{\pi \cdot 5} = \frac{\mu_0}{2\pi} \quad (\text{where } I_1 = 2.5 \text{ amp})$$

The direction of $\vec{B}_p$ is downward ⊙

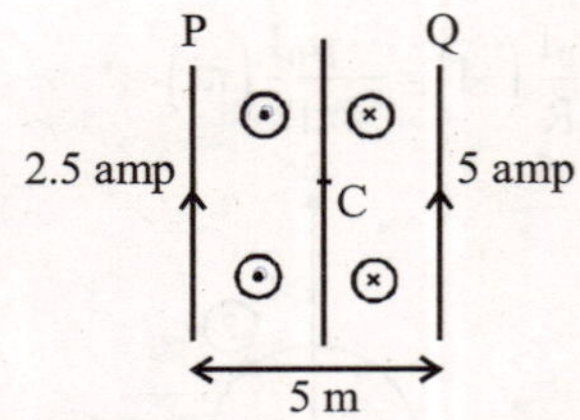

Magnetic field at half way due to wire Q

$$\vec{B}_Q = \frac{\mu_0 I_2}{2\pi\frac{5}{2}} = \frac{\mu_0}{\pi} \quad [\text{upward } ⊗]$$

[where $I_2 = 2.5$ amp.]

Net magnetic field at half way

$$\vec{B} = \vec{B}_P + \vec{B}_Q$$

$$= -\frac{\mu_0}{2\pi} + \frac{\mu_0}{\pi} = \frac{\mu_0}{2\pi} \quad (\text{upward})$$

Hence, net magnetic field at midpoint $= \frac{\mu_0}{2\pi}$

50. **(b)** We know that magnetic field at the centre of circular coil,

$$B = \frac{\mu_0 In}{2r} = \frac{4\pi \times 10^{-7} \times 2 \times 50}{2 \times 0.5} = 1.25 \times 10^{-4}\,N$$

51. **(b)** Inside a hollow pipe carrying current, the magnetic field is zero, since according to Ampere's law, $B_i.\,2\pi r = \mu_0 \times 0 \Rightarrow B_i = 0$.

But for external points, the current behaves as if it was concentrated at the axis only; so, outside, $B_0 = \frac{\mu_0 i}{2\pi r}$. Thus, the magnetic field is produced outside the pipe only.

52. **(b)** Let ℓ be length of wire.

Ist case : $\ell = 2\pi r \Rightarrow \quad r = \frac{\ell}{2\pi}$

$$B = \frac{\mu_0 In}{2r} = \frac{\mu_0 I \times 2\pi}{2\ell} = \frac{\mu_0 \pi I}{\ell} \; [\because n = 1] \ldots(1)$$

2nd Case : $\ell = 2(2\pi r') \Rightarrow r' = \frac{\ell}{4\pi}$

$$B' = \frac{\mu_0 In}{2\frac{\ell}{4\pi}} = \frac{2\mu_0 I\pi}{\frac{\ell}{2}} = 4\left(\frac{\mu_0 \pi I}{\ell}\right) = 4B,$$

using (1) (where $n = 2$)

NOTES

If a current carrying circular loop ($n = 1$) is turned into a coil having n identical turns then magnetic field at the centre of the coil becomes n^2 times the previous field *i.e.*, $B_{(n \text{ turn})} = n^2 B_{(\text{single turn})}$

53. **(d)** According to Biot Savart law,

$$d\vec{B}=\frac{\mu_0}{4\pi}\frac{i(d\vec{\ell}\times\vec{r})}{r^3}$$

54. **(a)**

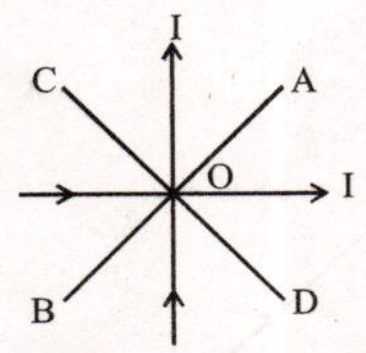

Net magnetic field on AB is zero because magnetic field due to both current carrying wires is equal in magnitude but opposite in direction.

55. **(b)** $B=\frac{\mu_0 i}{2\pi r}$ and so it is independent of thickness.
The current is same in both the wires, hence magnetic field induced will be same.

56. **(a)** Current (I) = 12 A and magnetic field (B) $= 3\times10^{-5}\ Wb/m^2$. Consider magnetic field $\vec{B}$ at distance r.

Magnetic field, $B=\frac{\mu_0 I}{2\pi r}$

$$\Rightarrow r=\frac{\mu_0 I}{2\pi B}=\frac{(4\pi\times10^{-7})\times12}{2\times\pi\times(3\times10^{-5})}=8\times10^{-2}\ m$$

57. **(a)** $B=\frac{\mu_0 i}{2\pi r}$ or $B\propto\frac{1}{r}$

When r is doubled, the magnetic field becomes half, *i.e.*, now the magnetic field will be 0.2 T.

58. **(a)** $B=\frac{\mu_0 I}{2\pi r}\Rightarrow B\propto\frac{1}{r}$

As the distance is increased to three times, the magnetic induction reduces to one third. Hence,

$$B=\frac{1}{3}\times10^{-3}\ \text{tesla}=3.33\times10^{-4}\ \text{tesla}$$

59. **(b)** Energy is stored in magnetic field.

60. **(b)** Tesla is the unit of magnetic field.

61. **(c)** From figure, for equilibrium,
mg sin 30° = I/B cos 30°

$$\Rightarrow I=\frac{mg}{\ell B}\tan 30°$$

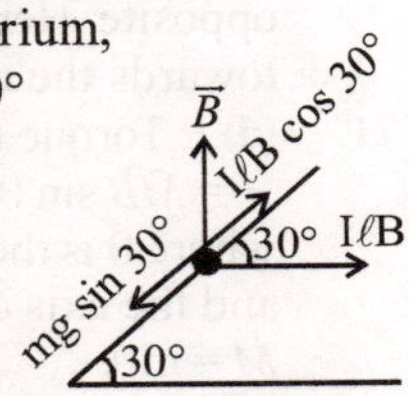

$$=\frac{0.5\times9.8}{0.25\times\sqrt{3}}=11.32\ A$$

62. **(d)** Work done, $W=MB(\cos\theta_1-\cos\theta_2)$
When it is rotated by angle 180° then

$W=MB(\cos0°-\cos180°)=MB(1+1)$

$W=2MB$

$W=2(NIA)B$

$=2\times250\times85\times10^{-6}[1.25\times2.1\times10^{-4}]\times85\times10^{-2}$

$=9.1\ \mu J$

63. **(c)** Force per unit length between two parallel current carrying conductors,

$$F=\frac{\mu_0 i_1 i_2}{2\pi d}$$

Since same current flowing through both the wires $i_1=i_2=i$

$$\text{so } F_1=\frac{\mu_0 i^2}{2\pi d}=F_2$$

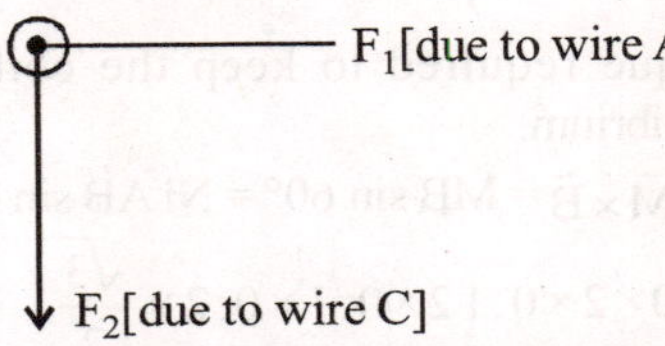

∴ Magnitude of force per unit length on the middle wire 'B'

$$F_{net}=\sqrt{F_1^2+F_2^2}=\frac{\mu_0 i^2}{\sqrt{2}\pi d}$$

64. **(a)** The direction of current in conductor

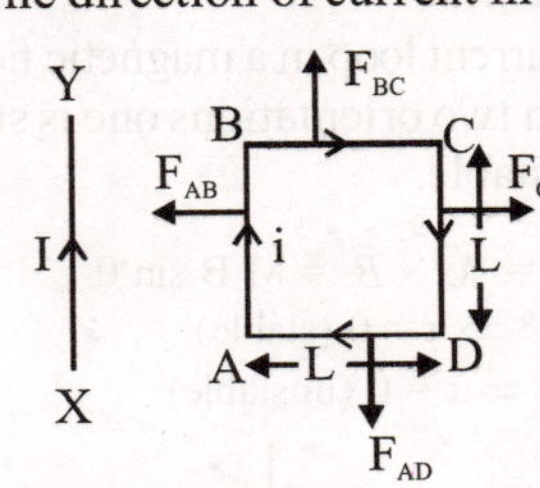

XY and AB is same

∴ $F_{AB}=i\ell B$ (attractive)

$$F_{AB}=i(L).\frac{\mu_0 I}{2\pi\left(\frac{L}{2}\right)}(\leftarrow)=\frac{\mu_0 iI}{\pi}(\leftarrow)$$

F_{BC} opposite to F_{AD}

$F_{BC}(\uparrow)$ and $F_{AD}(\downarrow)$

⇒ cancels each other

$F_{CD}=i\ell B$ (repulsive)

$$F_{CD}=i(L)\frac{\mu_0 I}{2\pi\left(\frac{3L}{2}\right)}(\rightarrow)=\frac{\mu_0 iI}{3\pi}(\rightarrow)$$

Therefore the net force on the loop

$F_{net} = F_{AB} + F_{BC} + F_{CD} + F_{AD}$

$\Rightarrow \quad F_{net} = \dfrac{\mu_o iI}{\pi} - \dfrac{\mu_o iI}{3\pi} = \dfrac{2\mu_o iI}{3\pi}$

65. (a) Here, number of turns of coil, N = 50

Current through the coil I = 2A

Area $A = l \times b = 0.12 \times 0.1 m^2 = 0.012\ m^2$

Magnetic field $\vec{B} = 0.2\ W/m^2$

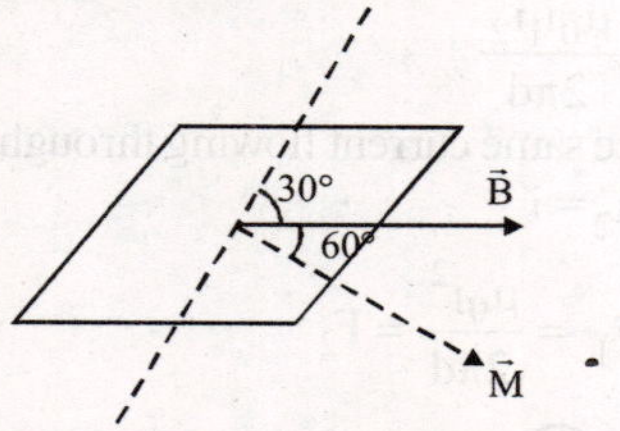

Torque required to keep the coil in stable equilibrium.

$\tau = \vec{M} \times \vec{B} = MB \sin 60° = Ni AB \sin 60°$

$= 50 \times 2 \times 0.12 \times 0.1 \times 0.2 \times \dfrac{\sqrt{3}}{2}$

$= 12\sqrt{3} \times 10^{-2} = 0.20784\ Nm$

NOTES

Torque $\tau = nIBA \sin \theta$

Here, θ is the angle which a normal drawn on the plane of the coil makes with the direction of magnetic field.

66. (c) A current loop in a magnetic field is in equilibrium in two orientations one is stable and another unstable.

NOTES

$\because \quad \vec{\tau} = \vec{M} \times \vec{B} = MB \sin \theta$

If $\theta = 0° \Rightarrow \tau = 0$ (stable)

If $\theta = \pi \Rightarrow \tau = 0$ (unstable)

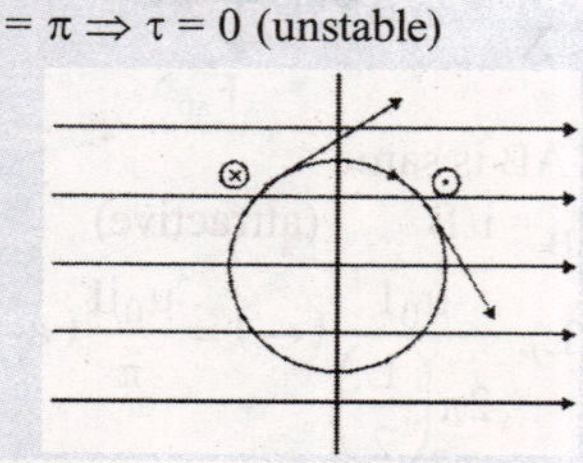

67. (c) Given:

Magnetic field $B = 2 \times 10^{-4}$ weber/m²

Velocity of electron, $v = 10^7$ m/s

Lorentz force $F = qvB \sin \theta$

$= 1.6 \times 10^{-19} \times 10^7 \times 2 \times 10^{-4}$ ($\because \theta = 90°$)

$= 3.2 \times 10^{-16}$ N

68. (b) Here, $\vec{F}_{AB} + \vec{F}_{BCDA} = \vec{0}$

$\Rightarrow \vec{F}_{BCDA} = -\vec{F}_{AB} = -\vec{F} \quad (\because \vec{F}_{AB} = \vec{F})$

69. (b) Let a current i be flowing in the loop ABC in the direction shown in the figure. If the length of each of the sides AB and BC be x then

$|\vec{F}| = i \times B$

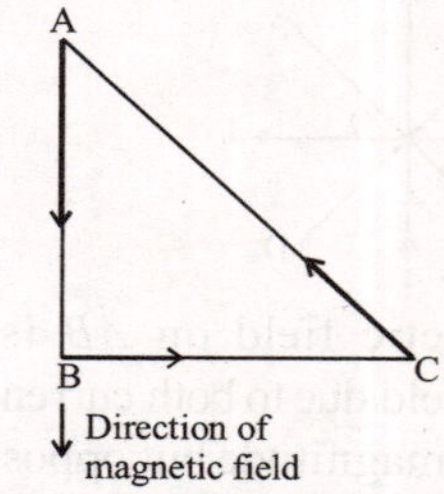

where B is the magnitude of the magnetic force. The direction of $\vec{F}$ will be in the direction perpendicular to the plane of the paper and going into it.

By Pythagorus theorem,

$AC = \sqrt{x^2 + x^2} = \sqrt{2}x$

∴ Magnitude of force on AC

$= i\sqrt{2} \times B \sin 45°$

$= i\sqrt{2}xB \times \dfrac{1}{\sqrt{2}}$

$= ixB = |\vec{F}|$

The direction of the force on AC is perpendicular to the plane of the paper and going out of it.

Hence, force on AC = $-\vec{F}$

70. (d)

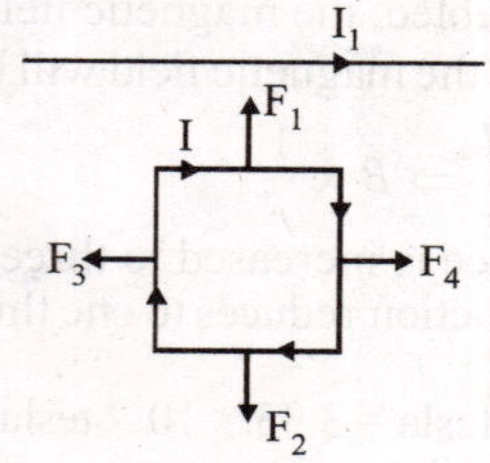

$F_1 > F_2$ as $F \propto \dfrac{1}{d}$, and F_3 and F_4 are equal and opposite. Hence, the net attraction force will be towards the conductor.

71. (d) Torque on the solenoid is given by

$\tau = MB \sin \theta$

where θ is the angle between the magnetic field and the axis of solenoid.

$M = niA$

$\therefore \quad \tau = niA B \sin 30°$

$= 2000 \times 2 \times 1.5 \times 10^{-4} \times 5 \times 10^{-2} \times \dfrac{1}{2}$

$= 1.5 \times 10^{-2}\ N{-}m$

72. **(b)** According to the figure the magnitude of force on the segment QM is $F_3 - F_1$ and PM is F_2.

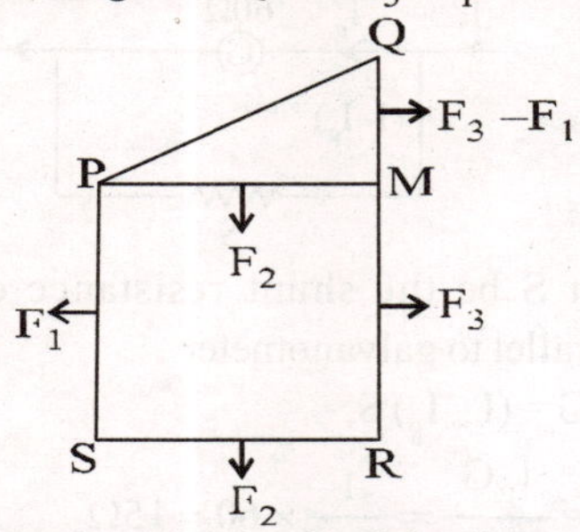

Therefore, the magnitude of the force on segment PQ is $\sqrt{(F_3 - F_1)^2 + F_2^2}$

73. **(b)** Force on a particle moving with velocity v in a magnetic field B is $F = q(\vec{v} \times \vec{B})$

If angle between $\vec{v}$ & $\vec{B}$ is either zero or 180°, then value of F will be zero as cross product of $\vec{v}$ and $\vec{B}$ will be zero.

So option (b) is correct.

74. **(a)** The direction of $\vec{B}$ is along $(-\hat{k})$

$\therefore$ The magnetic force

$\vec{F} = Q(\vec{v} \times \vec{B}) = Q(v\hat{i}) \times B(-\hat{k}) = QvB\hat{j}$

$\Rightarrow \vec{F}$ is along OY.

75. **(d)** Force due to electric field = $q\vec{E}$

Force due to magnetic field = $q(\vec{v} \times \vec{B})$

Net force experienced = $q\vec{E} + q(\vec{v} \times \vec{B})$

76. **(a)** $F = \dfrac{\mu_0}{4\pi} \times \dfrac{2 i_1 i_2 l}{r} = 10^{-7} \times \dfrac{2 \times 1 \times 1 \times 1}{1}$

$= 2 \times 10^{-7}$ N/m.

[This relates to the definition of ampere]

77. **(a)** A current carrying coil has magnetic dipole moment. Hence, a torque acts on it in magnetic field.

78. **(b)** $F = Bi\ell = 2 \times 1.2 \times 0.5 = 1.2$ N

79. **(d)** Current sensitivity of moving coil galvanometer

$$I_s = \frac{NBA}{C} \quad \text{...(i)}$$

Voltage sensitivity of moving coil galvanometer,

$$V_s = \frac{NBA}{CR_G} \quad \text{...(ii)}$$

Dividing eqn. (i) by (ii)

Resistance of galvanometer

$$R_G = \frac{I_s}{V_s} = \frac{5 \times 1}{20 \times 10^{-3}} = \frac{5000}{20} = 250\,\Omega$$

80. **(d)** From circuit diagram

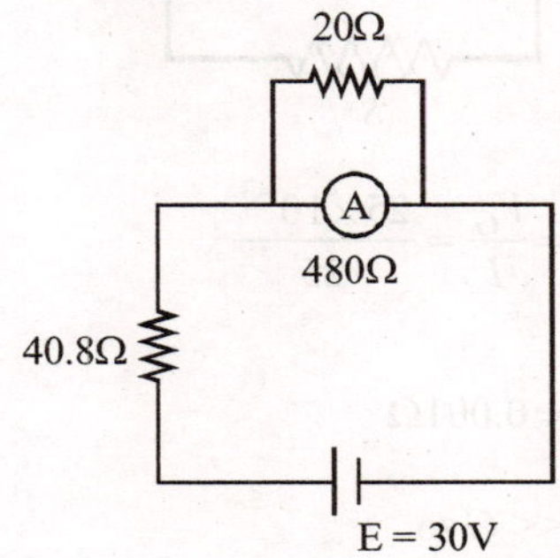

Resistance of ammeter = $\dfrac{480 \times 20}{480 + 20} = 19.2\Omega$.

Total resistance $R = 40.8 + 19.2 = 60\Omega$

Reading in the ammeter $i = \dfrac{V}{R}$

$= \dfrac{30}{40.8 + 19.2} = 0.5$A

81. **(c)** As 0.2% of main current passes through the galvanometer hence $\dfrac{998}{1000}$I current through the shunt.

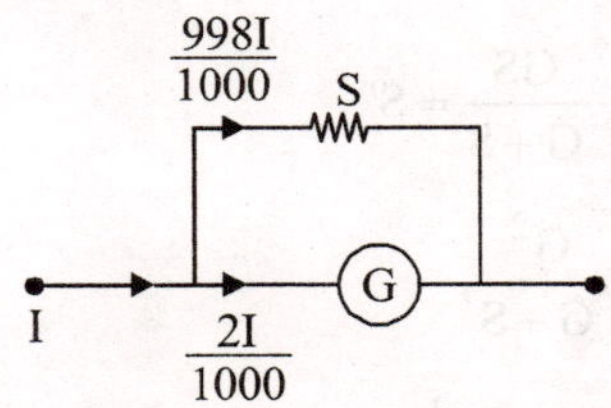

$$\left(\frac{2I}{1000}\right)G = \left(\frac{998I}{1000}\right)S \Rightarrow S = \frac{G}{499}$$

Total resistance of Ammeter

$$R = \frac{SG}{S+G} = \frac{\left(\frac{G}{499}\right)G}{\left(\frac{G}{499}\right) + G} = \frac{G}{500}$$

82. **(a)** Galvanometer is converted into ammeter, by connected a shunt, in parallel with it.

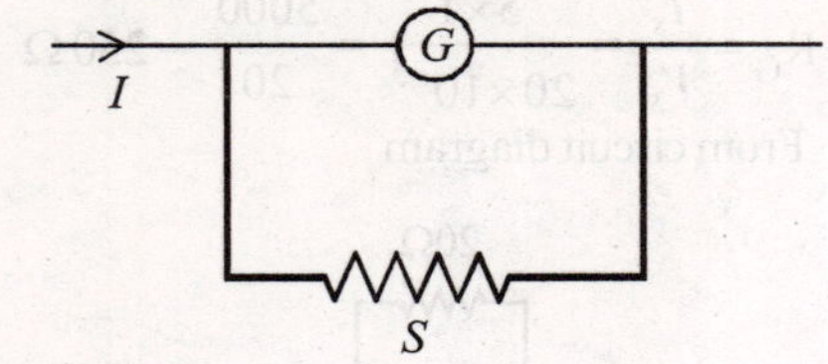

$$\frac{GS}{G+S} = \frac{V_G}{I} = \frac{25\times10^{-3}}{25}$$

$$\frac{GS}{G+S} = 0.001\Omega$$

Here $S << G$ so

$S = 0.001\ \Omega$

NOTES

In order to increase the range of ammeter n times, the value of shunt resistance to be connected in parallel to galvanometer is

$$S = \frac{G}{n-1}$$

Here, G = resistance of galvanometer

83. **(c)** To keep the main current in the circuit unchanged, the resistance of the galvanometer should be equal to the net resistance.

$$\therefore G = \left(\frac{GS}{G+S}\right) + S'$$

$$\Rightarrow G - \frac{GS}{G+S} = S'$$

$$\therefore S' = \frac{G^2}{G+S}.$$

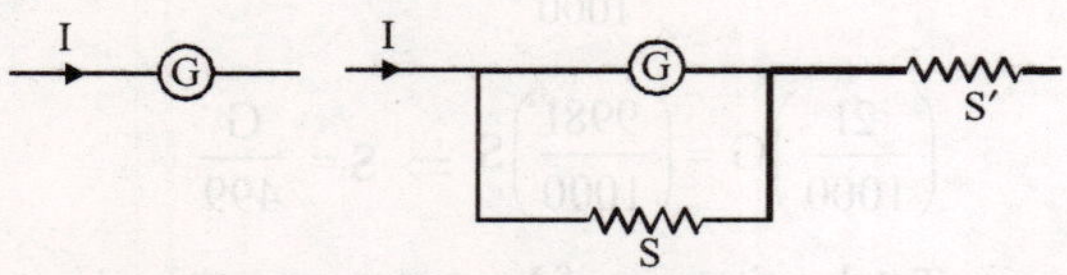

84. **(a)** Let the resistance to be added be R, then

$30 = I_g(r+R)$

$$\therefore \quad R = \frac{30}{I_g} - r = \frac{30}{30\times10^{-3}} - 100$$

$= 1000 - 100 = 900\ \Omega$

85. **(c)** $G = 60\Omega, I_g = 1.0A, I = 5A.$

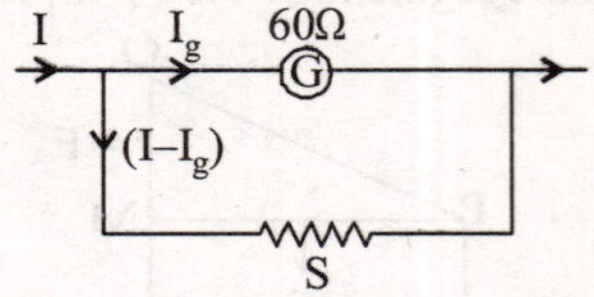

Let S be the shunt resistance connected in parallel to galvanometer

$I_g G = (I - I_g)S,$

$$S = \frac{I_g G}{I - I_g} = \frac{1}{5-1}\times 60 = 15\Omega$$

Thus by putting $15\ \Omega$ in parallel, the galvanometer can be converted into an ammeter.

86. **(d)** Total internal resistance $= (50 + 2950)\ \Omega$

$= 3000\ \Omega$

Emf of the cell, $\varepsilon = 3V$

$$\therefore \text{Current} = \frac{\varepsilon}{R} = \frac{3}{3000} = 1\times10^{-3}\ A = 1.0\ mA$$

$\therefore$ Current for full scale deflection of 30 divisions is 1.0 mA.

$\therefore$ Current for a deflection of 20 divisions,

$$I = \left(\frac{20}{30}\times1\right) mA \text{ or } I = \frac{2}{3} mA$$

Let the resistance be $x\ \Omega$. Then

$$x = \frac{\varepsilon}{I} = \frac{3V}{\left(\frac{2}{3}\times10^{-3} A\right)} = \frac{3\times3\times10^3}{2}\Omega$$

$= 4500\ \Omega$

But the resistance of the galvanometer is 50Ω.

$\therefore$ Resistance to be added

$= (4500 - 50)\ \Omega = 4450\ \Omega$

87. **(a)** We know

$$\frac{I}{I_S} = 1 + \frac{G}{S}$$

$$\frac{750}{100} = 1 + \frac{13}{S}$$

$S \Rightarrow 2\Omega$

88. **(c)** A galvanometer can be converted into a voltmeter by connecting *the high ressistance in series* with the galvanometer so that only a small amount of current passes through it.

89. **(a)** $R_g = 50\Omega, I_g = 25 \times 4 \times 10^{-A}\Omega = 10^{-2}$ A

Range of V = 25 volts

$$V = I_g(R_e + R_g)$$

$$\therefore R_e = \frac{V}{I_g} - R_g = 2450\Omega$$

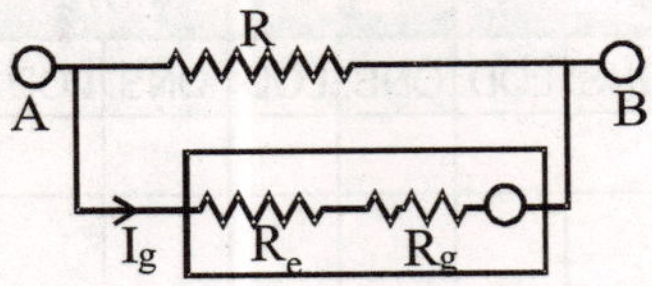

> **NOTES**
> In order to increase the range of voltmeter n times the value of resistance to be connected in series with galvanometer is
> $R = (n - 1)G$
> Here, G = resistance of galvanometer

90. **(b)** $I_s = I \times \frac{G}{S+G} = 1 \times \frac{8}{2+8}$

$$= \frac{8}{10} = 0.8 \text{ amp}.$$

91. **(c)** Maximum current which can pass through galvanometer, $I_g = 0.004$A

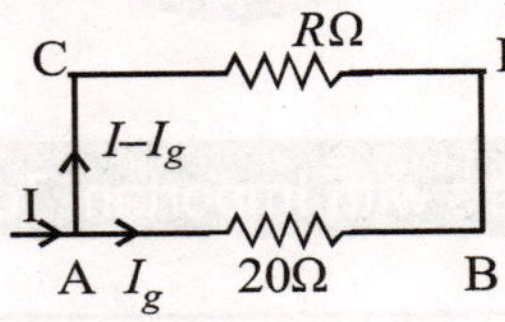

Let R be the resistance of shunt.

We know potential drop across AB

= Potential drop across CD

$R(I - I_g) = I_g(20)$

$\Rightarrow R(1 - 0.004) = 0.004 \times 20 \Rightarrow R = 0.08\,\Omega$

92. **(a)** To convert a galvanometer into an ammeter, one needs to connect a low resistance in parallel so that maximum current passes through the shunt wire and ammeter remains protected.

19 Magnetism and Matter

Trend Analysis with Important Topics & Sub-Topics

Topic Name	Sub-Topic	2021		2020		2019		2018		2017		2016	
		QNS.	LOD	QNS	LOD	QNS.	LOD	QNS.	LOD	QNS.	LOD	QNS.	LOD
Magnetism, Gauss's Law, Magnetic Moment & Properties of Magnet													
The Earth's Magnetism, Magnetic Materials and their Properties	Magnetic susceptibity and permeability diamagnetic material			1	A			1	E			1	E
	Angle of dip.					1	A			1	A		
Magnetic Equipments	Time period of oscillations of freely suspended magnet												

LOD - Level of Difficulty | E - Easy | A - Average | D - Difficult | Qns - No. of Questions

Topic 1 Magnetism, Gauss's Law, Magnetic Moment & Properties of Magnet

1. Following figures show the arrangement of bar magnets in different configurations. Each magnet has magnetic dipole moment $\vec{m}$. Which configuration has highest net magnetic dipole moment? *[2014]*

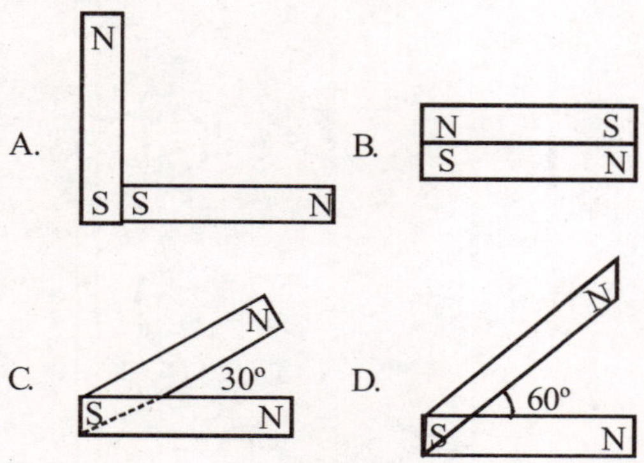

(a) A (b) B
(c) C (d) D

2. A bar magnet of length 'ℓ' and magnetic dipole moment 'M' is bent in the form of an arc as shown in figure. The new magnetic dipole moment will be *[2013]*

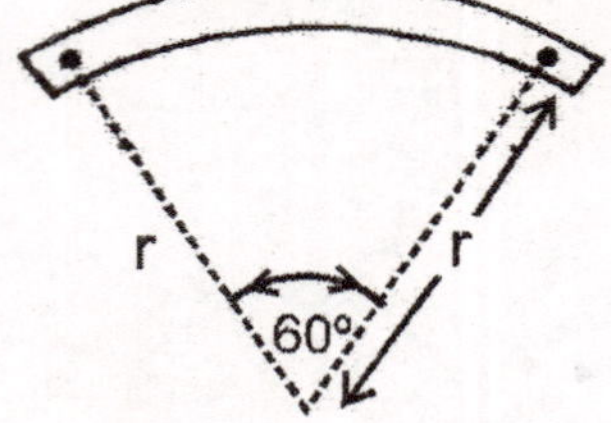

(a) $\frac{3}{\pi}$M (b) $\frac{2}{\pi}$ M
(c) $\frac{M}{2}$ (d) M

3. A bar magnet of magnetic moment M is placed at right angles to a magnetic induction B. If a force F is experienced by each pole of the magnet, the length of the magnet will be *[NEET Kar. 2013]*

(a) F/MB (b) MB/F
(c) BF/M (d) MF/B

4. A magnetic needle suspended parallel to a magnetic field requires $\sqrt{3}$ J of work to turn it through 60°. The torque needed to maintain the needle in this position will be : *[2012M]*

(a) $2\sqrt{3}J$ (b) 3J

(c) $\sqrt{3}J$ (d) $\frac{3}{2}J$

5. A short bar magnet of magnetic moment $0.4 J T^{-1}$ is placed in a uniform magnetic field of 0.16 T. The magnet is in stable equilibrium when the potential energy is *[2011M]*

(a) –0.064 J (b) zero

(c) –0.082 J (d) 0.064 J

6. A bar magnet having a magnetic moment of $2\times 10^4 JT^{-1}$ is free to rotate in a horizontal plane. A horizontal magnetic field $B = 6\times 10^{-4}$ T exists in the space. The work done in taking the magnet slowly from a direction parallel to the field to a direction 60° from the field is *[2009]*

(a) 12 J (b) 6 J

(c) 2 J (d) 0.6 J

7. A coil in the shape of an equilateral triangle of side l is suspended between the pole pieces of a permanent magnet such that $\vec{B}$ is in the plane of the coil. If due to a current i in the triangle a torque τ acts on it, the side l of the triangle is *[2005]*

(a) $\frac{2}{\sqrt{3}}\left(\frac{\tau}{B.i}\right)^{\frac{1}{2}}$ (b) $2\left(\frac{\tau}{\sqrt{3}B.i}\right)^{\frac{1}{2}}$

(c) $\frac{2}{\sqrt{3}}\left(\frac{\tau}{B.i}\right)$ (d) $\frac{1}{\sqrt{3}}\frac{\tau}{B.i}$

8. Current i is flowing in a coil of area A and number of turns N, then magnetic moment of the coil, M is *[2001]*

(a) NiA (b) $\frac{Ni}{A}$

(c) $\frac{Ni}{\sqrt{A}}$ (d) N^2Ai

9. A bar magnet, of magnetic moment $\vec{M}$, is placed in a magnetic field of induction $\vec{B}$. The torque exerted on it is *[1999]*

(a) $\vec{M}\cdot\vec{B}$ (b) $-\vec{M}\cdot\vec{B}$

(c) $\vec{M}\times\vec{B}$ (d) $\vec{B}\times\vec{M}$

10. The work done in turning a magnet of magnetic moment M by an angle of 90° from the meridian, is n times the corresponding work done to turn it through an angle of 60°. The value of n is given by *[1995]*

(a) 2 (b) 1

(c) 0.5 (d) 0.25

Topic 2 The Earth's Magnetism, Magnetic Materials and their Properties

11. An iron rod of susceptibility 599 is subjected to a magnetising field of 1200 A m^{-1}. The permeability of the material of the rod is : ($\mu_0 = 4\pi\times 10^{-7} T m A^{-1}$) *[2020]*

(a) $8.0\times 10^{-5} T m A^{-1}$

(b) $2.4\pi\times 10^{-5} T m A^{-1}$

(c) $2.4\pi\times 10^{-7} T m A^{-1}$

(d) $2.4\pi\times 10^{-4} T m A^{-1}$

12. At a point A on the earth's surface the angle of dip, δ = +25°. At a point B on the earth's surface the angle of dip, δ = –25°. We can interpret that: *[2019]*

(a) A and B are both located in the northern hemisphere.

(b) A is located in the southern hemisphere and B is located in the northern hemisphere.

(c) A is located in the northern hemisphere and B is located in the southern hemisphere.

(d) A and B are both located in the southern hemisphere.

13. The relations amongst the three elements of earth's magnetic field, namely horizontal component H, vertical component V and dip δ are, (B_E = total magnetic field) *[NEET Odisha 2019]*

(a) $V = B_E, H = B_E \tan\delta$

(b) $V = B_E \tan\delta, H = B_E$

(c) $V = B_E \sin\delta, H = B_E \cos\delta$

(d) $V = B_E \cos\delta, H = B_E \sin\delta$

14. A thin diamagnetic rod is placed vertically between the poles of an electromagnet. When the current in the electromagnet is switched on, then the diamagnetic rod is pushed up, out of the horizontal magnetic field. Hence the rod gains gravitational potential energy. The work required to do this comes from *[2018]*

(a) the current source
(b) the magnetic field
(c) the induced electric field due to the changing magnetic field
(d) the lattice structure of the material of the rod

15. If θ_1 and θ_2 be the apparent angles of dip observed in two vertical planes at right angles to each other, then the true angle of dip θ is given by :- *[2017]*
(a) $\tan^2\theta = \tan^2\theta_1 + \tan^2\theta_2$
(b) $\cot^2\theta = \cot^2\theta_1 - \cot^2\theta_2$
(c) $\tan^2\theta = \tan^2\theta_1 - \tan^2\theta_2$
(d) $\cot^2\theta = \cot^2\theta_1 + \cot^2\theta_2$

16. The magnetic susceptibility is negative for : *[2016]*
(a) diamagnetic material only
(b) paramagnetic material only
(c) ferromagnetic material only
(d) paramagnetic and ferromagnetic materials

17. There are four light–weight–rod samples A,B,C,D separately suspended by threads. A bar magnet is slowly brought near each sample and the following observations are noted *[2011]*
(i) A is feebly repelled
(ii) B is feebly attracted
(iii) C is strongly attracted
(iv) D remains unaffected
Which one of the following is true ?
(a) B is of a paramagnetic material
(b) C is of a diamagnetic material
(c) D is of a ferromagnetic material
(d) A is of a non–magnetic material

18. Electromagnets are made of soft iron because soft iron has *[2010]*
(a) low retentivity and high coercive force
(b) high retentivity and high coercive force
(c) low retentivity and low coercive force
(d) high retentivity and low coercive force

19. The magnetic moment of a diamagnetic atom is *[2010]*
(a) equal to zero
(b) much greater than one
(c) 1
(d) between zero and one

20. If a diamagnetic substance is brought near the north or the south pole of a bar magnet, it is: *[2009, 1999]*
(a) repelled by the north pole and attracted by the south pole
(b) attracted by the north pole and repelled by the south pole
(c) attracted by both the poles
(d) repelled by both the poles

21. Curie temperature is the temperature above which *[2008, 2006]*
(a) ferromagnetic material becomes paramagnetic material
(b) paramagnetic material becomes diamagnetic material
(c) paramagnetic material becomes ferromagnetic material
(d) ferromagnetic material becomes diamagnetic material.

22. Nickel shows ferromagnetic property at room temperature. If the temperature is increased beyond Curie temperature, then it will show *[2007]*
(a) anti ferromagnetism
(b) no magnetic property
(c) diamagnetism
(d) paramagnetism

23. If the magnetic dipole moment of an atom of diamagnetic material, paramagnetic material and ferromagnetic material are denoted by μ_d, μ_p and μ_f respectively, then *[2005]*
(a) $\mu_d = 0$ and $\mu_p \neq 0$ (b) $\mu_d \neq 0$ and $\mu_p = 0$
(c) $\mu_p = 0$ and $\mu_f \neq 0$ (d) $\mu_d \neq 0$ and $\mu_f \neq 0$

24. According to Curie's law, the magnetic susceptibility of a substance at an absolute temperature T is proportional to *[2003]*
(a) T^2 (b) $1/T$
(c) T (d) $1/T^2$

25. A diamagnetic material in a magnetic field moves *[2003]*
(a) perpendicular to the field
(b) from stronger to the weaker parts of the field
(c) from weaker to the stronger parts of the field
(d) in none of the above directions

Topic 3 Magnetic Equipments

26. A compass needle which is allowed to move in a horizontal plane is taken to a geomagnetic pole. It : *[2012]*
(a) will become rigid showing no movement
(b) will stay in any position
(c) will stay in north-south direction only
(d) will stay in east-west direction only

27. A vibration magnetometer placed in magnetic meridian has a small bar magnet. The magnet executes oscillations with a time period of 2 sec in earth's horizontal magnetic field of 24 microtesla. When a horizontal field of 18 microtesla is produced opposite to the earth's field by placing a current carrying wire, the new time period of magnet will be *[2010]*

(a) 1 *s* (b) 2 *s*
(c) 3 *s* (d) 4 *s*

28. Two magnets of magnetic moments M and 2M are placed in a vibration magnetometer, with the identical poles in the same direction. The time period of vibration is T_1. If the magnets are placed with opposite poles together and vibrate with time period T_2, then *[2002]*

(a) T_2 is infinite. (b) $T_2 = T_1$
(c) $T_2 > T_1$ (d) $T_2 < T_1$

29. For protecting a sensitive equipment from the external electric arc, it should be *[1998]*

(a) wrapped with insulation around it when a current is passing through it
(b) placed inside an iron can
(c) surrounded with fine copper sheet
(d) placed inside an aluminium can

30. A bar magnet is oscillating in the earth's magnetic field with a period T. What happens to its period of motion, if its mass is quadrupled *[1994]*

(a) motion remains simple harmonic with new period = T/2
(b) motion remains simple harmonic with new period = 2 T
(c) motion remains simple harmonic with new period = 4T
(d) motion remains simple harmonic and the period stays nearly constant

ANSWER KEY

1	(c)	5	(a)	9	(c)	13	(c)	17	(a)	21	(a)	25	(b)	29	(b)
2	(a)	6	(b)	10	(a)	14	(a)	18	(d)	22	(d)	26	(b)	30	(b)
3	(b)	7	(b)	11	(d)	15	(d)	19	(a)	23	(a)	27	(d)		
4	(b)	8	(a)	12	(c)	16	(a)	20	(d)	24	(b)	28	(c)		

Hints & Solutions

1. **(c)** Net magnetic dipole moment $= 2\,M\cos\dfrac{\theta}{2}$

As value of $\cos\dfrac{\theta}{2}$ is maximum in case (c) hence net magnetic dipole moment is maximum for option (c).

2. **(a)** Magnetic dipole moment

$M = m \times \ell \qquad M' = m \times r$

From figure

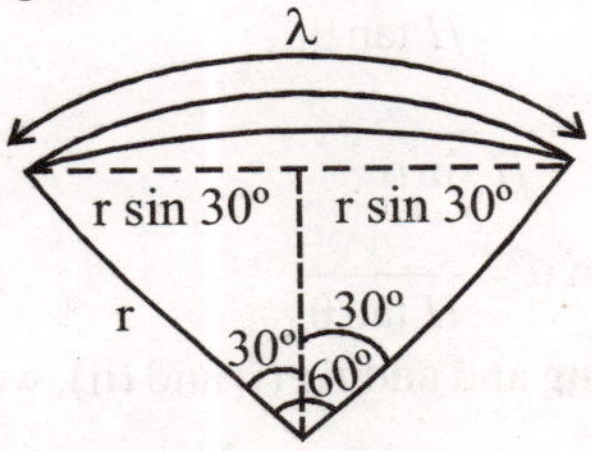

$\ell = \dfrac{\pi r}{3}$ or $r = \dfrac{3\ell}{\pi}$

so, $M' = m \times r = \dfrac{m \times 3\ell}{\pi} = \dfrac{3}{\pi} M$

3. **(b)** Torque $FL = MB \Rightarrow L = \dfrac{MB}{F}$

4. **(b)** According to work energy theorem

$W = U_{final} - U_{initial} = MB\,(\cos 0 - \cos 60°)$

$$W = \frac{MB}{2} = \sqrt{3}J \quad \text{...(i)}$$

$$\tau = \vec{M} \times \vec{B} = MB\sin 60° = \left(\frac{MB\sqrt{3}}{2}\right) \quad \text{...(ii)}$$

From equation (i) and (ii)

$$\tau = \frac{2\sqrt{3} \times \sqrt{3}}{2} = 3J$$

5. **(a)** For stable equilibrium

U= –MB

$= -(0.4)(0.16)$
$= -0.064$ J

NOTES When the magnetic dipole is aligned along the magnetic field it is stable equilibrium having minimum potential energy

$U = MB\cos\theta = -MB\cos\theta°$
$= -MB$ i.e., $\theta = 0°$

When $\theta = 180°$

$U = -MB\cos\theta = -MB\cos 180° = MB$ (Maximum)

This is the position of unstable equilibrium

6. (b) Work done
$= MB(\cos\theta_1 - \cos\theta_2)$
$= MB(\cos 0° - \cos 60°)$

$$= MB\left(1 - \frac{1}{2}\right) = \frac{2\times10^4\times6\times10^{-4}}{2} = 6\text{ J}$$

7. (b) $\tau = MB\sin\theta$
$\tau = iAB\sin 90°$

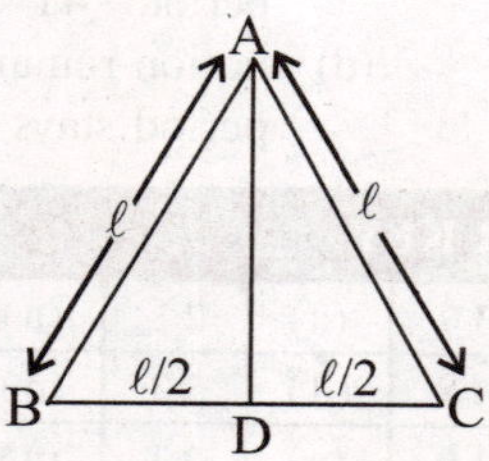

$\therefore A = \frac{\tau}{iB}$

Also, $A = 1/2(BC)(AD)$

But $\frac{1}{2}(BC)(AD) = \frac{1}{2}(l)\sqrt{l^2 - \left(\frac{l}{2}\right)^2}$

$= \frac{\sqrt{3}}{4}l^2 \quad \Rightarrow \quad \frac{\sqrt{3}}{4}(l)^2 = \frac{\tau}{Bi}$

$\therefore \quad l = 2\left(\frac{\tau}{\sqrt{3}\,B.i}\right)^{\frac{1}{2}}$

8. (a) Magnetic moment linked with one turn $= iA$
Magnetic moment linked with,
N turns $= iNA$ amp-m^2. Here, A = Area of current loop.

9. (c) We know that when a bar magnet is placed in the magnetic field at an angle θ, then torque acting on the bar magnet (τ)
$= MB\sin\theta = \vec{M}\times\vec{B}$.

NOTES The direction of torque (τ) is perpendicular to the plane containing M and B and is given by right hand screw rule.

10. (a) Magnetic moment = M; Initial angle through which magnet is turned $(\theta_1) = 90°$ and final angle through which magnet is turned $(\theta_2) = 60°$. Work done in turning the magnet through $90°(W_1) = MB(\cos 0° - \cos 90°)$
$= MB(1-0) = MB$.

Similarly, $W_2 = MB(\cos 0° - \cos 60°)$

$= MB\left(1 - \frac{1}{2}\right) = \frac{MB}{2}$

$\therefore W_1 = 2W_2$ or $n = 2$.

11. (d) Given : Magnetic susceptibility of iron, $\chi_m = 599$
Using, $\mu_r = 1 + \chi_m = 600$
$\mu = \mu_r\mu_0 = 600\times4\pi\times10^{-7} = 2400\pi\times10^{-7}$
Hence, permeability of the material of the rod,
$\mu = 2.4\pi\times10^{-4}$ T m A^{-1}

12. (c) As we know that the angle of dip is the angle between earth's resultant magnetic field from horizontal.

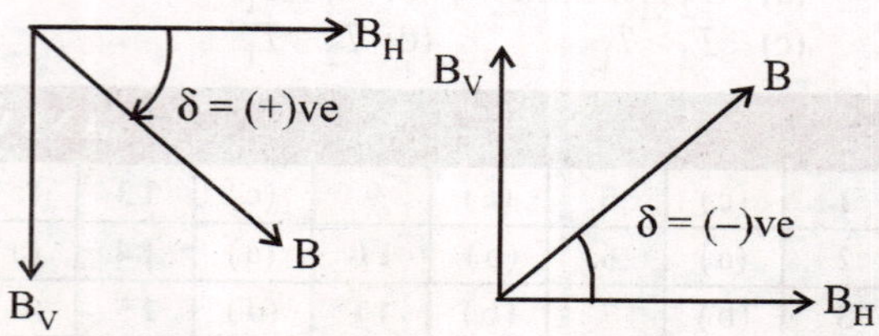

At equator, dip is zero. At Northern hemisphere, dip is positive. At southern hemisphere, dip is negative.

13. (c) $H = B_E\cos\delta$
$V = B_E\sin\delta$

14. (a) Rod gains gravitational potential energy which comes from energy of current source.

15. (d) If θ_1 and θ_2 are opparent angles of dip Let α be the angle which one of the plane make with the magnetic meridian.

$\tan\theta_1 = \frac{v}{H\cos\alpha}$

i.e., $\cos\alpha = \frac{v}{H\tan\theta_1}$ ...(i)

$\tan\theta_2 = \frac{v}{H\sin\alpha}$,

i.e., $\sin\alpha = \frac{v}{H\tan\theta_2}$ ...(ii)

Squaring and adding (i) and (ii), we get

$$\cos^2\alpha + \sin^2\alpha = \left(\frac{V}{H}\right)^2\left(\frac{1}{\tan^2\theta_1} + \frac{1}{\tan^2\theta_2}\right)$$

i.e., $1=\frac{V^2}{H^2}\left[\cot^2\theta_1+\cot^2\theta_2\right]$

or $\frac{H^2}{V^2}=\cot^2\theta_1+\cot^2\theta_2$

i.e., $\cot^2\theta=\cot^2\theta_1+\cot^2\theta_2$

16. **(a)** Magnetic susceptibility χ for dia-magnetic materials only is negative and low $|\chi| = -1$; for paramagnetic substances low but positive $|\chi| = 1$ and for ferromagnetic substances positive and high $|\chi| = 10^2$.

17. **(a)** A $\rightarrow$ diamagnetic
B $\rightarrow$ paramagnetic
C $\rightarrow$ Ferromagnetic
D $\rightarrow$ Non magnetic

18. **(d)** Soft iron is a soft magnetic material soft iron has low retentivity and low coercive force or coercivity suitable for electromagnets.
Retentivity is the residual magnetism i.e., the magnetism remains in a materials even on the removal of the magnetising field.
Magnetic hard substance like steel has high coercivity and magnetic soft substance-soft iron has low coercivity.

19. **(a)** The magnetic moment of a diamagnetic atom is equal to zero.

20. **(d)** Diamagnetic substances do not have any unpaired electron. And they are magnetised in direction opposite to that of magnetic field. Hence, when they are brought to north or south pole of a bar magnet, they are repelled by poles.

21. **(a)** Curie temperature is the temperature above which ferromagnetic material becomes paramagnetic material.

22. **(d)** Beyond Curie temperature, ferromagnetic substances behaves like a paramagnetic substance.

23. **(a)** The magnetic dipole moment of diamagnetic material is zero as each of its pair of electrons have opposite spins, i.e., $\mu_d = 0$.
Paramagnetic substances have dipole moment > 0, i.e. $\mu_p \neq 0$, because of excess of electrons in its molecules spinning in the same direction.
Ferro-magnetic substances are very strong magnets and they also have permanent magnetic moment, i.e. $\mu_p \neq 0$.

24. **(b)** According to Curie's law, $\chi_m \propto \frac{1}{T}$

25. **(b)** A diamagnetic material in a magnetic field moves from stronger to the weaker parts of the field.

26. **(b)** Since magnetic field is in vertical direction and needle is free to totate in horizontal plane only so magnetic force cannot rotate the needle in horizontal plane so needle can stay in any position.

27. **(d)** Time period of a vibration magnetometer,

$$T\propto\frac{1}{\sqrt{B}} \Rightarrow \frac{T_1}{T_2}=\sqrt{\frac{B_2}{B_1}}$$

$$\Rightarrow T_2=T_1\sqrt{\frac{B_1}{B_2}}=2\sqrt{\frac{24\times10^{-6}}{6\times10^{-6}}}=4\,s$$

28. **(c)** $T_1=2\pi\sqrt{\frac{I_1+I_2}{(M+2M)H}}=2\pi\sqrt{\frac{I}{3MH}}$

$$T_2=2\pi\sqrt{\frac{I_1+I_2}{(2M-M)H}}=2\pi\sqrt{\frac{I}{MH}}$$

Obviously, $T_2 > T_1$

29. **(b)** The iron can produces a magnetic screening for the equipment as lines of magnetic force can not enter iron enclosure.

NOTES

When two magnets are placed in a vibration magnetometer such that identical poles in same direction then net magnetic moment $M_s = M_1 + M_2$

And when two magnets placed such that north facing south of other and vice versa then net magnetic moment $M_d = M_1 - M_2$

30. **(b)** The time period of a bar magnet in a magnetic field is given by.

$$T=2\pi\sqrt{\frac{I}{MB}};$$

Here, I = moment of inertia $\propto m$, M = moment of magnet, B = magnetic field.

$T\propto\sqrt{I}\propto\sqrt{m}$; so, T becomes twice as mass becomes four times

20 Electromagnetic Induction

Trend Analysis with Important Topics & Sub-Topics

Topic Name	Sub-Topic	2021		2020		2019		2018		2017		2016	
		QNS.	LOD	QNS	LOD	QNS.	LOD	QNS.	LOD	QNS.	LOD	QNS.	LOD
Magnetic Flux, Faraday's & Lenz's Law	$\phi = B.A. = BA\cos\theta$, $q = \frac{\lvert\Delta\phi\rvert}{R}$					1	A		1	A			
Motional and Static EMI & Applications of EMI	Eddy Current					1	E						
	Energy stored in inductor							1	E				
	Self inductance, $L = \frac{\phi}{I}$											1	E
	Mutual inductance $M = \frac{\phi}{I}$	1	A										

LOD - Level of Difficulty | E - Easy | A - Average | D - Difficult | Qns - No. of Questions

Topic 1 Magnetic Flux, Faraday's & Lenz's Law

1. A 800 turn coil of effective area 0.05 m^2 is kept perpendicular to a magnetic field 5×10^{-5} T. When the plane of the coil is rotated by 90° around any of its coplanar axis in 0.1 s, the emf induced in the coil will be : *[2019]*
   (a) 2 V (b) 0.2 V
   (c) 2×10^{-3} V (d) 0.02 V

2. A long solenoid of diameter 0.1 m has 2×10^4 turns per meter. At the centre of the solenoid, a coil of 100 turns and radius 0.01 m is placed with its axis coinciding with the solenoid axis. The current in the solenoid reduces at a constant rate to 0A from 4 A in 0.05 s. If the resistance of the coil is $10\pi^2\Omega$. the total charge flowing through the coil during this time is :- *[2017]*
   (a) 16 μC (b) 32 μC
   (c) 16 π μC (d) 32 π μC

3. An electron moves on a straight line path XY as shown. The abcd is a coil adjacent to the path of electron. What will be the direction of current if any, induced in the coil?

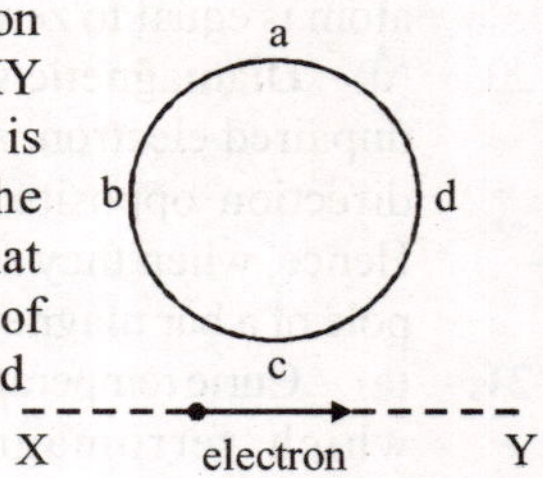

[2015 RS]
   (a) adcb
   (b) The current will reverse its direction as the electron goes past the coil
   (c) No current induced
   (d) abcd

4. A thin semicircular conducting ring (PQR) of radius 'r' is falling with its plane vertical in a horizontal magnetic field B, as shown in figure. The potential difference developed across the ring when its speed is v, is : *[2014]*

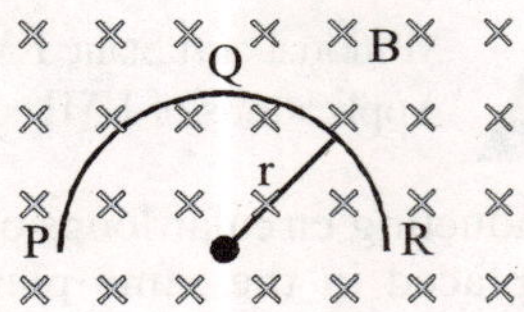

(a) Zero
(b) $Bv\pi r^2/2$ and P is at higher potnetial
(c) πrBv and R is at higher potnetial
(d) 2rBv and R is at higher potential

5. A wire loop is rotated in a magnetic field. The frequency of change of direction of the induced e.m.f. is *[2013]*
(a) twice per revolution
(b) four times per revolution
(c) six times per revolution
(d) once per revolution

6. A current of 2.5 A flows through a coil of inductance 5 H. The magnetic flux linked with the coil is *[NEET Kar. 2013]*
(a) 2 Wb (b) 0.5 Wb
(c) 12.5 Wb (d) Zero

7. A coil of resistance 400Ω is placed in a magnetic field. If the magnetic flux φ (wb) linked with the coil varies with time *t* (sec) as $\phi = 50t^2 + 4$. The current in the coil at $t = 2$ sec is : *[2012]*
(a) 0.5 A (b) 0.1 A
(c) 2 A (d) 1 A

8. In a coil of resistance 10 Ω, the induced current developed by changing magnetic flux through it, is shown in figure as a function of time. The magnitude of change in flux through the coil in Weber is : *[2012M]*

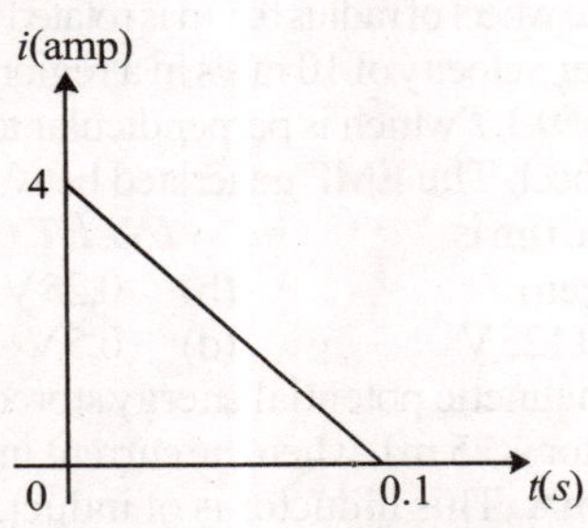

(a) 8 (b) 2
(c) 6 (d) 4

9. The current i in a coil varies with time as shown in the figure. The variation of induced emf with time would be *[2011]*

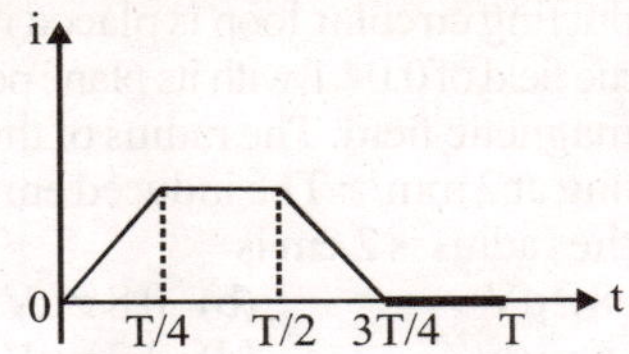

(a)
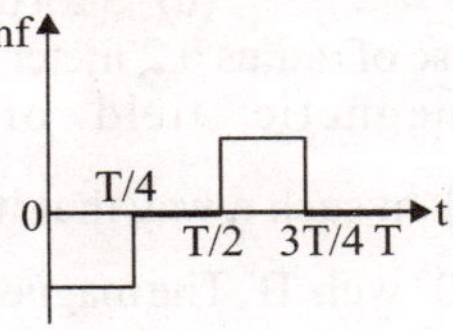

(b)
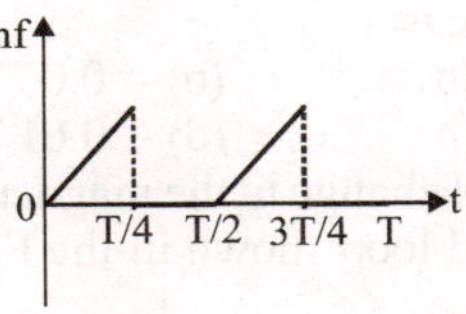

(c)
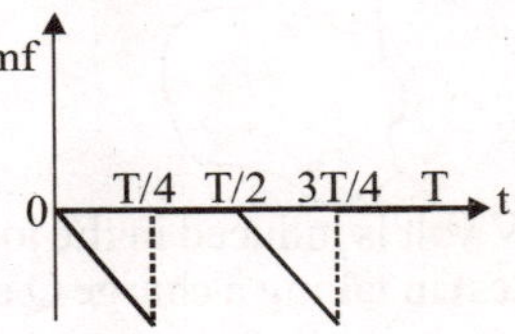

(d)
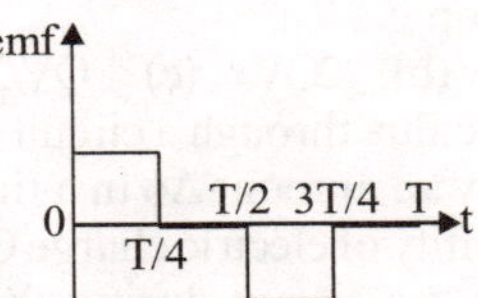

10. A conducting circular loop is placed in a uniform magnetic field, $B = 0.025\ T$ with its plane perpendicular to the loop. *[2010]*
The radius of the loop is made to shrink at a constant rate of 1 mm s^{-1}. The induced e.m.f. when the radius is 2 cm, is
(a) $2\pi\mu V$ (b) $\pi\mu V$
(c) $\frac{\pi}{2}\mu V$ (d) $2\mu V$

11. A rectangular, a square, a circular and an elliptical loop, all in the (x – y) plane, are moving out of a uniform magnetic field with a constant velocity, $\vec{V} = v\hat{i}$. The magnetic field is directed along the negative z axis direction. The induced emf, during the passage of these loops, out of the field region, will not remain constant for *[2009]*
(a) the circular and the elliptical loops.
(b) only the elliptical loop.
(c) any of the four loops.
(d) the rectangular, circular and elliptical loops.

12. A conducting circular loop is placed in a uniform magnetic field of 0.04 T with its plane perpendicular to the magnetic field. The radius of the loop starts shrinking at 2 mm/s. The induced emf in the loop when the radius is 2 cm is *[2009]*
(a) $4.8\pi\mu V$ (b) $0.8\pi\mu V$
(c) $1.6\pi\mu V$ (d) $3.2\pi\mu V$

13. A circular disc of radius 0.2 meter is placed in a uniform magnetic field of induction $\frac{1}{\pi}\left(Wb/m^2\right)$ in such a way that its axis makes an angle of 60° with $\vec{B}$. The magnetic flux linked with the disc is: *[2008]*
(a) 0.02 Wb (b) 0.06 Wb
(c) 0.08 Wb (d) 0.01 Wb

14. As a result of change in the magnetic flux linked to the closed loop shown in the Fig,

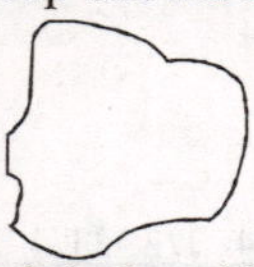

an e.m.f. V volt is induced in the loop. The work done (joules) in taking a charge Q coulomb once along the loop is *[2005]*
(a) QV (b) 2QV (c) QV/2 (d) Zero

15. The magnetic flux through a circuit of resistance R changes by an amount $\Delta\phi$ in a time Δt. Then the total quantity of electric charge Q that passes any point in the circuit during the time Δt is represented by *[2004]*
(a) $Q = R.\frac{\Delta\phi}{\Delta t}$ (b) $Q = \frac{1}{R}.\frac{\Delta\phi}{\Delta t}$
(c) $Q = \frac{\Delta\phi}{R}$ (d) $Q = \frac{\Delta\phi}{\Delta t}$

16. A magnetic field of 2×10^{-2} T acts at right angles to a coil of area 100 cm^2, with 50 turns. The average e.m.f. induced in the coil is 0.1 V, when it is removed from the field in t sec. The value of t is *[1991]*
(a) 10 s (b) 0.1 s
(c) 0.01 s (d) 1 s

17. In the circuit of Fig, the bulb will become suddenly bright if *[1989]*

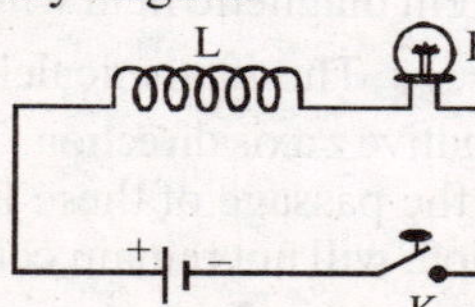

(a) contact is made or broken
(b) contact is made
(c) contact is broken
(d) won't become bright at all

Topic 2 Motional and Static EMI & Applications of EMI

18. Two conducting circular loops of radii R_1 and R_2 are placed in the same plane with their centres coinciding. If $R_1 >> R_2$, the mutual inductance M between them will be directly proportional to *[2021]*
(a) $\frac{R_2^2}{R_1}$ (b) $\frac{R_1}{R_2}$
(c) $\frac{R_2}{R_1}$ (d) $\frac{R_1^2}{R_2}$

19. In which of the following devices, the eddy current effect is not used ? *[2019]*
(a) induction furnace
(b) magnetic braking in train
(c) electromagnet
(d) electric heater

20. The variation of EMF with time for four types of generators are shown in the figures. Which amongst them can be called AC? *[NEET Odisha 2019]*

(a) 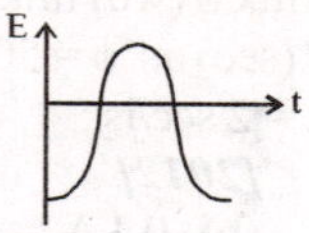

(b) 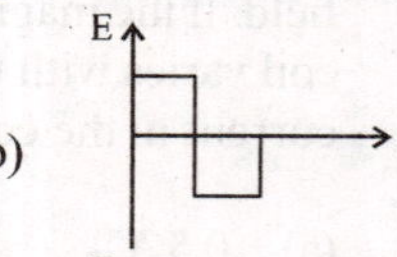

(c) 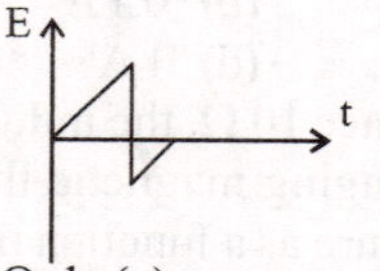

(d) 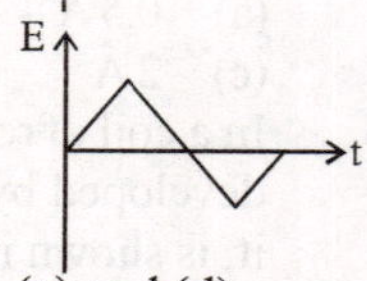

(a) Only (a) (b) (a) and (d)
(c) (a), (b), (c), (d) (d) (a) and (b)

21. A cycle wheel of radius 0.5 m is rotated with constant angular velocity of 10 rad/s in a region of magnetic field of 0.1 *T* which is perpendicular to the plane of the wheel. The EMF generated between its centre and the rim is *[NEET Odisha 2019]*
(a) zero (b) 0.25 V
(c) 0.125 V (d) 0.5 V

22. The magnetic potential energy stored in a certain inductor is 25 mJ, when the current in the inductor is 60 mA. This inductor is of inductance *[2018]*
(a) 0.138 H (b) 138.88 H
(c) 13.89 H (d) 1.389 H

23. A long solenoid has 1000 turns. When a current of 4A flows through it, the magnetic flux linked with each turn of the solenoid is 4×10^{-3} Wb. The self inductance of the solenoid is : *[2016]*
(a) 4H (b) 3H (c) 2H (d) 1H

24. A conducting square frame of side 'a' and a long staight wire carrying current I are located in the same plane as shown in the figure. The frame moves to the right with a constant velocity 'V'. The emf induced in the frame will be proportional to *[2015]*

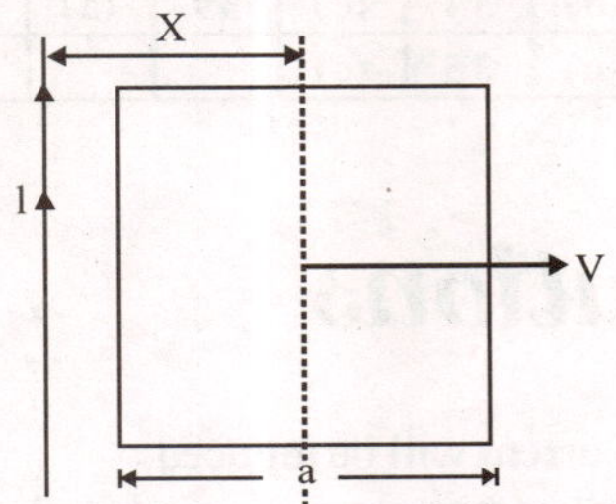

(a) $\frac{1}{(2x-a)^2}$ (b) $\frac{1}{(2x+a)^2}$

(c) $\frac{1}{(2x-a)(2x+a)}$ (d) $\frac{1}{x^2}$

25. A long solenoid has 500 turns. When a current of 2 ampere is passed through it, the resulting magnetic flux linked with each turn of the solenoid is 4×10^{-3} Wb. The self- inductance of the solenoid is *[2008]*

(a) 2.5 henry (b) 2.0 henry
(c) 1.0 henry (d) 40 henry

26. Two coils of self inductances 2 mH and 8 mH are placed so close together that the effective flux in one coil is completely linked with the other. The mutual inductance between these coils is *[2006]*

(a) 6mH (b) 4mH
(c) 16mH (d) 10mH

27. In an inductor of self-inductance L = 2 mH, current changes with time according to relation $i = t^2 e^{-t}$. At what time emf is zero? *[2001]*

(a) 4s (b) 3s (c) 2s (d) 1s

28. Two coils have a mutual inductance 0.005 H. The current changes in the first coil according to equation $I = I_0 \sin \omega t$, where I_0 = 10A and $\omega = 100\pi$ radian/sec. The maximum value of e.m.f. in the second coil is *[1998]]*

(a) 2p (b) 5p (c) p (d) 4p

29. A conductor of length 0.4 m is moving with a speed of 7 m/s perpendicular to a magnetic field of intensity 0.9 Wb/m^2. The induced e.m.f. across the conductor is *[1995]*

(a) 1.26V (b) 2.52V
(c) 5.04V (d) 25.2V

30. A varying current in a coil changes from 10A to zero in 0.5 sec. If the average e.m.f induced in the coil is 220V, the self-inductance of the coil is *[1995]*

(a) 5 H (b) 6 H
(c) 11 H (d) 12 H

31. What is the self-inductance of a coil which produces 5V when the current changes from 3 ampere to 2 ampere in one millisecond? *[1993]*

(a) 5000 henry (b) 5 milli-henry
(c) 50 henry (d) 5 henry

32. If N is the number of turns in a coil, the value of self inductance varies as *[1993]*

(a) N^0 (b) N (c) N^2 (d) N^{-2}

33. A rectangular coil of 20 turns and area of cross-section 25 sq. cm has a resistance of 100Ω. If a magnetic field which is perpendicular to the plane of coil changes at a rate of 1000 tesla per second, the current in the coil is *[1992]*

(a) 1 A (b) 50 A (c) 0.5 A (d) 5 A

34. The total charge induced in a conducting loop when it is moved in a magnetic field depend on *[1992]*

(a) the rate of change of magnetic flux
(b) initial magnetic flux only
(c) the total change in magnetic flux
(d) final magnetic flux only

35. A 100 millihenry coil carries a current of 1A. Energy stored in its magnetic field is *[1991]*

(a) 0.5 J (b) 1 A (c) 0.05 J (d) 0.1 J

36. If the number of turns per unit length of a coil of solenoid is doubled, the self-inductance of the solenoid will *[1991]*

(a) remain unchanged (b) be halved
(c) be doubled (d) become four times

37. An inductor may store energy in *[1990]*

(a) its electric field
(b) its coils
(c) its magnetic field
(d) both in electric and magnetic fields

38. The total charge induced in a conducting loop when it is moved in a magnetic field depends on *[1990]*

(a) the rate of change of magnetic flux
(b) initial magnetic flux only
(c) the total change in magnetic flux
(d) final magnetic flux only

39. The current in self inductance L = 40 mH is to be increased uniformly from 1 amp to 11 amp in 4 milliseconds. The e.m.f. induced in the inductor during the process is *[1990]*

(a) 100 volt (b) 0.4 volt
(c) 4.0 volt (d) 440 volt

ANSWER KEY															
1	(d)	6	(c)	11	(a)	16	(b)	21	(c)	26	(b)	31	(b)	36	(d)
2	(b)	7	(a)	12	(d)	17	(c)	22	(c)	27	(c)	32	(c)	37	(c)
3	(b)	8	(b)	13	(a)	18	(a)	23	(d)	28	(b)	33	(c)	38	(c)
4	(d)	9	(a)	14	(a)	19	(d)	24	(c)	29	(b)	34	(c)	39	(a)
5	(a)	10	(b)	15	(c)	20	(c)	25	(c)	30	(c)	35	(c)		

Hints & Solutions

1. **(d)** Given :

Magnetic field $B = 5 \times 10^{-5}$ T

Number of turns in coil N = 800

Area of coil $A = 0.05\ m^2$

Time taken to rotate $= \Delta t = 0.1$ s

Initial angle $\theta_1 = 0°$

Final angle $\theta_2 = 90°$

Change in magnetic flux $\Delta\phi$

$= NBA \cos 90° - BA \cos 0°$

$= -NBA$

$= -800 \times 5 \times 10^{-5} \times 0.05$

$= -2 \times 10^{-2}$ weber

$$\theta = -\frac{\Delta\phi}{\Delta t} = \frac{-(-)2 \times 10^{-3}\ \text{Wb}}{0.1\ \text{s}} = 0.02\ \text{V}$$

2. **(b)** Given, no. of turns N = 100

radius, r = 0.01 m

resistance, $R = 10\pi^2\ \Omega$, $n = 2 \times 10^4$

As we know,

$$\varepsilon = -N\frac{d\phi}{dt} \Rightarrow \frac{\varepsilon}{R} = -\frac{N}{R}\frac{d\phi}{dt}$$

$$\Delta I = -\frac{N}{R}\frac{d\phi}{dt} \Rightarrow \frac{\Delta q}{\Delta t} = -\frac{N}{R}\frac{\Delta\phi}{\Delta t}$$

$$\Delta q = -\left[\frac{N}{R}\left(\frac{\Delta\phi}{\Delta t}\right)\right]\Delta t$$

'–' ve sign shows that induced emf opposes the change of flux.

$$\Delta q = \left[\mu_0 nN\pi r^2\left(\frac{\Delta i}{\Delta t}\right)\right]\frac{1}{R}\Delta t = \frac{\mu_0 nN\pi r^2 \Delta i}{R}$$

$$\Delta q = \frac{4\pi \times 10^{-7} \times 100 \times 4 \times \pi \times (0.01)^2 \times 2 \times 10^4}{10\pi^2}$$

$\Delta q = 32\mu C$

3. **(b)** Current will be induced,

when e^- comes closer the induced current will be anticlockwise

when e^- comes further induced current will be clockwise

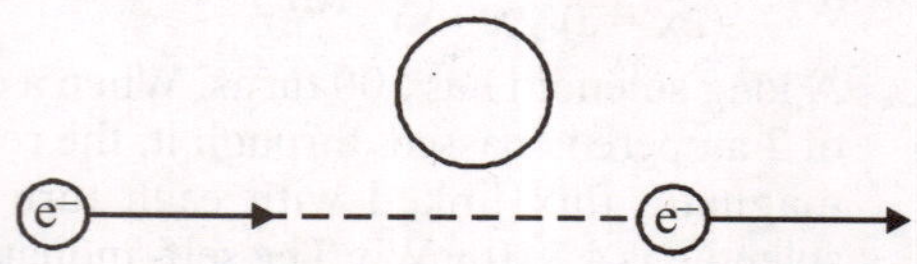

NOTES Presence of magnetic flux is not enough. The amount of magnetic flux linked with the coil must change in order to produce any induced emf in the coil.

4. **(d)** Rate of decrease of area of semicircular ring

$$= \frac{dA}{dt} = (2r)V$$

From Faraday's law of electromagnetic induction

$$e = -\frac{d\phi}{dt} = -B\frac{dA}{dt} = -B(2rV)$$

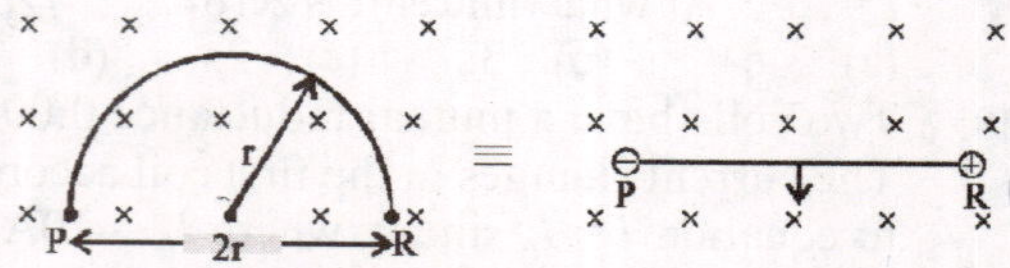

As induced current in ring produces magnetic field in upward direction hence R is at higher potential.

5. **(a)** This is the case of periodic EMI

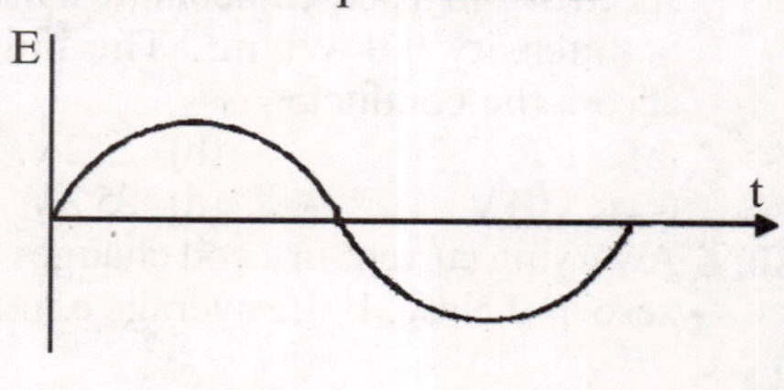

From graph, it is clear that direction is changing once in $\frac{1}{2}$ cycle.

6. (c) Given: current I = 2.5 A
Inductance, L = 5H
Magnatic flux, ϕ = ?
We know, $\phi = LI \Rightarrow 5 \times 2.5$ Wb = 12.5 Wb

7. (a) According, to Faraday's law of induction

Induced e.m.f. $\varepsilon = -\frac{d\phi}{dt} = -(100t)$

Induced current i at $t = 2$ sec.

$$= \left|\frac{\varepsilon}{R}\right| = +\frac{100 \times 2}{400} = +0.5\,\text{A}$$

NOTES The magnetic flux linked with a loop does not change when magnet and loop are moving with the same velocity.

8. (b) The charge through the coil = area of current-time($i-t$) graph

$$q = \frac{1}{2} \times 0.1 \times 4 = 0.2\,\text{C}$$

$q = \frac{\Delta\phi}{R}$ $\because$ Change in flux $(\Delta\phi) = q \times R$

$$q = 0.2 = \frac{\Delta\phi}{10}$$

$\Delta\phi = 2$ Weber

9. (a) $e = -L\frac{di}{dt}$

During 0 to $\frac{T}{4}, \frac{di}{dt}$ = const.
$\therefore$ e = – ve

During $\frac{T}{4}$ to $\frac{T}{2}, \frac{di}{dt} = 0$
$\therefore$ e = 0

During $\frac{T}{2}$ to $\frac{3T}{4}, \frac{di}{dt}$ = const.
$\therefore$ e = +ve

Thus graph given in option (a) represents the variation of induced emf with time.

NOTES Inductance does not play any role till there is a constant current flowing in the circuit. It opposes any change in the current in the circuit.

10. (b) Magnetic flux linked with the loop is $\phi = B\pi r^2$

$$|e| = \frac{d\phi}{dt} = B\pi \cdot 2r\frac{dr}{dt}$$

When $r = 2$ cm, $\frac{dr}{dt} = 1\,\text{mm}\,s^{-1}$

$e = 0.025 \times \pi \times 2 \times 2 \times 10^{-2} \times 10^{-3}$
$= 0.100 \times \pi \times 10^{-5} = \pi \times 10^{-6}\,\text{V} = \pi\mu\text{V}$

11. (a) The induced emf will remain constant only in the case of rectangular and square loops. In case of the circular and the elliptical loops, the rate of change of area of the loops during their passage out of the field is not constant, hence induced emf will not remain constant for them.

12. (d) Induced emf in the loop is given by

$e = -B.\frac{dA}{dt}$ where A is the area of the loop.

$$e = -B.\frac{d}{dt}(\pi r^2) = -B\pi 2r\frac{dr}{dt}$$

r = 2cm = 2×10^{-2} m
dr = 2 mm = 2×10^{-3} m
dt = 1 s

$$e = -0.04 \times 3.14 \times 2 \times 2 \times 10^{-2} \times \frac{2 \times 10^{-3}}{1}\text{V}$$

$= 0.32\,\pi \times 10^{-5}$V
$= 3.2\,\pi \times 10^{-6}$V
$= 3.2\,\pi\mu$V

13. (a) Here, $B = \frac{1}{\pi}$ (Wb/m^2)
$\theta = 60°$
Area normal to the plane of the disc

$$= \pi r^2 \cos 60° = \frac{\pi r^2}{2}$$

Flux = B × normal area

$$= \frac{0.2 \times 0.2}{2} = 0.02\,\text{Wb}$$

14. (a) $\xi = \frac{W}{Q} \Rightarrow V = \frac{W}{Q} \Rightarrow W = QV$

15. (c) $\frac{\Delta\phi}{\Delta t} = \varepsilon = iR \Rightarrow \Delta\phi = (i\Delta t)R = \text{QR}$

$$\Rightarrow Q = \frac{\Delta\phi}{R}$$

16. (b) $e = \frac{-(\phi_2 - \phi_1)}{t} = \frac{-(0 - NBA)}{t} = \frac{NBA}{t}$

$$t = \frac{NBA}{e} = \frac{50 \times 2 \times 10^{-2} \times 10^{-2}}{0.1} = 0.1\,s$$

17. (c) When a circuit is broken, the induced e.m.f. is largest. So the answer is (c).

18. (a) Let i be the current in outer loop

Magnetic field at centre $B = \frac{\mu_0 i}{2R_1}$

Magnetic flux through inner coil $\phi_i = B \times \pi R_2^2$

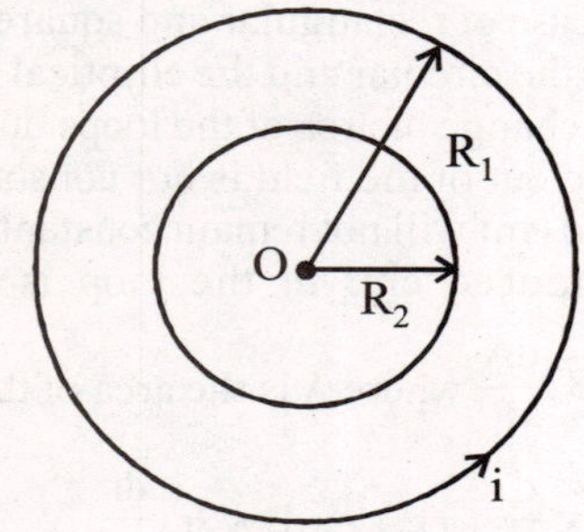

$$\phi_i = \frac{\mu_0 i}{2R_1} \times \pi R_2^2$$

$$\phi_i = \frac{\mu_0 i}{2} \times \frac{\pi R_2^2}{R_1}$$

$\therefore \phi = Mi$

$$\therefore M = \left(\frac{\mu_0 \pi}{2}\right)\frac{R_2^2}{R_1} \text{ or, } M \propto \frac{R_2^2}{R_1}$$

19. (d) An electric heater works on the principle of Joule's heating effect and it does not involve Eddy currents.

20. (c) A current which changes its direction periodically is called alternating current.

21. (c) Using Faraday's law,

$$e = \frac{Bl^2\omega}{2} = \frac{1}{2} \times 0.1 \times \left(\frac{1}{2}\right)^2 \times 10 = \frac{1}{8} = 0.125 \text{ V}$$

22. (c) From question energy stored in inductor, $U = 25 \times 10^{-3}$ J
Current, I = 60 mA

Energy stored in inductor $U = \frac{1}{2}LI^2$

$$25 \times 10^{-3} = \frac{1}{2} \times L \times (60 \times 10^{-3})^2$$

$$L = \frac{25 \times 2 \times 10^6 \times 10^{-3}}{3600} = 13.89 \text{ Hm}$$

23. (d) Here, number of turns n = 1000; current through the solenoid i = 4A; flux linked with each turn $= 4 \times 10^{-3}$ Wb

$\therefore$ Total flux linked,
$= 1000[4 \times 10^{-3}] = 4$ Wb

$\phi_{total} = 4 \Rightarrow Li = 4$

$\Rightarrow L = 1$ H

NOTES
Number of flux linkages with the coil $\propto$ current (I)
$N\phi \propto I \Rightarrow N\phi = L_I \Rightarrow L = \frac{N\phi}{I}$.

24. (c) Emf induced in side 1 of frame $e_1 = B_1 V\ell$

$$B_1 = \frac{\mu_0 I}{2\pi (x - a/2)}$$

Emf induced in side 2 of frame $e_2 = B_2 V\ell$

$$B_2 = \frac{\mu_0 I}{2\pi (x + a/2)}$$

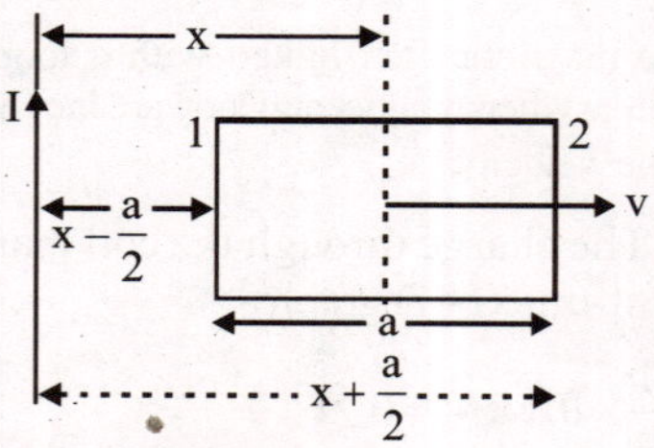

Emf induced in square frame
$e = B_1 V\ell - B_2 V\ell$

$$= \frac{\mu_0 I}{2\pi (x - a/2)}\ell V - \frac{\mu_0 I}{2\pi (x + a/2)}\ell V$$

$$\text{or, } e \propto \frac{1}{(2x - a)(2x + a)}$$

25. (c) Total number of turns in the solenoid, N = 500
Current, I = 2A.
Magnetic flux linked with each turn
$= 4 \times 10^{-3}$ Wb

As, $\phi = LI$ or $N\phi = LI \Rightarrow L = \frac{N\phi}{1}$

$$= \frac{500 \times 4 \times 10^{-3}}{2} \text{ henry} = 1 \text{ H.}$$

26. (b) Mutual Inductance of two coils

$$M = \sqrt{M_1 M_2} = \sqrt{2mH \times 8mH} = 4mH$$

NOTES
It is not possible to have mutual inductance without self inductance but if may or may not be possible self inductance without mutual inductance.

27. (c) $L = 2$mH, $i = t^2 e^{-t}$

$$E = -L\frac{di}{dt} = -L[-t^2 e^{-t} + 2te^{-t}]$$

when $E=0$,
$-e^{-t}t^2+2te^{-t}=0$
or, $2te^{-t}=e^{-t}t^2$
$\Rightarrow t=2$ sec.

28. (b) $e=-M\dfrac{di}{dt}=-0.005\times\dfrac{d(i_0\sin\omega t)}{dt}$
$=-0.005\times i_0\times(\omega\cos\omega t)$
$e_{max.}=0.005\times i_0\times\omega$ (when $\cos\omega t=-1$)
$=0.005\times10\times100\pi=5\pi$ V

29. (b) Length of conductor (l) = 0.4 m; Speed (v) = 7 m/s and magnetic field (B) = 0.9 Wb/ m^2. Induced e.m.f. (V) = $Blv\sin\theta=0.9\times0.4\times7\times\sin 90°=2.52$ V.

30. (c) Initial current (I_1) = 10 A; Final current (I_2) = 0; Time (t) = 0.5 sec and induced e.m.f. (ε) = 220 V.
Induced e.m.f. (ε)
$$=-L\frac{dI}{dt}=-L\frac{(I_2-I_1)}{t}=-L\frac{(0-10)}{0.5}=20L$$
or, $L=\dfrac{220}{20}=11$H
[where L = Self inductance of coil]

31. (b) $L=\dfrac{e}{\frac{di}{dt}}=\dfrac{e\,dt}{di}=\dfrac{5\times10^{-3}}{(3-2)}\text{H}=5\text{mH}$

32. (c) $L=\dfrac{N\phi}{i};\ \phi=BA;\ B=\dfrac{\mu_0Ni}{2R}$
$$\Rightarrow L=\frac{N}{i}\left(\frac{\mu_0Ni}{2R}\right)A=\frac{\mu_0N^2}{2R}A\Rightarrow L\propto N^2$$

33. (c) $i=\dfrac{e}{R}=\dfrac{\frac{nAdB}{dt}}{R}$
$$=\frac{20\times(25\times10^{-4})\times1000}{100}=0.5\text{A}$$

34. (c) $q=\int idt=\dfrac{1}{R}\int edt=\dfrac{1}{R}\int\left(\dfrac{-d\phi}{dt}\right)dt$
$=\dfrac{1}{R}\int d\phi$ (taking only magnitude of e)
Hence, total charge induced in the conducting loop depends upon the total change in magnetic flux.

35. (c) $E=\dfrac{1}{2}Li^2=\dfrac{1}{2}\times(100\times10^{-3})\times1^2=0.05$ J

NOTES: In building a steady current in the circuit, the source emf has to do work against self inductance of coil and whatever energy consumed for tis work stored in magnetic field of coil-called magnetic potential energy (u) of coil $u=\dfrac{1}{2}LI^2=\dfrac{N\phi i}{2}$

36. (d) Self inductance of a solenoid = $\dfrac{\mu n^2A}{\ell}$
So, self inductance $\propto n^2$
So, inductance becomes 4 times when n is doubled.

37. (c) An inductor stores energy in its magnetic field.

38. (c) $e=\dfrac{d\phi}{dt};\ i=\dfrac{e}{R}=\dfrac{1}{R}\dfrac{d\phi}{dt}$
Total charge induced = $\int i\,dt=\int\dfrac{1}{R}\dfrac{d\phi}{dt}.dt$
$$=\frac{1}{R}\int_{\phi_1}^{\phi_2}d\phi=\frac{1}{R}(\phi_2-\phi_1)$$

39. (a) $e=L\dfrac{di}{dt}$
Given that $L=40\times10^{-3}$ H,
di = 11 A – 1 A = 10 A
and $dt=4\times10^{-3}$ s
$\therefore e=40\times10^{-3}\times\left(\dfrac{10}{4\times10^{-3}}\right)=100$V

21 Alternating Current

Trend Analysis with Important Topics & Sub-Topics

Topic Name	Sub-Topic	2021 QNS.	2021 LOD	2020 QNS.	2020 LOD	2019 QNS.	2019 LOD	2018 QNS.	2018 LOD	2017 QNS.	2017 LOD	2016 QNS.	2016 LOD
Alternating Current, Voltage & Power	Alternating Current, Voltage & Power			1	E								
A.C. Circuit, LCR Circuit, Quality & Power Factor	Current, Power dissipated in LCR							1	A	1	A	2	A
	Impedance	1	A										
	LCR, Quality & Power Factor, Resonant Frequency	1	A	1	A								
Transformers & LC Oscillations	Step down Transformers	1	E										

LOD - Level of Difficulty | E - Easy | A - Average | D - Difficult | Qns - No. of Questions

Topic 1 Alternating Current, Voltage & Power

1. A 40 μF capacitor is connected to a 200 V, 50 Hz ac supply. The rms value of the current in the circuit is, nearly : *[2020]*
   (a) 2.05 A (b) 2.5 A
   (c) 25.1 A (d) 1.7 A

2. A small signal voltage $V(t) = V_0 \sin \omega t$ is applied across an ideal capacitor C : *[2016]*
   (a) Current I (t), lags voltage V(t) by 90°.
   (b) Over a full cycle the capacitor C does not consume any energy from the voltage source.
   (c) Current I (t) is in phase with voltage V(t).
   (d) Current I (t) leads voltage V(t) by 180°.

3. The current (I) in the inductance is varying with time according to the plot shown in figure. *[2012]*

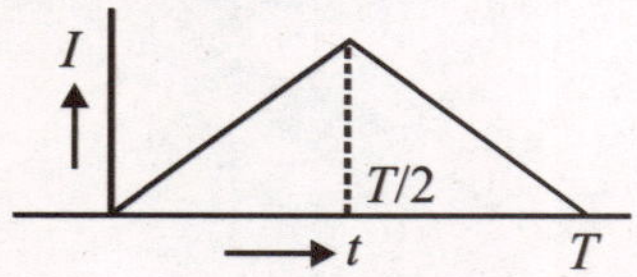

Which one of the following is the correct variation of voltage with time in the coil?

(a)
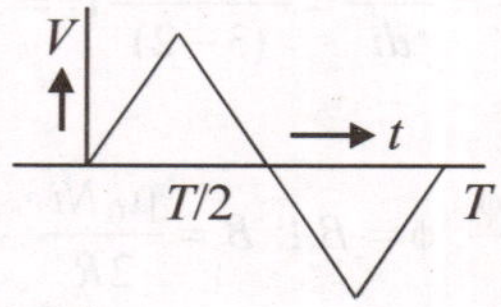

(b)
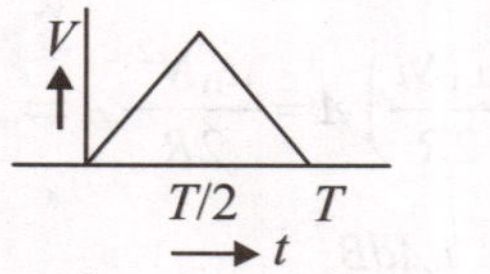

(c)
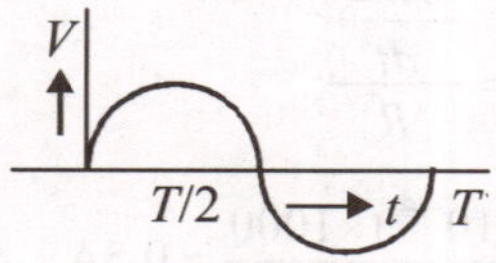

(d)
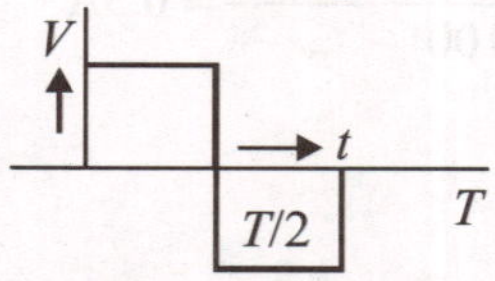

4. The instantaneous values of alternating current and voltages in a circuit are given as

$$i = \frac{1}{\sqrt{2}} \sin(100\pi t) \text{ amper}$$

$$e = \frac{1}{\sqrt{2}} \sin(100\pi t + \pi/3) \text{ Volt}$$

The average power in Watts consumed in the circuit is : *[2012M]*

(a) $\frac{1}{4}$ (b) $\frac{\sqrt{3}}{4}$ (c) $\frac{1}{2}$ (d) $\frac{1}{8}$

5. In an ac circuit an alternating voltage $e = 200\sqrt{2}\sin 100t$ volts is connected to a capacitor of capacity 1 μF. The r.m.s. value of the current in the circuit is *[2011]*

(a) 10 mA (b) 100 mA
(c) 200 mA (d) 20 mA

6. The r.m.s. value of potential difference V shown in the figure is *[2011M]*

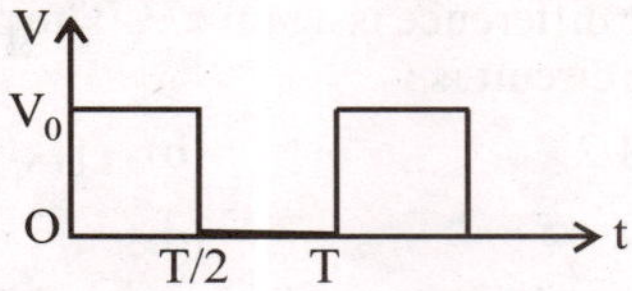

(a) V_0 (b) $V_0/\sqrt{2}$
(c) $V_0/2$ (d) $V_0/\sqrt{3}$

7. In an a.c circuit the e.m.f. (e) and the current (i) at any instant are given respectively by
$e = E_0 \sin \omega t$ *[2008]*
$i = I_0 \sin(\omega t - \phi)$
The average power in the circuit over one cycle of a.c. is

(a) $\frac{E_0 I_0}{2}$ (b) $\frac{E_0 I_0}{2}\sin\phi$
(c) $\frac{E_0 I_0}{2}\cos\phi$ (d) $E_0 I_0$

8. In an A.C. circuit with voltage V and current I the power dissipated is *[1997]*

(a) dependent on the phase between V and I
(b) $\frac{1}{\sqrt{2}}VI$
(c) $\frac{1}{2}VI$
(d) VI

9. In an a.c. circuit, the r.m.s. value of current, i_{rms} is related to the peak current, i_0 by the relation *[1994]*

(a) $I_{rms} = \sqrt{2}\, I_0$ (b) $I_{rms} = \pi I_0$
(c) $I_{rms} = \frac{1}{\pi} I_0$ (d) $I_{rms} = \frac{1}{\sqrt{2}} I_0$

10. In a region of uniform magnetic induction $B = 10^{-2}$ tesla, a circular coil of radius 30 cm and resistance π^2 ohm is rotated about an axis which is perpendicular to the direction of B and which forms a diameter of the coil. If the coil rotates at 200 rpm the amplitude of the alternating current induced in the coil is *[1988]*

(a) $4\pi^2$ mA (b) 30 mA
(c) 6 mA (d) 200 mA

Topic 2 A.C. Circuit, LCR Circuit, Quality & Power Factor

11. An inductor of inductance *L*, a capacitor of capacitance *C* and a resistor of resistance '*R*' are connected in series to an ac source of potential difference '*V*' volts as shown in figure. Potential difference across *L*, *C* and *R* is 40 *V*, 10 *V* and 40 *V*, respectively. The amplitude of current flowing through *LCR* series circuit is $10\sqrt{2}$ *A*. The impedance of the circuit is *[2021]*

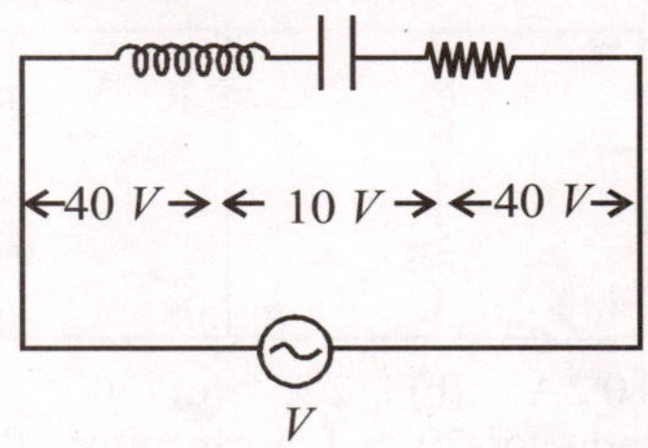

(a) 5 Ω (b) $4\sqrt{2}$ Ω
(c) $5\sqrt{2}$ Ω (d) 4 Ω

12. A series LCR circuit containing 5.0 H inductor, 80μ F capacitor and 40 Ω resistor is connected to 230V variable frequency ac source. The angular frequencies of the source at which power transferred to the circuit is half the power at the resonant angular frequency are likely to be *[2021]*

(a) 42 rad/s and 58 rad/s
(b) 25 rad/s and 75 rad/s
(c) 50 rad/s and 25 rad/s
(d) 46 rad/s and 54 rad/s

13. A series LCR circuit is connected to an ac voltage source. When L is removed from the circuit, the phase difference between current and voltage is $\frac{\pi}{3}$. If instead C is removed from the circuit, the phase difference is again $\frac{\pi}{3}$ between current and voltage. The power factor of the circuit is : *[2020]*

(a) 0.5 (b) 1.0 (c) –1.0 (d) zero

14. A circuit when connected to an AC source of 12 V gives a current of 0.2 A. The same circuit when connected to a DC source of 12 V, gives a current of 0.4 A. The circuit is *[NEET Odisha 2019]*

(a) Series LCR (b) Series LR
(c) Series RC (d) Series LC

15. An inductor 20 mH, a capacitor 100 μF and a resistor 50Ω are connected in series across a source of emf, V = 10 sin 314 t. The power loss in the circuit is *[2018]*

(a) 0.79 W (b) 0.43 W
(c) 1.13 W (d) 2.74 W

16. Figure shows a circuit that contains three identical resistors with resistance R = 9.0 Ω each, two identical inductors with inductance L = 2.0 mH each, and an ideal battery with emf ε = 18 V. The current 'i' through the battery just after the switch closed is *[2017]*

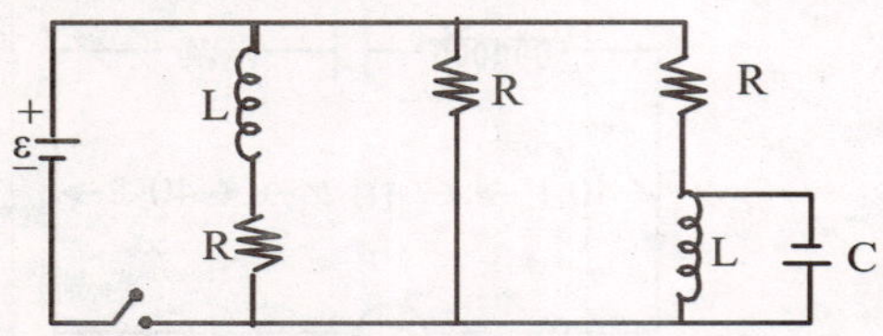

(a) 0.2 A (b) 2 A (c) 0 (d) 2 mA

17. An inductor 20 mH, a capacitor 50 μF and a resistor 40Ω are connected in series across a source of emf V = 10 sin 340 t. The power loss in A.C. circuit is : *[2016]*

(a) 0.51 W (b) 0.67 W
(c) 0.76 W (d) 0.89 W

18. A resistance 'R' draws power 'P' when connected to an AC source. If an inductance is now placed in series with the resistance, such that the impedance of the circuit becomes 'Z', the power drawn will be *[2015]*

(a) $P\sqrt{\frac{R}{Z}}$ (b) $P\left(\frac{R}{Z}\right)$

(c) P (d) $P\left(\frac{R}{Z}\right)^2$

19. A series R-C circuit is connected to an alternating voltage source. Consider two situations: *[2015 RS]*

(A) When capacitor is air filled.
(B) When capacitor is mica filled.

Current through resistor is i and voltage across capacitor is V then :

(a) $V_a > V_b$ (b) $i_a > i_b$
(c) $V_a = V_b$ (d) $V_a < V_b$

20. A coil of self-inductance L is connected in series with a bulb B and an AC source. Brightness of the bulb decreases when *[2013]*

(a) number of turns in the coil is reduced
(b) a capacitance of reactance $X_C = X_L$ is included in the same circuit
(c) an iron rod is inserted in the coil
(d) frequency of the AC source is decreased

21. In an electrical circuit *R*, *L*, *C* and an a.c. voltage source are all connected in series. When *L* is removed from the circuit, the phase difference between the voltage the current in the circuit is π/3. If instead, *C* is removed from the circuit, the phase difference is again π/3. The power factor of the circuit is : *[2012]*

(a) 1/2 (b) $1/\sqrt{2}$
(c) 1 (d) $\sqrt{3}/2$

22. An ac voltage is applied to a resistance R and an inductor L in series. If R and the inductive reactance are both equal to 3Ω, the phase difference between the applied voltage and the current in the circuit is *[2011]*

(a) π/6 (b) π/4 (c) π/2 (d) zero

23. A coil has resistance 30 ohm and inductive reactance 20 ohm at 50 Hz frequency. If an ac source, of 200 volt, 100 Hz, is connected across the coil, the current in the coil will be *[2011M]*

(a) 4.0A (b) 8.0A

(c) $\frac{20}{\sqrt{13}}$ A (d) 2.0A

24. In the given circuit the reading of voltmeter V_1 and V_2 are 300 volts each. The reading of the voltmeter V_3 and ammeter *A* are respectively *[2010]*

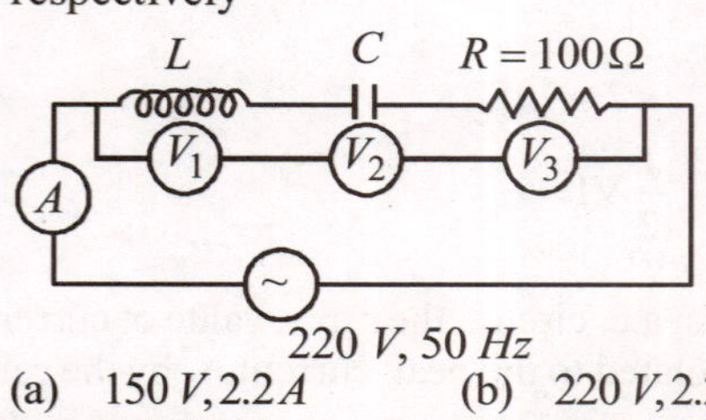

(a) 150 V, 2.2 A (b) 220 V, 2.2 A
(c) 220 V, 2.0 A (d) 100 V, 2.0 A

25. A condenser of capacity C is charged to a potential difference of V_1. The plates of the condenser are then connected to an ideal inductor of inductance L. The current through the inductor when the potential difference across the condenser reduces to V_2 is *[2010]*

(a) $\left(\frac{C(V_1^2-V_2^2)}{L}\right)^{1/2}$ (b) $\left(\frac{C(V_1-V_2)^2}{L}\right)^{1/2}$

(c) $\frac{C(V_1^2-V_2^2)}{L}$ (d) $\frac{C(V_1-V_2)}{L}$

26. Power dissipated in an LCR series circuit connected to an a.c source of emf ε is [2009]

(a) $\frac{\varepsilon^2\sqrt{R^2+\left(L\omega-\frac{1}{C\omega}\right)^2}}{R}$

(b) $\frac{\varepsilon^2\left[R^2+\left(L\omega-\frac{1}{C\omega}\right)^2\right]}{R}$

(c) $\frac{\varepsilon^2 R}{\sqrt{R^2+\left(L\omega-\frac{1}{C\omega}\right)^2}}$

(d) $\frac{\varepsilon^2 R}{\left[R^2+\left(L\omega-\frac{1}{C\omega}\right)^2\right]}$

27. What is the value of inductance L for which the current is maximum in a series LCR circuit with $C = 10\,\mu F$ and $\omega = 1000 s^{-1}$? [2007]
(a) 1 mH
(b) cannot be calculated unless R is known
(c) 10 mH
(d) 100 mH

28. A coil of inductive reactance 31 Ω has a resistance of 8 Ω. It is placed in series with a condenser of capacitative reactance 25Ω. The combination is connected to an a.c. source of 110 volt. The power factor of the circuit is [2006]
(a) 0.64 (b) 0.80 (c) 0.33 (d) 0.56

29. In a circuit, L, C and R are connected in series with an alternating voltage source of frequency f. The current leads the voltage by 45°. The value of C is [2005]

(a) $\frac{1}{\pi f(2\pi fL-R)}$ (b) $\frac{1}{2\pi f(2\pi fL-R)}$

(c) $\frac{1}{\pi f(2\pi fL+R)}$ (d) $\frac{1}{2\pi f(2\pi fL+R)}$

30. A coil of 40 henry inductance is connected in series with a resistance of 8 ohm and the combination is joined to the terminals of a 2 volt battery. The time constant of the circuit is [2004]
(a) 20 seconds (b) 5 seconds
(c) 1/5 seconds (d) 40 seconds

31. In a series resonant circuit, having *L*, *C* and *R* as its elements, the resonant current is *i*. The power dissipated in the circuit at resonance is

(a) $\frac{i^2R}{\left(\omega L-\frac{1}{\omega C}\right)}$ (b) zero [2002]

(c) $i^2\omega L$ (d) i^2R

where ω is the angular resonance frequency.

32. A capacitor has capacity *C* and reactance *X*. If capacitance and frequency become double, then reactance will be [2001]
(a) 4*X* (b) *X*/2
(c) *X*/4 (d) 2*X*

33. An inductance L having a resistance R is connected to an alternating source of angular frequency ω. The quality factor Q of the inductance is [2000]

(a) $\frac{R}{\omega L}$ (b) $\left(\frac{\omega L}{R}\right)^2$

(c) $\left(\frac{R}{\omega L}\right)^{1/2}$ (d) $\frac{\omega L}{R}$

34. In an experiment, 200 V A.C. is applied at the ends of an LCR circuit. The circuit consists of an inductive reactance $(X_L) = 50\,\Omega$, capacitive reactance $(X_C) = 50\,\Omega$ and ohmic resistance (R) = 10 Ω. The impedance of the circuit is [1996]
(a) 10Ω (b) 20Ω (c) 30Ω (d) 40Ω

35. An LCR series circuit is connected to a source of alternating current. At resonance, the applied voltage and the current flowing through the circuit will have a phase difference of [1994]

(a) π (b) $\frac{\pi}{2}$

(c) $\frac{\pi}{4}$ (d) 0

36. The time constant of C–R circuit is [1992]
(a) 1/CR (b) C/R
(c) CR (d) R/C

Topic 3 Transformers & LC Oscillations

37. A step down transformer connected to an ac mains supply of 220 V is made to operate at 11 V, 44 W lamp. Ignoring power losses in the transformer, what is the current in the primary circuit? *[2021]*

(a) 4 A (b) 0.2 A (c) 0.4 A (d) 2 A

38. A transformer having efficiency of 90% is working on 200V and 3kW power supply. If the current in the secondary coil is 6A, the voltage across the secondary coil and the current in the primary coil respectively are : *[2014]*

(a) 300 V, 15A (b) 450 V, 15A
(c) 450V, 13.5A (d) 600V, 15A

39. The primary of a transformer when connected to a dc battery of 10 volt draws a current of 1 mA. The number of turns of the primary and secondary windings are 50 and 100 respectively. The voltage in the secondary and the current drawn by the circuit in the secondary are respectively *[NEET Kar. 2013]*

(a) 20 V and 0.5 mA
(b) 20 V and 2.0 mA
(c) 10 V and 0.5 mA
(d) Zero and therefore no current

40. A 220 volts input is supplied to a transformer. The output circuit draws a current of 2.0 ampere at 440 volts. If the efficiency of the transformer is 80%, the current drawn by the primary windings of the transformer is *[2010]*

(a) 3.6 ampere (b) 2.8 ampere
(c) 2.5 ampere (d) 5.0 ampere

41. The primary and secondary coil of a transformer have 50 and 1500 turns respectively. If the magnetic flux ϕ linked with the primary coil is given by $\phi = \phi_0 + 4t$, where ϕ is in webers, t is time in seconds and ϕ_0 is a constant, the output voltage across the secondary coil is *[2007]*

(a) 120 volts (b) 220 volts
(c) 30 volts (d) 90 volts

42. A transformer is used to light a 100 W and 110 V lamp from a 220 V mains. If the main current is 0.5 amp, the efficiency of the transformer is approximately *[2007]*

(a) 50% (b) 90% (c) 10% (d) 30%.

43. The core of a transformer is laminated because *[2006]*

(a) the weight of the transformer may be reduced
(b) rusting of the core may be prevented
(c) ratio of voltage in primary and secondary may be increased
(d) energy losses due to eddy currents may be minimised

44. A transistor-oscillator using a resonant circuit with an inductor L (of negligible resistance) and a capacitor C in series produce oscillations of frequency f. If L is doubled and C is changed to 4C, the frequency will be *[2006]*

(a) 8f (b) $f/2\sqrt{2}$
(c) f/2 (d) f/4

45. A step-up transformer operates on a 230 V line and supplies a load of 2 ampere. The ratio of the primary and secondary windings is 1 : 25. The current in the primary is *[1998]*

(a) 25 A (b) 50 A
(c) 15 A (d) 12.5 A

46. The primary winding of a transformer has 500 turns whereas its secondary has 5000 turns. The primary is connected to an A.C. supply of 20 V, 50 Hz. The secondary will have an output of *[1997]*

(a) 2 V, 5 Hz (b) 200 V, 500 Hz
(c) 2V, 50 Hz (d) 200 V, 50 Hz

47. Eddy currents are produced when *[1988]*

(a) a metal is kept in varying magnetic field
(b) a metal is kept in steady magnetic field
(c) a circular coil is placed in a magnetic field
(d) through a circular coil, current is passed

ANSWER KEY

1	(b)	7	(c)	13	(b)	19	(a)	25	(a)	31	(d)	37	(b)	43	(d)				
2	(b)	8	(a)	14	(b)	20	(c)	26	(d)	32	(c)	38	(b)	44	(b)				
3	(d)	9	(d)	15	(a)	21	(c)	27	(d)	33	(d)	39	(d)	45	(b)				
4	(d)	10	(c)	16	(b)	22	(b)	28	(b)	34	(a)	40	(d)	46	(d)				
5	(d)	11	(a)	17	(a)	23	(a)	29	(d)	35	(d)	41	(a)	47	(a)				
6	(b)	12	(d)	18	(d)	24	(b)	30	(b)	36	(c)	42	(b)						

Hints & Solutions

1. (b) Given:
Capacitance, $C = 40\ \mu F = 40 \times 10^{-6}$ F
Frequency, $f = 50$ Hz $\quad \therefore \omega = 2\pi f = 100\pi$

$$\varepsilon_{rms} = 200 \text{ V} \quad \therefore I_{rms} = \frac{\varepsilon_{rms}}{X_C} = \frac{\varepsilon_{rms}}{\frac{1}{C\omega}}$$

$= 200 \times 40 \times 10^{-6} \times 2\pi \times 50 = 2.5$ A.

2. (b) As we know, power $P = V_{rms} \cdot I_{rms} \cos\phi$
as $\cos\phi = 0 \qquad (\because \ \phi = 90°)$
$\therefore$ Power consumed = 0 (in one complete cycle)

3. (d) $V = -L\dfrac{di}{dt}$

Here $\dfrac{di}{dt}$ is + ve for $\dfrac{T}{2}$ time and

$\dfrac{di}{dt}$ is – ve for next $\dfrac{T}{2}$ time so

4. (d) The average power in the circuit where $\cos\phi$ = power factory
$<P> = V_{rms} \times I_{rms} \cos\phi$

$$\phi = \pi/3 = \text{phase difference} = \frac{180}{3} = 60$$

$$V_{rms} = \frac{\frac{1}{\sqrt{2}}}{\sqrt{2}} = \frac{1}{2}\text{volt}$$

$$I_{rms} = \frac{\frac{1}{\sqrt{2}}}{\sqrt{2}} = \left(\frac{1}{2}\right)A$$

$$\cos\phi = \cos\frac{\pi}{3} = \frac{1}{2}$$

$$<P> = \frac{1}{2} \times \frac{1}{2} \times \frac{1}{2} = \frac{1}{8}W$$

NOTES
The instantaneous power is the power in circuit at any instant of time. It is equal to the product of values of alternating voltage and alternating current at that time
$P_{in} = EI = (E_0 \sin \omega t)(I_0 \sin \omega t)$
(in non-inductive circuit)
$P_{in} = (E_0 \sin \omega t)(I_0 \sin \omega t \pm \theta)$
(in inductive circuit)

5. (d) $V_{rms} = \dfrac{200\sqrt{2}}{\sqrt{2}} = 200\text{V}$

$$I_{rms} = \frac{V_{rms}}{X_C} = \frac{\frac{200}{1}}{100 \times 10^{-6}} = 2 \times 10^{-2} = 20\text{mA}$$

6. (b) $V_{rms} = \sqrt{\dfrac{(T/2)V_0^{\,2} + 0}{T}} = \dfrac{V_0}{\sqrt{2}}$.

7. (c) The average power in the circuit over one cycle of a.c. is given by
$P_{av} = e_{rms} \times i_{rms} \times \cos\phi$
$= \dfrac{E_0}{\sqrt{2}} \times \dfrac{I_0}{\sqrt{2}} \times \cos\phi = \dfrac{E_0 I_0}{2}\cos\phi$

8. (a) Power dissipated $= E_{rms} \cdot I_{rms}$
$= (E_{rms})(I_{rms}) \cos\phi$
Hence, power dissipated depends upon phase difference.

9. (d) $I_{rms} = \dfrac{I_0}{\sqrt{2}}$

10. (c) $I_0 = \dfrac{E_0}{R} = \dfrac{nBA\omega}{R}$
Given, $n = 1, B = 10^{-2}$ T, $A = \pi(0.3)^2 m^2, R = \pi^2$
$f = (200/60)$ and $\omega = 2\pi(200/60)$
Substituting these values and solving, we get
$I_0 = 6 \times 10^{-3}$ A = 6mA

11. (a) Impedance $z = \dfrac{V_{RMS}}{I_{RMS}}$

$$V_{RMS} = \sqrt{V_R^2 + (V_L - V_C)^2}$$

$$= \sqrt{(40)^2 + (40-10)^2} = 50 \text{ V}$$

$$I_{RMS} = \frac{I_0}{\sqrt{2}} = \frac{10\sqrt{2}}{\sqrt{2}} = 10\text{A}$$

$$\therefore Z = \frac{V_{RMS}}{I_{RMS}} = \frac{50}{10} = 5\,\Omega$$

NOTES
NOTE : Impedance (z) is the oppsition offered by ac circuits to the flow of ac. For LCR series circuit.

$$\text{Voltage } V = \sqrt{V_R^2 + (V_L - V_C)^2}$$

$$\text{and } Z = \sqrt{R^2 + (X_L - X_C)^2}$$

12. (d) For LCR series circuit resonance frequency

$$\omega_0 = \frac{1}{\sqrt{LC}} = \frac{1}{\sqrt{5\times 80\times 10^{-6}}} = 50 \text{ rad/s}$$

Half power frequency, $\omega = \omega_0 \pm \frac{R}{2L}$

i.e. $\omega_L = 50 - \frac{40}{2\times 5} = 46$ rad / s and

$$\omega_H = 50 + \frac{40}{2\times 5} = 54 \text{ rad/s}$$

13. (b) When L is removed,

Phase difference

$$\tan\phi = \frac{|X_C|}{R} = \tan\frac{\pi}{3} = \frac{X_C}{R} \quad ...(1)$$

When C is removed,

Phase difference

$$\tan\phi = \frac{|X_L|}{R} = \tan\frac{\pi}{3} = \frac{X_L}{R} \quad ...(2)$$

From eqs. (1) and (2), $X_L = X_C$

Since, $X_L = X_C$, the circuit is in resonance.

In this case, $Z = R$

$\therefore$ Power factor, $\cos\phi = \frac{R}{Z} = 1.$

14. (b) $I_1 = \frac{V}{Z}$

$$I_1 = \frac{12}{\sqrt{R^2 + (X_L - X_C)^2}} = 0.2\text{ A}$$

In second case, the capacitor would provide infinite resistance but current is present in circuit, it means resistor and inductor can be present in the circuit.

Since current with AC source and DC source are different, inductor must be present with resistance.

15. (a) Power dissipated in an LCR series circuit connected to an a.c. source of emf E

$$P = E_{rms}\, i_{rms} \cos\phi = \frac{E_{rms}^2 R}{Z^2} = \frac{E_{rms}^2 R}{R^2 + \left(\omega L - \frac{1}{C\omega}\right)^2}$$

$$= \frac{\left(\frac{10}{\sqrt{2}}\right)^2 \times 50}{(50)^2 + \left(314\times 20\times 10^{-3} - \frac{1}{314\times 100\times 10^{-6}}\right)^2}$$

Solving we get, P = 0.79 W

16. (b)

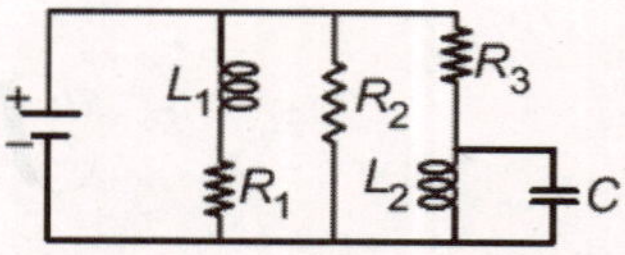

At t = 0, no current flows through R_1 and R_3

$\therefore$

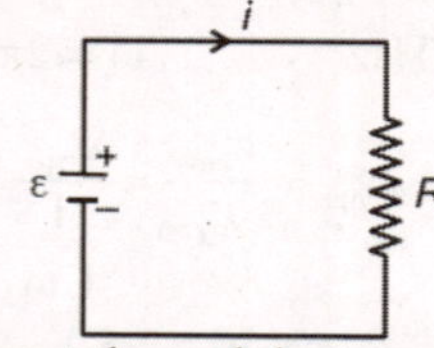

Current through battery just after the switch closed is

$$i = \frac{\varepsilon}{R_2} = \frac{18}{9} = 2\text{A}$$

17. (a) Given: L = 20 mH; C = 50 μF; R = 40 Ω

V = 10 sin 340 t

$$\therefore \quad V_{runs} = \frac{10}{\sqrt{2}}$$

$$X_C = \frac{1}{\omega C} = \frac{1}{340\times 50\times 10^{-6}} = 58.8\,\Omega$$

$$X_L = \omega L = 340 \times 20 \times 10^{-3} = 6.8\,\Omega$$

Impedance, $Z = \sqrt{R^2 + (X_C - X_L)^2}$

$$= \sqrt{40^2 + (58.8 - 6.8)^2} = \sqrt{4304}\,\Omega$$

Power loss in A.C. circuit,

$$P = i_{rms}^2 R = \left(\frac{V_{rms}}{Z}\right)^2 R$$

$$= \left(\frac{10/\sqrt{2}}{\sqrt{4304}}\right)^2 \times 40 = \frac{50\times 40}{4304} \simeq 0.51\text{ W}$$

18. (d)

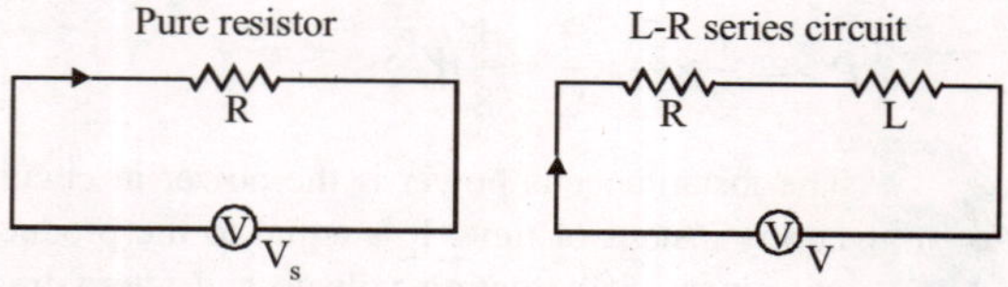

Phasor diagram

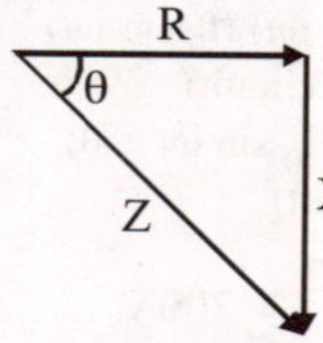

$\cos\theta = \frac{R}{Z}$

Z = impedance

For pure resistor circuit, power

$$P = \frac{V^2}{R} \Rightarrow V^2 = PR$$

For L-R series circuit, power

$$P^{l} = \frac{V^2}{Z}\cos\theta = \frac{V^2}{Z}.\frac{R}{Z} = \frac{PR}{Z^2}.R = P\left(\frac{R}{Z}\right)^2$$

19. (a) For series R – C circuit, capacitive reactance,

$$Z_c = \sqrt{R^2 + \left(\frac{1}{C\omega}\right)^2}$$

$X_c = 1/c\omega$

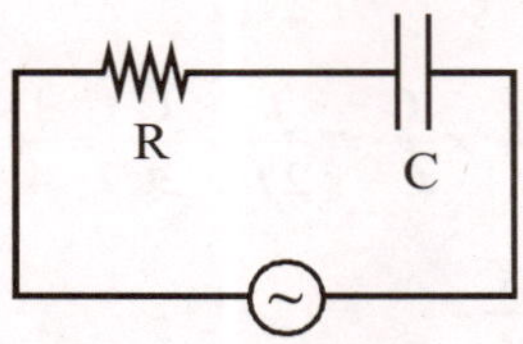

AC Source

$$\text{Current } i = \frac{V}{Z_c} = \frac{V}{\sqrt{R^2 + \left(\frac{1}{C\omega}\right)^2}}$$

$$V_c = iX_c = \frac{V}{\sqrt{R^2 + \left(\frac{1}{C\omega}\right)^2}} \times \frac{1}{C\omega}$$

$$V_c = \frac{V}{\sqrt{(RC\omega)^2 + 1}}$$

If we fill a di-electric material like mica instead of air then capacitance $C\uparrow \Rightarrow V_c\downarrow$

So, $V_a > V_b$

20. (c) By inserting iron rod in the coil, $L\uparrow Z\uparrow I\downarrow$ so brightness $\downarrow$

NOTES

Inductive reactance, $X_L = \omega L$. On inserting iron rod in the coil, L increases. As a result Impedance $Z = \sqrt{X_L^2 + R^2} = \sqrt{(\omega L)^2 + R^2}$ also increases.

So, Current, $I = \frac{V}{Z}$ decreases.

21. (c) 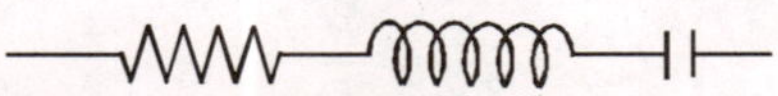

when L is removed from the circuit

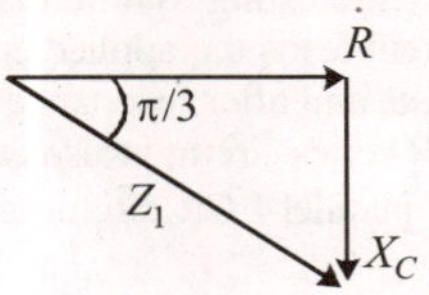

$$\frac{X_C}{R} = \tan\frac{\pi}{3}$$

$$X_C = R\tan\frac{\pi}{3} \quad \text{....(1)}$$

when C is remove from the circuit

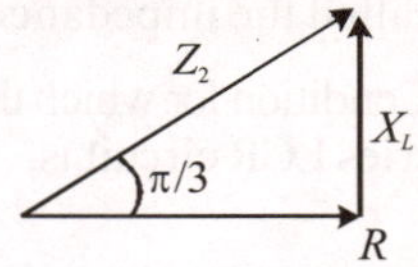

$$\frac{X_L}{R} = \tan\frac{\pi}{3}$$

$$X_C = R\tan\frac{\pi}{3} \quad \text{...(2)}$$

net impedence $Z = \sqrt{R^2 + (X_L - X_C)^2} = R$

power factor $\cos\phi = \frac{R}{Z} = 1$

22. (b) The phase difference ϕ is given by

$$\tan\phi = \frac{X_L}{R} = \frac{3}{3} = 1 \quad \Rightarrow \phi = \frac{\pi}{4}.$$

23. (a) If $\omega = 50 \times 2\pi$ then $\omega L = 20\Omega$

If $\omega' = 100 \times 2\pi$ then $\omega' L = 40\Omega$

Current flowing in the coil is

$$I = \frac{200}{Z} = \frac{200}{\sqrt{R^2 + (\omega' L)^2}} = \frac{200}{\sqrt{(30)^2 + (40)^2}}$$

$I = 4A$.

24. (b) As $V_L = V_C = 300\ V$, resonance will take place

$\therefore \quad V_R = 220\ V$

$$\text{Current, } I = \frac{220}{100} = 2.2A$$

$\therefore$ reading of $V_3 = 220\ V$

and reading of $A = 2.2\ A$

25. (a) $q = CV_1\cos\omega \Rightarrow i = \frac{dq}{dt} = -\omega Cv_1\sin\omega t$

Also, $\omega^2 = \frac{1}{LC}$ and $V = V_1\cos\omega t$

At $t = t_1$, $V = V_2$ and $i = -\omega CV_1\sin\omega t_1$

$\therefore \cos\omega t_1 = \frac{V_2}{V_1}$ (–ve sign gives direction)

Hence,

$$i = V_1\sqrt{\frac{C}{L}}\left(1 - \frac{V_2^2}{V_1^2}\right)^{1/2} = \left(\frac{C(V_1^2 - V_2^2)}{L}\right)^{1/2}$$

26. (d) Power dissipated in series LCR;

$$P = I^2R = \frac{\varepsilon^2}{(Z)^2}R = \frac{\varepsilon^2 R}{\left[R^2 + \left(\omega L - \frac{1}{\omega C}\right)^2\right]}$$

NOTES

$$Z = \sqrt{R^2 + \left(\omega L - \frac{1}{\omega C}\right)^2}$$

is called the impedance of the circuit.

27. (d) Condition for which the current is maximum in a series LCR circuit is,

$$\omega = \frac{1}{\sqrt{LC}} \Rightarrow 1000 = \frac{1}{\sqrt{L(10\times10^{-6})}}$$

$\Rightarrow L = 100\,\text{mH}$

28. (b) Power factor, $\cos\phi = \dfrac{R}{\sqrt{\left(\omega L - \dfrac{1}{\omega C}\right)^2 + R^2}}$

$$= \frac{8}{\sqrt{(31-25)^2+8^2}} = \frac{8}{\sqrt{6^2+8^2}} = \frac{8}{10} = 0.8$$

NOTES

Power factor = $\cos\theta = \dfrac{R}{Z}$

For purely inductive and purely capacitive circuits, $\theta = 90°$

Power factor = $\cos\theta = \cos 90° = \theta$

For non-inductive circuit, $\theta = 0°$

$\cos\theta = \cos 0° = 1$

29. (d) From figure,

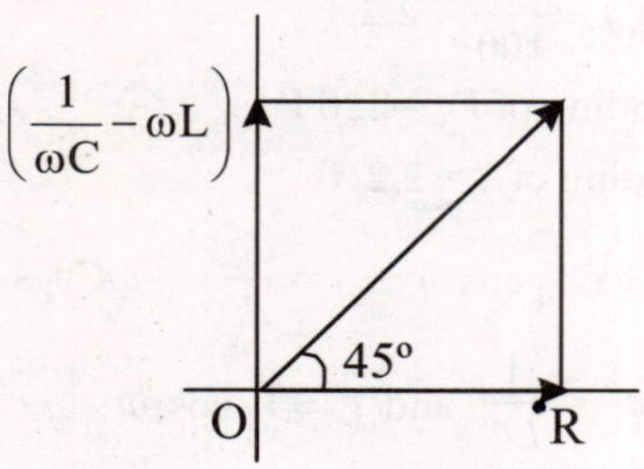

$$\tan 45° = \frac{\dfrac{1}{\omega C} - \omega L}{R} \Rightarrow \frac{1}{\omega C} - \omega L = R$$

$$\Rightarrow \frac{1}{\omega C} = R + \omega L \quad \left(\because \omega = \frac{2\pi}{T} = 2\mu f\right)$$

$$C = \frac{1}{\omega(R+\omega L)} = \frac{1}{2\pi f(R + 2\pi f L)}$$

30. (b) Time constant is L/R

Given, $L = 40\text{H}$ & $R = 8\Omega$

$\therefore \tau = 40/8 = 5$ sec.

31. (d) At resonance $L\omega = \dfrac{1}{C\omega}$, $\omega = \dfrac{1}{\sqrt{LC}}$

Current through circuit $i = \dfrac{E}{R}$

Power dissipated at Resonance = i^2R

32. (c) Capacitive reactance, $X = \dfrac{1}{\omega C} = \dfrac{1}{2\pi f C}$

$$\Rightarrow X \propto \frac{1}{fC}$$

$$\therefore \frac{X'}{X} = \frac{f}{f'} \times \frac{C}{C'} = \frac{f}{2f} \times \frac{C}{2C} = \frac{1}{4}$$

$$\Rightarrow X' = \frac{X}{4}$$

33. (d) Quality factor

$$= \frac{\text{Potential drop across capacitor or inductor}}{\text{Potential drop across R}}$$

$$= \frac{I\omega L}{IR} = \frac{\omega L}{R}$$

34. (a) Given : Supply voltage $(V_{ac}) = 200$ V

Inductive reactance $(X_L) = 50\,\Omega$

Capacitive reactance $(X_C) = 50\,\Omega$

Ohmic resistance $(R) = 10\,\Omega$.

We know that impedance of the LCR circuit (Z)

$$= \sqrt{\{(X_L - X_C)^2 + R^2\}}$$

$$= \sqrt{\{(50-50)^2 + (10)^2\}} = 10\Omega$$

35. (d) At resonance, $\omega L = \dfrac{1}{\omega C}$. The circuit behaves as if it contains R only. So, phase difference = 0

NOTES

At resonance, impedance is minimum $Z_{min} = R$ and current is maximum, given by

$$I_{max} = \frac{E}{Z_{min}} = \frac{E}{R}$$

It is interesting to note that before resonance the current leads the applied emf, at resonance it is in phase, and after resonance it lags behind the emf. LCR series circuit is also called as acceptor circuit and parallel LCR circuit is called rejector circuit.

36. (c) The time constant for resonance circuit,

=CR

> **NOTES** Growth of charge in a circuit containing capacitance and resistance is given by the formula, $q = q_0 (1 - e^{-t/CR})$
> CR is known as time constant in this formula.

37. (b) Power P = V × I

For an ideal transformer

Input power = Output power

$V_P I_P = V_S I_S$

or, $220 \times I_P = 44 \therefore I_P = 0.2$ A

38. (b) Efficiency $\eta = \frac{V_s I_s}{V_p I_p} \Rightarrow 0.9 = \frac{V_s(6)}{3 \times 10^3}$

$\Rightarrow V_s = 450$ V

As $V_p I_p = 3000$ so

$$I_p = \frac{3000}{V_p} = \frac{3000}{200} A = 15 A$$

39. (d) A transformer is essentially an AC device. DC source so no mutual induction between coils

$\Rightarrow E_2 = 0$ and $I_2 = 0$

40. (d) $\frac{V_2}{V_1} = 0.8 \frac{I_1}{I_2} \Rightarrow \frac{V_2 I_2}{V_1 I_1} = 0.8$

$V_1 = 220\ V, I_2 = 2.0\ A, V_2 = 440\ V$

$$I_1 = \frac{V_2 I_2}{V_1} \times \frac{10}{8} = \frac{440 \times 2 \times 10}{220 \times 8} = 5\ A$$

41. (a) Since $\frac{V_s}{V_p} = \frac{N_s}{N_p}$

Where

N_s = No. of turns across primary coil = 50

N_p = No. of turns across secondary coil = 1500

and $V_p = \frac{d\phi}{dt} = \frac{d}{dt}(\phi_0 + 4t) = 4$

$$\Rightarrow V_s = \frac{1500}{50} \times 4 = 120 \text{ V}$$

42. (b) Efficiency of the transformer

$$\eta = \frac{P_{output}}{P_{input}} \times 100 = \frac{100}{220 \times 0.5} \times 100 = 90.9\%$$

43. (d) When there is change of flux in the core of a transformer due to change in current around it, eddy current is produced. The direction of this current is opposite to the current which produces it, so it will reduce the main current. We laminate the core so that flux is reduced resulting in the reduced production of eddy current.

44. (b) We know that frequency of electrical oscillation in L.C. circuit is

$$f = \frac{1}{2\pi}\sqrt{\frac{1}{LC}}$$

Now, $L = 2L$ & $C = 4C$

$$f' = \frac{1}{2\pi}\sqrt{\frac{1}{2L \cdot 4C}} = \frac{1}{2\pi}\sqrt{\frac{1}{LC}} \times \frac{1}{2\sqrt{2}}$$

$$\Rightarrow f' = \frac{1}{2\sqrt{2}} \times f$$

45. (b) $\frac{N_P}{N_s} = \frac{E_P}{E_s} = \frac{1}{25}$

$\therefore E_s = 25 E_p$

But $E_s I_s = E_p I_p$

$25 E_p \times 2 = E_p \times I_p \Rightarrow I_p = 50$ A

> **NOTES** If n_p is number of turns in primary coil and n_s is number of turns in secondary coil, then
> $$\frac{E_s}{E_p} = \frac{I_p}{I_s} = \frac{n_s}{n_p} = k$$
> Here, I_p = current in primary
> k = Transformation ratio
> I_s = current is secondary
> For a step up transformer, $k > 1$
> For step down transformer, $k < 1$

46. (d) The transformer converts A.C. high voltage into A.C. low voltage, but it does not cause any change in frequency. The formula for voltage is

$$\frac{E_s}{E_p} = \frac{N_s}{N_p} \Rightarrow E_s = \frac{N_s}{N_p} \times E_p$$

$$= \frac{5000}{500} \times 20 = 200 \text{V}$$

Thus, output has voltage 200 V and frequency 50 Hz.

47. (a) Eddy currents are produced when a metal is kept in a varying magnetic field.

22 Electromagnetic Waves

Trend Analysis with Important Topics & Sub-Topics

Topic Name	Sub-Topic	2021 QNS.	2021 LOD	2020 QNS.	2020 LOD	2019 QNS.	2019 LOD	2018 QNS.	2018 LOD	2017 QNS.	2017 LOD	2016 QNS.	2016 LOD
Electromagnetic Waves, Conduction & Displacement Current	Electromagnetic waves $\vec{E}\times\vec{B}=\vec{C}$	1	A	1	E							1	E
	Displacement Current	1	A										
	E/B = C & F = QE							1	E	1	E		
Electromagnetic Spectrum						1	E						

LOD - Level of Difficulty | E - Easy | A - Average | D - Difficult | Qns - No. of Questions

Topic 1 Electromagnetic Waves, Conduction & Displacement Current

1. A capacitor of capacitance 'C', is connected across an ac source of voltage V, given by $V = V_0 \sin \omega t$
The displacement current between the plates of the capacitor, would then be given by *[2021]*
(a) $I_d = V_0 \omega C \sin \omega t$ (b) $I_d = V_0 \omega C \cos \omega t$
(c) $I_d = \dfrac{V_0}{\omega C}\cos \omega t$ (d) $I_d = \dfrac{V_0}{\omega C}\sin \omega t$

2. For a plane electromagnetic wave propagating in x-direction, which one of the following combination gives the correct possible directions for electric field (E) and magnetic field (B) respectively? *[2021]*
(a) $-\hat{j}+\hat{k}, -\hat{j}+\hat{k}$ (b) $\hat{j}+\hat{k}, \hat{j}+\hat{k}$
(c) $-\hat{j}+\hat{k}, -\hat{j}-\hat{k}$ (d) $\hat{j}+\hat{k}, -\hat{j}-\hat{k}$

3. The ratio of contributions made by the electric field and magnetic field components to the intensity of an electromagnetic wave is : (c = speed of electromagnetic waves) *[2020]*
(a) 1 : 1 (b) 1 : c
(c) $1 : c^2$ (d) c : 1

4. An em wave is propagating in a medium with a velocity $\vec{V} = V\hat{i}$. The instantaneous oscillating electric field of this em wave is along +y axis. Then the direction of oscillating magnetic field of the em wave will be along *[2018]*
(a) –z direction (b) +z direction
(c) –x direction (d) –y direction

5. In an electromagnetic wave in free space the root mean square value of the electric field is E_{rms} = 6V/m. The peak value of the magnetic field is :- *[2017]*
(a) 2.83×10^{-8} T (b) 0.70×10^{-8} T
(c) 4.23×10^{-8} T (d) 1.41×10^{-8} T

6. Out of the following options which one can be used to produce a propagating electromagnetic wave ? *[2016]*
(a) A charge moving at constant velocity
(b) A stationary charge
(c) A chargeless particle
(d) An accelerating charge

7. A radiation of energy 'E' falls normally on a perfectly reflecting surface. The momentum transferred to the surface is (C = Velocity of light) *[2015]*
(a) $\dfrac{2E}{C}$ (b) $\dfrac{2E}{C^2}$
(c) $\dfrac{E}{C^2}$ (d) $\dfrac{E}{C}$

8. Light with an energy flux of 25×10^4 Wm^{-2} falls on a perfectly reflecting surface at normal incidence. If the surface area is 15 cm^2, the average force exerted on the surface is : [2014]
(a) 1.25×10^{-6} N (b) 2.50×10^{-6} N
(c) 1.20×10^{-6} N (d) 3.0×10^{-6} N

9. An electromagnetic wave of frequency $\nu = 3.0$ MHz passes from vacuum into a dielectric medium with relative permittivity $\varepsilon = 4.0$. Then [NEET Kar. 2013]
(a) wavelength is doubled and frequency is unchanged
(b) wavelength is doubled and frequency becomes half
(c) wavelength is halved and frequency remains unchanged
(d) wavelength and frequency both remain unchanged

10. The electric field associated with an e.m. wave in vacuum is given by $\vec{E} = \hat{i}\, 40 \cos (kz - 6 \times 10^8 t)$, where E, z and t are in volt/m, meter and seconds respectively. The value of wave vector k is : [2012]
(a) $2\ m^{-1}$ (b) $0.5\ m^{-1}$
(c) $6\ m^{-1}$ (d) $3\ m^{-1}$

11. The ratio of amplitude of magnetic field to the amplitude of electric field for an electromagnetic wave propagating in vacuum is equal to :
(a) the speed of light in vacuum [2012M]
(b) reciprocal of speed of light in vacuum
(c) the ratio of magnetic permeability to the electric susceptibility of vacuum
(d) unity

12. The electric and the magnetic field associated with an E.M. wave, propagating along the +z-axis, can be represented by [2011]
(a) $\left[\vec{E} = E_0\hat{i}, \vec{B} = B_0\hat{j}\right]$ (b) $\left[\vec{E} = E_0\hat{k}, \vec{B} = B_0\hat{i}\right]$
(c) $\left[\vec{E} = E_0\hat{j}, \vec{B} = B_0\hat{i}\right]$ (d) $\left[\vec{E} = E_0\hat{j}, \vec{B} = B_0\hat{k}\right]$

13. Which of the following statement is false for the properties of electromagnetic waves? [2010]
(a) Both electric and magnetic field vectors attain the maxima and minima at the same place and same time.
(b) The energy in electromagnetic wave is divided equally between electric and magnetic vectors
(c) Both electric and magnetic field vectors are parallel to each other and perpendicular to the direction of propagation of wave
(d) These waves do not require any material medium for propagation.

14. The electric field part of an electromagnetic wave in a medium is represented by $E_x = 0$;
$$E_y = 2.5\frac{N}{C}\cos\left[\left(2\pi\times10^6\ \frac{rad}{m}\right)t - \left(\pi\times10^{-2}\ \frac{rad}{s}\right)x\right];$$
$E_z = 0$. The wave is : [2009]
(a) moving along x direction with frequency 10^6 Hz and wave length 100 m.
(b) moving along x direction with frequency 10^6 Hz and wave length 200 m.
(c) moving along – x direction with frequency 10^6 Hz and wave length 200 m.
(d) moving along y direction with frequency $2\pi \times 10^6$ Hz and wave length 200 m.

15. The velocity of electromagnetic radiation in a medium of permittivity ε_0 and permeability μ_0 is given by [2008]
(a) $\sqrt{\frac{\epsilon_0}{\mu_0}}$ (b) $\sqrt{\mu_0 \epsilon_0}$
(c) $\frac{1}{\sqrt{\mu_0 \epsilon_0}}$ (d) $\sqrt{\frac{\mu_0}{\epsilon_0}}$

16. The electric and magnetic field of an electromagnetic wave are [2007]
(a) in opposite phase and perpendicular to each other
(b) in opposite phase and parallel to each other
(c) in phase and perpendicular to each other
(d) in phase and parallel to each other.

17. If $\vec{E}$ and $\vec{B}$ represent electric and magnetic field vectors of the electromagnetic waves, then the direction of propagation of the waves will be along [2002]
(a) $\vec{B} \times \vec{E}$ (b) $\vec{E}$
(c) $\vec{B}$ (d) $\vec{E} \times \vec{B}$

18. The electromagnetic radiations are caused by
(a) a stationary charge [1999]
(b) uniformly moving charges
(c) accelerated charges
(d) all of above

19. If ε_0 and μ_0 are the electric permittivity and magnetic permeability in vacuum, ε and μ are corresponding quantities in medium, then refractive index of the medium is [1997]
(a) $\sqrt{\frac{\varepsilon}{\varepsilon_0}}$ (b) $\sqrt{\frac{\varepsilon_0\mu}{\varepsilon\mu_0}}$
(c) $\sqrt{\frac{\varepsilon_0\mu_0}{\varepsilon\mu}}$ (d) $\sqrt{\frac{\varepsilon\mu}{\varepsilon_0\mu_0}}$

20. The oscillating electric and magnetic field vectors of electromagnetic wave are oriented along [1994]
(a) the same direction and in phase
(b) the same direction but have a phase difference of 90º
(c) mutually perpendicular directions and are in phase
(d) mutually perpendicular directions but has a phase difference of 90º

21. The frequency of electromagnetic wave, which is best suited to observe a particle of radius 3×10^{-4} cm is of the order of [1991]
(a) 10^{15} (b) 10^{14}
(c) 10^{13} (d) 10^{12}

Topic 2 Electromagnetic Spectrum

22. Which colour of the light has the longest wavelength ? [2019]
(a) red (b) blue
(c) green (d) violet

23. The energy of the em waves is of the order of 15 keV. To which part of the spectrum does it belong? [2015 RS]
(a) Infra-red rays (b) Ultraviolet rays
(c) γ-rays (d) X-rays

24. The condition under which a microwave oven heats up a food item containing water molecules most efficiently is [2013]
(a) the frequency of the microwaves has no relation with natural frequency of water molecules.
(b) microwaves are heat waves, so always produce heating.
(c) infra-red waves produce heating in a microwave oven.
(d) the frequency of the microwaves must match the resonant frequency of the water molecules.

25. The decreasing order of wavelength of infrared, microwave, ultraviolet and gamma rays is [2011]
(a) microwave, infrared, ultraviolet, gamma rays
(b) gamma rays, ultraviolet, infrared, micro-waves
(c) microwaves, gamma rays, infrared, ultraviolet
(d) infrared, microwave, ultraviolet, gamma rays

26. If λ_v, λ_x and λ_m represent the wavelengths of visible light, X–rays and microwaves respectively, then [2005]
(a) $\lambda_m > \lambda_x > \lambda_v$ (b) $\lambda_m > \lambda_v > \lambda_x$
(c) $\lambda_v > \lambda_x > \lambda_m$ (d) $\lambda_v > \lambda_m > \lambda_x$

27. Which one of the following rays is not electromagnetic wave ? [2003]
(a) heat rays (b) X-rays
(c) γ-rays (d) β-rays

28. We consider the radiation emitted by the human body. Which of the following statements is true? [2003]
(a) the radiation emitted lies in the ultraviolet region and hence is not visible.
(b) the radiation emitted is in the infra-red region.
(c) the radiation is emitted only during the day.
(d) the radiation is emitted during the summers and absorbed during the winters.

29. Which of the following electromagnetic radiations has the least wavelength? [2002]
(a) gamma rays (b) infra-red
(c) ultraviolet (d) X-rays

30. Green-house effect is the heating up of earth's atmosphere due to [2002]
(a) green plants (b) infra-red rays
(c) X-rays (d) ultraviolet rays

31. Which of the following is positively charged?
(a) α-particle (b) β-particle [2001]
(c) γ-rays (d) X-rays

32. The frequencies of X-rays, γ-rays and ultraviolet rays are respectively a, b, and c. Then [2000]
(a) $a < b, b < c$ (b) $a < b, b > c$
(c) $a > b, b > c$ (d) $a > b, b < c$

33. Which one of the following electromagnetic radiations has the smallest wavelength? [1994]
(a) ultraviolet waves (b) X-rays
(c) γ-rays (d) microwaves

34. The structure of solids is investigated by using
(a) cosmic rays [1992]
(b) X-rays
(c) γ-rays
(d) infra-red radiations

35. Pick out the longest wavelength from the following types of radiation. [1990]
(a) blue light (b) gamma rays
(c) X-rays (d) red light

36. Which of the following, has the longest wavelength? [1989]
(a) X-rays (b) γ-rays
(c) microwaves (d) radio waves

ANSWER KEY

1	(b)	5	(a)	9	(c)	13	(c)	17	(d)	21	(b)	25	(a)	29	(a)	33	(c)
2	(c)	6	(d)	10	(a)	14	(b)	18	(c)	22	(a)	26	(b)	30	(b)	34	(b)
3	(a)	7	(a)	11	(b)	15	(c)	19	(d)	23	(d)	27	(d)	31	(a)	35	(d)
4	(b)	8	(b)	12	(a)	16	(c)	20	(c)	24	(d)	28	(b)	32	(b)	36	(d)

Hints & Solutions

1. (b) Displacement current

$$I_d = C\frac{dV}{dt}$$

$$= C\frac{d}{dt}(V_0 \sin\omega t)\ (\because V = V_0 \sin\omega t \text{ given})$$

$$= C(V_0\omega)\cos\omega t$$

$$\therefore I_d = V_0\omega C\cos\omega t$$

2. (c) Direction of propagation along x-axis.

$$\vec{E}\times\vec{B} = (-\hat{j}-\hat{k})\times(-\hat{j}-\hat{k}) = 2\hat{i}$$

NOTES: $\vec{E}\times\vec{B}$ is always in the direction of propagation of wave. The EM wave propagating in the positive x-direction $E = E_y = E_0 \sin(kx - \omega t)$ and $B = B_z = B_0 \sin(kx - \omega t)$

3. (a) The energy in electromagnetic wave is divided equally between the electric and magnetic field.
So, in an electromagnetic wave, half of the intensity is provided by the electric field and half by the magnetic field.
Hence, required ratio should be 1 : 1.

4. (b) As we know, $\vec{E}\times\vec{B} = \vec{V}$

$$(E\hat{j})\times(\vec{B}) = V\hat{i}$$

($\because$ Electric field vector is along +y axis)

So, $\vec{B} = B\hat{k}$
i.e., direction of magnetic field vector is along +z direction.

5. (a) Given, $E_{rms} = 6$ V/m

$$\frac{E_{rms}}{B_{rms}} = c \Rightarrow B_{rms} = \frac{E_{rms}}{c} \quad ...(i)$$

$$B_{rms} = \frac{B_0}{\sqrt{2}} \Rightarrow B_0 = \sqrt{2}\,B_{rms}$$

$$B_0 = \sqrt{2}\times\frac{E_{rms}}{c} \text{ From equation (i)}$$

$$= \frac{\sqrt{2}\times 6}{3\times10^8} = 2.83\times10^{-8}\text{ T}$$

6. (d) To generate electromagnetic waves we need accelerating charge particle.

7. (a) Momentum of light falling on reflecting surface $p = \frac{E}{C}$

As surface is perfectly reflecting so momentum reflect $p' = -\frac{E}{C}$

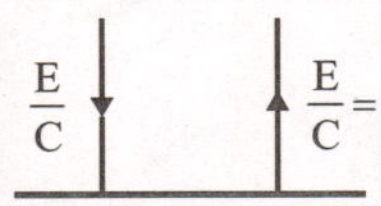

So, momentum transferred

$$= P - P' = \frac{E}{C} - \left(-\frac{E}{C}\right) = \frac{2E}{C}$$

8. (b) Average force $F_{av} = \frac{\Delta p}{\Delta t} = \frac{2IA}{c}$ ($\because$ Power = F.V)

$$= \frac{2\times25\times10^4\times15\times10^{-4}}{3\times10^8} = 2.50\times10^{-6}\text{N}$$

NOTES: Pressure $= \frac{\text{Force}}{\text{Area}} = \frac{F}{A}$

Force is the rate of change of momentum $F = \frac{dp}{dt}$

But energy, $U = pc$

$\therefore$ Pressure $= \frac{1}{A}\frac{U}{c\,.\,dt} = \frac{I}{c}$ $\left[\because \text{Intensity} = \frac{U}{Adt}\right]$

For perfectly reflecting surface, $P_r = \frac{2I}{c}$

$\therefore$ For perfectly reflecting surface, Force $= \frac{2IA}{c}$

9. **(c)** Given: frequency f = 2MHz, relative permittivity $\in_r = 4$

From formula,

velocity $v = \frac{c}{\sqrt{\in_r}} = \frac{c}{2} \Rightarrow \lambda' = \frac{\lambda}{2}$

[Since frequency remains unchanged]

10. **(a)** On comparing the given equation to

$\vec{E} = a_0 \hat{i} \cos(\omega t - kz)$

$\omega = 6 \times 10^8$

Wave number,

$k = \frac{2\pi}{\lambda} = \frac{\omega}{c}$

$k = \frac{\omega}{c} = \frac{6 \times 10^8}{3 \times 10^8} = 2\,\mathbf{m}^{-1}$

11. **(b)** The average energy stored in the electric field, $U_E = \frac{1}{2}\varepsilon_0 E^2$

The average energy stored in the magnetic field

$= U_B = \frac{1}{2}\frac{B^2}{\mu_0}$,

According to conservation of energy $U_E = U_B$

$\varepsilon_0 \mu_0 = \frac{B^2}{E^2}$

$\frac{B}{E} = \sqrt{\varepsilon_0 \mu_0} = \frac{1}{c}$

NOTES

The average energy density of electric field

$\mu_E = \frac{1}{4}\in_0 E_0^2$

$\Rightarrow \mu_E = \frac{1}{4}\in_0 (c^2 B_0^2) \quad \left[\because \frac{E_0}{B_0} = c\right]$

$\Rightarrow \mu_E = \frac{1}{4} E_0 B_0^2 \times \frac{1}{\mu_0 \in_0} \quad \left[\because c = \frac{1}{\sqrt{\mu_0 \in_0}}\right]$

$\Rightarrow \mu_E = \frac{1}{4}\frac{B_0^2}{\mu_0} = \mu_B$

Thus, the average energy density of electric field equal to the average density of magnetic field.

12. **(a)** E.M. wave always propagates in a direction perpendicular to both electric and magnetic fields. So, electric and magnetic fields should be along + X and + Y–directions respectively. Therefore, option (a) is the correct option.

13. **(c)** Electromagnetic waves are the combination of mutually perpendicular electric and magnetic fields. So, option (c) is false.

14. **(b)** Comparing with the equation of wave.

$E_y = E_0 \cos(\omega t - kx)$

$\omega = 2\pi f = 2\pi \times 10^6 \quad \therefore \quad f = 10^6$ Hz

$\frac{2\pi}{\lambda} = k = \pi \times 10^{-2}\,\text{m}^{-1}, \lambda = 200\,\text{m}$

15. **(c)** The velocity of electromagnetic radiation in a medium of permittivity $\in_0$ and permeability μ_0 is is equal to $\frac{1}{\sqrt{\mu_0 \in_0}}$.

16. **(c)** Variation in magnetic field causes electric field and vice-versa.

In electromagnetic waves, $\vec{E} \perp \vec{B}$. Both $\vec{E}$ and $\vec{B}$ are in the same phase.

In electromagnetic waves

$E = E_0 \sin(\omega t - kx)$

$B = B_0 \sin(\omega t - kx)$

The electromagnetic waves travel in the direction of $(\vec{E} \times \vec{B})$.

17. **(d)** Direction of propagation of electromagnetic waves is perpendicular to Electric field and Magnetic field. Hence, direction is given by vector $\vec{S} = \vec{E} \times \vec{H} = \frac{\vec{E} \times \vec{B}}{\mu_0}$.

NOTES

In electromagnetic waves, the rate of flow of energy crossing a unit area is described by Poynting vector. The magnitude of poynting vector

$|\vec{S}| = \frac{1}{\mu_0}$ EB sin 90° $= \frac{EB}{\mu_0}$

$= \frac{E^2}{\mu_0 c} \quad \left(\because B = \frac{E}{c}\right)$

18. **(c)** A stationary charge produces electric field only; an uniformly moving charge produces localised electromagnetic field; an accelerated charge produces electromagnetic radiations.

19. **(d)** We know that velocity of electromagnetic wave in vacuum

$(v_0) = \frac{1}{\sqrt{\mu_0 \varepsilon_0}}$ and velocity of

electromagnetic wave in medium is

$$(v) = \frac{1}{\sqrt{\mu\varepsilon}}.$$

Therefore refractive index of the medium

$$(\mu) = \frac{\text{Vel. of E.M. wave in vacuum}(v_0)}{\text{Vel. of E.M. wave in medium}(v)}$$

$$= \frac{1/\sqrt{\mu_0\varepsilon_0}}{1/\sqrt{\mu\varepsilon}} = \sqrt{\frac{\mu\varepsilon}{\mu_0\varepsilon_0}}$$

20. **(c)** The direction of oscillations of E and B fields are perpendicular to each other as well as to the direction of propagation. So, electromagnetic waves are transverse in nature.
The electric and magnetic fields oscillate in same phase.

21. **(b)** Size of particle $= \lambda = \frac{c}{\nu}$

$$3 \times 10^{-6} = \frac{3 \times 10^8}{\nu}$$

$\nu = 10^{14}$ Hz

NOTES: When frequency is higher than this, wavelength is still smaller. Resolution becomes better.

22. **(a)** $\lambda_{Red} > \lambda_{Green} > \lambda_{Blue} > \lambda_{Violet.}$ Red has the longest wavelength.

23. **(d)** Energy of x-ray is (100 ev to 100 kev)
Hence energy of the order of 15 keV belongs to x-rays.

24. **(d)** Required condition : Frequency of microwaves = Resonant frequency of water molecules.

25. **(a)** The decreasing order of the wavelengths is as given below :
microwave, infrared, ultraviolet, gamma rays.

26. **(b)** We know $E = \frac{hc}{\lambda} \Rightarrow E \propto \frac{1}{\lambda}$

$\Rightarrow E_m < E_v < E_x$

$\therefore \lambda_m > \lambda_v > \lambda_x$

NOTES: Energy of electromagnetic wave, $E = \frac{hc}{\lambda}$

For *em* wave of lower energy, (ν) is small and λ is larger. For *em* wave of higher energy (ν) is large and λ is small.

27. **(d)** β ray is not electromagnetic ray

28. **(b)** Depends on the magnitude of frequency

29. **(a)** Gamma ray has highest frequency and lowest wavelength.

30. **(b)** Infrared rays is the cause of Green house effect. The glass transmits visible light and short infrared rays which are absorbed by plants. Then it emits long infrared rays, which are reflected back by glass.

31. **(a)** α rays contain Helium nuclei which contains 2 unit of positive charge.

32. **(b)** γ rays has lowest wavelength and highest frequency among them while ultraviolet ray has highest wavelength and lowest frequency.
Order of frequency : $b > a > c$

33. **(c)**

Rays	Wavelength [Range in m]
X-rays	1×10^{-11} to 3×10^{-8}
γ-rays	6×10^{-14} to 1×10^{-11}
Microwaves	10^{-3} to 0.3
Radiowaves	10 to 10^4

Wavelength of U.V. Rays ranges from 6×10^{-8} to 4×10^{-7}.

34. **(b)** X-rays are used for the investigation of structure of solids.

35. **(d)** Wavelength of red light is longest.

36. **(d)**

Rays	Wavelength [Range in m]
X-rays	1×10^{-11} to 3×10^{-8}
γ-rays	6×10^{-14} to 1×10^{-11}
Microwaves	10^{-3} to 0.3
Radio waves	10 to 10^4

23 Ray Optics and Optical Instruments

Trend Analysis with Important Topics & Sub-Topics

Topic Name	Sub-Topic	2021		2020		2019		2018		2017		2016	
		QNS.	LOD	QNS.	LOD	QNS.	LOD	QNS.	LOD	QNS.	LOD	QNS.	LOD
Plane Mirror, Spherical Mirror & Reflection of Light	Mirror & Refl of Light									1	A	1	A
Refraction of Light at Plane Surface & Total Internal Reflection						1	A						
Refraction at Curved Surface, Lenses & Power of Lens	Combination of Lenses & Lens formula	1	A			1	E						
	Combination of Lens and mirror	1	A					1	A	1	A	1	A
Prism & Dispersion of Light	Prism & Dispersion, Refraction through a Prism	1	A										
Optical Instruments	Rainbow					1	A						
	Magnification of Telescope & Microscope							1	E			1	A

LOD - Level of Difficulty | E - Easy | A - Average | D - Difficult | Qns - No. of Questions

Topic 1 Plane Mirror, Spherical Mirror & Reflection of Light

1. A beam of light from a source L is incident normally on a plane mirror fixed at a certain distance x from the source. The beam is reflected back as a spot on a scale placed just above the source I. When the mirror is rotated through a small angle θ, the spot of the light is found to move through a distance y on the scale. The angle θ is given by *[2017]*

   (a) $\frac{y}{x}$ (b) $\frac{x}{2y}$

   (c) $\frac{x}{y}$ (d) $\frac{y}{2x}$

2. Match the corresponding entries of column-1 with column-2 (Where m is the magnification produced by the mirror): *[2016]*

Column-1	Column-2
(P) $m = -2$	(A) Convex mirror
(Q) $m = -\frac{1}{2}$	(B) Concave mirror
(R) $m = +2$	(C) Real image
(S) $m = +\frac{1}{2}$	(D) Virtual image

   (a) P → B and C, Q → B and C, R → B and D, S → A and D.
   (b) P → A and C, Q → A and D, R → A and B, S → C and D
   (c) P → A and D, Q → B and C, R → B and D, S → B and C
   (d) P → C and D, Q → B and D, R → B and C, S → A and D

3. Two plane mirrors are inclined at 70°. A ray incident on one mirror at angle θ after reflection falls on second mirror and is reflected from there parallel to first mirror. The value of θ is *[NEET Kar. 2013]*

(a) 50° (b) 45°
(c) 30° (d) 55°

4. A rod of length 10 cm lies along the principal axis of a concave mirror of focal length 10 cm in such a way that its end closer to the pole is 20 cm away from the mirror. The length of the image is : *[2012M]*

(a) 10 cm (b) 15 cm
(c) 2.5 cm (d) 5 cm

5. A person is six feet tall. How tall must a vertical mirror be if he is able to see his entire length? *[2000]*

(a) 3 ft (b) 4.5 ft
(c) 7.5 ft (d) 6 ft

6. If two mirrors are kept inclined at 60° to each other and a body is placed at the middle, then total number of images formed is *[1995]*

(a) six (b) five
(c) four (d) three

7. Ray optics is valid, when characteristic dimensions are *[1989]*

(a) of the same order as the wavelength of light
(b) much smaller than the wavelength of light
(c) of the order of one millimetre
(d) much larger than the wavelength of light

Topic 2 Refraction of Light at Plane Surface & Total Internal Reflection

8. In total internal reflection when the angle of incidence is equal to the critical angle for the pair of media in contact, what will be angle of refraction ? *[2019]*

(a) 180°
(b) 0°
(c) equal to angle of incidence
(d) 90°

9. Which of the following is not due to total internal reflection? *[2011]*

(a) Working of optical fibre
(b) Difference between apparent and real depth of pond
(c) Mirage on hot summer days
(d) Brilliance of diamond

10. A ray of light travelling in a transparent medium of refractive index μ, falls on a surface separating the medium from air at an angle of incidence of 45°. For which of the following value of μ the ray can undergo total internal reflection? *[2010]*

(a) $\mu = 1.33$ (b) $\mu = 1.40$
(c) $\mu = 1.50$ (d) $\mu = 1.25$

11. The frequency of a light wave in a material is 2×10^{14} Hz and wavelength is 5000 Å. The refractive index of material will be *[2007]*

(a) 1.50 (b) 3.00
(c) 1.33 (d) 1.40

12. A small coin is resting on the bottom of a beaker filled with liquid. A ray of light from the coin travels upto the surface of the liquid and moves along its surface. How fast is the light travelling in the liquid? *[2007]*

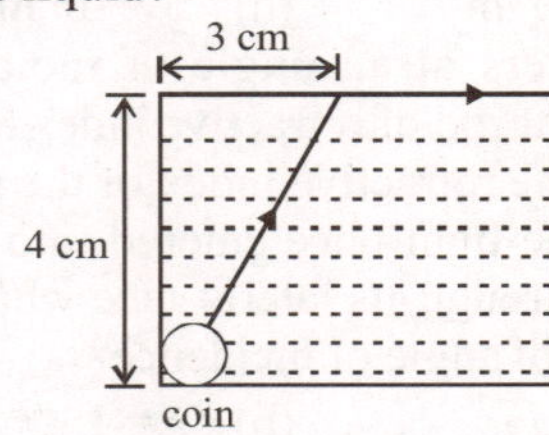

(a) 2.4×10^8 m/s (b) 3.0×10^8 m/s
(c) 1.2×10^8 m/s (d) 1.8×10^8 m/s

13. A beam of light composed of red and green rays is incident obliquely at a point on the face of rectangular glass slab. When coming out on the opposite parallel face, the red and green rays emerge from *[2004]*

(a) one point propagating in the same direction
(b) two points propagating in two different non-parallel directions
(c) two points propagating in two different parallel directions
(d) one point propagating in two different directions

14. A light ray falls on a rectangular glass slab as shown. The index of refraction of the glass, if total internal reflection is to occur at the vertical face, is *[2002]*

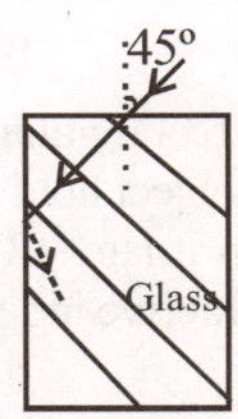

(a) $\sqrt{3/2}$ (b) $\frac{(\sqrt{3}+1)}{2}$
(c) $\frac{(\sqrt{2}+1)}{2}$ (d) $\sqrt{5/2}$

15. The reddish appearance of the sun at sunrise and sunset is due to *[2000]*
(a) the colour of the sky
(b) the scattering of light
(c) the polarisation of light
(d) the colour of the sun

16. An air bubble in a glass slab ($\mu = 1.5$) is 5 cm deep when viewed from one face and 2 cm deep when viewed from the opposite face. The thickness of the slab is *[2000]*
(a) 7.5 cm (b) 10.5 cm
(c) 7 cm (d) 10 cm

17. Wavelength of light of frequency 100 Hz *[1999]*
(a) 2×16^6 m (b) 3×10^6 m
(c) 4×10^6 m (d) 5×10^6 m

18. Light enters at an angle of incidence in a transparent rod of refractive index n. For what value of the refractive index of the material of the rod the light once entered into it will not leave it through its lateral face whatsoever be the value of angle of incidence? *[1998]*
(a) $n > \sqrt{2}$ (b) $n = 1$
(c) $n = 1.1$ (d) $n = 1.3$

19. An electromagnetic radiation of frequency n, wavelength λ, travelling with velocity v in air enters in a glass slab of refractive index (μ). The frequency, wavelength and velocity of light in the glass slab will be respectively *[1997]*
(a) $n, \frac{\lambda}{\mu}$ and $\frac{v}{\mu}$ (b) $n, 2\lambda$ and $\frac{v}{\mu}$
(c) $\frac{n}{\mu}, \frac{\lambda}{\mu}$ and $\frac{v}{\mu}$ (d) $\frac{2\pi}{\mu}, \frac{\lambda}{\mu}$ and v

20. Light travels through a glass plate of thickness t and refractive index μ. If c is the speed of light in vacuum, the time taken by light to travel this thickness of glass is *[1996]*
(a) μtc (b) $\frac{tc}{\mu}$
(c) $\frac{t}{\mu c}$ (d) $\frac{\mu t}{c}$

21. One face of a rectangular glass plate 6 cm thick is silvered. An object held 8 cm in front of the first face forms an image 12 cm behind the silvered face. The refractive index of the glass is *[1996]*
(a) 0.4 (b) 0.8
(c) 1.2 (d) 1.6

22. A point source of light is placed 4 m below the surface of water of refractive index $\frac{5}{3}$. The minimum diameter of a disc, which should be placed over the source, on the surface of water to cut off all light coming out of water is *[1994]*
(a) ∞ (b) 6m
(c) 4m (d) 3m

23. Time taken by sunlight to pass through a window of thickness 4 mm whose refractive index is $\frac{3}{2}$ is *[1993]*
(a) 2×10^{-4} sec (b) 2×10^{8} sec
(c) 2×10^{-11} sec (d) 2×10^{11} sec

24. A beam of monochromatic light is refracted from vacuum into a medium of refractive index 1.5, the wavelength of refracted light will be *[1991]*
(a) dependent on intensity of refracted light
(b) same
(c) smaller
(d) larger

25. Green light of wavelength 5460 Å is incident on an air-glass interface. If the refractive index of glass is 1.5, the wavelength of light in glass would be ($c = 3 \times 10^8\ ms^{-1}$) *[1991]*
(a) 3640 Å (b) 5460 Å
(c) 4861 Å (d) none of the above

Topic 3 Refraction at Curved Surface, Lenses & Power of Lens and Combination of Lens & Mirror

26. A convex lens 'A' of focal length 20 cm and a concave lens 'B' of focal length 5 cm are kept along the same axis with a distance 'd' between them. If a parallel beam of light falling on 'A' leaves 'B' as a parallel beam, then the distance 'd' in cm will be *[2021]*
(a) 30 (b) 25
(c) 15 (d) 50

27. A point object is placed at a distance of 60 cm from a convex lens of focal length 30 cm. If a plane mirror were put perpendicular to the principal axis of the lens and at a distance of 40 cm from it, the final image would be formed at a distance of *[2021]*

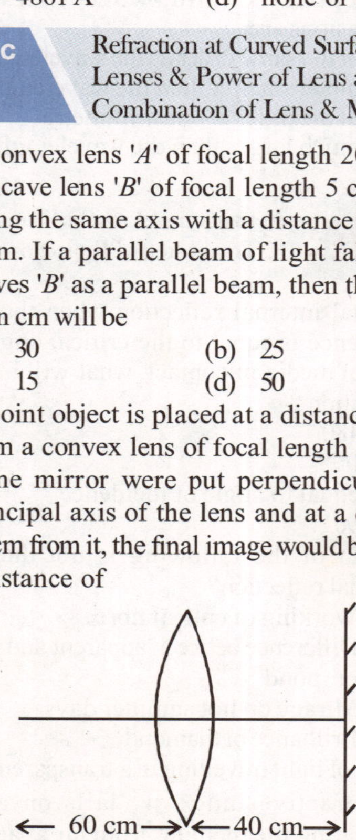

(a) 20 cm from the plane mirror, it would be a virtual image
(b) 20 cm from the lens, it would be a real image
(c) 30 cm from the lens, it would be a real image
(d) 30 cm from the plane mirror, it would be a virtual image

28. Two similar thin equi-convex lenses, of focal length f each, are kept coaxially in contact with each other such that the focal length of the combination is F_1. When the space between the two lenses is filled with glycerin (which has the same refractive index (= 1.5) as that of glass) then the equivalent focal length is F_2. The ratio $F_1 : F_2$ will be : *[2019]*
(a) 2 : 1 (b) 1 : 2
(c) 2 : 3 (d) 3 : 4

29. An equiconvex lens has power *P*. It is cut into two symmetrical halves by a plane containing the principal axis. The power of one part will be, *[NEET Odisha 2019]*
(a) P (b) 0
(c) $\frac{P}{2}$ (d) $\frac{P}{4}$

30. A double convex lens has focal length 25 cm. The radius of curvature of one of the surfaces is double of the other. Find the radii if the refractive index of the material of the lens is 1.5. *[NEET Odisha 2019]*
(a) 50 cm, 100 cm (b) 100 cm, 50 cm
(c) 25 cm, 50 cm (d) 18.75 cm, 37.5 cm

31. Two identical thin plano-convex glass lenses (refractive index 1.5) each having radius of curvature of 20 cm are placed with their convex surfaces in contact at the centre. The intervening space is filled with oil of refractive index 1.7. The focal length of the combination is *[2015]*
(a) –25 cm (b) –50 cm
(c) 50 cm (d) –20 cm

32. A plano convex lens fits exactly into a plano concave lens. Their plane surfaces are parallel to each other. If lenses are made of different materials of refractive indices μ_1 and μ_2 and R is the radius of curvature of the curved surface of the lenses, then the focal length of the combination is *[2013]*
(a) $\frac{R}{2(\mu_1 - \mu_2)}$ (b) $\frac{R}{(\mu_1 - \mu_2)}$
(c) $\frac{2R}{(\mu_2 - \mu_1)}$ (d) $\frac{R}{2(\mu_1 + \mu_2)}$

33. When a biconvex lens of glass having refractive index 1.47 is dipped in a liquid, it acts as a plane sheet of glass. This implies that the liquid must have refractive index. *[2012]*
(a) equal to that of glass
(b) less then one
(c) greater than that of glass
(d) less then that of glass

34. A concave mirror of focal length 'f_1' is placed at a distance of '*d*' from a convex lens of focal length 'f_2'. A beam of light coming from infinity and falling on this convex lens-concave mirror combination returns to infinity. *[2012]*
The distance '*d*' must equal :
(a) $f_1 + f_2$ (b) $-f_1 + f_2$
(c) $2f_1 + f_2$ (d) $-2f_1 + f_2$

35. A biconvex lens has a radius of curvature of magnitude 20 cm. Which one of the following options best describe the image formed of an object of height 2 cm placed 30 cm from the lens? *[2011]*
(a) Virtual, upright, height = 1 cm
(b) Virtual, upright, height = 0.5 cm
(c) Real, inverted, height = 4 cm
(d) Real, inverted, height = 1cm

36. A converging beam of rays is incident on a diverging lens. Having passed through the lens the rays intersect at a point 15 cm from the lens on the opposite side. If the lens is removed the point where the rays meet will move 5 cm closer to the lens. The focal length of the lens is *[2011M]*
(a) – 10 cm (b) 20 cm
(c) –30 cm (d) 5 cm

37. A lens having focal length *f* and aperture of diameter *d* forms an image of intensity *I*. Aperture of diameter $\frac{d}{2}$ in central region of lens is covered by a black paper. Focal length of lens and intensity of image now will be respectively: *[2010]*
(a) f and $\frac{I}{4}$ (b) $\frac{3f}{4}$ and $\frac{I}{2}$
(c) f and $\frac{3I}{4}$ (d) $\frac{f}{2}$ and $\frac{I}{2}$

38. Two thin lenses of focal lengths f_1 and f_2 are in contact and coaxial. The power of the combination is: *[2008]*
(a) $\sqrt{\frac{f_1}{f_2}}$ (b) $\sqrt{\frac{f_2}{f_1}}$
(c) $\frac{f_1 + f_2}{2}$ (d) $\frac{f_1 + f_2}{f_1 f_2}$

39. A boy is trying to start a fire by focusing sunlight on a piece of paper using an equiconvex lens of focal length 10 cm. The diameter of the Sun is 1.39×10^9 m and its mean distance from the earth is 1.5×10^{11} m. What is the diameter of the Sun's image on the paper? *[2008]*

(a) 9.2×10^{-4} m (b) 6.5×10^{-4} m
(c) 6.5×10^{-5} m (d) 12.4×10^{-4} m

40. A convex lens and a concave lens, each having same focal length of 25 cm, are put in contact to form a combination of lenses. The power in diopters of the combination is *[2006]*

(a) 50 (b) infinite
(c) zero (d) 25

41. An equiconvex lens is cut into two halves along (i) XOX' and (ii) YOY' as shown in the figure. Let f, f', f'' be the focal lengths of the complete lens, of each half in case (i), and of each half in case (ii), respectively. *[2003]*

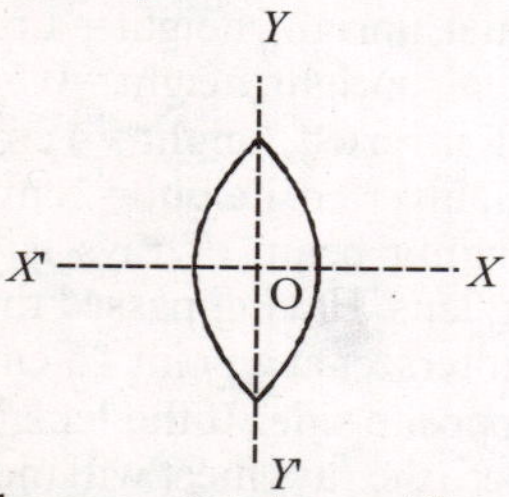

Choose the correct statement from the following

(a) $f' = 2f, f'' = 2f$ (b) $f' = f, f'' = 2f$
(c) $f' = 2f, f'' = f$ (d) $f' = f, f'' = f$

42. A convex lens is dipped in a liquid whose refractive index is equal to the refractive index of the lens. Then its focal length will *[2003]*

(a) remain unchanged
(b) become zero
(c) become infinite
(d) become small, but non-zero

43. A body is located on a wall. Its image of equal size is to be obtained on a parallel wall with the help of a convex lens. The lens is placed at a distance 'd' ahead of second wall, then the required focal length will be *[2002]*

(a) only $\frac{d}{4}$

(b) only $\frac{d}{2}$

(c) more than $\frac{d}{4}$ but less than $\frac{d}{2}$

(d) less than $\frac{d}{4}$

44. The radius of curvature of a thin plano-convex lens is 10 cm (of curved surface) and the refractive index is 1.5. If the plane surface is silvered, then it behaves like a concave mirror of focal length *[2000]*

(a) 10 cm (b) 15 cm
(c) 20 cm (d) 5 cm

45. A plano-convex lens is made of material of refractive index 1.6. The radius of curvature of the curved surface is 60 cm. The focal length of the lens is *[1999]*

(a) 50 cm (b) 100 cm
(c) 200 cm (d) 400 cm

46. A luminous object is placed at a distance of 30 cm from the convex lens of focal length 20 cm. On the other side of the lens, at what distance from the lens a convex mirror of radius of curvature 10 cm be placed in order to have an upright image of the object coincident with it? *[1998]*

(a) 12 cm (b) 30 cm
(c) 50 cm (d) 60 cm

47. The focal length of converging lens is measured for violet, green and red colours. It is respectively f_v, f_g, f_r. We will get *[1997]*

(a) $f_v = f_g$ (b) $f_g > f_r$
(c) $f_v < f_r$ (d) $f_v > f_r$

48. A convex lens of focal length 80 cm and a concave lens of focal length 50 cm are combined together. What will be their resulting power? *[1996]*

(a) $+6.5$ D (b) -6.5 D
(c) $+7.5$ D (d) -0.75 D

49. If f_V and f_R are the focal lengths of a convex lens for violet and red light respectively and F_V and F_R are the focal lengths of concave lens for violet and red light respectively, then we have *[1996]*

(a) $f_V < f_R$ and $F_V > F_R$
(b) $f_V < f_R$ and $F_V < F_R$
(c) $f_V > f_R$ and $F_V > F_R$
(d) $f_V > f_R$ and $F_V < F_R$

50. An achromatic combination of lenses is formed by joining *[1995]*

(a) 2 convex lenses
(b) 2 concave lenses
(c) 1 convex and 1 concave lens
(d) 1 convex and 1 plane mirror

51. A lens is placed between a source of light and a wall. It forms images of area A_1 and A_2 on the wall for its two different positions. The area of the source of light is *[1995]*

(a) $\sqrt{A_1 A_2}$ (b) $\frac{A_1 + A_2}{2}$

(c) $\frac{A_1 - A_2}{2}$ (d) $\frac{1}{A_1} + \frac{1}{A_2}$

52. Focal length of a convex lens will be maximum for [1994]
(a) blue light (b) yellow light
(c) green light (d) red light

53. Focal length of a convex lens of refractive index 1.5 is 2 cm. Focal length of the lens when immersed in a liquid of refractive index of 1.25 will be [1988]
(a) 10 cm (b) 2.5 cm
(c) 5 cm (d) 7.5 cm

Topic 4 Prism & Dispersion of Light

54. Find the value of the angle of emergence from the prism. Refractive index of the glass is $\sqrt{3}$. [2021]

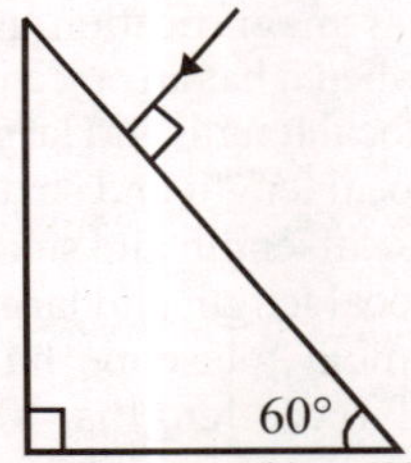

(a) 90° (b) 60°
(c) 30° (d) 45°

55. A ray is incident at an angle of incidence i on one surface of a small angle prism (with angle of prism A) and emerges normally from the opposite surface. If the refractive index of the material of the prism is μ, then the angle of incidence is nearly equal to : [2020]
(a) $\frac{2A}{\mu}$ (b) μA
(c) $\frac{\mu A}{2}$ (d) $\frac{A}{2\mu}$

56. Pick the wrong answer in the context with rainbow. [2019]
(a) When the light rays undergo two internal reflections in a water drop, a secondary rainbow is formed.
(b) The order of colours is reversed in the secondary rainbow.
(c) An observer can see a rainbow when his front is towards the sun.
(d) Rainbow is a combined effect of dispersion, refraction and reflection of sunlight.

57. The refractive index of the material of a prism is $\sqrt{2}$ and the angle of the prism is 30°. One of the two refracting surfaces of the prism is made a mirror inwards, by silver coating. A beam of monochromatic light entering the prism from the other face will retrace its path (after reflection from the silvered surface) if its angle of incidence on the prism is [2018]
(a) 60° (b) 45°
(c) Zero (d) 30°

58. A thin prism having refracting angle 10° is made of glass of refractive index 1.42. This prism is combined with another thin prism of glass of refractive index 1.7. This combination produces dispersion without deviation. The refracting angle of second prism should be [2017]
(a) 6° (b) 8°
(c) 10° (d) 4°

59. The angle of incidence for a ray of light at a refracting surface of a prism is 45°. The angle of prism is 60°. If the ray suffers minimum deviation through the prism, the angle of minimum deviation and refractive index of the material of the prism respectively, are : [2016]
(a) $45°, \frac{1}{\sqrt{2}}$ (b) $30°, \sqrt{2}$
(c) $45°, \sqrt{2}$ (d) $30°, \frac{1}{\sqrt{2}}$

60. The refracting angle of a prism is 'A', and refractive index of the material of the prism is cot(A/2). The angle of minimum deviation is : [2015]
(a) 180° – 2A (b) 90° – A
(c) 180° + 2A (d) 180° – 3A

61. A beam of light consisting of red, green and blue colours is incident on a right angled prism. The refractive index of the material of the prism for the above red, green and blue wavelengths are 1.39, 1.44 and 1.47, respectively. [2015 RS]

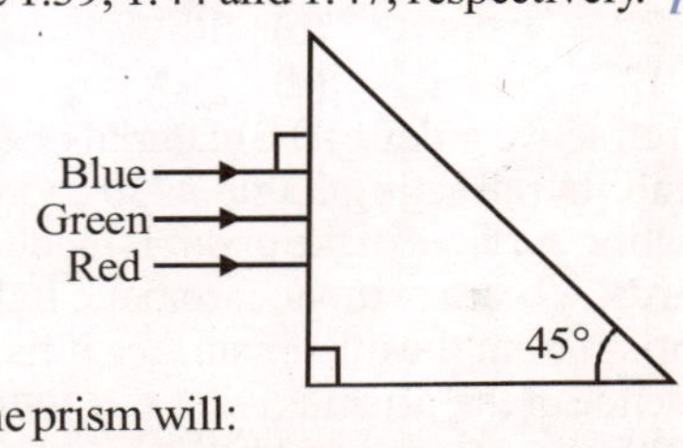

The prism will:
(a) separate all the three colours from one another

(b) not separate the three colours at all
(c) separate the red colour part from the green and blue colours
(d) separate the blue colour part from the red and green colours

62. The angle of a prism is 'A'. One of its refracting surfaces is silvered. Light rays falling at an angle of incidence 2A on the first surface returns back through the same path after suffering reflection at the silvered surface. The refractive index μ, of the prism is : *[2014]*
(a) 2 sin A (b) 2 cos A
(c) $\frac{1}{2}\cos A$ (d) tan A

63. Rainbow is formed due to a combination of *[2000]*
(a) dispersion and total internal reflection
(b) refraction and absorption
(c) dispersion and focussing
(d) refraction and scattering

64. A ray of light is incident at an angle of incidence, *i*, on one face of prism of angle *A* (assumed to be small) and emerges normally from the opposite face. If the refractive index of the prism is μ, the angle of incidence *i*, is nearly equal to: *[2012]*
(a) μA (b) $\frac{\mu A}{2}$
(c) $\frac{A}{\mu}$ (d) $\frac{A}{2\mu}$

65. For the angle of minimum deviation of a prism to be equal to its refracting angle, the prism must be made of a material whose refractive index : *[2012M]*
(a) lies between $\sqrt{2}$ and 1
(b) lies between 2 and $\sqrt{2}$
(c) is less than 1
(d) is greater than 2

66. A thin prism of angle 15° made of glass of refractive index $\mu_1 = 1.5$ is combined with another prism of glass of refractive index $\mu_2 = 1.75$. The combination of the prism produces dispersion without deviation. The angle of the second prism should be *[2011M]*
(a) 7° (b) 10°
(c) 12° (d) 5°

67. The refractive index of the material of a prism is $\sqrt{2}$ and its refracting angle is 30°. One of the refracting surfaces of the prism is made a mirror inwards. A beam of monochromatic light enters the prism from the mirror surface if its angle of incidence of the prism is *[2004, 1992]*
(a) 30° (b) 45°
(c) 60° (d) 0°

68. The refractive index of the material of the prism is $\sqrt{3}$; then the angle of minimum deviation of the prism is *[1999]*
(a) 30° (b) 45°
(c) 60° (d) 75°

69. Angle of deviation (δ) by a prism (refractive index = μ and supposing the angle of prism A to be small) can be given by *[1994]*
(a) $\delta = (\mu - 1)A$ (b) $\delta = (\mu + 1)A$
(c) $\delta = \frac{\sin\frac{A+\delta}{2}}{\sin\frac{A}{2}}$ (d) $\delta = \frac{\mu - 1}{\mu + 1}A$

Topic 5 Optical Instruments

70. An astronomical refracting telescope will have large angular magnification and high angular resolution, when it has an objective lens of *[2018]*
(a) small focal length and large diameter
(b) large focal length and small diameter
(c) small focal length and small diameter
(d) large focal length and large diameter

71. A astronomical telescope has objective and eyepiece of focal lengths 40 cm and 4 cm respectively. To view an object 200 cm away from the objective, the lenses must be separated by a distance : *[2016]*
(a) 37.3 cm (b) 46.0 cm
(c) 50.0 cm (d) 54.0 cm

72. In an astronomical telescope in normal adjustment a straight black line of lenght L is drawn on inside part of objective lens. The eye-piece forms a real image of this line. The length of this image is *l*. The magnification of the telescope is : *[2015 RS]*
(a) $\frac{L}{I} - 1$ (b) $\frac{L+I}{L-I}$
(c) $\frac{L}{I}$ (d) $\frac{L}{I} + 1$

73. If the focal length of objective lens is increased then magnefying power of : *[2014]*
(a) microscope will increase but that of telescope decrease.
(b) microscope and telescope both will increase.
(c) microscope and telescope both will decrease
(d) microscope will decrease but that of telescope increase.

74. For a normal eye, the cornea of eye provides a converging power of 40D and the least converging power of the eye lens behind the cornea is 20D. Using this information, the distance between the retina and the eye lens of the eye can be estimated to be *[2013]*

(a) 2.5 cm (b) 1.67 cm
(c) 1.5 cm (d) 5 cm

75. The magnifying power of a telescope is 9. When it is adjusted for parallel rays the distance between the objective and eyepiece is 20 cm. The focal length of lenses are : *[2012]*

(a) 10 cm, 10 cm (b) 15 cm, 5 cm
(c) 18 cm, 2 cm (d) 11 cm, 9 cm

76. A microscope is focussed on a mark on a piece of paper and then a slab of glass of thickness 3 cm and refractive index 1.5 is placed over the mark. How should the microscope be moved to get the mark in focus again ? *[2006]*

(a) 4.5 cm downward (b) 1 cm downward
(c) 2 cm upward (d) 1 cm upward

77. A telescope has an objective lens of 10 cm diameter and is situated at a distance of one kilometer from two objects. The minimum distance between these two objects, which can be resolved by the telescope, when the mean wavelength of light is 5000 Å, is of the order of *[2004]*

(a) 5 cm (b) 0.5 m
(c) 5 m (d) 5 mm

78. An astronomical telescope has a length of 44 cm and tenfold magnification. The focal length of the objective lens is *[1997]*

(a) 4 cm (b) 40 cm
(c) 44 cm (d) 440 cm

79. The hypermetropia is a *[1995]*

(a) short-sight defect
(b) long-sight defect
(c) bad vision due to old age
(d) none of these

ANSWER KEY																	
1	(d)	10	(c)	19	(a)	28	(b)	37	(c)	46	(c)	55	(b)	64	(a)	73	(d)
2	(a)	11	(b)	20	(d)	29	(a)	38	(d)	47	(c)	56	(c)	65	(b)	74	(b)
3	(a)	12	(d)	21	(c)	30	(d)	39	(a)	48	(d)	57	(b)	66	(b)	75	(c)
4	(d)	13	(c)	22	(b)	31	(b)	40	(c)	49	(a)	58	(a)	67	(b)	76	(d)
5	(a)	14	(a)	23	(c)	32	(b)	41	(b)	50	(c)	59	(b)	68	(c)	77	(d)
6	(b)	15	(b)	24	(c)	33	(a)	42	(c)	51	(a)	60	(a)	69	(a)	78	(b)
7	(d)	16	(b)	25	(a)	34	(c)	43	(b)	52	(d)	61	(c)	70	(d)	79	(b)
8	(d)	17	(b)	26	(c)	35	(c)	44	(a)	53	(c)	62	(b)	71	(d)		
9	(b)	18	(a)	27	(a)	36	(c)	45	(b)	54	(b)	63	(a)	72	(c)		

Hints & Solutions

1. **(d)** When mirror is rotated by angle θ reflected ray will be rotated by 2θ.
According to the condition,

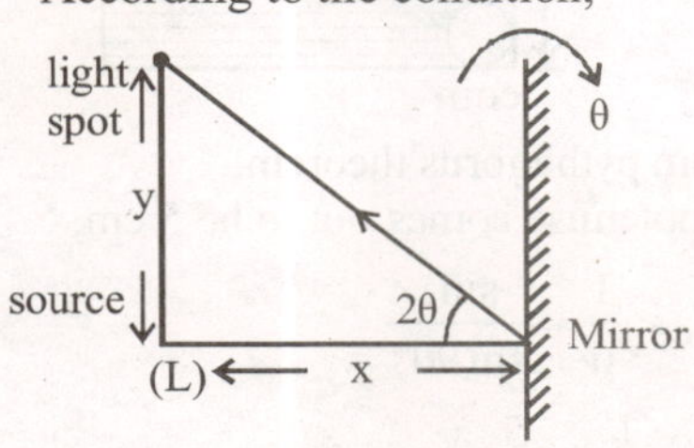

$$\frac{y}{x} = 2\theta \Rightarrow \theta = \frac{y}{2x}$$

2. **(a)** Magnitude m = +ve ⇒ virtual image
m = –ve ⇒ real image
magnitude of magnification,
$|m| > 1$ ⇒ magnified image
$|m| < 1$ ⇒ diminished image

3. **(a)**

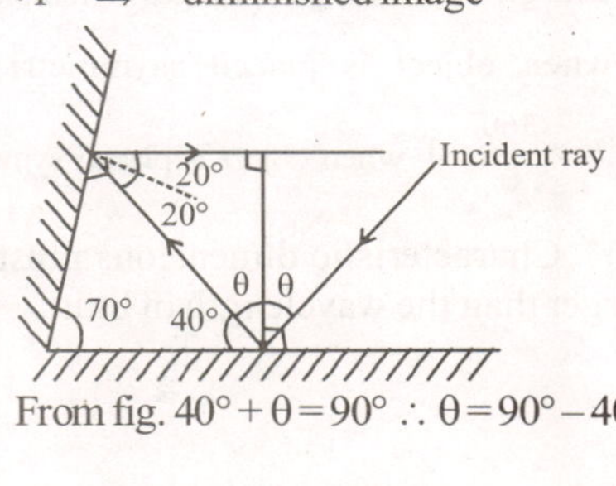

From fig. $40° + \theta = 90°$ ∴ $\theta = 90° - 40° = 50°$

4. (d) According to the condition,

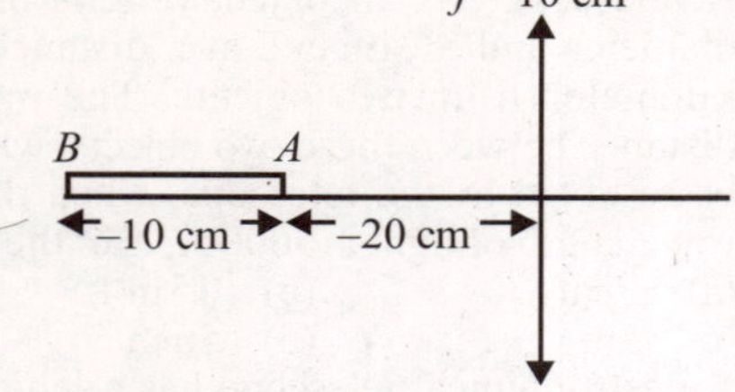

The focal length of the mirror

$$-\frac{1}{f}=\frac{1}{v}+\frac{1}{u}$$

For A end of the rod the image distance
When $u_1 = -20$ cm

$$\Rightarrow \frac{-1}{10}=\frac{1}{v_1}-\frac{1}{20}$$

$$\frac{1}{v_1}=\frac{-1}{10}+\frac{1}{20}=\frac{-2+1}{20}$$

$v_1 = -20$ cm
For when $u_2 = -30$ cm

$$\frac{1}{f}=\frac{1}{v_2}-\frac{1}{30}$$

$$\frac{1}{v_2}=\frac{-1}{10}+\frac{1}{30}=\frac{-30+10}{300}=\frac{-20}{300}$$

$v_2 = -15$ cm
$L = v_2 - v_1 = -15 - (-20)$
$L = 5$ cm

5. (a) To see his full image in a plane mirror a person requires a mirror of at least half of his height.

6. (b) Angle between two mirrors $(\theta) = 60°$. Number of images formed by the inclined mirror

$$(n)=\frac{360°}{\theta}-1=\frac{360°}{60°}-1=6-1=5$$

NOTES When two plane mirrors are inclined to each other at an angle θ then number of images (n) formed $n=\frac{360}{\theta}$ when object is placed asymmetrically, and $n=\frac{360}{\theta}-1$ when object is placed symmetrically.

7. (d) Characteristic dimensions must be much larger than the wavelength of light.

8. (d) For total internal reflection when $i = i_c$, then

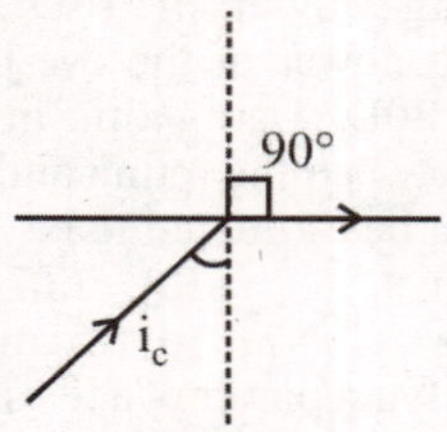

refracted ray grazes with the surface. That means the angle of refraction $r = 90°$.

9. (b) Difference between apparent and real depth of a pond is due to the refraction of light, not due to the total internal reflection. Other three phenomena are due to the total internal reflection.

10. (c) For total internal reflection,

$$\mu \geq \frac{1}{\sin C} \geq \sqrt{2} \geq 1.414$$

$\Rightarrow \mu = 1.50$

11. (b) By using $v = n\lambda$

Here, $n = 2\times10^{14}$ Hz

$\lambda = 5000 Å = 5000\times10^{-10}$ m

$v = 2\times10^{14}\times5000\times10^{-10} = 10^8$ m/s

Refractive index of the material,

$$\mu = c/v = \frac{3\times10^8}{10^8}=3$$

NOTES In case of refraction of light frequency (and hence colour) and phase do not change while wavelength and velocity will change.

12. (d)

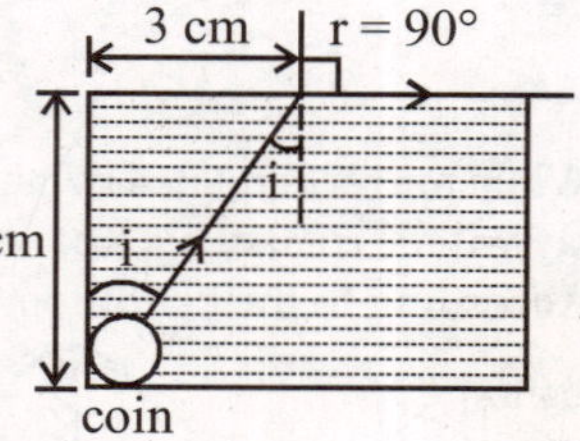

From pythagorus theorem,
Hypotenuse comes out to be 5 cm.

Since, $\frac{1}{\mu}=\frac{\sin i}{\sin 90°}$

$$\mu=\frac{1}{\sin i}=\frac{5}{3}$$

Speed, $v = \frac{c}{\mu} = \frac{3\times10^8}{5/3} = 1.8\times10^8$ m/s

13. (c) Since refractive index for both the light are different, so they emerge out moving in two different parallel directions.

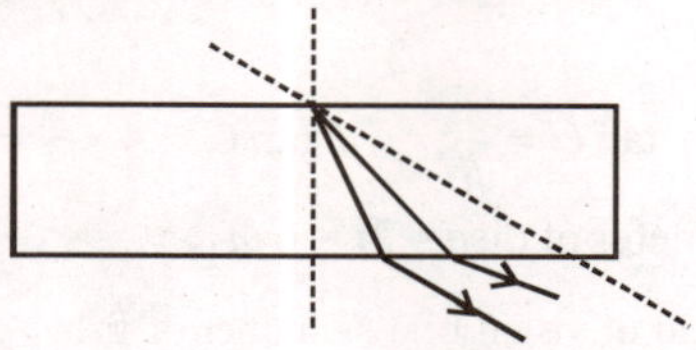

14. (a) For point A, ${}_a\mu_g = \frac{\sin 45°}{\sin r}$

$$\Rightarrow \sin r = \frac{1}{\sqrt{2}\,{}_a\mu_g}$$

For point B, $\sin(90° - r) = {}_g\mu_a$ where, $(90° - r)$ is critical angle.

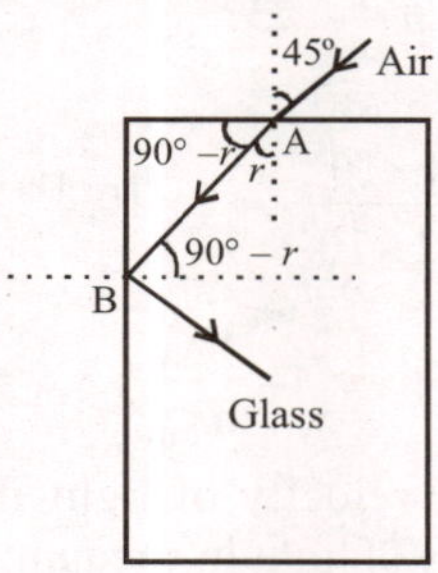

$$\therefore \cos r = {}_g\mu_a = \frac{1}{{}_a\mu_g} \Rightarrow {}_a\mu_g = \frac{1}{\cos r}$$

$$= \frac{1}{\sqrt{1-\sin^2 r}} = \frac{1}{\sqrt{1-\frac{1}{2\,{}_a\mu_g^2}}}$$

$$\Rightarrow {}_a\mu_g^2 = \frac{1}{1-\frac{1}{2\,{}_a\mu_g^2}} = \frac{2\,{}_a\mu_g^2}{2\,{}_a\mu_g^2 - 1}$$

$$\Rightarrow 2\,{}_a\mu_g^2 - 1 = 2 \Rightarrow {}_a\mu_g = \sqrt{\frac{3}{2}}$$

15. (b) It is due to scattering of light.

Scattering $\propto \frac{1}{\lambda^4}$. Hence the light reaches us is rich in red.

16. (b) $1.5 = \frac{\text{Real depth } (R_1)}{\text{Apparent depth } (5\text{ cm})}$

$\therefore R_1 = 1.5\times5 = 7.5$ cm

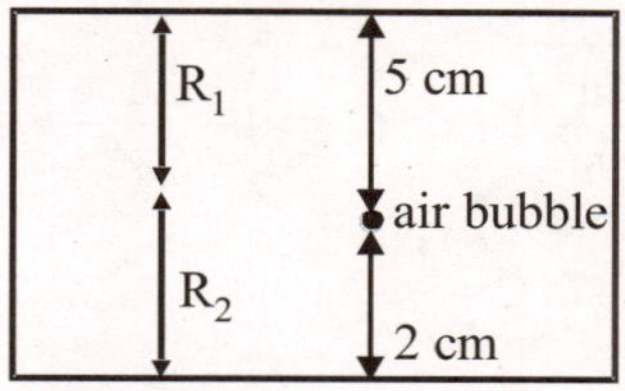

For opposite face,

$$1.5 = \frac{R_2}{2} \Rightarrow R_2 = 3.0 \text{ cm}$$

$\therefore$ Thickness of the slab $= R_1 + R_2$
$= 7.5 + 3 = 10.5$ cm

> **NOTES**
> If object and observer are situated in different medium then due to refraction, object appears to be displaced
>
> Refractive index, $\mu = \frac{\text{Real depth}(h)}{\text{Apparent depth}(h')}$
>
> Real depth > Apparent depth
>
> In case object is in rarer medium and object is in denser medium then, $\mu = \frac{h'}{h}$
>
> Real depth < apparent depth

17. (b) Frequency (n) = 100 Hz
$v = n\lambda$

$$\lambda = \frac{3\times10^8}{100}$$

[where, velocity of light (v) = 3×10^8 m/s]

$$\lambda = 3\times10^6 \text{ m}\cdot$$

18. (a) Let a ray of light enter at A and the refracted beam is AB. This is incident at an angle θ. For no refraction at the lateral face, $\theta > C$

or, $\sin\theta > \sin C$ But $\theta + r = 90° \Rightarrow \theta = (90° - r)$

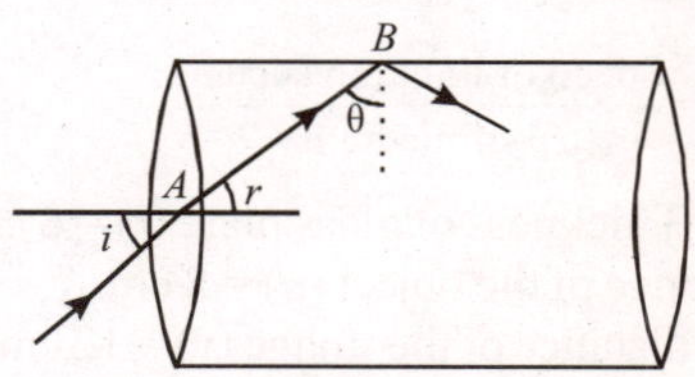

$\therefore \sin(90° - r) > \sin C$

or $\cos r > \sin C$ ..(1)

From Snell's law, $n = \frac{\sin i}{\sin r} \Rightarrow \sin r = \frac{\sin i}{n}$

$$\therefore \cos r = \sqrt{1-\sin^2 r} = \sqrt{\left(1-\frac{\sin^2 i}{n^2}\right)}$$

∴ equation (1) gives

$$\sqrt{1-\frac{\sin^2 i}{n^2}} > \sin C \Rightarrow 1-\frac{\sin^2 i}{n^2} > \sin^2 C$$

Also, $\sin C = \frac{1}{n}$

$$\therefore 1-\frac{\sin^2 i}{n^2} > \frac{1}{n^2} \text{ or } 1 > \frac{\sin^2 i}{n^2}+\frac{1}{n^2}$$

$$\text{or } \frac{1}{n^2}(\sin^2 i+1) < 1 \text{ or } n^2 > (\sin^2 i+1)$$

Maximum value of sin i = 1

$\therefore n^2 > 2 \Rightarrow n > \sqrt{2}$

19. **(a)** For e.m. wave entering from air to glass slab (μ), frequency remains n,

wavelength, $\lambda' = \frac{\lambda}{\mu}$ and velocity, $v' = \frac{v}{\mu}$

> **NOTES** When electromagnetic wave enters in other medium, frequency remains unchanged while wavelength and velocity become $\frac{1}{\mu}$ times.

20. **(d)** Total thickness = t; Refrative index = μ

Speed of light in Glass plate = $\frac{c}{\mu}$

Time taken = $\frac{t}{\left(\frac{c}{\mu}\right)} = \frac{\mu t}{c}$.

where, t = thickness of glass plate

> **NOTES** Speed of light in medium
>
> $v = \frac{\text{Speed of light in vaccum}}{\mu \text{ of medium}}$

21. **(c)** Thickness of glass plate (t) = 6 cm;
Distance of the object (u) = 8 cm.
And distance of the image (v) = 12 cm.
Let x = Apparent position of the silvered surface in cm.
Since the image is formed due to reflection at the silvered face and by the property of mirror image
Distance of object from the mirror = Distance of image from the mirror
or, $x+8 = 12+6-x \Rightarrow x = 5$ cm.

Therefore, refractive index of glass

$$= \frac{\text{Real depth}}{\text{Apparent depth}} = \frac{6}{5} = 1.2.$$

22. **(b)** $\sin C = \frac{1}{\mu} = \frac{1}{\frac{5}{3}} = \frac{3}{5} \Rightarrow$ So, $\tan C = \frac{3}{4}$

Now, $\tan C = \frac{r}{h}; r = h\tan C = 4\times\frac{3}{4} = 3m$

Diameter of disc = $2r = 6\,m$

> **NOTES** Field of vision of a swimmer or fish,
>
> Radius, $r = h \tan c = \frac{h}{\sqrt{\mu^2-1}}$
>
> Area, $A = \pi r^2 = \frac{\pi h^2}{(\mu^2-1)}$

23. **(c)** $v_g = \frac{c}{\mu} = \frac{3\times10^8}{\frac{3}{2}} = 2\times10^8$ m/s

$$t = \frac{x}{v_g} = \frac{4\times10^{-3}}{2\times10^8} = 2\times10^{-11} \text{ s}$$

24. **(c)** From $\mu = \frac{c}{v} = \frac{n\lambda_v}{n\lambda_m}$, $\lambda_m = \frac{\lambda_v}{\mu}$

Here, c = velocity of light in medium and v = velocity of light in vacuum;
μ = refractive index of the medium.
Hence, wavelength in medium $(\lambda_m) < \lambda_a$
($\because \mu > 1$, given)
So, the required wavelength decreases.

25. **(a)** $\lambda_g = \frac{\lambda_a}{\mu} = \frac{5460}{1.5} = 3640$Å

26. **(c)**

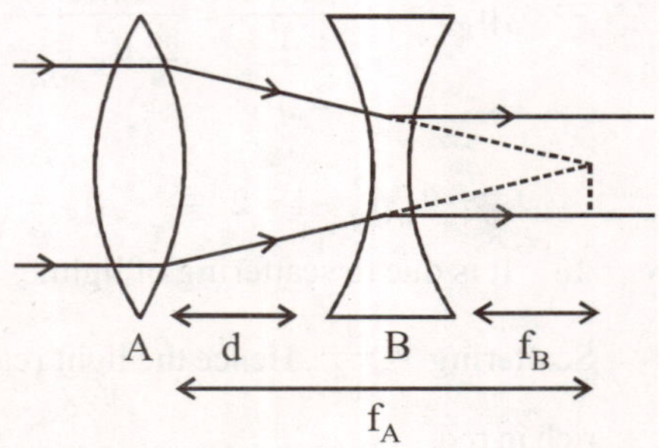

From the ray diagram light refracted from convex lens virtually meet at focus of concave lens

$\therefore d = f_A - f_B$
$= 20 - 5 = 15$ cm

27. (a) For first refraction from convex lens
$v_1 = ?$, $u = -60$ cm, $f = 30$ cm
From lens formula,

$$\frac{1}{v_1} - \frac{1}{u} = \frac{1}{f}$$

$$\Rightarrow \frac{1}{v_1} + \frac{1}{60} = \frac{1}{30} \Rightarrow v_1 = 60\,\text{cm}$$

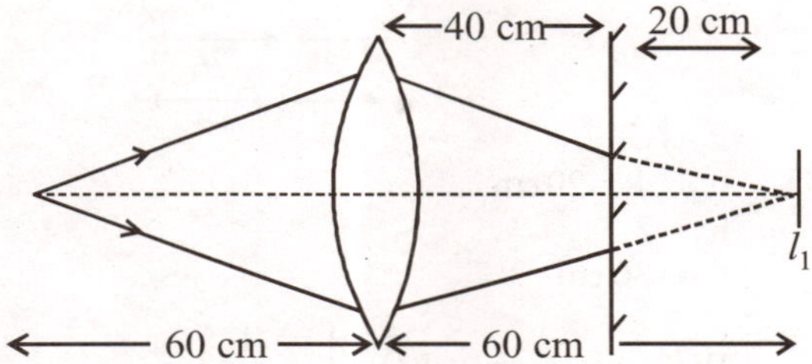

I_1 = first image by lens
Image produced by plane mirror at distance 20 cm left of it.
Therefore for second refraction from convex lens,
$u = -20$ cm, $v = ?$, $f = 30$ cm

$$\frac{1}{v} - \frac{1}{u} = \frac{1}{f} \Rightarrow \frac{1}{v} + \frac{1}{20} = \frac{1}{30}$$

$$\Rightarrow \frac{1}{v} = \frac{1}{30} - \frac{1}{20} \Rightarrow v = -60\text{cm}$$

Hence the final image which is virtual formed at a distance,
$60 - 40 = 20$ cm from plane mirror.

28. (b)

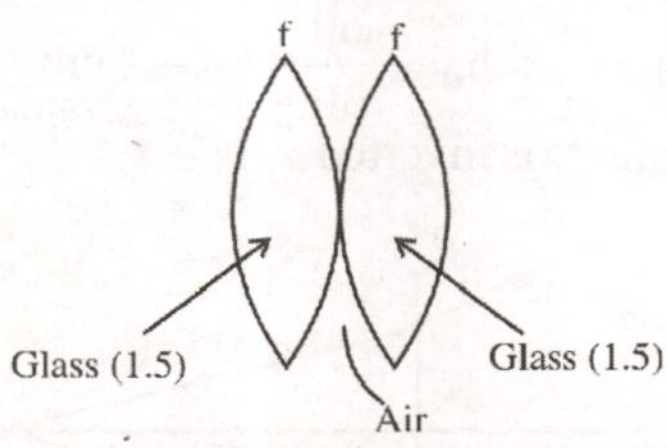

Equivalent focal length in air

$$\frac{1}{F_1} = \frac{1}{f} + \frac{1}{f} = \frac{2}{f}$$

$$\Rightarrow F_1 = \frac{f}{2} \qquad \text{...(i)}$$

When glycerin is filled inside, it behaves like a concave lens of focal length $(-f)$

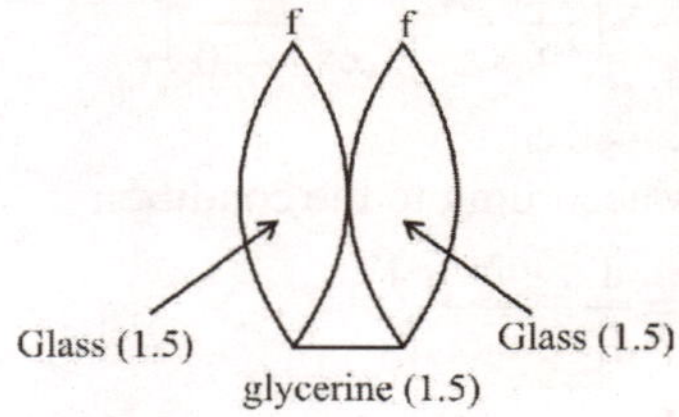

$$\therefore \quad \frac{1}{F_2} = \frac{1}{f} + \frac{1}{f} - \frac{1}{f} \Rightarrow F_2 = f \qquad \text{...(ii)}$$

Dividing (i) by (ii), we get

$$\frac{F_1}{F_2} = \frac{1}{2}$$

29. (a) When lens is cut in two half then power of one part will be same. i.e., P, as focal length remains same.

30. (d) $\frac{1}{f} = (\mu - 1)\left(\frac{1}{R_1} - \frac{1}{R_2}\right)$

$$\frac{1}{25} = (1.5 - 1)\left(\frac{1}{R} + \frac{1}{2R}\right) \quad \frac{1}{25} = 0.5\left(\frac{3}{2R}\right)$$

$2R = 37.5$ cm $\qquad R = 18.75$ cm

31. (b) Using lens maker's formula,

$$\frac{1}{f} = (\mu - 1)\left(\frac{1}{R_1} - \frac{1}{R_2}\right)$$

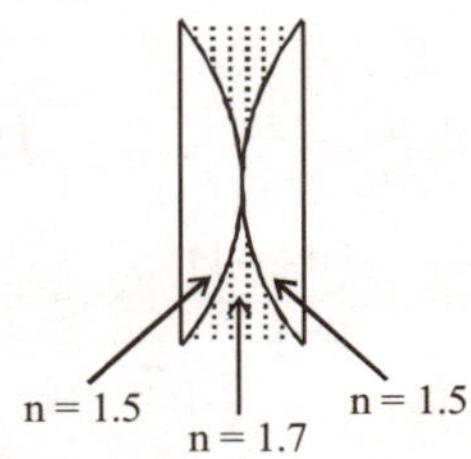

$$\Rightarrow \frac{1}{f_1} = \left(\frac{1.5}{1} - 1\right)\left(\frac{1}{\infty} - \frac{1}{-20}\right)$$

$$\Rightarrow f_1 = 40\text{cm}$$

$$\Rightarrow \frac{1}{f_2} = \left(\frac{1.7}{1} - 1\right)\left(\frac{1}{-20} - \frac{1}{+20}\right)$$

$$\Rightarrow f_2 = -\frac{100}{7}\,\text{cm}$$

Similarly,

$$\Rightarrow \frac{1}{f_3} = \left(\frac{1.5}{1} - 1\right)\left(\frac{1}{\infty} - \frac{1}{-20}\right)$$

$\Rightarrow f_3 = 40$ cm

Now according to the condition

$$\frac{1}{f_{eq}} = \frac{1}{f_1} + \frac{1}{f_2} + \frac{1}{f_3}$$

$$\Rightarrow \frac{1}{f_{eq}} = \frac{1}{40} + \frac{1}{-100/7} + \frac{1}{40}$$

$\therefore \quad f_{eq} = -50$ cm

Therefore, the focal length of the combination is – 50 cm.

NOTES For a system of lenses net focal length $F_{net} = \frac{1}{f_1} + \frac{1}{f_2} + ... + \frac{1}{f_n}$

Net power $P_{net} = P_1 + P_2 + ... + P_n$

32. (b) From the question,

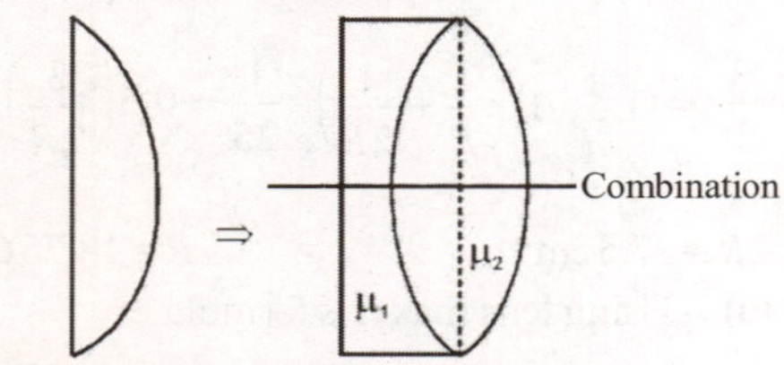

$$\frac{1}{f} = \frac{1}{f_1} + \frac{1}{f_2}$$

$$= (\mu_1 - 1)\left(\frac{1}{\infty} - \frac{1}{-R}\right) + (\mu_2 - 1)\left(\frac{1}{\infty} - \frac{1}{R}\right)$$

$$= \frac{(\mu_1 - 1)}{R} - \frac{(\mu_2 - 1)}{R} \Rightarrow \frac{1}{f} = \frac{\mu_1 - \mu_2}{R}$$

$$\Rightarrow f = \frac{R}{\mu_1 - \mu_2}$$

Hence, focal length of the combination is $\frac{R}{\mu_1 - \mu_2}$.

33. (a) $\frac{1}{f} = \left(\frac{\mu_g}{\mu_m} - 1\right)\left(\frac{1}{R_1} - \frac{1}{R_2}\right)$

If $\mu_g = \mu_m$, then $\frac{1}{f} = (1-1)\left(\frac{1}{R_1} - \frac{1}{R_2}\right)$

$$\Rightarrow \frac{1}{f} = 0 \qquad \boxed{f = \frac{1}{0} = \infty}$$

This implies that the liquid must have refractive index equal to glass.

NOTES A transparent solid is invisible in a liquid of same refractive index because of no refraction.

34. (c)

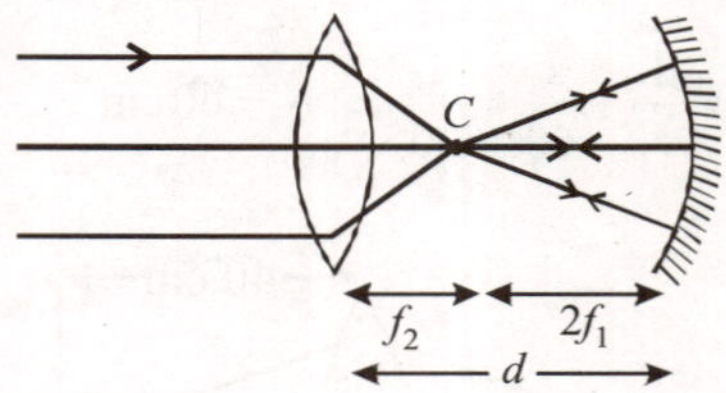

$d = f_2 + 2f_1$

35. (c) R = 20 cm

$h_0 = 2$

u = –30 cm

We have, $\frac{1}{f} = (\mu - 1)\left(\frac{1}{R_1} - \frac{1}{R_2}\right)$

$$= \left(\frac{3}{2} - 1\right)\left[\frac{1}{20} - \left(-\frac{1}{20}\right)\right]$$

$$\Rightarrow \frac{1}{f} = \left(\frac{3}{2} - 1\right) \times \frac{2}{20}$$

$\therefore$ f = 20 cm

$$\frac{1}{f} = \frac{1}{v} - \frac{1}{u} \quad \Rightarrow \frac{1}{20} = \frac{1}{v} + \frac{1}{30}$$

$$\frac{1}{v} = \frac{1}{20} - \frac{1}{30} = \frac{10}{600}$$

v = 60 cm

$$m = \frac{h_i}{h_0} = \frac{v}{u}$$

$$\Rightarrow h_i = \frac{v}{u} \times h_0 = \frac{60}{30} \times 2 = -4 \text{ cm}$$

So, image is inverted.

36. (c)

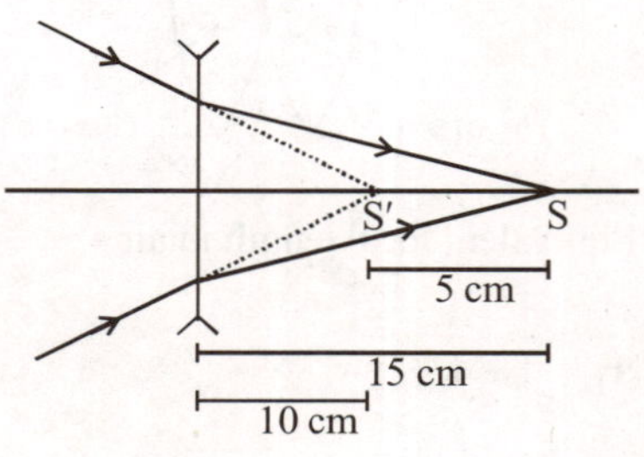

By lens formula,

$$\frac{1}{v}-\frac{1}{u}=\frac{1}{f}$$

u = 10 cm

v = 15 cm

f = ?

Putting the values, we get

$$\frac{1}{15}-\frac{1}{10}=\frac{1}{f}$$

$$\frac{10-15}{150}=\frac{1}{f} \quad \therefore f=-\frac{150}{3}=-30 \text{ cm}$$

37. **(c)** By covering aperture, focal length does not change. But intensity is reduced by $\frac{1}{4}$ times, as aperture diameter $\frac{d}{2}$ is covered.

$$\therefore \quad I' = I - \frac{I}{4} = \frac{3I}{4}$$

$\therefore$ New focal length = f and intensity = $\frac{3I}{4}$.

38. **(d)** The focal length of the combination

$$\frac{1}{f}=\frac{1}{f_1}+\frac{1}{f_2}$$

$\therefore$ Power of the combinations,

$$P=\frac{f_1+f_2}{f_1 f_2} \quad \left(\because P=\frac{1}{f}\right)$$

39. **(a)** We have, $\left|\frac{v}{u}\right| = \frac{\text{Size of image}}{\text{Size of object}}$

or, Size of image = $\left|\frac{v}{u}\right| \times$ Size of object

$$=\left(\frac{10^{-1}}{1.5\times10^{11}}\right)\times(1.39\times10^{9})$$

$= 0.92 \times 10^{-3}$ m $= 9.2 \times 10^{-4}$ m

$\therefore$ Diameter of the sun's image $= 9.2 \times 10^{-4}$ m.

40. **(c)** From the formula,

$$\frac{1}{f}=\frac{1}{f_1}+\frac{1}{f_2}=\frac{1}{25}-\frac{1}{25}=0$$

Power of combination $=\frac{1}{f}=0$

41. **(b)**

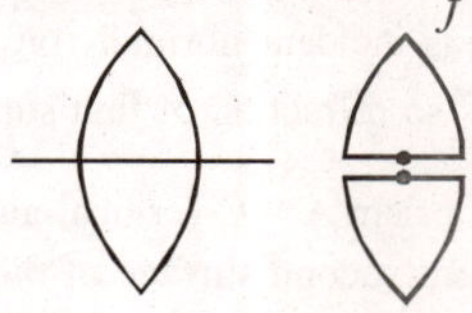

$$\frac{1}{f}=(\mu-1)\left(\frac{1}{R_1}-\frac{1}{R_2}\right)$$

In this case, R_1 and R_2 are unchanged

So, f will remain unchanged for both pieces of the lens

$\therefore f = f'$

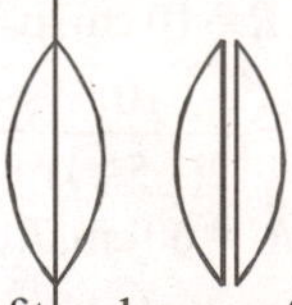

$$\frac{1}{f}=\frac{1}{f_1}+\frac{1}{f_2}$$

This is combination of two lenses of equal focal lengths

$$\therefore \quad \frac{1}{f}=\frac{1}{f''}+\frac{1}{f''}=\frac{2}{f''} \Rightarrow f''=2f.$$

> **NOTES** A symmetric lens is cut along optical axis in two equal parts. Intensity of image formed by each part will be same as that of complete lens. Focal length is double the original for each part.

42. **(c)** $\frac{1}{f}=\left({}_{\ell}\mu_g-1\right)\left(\frac{1}{R_1}-\frac{1}{R_2}\right)$

where, ${}_{\ell}\mu_g = 1$ is given.

$$\Rightarrow \frac{1}{f}=(1-1)\left(\frac{1}{R_1}-\frac{1}{R_2}\right)=0 \quad \Rightarrow \quad f=\infty$$

43. **(b)** Using the lens formula $\frac{1}{f}=\frac{1}{v}-\frac{1}{u}$

Given $v = d$, for equal size image $|v| = |u| = d$

By sign convention $u = -d$

$$\therefore \quad \frac{1}{f}=\frac{1}{d}+\frac{1}{d} \quad \text{or } f=\frac{d}{2}$$

44. **(a)** The silvered plano convex lens behaves as a concave mirro; whose focal length is given by

$$\frac{1}{F}=\frac{2}{f_1}+\frac{1}{f_m}$$

If plane surface is silvered

$$f_m=\frac{R_2}{2}=\frac{\infty}{2}=\infty$$

$$\therefore \quad \frac{1}{f_1}=(\mu-1)\left(\frac{1}{R_1}-\frac{1}{R_2}\right)$$

$$=(\mu-1)\left(\frac{1}{R}-\frac{1}{\infty}\right)=\frac{\mu-1}{R}$$

$\therefore \frac{1}{F} = \frac{2(\mu-1)}{R} + \frac{1}{\infty} = \frac{2(\mu-1)}{R}$

$F = \frac{R}{2(\mu-1)}$

Here, $R = 10$ cm, $\mu = 1.5$

$\therefore F = \frac{10}{2(1.5-1)} = 10\text{cm}$

45. **(b)** $R_1 = 60$ cm, $R_2 = \infty$, $\mu = 1.6$

$\frac{1}{f} = (\mu-1)\left(\frac{1}{R_1} - \frac{1}{R_2}\right)$

$\frac{1}{f} = (1.6-1)\left(\frac{1}{60}\right) \Rightarrow f = 100$ cm.

46. **(c)**

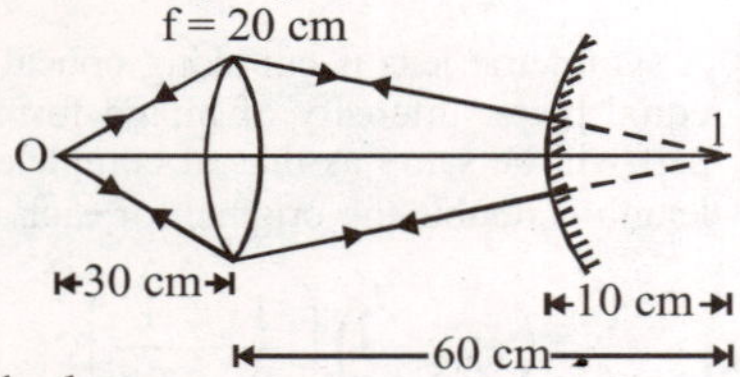

For the lens,

$\frac{1}{v} - \frac{1}{u} = \frac{1}{f}$; $\frac{1}{v} - \frac{1}{-30} = \frac{1}{20} \Rightarrow v = 60$ cm

Coincidence is possible when the image is formed at the centre of curvature of the mirror. Only then the rays refracting through the lens will fall normally on the convex mirror and retrace their path to form the image at O. So, the distance between lens and mirror = 60 – 10 = 50 cm.

47. **(c)** Red light has wavelength greater than violet light. Therefore focal length of lens for red is greater than for violet. $(f_r > f_v)$

> **NOTES**
> Cauchy relation
> $\mu = A + \frac{B}{\lambda^2} + \frac{C}{\lambda^4} \ldots$ Hence $f \propto \lambda$.

48. **(d)** We know that $\frac{1}{f} = \sum_{i=1}^{n} \frac{1}{f_i}$

$\frac{1}{f} = \frac{1}{f_1} + \frac{1}{f_2}$; $f_1 = 80$ cm, $f_2 = -50$ cm

$\frac{1}{f} = \frac{1}{\frac{80}{100}} - \frac{1}{\frac{50}{100}}$

$\Rightarrow P = \frac{1}{f} = 1.25 - 2 = -0.75\text{D}$

49. **(a)** $\frac{1}{f} = (\mu-1)\left(\frac{1}{R_1} - \frac{1}{R_2}\right)$

According to Cauchy relation

$\mu = A + \frac{B}{\lambda^2} + \frac{C}{\lambda^4} \ldots$ Hence $f \propto \lambda$.

Hence, red light having maximum wavelength has maximum focal length.

$\therefore f_v < f_r$ and also $F_v > F_r$ as focal length is negative for a concave lens.

50. **(c)** If two or more lenses are combined together in such a way that this combination produces images of different colours at the same point and of the same size, then this property is called 'achromatism'. Concave and convex type of lenses are used for this combination.

51. **(a)** Size of images = A_1 and A_2. From the displacement method, the area of the source of light = $\sqrt{A_1 A_2}$.

52. **(d)** For red light, focal length of lens is maximum because $f \propto \lambda$ and λ is maximum for red light.

53. **(c)** $\frac{f_a}{f_l} = \frac{\left(\frac{\mu_g}{\mu_l} - 1\right)}{(\mu_g - 1)} = \frac{\left(\frac{1.5}{1.25} - 1\right)}{1.5-1} = \frac{\frac{1}{5}}{\frac{1}{2}} = \frac{2}{5}$

$f_l = \frac{5}{2} f_a = \frac{5}{2} \times 2 = 5$ cm

54. **(b)** From Snell's law at point P.

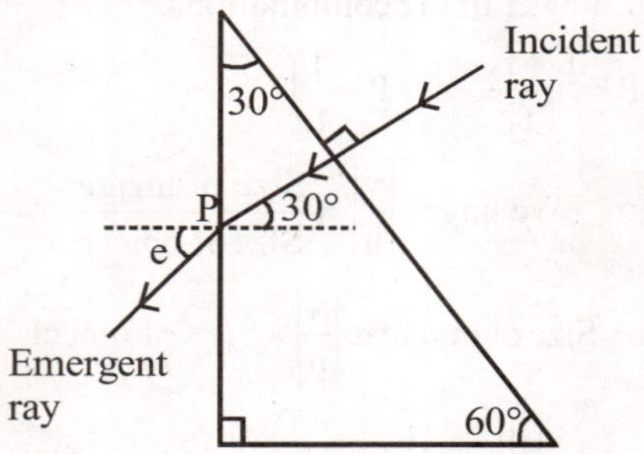

$\frac{\sin i}{\sin r} = \frac{\mu_{air}}{\mu_{Prism}}$

$\Rightarrow \frac{\sin 30°}{\sin e} = \frac{1}{\sqrt{3}}$ ($\angle r = \angle e$ emergent angle)

$\Rightarrow \sin e = \sqrt{3} . \frac{1}{2}$ $\therefore \angle e = 60°$

> **NOTES**
> If a light ray is incident normally on first surface i.e., $\angle i = 0°$ so refraction of first surface $\angle r_1 = 0°$.
> When angle of prism A > C (critical angle) then no ray emerges from second surface of the prism.

55. (b) Light ray emerges normally from another surface, hence, e (angle of emergence) = 0

$\therefore r_2 = 0 \qquad r_1 + r_2 = A \Rightarrow r_1 = A$

$\mu_1 \cdot \sin i = \mu_2 \cdot \sin r$

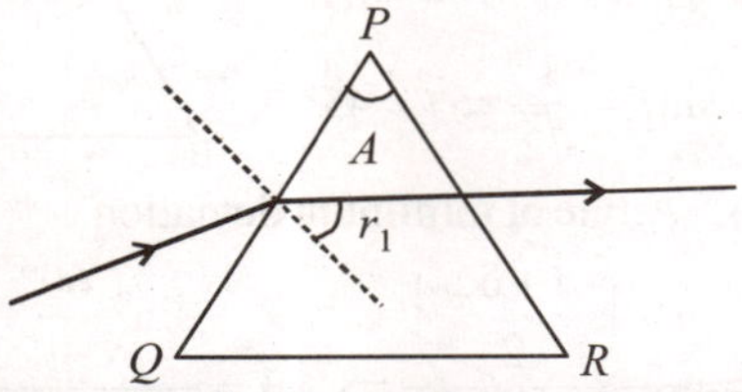

Applying Snell's law on first surface PQ

$\Rightarrow \mu_1 \cdot \sin i = \mu_2 \cdot \sin r_1 \Rightarrow \sin i = \mu \sin A$

For small angles $(\sin\theta \approx \theta)$

$\therefore \ i = \mu A$

56. (c) Rainbow will be observed only when the sun is at the back side of observer.

57. (b) For retracing the path, light ray should be normally incident on silvered face.

$A = r + O \Rightarrow r = 30°$

Applying Snell's law at point M,

$$\frac{\sin i}{\sin 30°} = \frac{\sqrt{2}}{1}$$

$$\Rightarrow \ \sin i = \sqrt{2} \times \frac{1}{2}$$

or, $\sin i = \dfrac{1}{\sqrt{2}}$ i.e., $i = 45°$

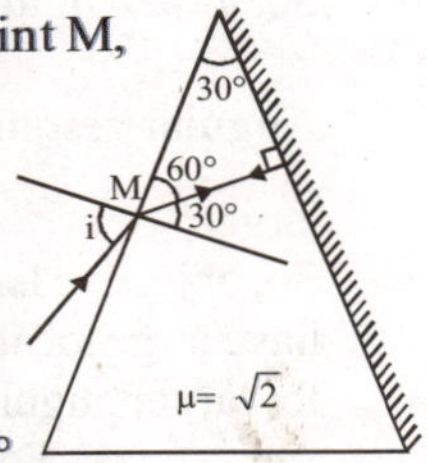

58. (a) For dispersion without deviation

$(\mu - 1)A_1 + (\mu' - 1)A_2 = 0$

$|(\mu - 1)A_1| = |(\mu' - 1)A_2|$

$(1.42 - 1) \times 10° = (1.7 - 1)A_2$

$4.2 = 0.7A_2$

$A_2 = 6°$

NOTES: In case of deviation without dispersion

$$\frac{A'}{A'} = \frac{(\mu_V - \mu_R)}{(\mu_V' - \mu_R')}$$

59. (b) Given: Angle of incidence $i = 45°$

angle of prism, $A = 60°$

Angle of minimum deviation,

$\delta_m = 2i - A = 30°$

Refractive index of material of prism.

$$\mu = \frac{\sin\left(\dfrac{A + \delta_m}{2}\right)}{\sin(A/2)}$$

$$= \frac{\sin 45°}{\sin 30°} = \frac{1}{\sqrt{2}} \cdot \left(\frac{2}{1}\right) = \sqrt{2}$$

60. (a) As we know, the refractive index of the material of the prism

$$\mu = \frac{\sin\left(\dfrac{\delta_m + A}{2}\right)}{\sin(A/2)}$$

$$\cot A/2 = \frac{\sin\left(\dfrac{A + \delta_m}{2}\right)}{\sin A/2} = \frac{\cos(A/2)}{\sin(A/2)}$$

$[\because \mu = \cot(A/2)]$

$$\Rightarrow \ \sin\left(\frac{\delta_m + A}{2}\right) = \sin(90° + A/2)$$

$$\Rightarrow \ \delta_{min} = 180° - 2A$$

61. (c) For total internal reflection,

incident angle(i) > critical angle(i_c)

So, $\sin i > \sin i_c$

$$\sin 45° > \frac{1}{\mu}$$

$\Rightarrow \mu > \sqrt{2} \Rightarrow 1.414$

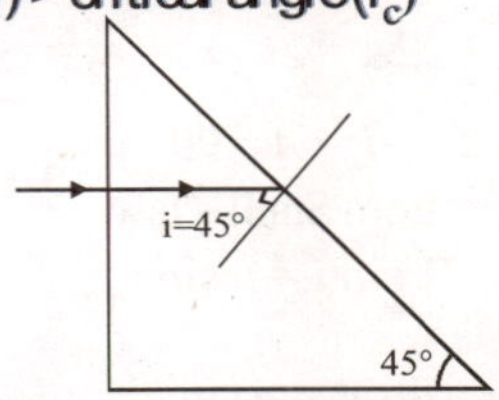

Since refractive index μ of green and voilet are greater than 1.414 so they will total internal reflected. But red colour will be refracted.

62. (b)

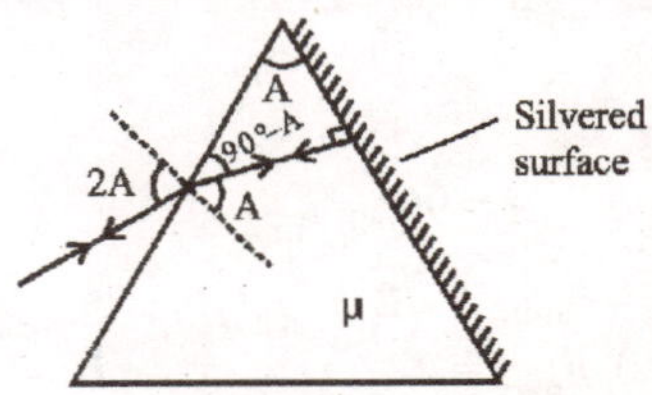

According to Snell's law $\mu = \dfrac{\sin i}{\sin r}$

$\Rightarrow (1) \sin 2A = (\mu) \sin A \Rightarrow \mu = 2\cos A$

63. (a) Rainbow is formed due to combination of total internal reflection and dispersion.

NOTES: Rainbow is formed due to the dispersion of light suffering refraction and TIR in the droplets present in the atmosphere. To observe rainbow observer should stand its back towards sun.

64. **(a)** For normally emerge $e = 0$

Therefore $r_2 = 0$ and $r_1 = A$

Snell's Law for incident ray's

$1 \sin i = \mu \sin r_1 = \mu \sin A$

For small angle

$i = \mu A$

65. **(b)**

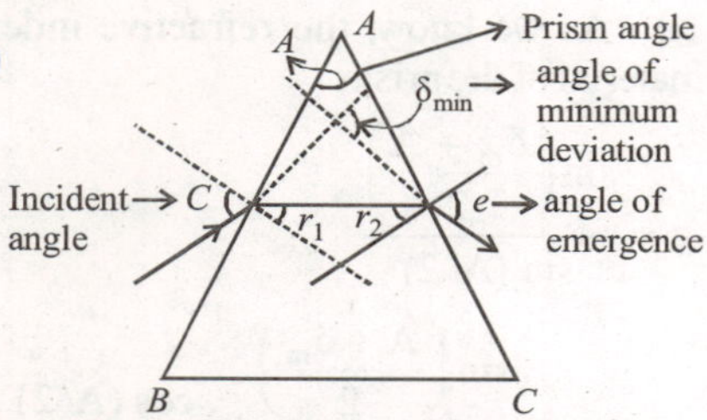

The angle of minimum deviation is given as

$\delta_{min} = i + e - A$

for minimum deviation

$\delta_{min} = A$ then

$2A = i + e$

in case of δ_{min}, $i = e$

$2A = 2i \quad r_1 = r_2 = \frac{A}{2}$

$i = A = 90°$

from Snell's law

$1 \sin i = n \sin r_1$

$\sin A = n \sin \frac{A}{2}$

$2 \sin\frac{A}{2} \cos\frac{A}{2} = n \sin\frac{A}{2}$

$2 \cos\frac{A}{2} = n$

when $\quad A = 90° = i_{min}$

then $\quad n_{min} = \sqrt{2}$

$i = A = 0 \; n_{max} = 2$

66. **(b)** Deviation = zero

So, $\delta = \delta_1 + \delta_2 = 0$

$\Rightarrow (\mu_1 - 1)A_1 + (\mu_2 - 1)A_2 = 0$

$\Rightarrow A_2(1.75 - 1) = -(1.5 - 1)\, 15°$

$\Rightarrow A_2 = -\frac{0.5}{0.75} \times 15°$

or $A_2 = -10°$.

Negative sign shows that the second prism is inverted with respect to the first.

67. **(b)** $\angle r = 30°$ (using law of triangle)

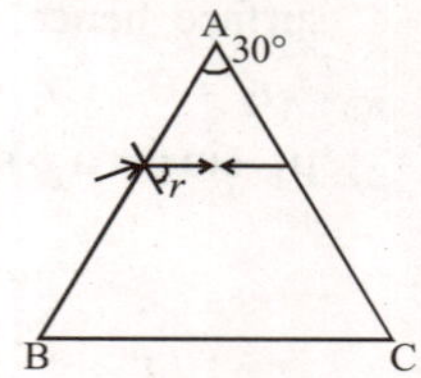

$\Rightarrow \mu = \frac{\sin i}{\sin r}$

$\Rightarrow \sqrt{2} \times \sin 30° = \sin i$

$\Rightarrow \sin i = \frac{1}{\sqrt{2}} \Rightarrow i = 45°.$

68. **(c)** Angle of minimum deviation

$$\mu = \frac{\sin\left(\frac{A+\delta_m}{2}\right)}{\sin\left(\frac{A}{2}\right)} \Rightarrow \sqrt{3} = \frac{\sin\left(\frac{60° + \delta_m}{2}\right)}{\sin\left(\frac{60°}{2}\right)}$$

$$\Rightarrow \sin\left(30° + \frac{\delta_m}{2}\right) = \frac{\sqrt{3}}{2} \Rightarrow 30° + \frac{\delta_m}{2} = 60°$$

$\Rightarrow \delta_m = 60°.$

69. **(a)** When the angle of prism is small, $\delta = (\mu - 1)A$

70. **(d)** For telescope, angular magnification $= \frac{f_0}{f_E}$

So, focal length of objective lens should be large.

Angular resolution $= \frac{D}{1.22\lambda}$ So, D should be large.

So, objective lens of refracting telescope should have large focal length (f_0) and large diameter D for larger angular magnification.

71. **(d)** **Given:** Focal length of objective, $f_0 = 40$cm

Focal length of eye – piece $f_e = 4$ cm

image distance, $v_0 = 200$ cm

Using lens formula for objective lens

$$\frac{1}{v_0} - \frac{1}{u_0} = \frac{1}{f_0} \quad \Rightarrow \quad \frac{1}{v_0} = \frac{1}{f_0} + \frac{1}{u_0}$$

$$\Rightarrow \frac{1}{v_0} = \frac{1}{40} + \frac{1}{-200} = \frac{+5-1}{200}$$

$\Rightarrow v_0 = 50$ cm

Tube length $\ell = |v_0| + f_e = 50 + 4 = 54$ cm.

NOTES

In a telescope, if field and eye lenses are interchanged magnification will change from $\left(\frac{f_o}{f_e}\right)$ to $\left(\frac{f_e}{f_o}\right)$ i.e., it will change from m to $\frac{1}{m}$ i.e., will become $\left(\frac{1}{m^2}\right)$ times of its initial value.

72. **(c)**

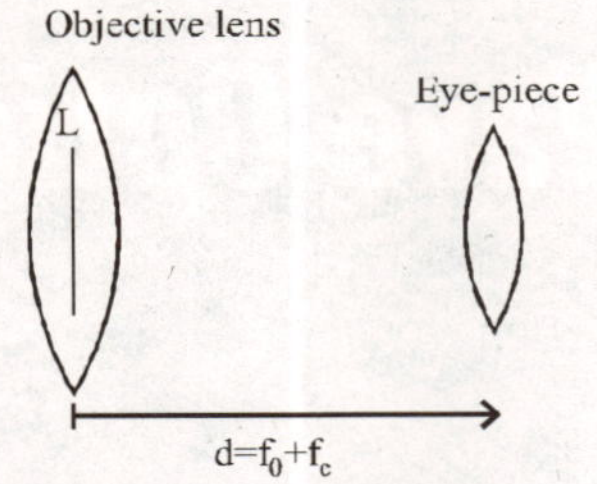

Magnification by eye piece

$$m=\frac{f}{f+u}$$

$$-\frac{I}{L}=\frac{f_e}{f_e+[-(f_0+f_e)]}=-\frac{f_e}{f_0} \text{ or, } \frac{I}{L}=\frac{f_e}{f_0}$$

$$\text{Magnification, } M=\frac{f_0}{f_e}=\frac{L}{I}$$

73. **(d)** Magnifying power of microscope

$$=\frac{LD}{f_0 f_e}\propto\frac{1}{f_0}$$

Hence with increase f_0 magnifyig power of microscope decreases.

$$\text{Magnifying power of telescope}=\frac{f_0}{f_e}\propto f_0$$

Hence with increase f_0 magnifying power of telescope increases.

74. **(b)** $P_{cornea}=+40\text{ D}$

$P_e=+20\text{ D}$

Total power of combination = 40 + 20 = 60 D

$$\text{Focal length of combination }=\frac{1}{60}\times 100\text{ cm}$$

$$=\frac{5}{3}\text{ cm}$$

For minimum converging state of eye lens,

$$u=-\infty \quad v=? \quad f=\frac{5}{3}$$

From lens formula,

$$\frac{1}{f}=\frac{1}{v}-\frac{1}{u}\Rightarrow v=\frac{5}{3}\text{ cm}$$

Distance between retina and cornea-eye lens

$$=\frac{5}{3}=1.67\text{ m}$$

75. **(c)** $\text{M.P.}=9=\dfrac{f_0}{f_e}$

$\Rightarrow f_0=9f_e$ ...(1)

$f_0+f_e=20$ ...(2)

on solving

$f_0=18$ cm = focal length of the objective

$f_e=2$ cm = focal length of the eyepiece

76. **(d)** In the later case microscope will be focussed for O'. So, it is required to be lifted by distance OO'.

OO' = real depth of O – apparent depth of O.

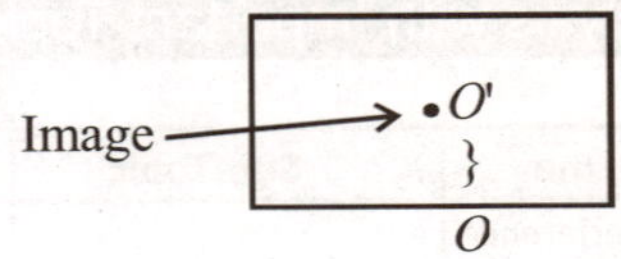

$$=3-\frac{3}{1.5}\quad\left[\mu=\frac{\text{real depth}}{\text{apparent depth}}\right]$$

$$=3\left[\frac{1.5-1}{1.5}\right]=\frac{3\times .5}{1.5}=1\text{ cm}$$

77. **(d)** Here, $\dfrac{x}{1000}=\dfrac{1.22\lambda}{D}$

$$\text{or, } x=\frac{1.22\times 5\times 10^3\times 10^{-10}\times 10^3}{10\times 10^{-2}}$$

or, $x=1.22\times 5\times 10^{-3}\text{ m}=6.1$ mm

$\therefore$ x is of the order of 5 mm.

78. **(b)** Given : Length of astronomical telescope

$(f_0+f_e)=44$ cm and magnification $\left(\dfrac{f_0}{f_e}\right)=10$.

From the given magnification, we find that $f_0=10f_e$. Therefore, $10f_e+f_e=44$ or $11f_e=44$ or f_e = 4. And focal length of the objective $(f_0)=44-f_e=44-4=40$ cm.

> **NOTES** In case if objective and eye lens of a telescope are interchanged, it will not behave as a microscope but object appears very small.

79. **(b)** A person suffering from hyper metropia can see objects beyond a particular point called the near point. If the object lies at a point nearer than this point, then image is not formed at the retina. This is also known as long-sight defect.

24 Wave Optics

Trend Analysis with Important Topics & Sub-Topics

Topic Name	Sub-Topic	2021		2020		2019		2018		2017		2016	
		QNS.	LOD	QNS	LOD	QNS.	LOD	QNS.	LOD	QNS.	LOD	QNS.	LOD
Wavefront, Interference of Light, Coherent & Incoherent Sources													
Young's Double Slit Experiment	YDSE Fringe width			1	A							1	A
Diffraction, Polarization of Light & Resolving Power	Angular fringe width in water					1	A	1	A				
	Diffraction & Polarisation of light			1	E			1	A	2	A	1	A
	Resolution of Telescope & Microscope	1	A	1	E					1	E		

LOD - Level of Difficulty | E - Easy | A - Average | D - Difficult | Qns - No. of Questions

Topic 1 Wavefront, Interference of Light, Coherent & Incoherent Sources

1. The periodic waves of intensities I_1 and I_2 pass through a region at the same time in the same direction. The sum of the maximum and minimum intensities is: *[2008]*
   (a) $I_1 + I_2$ (b) $\left(\sqrt{I_1}+\sqrt{I_2}\right)^2$
   (c) $\left(\sqrt{I_1}-\sqrt{I_2}\right)^2$ (d) $2(I_1+I_2)$
2. Colours appear on a thin soap film and on soap bubbles due to the phenomenon of *[1999]*
   (a) refraction (b) dispersion
   (c) interference (d) diffraction
3. Interference was observed in interference chamber where air was present, now the chamber is evacuated, and if the same light is used, a careful observer will see *[1993]*
   (a) no interference
   (b) interference with brighter bands
   (c) interference with dark bands
   (d) interference fringe with larger width
4. Ratio of intensities of two waves are given by 4 : 1. Then the ratio of the amplitudes of the two waves is *[1991]*
   (a) 2 : 1 (b) 1 : 2
   (c) 4 : 1 (d) 1 : 4
5. Interference is possible in *[1989]*
   (a) light waves only
   (b) sound waves only
   (c) both light and sound waves
   (d) neither light nor sound waves
6. Which one of the following phenomena is not explained by Huygens construction of wavefront? *[1988]*
   (a) Refraction (b) Reflection
   (c) Diffraction (d) Origin of spectra

Topic 2 Young's Double Slit Experiment

7. In Young's double slit experiment, if the separation between coherent sources is halved and the distance of the screen from the coherent sources is doubled, then the fringe width becomes : *[2020]*
(a) half (b) four times
(c) one-fourth (d) double

8. In a double slit experiment, when light of wavelength 400 nm was used, the angular width of the first minima formed on a screen placed 1 m away, was found to be 0.2°. What will be the angular width of the first minima, if the entire experimental apparatus is immersed in water? ($\mu_{water} = 4/3$) *[2019]*
(a) 0.266° (b) 0.15°
(c) 0.05° (d) 0.1°

9. In a Young's double slit experiment, if there is no initial phase difference between the light from the two slits, a point on the screen corresponding to the fifth minimum has path difference *[NEET Odisha 2019]*
(a) $11\frac{\lambda}{2}$ (b) $5\frac{\lambda}{2}$
(c) $10\frac{\lambda}{2}$ (d) $9\frac{\lambda}{2}$t

10. In Young's double slit experiment the separation d between the slits is 2 mm, the wavelength λ of the light used is 5896 Å and distance D between the screen and slits is100 cm. It is found that the angular width of the fringes is 0.20°. To increase the fringe angular width to 0.21° (with same λ and D) the separation between the slits needs to be changed to *[2018]*
(a) 1.8mm (b) 1.9mm
(c) 1.7mm (d) 2.1mm

11. Young's double slit experment is first performed in air and then in a medium other than air. It is found that 8th bright fringe in the medium lies where 5th dark fringe lies in air. The refractive index of the medium is nearly *[2017]*
(a) 1.59 (b) 1.69
(c) 1.78 (d) 1.25

12. The intensity at the maximum in a Young's double slit experiment is I_0. Distance between two slits is $d = 5\lambda$, where λ is the wavelength of light used in the experiment. What will be the intensity in front of one of the slits on the screen placed at a distance D = 10 d ? *[2016]*
(a) I_0 (b) $\frac{I_0}{4}$
(c) $\frac{3}{4}I_0$ (d) $\frac{I_0}{2}$

13. In a double slit experiment, the two slits are 1 mm apart and the screen is placed 1 m away. A monochromatic light wavelength 500 nm is used. What will be the width of each slit for obtaining ten maxima of double slit within the central maxima of single slit pattern ? *[2015]*
(a) 0.1 mm (b) 0.5 mm
(c) 0.02 mm (d) 0.2 mm

14. Two slits in Young's experiment have widths in the ratio 1 : 25. The ratio of intensity at the maxima and minima in the interference pattern, $\frac{I_{max}}{I_{min}}$ is: *[2015 RS]*
(a) $\frac{121}{49}$ (b) $\frac{49}{121}$
(c) $\frac{4}{9}$ (d) $\frac{9}{4}$

15. In the Young's double-slit experiment, the intensity of light at a point on the screen where the path difference is λ is K, (λ being the wave length of light used). The intensity at a point where the path difference is $\lambda/4$, will be: *[2014]*
(a) K (b) K/4
(c) K/2 (d) Zero

16. In Young's double slit experiment, the slits are 2 mm apart and are illuminated by photons of two wavelengths $\lambda_1 = 12000$Å and $\lambda_2 = 10000$Å. At what minimum distance from the common central bright fringe on the screen 2 m from the slit will a bright fringe from one interference pattern coincide with a bright fringe from the other ? *[2013]*
(a) 6mm (b) 4mm
(c) 3mm (d) 8mm

17. In Young's double slit experiment the distance between the slits and the screen is doubled. The separation between the slits is reduced to half. As a result the fringe width *[NEET Kar. 2013]*
(a) is doubled
(b) is halved
(c) becomes four times
(d) remains unchanged

18. In Young's double slit experiment carried out with light of wavelength (λ) = 5000Å, the distance between the slits is 0.2 mm and the screen is at 200 cm from the slits. The central maximum is at $x = 0$. The third maximum (taking the central maximum as zeroth maximum) will be at x equal to *[1992]*
(a) 1.67 cm (b) 1.5 cm
(c) 0.5 cm (d) 5.0 cm

19. If yellow light emitted by sodium lamp in Young's double slit experiment is replaced by a monochromatic blue light of the same intensity *[1992]*
(a) fringe width will decrease
(b) finge width will increase
(c) fringe width will remain unchanged
(d) fringes will become less intense

20. In Young's experiment, two coherent sources are placed 0.90 mm apart and fringe are observed one metre away. If it produces second dark fringe at a distance of 1 mm from central fringe, the wavelength of monochromatic light used would be *[1991, 1992]*
(a) 60×10^{-4} cm (b) 10×10^{-4} cm
(c) 10×10^{-5} cm (d) 6×10^{-5} cm

21. The Young's double slit experiment is performed with blue and with green light of wavelengths 4360Å and 5460Å respectively. If x is the distance of 4th maxima from the central one, then *[1990]*
(a) x (blue) = x (green)
(b) x (blue) > x (green)
(c) x (blue) < x (green)
(d) $\frac{x(\text{blue})}{x(\text{green})} = \frac{5460}{4360}$

22. In Young's double slit experiment, the fringe width is found to be 0.4 mm. If the whole apparatus is immersed in water of refrative index $\frac{4}{3}$, without disturbing the geometrical arrangement, the new fringe width will be *[1990]*
(a) 0.30 mm (b) 0.40 mm
(c) 0.53 mm (d) 450 microns

Topic 3 Diffraction, Polarization of Light & Resolving Power

23. A lens of large focal length and large aperture is best suited as an objective of an astronomical telescope since *[2021]*
(a) a large aperture contributes to the quality and visibility of the images.
(b) a large area of the objective ensures better light gathering power.
(c) a large aperture provides a better resolution.
(d) all of the above

24. The Brewsters angle i_b for an interface should be *[2020]*
(a) $30° < i_b < 45°$ (b) $45° < i_b < 90°$
(c) $i_b = 90°$ (d) $0° < i_b < 30°$

25. Assume that light of wavelength 600 nm is coming from a star. The limit of resolution of telescope whose objective has a diameter of 2 m is: *[2020]*
(a) 1.83×10^{-7} rad (b) 7.32×10^{-7} rad
(c) 6.00×10^{-7} rad (d) 3.66×10^{-7} rad

26. Angular width of the central maxima in the Fraunhofer diffraction for λ = 6000Å is θ_0. When the same slit is illuminated by another monochromatic light, the angular width decreases by 30%. The wavelength of this light is *[NEET Odisha 2019]*
(a) 420 Å (b) 1800 Å
(c) 4200 Å (d) 6000 Å

27. Unpolarised light is incident from air on a plane surface of a material of refractive index 'μ'. At a particular angle of incidence 'i', it is found that the reflected and refracted rays are perpendicular to each other. Which of the following options is correct for this situation? *[2018]*
(a) Reflected light is polarised with its electric vector parallel to the plane of incidence
(b) Reflected light is polarised with its electric vector perpendicular to the plane of incidence
(c) $i = \tan^{-1}\left(\frac{1}{\mu}\right)$ (d) $i = \sin^{-1}\left(\frac{1}{\mu}\right)$

28. Two Polaroids P_1 and P_2 are placed with their axis perpendicular to each other. Unpolarised light I_0 is incident on P_1. A third polaroid P_3 is kept in between P_1 and P_2 such that its axis makes an angle 45° with that of P_1. The intensity of transmitted light through P_2 is *[2017]*
(a) $\frac{I_0}{4}$ (b) $\frac{I_0}{8}$
(c) $\frac{I_0}{16}$ (d) $\frac{I_0}{2}$

29. The ratio of resolving powers of an optical microscope for two wavelengths $\lambda_1 = 4000$ Å and $\lambda_2 = 6000$ Å is *[2017]*
(a) 9 : 4 (b) 3 : 2
(c) 16 : 81 (d) 8 : 27

30. In a diffraction pattern due to a single slit of width 'a', the first minimum is observed at an angle 30° when light of wavelength 5000 Å is incident on the slit. The first secondary maximum is observed at an angle of : *[2016]*
(a) $\sin^{-1}\left(\frac{1}{4}\right)$ (b) $\sin^{-1}\left(\frac{2}{3}\right)$
(c) $\sin^{-1}\left(\frac{1}{2}\right)$ (d) $\sin^{-1}\left(\frac{3}{4}\right)$

31. For a parallel beam of monochromatic light of wavelength 'λ', diffraction is produced by a single slit whose width 'a' is of the wavelength of the light. If 'D' is the distance of the screen from the slit, the width of the central maxima will be : *[2015]*
(a) $\frac{D\lambda}{a}$ (b) $\frac{Da}{\lambda}$
(c) $\frac{2Da}{\lambda}$ (d) $\frac{2D\lambda}{a}$

32. At the first minimum adjacent to the central maximum of a single-slit diffraction pattern, the phase difference between the Huygen's wavelet from the edge of the slit and the wavelet from the midpoint of the slit is : *[2015 RS]*
(a) $\frac{\pi}{2}$ radian (b) π radian
(c) $\frac{\pi}{8}$ radian (d) $\frac{\pi}{4}$ radian

33. A beam of light of $\lambda = 600$ nm from a distant source falls on a single slit 1 mm wide and the resulting diffraction pattern is observed on a screen 2 m away. The distance between first dark fringes on either side of the central bright fringe is: *[2014]*
(a) 1.2 cm (b) 1.2 mm
(c) 2.4 cm (d) 2.4 mm

34. A parallel beam of fast moving electrons is incident normally on a narrow slit. A fluorescent screen is placed at a large distance from the slit. If the speed of the electrons is increased, which of the following statements is correct ? *[2013]*
(a) The angular width of the central maximum of the diffraction pattern will increase.
(b) The angular width of the central maximum will decrease.
(c) The angular width of the central maximum will be unaffected.
(d) Diffraction pattern is not observed on the screen in case of electrons.

35. A parallel beam of light of wavelength λ is incident normally on a narrow slit. A diffraction pattern is formed on a screen placed perpendicular to the direction of the incident beam. At the second minimum of the diffraction pattern, the phase difference between the rays coming from the two edges of slit is *[NEET Kar. 2013]*
(a) $\pi\lambda$ (b) 2π
(c) 3π (d) 4π

36. The angular resolution of a 10 cm diameter telescope at a wavelength of 5000 Å is of the order of *[2005]*
(a) 10^6 rad (b) 10^{-2} rad
(c) 10^{-4} rad (d) 10^{-6} rad

37. A paper, with two marks having separation d, is held normal to the line of sight of an observer at a distance of 50m. The diameter of the eye-lens of the observer is 2 mm. Which of the following is the least value of d, so that the marks can be seen as separate ? The mean wavelength of visible light may be taken as 5000 Å. *[2002]*
(a) 1.25 m (b) 12.5 cm
(c) 1.25 cm (d) 2.5 mm

38. In a Fresnel biprism experiment, the two positions of lens give separation between the slits as 16 cm and 9 cm respectively. What is the actual distance of separation? *[1996]*
(a) 12.5 cm (b) 12 cm
(c) 13 cm (d) 14 cm

39. A parallel beam of monochromatic light of wavelength 5000Å is incident normally on a single narrow slit of width 0.001 mm. The light is focussed by a convex lens on a screen placed in focal plane. The first minimum will be formed for the angle of diffraction equal to *[1993]*
(a) 0° (b) 15°
(c) 30° (d) 50°

40. Which of the following phenomenon is not common to sound and light waves ? *[1988]*
(a) Interference (b) Diffraction
(c) Coherence (d) Polarisation

ANSWER KEY															
1	(d)	6	(d)	11	(c)	16	(a)	21	(c)	26	(c)	31	(d)	36	(d)
2	(c)	7	(b)	12	(d)	17	(c)	22	(a)	27	(b)	32	(b)	37	(b)
3	(d)	8	(b)	13	(d)	18	(b)	23	(d)	28	(b)	33	(d)	38	(b)
4	(a)	9	(d)	14	(d)	19	(a)	24	(b)	29	(b)	34	(b)	39	(c)
5	(c)	10	(b)	15	(c)	20	(d)	25	(d)	30	(d)	35	(d)	40	(d)

Hints & Solutions

1. **(d)** The resultant intensity of two periodic waves at a point is given by
$I = I_1 + I_2 + 2\sqrt{I_1 I_2}.\cos\phi.$
Resultant intensity is maximum if
$\cos\phi = -1$
i.e. $I_{max} = I_1 + I_2 + 2\sqrt{I_1 I_2}$
Resultant intensity is minimum if
$\cos\phi = +1$
i.e, $I_{min} = I_1 + I_2 - 2\sqrt{I_1 I_2}$
Therefore, the sum of the maximum and minimum intensities is $I_{max} + I_{min}$
$= I_1 + I_2 + 2\sqrt{I_1 I_2} + I_1 + I_2 - 2\sqrt{I_1 I_2}$
$= 2(I_1 + I_2)$

2. **(c)** We know that the colours for which the condition of constructive interference is satisfied are observed in a given region of the film. The path difference between the light waves reaching the eye changes when the position of the eye is changed. Therefore, colours appear on a thin soap film or soap bubbles due to the phenomenon of interference.

3. **(d)** In vaccum, λ increases very slightly compared to that in air. As $\beta \propto \lambda$, therefore, width of interference fringe increases slightly.

4. **(a)** $\frac{I_1}{I_2} = \frac{a_1^2}{a_2^2} = \frac{4}{1} \quad \therefore \frac{a_1}{a_2} = \frac{2}{1}$

NOTES

If w_1 and w_2 are widths of two slits from which intensities of light I_1 and I_2 emanate, then $\frac{I_1}{I_2} = \frac{a^2}{b^2} = \frac{w_1}{w_2}$; a and b are amplitudes.

5. **(c)** Interference is a wave phenomenon shown by both the light waves and sound waves.

6. **(d)** Huygen's construction of wavefront does not apply to origin of spectra which is explained by quantum theory.

7. **(b)** Fringe width $\beta = \frac{\lambda D}{d}$
Here, λ = wavelength of light from coherent sources,
D = distance of screen from the coherent sources,
d = separation between coherent sources
When, $d' = \frac{d}{2}$ and $D' = 2D$
New Fringe width, $\beta' = \frac{\lambda(2D)}{d/2} = \frac{4\lambda D}{d}$
$\Rightarrow \beta' = 4\beta$
Fringe width becomes 4 times.

8. **(b)** For double slit experiment angular fringe width $\theta_0 = \frac{\beta}{D}$
Angular fringe width (in water)
$$\theta_w = \frac{\beta}{\mu D} = \frac{\theta_0}{\mu}$$
$$= \frac{0.2°}{\left(\frac{4}{3}\right)} = 0.15°$$

9. **(d)** Path difference for destructive interference in YDSE
$$\Rightarrow \Delta X_n = \frac{(2n-1)}{2}\lambda \qquad n = 1, 2, 3.....$$
$$\Delta X_{5th} = \frac{9\lambda}{2}$$

10. **(b)** Angular width $= \frac{\lambda}{d}$

So, $0.20° = \frac{\lambda}{2\text{mm}}$

$\Rightarrow \lambda = 0.20° \times 2$

Again, $0.21° = \frac{\lambda}{d}$

Now putting the value of λ

$$d = \frac{0.20° \times 2\text{mm}}{0.21°}$$

$\therefore \quad d = 1.9\,\text{mm}$

11. **(c)** According to question

8th bright fringe in medium = 5th dark fringe in air

$$Y_{8\text{th bright}} = 8\frac{\lambda D}{\mu d}$$

$$Y_{5\text{th dark}} = (2 \times 5 - 1)\frac{\lambda D}{2d} = \frac{9}{2}\frac{\lambda D}{d}$$

$$\Rightarrow \quad \frac{9}{2}\frac{\lambda D}{d} = 8\frac{\lambda D}{\mu d}$$

or, refractive index $\mu = \frac{16}{9} = 1.78$

> **NOTES**
> In YDSE for constructive or bright fringes. $Y_n = \frac{nD\lambda}{d}$ for nth bright fringe and for destructive or dark fringes. $y_n = (2n-1)\frac{D\lambda}{2d}$ for nth dark fringe.

12. **(d)** Let P is a point infront of one slit at which intensity is to be calculated. From figure,

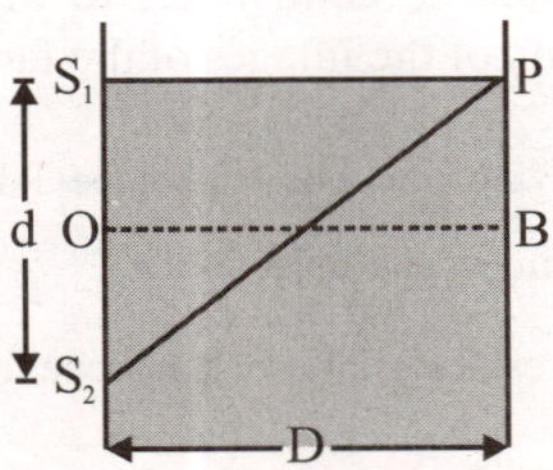

Path difference $= S_2P - S_1P$

$$= \sqrt{D^2 + d^2} - D = D\left(1 + \frac{1}{2}\frac{d^2}{D^2}\right) - D$$

$$= D\left[1 + \frac{d^2}{2D^2} - 1\right] = \frac{d^2}{2D}$$

$$\Delta x = \frac{d^2}{2 \times 10d} = \frac{d}{20} = \frac{5\lambda}{20} = \frac{\lambda}{4}$$

Phase difference,

$$\Delta\phi = \frac{2\pi}{\lambda} \times \frac{\lambda}{4} = \frac{\pi}{2}$$

So, resultant intensity at the desired point 'p' is

$$I = I_0\cos^2\frac{\phi}{2} = I_0\cos^2\frac{\pi}{4} = \frac{I_0}{2}$$

13. **(d)** Here, distance between two slits,

$d = 1\text{mm} = 10^{-3}\text{m}$

distance of screen from slits, D = 1 m

wavelength of monochromatic light used,

$\lambda = 500\text{nm} = 500 \times 10^{-9}\text{m}$

width of each slit a = ?

Width of central maxima in single slit pattern

$$= \frac{2\lambda D}{a}$$

Fringe width in double slit experiment $\beta = \frac{\lambda D}{d}$

So, required condition $\frac{10\lambda D}{d} = \frac{2\lambda D}{a}$

$$\Rightarrow \quad a = \frac{d}{5D} = \frac{1}{5} \times 10^{-3}\text{m} = 0.2\,\text{mm}$$

14. **(d)** The ratio of slits width $= \frac{1}{25}$ (given)

$$\therefore \frac{I_1}{I_2} = \frac{25}{1}$$

$$I \propto A^2 \Rightarrow \frac{I_1}{I_2} = \frac{A_1^2}{A_2^2} = \frac{25}{1} \text{ or } \frac{A_1}{A_2} = \frac{5}{1}$$

As we know that,

$$\frac{A_{max}}{A_{min}} = \frac{A_1 + A_2}{A_1 - A_2} = \frac{5+1}{5-1} = \frac{6}{4} = \frac{3}{2}$$

$$\therefore \frac{I_{max}}{I_{min}} = \frac{A_{max}^2}{A_{min}^2} = \left(\frac{3}{2}\right)^2 = \frac{9}{4}$$

> **NOTES**
> In YDSE, the radio of $\frac{I_{max}}{I_{min}}$ is maximum when both the sources have same intensity.

15. (c) For path difference λ, phase difference $= 2\pi$ rad.

For path difference $\frac{\lambda}{4}$, phase difference $= \frac{\pi}{2}$ rad.

As $K = 4I_0$ so intensity at given point where path difference is $\frac{\lambda}{4}$

$$K' = 4I_0 \cos^2\left(\frac{\pi}{4}\right)\left(\cos\frac{\pi}{4} = \cos 45^\circ\right) = 2I_0 = \frac{K}{2}$$

16. (a) $\because Y = \frac{n\lambda D}{d}$

$\therefore n_1\lambda_1 = n_2\lambda_2$

$\Rightarrow n_1 \times 12000 \times 10^{-10} = n_2 \times 10000 \times 10^{-10}$

or, $n(12000 \times 10^{-10}) = (n+1)(10000 \times 10^{-10})$

$\Rightarrow n = 5$

$$(\because \lambda_1 = 12000 \times 10^{-10} m;\ \lambda_2 = 10000 \times 10^{-10} m)$$

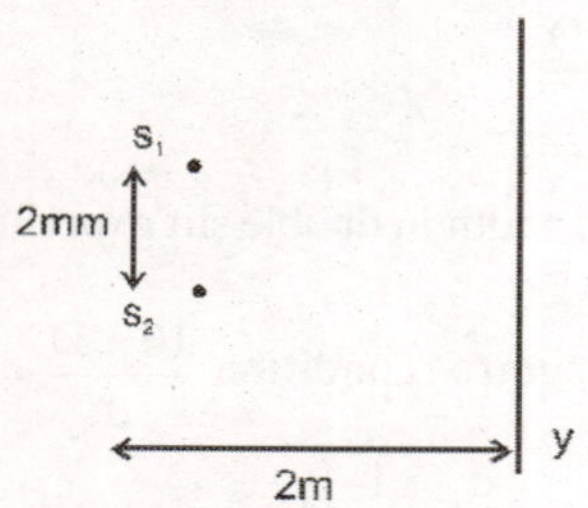

Hence, $Y_{common} = \frac{n\lambda_1 D}{d}$

$$= \frac{5(12000 \times 10^{-10}) \times 2}{2 \times 10^{-3}}$$

$(\because d = 2$ mm and $D = 2m)$

$= 5 \times 12 \times 10^{-4}$ m

$= 60 \times 10^{-4}$ m

$= 6 \times 10^{-3}$ m $= 6$ mm

17. (c) Fringe width $\beta = \frac{\lambda D}{d}$;

From question $D' = 2D$ and $d' = \frac{d}{2}$

$\therefore \beta' = \frac{\lambda D'}{d'} = 4\beta$

18. (b) $x = (n)\lambda\frac{D}{d} = 3 \times 5000 \times 10^{-10} \times \frac{2}{0.2 \times 10^{-3}}$

$= 1.5 \times 10^{-2}$ m $= 1.5$ cm

19. (a) As $\beta = \frac{\lambda D}{d}$ and $\lambda_b < \lambda_y$,

Fringe width β will decrease

20. (d) For dark fringe

$$x = (2n-1)\frac{\lambda D}{2d}$$

$$\therefore \lambda = \frac{2xd}{(2n-1)D} = \frac{2 \times 10^{-3} \times 0.9 \times 10^{-3}}{(2 \times 2 - 1) \times 1}$$

$\lambda = 0.6 \times 10^{-6}$ m $= 6 \times 10^{-5}$ cm

21. (c) Distance of nth maxima, $x = n\lambda\frac{D}{d} \propto \lambda$

As $\lambda_b < \lambda_g \quad \therefore x_{blue} < x_{green}$

> **NOTES**
> Fringes with red light are thicker than those for blue light.
> $\therefore \lambda_{red} > \lambda_{blue}$

22. (a) $\beta' = \frac{\beta}{\mu} = \frac{0.4}{\frac{4}{3}} = 0.3$ mm

> **NOTES**
> If the whole YDSE set up is taken in another medium then λ changes. So B changes.
> In water $\lambda\omega = \frac{\lambda_a}{\mu_w}$
> $\Rightarrow \beta\omega = \frac{B_a}{\mu_w} = \frac{3}{4}B_a$

23. (d) A large focal length and large aperture of an objective of an astronomical telescope, ensures better light gathering power, and resolution so contributes to the quality and visibility of the images of the far-off objects.

> **NOTES**
> Resolving power of a telescope $RP = \frac{a}{1.22\lambda}$, a = aperture of the objective.
> And resolving limit of telescope $d\theta = \frac{1.22\lambda}{a}$

24. (b) According to Brewster's law, when a beam of unpolarised light is reflected from a transparent medium of refractive index (μ_2), the reflected light is completely polarised at certain

angle of incidence called the angle of polarisation (i_b).

$$\tan i_b = \frac{\mu_2}{\mu_1}$$

For air, $\mu_1 = 1$

$\therefore \tan i_b = \mu_2 > 1$

$\Rightarrow \tan i_b > 1 \;\Rightarrow\; 90° > i_b > 45°$

25. **(d)** Given :

Wavelength, $\lambda = 600 \text{ nm} = 600 \times 10^{-9}$ m

and diameter of objective, $d = 2$ m

Limit of resolution of telescope,

$$\theta = \frac{1.22\lambda}{d} = \frac{1.22 \times 600 \times 10^{-9}}{2} = 3.66 \times 10^{-7} \text{ rad}$$

26. **(c)** As $\theta = \frac{2\lambda}{a}$

$$\theta_0 = \frac{2 \times 6000}{a} \quad ...(i)$$

$$\frac{\theta_1}{\theta_0} = \frac{\lambda_1}{6000} \quad ...(ii)$$

$\Rightarrow \lambda_1 = 0.7 \times 6000$ (as $\theta_1 = 0.7\,\theta_0$)

$\Rightarrow$ 4200 Å

27. **(b)** When reflected light rays and refracted rays are perpendicular, reflected light is polarised with electric field vector perpendicular to the plane of incidence.

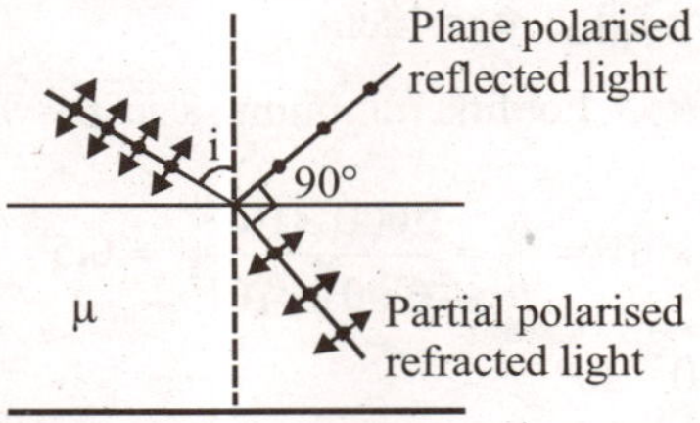

Also, tan i = μ (i = Brewester angle)

28. **(b)**

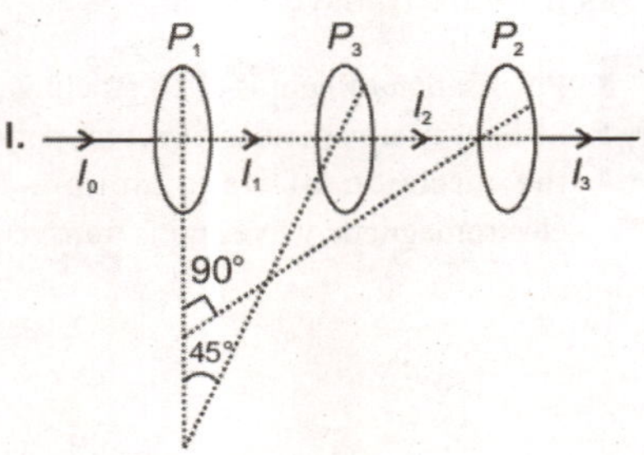

According to Malus law, $I = I_0 \cos^2 \theta$

$$I_1 = \frac{I_0}{2}$$

$$I_2 = \frac{I_0}{2}\cos^2 45° = \frac{I_0}{2} \times \frac{1}{2} = \frac{I_0}{4}$$

$$I_3 = \frac{I_0}{4}\cos^2 45° = \frac{I_0}{8}$$

29. **(b)** Resolving power of a microscope $= \frac{2\mu \sin\theta}{\lambda}$

i.e., $R \propto \frac{1}{\lambda}$ or, $\frac{R_1}{R_2} = \frac{\lambda_2}{\lambda_1}$

Given that the two wavelengths,

$\lambda_1 = 4000$Å and $\lambda_2 = 6000$Å

$$\therefore \frac{R_1}{R_2} = \frac{6000\text{Å}}{4000\text{Å}} = \frac{3}{2}$$

30. **(d)** For the first minima,

$$\theta = \frac{\eta\lambda}{a} \;\Rightarrow\; \sin 30° = \frac{\lambda}{a} = \frac{1}{2}$$

First secondary maxima will be at

$$\sin\theta = \frac{3\lambda}{2a} = \frac{3}{2}\left(\frac{1}{2}\right) \;\Rightarrow\; \theta = \sin^{-1}\left(\frac{3}{4}\right)$$

31. **(d)** Linear width of central maxima Y

$$= D(2\theta) = 2D\theta = \frac{2D\lambda}{a} \qquad \therefore \theta = \frac{\lambda}{a}$$

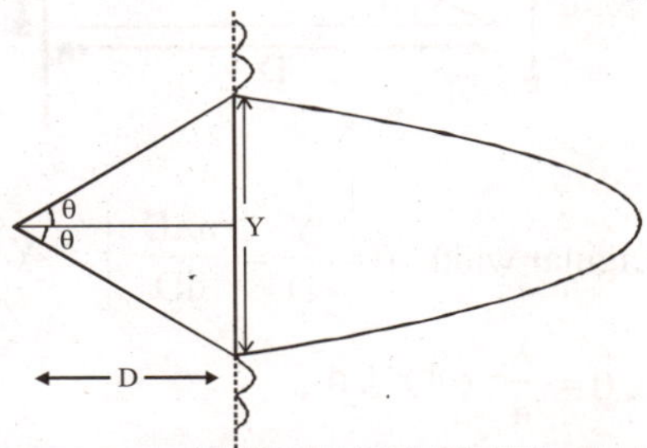

NOTES

The central maxima lies between the first minima on both sides.

The angular width 'd' central maxima $= 2\theta = \frac{2\lambda}{b}$

Linear width of central maxima $= 2D\theta = \frac{2D\lambda}{b}$

b = width of slits and D = distance between slit and screen.

32. **(b)** For first minima at P

$AP - BP = \lambda$

$AP - MP = \frac{\lambda}{2}$

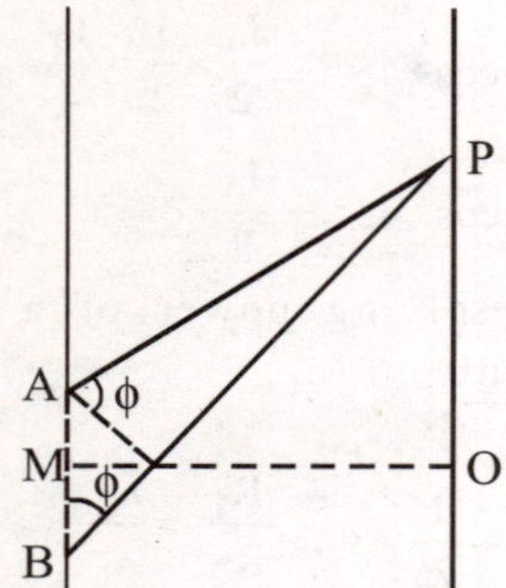

So phase difference, $\phi = \frac{2\pi}{\lambda} \times \frac{\lambda}{2} = \pi$ radian

33. **(d)** Given: $D = 2m$; $d = 1\ mm = 1 \times 10^{-3}\ m$
$\lambda = 600\ nm = 600 \times 10^{-6}\ m$
Width of central bright fringe $(= 2\beta)$

$$= \frac{2\lambda D}{d} = \frac{2 \times 600 \times 10^{-6} \times 2}{1 \times 10^{-3}} m$$

$= 2.4 \times 10^{-3}\ m = 2.4\ mm$

34. **(b)**

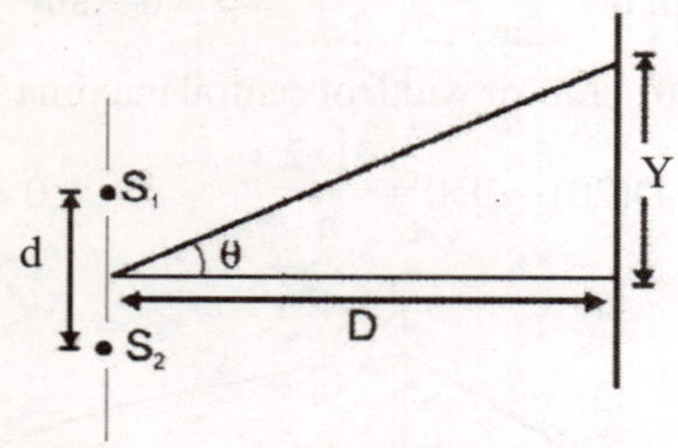

Angular width, $\theta = \frac{Y}{D} = \frac{n\lambda D}{dD}$ $\left[\because Y = \frac{D\lambda}{d}\right]$

so, $\theta = \frac{\lambda}{d}$, $v \uparrow \lambda \downarrow \theta \downarrow$

[For central maxima n = 1]

Hence, with increase in speed of electrons angular width of central maximum decreases.

35. **(d)** Conditions for diffraction minima are

Path diff. $\Delta x = n\lambda$ and Phase diff. $\delta\phi = 2n\pi$

Path diff. $= n\lambda = 2\lambda$

Phase diff. $= 2n\pi = 4\pi$ $(\because n = 2)$

36. **(d)** $\delta\phi = 1.22\frac{\lambda}{D} = 1.22\frac{5000 \times 10^{-10}}{10 \times 10^{-2}} = 6.1 \times 10^{-6}$

$\therefore$ Order $= 10^{-6}$

37. **(b)** Angular limit of resolution of eye, $\theta = \frac{\lambda}{d}$,

where, d is diameter of eye lens.

Also, if Y is the minimum separation between two objects at distance D from eye then,

$$\theta = \frac{Y}{D}$$

$$\Rightarrow \frac{Y}{D} = \frac{\lambda}{d} \Rightarrow Y = \frac{\lambda D}{d} \qquad ...(1)$$

Here, wavelength $\lambda = 5000Å = 5 \times 10^{-7}$ m

$D = 50$ m

Diameter of eye lens $= 2$ mm $= 2 \times 10^{-3}$ m

From eq. (1), minimum separation is

$$Y = \frac{5 \times 10^{-7} \times 50}{2 \times 10^{-3}} = 12.5 \times 10^{-3}\ m = 12.5\ cm$$

38. **(b)** Separation between slits are $(r_1 =)$ 16 cm and $(r_2 =)$ 9 cm.

Actual distance of separation $= \sqrt{r_1 r_2}$

$= \sqrt{16 \times 9} = 12$cm

39. **(c)** For first minimum, $a \sin\theta = n\lambda = 1\lambda$

$$\sin\theta = \frac{\lambda}{a} = \frac{5000 \times 10^{-10}}{0.001 \times 10^{-3}} = 0.5$$

$\theta = 30°$

40. **(d)** Sound waves can not be polarised as they are longitudinal. Light waves can be polarised as they are transverse.

NOTES

Polarisation of light is the restricting the vibration of light in a particular direction perpendicular to the direction of wave motion. Light is an electromagnetic wave and is transverse in nature.

25 Dual Nature of Radiation and Matter

Trend Analysis with Important Topics & Sub-Topics

Topic Name	Sub-Topic	2021		2020		2019		2018		2017		2016	
		QNS.	LOD	QNS.	LOD	QNS.	LOD	QNS.	LOD	QNS.	LOD	QNS.	LOD
Matter Waves, Cathode & Positive Rays	de-Broglie wavelength of electron			1	E	1	E	1	A			1	A
	De-Brglie of neutron									1	A		
Electron Emission, Photon Photoelectric Effect & X-ray	Einstein's photoelectric equation	1	E	1	E					1	A	1	A
	No. of photons/s $n = \frac{P\lambda}{hc}$	1	E										

LOD - Level of Difficulty | E - Easy | A - Average | D - Difficult | Qns - No. of Questions

Topic 1 Matter Waves, Cathode & Positive Rays

1. An electron is accelerated from rest through a potential difference of V volt. If the de Broglie wavelength of the electron is 1.227×10^{-2} nm, the potential difference is : *[2020]*
   (a) 10^2 V (b) 10^3 V
   (c) 10^4 V (d) 10 V

2. An electron is accelerated through a potential difference of 10,000 V. Its de Broglie wavelength is, (nearly): ($m_e = 9\ 10^{-31}$ kg) *[2019]*
   (a) $12.2\ 10^{-13}$ m (b) $12.2\ 10^{-12}$ m
   (c) $12.2\ 10^{-14}$ m (d) 12.2 nm

3. A proton and an α–particle are accelerated from rest to the same energy. The de Broglie wavelengths λ_p and λ_a are in the ratio, *[NEET Odisha 2019]*
   (a) 4 : 1 (b) 2 : 1
   (c) 1 : 1 (d) $\sqrt{2}$: 1

4. An electron of mass m with an initial velocity $\vec{V} = V_0\hat{i}$ ($V_0 > 0$) enters an electric field $\vec{E} = -E_0\hat{i}$ (E_0 = constant > 0) at t = 0. If λ_0 is its de-Broglie wavelength initially, then its de-Broglie wavelength at time t is *[2018]*
   (a) $\dfrac{\lambda_0}{\left(1+\dfrac{eE_0}{mV_0}t\right)}$ (b) $\lambda_0\left(1+\dfrac{eE_0}{mV_0}t\right)$
   (c) λ_0 (d) $\lambda_0 t$

5. The de-Broglie wavelength of a neutron in thermal equilibrium with heavy water at a temperature T (Kelvin) and mass m, is :- *[2017]*
   (a) $\dfrac{h}{\sqrt{3mkT}}$ (b) $\dfrac{2h}{\sqrt{3mkT}}$
   (c) $\dfrac{2h}{\sqrt{mkT}}$ (d) $\dfrac{h}{\sqrt{mkT}}$

6. An electron of mass m and a photon have same energy E. The ratio of de-Broglie wavelengths associated with them is : *[2016]*
   (a) $\dfrac{1}{c}\left(\dfrac{E}{2m}\right)^{\frac{1}{2}}$ (b) $\left(\dfrac{E}{2m}\right)^{\frac{1}{2}}$
   (c) $c(2mE)^{\frac{1}{2}}$ (d) $\dfrac{1}{xc}\left(\dfrac{2m}{E}\right)^{\frac{1}{2}}$

7. Which of the following figures represents the variation of particle momentum and the associated de-Broglie wavelength? *[2015]*

(a) 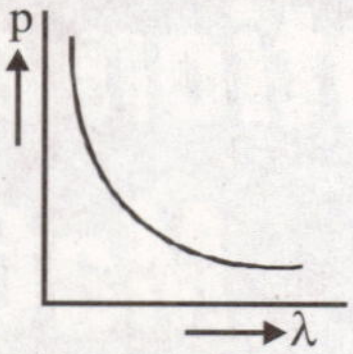

(b) 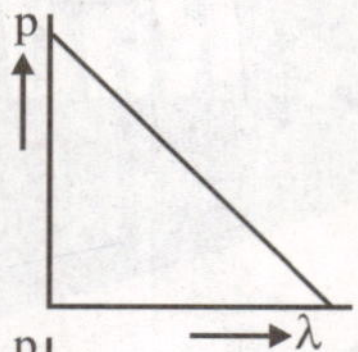

(c) 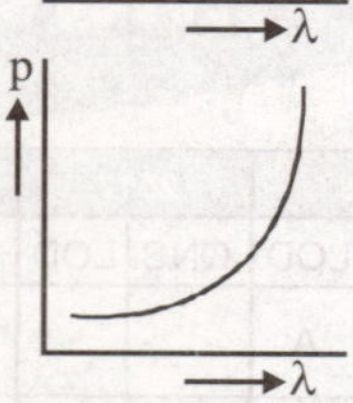

(d) 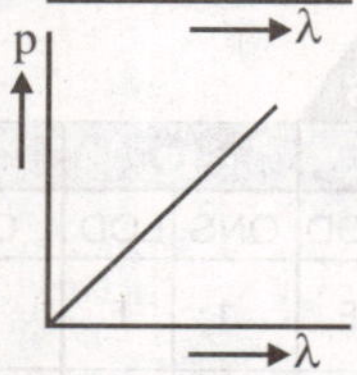

8. If the kinetic energy of the particle is increased to 16 times its previous value, the percentage change in the de-Broglie wavelength of the particle is : *[2014]*
(a) 25 (b) 75
(c) 60 (d) 50

9. The wavelength λ_e of an electron and λ_p of a photon are of same energy E are related by *[2013]*
(a) $\lambda_p \propto \lambda_e$ (b) $\lambda_p \propto \sqrt{\lambda_e}$
(c) $\lambda_p \propto \frac{1}{\sqrt{\lambda_e}}$ (d) $\lambda_p \propto \lambda_e^2$

10. The de-Broglie wavelength of neutron in thermal equilibrium at temperature T is *[NEET Kar. 2013]*
(a) $\frac{30.8}{\sqrt{T}}$ Å (b) $\frac{3.08}{\sqrt{T}}$ Å
(c) $\frac{0.308}{\sqrt{T}}$ Å (d) $\frac{0.0308}{\sqrt{T}}$ Å

11. If the momentum of electron is changed by P, then the de Broglie wavelength associated with it changes by 0.5%. The initial momentum of electron will be : *[2012M]*
(a) $200P$ (b) $400P$
(c) $\frac{P}{200}$ (d) $100P$

12. Electrons used in an electron microscope are accelerated by a voltage of 25 kV. If the voltage is increased to 100kV then the de–Broglie wavelength associated with the electrons would *[2011]*
(a) increase by 2 times
(b) decrease by 2 times
(c) decrease by 4 times
(d) increase by 4 times

13. In the Davisson and Germer experiment, the velocity of electrons emitted from the electron gun can be increased by *[2011]*
(a) increasing the potential difference between the anode and filament
(b) increasing the filament current
(c) decreasing the filament current
(d) decreasing the potential difference between the anode and filament

14. In the phenomenon of electric discharge through gases at low pressure, the coloured glow in the tube appears as a result of: *[2008]*
(a) excitation of electrons in the atoms
(b) collision between the atoms of the gas
(c) collisions between the charged particles emitted from the cathode and the atoms of the gas
(d) collision between different electrons of the atoms of the gas

15. A particle of mass 1 mg has the same wavelength as an electron moving with a velocity of 3×10^6 ms^{-1}. The velocity of the particle is: *[2008]*
(a) 2.7×10^{-18} ms^{-1} (b) 9×10^{-2} ms^{-1}
(c) 3×10^{-31} ms^{-1} (d) 2.7×10^{-21} ms^{-1}
(mass of electron $= 9.1 \times 10^{-31}$kg)

16. In a discharge tube ionization of enclosed gas is produced due to collisions between *[2006]*
(a) negative electrons and neutral atoms / molecules
(b) photons and neutral atoms /molecules
(c) neutral gas atoms/molecules
(d) positive ions and neutral atoms/molecules

17. Which of the following moving particles (moving with same velocity) has largest wavelength of matter waves? *[2002]*
(a) Electron (b) α-particle
(c) Proton (d) Neutron

18. Which one of the following statements is not true for cathode rays? *[2002]*
(a) Cathode rays produce heat when incident on metals
(b) Cathode rays travel in a straight line
(c) Cathode rays do not deflect in electric field
(d) Cathode rays produce fluorescence when they fall on certain meterials

19. The wavelength associated with an electron, accelerated through a potential difference of 100 V, is of the order of *[1996]*
(a) 1000 Å (b) 100 Å
(c) 10.5 Å (d) 1.2 Å

20. Gases begin to conduct electricity at low pressure because [1994]
(a) at low pressures gases turn of plasma
(b) colliding electrons can acquire higher kinetic energy due to increased mean free path leading to ionisation of atoms
(c) atoms break up into electrons and protons
(d) the electrons in atoms can move freely at low pressures

21. An ionization chamber with parallel conducting plates as anode and cathode has 5×10^7 electrons and the same number of singly charged positive ions per cm^3. The electrons are moving towards the anode with velocity 0.4 m/s. The current density from anode to cathode is $4 \mu A/m^2$. The velocity of positive ions moving towards cathode is [1992]
(a) 0.4 m/s (b) 1.6 m/s
(c) zero (d) 0.1 m/s

Topic 2 Electron Emission, Photon Photoelectric Effect & X-ray

22. The number of photons per second on an average emitted by the source of monochromatic light of wavelength 600 nm, when it delivers the power of 3.3×10^{-3} watt will be ($h = 6.6 \times 10^{-34}$ J s) [2021]
(a) 10^{15} (b) 10^{18}
(c) 10^{17} (d) 10^{16}

23. An electromagnetic wave of wavelength 'λ' is incident on a photosensitive surface of negligible work function. If 'm' mass is of photoelectron emitted from the surface has de-Broglie wavelength λ_d, then [2021]
(a) $\lambda = \left(\frac{2h}{mc}\right)\lambda_d^2$ (b) $\lambda = \left(\frac{2m}{hc}\right)\lambda_d^2$
(c) $\lambda_d = \left(\frac{2mc}{h}\right)\lambda^2$ (d) $\lambda = \left(\frac{2mc}{h}\right)\lambda_d^2$

24. Light of frequency 1.5 times the threshold frequency is incident on a photosensitive material. What will be the photoelectric current if the frequency is halved and intensity is doubled? [2020]
(a) four times (b) one-fourth
(c) zero (d) doubled

25. The work function of a photosensitive material is 4.0 eV. The longest wavelength of light that can cause photon emission from the substance is (approximately) [NEET Odisha 2019]
(a) 310 nm (b) 3100 nm
(c) 966 nm (d) 31 nm

26. The photoelectric threshold wavelength of silver is 3250×10^{-10}m. The velocity of the electron ejected from a silver surface by ultraviolet light of wavelength 2536×10^{-10} m is [2017]
(Given $h = 4.14 \times 10^{-15}$ eVs and $c = 3 \times 10^8$ ms^{-1})
(a) $\approx 0.6 \times 10^6$ ms^{-1} (b) $\approx 61 \times 10^3$ ms^{-1}
(c) $\approx 0.3 \times 10^6$ ms^{-1} (d) $\approx 6 \times 10^5$ ms^{-1}

27. When a metallic surface is illuminated with radiation of wavelength λ, the stopping potential is V. If the same surface is illuminated with radiation of wavelength 2λ, the stopping potential is $\frac{V}{4}$. The threshold wavelength for the metallic surface is : [2016]
(a) 4λ (b) 5λ
(c) $\frac{5}{2}\lambda$ (d) 3λ

28. A certain metallic surface is illuminated with monochromatic light of wavelength λ. The stopping potential for photo-electric current for this light is $3V_0$. If the same surface is illuminated with light of wavelength 2λ, the stopping potential is V_0. The threshold wavelength for this surface for photo-electric effect is [2015]
(a) 4λ (b) $\frac{\lambda}{4}$
(c) $\frac{\lambda}{6}$ (d) 6λ

29. Light of wavelength 500 nm is incident on a metal with work function 2.28 eV. The wavelength of the emitted electron is: [2015 RS]
(a) $< 2.8 \times 10^{-9}$ m (b) $\geq 2.8 \times 10^{-9}$ m
(c) $\leq 2.8 \times 10^{-12}$ m (d) $< 2.8 \times 10^{-10}$ m

30. A photoelectric surface is illuminated successively by monochromatic light of wavelength λ and $\frac{\lambda}{2}$. If the maximum kinetic energy of the emitted photoelectrons in the second case is 3 times that in the first case, the work function of the surface of the material is : (h = Planck's constant, c = speed of light) [2015 RS]
(a) $\frac{hc}{\lambda}$ (b) $\frac{2hc}{\lambda}$ (c) $\frac{hc}{3\lambda}$ (d) $\frac{hc}{2\lambda}$

31. When the energy of the incident radiation is increased by 20%, the kinetic energy of the photoelectrons emitted from a metal surface increased from 0.5 eV to 0.8 eV. The work function of the metal is : [2014]
(a) 0.65 eV (b) 1.0 eV
(c) 1.3 eV (d) 1.5 eV

32. For photoelectric emission from certain metal the cut-off frequency is ν. If radiation of frequency 2ν impinges on the metal plate, the maximum possible velocity of the emitted electron will be (m is the electron mass) [2013]
(a) $\sqrt{h\nu / m}$ (b) $\sqrt{2h\nu / m}$
(c) $2\sqrt{h\nu / m}$ (d) $\sqrt{h\nu / (2m)}$

33. A source of light is placed at a distance of 50 cm from a photocell and the stopping potential is found to be V_0. If the distance between the light source and photocell is made 25 cm, the new stopping potential will be [NEET Kar. 2013]
(a) $2V_0$ (b) $V_0/2$
(c) V_0 (d) $4V_0$

34. A 200 W sodium street lamp emits yellow light of wavelength 0.6 μm. Assuming it to be 25% efficient in converting electrical energy to light, the number of photons of yellow light it emits per second is [2012]
(a) 1.5×10^{20} (b) 6×10^{18}
(c) 62×10^{20} (d) 3×10^{19}

35. Monochromatic radiation emitted when electron on hydrogen atom jumps from first excited to the ground state irradiates a photosensitive material. The stopping potential is measured to be 3.57 V. The threshold frequency of the materials is : [2012]
(a) 4×10^{15} Hz (b) 5×10^{15} Hz
(c) 1.6×10^{15} Hz (d) 2.5×10^{15} Hz

36. Two radiations of photons energies 1 eV and 2.5 eV, successively illuminate a photosensitive metallic surface of work function 0.5 eV. The ratio of the maximum speeds of the emitted electrons is : [2012M]
(a) 1 : 4 (b) 1 : 2
(c) 1 : 1 (d) 1 : 5

37. Photoelectric emmision occurs only when the incident light has more than a certain minimum [2011]
(a) power (b) wavelength
(c) intensity (d) frequency

38. The momentum of a photon of energy hν will be [2011]
(a) hν/c (b) c/hν
(c) hν (d) $h\nu/c^2$

39. Light of two different frequencies whose photons have energies 1 eV and 2.5 eV respectively illuminate a metallic surface whose work function is 0.5 eV successively. Ratio of maximum speeds of emitted electrons will be [2011]
(a) 1 : 4 (b) 1 : 2
(c) 1 : 1 (d) 1 : 5

40. In photoelectric emission process from a metal of work function 1.8 eV, the kinetic energy of most energetic electrons is 0.5 eV. The corresponding stopping potential is [2011]
(a) 1.8 V (b) 1.2 V
(c) 0.5 V (d) 2.3 V

41. The threshold frequency for a photosensitive metal is 3.3×10^{14} Hz. If light of frequency 8.2×10^{14} Hz is incident on this metal, the cut-off voltage for the photoelectric emission is nearly [2011M]
(a) 2 V (b) 3 V
(c) 5 V (d) 1 V

42. A source S_1 is producing, 10^{15} photons per second of wavelength 5000 Å. Another source S_2 is producing 1.02×10^{15} photons per second of wavelength 5100Å Then, (power of S_2) (power of S_1) is equal to : [2010]
(a) 1.00 (b) 1.02
(c) 1.04 (d) 0.98

43. The potential difference that must be applied to stop the fastest photoelectrons emitted by a nickel surface, having work function 5.01 eV, when ultraviolet light of 200 nm falls on it, must be: [2010]
(a) 2.4 V (b) −1.2 V
(c) −2.4 V (d) 1.2 V

44. The number of photo electrons emitted for light of a frequency ν (higher than the threshold frequency ν_0) is proportional to: [2009]
(a) Threshold frequency (ν_0)
(b) Intensity of light
(c) Frequency of light (ν)
(d) $\nu - \nu_0$

45. The figure shows a plot of photo current versus anode potential for a photo sensitive surface for three different radiations. Which one of the following is a **correct** statement? [2009]

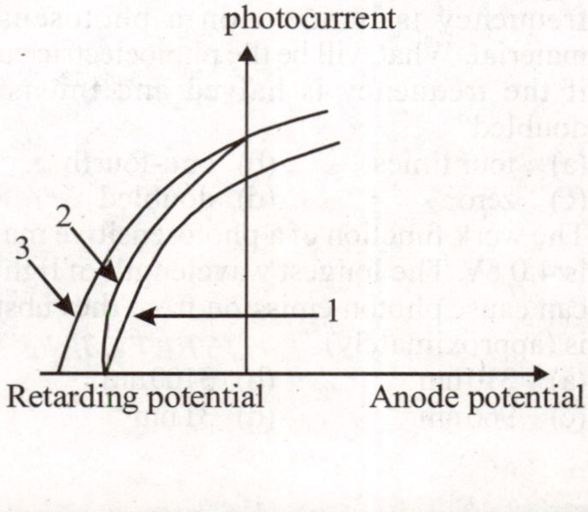

(a) Curves (1) and (2) represent incident radiations of same frequency but of different intensities.
(b) Curves (2) and (3) represent incident radiations of different frequencies and different intensities.
(c) Curves (2) and (3) represent incident radiations of same frequency having same intensity.
(d) Curves (1) and (2) represent incident radiations of different frequencies and different intensities.

46. Monochromatic light of wavelength 667 nm is produced by a helium neon laser. The power emitted is 9 mW. The number of photons arriving per sec on the average at a target irradiated by this beam is: *[2009]*
(a) 3×10^{16} (b) 9×10^{15}
(c) 3×10^{19} (d) 9×10^{17}

47. The work function of a surface of a photosensitive material is 6.2 eV. The wavelength of incident radiation for which the stopping potential is 5 V lies in the: *[2008]*
(a) Ultraviolet region
(b) Visible region
(c) Infrared region
(d) X-ray region

48. Monochromatic light of frequency 6.0×10^{14} Hz is produced by a laser. The power emitted is 2×10^{-3} W. The number of photons emitted, on the average, by the sources per second is *[2007]*
(a) 5×10^{16} (b) 5×10^{17}
(c) 5×10^{14} (d) 5×10^{15}

49. A 5 watt source emits monochromatic light of wavelength 5000 Å. When placed 0.5 m away, it liberates photoelectrons from a photosensitive metallic surface. When the source is moved to a distance of 1.0 m, the number of photoelectrons liberated will be reduced by a factor of *[2007]*
(a) 8 (b) 16 (c) 2 (d) 4

50. When photons of energy $h\nu$ fall on an aluminium plate (of work function E_0), photoelectrons of maximum kinetic energy K are ejected. If the frequency of the radiation is doubled, the maximum kinetic energy of the ejected photoelectrons will be *[2006]*
(a) $2K$ (b) K
(c) $K + h\nu$ (d) $K + E_0$

51. A photo-cell employs photoelectric effect to convert *[2006]*
(a) change in the intensity of illumination into a change in photoelectric current
(b) change in the intensity of illumination into a change in the work function of the photocathode
(c) change in the frequency of light into a change in the electric current
(d) change in the frequency of light into a change in electric voltage

52. The momentum of a photon of energy 1 MeV in kg m/s, will be *[2006]*
(a) 7×10^{-24} (b) 10^{-22}
(c) 5×10^{-22} (d) 0.33×10^{6}

53. A photosensitive metallic surface has work function, $h\nu_0$. If photons of energy $2h\nu_0$ fall on this surface, the electrons come out with a maximum velocity of 4×10^6 m/s. When the photon energy is increased to $5h\nu_0$, then maximum velocity of photoelectrons will be
(a) 2×10^7 m/s (b) 2×10^6 m/s *[2005]*
(c) 8×10^6 m/s (d) 8×10^5 m/s

54. The work functions for metals A, B and C are respectively 1.92 eV, 2.0 eV and 5 eV. According to Einstein's equation, the metals which will emit photoelectrons for a radiation of wavelength 4100 Å is/are *[2005]*
(a) none (b) A only
(c) A and B only (d) all three metals

55. According to Einstein's photoelectric equation, the graph between the kinetic energy of photoelectrons ejected and the frequency of incident radiation is *[2004]*

(a) 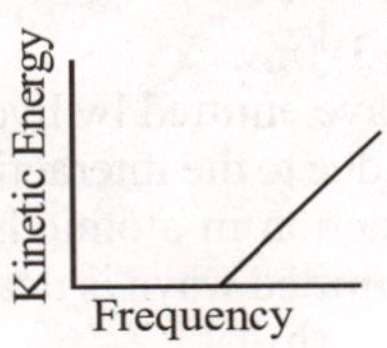

(b) 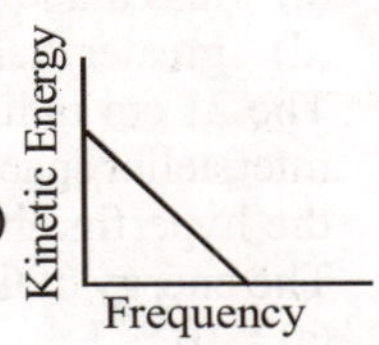

(c) 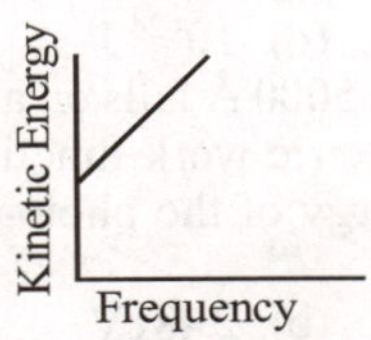

(d) 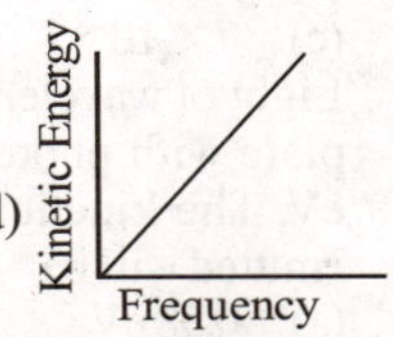

56. A photoelectric cell is illuminated by a point source of light 1m away. When the source is shifted to 2m then *[2003]*
(a) number of electrons emitted is a quarter of the initial number
(b) each emitted electron carries one quarter of the initial energy
(c) number of electrons emitted is half the initial number
(d) each emitted electron carries half the initial energy

57. When ultraviolet radiation is incident on a surface, no photoelectrons are emitted. If a second beam causes photoelectrons to be ejected, it may consists of [2002]
(a) infra-red waves (b) X-rays
(c) visible light rays (d) radio waves

58. Einstein work on the photoelectric effect provided support for the equation [2000]
(a) $E = h\nu$ (b) $E = mc^2$
(c) $E = \frac{Rhc}{n^2}$ (d) $K.E. = \frac{1}{2}mv^2$

59. As the intensity of incident light increases [1999]
(a) photoelectric current increases
(b) K. E. of emitted photelectrons increases
(c) photoelectric current decreases
(d) K.E. of emitted photoelectrons decreases

60. The photoelectric work function for a metal surface is 4.125 eV. The cut off wavelength for this surface is [1999]
(a) 4125 Å (b) 3000 Å
(c) 6000 Å (d) 2062.5 Å

61. In a photo-emissive cell, with exciting wavelength λ, the fastest electron has speed v. If the exciting wavelength is changed to $\frac{3\lambda}{4}$, the speed of the fastest emitted electron will be [1998]
(a) $(3/4)^{1/2}$. v
(b) $(4/3)^{1/2}$. v
(c) less than $(4/3)^{1/2}$. v
(d) greater than $(4/3)^{1/2}$. v

62. The 21 cm radio wave emitted by hydrogen in interstellar space is due to the interaction called the hyperfine interaction in atomic hydrogen. The energy of the emitted wave is nearly [1998]
(a) 10^{-17} J (b) 1 J
(c) 7×10^{-8} J (d) 10^{-24} J

63. Light of wavelength 5000 Å falls on a sensitive plate with photo-electric work function of 1.9 eV. The kinetic energy of the photo-electrons emitted will be [1998]
(a) 0.58 eV (b) 2.48 eV
(c) 1.24 eV (d) 1.16 eV

64. Which of the following statement is correct? [1997]
(a) Photocurrent increases with intensity of light
(b) Photocurrent is proportional to the applied voltage
(c) Current in photocell increases with increasing frequency
(d) Stopping potential increases with increase of incident light

65. The X-rays cannot be diffracted by means of an ordinary grating because of [1997]
(a) high speed (b) short wavelength
(c) large wavelength (d) none of these

66. An electron of mass m and charge e is accelerated from rest through a potential difference of V volt in vacuum. Its final speed will be [1996]
(a) $\frac{eV}{2m}$ (b) $\frac{eV}{m}$
(c) $\sqrt{\frac{2eV}{m}}$ (d) $\sqrt{\frac{eV}{2m}}$

67. The nature of ions knocked out from hot surfaces is [1995]
(a) Protons (b) Neutrons
(c) Electrons (d) Nuclei

68. If the threshold wavelength for a certain metal is 2000 Å, then the work-function of the metal is [1995]
(a) 6.2 J (b) 6.2 eV
(c) 6.2 MeV (d) 6.2 keV

69. Kinetic energy of an electron, which is accelerated in a potential difference of 100 V is [1995]
(a) 1.6×10^{-17} J (b) 1.6×10^{-19} J
(c) 1.6×10^{-21} J (d) 1.6×10^{-25} J

70. In photoelectric effect the work function of a metal is 3.5 eV. The emitted electrons can be stopped by applying a potential of –1.2 V. Then [1994]
(a) the energy of the incident photon is 4.7 eV
(b) the energy of the incident photon is 2.3 eV
(c) if higher frequency photon be used, the photoelectric current will rise
(d) when the energy of photon is 3.5 eV, the photoelectric current will be maximum

71. Doubly ionised helium atoms and hydrogen ions are accelerated from rest through the same potential drop. The ratio of the final velocities of the helium and the hydrogen ion is [1994]
(a) 1/2 (b) 2
(c) $1/\sqrt{2}$ (d) $\sqrt{2}$

72. When light of wavelength 300 nm (nanometer) falls on a photoelectric emitter, photoelectrons are liberated. For another emitter, however, light of 600 nm wavelength is sufficient for creating photoemission. What is the ratio of the work functions of the two emitters? [1993]
(a) 1 : 2 (b) 2 : 1
(c) 4 : 1 (d) 1 : 4

73. Number of ejected photoelectron increases with increase *[1993]*
(a) in intensity of light
(b) in wavelength of light
(c) in frequency of light
(d) never

74. Momentum of a photon of wavelength λ is
(a) $\frac{h}{\lambda}$ (b) zero *[1993]*
(c) $\frac{h\lambda}{c^2}$ (d) $\frac{h\lambda}{c}$

75. The cathode of a photoelectric cell is changed such that the work function changes from W_1 to W_2 ($W_2 > W_1$). If the current before and after changes are I_1 and I_2, all other conditions remaining unchanged, then (assuming $h\nu > W_2$)
(a) $I_1 = I_2$ (b) $I_1 < I_2$ *[1992]*
(c) $I_1 > I_2$ (d) $I_1 < I_2 < 2I_1$

76. The wavelength of a 1 keV photon is 1.24×10^{-9} m. What is the frequency of 1 MeV photon? *[1991]*
(a) 1.24×10^{15} (b) 2.4×10^{20}
(c) 1.24×10^{18} (d) $2 \times 4 \times 10^{23}$

77. Photoelectric work function of a metal is 1eV. Light of wavelength $\lambda = 3000$ Å falls on it. The photo electrons come out with velocity *[1991]*
(a) 10 metres/sec (b) 10^2 metres/sec
(c) 10^4 metres/sec (d) 10^6 metres/sec

78. Energy levels *A*, *B*, *C* of a certain atom correspond to increasing values of energy i.e., $E_A < E_B < E_C$. If λ_1, λ_2, λ_3 are the wavelengths of radiation corresponding to the transitions *C* to *B*, *B* to *A* and *C* to *A* respectively, which of the following relation is correct? *[1990, 2005]*
(a) $\lambda_3 = \lambda_1 + \lambda_2$ (b) $\lambda_3 = \frac{\lambda_1\lambda_2}{\lambda_1 + \lambda_2}$
(c) $\lambda_1 + \lambda_2 + \lambda_3 = 0$ (d) $\lambda_3^2 = \lambda_1^2 + \lambda_2^2$

79. A radio transmitter operates at a freqency 880 kHz and a power of 10 kW. The number of photons emitted per second is *[1990]*
(a) 1.72×10^{31} (b) 1.327×10^{25}
(c) 1.327×10^{37} (d) 1.327×10^{45}

80. The momentum of a photon of an electromagnetic radiation is 3.3×10^{-29} kgms^{-1}. What is the frequency of the associated waves ? *[1990]*
[$h = 6.6 \times 10^{-34}$ Js; $c = 3 \times 10^8$ ms^{-1}]
(a) 1.5×10^{13} Hz (b) 7.5×10^{12} Hz
(c) 6.0×10^3 Hz (d) 3.0×10^3 Hz

81. Ultraviolet radiations of 6.2 eV falls on an aluminium surface. K.E. of fastest electron emitted is (work function = 4.2 eV) *[1989]*
(a) 3.2×10^{-21} J (b) 3.2×10^{-19} J
(c) 7×10^{-25} J (d) 9×10^{-32} J

82. The energy of a photon of wavelength λ is
(a) $hc\lambda$ (b) $\frac{hc}{\lambda}$ *[1988]*
(c) $\frac{\lambda}{hc}$ (d) $\frac{\lambda h}{c}$

83. The threshold frequency for photoelectric effect on sodium corresponds to a wavelength of 5000 Å. Its work function is *[1988]*
(a) 4×10^{-19} J (b) 1 J
(c) 2×10^{-19} J (d) 3×10^{-19} J

ANSWER KEY

1	(c)	10	(a)	19	(d)	28	(a)	37	(d)	46	(a)	55	(a)	64	(a)	73	(a)	82	(b)
2	(b)	11	(a)	20	(b)	29	(b)	38	(a)	47	(a)	56	(a)	65	(b)	74	(a)	83	(a)
3	(b)	12	(b)	21	(d)	30	(d)	39	(b)	48	(d)	57	(b)	66	(c)	75	(a)		
4	(a)	13	(a)	22	(d)	31	(b)	40	(c)	49	(d)	58	(a)	67	(c)	76	(b)		
5	(a)	14	(c)	23	(d)	32	(b)	41	(a)	50	(c)	59	(a)	68	(b)	77	(d)		
6	(a)	15	(a)	24	(c)	33	(c)	42	(a)	51	(a)	60	(b)	69	(a)	78	(b)		
7	(a)	16	(a)	25	(a)	34	(a)	43	(b)	52	(c)	61	(d)	70	(a)	79	(a)		
8	(b)	17	(a)	26	(a, d)	35	(c)	44	(b)	53	(c)	62	(d)	71	(d)	80	(a)		
9	(d)	18	(c)	27	(d)	36	(b)	45	(a)	54	(c)	63	(a)	72	(b)	81	(b)		

Hints & Solutions

1. (c) Using, $\lambda = \frac{12.27}{\sqrt{V}}$ Å

$$\Rightarrow \sqrt{V} = \frac{12.27 \times 10^{-10}}{1.227 \times 10^{-11}} = 10^2$$

[Given, $\lambda = 1.227 \times 10^{-11}$ m]

$\therefore V = 10^4$ volt

2. (b) de-Broglie wavelength of electron

$$\lambda = \frac{12.27}{\sqrt{V}} \text{ Å} = \frac{12.27 \times 10^{-10}}{\sqrt{10000}} = 12.27 \times 10^{-12} \text{ m}$$

3. (b) As $\lambda = \frac{h}{p} = \frac{h}{\sqrt{2mK}}$

$\therefore \quad \lambda \propto \frac{1}{\sqrt{m}}$ (Kinetic energies are same)

$$\frac{\lambda_p}{\lambda_\alpha} = \sqrt{\frac{m_\alpha}{m_p}}$$

$$\therefore \quad \frac{\lambda_p}{\lambda_\alpha} = \frac{2}{1}$$

4. (a) Initial de-Brogile wavelength

$$\lambda_0 = \frac{h}{mV_0} \quad \text{...(i)}$$

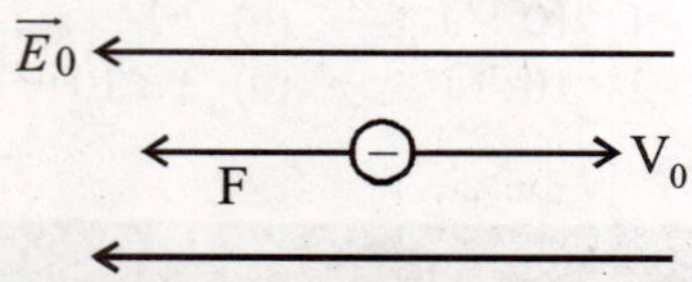

Acceleration of electron

$$a = \frac{eE_0}{m} \qquad (\because F = ma = eE_0)$$

Velocity after time 't'

$$V = \left(V_0 + \frac{eE_0}{m}t\right)$$

$$\text{So, } \lambda = \frac{h}{mV} = \frac{h}{m\left(V_0 + \frac{eE_0}{m}t\right)}$$

$$= \frac{h}{mV_0\left[1 + \frac{eE_0}{mV_0}t\right]} = \frac{\lambda_0}{\left[1 + \frac{eE_0}{mV_0}t\right]} \quad \text{...(ii)}$$

Dividing eqs. (ii) by (i),

$$\text{de-Broglie wavelength } \lambda = \frac{\lambda_0}{\left[1 + \frac{eE_0}{mV_0}t\right]}$$

5. (a) We know that,

$$\text{de-Broglie wavelength } \lambda = \frac{\hbar}{p} = \frac{h}{\sqrt{2m(KE)}}$$

$$\text{K.E. of thermal neutron} = \frac{3}{2}kT$$

$$= \frac{h}{\sqrt{2m\left(\frac{3}{2}kT\right)}}$$

$$\lambda = \frac{h}{\sqrt{3mkT}}$$

6. (a) For electron De-Broglie wavelength,

$$\lambda_e = \frac{h}{\sqrt{2mE}}$$

For photon $E = pc$

$\Rightarrow$ De-Broglie wavelength, $\lambda_{Ph} = \frac{hc}{E}$

$$\therefore \quad \frac{\lambda_e}{\lambda_{Ph}} = \frac{h}{\sqrt{2mE}} \times \frac{E}{hc} = \left(\frac{E}{2m}\right)^{1/2} \frac{1}{c}$$

7. (a) According to De-broglie $p = \frac{h}{\lambda}$ or $P \propto \frac{1}{\lambda}$

where
P = particle momentum;
λ = de-Broglie wavelength
h = Plank's constant

$P \propto \frac{1}{\lambda}$ represents rectangular hyperbola.

8. (b) As we know

$$\lambda = \frac{h}{P} = \frac{h}{\sqrt{2mK}} \qquad (\because P = \sqrt{2mKE})$$

or $\frac{\lambda_1}{\lambda_2} = \sqrt{\frac{K_2}{K_1}} = \sqrt{\frac{16K}{K}} = \frac{4}{1}$

Therefore the percentage change in de-Broglie wavelength $= \frac{1-4}{4} \times 100 = -75\%$

9. **(d)** As $P = \frac{E}{c}$

$$\lambda_p = \frac{hc}{E} \quad ...(i)$$

$$\lambda_e = \frac{h}{\sqrt{2mE}} \Rightarrow \lambda_e^2 = \frac{h^2}{2mE} \quad ...(ii)$$

From equations (i) and (ii)

$\lambda_p \propto \lambda_e^2$

NOTES

de-Broglie wavelength, $\lambda = \frac{h}{p}$

Here, h = plank's constant

p = momentum

Momentum, $p = \sqrt{2mk}$

$\therefore \lambda = \frac{h}{\sqrt{2mE}}$ (Here, E = kinetic energy)

$\therefore E = \frac{h^2}{2m\lambda^2} \Rightarrow \lambda^2 = \frac{h^2}{2mE}$

10. **(a)** From formula $\lambda = \frac{h}{\sqrt{2mKT}}$

$$= \frac{6.63 \times 10^{-34}}{\sqrt{2 \times 1.67 \times 10^{-27} \times 1.38 \times 10^{-23} T}} m$$

[By placing value of h, m and k]

$$= \frac{30.8}{\sqrt{T}} \text{Å}$$

11. **(a)** The de-Broglie's wavelength associated with the moving electron $\lambda = \frac{h}{P}$

Now, according to problem

$$\frac{d\lambda}{\lambda} = -\frac{dp}{P}$$

$$\frac{0.5}{100} = \frac{P}{P'}$$

$P' = 200P$

12. **(b)** $\lambda \propto \frac{1}{\sqrt{V}}$

$$\Rightarrow \frac{\lambda_1}{\lambda_2} = \sqrt{\frac{V_2}{V_1}} = \sqrt{\frac{100\text{keV}}{25\text{keV}}} = 2$$

$$\Rightarrow \lambda_2 = \frac{\lambda_1}{2}$$

13. **(a)** In the Davisson and Germer experiment, the velocity of electrons emitted from the electron gun can be increased by increasing the potential difference between the anode and filament.

14. **(c)** In discharge tube, collisions between the charged particles emitted from the cathode and atoms of the gas results to the coloured glow in the tube.

15. **(a)** Wavelength of particle

$$(\lambda_1) = \frac{h}{mv} = \frac{h}{(1\times10^{-6})\times v}$$

where v is the velocity of the particle.

Wave length of electron,

$$(\lambda_2) = \frac{h}{(9.1\times10^{-31})\times(3\times10^6)}$$

But $\lambda_1 = \lambda_2$ (Given)

$$\frac{h}{(1\times10^{-6})\times v} = \frac{h}{(9.1\times10^{-31})\times(3\times10^6)}$$

$$\Rightarrow v = \frac{9.1\times10^{-31}\times3\times10^6}{10^{-6}} = 2.73\times10^{-18}\ ms^{-1}$$

16. **(a)** When electrons emitted from cathode collide with gas molecules or atoms, they knock out outer electrons and produce positively charged ions. They become part of positive rays.

17. **(a)** de-Broglie wavelength $\lambda = \frac{h}{mv}$

For same velocity, $\lambda \propto \frac{1}{m}$

Out of given particles, the mass of electron is minimum, so the associated de-Broglie wavelength is maximum for electron.

18. **(c)** Cathode rays are negatively charged rays. Hence, when electric field is applied to them, then they deflect in direction opposite to electric field.

19. **(d)** Potential difference = 100 V

K.E. acquired by electron = e (100)

$$\frac{1}{2}mv^2 = e(100) \Rightarrow v = \sqrt{\frac{2e(100)}{m}}$$

According to de Broglie's concept

$$\lambda = \frac{h}{mv} \quad \Rightarrow \lambda = \frac{h}{m\sqrt{\frac{2e(100)}{m}}}$$

$$= \frac{h}{\sqrt{2me(100)}} = 1.2 \times 10^{-10} = 1.2 \text{Å}$$

NOTES

de-Broglie wavelength, $\lambda_e = \frac{h}{p} = \frac{h}{\sqrt{2mE}}$

Here, h = planck's constant

m = mass of electron

Kinetic energy, $E = eV$

Here, V = potential difference

$$\therefore \lambda_e \frac{h}{\sqrt{2m_e eV}}$$

Substituting h = 6.63×10^{-34} Js

$e = 1.6 \times 10^{-19}$ C

$m_e = 9.1 \times 10^{-31}$ kg

we get $\lambda_e = \frac{12.27}{\sqrt{V}} \mathring{A}$

20. **(b)** The ionisation requires high energy electrons.

21. **(d)** Current $= I_e + I_p$ ....(i)

I_e and I_p are current due to electrons and positively charged ions.

$$I = neAV_d$$

where,

$$n = 5 \times 10^7 / \text{cm}^3 = 5 \times 10^7 \times 10^6 / \text{m}^3 = 5 \times 10^{13} / \text{m}^3$$

$$I_e = 5 \times 10^{13} \times 1.6 \times 10^{-19} \times A \times 0.4$$

$$I_p = 5 \times 10^{13} \times 1.6 \times 10^{-19} \times A \times v$$

$I = I_e + I_p$ (from equation (i))

$$= 5 \times 10^{13} \times 1.6 \times 10^{-19} \times A(v + 0.4)$$

Given, $I / A = 4 \times 10^{-6}\ A / m^2$

$$4 \times 10^{-6} \times A = 5 \times 10^{-6} \times 1.6 \times A(v + 0.4)$$

$$\frac{4}{8} = v + 0.4 \Rightarrow 0.5 = v + 0.4 \Rightarrow v = 0.1 \text{ m/s}$$

22. **(d)** Power, $P = \frac{E}{t} = \frac{n}{t}\left(\frac{hc}{\lambda}\right)$

$\therefore$ No. of photons / second, $\frac{n}{t} = \frac{P}{\left(\frac{hc}{\lambda}\right)}$

$$\frac{n}{t} = \frac{3.3 \times 10^{-3} \times 6 \times 10^{-7}}{6.6 \times 10^{-34} \times 3 \times 10^8} = 10^{16}$$

NOTES

The number of photons emitted per second from a source of monochromatic radiation of wavelength λ and power p and if E = energy of each photon then,

$$\frac{n}{t} = \frac{P}{E} = \frac{P}{h\nu} = \frac{P}{hc/\lambda}$$

23. **(d)** From Einstein's photoelectric equation

$$\frac{hc}{\lambda} = \phi_0 + k$$

Where ϕ_0 = work function

k = maximum kinetic energy of photoelectrons

$$\therefore \frac{hc}{\lambda} = k = \frac{P^2}{2m} \Rightarrow P = \sqrt{\frac{2mhc}{\lambda}} \quad [\because \phi = \text{negligible}]$$

de-Broglie wavelength,

$$\lambda_d = \frac{h}{P} = \frac{h}{\sqrt{2mhc/\lambda}} \Rightarrow \sqrt{\lambda} = \lambda_d\sqrt{\frac{2mc}{h}}$$

$$\therefore \lambda = \left(\frac{2mc}{h}\right)\lambda_d^2$$

24. **(c)** For photoelectric emission, photoelectric current, incident light frequency should be greater than threshold frequency.

Light of frequency 1.5 times the threshold frequency ν_0 incident.

$$\nu = \frac{3}{2}\nu_0$$

If frequency is halved,

$$\therefore \nu' = \frac{\nu}{2} = \frac{3}{4}\nu_0$$

$$\because \nu' < \nu_0$$

$\therefore$ No photoelectric emission will take place.

25. **(a)** As $E = \frac{12400}{\lambda}$

$$\lambda = \frac{12400}{4} = 3100 \text{ Å}$$

$$\therefore \quad \lambda = 310 \text{nm}$$

26. **(a, d)** Both answers are correct

Given,

$\lambda_0 = 3250 \times 10^{-10}$ m

$\lambda = 2536 \times 10^{-10}$ m

$$\phi = \frac{hc}{\lambda_0} = \frac{4.14\times10^{-15}\times3\times10^{8}}{3250\times10^{-10}} = 3.82\text{eV}$$

$$h\nu = \frac{hc}{\lambda} = \frac{4.14\times10^{-15}\times3\times10^{8}}{2536\times10^{-10}} = 4.89\text{eV}$$

According to Einstein's photoelectric equation,

$K_{max} = h\nu - \phi$

$KE_{max} = (4.89–3.82)\text{eV} = 1.077\text{ eV}$

$$\frac{1}{2}mv^2 = 1.077 \times 1.6 \times 10^{-19}$$

$$\Rightarrow v = \sqrt{\frac{2\times1.077\times1.6\times10^{-19}}{9.1\times10^{-31}}}$$

or, $v = 0.6 \times 10^6$ m/s or 6×10^5 m/s

27. **(d)** According to Einstein's photoelectric effect,

$$eV = \frac{hc}{\lambda} - \frac{hc}{\lambda_0} \quad \text{...(i)}$$

$$eV/4 = \frac{hc}{2\lambda} - \frac{hc}{\lambda_0} \quad \text{...(ii)}$$

Dividing equation (i) by (ii) by

$$\Rightarrow \quad 4 = \frac{\frac{1}{\lambda} - \frac{1}{\lambda_0}}{\frac{1}{2\lambda} - \frac{1}{\lambda_0}} \text{ on solving we get,}$$

$\lambda_0 = 3\lambda$

NOTES

Threshold wavelength (λ_0) is the maximum wavelength of incident radiations required to eject the electrons from a metallic surface.

If incident wavelength $\lambda > \lambda_0$

$\Rightarrow$ No photoelectron emission occur

28. **(a)** As we know,

$$eV_s = \frac{hc}{\lambda} - \Psi$$

$$3eV_o = \frac{hc}{\lambda} - \Psi \quad \text{...(1)}$$

$$eV_o = \frac{hc}{2\lambda} - \Psi \quad \text{...(2)}$$

$$3eV_o = \frac{3hc}{2\lambda} - 3\Psi \quad \text{...(3)}$$

Multiplying eqn. (2) by (3) and subtracting it from eqn (1)

$$\Psi = \frac{hc}{4\lambda}$$

So, threshold wavelength,

$$\lambda_{th} = \frac{hc}{\Psi} = \frac{hc}{hc/4\lambda} = 4\lambda$$

29. **(b)** Given: Work function ϕ of metal = 2.28 eV

Wavelength of light $\lambda = 500$ nm $= 500 \times 10^{-9}$m

$$KE_{max} = \frac{hc}{\lambda} - \phi$$

$$KE_{max} = \frac{6.6\times10^{-34}\times3\times10^{8}}{5\times10^{-7}\times1.6\times10^{-19}} - 2.28$$

$KE_{max} = 2.48 - 2.28 = 0.2$ ev

$$\lambda_{min} = \frac{h}{p} = \frac{h}{\sqrt{2m(KE)_{max}}}$$

$$= \frac{\frac{20}{3}\times10^{-34}}{\sqrt{2\times9\times10^{-31}\times0.2\times1.6\times10^{-19}}}$$

$$\lambda_{min} = \frac{25}{9} \times 10^{-9}$$

$= 2.80 \times 10^{-9}$ nm $\quad \therefore \lambda \geq 2.8 \times 10^{-9}$ m

30. **(d)** Photoelectric equations

$$Ek_{1max} = \frac{hc}{\lambda} - \phi \quad \text{...(i)}$$

$$\text{and } Ek_{2max} = \frac{hc}{\lambda/2} - \phi$$

$$EK_{2max} = \frac{2hc}{\lambda} - \phi \quad \text{...(ii)}$$

From question, $Ek_{2max} = 3Ek_{1max}$

Multiplying equation (i) by 3

$$3Ek_{1max} = 3\left(\frac{hc}{\lambda} - \phi\right) \quad \text{...(iii)}$$

From equation (*ii*) and (*iii*)

$$\frac{3hc}{\lambda} - 3\phi = \frac{2hc}{\lambda} - \phi \quad \therefore \phi\text{ (work function)} = \frac{hc}{2\lambda}$$

31. **(b)** According to Einstein's photoelectric equation, $h\nu = \phi_0 + K_{max}$

We have

$h\nu = \phi_0 + 0.5$ ...(i)

and $1.2h\nu = \phi_0 + 0.8$ ...(ii)

Therefore, from above two equations $\phi_0 = 1.0$ eV.

NOTES

Work function (ϕ_0) varies from metal to metal. The material is better for photoelectric emission whose work function is least. As caesium has least work function, hence it is best metal for photoelectric emission.

32. **(b)** From photo-electric equation,

$$hv' = hv + K_{max} \quad ...(i)$$

$$h.2v = hv + \frac{1}{2} mV^2_{max} \quad [\therefore v' = 2v]$$

$$\Rightarrow hv = \frac{1}{2} mV^2_{max} \Rightarrow V_{max} = \sqrt{\frac{2hv}{m}}$$

33. **(c)** Since, stopping potential is independent of distance hence new stopping potential will remain unchanged i.e., new stopping potential = V_0.

34. **(a)** Give that, only 25% of 200W converter electrical energy into light of yellow colour

$$\left(\frac{hc}{\lambda}\right) \times N = 200 \times \frac{25}{100}$$

Where N is the No. of photons emitted per second, h = plank's constant, c, speed of light.

$$N = \frac{200 \times 25}{100} \times \frac{\lambda}{hc}$$

$$= \frac{200 \times 25 \times 0.6 \times 10^{-6}}{100 \times 6.2 \times 10^{-34} \times 3 \times 10^8} = 1.5 \times 10^{20}$$

35. **(c)** $n \to 2-1$

$E = 10.2$ eV

$kE = E - \phi$

$Q = 10.20 - 3.57$

$h\upsilon_0 = 6.63$ eV

$$\upsilon_0 = \frac{6.63 \times 1.6 \times 10^{-19}}{6.67 \times 10^{-34}} = 1.6 \times 10^{15}\text{ Hz}$$

36. **(b)** According to Einsten's photoelectric effect, the K.E. of the radiated electrons

$K.E_{max} = E - W$

$$\frac{1}{2} m v_1^{\,2} = (1 - 0.5)\text{ eV} = 0.5\text{ eV}$$

$$\frac{1}{2} m v_2^{\,2} = (2.5 - 0.5)\text{ eV} = 2\text{ eV}$$

$$\frac{v_1}{v_2} = \sqrt{\frac{0.5}{2}} = \frac{1}{\sqrt{4}} = 1:2$$

37. **(d)** For occurence of photoelectric effect, the incident light should have frequency more than a certain minimum which is called the threshold frequency (v_0).

NOTES

We have, $\frac{1}{2} mv^2 = hv - hv_0$

For photoelectric effect emission $n > n_0$ where n is the frequency of the incident light.

38. **(a)**

39. **(b)** The maximum kinetic energy of emitted electrons is given by

$K.E = \phi - \phi_0$

$K.E_1 = 1\text{ eV} - 0.5\text{ eV} = 0.5\text{ eV}$

$K.E_2 = 2.5\text{ eV} - 0.5\text{ eV} = 2\text{ eV}$

$$\therefore \frac{K.E_1}{K.E_2} = \frac{0.5\text{eV}}{2\text{eV}} = \frac{1}{4}$$

$$\Rightarrow \frac{v_1}{v_2} = \sqrt{\frac{1}{4}} = \frac{1}{2}$$

40. **(c)** The stopping potential is equal to maximum kinetic energy.

41. **(a)** $K.E. = hv - hv_{th} = eV_0$ (V_0 = cut off voltage)

$$\Rightarrow V_0 = \frac{h}{e}(8.2 \times 10^{14} - 3.3 \times 10^{14})$$

$$= \frac{6.6 \times 10^{-34} \times 4.9 \times 10^{14}}{1.6 \times 10^{-19}} \approx 2\text{V}.$$

42. **(a)** Energy emitted/sec by $S_1, P_1 = n_1 \frac{hc}{\lambda_1}$

Energy emitted/sec by $S_2, P_2 = n_2 \frac{hc}{\lambda_2}$

$$\therefore \frac{P_2}{P_1} = \frac{n_2}{n_1} \cdot \frac{\lambda_1}{\lambda_2}$$

$$= \frac{1.02 \times 10^{15}}{10^{15}} \cdot \frac{5000}{5100} = 1.0$$

43. **(b)** $K_{max} = \frac{hc}{\lambda} - W = \frac{hc}{\lambda} - 5.01$

$$= \frac{12375}{\lambda(\text{in Å})} - 5.01$$

$$= \frac{12375}{2000} - 5.01 = 6.1875 - 5.01 = 1.17775$$

$\simeq 1.2$ V

The potential difference that must be applied to stop photoelectrons = –1.2V

44. **(b)** The number of photoelectrons emitted is proportional to the intensity of incident light. Saturation current $\propto$ intensity.

45. (a) Retarding potential depends on the frequency of incident radiation but is independent of intensity.

46. (a) $\lambda = 667 \times 10^{-9}$ m, $P = 9 \times 10^{-3}$ W

$P = \frac{Nhc}{\lambda}$, N= No. of photons emitted/sec.

$$N = \frac{9 \times 10^{-3} \times 667 \times 10^{-9}}{6.6 \times 10^{-34} \times 3 \times 10^{8}}$$

$$= \frac{9 \times 6.67 \times 10^{-10}}{3 \times 6.6 \times 10^{-26}} = 3 \times 10^{16} \text{ / sec}$$

47. (a) Work function, $\phi_0 = 6.2$eV.

Stopping potential $V_0 = 5$V.

As, $eV_0 = h\nu - \phi_0$

or $eV_0 + \phi_0 = \frac{hc}{\lambda}$

or $\lambda = \frac{hc}{eV_0 + \phi_0}$

$$\lambda = \frac{6.62 \times 10^{-34} Js \times 3 \times 10^{8} ms^{-1}}{1.6 \times 10^{-19} \times 5J + 6.2 \times 1.6 \times 10^{-19} J}$$

$$= \frac{6.62 \times 10^{-34} \times 3 \times 10^{8}}{1.6 \times 10^{-19}(5 + 6.2)} m$$

$$= \frac{19.86 \times 10^{-26}}{17.92 \times 10^{-19}} = 1.10 \times 10^{-7} \text{ m.}$$

Thus the wavelength of the incident radiation lies in the ultraviolet region.

48. (d) Since $p = nh\nu$

$$\Rightarrow n = \frac{p}{h\nu} = \frac{2 \times 10^{-3}}{6.6 \times 10^{-34} \times 6 \times 10^{14}} = 5 \times 10^{15}$$

49. (d) Number of emitted electrons $N_E \propto$ Intensity

$$\propto \frac{1}{(\text{Distance})^2}$$

Therefore, as distance is doubled, N_E decreases by (1/4) times.

50. (c) Applying Einstein's formula for photo-electricity

$$h\nu = \phi + \frac{1}{2}mv^2; \quad h\nu = \phi + K$$

$$\phi = h\nu - K$$

If we use 2ν frequency then let the kinetic energy becomes K'

So, $h.2\nu = \phi + K'$

$2h\nu = h\nu - K + K'$

$K' = h\nu + K$

51. (a) A photo-cell employs photoelectric effect to convert light energy into photoelectric current.

52. (c) 1 MeV $= 10^6 \times 1.6 \times 10^{-19}$ joule

Momentum of photon,

$$p = \frac{E}{c} = \frac{1.6 \times 10^{-13}}{3 \times 10^{8}}$$

$$p = \frac{1.6}{3} \times 10^{-21} = \frac{16}{3} \times 10^{-22}$$

$p = 5 \times 10^{-22}$ kg m/sec

NOTES

Rest mass of photon is zero. But its effective mass is given by

$E = mc^2 = h\nu$

$\Rightarrow m = \frac{E}{c^2}$

∴ Momentum of photon, $p = mc = \frac{E}{c^2} \times C$

$\Rightarrow p = \frac{E}{c}$

53. (c) We know that

$$h\nu - \phi = K_{max} = \frac{1}{2}mv_{max}^2$$

According to question

$$\frac{5h\nu_0 - h\nu_0}{2h\nu_0 - h\nu_0} = \frac{v_2^2}{v_1^2}$$

$v_2 = 2v_1 = 2 \times 4 \times 10^6 = 8 \times 10^6$ m/s.

54. (c) $E = \frac{hc}{\lambda}$

$$\Rightarrow E = \frac{12375}{\lambda(in\ \text{Å})} \text{eV} \Rightarrow E = \frac{12375}{4100} \text{eV}$$

$\Rightarrow E \approx 3$ eV

So, only metals having work function less than 3eV can emit photoelectrons for the incident radiation of wavelength 4100Å.

55. (a) K.E. $= h\nu - \phi$

∴ graph of K.E. versus ν is a straight line with +ve slope h and K.E. > 0 for ν such that $h\nu > \phi$.

56. (a) Intensity $\propto$ No. of electrons emitted (N)

$$I \propto \frac{1}{r^2} \Rightarrow N \propto \frac{1}{r^2}$$

NOTES

Energy crossing per unit area normally per second is called Intensity.

$$I = \frac{E}{At} = \frac{P}{A} \left(\text{Here, } P = \frac{E}{t} = \text{radiation power}\right)$$

At a distance r from a point source of power P intensity is given by $I = \frac{P}{4\pi r^2}$

$$\Rightarrow I \propto \frac{1}{r^2}$$

57. (b) Energy of photon of X-rays is more than energy of photon of ultraviolet rays. Because frequency of X rays is more than ultraviolet rays.

58. (a) Einstein work on photoelectric effect supports the equation $E = h\nu$. It is based on quantum theory of light.

59. (a) K.E. of electrons emitted depends upon the frequency of incident rays rather than the intensity. While number of photo electrons emitted depends upon intensity of radiation.

60. (b) Let λ_0 be cut off wavelength.

$$\text{Work function} = \frac{hc}{\lambda_0} = 4.125 \times 1.6 \times 10^{-19}$$

$$\lambda_0 = \frac{6.626 \times 10^{-34} \times 3 \times 10^8}{4.125 \times 1.6 \times 10^{-19}} = 3000\text{Å}$$

61. (d) $\frac{1}{2}mv^2 = \frac{hc}{\lambda} - W_0$ or $\frac{hc}{\lambda} = \frac{1}{2}mv^2 + W_0$

and

$$\frac{1}{2}mv_1^2 = \frac{hc}{(3\lambda/4)} - W_0$$

$$= \frac{4}{3}\left(\frac{1}{2}mv^2 + W_0\right) - W_0$$

So, v_1 is greater than $v\left(\frac{4}{3}\right)^{1/2}$.

62. (d) $E = h\nu = \frac{hc}{\lambda} = \frac{6.6 \times 10^{-34} \times 3 \times 10^8}{0.21}$

$$= \frac{6.6}{7} \times 10^{-24} = 0.94 \times 10^{-24}\ \text{J}$$

63. (a) K.E. $= h\nu - h\nu_0$

$$= \frac{hc}{\lambda} - h\nu_0 = \frac{12375}{5000} - 1.9$$

$= 2.48 - 1.9 = 0.58$ eV

64. (a) According to photoelectric effect, speed of electron (kinetic energy) emitted depends upon frequency of incident light while number of photoelectrons emitted depends upon intensity of incident light. Hence, as the intensity of light increases, the photocurrent increases.

NOTES

In a photo-cell, the photocurrent has no relation with the applied voltage.

Stopping potential is the (negative) potential at which the current is just reduced to zero. It is independent of intensity of light but depends on the frequency of light similar to K.E.

65. (b) We know that the X-rays are of short wavelength as compared to grating constant of optical grating. As a result of this, it makes difficult to observe X-rays diffraction with ordinary grating.

66. (c) Kinetic energy of electron accelerated through a potential V= eV

$$\Rightarrow \frac{1}{2}mv^2 = \text{eV}$$

$$\Rightarrow v^2 = \frac{2\text{eV}}{m} \Rightarrow v = \sqrt{\frac{2\text{eV}}{m}}$$

67. (c)

68. (b) Threshold wavelength $(\lambda) = 2000$ Å $= 2000 \times 10^{-10}$ m.

$$\text{Work function } (W) = \frac{hc}{\lambda}$$

$$= \frac{(6.6 \times 10^{34}) \times (3 \times 10^8)}{2000 \times 10^{-10}}$$

$$= 9.9 \times 10^{-19}\ \text{J} = \frac{9.9 \times 10^{-19}}{1.6 \times 10^{-19}} = 6.2\text{eV}$$

69. (a) Potential difference (V) = 100 V. The kinetic energy of an electron = 1 eV = $1 \times (1.6 \times 10^{-19})$ = 1.6×10^{-19} J. Therefore kinetic energy in 100 volts = $(1.6 \times 10^{-19}) \times 100 = 1.6 \times 10^{-17}$J.

NOTES

K.E. = qV

$= 1.6 \times 10^{-19} \times 100$ J $= 1.6 \times 10^{-17}$

70. (a) $h\nu = W_0 + E_k = 3.5 + 1.2 = 4.7$ eV

71. (d) $\frac{1}{2}mv^2 = \text{eV}$, or, $v = (2\text{eV}/m)^{1/2}$

The mass of helium ion is four times that of hydrogen ion. Further, the charge on helium ion is twice that of hydrogen ion.

$$\therefore\ v_{He}/v_H = \sqrt{2}$$

72. **(b)** $W_0 = \frac{hc}{\lambda_0}$ or $W_0 \propto \frac{1}{\lambda_0}$;

$$\Rightarrow \frac{W_1}{W_2} = \frac{\lambda_2}{\lambda_1} = \frac{600}{300} = 2$$

73. **(a)** Photoelectric current is directly proportional to the intensity of incident light.

NOTES: Kinetic energy of emitted photoelectrons varies with frequency of incident radiation and it does not depend upon the intensity of incident radiations. But, the number of photons emitted from the surface varies directly with intensity of incident light.

74. **(a)** According to de Brogie wave equation,

$$\lambda = \frac{h}{mv} = \frac{h}{p} \qquad \therefore\ p = \frac{h}{\lambda}$$

75. **(a)** The work function has no effect on photoelectric current so long as $h\nu > W_0$. The photoelectric current is proportional to the intensity of incident light. Since there is no change in the intensity of light, hence $I_1 = I_2$.

76. **(b)** Here, $\frac{hc}{\lambda} = 10^3$ eV and $h\nu = 10^6$ eV

Hence, $\nu = \frac{10^3 c}{\lambda} = \frac{10^3 \times 3 \times 10^8}{1.24 \times 10^{-9}}$

$= 2.4 \times 10^{20}$ Hz

77. **(d)** $h\nu = W + \frac{1}{2}mv^2$ or

$$\frac{hc}{\lambda} = W + \frac{1}{2}mv^2$$

Here $\lambda = 3000$ Å $= 3000 \times 10^{-10}$ m and W = 1 eV $= 1.6 \times 10^{-19}$ joule

$$\therefore \frac{(6.6\times10^{-34})(3\times10^8)}{3000\times10^{-10}}$$

$$= (1.6\times10^{-19}) + \frac{1}{2}\times(9.1\times10^{-31})v^2$$

Solving we get, $v \cong 10^6$ m/s

78. **(b)** $(E_2 - E_1) = h\nu = \frac{hc}{\lambda}$

$$\therefore \frac{hc}{\lambda_1} = (E_C - E_B),\ \frac{hc}{\lambda_2} = (E_B - E_A)$$

and $\frac{hc}{\lambda_3} = (E_C - E_A)$

Now,

$$(E_C - E_A) = (E_C - E_B) + (E_B - E_A)$$

or, $\frac{hc}{\lambda_3} = \frac{hc}{\lambda_1} + \frac{hc}{\lambda_2}$ or $\frac{1}{\lambda_3} = \frac{1}{\lambda_1} + \frac{1}{\lambda_2}$

$$\therefore \frac{1}{\lambda_3} = \frac{\lambda_1 + \lambda_2}{\lambda_1\lambda_2} \text{ or } \lambda_3 = \frac{\lambda_1\lambda_2}{\lambda_1 + \lambda_2}$$

79. **(a)** No. of photons emitted per sec,

$$n = \frac{\text{Power}}{\text{Energy of photon}}$$

$$= \frac{P}{h\nu} = \frac{10000}{6.6\times10^{-34}\times880\times10^3}$$

$= 1.72 \times 10^{31}$

80. **(a)** As $\lambda = \frac{h}{p}$ and $\lambda = \frac{c}{\nu}$; so

$$\nu = \frac{c}{\lambda} = \frac{cP}{h} = 3\times10^8 \times \frac{3.3\times10^{-29}}{6.6\times10^{-34}}$$

$= 1.5 \times 10^{13}$ Hz

81. **(b)** K.E. of fastest electron

$= E - W_0 = 6.2 - 4.2 = 2.0$ eV

$= 2 \times 1.6 \times 10^{-19} = 3.2 \times 10^{-19}$ J

82. **(b)** Energy of a photon $E = h\nu = \frac{hc}{\lambda}$

83. **(a)** $W_0 = \frac{hc}{\lambda_0}$

$$= \frac{6.63\times10^{-34}\times3\times10^8}{5000\times10^{-10}} = 4\times10^{-19}\text{ J}$$

26 Atoms

Trend Analysis with Important Topics & Sub-Topics

Topic	Sub-Topic	2020		2019		2018		2017		2016	
		QNS.	LOD	QNS.	LOD	QNS.	LOD	QNS.	LOD	QNS.	LOD
Atomic Structure, Rutherford's Nuclear Model of Atom	Distance of closest approach									1	A
Bohr Model & The Spectra of the Hydrogen Atom	Bohr's model, Balmer and Lyman Series	1	E					1	A		
	Kinetic energy and potential energy of electron in hydrogen atom			1	E	1	E				

LOD - Level of Difficulty | E - Easy | A - Average | D - Difficult | Qns - No. of Questions

Topic 1 Atomic Structure, Rutherford's Nuclear Model of Atom

1. When an α-particle of mass 'm' moving with velocity 'v' bombards on a heavy nucleus of charge 'Ze', its distance of closest approach from the nucleus depends on m as : *[2016]*

   (a) $\frac{1}{m}$ (b) $\frac{1}{\sqrt{m}}$

   (c) $\frac{1}{m^2}$ (d) m

2. An alpha nucleus of energy $\frac{1}{2}mv^2$ bombards a heavy nuclear target of charge Ze. Then the distance of closest approach for the alpha nucleus will be proportional to *[2010]*

   (a) $\frac{1}{Ze}$ (b) v^2

   (c) $\frac{1}{m}$ (d) $\frac{1}{v^4}$

3. In a Rutherford scattering experiment when a projectile of charge Z_1 and mass M_1 approaches a target nucleus of charge Z_2 and mass M_2, the distance of closest approach is r_0. The energy of the projectile is *[2009]*

   (a) directly proportional to $Z_1 Z_2$
   (b) inversely proportional to Z_1
   (c) directly proportional to mass M_1
   (d) directly proportional to $M_1 \times M_2$

4. J.J. Thomson's experiment demonstrated that *[2003]*

   (a) the e/m ratio of the cathode-ray particles changes when a different gas is placed in the discharge tube
   (b) cathode rays are streams of negatively charged ions
   (c) all the mass of an atom is essentially in the nucleus
   (d) the e/m of electrons is much greater than the e/m of protons

5. Who indirectly determined the mass of the electron by measuring the charge of the electron? *[2000]*

   (a) Millikan (b) Rutherford
   (c) Einstein (d) Thomson

6. In Rutherford scattering experiment, what will be the correct angle for α-scattering for an impact parameter, $b = 0$? [1994]
(a) 90° (b) 270°
(c) 0° (d) 180°

7. What is the radius of iodine atom (At. no. 53, mass no. 126) [1988]
(a) 2.5×10^{-11} m (b) 2.5×10^{-9} m
(c) 7×10^{-9} m (d) 7×10^{-6} m

Topic 2 Bohr Model & The Spectra of the Hydrogen Atom

8. For which one of the following, Bohr model is **not** valid? [2020]
(a) Singly ionised helium atom (He^+)
(b) Deuteron atom
(c) Singly ionised neon atom (Ne^+)
(d) Hydrogen atom

9. The total energy of an electron in an atom in an orbit is –3.4 eV. Its kinetic and potential energies are, respectively: [2019]
(a) –3.4 eV, –3.4 eV (b) –3.4 eV, –6.8 eV
(c) 3.4 eV, –6.8 eV (d) 3.4 eV, 3.4 eV

10. The radius of the first permitted Bohr orbit, for the electron, in a hydrogen atom equals 0.51 Å and its ground state energy equals –13.6 eV. If the electron in the hydrogen atom is replaced by muon (μ^-) [charge same as electron and mass 207 m_e], the first Bohr radius and ground state energy will be, [NEET Odisha 2019]
(a) 2.56×10^{-13} m, –13.6 eV
(b) 0.53×10^{-13} m, –3.6 eV
(c) 25.6×10^{-13} m, –2.8 eV
(d) 2.56×10^{-13} m, –2.8 eV

11. The ratio of kinetic energy to the total energy of an electron in a Bohr orbit of the hydrogen atom, is [2018]
(a) 1 : 1 (b) 1 : –1
(c) 1 : –2 (d) 2 : –1

12. The ratio of wavelengths of the last line of Balmer series and the last line of Lyman series is :- [2017]
(a) 1 (b) 4
(c) 0.5 (d) 2

13. Given the value of Rydberg constant is $10^7 m^{-1}$, the wave number of the last line of the Balmer series in hydrogen spectrum will be : [2016]
(a) $0.025 \times 10^4 m^{-1}$ (b) $0.5 \times 10^7 m^{-1}$
(c) $0.25 \times 10^7 m^{-1}$ (d) $2.5 \times 10^7 m^{-1}$

14. Consider 3rd orbit of He^+ (Helium), using non-relativistic approach, the speed of electron in this orbit will be [given K = 9×10^9 constant, Z = 2 and h (Plank's Constant) = 6.6×10^{-34} J s] [2015]
(a) 1.46×10^6 m/s (b) 0.73×10^6 m/s
(c) 3.0×10^8 m/s (d) 2.92×10^6 m/s

15. Two particles of masses m_1, m_2 move with initial velocities u_1 and u_2. On collision, one of the particles get excited to higher level, after absorbing energy ε. If final velocities of particles be v_1 and v_2 then we must have [2015]
(a) $\frac{1}{2}m_1u_1^2 + \frac{1}{2}m_2u_2^2 = \frac{1}{2}m_1v_1^2 + \frac{1}{2}m_2v_2^2 - \varepsilon$
(b) $\frac{1}{2}m_1u_1^2 + \frac{1}{2}m_2u_2^2 - \varepsilon = \frac{1}{2}m_1v_1^2 + \frac{1}{2}m_2v_2^2$
(c) $\frac{1}{2}m_1^2u_1^2 + \frac{1}{2}m_2^2u_2^2 + \varepsilon = \frac{1}{2}m_1^2v_1^2 + \frac{1}{2}m_2^2v_2^2$
(d) $m_1^2u_1 + m_2^2u_2 - \varepsilon = m_1^2v_1 + m_2^2v_2$

16. In the spectrum of hydrogen, the ratio of the longest wavelength in the Lyman series to the longest wavelength in the Balmer series is [2015 RS, 2013]
(a) $\frac{9}{4}$ (b) $\frac{27}{5}$
(c) $\frac{5}{27}$ (d) $\frac{4}{9}$

17. Hydrogen atom in ground state is excited by a monochromatic radiation of λ = 975 Å. Number of spectral lines in the resulting spectrum emitted will be [2014]
(a) 3 (b) 2
(c) 6 (d) 10

18. An electron in hydrogen atom makes a transition $n_1 \to n_2$ where n_1 and n_2 are principal quantum numbers of the two states. Assuming Bohr's model to be valid the time period of the electron in the initial state is eight times that in the final state. The possible values of n_1 and n_2 are [NEET Kar. 2013]
(a) $n_1 = 4$ and $n_2 = 2$
(b) $n_1 = 6$ and $n_2 = 2$
(c) $n_1 = 8$ and $n_2 = 1$
(d) $n_1 = 8$ and $n_2 = 2$

19. Electron in hydrogen atom first jumps from third excited state to second excited state and then from second excited to the first excited state. The ratio of the wavelength $\lambda_1 : \lambda_2$ emitted in the two cases is *[2012]*
(a) 7/5 (b) 27/20
(c) 27/5 (d) 20/7

20. An electron of a stationary hydrogen atom passes from the fifth energy level to the ground level. The velocity that the atom acquired as a result of photon emission will be : *[2012]*
(a) $\frac{24hR}{25m}$ (b) $\frac{25hR}{24m}$
(c) $\frac{25m}{24hR}$ (d) $\frac{24m}{25hR}$
(*m* is the mass of the electron, *R*, Rydberg constant and *h* Planck's constant)

21. The transition from the state $n = 3$ to $n = 1$ in a hydrogen like atom results in ultraviolet radiation. Infrared radiation will be obtained in the transition from : *[2012M]*
(a) $2 \rightarrow 1$ (b) $3 \rightarrow 2$
(c) $4 \rightarrow 2$ (d) $4 \rightarrow 3$

22. The wavelength of the first line of Lyman series for hydrogen atom is equal to that of the second line of Balmer series for a hydrogen like ion. The atomic number Z of hydrogen like ion is *[2011]*
(a) 3 (b) 4
(c) 1 (d) 2

23. An electron in the hydrogen atom jumps from excited state n to the ground state. The wavelength so emitted illuminates a photosensitive material having work function 2.75 eV. If the stopping potential of the photoelectron is 10 V, the value of n is *[2011M]*
(a) 3 (b) 4
(c) 5 (d) 2

24. Out of the following which one is not a possible energy for a photon to be emitted by hydrogen atom according to Bohr's atomic model? *[2011M]*
(a) 1.9 eV (b) 11.1 eV
(c) 13.6 eV (d) 0.65 eV

25. The energy of a hydrogen atom in the ground state is – 13.6 *eV*. The energy of a He^+ ion in the first excited state will be *[2010]*
(a) –13.6 *eV* (b) –27.2 *eV*
(c) –54.4 *eV* (d) –6.8 *eV*

26. The ionization energy of the electron in the hydrogen atom in its ground state is 13.6 eV. The atoms are excited to higher energy levels to emit radiations of 6 wavelengths. Maximum wavelength of emitted radiation corresponds to the transition between *[2009]*
(a) n = 3 to n = 1 states
(b) n = 2 to n = 1 states
(c) n = 4 to n = 3 states
(d) n = 3 to n = 2 states

27. The ground state energy of hydrogen atom is 13.6eV. When its electron is in the first excited state, its excitation energy is *[2008]*
(a) 3.4 eV (b) 6.8 eV
(c) 10.2 eV (d) 0

28. The total energy of electron in the ground state of hydrogen atom is – 13.6 eV. The kinetic energy of an electron in the first excited state is *[2007]*
(a) 6.8 eV (b) 13.6 eV
(c) 1.7 eV (d) 3.4 eV.

29. Ionization potential of hydrogen atom is 13.6eV. Hydrogen atoms in the ground state are excited by monochromatic radiation of photon energy 12.1 eV. According to Bohr's theory, the spectral lines emitted by hydrogen will be *[2006]*
(a) three (b) Four
(c) One (d) Two

30. The total energy of an electron in the first excited state of hydrogen atom is about –3.4 eV. Its kinetic energy in this state is *[2005]*
(a) 3.4 eV (b) 6.8 eV
(c) –3.4 eV (d) –6.8 eV

31. The Bohr model of atoms *[2004]*
(a) predicts the same emission spectra for all types of atoms
(b) assumes that the angular momentum of electrons is quantised
(c) uses Einstein's photoelectric equation
(d) predicts continuous emission spectra for atoms

32. Energy *E* of a hydrogen atom with principal quantum number n is given by $E = -13.6/n^2$ eV. The energy of photon ejected when the electron jumps from $n = 3$ state to $n = 2$ state of hydrogen is approximately *[2004]*
(a) 1.9 eV. (b) 1.5 eV.
(c) 0.85 eV (d) 3.4 eV

33. In which of the following systems will the radius of the first orbit ($n = 1$) be minimum ? [2003]
(a) Hydrogen atom
(b) Doubly ionized lithium
(c) Singly ionized helium
(d) Deuterium atom

34. An electron changes its position from orbit $n = 2$ to the orbit $n = 4$ of an atom. The wavelength of the emitted radiations is (R = Rydberg's constant) [2001]
(a) $\frac{16}{R}$ (b) $\frac{16}{3R}$
(c) $\frac{16}{5R}$ (d) $\frac{16}{7R}$

35. The energy of hydrogen atom in nth orbit is E_n, then the energy in nth orbit of single ionised helium atom will be [2001]
(a) $4E_n$ (b) $E_n/4$
(c) $2E_n$ (d) $E_n/2$

36. When an electron jumps from the fourth orbit to the second orbit, one gets the [2000]
(a) second line of Lyman series
(b) second line of Paschen series
(c) second line of Balmer series
(d) first line of Pfund series

37. Which of the following transitions in a hydrogen atom emits the photon of highest frequency? [2000]
(a) $n = 2$ to $n = 1$ (b) $n = 2$ to $n = 6$
(c) $n = 6$ to $n = 2$ (d) $n = 1$ to $n = 2$

38. In the Bohr model of a hydrogen atom, the centripetal force is furnished by the coulomb attraction between the proton and the electron. If a_0 is the radius of the ground state orbit, m is the mass, e is the charge on the electron and ε_0 is the vacuum permittivity, the speed of the electron is [1998]
(a) 0 (b) $\frac{e}{\sqrt{\varepsilon_0 a_0 m}}$
(c) $\frac{e}{\sqrt{4\pi\varepsilon_0 a_0 m}}$ (d) $\frac{\sqrt{4\pi\varepsilon_0 a_0 m}}{e}$

39. When hydrogen atom is in its first excited level, its radius is [1997]
(a) four times its ground state radius
(b) twice
(c) same
(d) half

40. The spectrum obtained from a sodium vapour lamp is an example of [1995]
(a) band spectrum
(b) continuous spectrum
(c) emission spectrum
(d) absorption spectrum

41. When a hydrogen atom is raised from the ground state to an excited state, [1995]
(a) P.E decreases and K.E. increases
(b) P.E. increases and K.E decreases
(c) both K.E. and P.E. decrease
(d) absorption spectrum

42. The radius of hydrogen atom in its ground state is 5.3×10^{-11} m. After collision with an electron it is found to have a radius of 21.2×10^{-11} m. What is the principal quantum number n of the final state of the atom [1994]
(a) $n = 4$ (b) $n = 2$
(c) $n = 16$ (d) $n = 3$

43. Which source is associated with a line emission spectrum ? [1993]
(a) Electric fire (b) Neon street sign
(c) Red traffic light (d) Sun

44. In terms of Bohr radius a_0, the radius of the second Bohr orbit of a hydrogen atom is given by [1992]
(a) $4\,a_0$ (b) $8\,a_0$
(c) $\sqrt{2}\,a_0$ (d) $2\,a_0$

45. The ionization energy of hydrogen atom is 13.6 eV. Following Bohr's theory, the energy corresponding to a transition between 3rd and 4th orbit is [1992]
(a) 3.40 eV (b) 1.51 eV
(c) 0.85 eV (d) 0.66 eV

46. The ground state energy of H-atom 13.6 eV. The energy needed to ionize H-atom from its second excited state. *[1991]*

(a) 1.51 eV (b) 3.4 eV
(c) 13.6 eV (d) 12.1 eV

47. To explain his theory, Bohr used *[1989]*

(a) conservation of linear momentum
(b) conservation of angular momentum
(c) conservation of quantum frequency
(d) conservation of energy

48. The ionisation energy of hydrogen atom is 13.6 eV, the ionisation energy of helium atom would be *[1988]*

(a) 13.6 eV (b) 27.2 eV
(c) 6.8 eV (d) 54.4 eV

ANSWER KEY

1	(a)	7	(a)	13	(c)	19	(d)	25	(a)	31	(b)	37	(a)	43	(b)
2	(c)	8	(c)	14	(a)	20	(a)	26	(c)	32	(a)	38	(c)	44	(a)
3	(a)	9	(c)	15	(b)	21	(d)	27	(c)	33	(b)	39	(a)	45	(d)
4	(b)	10	(d)	16	(c)	22	(d)	28	(d)	34	(b)	40	(c)	46	(a)
5	(d)	11	(b)	17	(c)	23	(b)	29	(a)	35	(a)	41	(d)	47	(b)
6	(d)	12	(b)	18	(a)	24	(b)	30	(a)	36	(c)	42	(b)	48	(d)

Hints & Solutions

1. **(a)** At closest distance of approach, the kinetic energy of the particle will convert completely into electrostatic potential energy.

Kinetic energy K.E. $= \frac{1}{2}mv^2$

Potential energy P.E. $= \frac{KQq}{r}$

$$\frac{1}{2}mv^2 = \frac{KQq}{r} \Rightarrow r \propto \frac{1}{m}$$

2. **(c)** Kinetic energy of alpha nucleus is equall to electrostatic potential energy of the system of the alpha particle and the heavy nucleus. That is,

$$\frac{1}{2}mv^2 = \frac{1}{4\pi\varepsilon_0}\frac{q_\alpha Ze}{r_0}$$

where r_0 is the distance of closest approach

$$r_0 = \frac{2}{4\pi\varepsilon_0}\frac{q_\alpha Ze}{mv^2}$$

$$\Rightarrow r_0 \propto Ze \propto q_\alpha \propto \frac{1}{m} \propto \frac{1}{v^2}$$

Hence, correct option is (c).

3. **(a)** The kinetic energy of the projectile is given by

$$\frac{1}{2}mv^2 = \frac{Ze(2e)}{4\pi\varepsilon_0 r_0}$$

$$= \frac{Z_1 Z_2}{4\pi\varepsilon_0 r_0}$$

Thus energy of the projectile is directly proportional to $Z_1 Z_2$.

4. **(b)** Cathode rays are streams of negatively charged ions

5. **(d)**

6. **(d)** Impact parameter for Rutherford scattering experiment,

$$b = \frac{Ze^2 \cot\left(\frac{\theta}{2}\right)}{4\pi\,\epsilon_0\,k_i} = 0 \Rightarrow \cot\left(\frac{\theta}{2}\right) = 0$$

$$\Rightarrow \frac{\theta}{2} = 90° \text{ or } \theta = 180°$$

NOTES For large value of b, alpha particle will go undeviated and for small b the alpha particle will suffer large scattering.

7. **(a)** 53 electrons in iodine atom are distributed as 2, 8, 18, 18, 7

$\therefore n = 5$

$$r_n = (0.53\times10^{-10})\frac{n^2}{Z}$$

$$= \frac{0.53\times10^{-10}\times5^2}{53} = 2.5\times10^{-11}\text{m}$$

8. **(c)** Bohr model is only valid for single electron species i.e., Hydrogen or hydrogen like atom – He^+, deuteron, etc. Singly ionised neon atom has more than one electron in orbit. Hence, Bohr model is not valid.

9. **(c)** According to Bohr's model of H-atom, the relation between kinetic energy, potential energy and total energy

$$\text{K.E.} = |\text{TE}| = \frac{|\text{P.E.}|}{2}$$

so, K.E. = – (T.E.)

$\therefore$ K.E. = 3.4 eV

and P.E. = 2 (T.E)

$\Rightarrow$ P.E. = – 6.8 eV

10. **(d)** $r \propto \frac{1}{m}$

$$R = \frac{0.51}{207} \quad \therefore\ 2.56\times10^{-13}\text{ m}$$

$E \propto m$

$\therefore\ (E)_\mu = -13.6\times207 = -2.8\text{ keV}$

11. **(b)** In a Bohr orbit of the hydrogen atom

Kinetic energy,

$$k = \frac{kze^2}{2r_n}$$

Total energy, $E = \frac{-kze^2}{2r_n}$

So, Kinetic energy : total energy = 1 : –1

12. **(b)** For last line of Balmer series :

$n_1 = 2$ and $n_2 = \infty$

$$\frac{1}{\lambda_B} = RZ^2\left[\frac{1}{n_1^2}-\frac{1}{n_2^2}\right] = R_1^2\left[\frac{1}{2^2}-\frac{1}{\infty^2}\right]$$

$$\lambda_B = \frac{4}{R} \quad \text{...(i)}$$

For last line of Lyman series : $n_1 = 1$ and $n_2 = \infty$

$$\frac{1}{\lambda_L} = RZ^2\left[\frac{1}{n_1^2}-\frac{1}{n_2^2}\right] = RZ^2\left[\frac{1}{1^2}-\frac{1}{\infty^2}\right]$$

$$\lambda_L = \frac{1}{R} \quad \text{...(ii)}$$

Dividing equation (i) by (ii)

$$\frac{\lambda_B}{\lambda_L} = \frac{\frac{4}{R}}{\frac{1}{R}}$$

Ratio of wavelengths is $\frac{\lambda_B}{\lambda_L} = 4$

13. **(c)** According to Bohr's theory, the wave number of the last line of the Balmer series in hydrogen spectrum,

For hydrogen atom Z = 1

$$\frac{1}{\lambda} = RZ^2\left(\frac{1}{n_2^2}-\frac{1}{n_1^2}\right)$$

$$= 10^7\times1^2\left(\frac{1}{2^2}-\frac{1}{\infty^2}\right)$$

$\Rightarrow$ wave number $\frac{1}{\lambda} = 0.25\times10^7\text{ m}^{-1}$

> **NOTES**
> Last line of the series is called series limit. For this line wavelength is minimum (λ_{min}). For minimum wavelength, $n_2 = \infty$, $n_1 = n$
>
> So, wavelength $\lambda_{min} = \frac{n^2}{R}$

14. **(a)** Speed of electron in nth orbit

$$v_n = \frac{2\pi\,KZe^2}{nh}$$

$$v = (2.19\times10^6\text{ m/s})\,\frac{Z}{n}$$

$$v = (2.19\times10^6)\,\frac{2}{3} \quad (Z = 2 \text{ and } n = 3)$$

$$v = 1.46\times10^6\text{ m/s}$$

15. **(b)** By law of conservation of energy,

$K.E_f = K.E_i$ – excitation energy (ε)

or $\frac{1}{2}mv_1^2 + \frac{1}{2}mv_2^2 = \frac{1}{2}m_1u_1^2 + \frac{1}{2}m_2u_2^2 - \varepsilon$

16. **(c)** For Lyman series ($2 \rightarrow 1$)

$$\frac{1}{\lambda_L} = R\left[1-\frac{1}{2^2}\right] = \frac{3R}{4}$$

For Balmer series $(3 \rightarrow 2)$

$$\frac{1}{\lambda_B} = R\left[\frac{1}{4} - \frac{1}{9}\right] = \frac{5R}{36}$$

$$\Rightarrow \frac{\lambda_L}{\lambda_B} = \frac{\frac{4}{3R}}{\frac{36}{5R}} = \frac{4}{36}\left(\frac{5}{3}\right) = \frac{5}{27}$$

NOTES

The wavelength of spectral lines increases with the increases of order of the series.

$\lambda_{PFund} > \lambda_{Brackett} > \lambda_{Paschen} > \lambda_{Balmer} > \lambda_{Lymen}$

17. **(c)** For the $\lambda = 975$ Å

$$\frac{1}{\lambda} = R\left(\frac{1}{n_1^2} - \frac{1}{n_2^2}\right)$$

where R is the Rydberg constant

Solving we get $n_2 = n = 4$

($\because n_1 = 1$ ground state)

Therefore number of spectral lines

$$= \frac{n(n-1)}{2} = \frac{4(4-1)}{2} = 6$$

18. **(a)** $\because T \propto n^3$

$$\frac{T_1}{T_2} = \frac{8T_2}{T_2} = \left(\frac{n_1}{n_2}\right)^3$$

Hence, $n_1 = 2n_2$

19. **(d)**

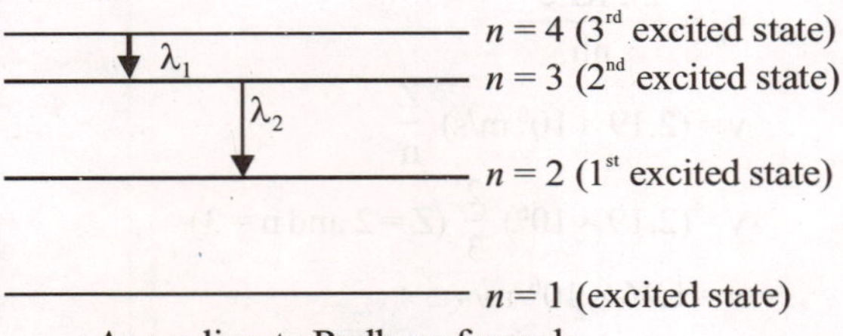

According to Rydberg formula

$$\frac{1}{\lambda} = R\left[\frac{1}{n_f^2} - \frac{1}{n_i^2}\right]$$

In first case, $n_f = 3, n_i = 4$

$$\therefore \frac{1}{\lambda_1} = R\left[\frac{1}{3^2} - \frac{1}{4^2}\right] = R\left[\frac{1}{9} - \frac{1}{16}\right] = \frac{7}{144}R \quad ...(i)$$

In second case, $n_f = 2, n_i = 3$

$$\therefore \frac{1}{\lambda_2} = R\left[\frac{1}{2^2} - \frac{1}{3^2}\right] = R\left[\frac{1}{4} - \frac{1}{9}\right] = \frac{5}{36}R \quad ...(ii)$$

Divide (ii) by (i), we get

$$\frac{\lambda_1}{\lambda_2} = \frac{5}{36} \times \frac{144}{7} = \frac{20}{7}$$

20. **(a)** For emission, the wave number of the radiation is given as

$$\frac{1}{\lambda} = Rz^2\left(\frac{1}{n_1^2} - \frac{1}{n_2^2}\right)$$

$R =$ Rydberg constant, $Z =$ atomic number

$$= R\left(\frac{1}{1^2} - \frac{1}{5^2}\right) = R\left(1 - \frac{1}{25}\right) \Rightarrow \frac{1}{\lambda} = R\frac{24}{25}$$

linear momentum

$$p = \frac{h}{\lambda} = h \times R \times \frac{24}{25} \text{ (de-Broglie hypothesis)}$$

$$\Rightarrow mv = \frac{24hR}{25} \Rightarrow v = \frac{24hR}{25m}$$

21. **(d)** $\because$ The frequency of the transition $v \propto \frac{1}{n^2}$, when $n = 1, 2, 3$.

22. **(d)** For first line of Lyman series of hydrogen

$$\frac{hc}{\lambda_1} = Rhc\left(\frac{1}{1^2} - \frac{1}{2^2}\right)$$

For second line of Balmer series of hydrogen like ion

$$\frac{hc}{\lambda_2} = Z^2 Rhc\left(\frac{1}{2^2} - \frac{1}{4^2}\right)$$

By question, $\lambda_1 = \lambda_2$

$$\Rightarrow \left(\frac{1}{1} - \frac{1}{2}\right) = Z^2\left(\frac{1}{4} - \frac{1}{16}\right) \text{ or } Z = 2$$

23. **(b)** $KE_{max} = 10eV$

$\phi = 2.75$ eV

Total incident energy

$E = \phi + KE_{max} = 12.75$ eV

$\therefore$ Energy is released when electron jumps from the excited state n to the ground state.

$\because E_4 - E_1 = \{-0.85 - (-13.6) \text{ ev}\}$

$= 12.75$eV

$\therefore$ value of n = 4

24. **(b)** Obviously, difference of 11.1eV is not possible.

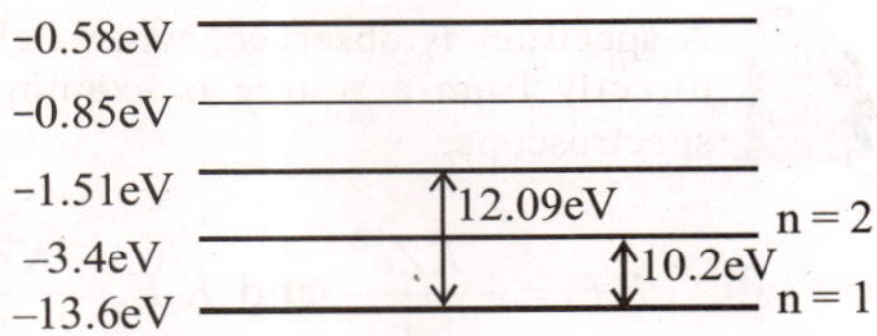

25. **(a)** Energy of a H-like atom in it's n^{th} state is given by

$$E_n = -Z^2 \times \frac{13.6}{n^2} eV$$

For, first excited state of He^+, n = 2, Z = 2

$$\therefore \quad E_{He^+} = -\frac{4}{2^2} \times 13.6 = -13.6 eV$$

26. **(c)** $\frac{n(n-1)}{2} = 6$

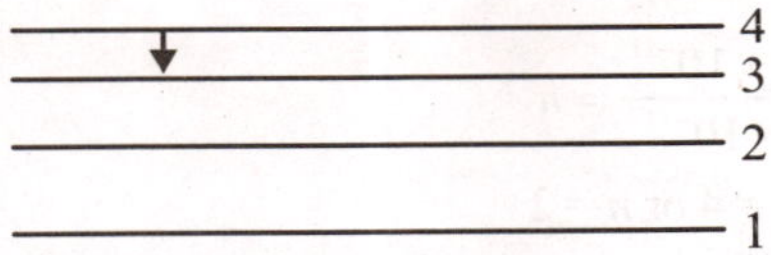

$n^2 - n - 12 = 0$

$(n-4)(n+3) = 0$ or $n = 4$

NOTES

If electron falls from orbit n_2 to n_1 then the number of spectral lines emitted is given by

$$N_E = \frac{(n_2 - n_1 + 1)(n_2 - n_1)}{2}$$

If an electron falls from nth orbit to ground state ($n = 1$) then number of spectral lines emitted, $N_E = \frac{n(n-1)}{2}$

27. **(c)** When the electron is in first excited state (n = 2), the excitation energy is given by

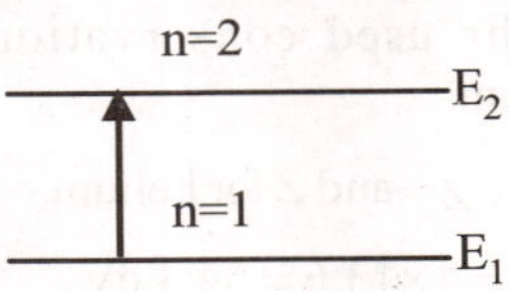

$\Delta E = E_2 - E_1$

We have, $E_n = -\frac{13.6}{n^2} eV$

$\therefore \; E_2 = -\frac{13.6}{2^2} eV = -3.4eV$

Given $E_1 = -13.6eV$

$\therefore \Delta E = (-3.4) - (-13.6) = 10.2$ eV.

28. **(d)** Energy in the first excited state

$$= \frac{-13.6}{n^2} = \frac{-13.6}{2^2} = -3.4 \text{ eV}$$

But K.E. = –(Total energy) = +3.4 eV.

29. **(a)** Energy of ground state 13.6 eV

Energy of first excited state

$$= -\frac{13.6}{4} = -3.4 \text{ eV}$$

Energy of second excited state

$$= -\frac{13.6}{9} = -1.5 \text{ eV}$$

Difference between ground state and 2nd excited state = 13.6 – 1.5 = 12.1 eV

So, electron can be excited upto 3rd orbit

No. of possible transition

$1 \to 2, 1 \to 3, 2 \to 3$

So, three lines are possible.

30. **(a)** $K.E. = \frac{Z^2}{n^2}(13.6 \text{ eV})$

Mechanical energy $= \frac{-Z^2}{n^2}(13.6 \text{ eV})$

$\therefore$ K.E. in 2nd orbital for hydrogen

= – Mechanical energy

$$= \frac{(1)^2}{(2)^2}(13.6) = +3.4 eV$$

31. **(b)** In Bohr's model, angular momentum is quantised i.e $\ell = n\left(\frac{h}{2\pi}\right)$

32. **(a)** $\Delta E = E_3 - E_2$

$$= \frac{-13.6}{3^2} + \frac{13.6}{2^2}$$

$$= 13.6\left\{\frac{1}{4} - \frac{1}{9}\right\} \text{eV} = 1.9 \text{ eV}$$

33. **(b)** $r \propto \frac{1}{Z}$; $Z(=3)$ is maximum for Li^{2+}.

34. **(b)** $\bar{\nu} = R\left(\frac{1}{n_1^2} - \frac{1}{n_2^2}\right)$, where $n_1 = 2, n_2 = 4$

$$\bar{\nu} = R\left(\frac{1}{4} - \frac{1}{16}\right)$$

$$\frac{1}{\lambda} = R\left(\frac{12}{4 \times 16}\right) \Rightarrow \lambda = \frac{16}{3R}$$

35. (a) We have $E_n = \frac{-2\pi^2 mK^2 Z^2 e^4}{n^2 h^2}$. For helium $Z = 2$. Hence requisite answer is $4E_n$

NOTES: The energy required to ionise an atom is called ionisation energy. It is energy required to make the electron Jump from the present orbit to the infinite orbit.

$$E_{ionisation} = E_\infty - E_n = 0 - \left(-13.6\frac{Z^2}{n^2}\right)$$

$$= +\frac{13.6z^2\,\text{eV}}{n^2}$$

36. (c) When the electron drops from any orbit to second orbit, then wavelength of line obtained belongs to Balmer series.

37. (a) $\bar{\nu} = R\left[\frac{1}{n_1^2} - \frac{1}{n_2^2}\right]$ or $\frac{1}{\lambda} = R\left[\frac{1}{n_1^2} - \frac{1}{n_2^2}\right]$

Frequency, $\nu = Rc\left[\frac{1}{n_1^2} - \frac{1}{n_2^2}\right]$

Note : See the greatest energy difference and also see that the transition is from higher to lower energy level. Hence, it is highest in case of $n = 2$ to $n = 1$.

38. (c) Centripetal force = Coulombian force

$$\frac{mv^2}{a_0} = \frac{1}{4\pi\varepsilon_0} \cdot \frac{e \times e}{a_0^2}$$

$$\Rightarrow v^2 = \frac{e^2}{4\pi\varepsilon_0 . a_0 . m} \Rightarrow v = \frac{e}{\sqrt{4\pi\varepsilon_0 . a_0 . m}}$$

39. (a) $R = \frac{R_0 n^2}{Z}$

Radius in ground state $= \frac{R_0}{Z}$

Radius in first excited state $= \frac{R_0 \times 4}{Z}$ $(\because n = 2)$

Hence, radius of first excited state is four times the radius in ground state.

40. (c) Spectrum obtained from a sodium vapour lamp is emission spectrum.

NOTES: A spectrum is observed, when light coming directly from a source is examined with a spectroscope.

41. (b) $P.E. = -\frac{KZe^2}{r}$ and $K.E. = \frac{KZe^2}{2r}$,

where, r is the radius of orbit which increases as we move from ground to an excited state. Therefore, when a hydrogen atom is raised from the ground state, it increases the value of r. As a result of this, P.E. increases (decreases in negative) and K.E. decreases.

42. (b) $r \propto n^2$

$$\therefore \frac{\text{radius of final state}}{\text{radius of initial state}} = n^2$$

$$\frac{21.2 \times 10^{-11}}{5.3 \times 10^{-11}} = n^2$$

$\therefore n^2 = 4$ or $n = 2$

43. (b) Neon street sign is a source of line emission spectrum.

44. (a) As $r \propto n^2$, therefore, radius of 2nd Bohr's orbit $= 4\,a_0$.

45. (d) $E = E_4 - E_3$

$$= -\frac{13.6}{4^2} - \left(-\frac{13.6}{3^2}\right) = -0.85 + 1.51$$

$= 0.66$ eV

46. (a) Second excited state corresponds to $n = 3$

$\therefore \quad E = \frac{13.6}{3^2}\text{eV} = 1.51\text{eV}$

47. (b) Bohr used conservation of angular momentum.

48. (d) $E \propto Z^2$ and Z for helium = 2

$\therefore (E)_{He} = 4 \times 13.6 = 54.4$ eV

27 Nuclei

Trend Analysis with Important Topics & Sub-Topics

Topic Name	Sub-Topic	2021		2020		2019		2018		2017		2016	
		QNS.	LOD	QNS.	LOD	QNS.	LOD	QNS.	LOD	QNS.	LOD	QNS.	LOD
Composition and Size of the Nucleus													
Mass-Energy & Nuclear Reactions	Nuclear reaction Binding energy	1	A	1	A								
	Mass – Energy			1	E								
Radioactivity	α particle composition					1	E						
	Emission of α & β particles	1	E										
	Rate of decay law & Half life period	1	E					1	A	1	D		

LOD - Level of Difficulty | E - Easy | A - Average | D - Difficult | Qns - No. of Questions

Topic 1 Composition and Size of the Nucleus

1. If radius of the $^{27}_{12}$Al nucleus is taken to be R_{Al}, then the radius of $^{125}_{53}$Te nucleus is nearly: *[2015]*

   (a) $\frac{5}{3}R_{Al}$ (b) $\frac{3}{5}R_{Al}$

   (c) $\left(\frac{13}{53}\right)^{1/3} R_{Al}$ (d) $\left(\frac{53}{13}\right)^{1/3} R_{Al}$

2. If the nuclear radius of ^{27}Al is 3.6 Fermi, the approximate nuclear radius of ^{64}Cu in Fermi is : *[2012]*

   (a) 2.4 (b) 1.2

   (c) 4.8 (d) 3.6

3. Two nuclei have their mass numbers in the ratio of 1 : 3. The ratio of their nuclear densities would be *[2008]*

   (a) 1 : 3 (b) 3 : 1

   (c) $(3)^{1/3}$: 1 (d) 1 : 1

4. If the nucleus $^{27}_{13}Al$ has nuclear radius of about 3.6 fm, then $^{125}_{32}Te$ would have its radius approximately as *[2007]*

   (a) 9.6 fm (b) 12.0 fm

   (c) 4.8 fm (d) 6.0 fm.

5. The radius of germanium (Ge) nuclide is measured to be twice the radius of $^{9}_{4}$Be. The number of nucleons in Ge are *[2006]*

   (a) 74 (b) 75

   (c) 72 (d) 73

6. The nuclei of which one of the following pairs of nuclei are isotones? *[2005]*

   (a) $_{34}Se^{74}$, $_{31}Ga^{71}$ (b) $_{38}Sr^{84}$, $_{38}Sr^{86}$

   (c) $_{42}Mo^{92}$, $_{40}Zr^{92}$ (d) $_{20}Ca^{40}$, $_{16}S^{32}$

7. A nucleus represented by the symbol $^{A}_{Z}X$ has *[2004]*

   (a) *A* protons and (*Z-A*) neutrons

   (b) *Z* neutrons and (*A-Z*) protons

   (c) *Z* protons and (*A-Z*) neutrons

   (d) *Z* protons and *A* neutrons

8. The mass number of a nucleus is *[2003]*

   (a) sometimes less than and sometimes more than its atomic number

   (b) always less than its atomic number

   (c) always more than its atomic number

   (d) sometimes equal to its atomic number

9. The volume occupied by an atom is greater than the volume of the nucleus by a factor of about *[2003]*
(a) 10^{15} (b) 10^{1}
(c) 10^{5} (d) 10^{10}

10. M_n and M_p represent mass of neutron and proton respectively. If an element having atomic mass M has N-neutron and Z-proton, then the correct relation will be *[2001]*
(a) $M < [NM_n + ZM_p]$
(b) $M > [NM_n + ZM_p]$
(c) $M = [NM_n + ZM_p]$
(d) $M = N[M_n + M_p]$

11. Atomic weight of Boron is 10.81 and it has two isotopes ${}_5B^{10}$ and ${}_5B^{11}$. Then the ratio ${}_5B^{10} : {}_5B^{11}$ in nature would be *[1998]*
(a) 19 : 81 (b) 10 : 11
(c) 15 : 16 (d) 81 : 19

12. The stable nucleus that has a radius half that of Fe^{56} is *[1997]*
(a) Li^7 (b) Na^{21}
(c) S^{16} (d) Ca^{40}

13. A nucleus ruptures into two nuclear parts, which have their velocity ratio equal to 2:1. What will be the ratio of their nuclear size (nuclear radius)? *[1996]*
(a) $2^{1/3} : 1$ (b) $1 : 2^{1/3}$
(c) $3^{1/2} : 1$ (d) $1 : 3^{1/2}$

14. The mass number of a nucleus is equal to the number of *[1995]*
(a) protons it contains
(b) nucleons it contains
(c) neutrons it contains
(d) electron it contains

15. The mass number of He is 4 and that for sulphur is 32. The radius of sulphur nuclei is larger than that of helium by *[1994]*
(a) $\sqrt{8}$ (b) 4
(c) 2 (d) 8

16. The nucleus which has radius one-third of the radius of Cs^{189} is *[1993]*
(a) ${}^{4}_{2}He$ (b) ${}^{7}_{3}Li$
(c) ${}^{141}_{56}Ba$ (d) ${}^{235}_{92}U$

17. The mass density of a nucleus varies with mass number A as *[1992]*
(a) A^2 (b) A
(c) constant (d) 1/A

18. The constituents of atomic nuclei are believed to be *[1991]*
(a) neutrons and protons
(b) protons only
(c) electrons and protons
(d) electrons, protons and neutrons.

19. In the nucleus of ${}_{11}Na^{23}$, the number of protons, neutrons and electrons are *[1991]*
(a) 11, 12, 0 (b) 23, 12, 11
(c) 12, 11, 0 (d) 23, 11, 12

20. The nuclei ${}_6C^{13}$ and ${}_7N^{14}$ can be described as *[1990]*
(a) isotones (b) isobars
(c) isotopes of carbon (d) isotopes of nitrogen

21. The ratio of the radii of the nuclei ${}_{13}Al^{27}$ and ${}_{52}Te^{125}$ is approximately *[1990]*
(a) 6 : 10 (b) 13 : 52
(c) 40 : 177 (d) 14 : 73

Topic 2 Mass-Energy & Nuclear Reactions

22. A nucleus with mass number 240 breaks into two fragments each of mass number 120, the binding energy per nucleon of unfragmented nuclei is 7.6 MeV while that of fragments is 8.5 MeV. The total gain in the binding energy in the process is *[2021]*
(a) 216 MeV (b) 0.9 MeV
(c) 9.4 MeV (d) 804 MeV

23. When a uranium isotope ${}^{235}_{92}U$ is bombarded with a neutron, it generates ${}^{89}_{36}kr$, three neutrons and : *[2020]*
(a) ${}^{91}_{40}Zr$ (b) ${}^{101}_{36}Kr$
(c) ${}^{103}_{36}Kr$ (d) ${}^{144}_{56}Ba$

24. The energy equivalent of 0.5 g of a substance is : *[2020]*
(a) 4.5×10^{13} J (b) 1.5×10^{13} J
(c) 0.5×10^{13} J (d) 4.5×10^{16} J

25. A nucleus of uranium decays at rest into nuclei of thorium and helium. Then : *[2015 RS]*
(a) the helium nucleus has less momentum than the thorium nucleus.
(b) the helium nucleus has more momentum than the thorium nucleus.
(c) the helium nucleus has less kinetic energy than the thorium nucleus.
(d) the helium nucleus has more kinetic energy than the thorium nucleus.

26. The Binding energy per nucleon of $^{7}_{3}Li$ and $^{4}_{2}He$ nuclei are 5.60 MeV and 7.06 MeV, respectively. In the nuclear reaction $^{7}_{3}Li + ^{1}_{1}H \rightarrow ^{4}_{2}He + Q$, the value of energy Q released is : [2014]
(a) 19.6 MeV (b) –2.4 MeV
(c) 8.4 MeV (d) 17.3 MeV

27. A certain mass of Hydrogen is changed to Helium by the process of fusion. The mass defect in fusion reaction is 0.02866 a.m.u. The energy liberated per a.m.u. is [2013]
(Given : 1 a.m.u = 931 MeV)
(a) 26.7 MeV (b) 6.675 MeV
(c) 13.35 MeV (d) 2.67 MeV

28. How does the binding energy per nucleon vary with the increase in the number of nucleons? [NEET Kar. 2013]
(a) Increases continuously with mass number
(b) Decreases continuously with mass number
(c) First decreases and then increases with increase in mass number
(d) First increases and then decreases with increase in mass number

29. The power obtained in a reactor using U^{235} disintegration is 1000 kW. The mass decay of U^{235} per hour is [2011]
(a) 10 microgram (b) 20 microgram
(c) 40 microgram (d) 1 microgram

30. A nucleus of mass M emits a photon of frequency ν and the nucleus recoils. The recoil energy will be [2011]
(a) $Mc^2 - h\nu$ (b) $h^2\nu^2 / 2Mc^2$
(c) zero (d) hν

31. Fusion reaction takes place at high temperature because [2011]
(a) nuclei break up at high temperature
(b) atoms get ionised at high temperature
(c) kinetic energy is high enough to overcome the coulomb repulsion between nuclei
(d) molecules break up at high temperature

32. The mass of a $^{7}_{3}Li$ nucleus is 0.042 u less than the sum of the masses of all its nucleons. The binding energy per nucleon of $^{7}_{3}Li$ nucleus is nearly [2010]
(a) 46 MeV (b) 5.6 MeV
(c) 3.9 MeV (d) 23 MeV

33. The binding energy per nucleon in deuterium and helium nuclei are 1.1 *MeV* and 7.0 *MeV*, respectively. When two deuterium nuclei fuse to form a helium nucleus the energy released in the fusion is : [2010]
(a) 30.2 *MeV* (b) 23.6 *MeV*
(c) 2.2 *MeV* (d) 28.0 *MeV*

34. If M (A; Z), M_p and M_n denote the masses of the nucleus $^{A}_{Z}X$, proton and neutron respectively in units of u (1u = 931.5 MeV/c²) and BE represents its bonding energy in MeV, then [2008]
(a) $M(A, Z) = ZM_p + (A - Z) M_n - BE/c^2$
(b) $M(A, Z) = ZM_p + (A-Z) M_n + BE$
(c) $M(A, Z) = ZM_p + (A - Z) M_n - BE$
(d) $M(A, Z) = ZM_p + (A - Z)M_n + BE/c^2$

35. A nucleus $^{A}_{Z}X$ has mass represented by $M(A, Z)$. If M_p and M_n denote the mass of proton and neutron respectively and B.E. the binding energy in MeV, then [2007]
(a) $B.E. = [ZM_p + (A-Z)M_n - M(A, Z)]c^2$
(b) $B.E. = [ZM_p + ZM_n - M(A, Z)]c^2$
(c) $B.E. = M(A, Z) - ZM_p - (A-Z)M_n$
(d) $B.E. = [M(A, Z) - ZM_p - (A-Z)M_n]c^2$

36. The binding energy of deuteron is 2.2 MeV and that of $^{4}_{2}He$ is 28 MeV. If two deuterons are fused to form one $^{4}_{2}He$, then the energy released is [2006]
(a) 23.6 MeV (b) 19.2 MeV
(c) 30.2 MeV (d) 25.8 MeV

37. In the reaction, $^{2}_{1}H + ^{3}_{1}H \longrightarrow ^{4}_{2}He + ^{1}_{0}n$, if the binding energies of $^{2}_{1}H$, $^{3}_{1}H$ and $^{4}_{2}He$ are respectively, *a*, *b* and *c* (in MeV), then the energy (in MeV) released in this reaction is [2005]
(a) $a + b + c$ (b) $a + b - c$
(c) $c - a - b$ (d) $c + a - b$

38. In any fission process, the ratio $\dfrac{\text{mass of fission products}}{\text{mass of parent nucleus}}$ is [2005]
(a) equal to 1
(b) greater than 1
(c) less than 1
(d) depends on the mass of the parent nucleus

39. Fission of nuclei is possible because the binding energy per nucleon in them [2005]
(a) increases with mass number at low mass numbers.
(b) decreases with mass number at low mass numbers.
(c) increases with mass number at high mass numbers.
(d) decreases with mass number at high mass numbers.

40. If in nuclear fusion process the masses of the fusing nuclei be m_1 and m_2 and the mass of the resultant nucleus be m_3, then [2004]
(a) $m_3 > (m_1 + m_2)$ (b) $m_3 = m_1 + m_2$
(c) $m_3 = |m_1 - m_2|$ (d) $m_3 < (m_1 + m_2)$

41. M_p denotes the mass of a proton and M_n that of a neutron. A given nucleus, of binding energy B, contains Z protons and N neutrons. The mass $M(N, Z)$ of the nucleus is given by (c is the velocity of light) [2004]
(a) $M(N, Z) = NM_n + ZM_p + B/c^2$
(b) $M(N, Z) = NM_n + ZM_p - Bc^2$
(c) $M(N, Z) = NM_n + ZM_p + Bc^2$
(d) $M(N, Z) = NM_n + ZM_p - B/c^2$

42. Solar energy is mainly caused due to [2003]
(a) gravitational contraction
(b) burning of hydrogen in the oxygen
(c) fission of uranium present in the Sun
(d) fusion of protons during synthesis of heavier elements

43. The mass of proton is 1.0073 u and that of neutron is 1.0087 u (u = atomic mass unit). The binding energy of ${}_2^4\text{He}$ is [2003]
(a) 0.061 u (b) 0.0305 j
(c) 0.0305 erg (d) 28.4 MeV

44. For a nuclear fusion process, the suitable nuclei are [2002]
(a) any nuclei
(b) heavy nuclei
(c) light nuclei
(d) nuclei lying in the middle of the periodic table

45. It is possible to understand nuclear fission on the basis of the [2000]
(a) liquid drop model of the nucleus
(b) meson theory of the nuclear forces
(c) proton-proton cycle
(d) independent particle model of the nucleus

46. In nuclear reactions, we have the conservation of [2000]
(a) mass only
(b) energy only
(c) momentum only
(d) mass, energy and momentum

47. The explosion of the atomic bomb takes place due to [1999]
(a) nuclear fission (b) nuclear fusion
(c) scattering (d) thermionic emission

48. Complete the equation for the following fission process : ${}_{92}U^{235} + {}_0\text{n}^1 \rightarrow {}_{38}Sr^{90} + ...$ [1998]
(a) ${}_{54}X^{143} + 3\ {}_0n^1$ (b) ${}_{54}X^{145} + 3\ {}_0n^1$
(c) ${}_{57}X^{142} + 3\ {}_0n^1$ (d) ${}_{54}X^{142} + {}_0n^1$

49. In a fission reaction
$${}_{92}^{236}U \rightarrow {}^{117}X + {}^{117}Y + n + n$$
the binding energy per nucleon of X and Y is 8.5 MeV whereas of ${}^{236}U$ is 7.6 MeV. The total energy liberated will be about [1997]
(a) 2000 MeV (b) 200 MeV
(c) 2 MeV (d) 200 keV

50. Which of the following is used as a moderator in nuclear reactors? [1997]
(a) Plutonium (b) Cadmium
(c) Heavy water (d) Uranium

51. If the binding energy per nucleon in ${}_3\text{Li}^7$ and ${}_2\text{He}^4$ nuclei are respectively 5.60 MeV and 7.06 MeV, then the energy of proton in the reaction ${}_3\text{Li}^7 + p \rightarrow 2\ {}_2\text{He}^4$ is [1994]
(a) 19.6 MeV (b) 2.4 MeV
(c) 8.4 MeV (d) 17.3 MeV

52. Heavy water is used as a moderator in a nuclear reactor. The function of the moderator is [1994]
(a) to control energy released in the reactor
(b) to absorb neutrons and stop chain reaction
(c) to cool the reactor
(d) to slow down the neutrons to thermal energies.

53. Energy released in the fission of a single ${}_{92}^{235}\text{U}$ nucleus is 200 MeV. The fission rate of a ${}_{92}^{235}\text{U}$ filled reactor operating at a power level of 5 W is [1993]
(a) $1.56 \times 10^{-10} s^{-1}$ (b) $1.56 \times 10^{11} s^{-1}$
(c) $1.56 \times 10^{-16} s^{-1}$ (d) $1.56 \times 10^{-17} s^{-1}$

54. Solar energy is due to [1992]
(a) fusion reaction
(b) fission reaction
(c) combustion reaction
(d) chemical reaction

55. The energy equivalent of one atomic mass unit is [1992]
(a) 1.6×10^{-19} J (b) 6.02×10^{23} J
(c) 931 MeV (d) 9.31 MeV

56. An electron with (rest mass m_0) moves with a speed of 0.8 c. Its mass when it moves with this speed is [1991]
(a) m_0 (b) $\frac{m_0}{6}$ (c) $\frac{5m_0}{3}$ (d) $\frac{3m_0}{5}$

57. If the nuclear force between two protons, two neutrons and between proton and neutron is denoted by F_{pp}, F_{nn} and F_{pn} respectively, then [1991]

(a) $F_{pp} \approx F_{nn} \approx F_{pn}$

(b) $F_{pp} \neq F_{nn}$ and $F_{pp} = F_{nn}$

(c) $F_{pp} = F_{nn} = F_{pn}$

(d) $F_{pp} \neq F_{nn} \neq F_{pn}$

58. Which of the following statements is true for nuclear forces? *[1990]*

(a) they obey the inverse square law of distance
(b) they obey the inverse third power law of distance
(c) they are short range forces
(d) they are equal in strength to electromagnetic forces.

59. The average binding energy of a nucleon inside an atomic nucleus is about *[1989]*

(a) 8 MeV (b) 8 eV
(c) 8 J (d) 8 ergs

Topic 3 Radioactivity

60. A radioactive nucleus $^A_Z X$ undergoes spontaneous decay in the sequence $^A_Z X \to _{Z-1} B \to _{Z-3} C \to _{Z-2} D$, where Z is the atomic number of element X. The possible decay particles in the sequence are *[2021]*

(a) β^-, α, β^+ (b) α, β^-, β^+
(c) α, β^+, β^- (d) β^+, α, β^-

61. The half-life of a radioactive nuclide is 100 hours. The fraction of original activity that will remain after 150 hours would be *[2021]*

(a) $\frac{2}{3\sqrt{2}}$ (b) $\frac{1}{2}$
(c) $\frac{1}{2\sqrt{2}}$ (d) $\frac{2}{3}$

62. α-particle consists of: *[2019]*

(a) 2 protons and 2 neutrons only
(b) 2 electrons, 2 protons and 2 neutrons
(c) 2 electrons and 4 protons only
(d) 2 protons only

63. The rate of radioactive disintegration at an instant for a radioactive sample of half life 2.2×10^9 s is 10^{10} s^{-1}. The number of radioactive atoms in that sample at that instant is, *[NEET Odisha 2019]*

(a) 3.17×10^{19} (b) 3.17×10^{20}
(c) 3.17×10^{17} (d) 3.17×10^{18}

64. For a radioactive material, half-life is 10 minutes. If initially there are 600 number of nuclei, the time taken (in minutes) for the disintegration of 450 nuclei is *[2018]*

(a) 20 (b) 10
(c) 15 (d) 30

65. Radioactive material 'A' has decay constant '8 λ' and material 'B' has decay constant 'λ'. Initially they have same number of nuclei. After what time, the ratio of number of nuclei of material 'B' to that 'A' will be $\frac{1}{e}$? *[2017]*

(a) $\frac{1}{7\lambda}$ (b) $\frac{1}{8\lambda}$ (c) $\frac{1}{9\lambda}$ (d) $\frac{1}{\lambda}$

66. A radio isotope 'X' with a half life 1.4×10^9 years decays to 'Y' which is stable. A sample of the rock from a cave was found to contain 'X' and 'Y' in the ratio 1 : 7. The age of the rock is : *[2014]*

(a) 1.96×10^9 years (b) 3.92×10^9 years
(c) 4.20×10^9 years (d) 8.40×10^9 years

67. The half life of a radioactive isotope 'X' is 20 years. It decays to another element 'Y' which is stable. The two elements 'X' and 'Y' were found to be in the ratio of 1 : 7 in a sample of a the given rock. The age of the rock is estimated to be *[2013]*

(a) 60 years (b) 80 years
(c) 100 years (d) 40 years

68. α-particles, β-particles and γ-rays are all having same energy. Their penetrating power in a given medium in increasing order will be *[NEET Kar. 2013]*

(a) β, γ, α (b) γ, α, β
(c) α, β, γ (d) β, α, γ

69. A mixture consists of two radioactive materials A_1 and A_2 with half lives of 20 s and 10 s respectively. Initially the mixture has 40 g of A_1 and 160 g of A_2. The amount of the two in the mixture will become equal after : ***[2012]***

(a) 60 s (b) 80 s
(c) 20 s (d) 40 s

70. The half life of a radioactive nucleus is 50 days. The time interval $(t_2 - t_1)$ between the time t_2 when $\frac{2}{3}$ of it has decayed and the time t_1 when $\frac{1}{3}$ of it had decayed is : *[2012M]*

(a) 30 days (b) 50 days
(c) 60 days (d) 15 days

71. The half life of a radioactive isotope 'X' is 50 years. It decays to another element 'Y' which is stable. The two elements 'X' and 'Y' were found to be in the ratio of 1 : 15 in a sample of a given rock. The age of the rock was estimated to be [2011]
(a) 150 years (b) 200 years
(c) 250 years (d) 100 years

72. A nucleus ${}^{m}_{n}X$ emits one α-particle and two β-particles. The resulting nucleus is [2011, 1998]
(a) ${}^{m-6}_{n-4}Z$ (b) ${}^{m-6}_{n}Z$ (c) ${}^{m-4}_{n}X$ (d) ${}^{m-4}_{n-2}Y$

73. Two radioactive nuclei P and Q, in a given sample decay into a stable nucleus R. At time t = 0, number of P species are 4 N_0 and that of Q are N_0. Half-life of P (for conversion to R) is 1 minute where as that of Q is 2 minutes. Initially there are no nuclei of R present in the sample. When number of nuclei of P and Q are equal, the number of nuclei of R present in the sample would be [2011M]
(a) $3N_0$ (b) $\frac{9N_0}{2}$ (c) $\frac{5N_0}{2}$ (d) $2N_0$

74. The activity of a radioactive sample is measured as N_0 counts per minute at $t = 0$ and N_0/e counts per minute at $t = 5$ minutes. The time (in minutes) at which the activity reduces to half its value is [2010]
(a) $\log_e 2/5$ (b) $\frac{5}{\log_e 2}$
(c) $5 \log_{10} 2$ (d) $5 \log_e 2$

75. The decay constant of a radio isotope is λ. If A_1 and A_2 are its activities at times t_1 and t_2 respectively, the number of nuclei which have decayed during the time $(t_1 - t_2)$: [2010]
(a) $\lambda(A_1 - A_2)$ (b) $A_1t_1 - A_2t_2$
(c) $A_1 - A_2$ (d) $(A_1 - A_2)/\lambda$

76. The number of beta particles emitted by a radioactive substance is twice the number of alpha particles emitted by it. The resulting daughter is an [2009]
(a) isomer of parent (b) isotone of parent
(c) isotope of parent (d) isobar of parent

77. In the nuclear decay given below: [2009]

$${}^{A}_{Z}X \longrightarrow {}^{A}_{Z+1}Y \longrightarrow {}^{A-4}_{Z-1}B^{*} \longrightarrow {}^{A-4}_{Z-1}B,$$

the particles emitted in the sequence are
(a) γ, β, α (b) β, γ, α (c) α, β, γ (d) β, α, γ

78. Two radioactive materials X_1 and X_2 have decay constants 5λ and λ respectively. If initially they have the same number of nuclei, then the ratio of the number of nuclei of X_1 to that of X_2 will be $\frac{1}{e}$ after a time [2008]
(a) λ (b) $\frac{1}{2}\lambda$ (c) $\frac{1}{4\lambda}$ (d) $\frac{e}{\lambda}$

79. Two radioactive substances A and B have decay constants 5λ and λ respectively. At $t = 0$ they have the same number of nuclei. The ratio of number of nuclei of A to those of B will be $(1/e)^2$ after a time interval [2007]
(a) 4λ (b) 2λ (c) 1/2λ (d) 1/4λ

80. In a radioactive decay process, the negatively charged emitted β -particles are [2007]
(a) the electrons produced as a result of the decay of neutrons inside the nucleus
(b) the electrons produced as a result of collisions between atoms
(c) the electrons orbiting around the nucleus
(d) the electrons present inside the nucleus

81. In a radioactive material the activity at time t_1 is R_1 and at a later time t_2, it is R_2. If the decay constant of the material is λ, then [2006]
(a) $R_1 = R_2 e^{\lambda(t_1 - t_2)}$ (b) $R_1 = R_2 e^{(t_2/t_1)}$
(c) $R_1 = R_2$ (d) $R_1 = R_2 e^{-\lambda(t_1 - t_2)}$

82. The half life of radium is about 1600 years. Of 100 g of radium existing now, 25 g will remain unchanged after [2004]
(a) 3200 years (b) 4800 years
(c) 6400 years (d) 2400 years

83. A nuclear reaction is given by [2003]
${}_{Z}X^{A} \rightarrow {}_{Z+1}Y^{A} + {}_{-1}e^{0} + \bar{\nu}$, represents
(a) fission (b) β-decay
(c) γ-decay (d) fusion

84. A sample of radioactive element has a mass of 10gm at an instant t=0. The approximate mass of this element in the sample after two mean lives is [2003]
(a) 6.30 gm (b) 1.35 gm
(c) 2.50 gm (d) 3.70 gm

85. A sample has 4×10^{16} radioactive nuclei of half life 10 days. The number of atoms decaying in 30 days is [2002]
(a) 3.9×10^{16} (b) 5×10^{15}
(c) 10^{16} (d) 3.5×10^{16}

86. A deuteron strikes ${}_8O^{16}$ nucleus with subsequent emission of an alpha particle. Identify the nucleus so produced [2002]
(a) ${}_3Li^7$ (b) ${}_5B^{10}$ (c) ${}_7N^{13}$ (d) ${}_7N^{14}$

87. The decay constant (λ) and the half-life (T) of a radioactive isotope are related as [2000]
(a) $\lambda = \frac{\log_e 2}{T}$ (b) $\lambda = \frac{1}{\log_e 2.T}$
(c) $\lambda = \frac{T}{\log_e 2}$ (d) $\lambda = \frac{2}{T}$

88. Atomic hydrogen has life period of [2000]
(a) one minute
(b) one day
(c) a fraction of a second
(d) one hour

89. Alpha-particles are [1999]
(a) protons (b) positron
(c) neutrally charged (d) ionized helium atoms

90. After 1α and $2\ \beta$-emissions [1999]
(a) mass number reduces by 4
(b) mass number reduces by 5
(c) mass number reduces by 6
(d) mass number increases by 4

91. Half-lives of two radioactive substances A and B are respectively 20 minutes and 40 minutes. Initially, the samples of *A* and *B* have equal number of nuclei. After 80 minutes the ratio of remaining numbers of A and B nuclei is [1998]
(a) 1 : 16 (b) 4 : 1 (c) 1 : 4 (d) 1 : 1

92. The activity of a radioactive sample is measured as 9750 counts per minute at $t = 0$ and as 975 counts per minute at $t = 5$ minutes. The decay constant is approximately [1997]
(a) 0.922 per minute (b) 0.691 per minute
(c) 0.461 per minute (d) 0.230 per minute

93. A free neutron decays into a proton, an electron and [1997]
(a) a beta particle (b) an alpha particle
(c) an anti-neutrino (d) a neutrino

94. The most penetrating radiation of the following is [1997]
(a) gamma-rays (b) alpha particles
(c) beta-rays (d) *X*-rays

95. What is the respective number of α and β-particles emitted in the following radioactive decay $^{200}X_{90} \rightarrow ^{168}Y_{80}$? [1995]
(a) 6 and 8 (b) 6 and 6
(c) 8 and 8 (d) 8 and 6

96. The count rate of a Geiger Muller counter for the radiation of a radioactive material of half-life 30 minutes decreases to 5 sec^{-1} after 2 hours. The initial count rate was [1995]
(a) 20 sec^{-1} (b) 25 sec^{-1}
(c) 80 sec^{-1} (d) 625 sec^{-1}

97. In a given reaction
$${}_zX^A \rightarrow {}_{z+1}Y^A \rightarrow {}_{z-1}K^{A-4} \rightarrow {}_{z-1}K^{A-4}$$
Radioactive radiations are emitted in the sequence of [1993]
(a) α, β, γ (b) γ, α, β
(c) β, α, γ (d) γ, β, α

98. The mass of α-particle is [1992]
(a) less than the sum of masses of two protons and two neutrons
(b) equal to mass of four protons
(c) equal to mass of four neutrons
(d) equal to sum of masses of two protons and two neutron

99. The half life of radium is 1600 years. The fraction of a sample of radium that would remain after 6400 years [1991]
(a) 1/4 (b) 1/2 (c) 1/8 (d) 1/16

100. The nucleus ${}_6C^{12}$ absorbs an energetic neutron and emits a beta particle (β). The resulting nucleus is [1990]
(a) ${}_7N^{14}$ (b) ${}_7N^{13}$ (c) ${}_5B^{13}$ (d) ${}_6C^{13}$

101. A radioactive element has half life period 800 years. After 6400 years what amount will remain? [1989]
(a) $\frac{1}{2}$ (b) $\frac{1}{16}$ (c) $\frac{1}{8}$ (d) $\frac{1}{256}$

102. An element A decays into element C by a two step process
$$A \rightarrow B + {}_2He^4$$
$$B \rightarrow C + 2e^-$$
Then [1989]
(a) *A* and *C* are isotopes
(b) *A* and *C* are isobars
(c) *A* and *B* are isotopes
(d) *A* and *B* are isobars

103. Curie is a unit of [1989]
(a) energy of gamma-rays
(b) half-life
(c) radioactivity
(d) intensity of gamma-rays

104. A radioactive sample with a half life of 1 month has the label : 'Activity = 2 micro curies on 1–8–1991. What would be its activity two months earlier ? [1988]
(a) 1.0 micro curie (b) 0.5 micro curie
(c) 4 micro curie (d) 8 micro curie

105. The nucleus ${}^{115}_{48}Cd$, after two successive β– decay will give [1988]
(a) ${}^{115}_{46}Pa$ (b) ${}^{114}_{49}In$ (c) ${}^{113}_{50}Sn$ (d) ${}^{115}_{50}Sn$

ANSWER KEY

1	(a)	12	(a)	23	(d)	34	(a)	45	(a)	56	(c)	67	(a)	78	(c)	89	(d)	100	(b)
2	(c)	13	(b)	24	(a)	35	(a)	46	(d)	57	(c)	68	(c)	79	(c)	90	(a)	101	(d)
3	(d)	14	(b)	25	(d)	36	(a)	47	(a)	58	(c)	69	(d)	80	(a)	91	(c)	102	(a)
4	(d)	15	(c)	26	(d)	37	(c)	48	(a)	59	(a)	70	(b)	81	(d)	92	(c)	103	(c)
5	(c)	16	(b)	27	(b)	38	(c)	49	(b)	60	(d)	71	(b)	82	(a)	93	(c)	104	(d)
6	(a)	17	(c)	28	(d)	39	(d)	50	(c)	61	(d)	72	(c)	83	(b)	94	(a)	105	(d)
7	(c)	18	(a)	29	(c)	40	(d)	51	(d)	62	(a)	73	(b)	84	(b)	95	(d)		
8	(d)	19	(a)	30	(b)	41	(d)	52	(d)	63	(a)	74	(d)	85	(d)	96	(c)		
9	(a)	20	(a)	31	(c)	42	(d)	53	(b)	64	(a)	75	(d)	86	(d)	97	(c)		
10	(a)	21	(a)	32	(b)	43	(d)	54	(a)	65	(a)	76	(c)	87	(a)	98	(a)		
11	(a)	22	(a)	33	(b)	44	(c)	55	(c)	66	(c)	77	(d)	88	(c)	99	(d)		

Hints & Solutions

1. **(a)** As we know, $R = R_0 (A)^{1/3}$
where A = mass number
$R_{Al} = R_0 (27)^{1/3} = 3R_0$
$R_{Te} = R_0 (125)^{1/3} = 5R_0 = \frac{5}{3} R_{Al}$

2. **(c)** The radius of the nuclears is directly proportional to cube root of atomic number i.e. $R \propto A^{1/3}$
$R = R_0 A^{1/3}$, where R_0 is a constant of proportionality

$$\frac{R_2}{R_1} = \left(\frac{A_2}{A_1}\right)^{1/3} \Rightarrow \left(\frac{64}{27}\right)^{1/3} = \frac{4}{3}$$

where R_1 = the radius of ^{27}Al, and A_1 = Atomic mass number of Al
R_2 = the radius of ^{64}Cu and A_2 = Atomic mass number of C4

$$R_2 = 3.6 \times \frac{4}{3} = 4.8\text{m}$$

3. **(d)** Requird ratio of nuclear densities = $\frac{\rho_1}{\rho_2}$

$$= \frac{\left(\frac{M_1}{V_1}\right)}{\left(\frac{M_2}{V_2}\right)} = \frac{M_1}{M_2} \times \frac{V_2}{V_1} = \frac{1}{3} \times \frac{\frac{4}{3}\pi R_2^3}{\frac{4}{3}\pi R_1^3}$$

$$= \frac{1}{3} \times \left(\frac{R_2}{R_1}\right)^3 = \frac{1}{3} \times \left(\frac{R_0 M_2^{1/3}}{R_0 M_1^{1/3}}\right)^3 \quad [\because R = R_0 M^{1/3}]$$

$$= \frac{1}{3} \times \left(\frac{M_2}{M_1}\right) = \frac{1}{3} \times \left(\frac{3}{1}\right) = 1:1$$

> **NOTES**
> Nuclear density, $\rho = \frac{3m}{4\pi R_0^3}$
> Here, $R_0 = 1.2 \times 10^{-15}$m m = Average of mass of a nuclean (mass of proton + mass of neutron) $= 1.66 \times 10^{-27}$ kg
> This formula suggest that density of nuclear matter is same for all nuclei.

4. **(d)** It has been known that a nucleus of mass number A has radius
$R = R_0 A^{1/3}$,
where $R_0 = 1.2 \times 10^{-15}$m
and A = mass number

In case of $^{27}_{13}Al$, let nuclear radius be R_1

and for $^{125}_{32}Te$, nuclear radius be R_2

For $^{27}_{13}Al$, $R_1 = R_0(27)^{1/3} = 3R_0$

For $^{125}_{32}Te$, $R_2 = R_0(125)^{1/3} = 5R_0$

$$\frac{R_2}{R_1} = \frac{5R_0}{3R_0} = \frac{5}{3} R_1 = \frac{5}{3} \times 3.6 = 6\text{fm}$$

5. **(c)** We use the formula,
$R = R_0 A^{1/3}$
This represents relation between atomic mass and radius of the nucleus.

For berillium, $R_1 = R_0 (9)^{1/3}$
For germanium, $R_2 = R_0 A^{1/3}$

$$\frac{R_1}{R_2} = \frac{(9)^{1/3}}{(A)^{1/3}} \Rightarrow \frac{1}{2} = \frac{(9)^{1/3}}{(A)^{1/3}}$$

$$\Rightarrow \frac{1}{8} = \frac{9}{A} \Rightarrow A = 8 \times 9 = 72.$$

6. (a) Isotones means equal number of neutrons i.e., $(A-Z) = 74 - 34 = 71 - 31 = 40$.

7. (c) ${}^{A}_{Z}X$ has Z protons and $(A - Z)$ neutrons

8. (d) In case of hydrogen atom,
Mass number = atomic number

9. (a) $V \propto r^3 \Rightarrow V_{atom} = \frac{r^3_{atom}}{r^3_{nuclers}} \times V_{nucleus}$

But $\frac{r_{nucleus}}{r_{atom}} = \left(\frac{1}{10^5}\right)^3$

Hence, $V_{atom} = 10^{15} \times V_{nucleus}$

10. (a) Given : Mass of neutron = M_n
Mass of proton = M_p; Atomic mass of the element = M; Number of neutrons in the element = N and number of protons in the element = Z.
We know that the atomic mass (M) of any stable nucleus is always less than the sum of the masses of the constituent particles.
Therefore, $M < [NM_n + ZM_p]$.
X is a neutrino, when β-particle is emitted.

11. (a) Suppose that,
The number of ${}^{10}B$ type atoms = x
and the number of ${}^{11}B$ type atoms = y
Weight of ${}^{10}B$ type atoms = $10x$
Weight of ${}^{11}B$ type atoms = $11y$
Total number of atoms = $x + y$

$$\therefore \text{Atomic weight} = \frac{10x + 11y}{x + y} = 10.81$$

$$\Rightarrow 10x + 11y = 10.81x + 10.81y$$

$$\Rightarrow 0.81x = 0.19y \Rightarrow \frac{x}{y} = \frac{19}{81}$$

NOTES

If relative abundance of isotopes in an element has ratio $n_1 : n_2$ whose atomic massers are m_1 and m_2 then atomic masses of elements is

$$M = \frac{n_1 m_1 + n_2 m_2}{n_1 + n_2}$$

12. (a) The nuclear radius $R = R_0 A^{1/3}$

or $R \propto A^{1/3}$

or, $R^3 \propto A \Rightarrow A \propto R^3$

$$\therefore \frac{A_2}{A_1} = \left(\frac{R_2}{R_1}\right)^3 = \left(\frac{R_1/2}{R_1}\right)^3 = \frac{1}{8}$$

$$A_2 = \frac{A_1}{8} = \frac{56}{8} = 7$$

13. (b) Applying law of conservation of momentum,

$$m_1 v_1 = m_2 v_2$$

$$\frac{v_1}{v_2} = \frac{m_2}{m_1}$$

As $m = \frac{4}{3}\pi r^3 \rho \Rightarrow m \propto r^3$

Hence, $\frac{m_2}{m_1} = \frac{r_2^3}{r_1^3}$

$$\therefore \frac{v_1}{v_2} = \frac{r_2^3}{r_1^3} \Rightarrow \frac{r_2}{r_1} = \left(\frac{1}{2}\right)^{\frac{1}{3}}$$

14. (b)

15. (c) $\frac{R_s}{R_{He}} = \left(\frac{A_s}{A_{He}}\right)^{1/3} = \left(\frac{32}{4}\right)^{1/3} = 2$

16. (b)

17. (c) The nuclear radius r varies with mass number A according to the relation

$r \propto A^{1/3}$ or $A \propto r^3$

$$\text{density} = \frac{\text{mass}}{\text{volume}}$$

Further, mass ∝ A and volume $\propto r^3$

$$\therefore \frac{\text{mass}}{\text{volume}} = \text{constant}$$

18. (a) Nucleus contains only neutrons and protons.

19. (a) $Z = 11$ i.e., number of protons = 11, $A = 23$
∴ Number of neutrons = $A - Z = 12$
Number of electron = 0 (No electron in nucleus)
Therefore 11, 12, 0.

20. (a) As ${}_6C^{13}$ and ${}_7N^{14}$ have same no. of neutrons (13 – 6 = 7 for C and 14 – 7 = 7 for N), so they are isotones.

21. (a) $R \propto (A)^{1/3}$

$\therefore R_{Al} \propto (27)^{1/3}$ and $R_{Te} \propto (125)^{1/3}$

$$\therefore \frac{R_{Al}}{R_{Te}} = \frac{3}{5} = \frac{6}{10}$$

22. **(a)** Gain in BE = (BE) of products – (BE) of reactants.
$= [120 + 120] \times 8.5 - [240] \times 7.6$
$= (240) \times 8.5 - 240 \times 7.6$
$= (2040 - 1824)$ MeV = 216 MeV

23. **(d)** It is a nuclear fission reaction.
$$^{235}_{92}U + ^{1}_{0}n \longrightarrow ^{89}_{36}Kr + ^{144}_{56}Ba + ^{1}_{3}n + Q$$

24. **(a)** From the Einstein's mass-energy equivalence, the relation between the mass of a substance m and its energy E is
$E = mc^2$
Here, c = speed of light
$\therefore E = 0.5 \times 10^{-3} \times (3 \times 10^8)^2 = 0.5 \times 10^{-3} \times 9 \times 10^{16}$
$= 4.5 \times 10^{13}$ J.

25. **(d)** In an explosion a body breaks up into two pieces of unequal masses both part will have numerically equal momentum and lighter part will have more velocity.
$U \to Th + He$
$$KE_{Th} = \frac{P^2}{2m_{Th}}, KE_{He} = \frac{P^2}{2m_{He}}$$
since m_{He} is less so KE_{He} will be more.

26. **(d)** BE of $_2He^4 = 4 \times 7.06 = 28.24$ MeV
BE of $^7_3Li = 7 \times 5.60 = 39.20$ MeV
$$^7_3Li + ^1_1H \to {}_2He^4 + {}_2He^4 + Q$$
39.20 $\quad$ 28.24×2(= 56.48 MeV)
Therefore, Q = 56.48 – 39.20 = 17.28 MeV.

27. **(b)** Mass defect $\Delta m = 0.02866$ a.m.u.
Energy = 0.02866 × 931 = 26.7 MeV
As $_1H^2 + {}_1H^2 \longrightarrow {}_2He^4$
Energy liberated per a.m.u = 13.35/2 MeV
= 6.675 MeV

28. **(d)** BE/A
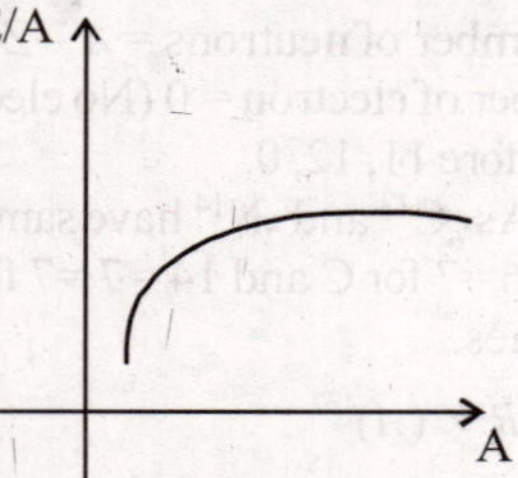

From the graph of BE/A versus mass number A it is clear that, BE/A first increases and then decreases with increase in mass number.

29. **(c)** $E = mc^2$
$$m = \frac{E}{c^2}$$
So, mass decay per second
$$\frac{dm}{dt} = \frac{1}{c^2}\frac{dE}{dt} = \frac{1}{c^2} \text{ (Power in watt)}$$
$$= \frac{1}{(3 \times 10^8)^2} \times 1000 \times 10^3$$
and mass decay per hour $= \frac{dm}{dt} \times 60 \times 60$
$$= \frac{1}{(3 \times 10^8)^2} \times 10^6 \times 3600 = 4 \times 10^{-8} \text{ kg}$$
= 40 microgram

30. **(b)** Momentum
$$Mu = \frac{E}{c} = \frac{h\nu}{c}$$
Recoil energy
$$\frac{1}{2}Mu^2 = \frac{1}{2}\frac{M^2u^2}{M} = \frac{1}{2M}\left(\frac{h\nu}{c}\right)^2 = \frac{h^2\nu^2}{2Mc^2}$$

31. **(c)** When the coulomb repulsion between the nuclei is overcome then nuclear fusion reaction takes place. This is possible when temperature is too high.

32. **(b)** B.E. = 0.042 × 931 ≃ 42 MeV
Number of nucleons in ^7_3Li is 7.
$$\therefore \quad \text{B.E./ nucleon} = \frac{42}{7} = 6 \text{ MeV} \simeq 5.6 \text{ MeV}$$

33. **(b)** Binding energy of two $_1H^2$ nuclei
$= 2(1.1 \times 2) = 4.4\,MeV$
Binding energy of one $_2He^4$ nucelus
$= 4 \times 7.0 = 28\,MeV$
$\therefore$ Energy released $= 28 - 4.4 = 23.6\,MeV$

34. **(a)** Mass defect = $ZM_p + (A-Z)M_n - M(A,Z)$
$$\text{or,} \quad \frac{B.E.}{c^2} = ZM_p + (A-Z)M_n - M(A,Z)$$
$$\therefore \quad M(A, Z) = ZM_p + (A-Z)M_n - \frac{B.E.}{c^2}$$

35. **(a)** The difference in mass of a nucleus and its constituents, ΔM, is called the mass defect and is given by
$$\Delta M = [ZM_p + (A-Z)M_n] - M$$
and binding energy = ΔMc^2
$$= [\{ZM_p + (A-Z)M_n\} - M]c^2$$

36. (a) $_1D^2 \longrightarrow {}_2He^4$

Energy released = 28 − 2 × 2.2 = 23.6 MeV

NOTES
Binding energy is energy released on formation of Nucleus

37. (c) $_1^2H$ and $_1^3H$ requires a and b amount of energies for their nucleons to be separated.

$_2^4$He releases c amount of energy in its formation i.e., in assembling the nucleons as nucleus.

Hence, Energy released $= c-(a+b) = c-a-b$

38. (c) Binding energy per nucleon for fission products is higher relative to Binding energy per nucleon for parent nucleus, i.e., more masses are lost and are obtained as kinetic energy of fission products. So, the given ratio < 1.

39. (d) B.E. per nucleon is smaller for lighter as well as heavier nucleus. But fusion reaction occurs for small mass number nuclei and fission reaction occurs for larger mass number nuclei to attain reaction binding energy per nucleon.

40. (d) $m_3 < (m_1 + m_2)$ $(\because m_1 + m_2 = m_3 + E]$

as $E = [m_1 + m_2 - m_3]\, C^2$

NOTES
Mass defect occur in both nuclear fission and fusion processes. Therefore, nuclear energy is released in both the processes as per Einstein mass energy relation $E = (\Delta m)c^2$

41. (d) Mass defect $= \dfrac{B.E}{c^2}$

Mass of nucleus = Mass of proton + mass of neutron − mass defect

42. (d) As a result of fusion, enormous amount of heat is liberated which is the main cause of source of solar energy.

43. (d) $\Delta m = (2 \times 1.0074 + 2 \times 1.0087 - 4.0015)$
$= 0.0307$

$E = (\Delta m) \times 931\ \text{MeV} = 0.0307 \times 931 = 28.5\ \text{MeV}$

44. (c) For nuclear fusion process the nuclei with low mass are suitable.

45. (a)

NOTES
According to liquid drop model of nucleus, an excited nucleus breaks into lighter nuclei just like an excited drop breaks into tiny drops.

46. (d)

47. (a)

48. (a) $_{92}U^{235} + {}_0n^1 \rightarrow {}_{38}Sr^{90} + {}_{54}Xe^{143} + 3\,{}_0n^1 + \text{energy}$

49. (b) Binding energy
$= 117 \times 8.5 + 117 \times 8.5 - 236 \times 7.6$
$= 234 \times 8.5 - 236 \times 7.6$
$= 1989 - 1793.6 = 200$ MeV

Thus, in per fission of Uranium nearly 200 MeV energy is liberated

50. (c) Moderator used in nuclear reactor are graphite and heavy water.

NOTES
Characteristics of moderator are
(i) Its atomic weight must be low
(ii) It should not absorb neutrons
(iii) It should undergo elastic collisions with neutrons and reduce their velocity.

51. (d) Applying principle of energy conservation, Energy of proton
= total B.E. of 2α − energy of Li^7
$= 8 \times 7.06 - 7 \times 5.6$
$= 56.48 - 39.2 = 17.28$ MeV

52. (d) Moderator slows down the neutrons to thermal energies.

53. (b) Fission rate

$$= \frac{\text{total power}}{\frac{\text{energy}}{\text{fission}}} = \frac{5}{200 \times 1.6 \times 10^{-13}}$$

$= 1.56 \times 10^{11}\ s^{-1}$

54. (a) Fusion reaction.

55. (c) 1 a.m.u = 931 MeV

56. (c) $$m = \frac{m_0}{\sqrt{1-\frac{v^2}{c^2}}} = \frac{m_0}{\sqrt{\frac{c^2-(0.8c)^2}{c^2}}} = \frac{5m_0}{3}$$

57. (c) Nuclear force is same between any two nucleons.

58. (c)

NOTES
Nuclear forces are short range attractive forces which balance the repulsive forces between the protons inside the nucleus.

59. (a) Average B.E./nucleon in nuclei is of the order of 8 MeV.

60. (d) Atomic number (z) decreases by 1 after one β^+ decay

Atomic number (z) increases by 1 after one

β^- decay

Atomic number (z) decreases by 2 after one α decay

$${}^{A}_{Z}X \xrightarrow{\beta^+ \text{ decay}} {}_{z-1}B \xrightarrow{\alpha \text{ decay}} {}_{z-3}C \xrightarrow{\beta^- \text{ decay}} {}_{z-2}D$$

$\therefore$ Correct order of decay is β^+, α, β^-

61. **(c)** Activity of a radioactive substance

$$A = A_0\left(\frac{1}{2}\right)^{\frac{t}{T_{1/2}}} \Rightarrow \frac{A}{A_0} = \left(\frac{1}{2}\right)^{\frac{t}{T_{1/2}}}$$

Given, $T_{1/2}$=100 hours & t = 150 hours

$$\Rightarrow \frac{A}{A_0} = \left(\frac{1}{2}\right)^{\frac{150}{100}} \Rightarrow \frac{A}{A_0} = \left(\frac{1}{2}\right)^{\frac{3}{2}}$$

$$\therefore \frac{A}{A_0} = \frac{1}{2\sqrt{2}}$$

NOTES

Activity of a radioactive substance is the rate of disintegration or the number of atoms of any material decaying per second.

$$\text{Activity} \propto \frac{1}{\text{half life}}$$

62. **(a)** α-particle is nucleus of helium He^{++} which has two protons and two neutrons.

63. **(a)** $t_{1/2} = \frac{ln 2}{\lambda} = \frac{0.693}{\lambda}$

$$\lambda = \frac{0.693}{2.2 \times 10^9} = 3.15 \times 10^{-10}$$

$$R = \lambda N$$

$$N = \frac{R}{\lambda} = \frac{10^{10}}{3.15 \times 10^{-10}} = 3.17 \times 10^{19}$$

64. **(a)** Number of nuclei remaining, N = 600 – 450 = 150 after time 't'

$$\frac{N}{N_0} = \frac{150}{600} = \frac{1}{4}$$

$$N = N_0 e^{-\lambda t} \Rightarrow \ell n\frac{N_0}{N} = \lambda t$$

$$\Rightarrow t = \frac{1}{\lambda}\ell n\frac{N_0}{N}$$

$$\Rightarrow t = \frac{2.303 \times T_{\frac{1}{2}}}{0.693}\log_{10}\frac{N_0}{N}$$

$$= \frac{2.303 \times 10}{0.693}\log_{10} 4$$

65. **(a)** Given, $\lambda_A = 8\lambda, \lambda_B = \lambda$

$$N_B = \frac{N_A}{e}$$

$$\Rightarrow N_o e^{-\lambda_B t} = N_o \frac{e^{-\lambda_A t}}{e}$$

$$e^{-\lambda t} = e^{-8\lambda t} e^{-1}$$

$$e^{-\lambda t} = e^{-8\lambda t - 1}$$

Comparing both side powers

$$-\lambda t = -8\lambda t - 1$$

$$-1 = 7\lambda t$$

$$t = -\frac{1}{7\lambda}$$

The best possible answer is $t = \frac{1}{7\lambda}$

66. **(c)** As $\frac{N_x}{N_y} = \frac{1}{7}$ (Given)

$$\Rightarrow \frac{N_x}{N_x + N_y} = \frac{1}{8} = \left(\frac{1}{2}\right)^3$$

Therefore, age of the rock

$$t = 3T_{1/2} = 3 \times 1.4 \times 10^9 \text{ yrs} = 4.2 \times 10^9 \text{ yrs.}$$

67. **(a)** The value of x is $\frac{1}{8}$

$$= \frac{x_0}{8} = \frac{x_0}{2^3} \Rightarrow t = 3T = 3 \times 20 = 60 \text{ years}$$

Hence the estimated age of the rock is 60 years

NOTES

	X →	Y_0
at t = 0	N_0	0
at t = t	N	$N_0 - N$

$$\frac{N}{N_0 - N} = \frac{1}{7} = \frac{N}{N_0} = \frac{1}{8}$$

t = 3T

= 3 × 20 = 60 years

68. **(c)** Increasing order of penetrating power : $\alpha < \beta < \gamma$.

NOTES

For same energy, lighter particle has higher penentrating power.

69. **(d)** Let, the amount of the two in the mixture will become equal after *t* years.

The amount of A_1, which remains after t years

$$N_1 = \frac{N_{01}}{(2)^{t/20}}$$

The amount of A_2, which remains, after t years

$$N_2 = \frac{N_{02}}{(2)^{t/10}}$$

According to the problem

$$N_1 = N_2$$

$$\frac{40}{(2)^{t/20}} = \frac{160}{(2)^{t/10}}$$

$$2^{t/20} = 2^{\left(\frac{t}{10}-2\right)}$$

$$\frac{t}{20} = \frac{t}{10} - 2$$

$$\frac{t}{20} - \frac{t}{10} = 2$$

$$\frac{t}{20} = 2$$

$$t = 40 \text{ s}$$

70. **(b)** $N_1 = N_0 e^{-\lambda t}$ $\quad N_1 = \frac{1}{3} N_0$

$$\frac{N_0}{3} = N_0 e^{-\lambda t_2}$$

$$\Rightarrow \frac{1}{3} = e^{-\lambda t^2} \quad \text{...(i)}$$

$$N_2 = \frac{2}{3} N_0$$

$$\frac{2}{3} N_0 = N_0 e^{-\lambda t_1}$$

$$\Rightarrow \frac{2}{3} = e^{-\lambda t_1} \quad \text{...(ii)}$$

Dividing equation (i) by equation (ii)

$$\frac{1}{2} = e^{-\lambda(t_2 - t_1)}$$

$$\lambda(t_2 - t_1) = \ln 2$$

$$t_2 - t_1 = \frac{\ln 2}{\lambda} = T_{1/2} = 50 \text{ days}$$

71. **(b)** Let number of atoms in $X = N_x$

Number of atoms in $Y = N_y$

By question

$$\frac{N_x}{N_y} = \frac{1}{15}$$

$$\therefore \text{Part of } N_x = \frac{1}{16}(N_x + N_y)$$

$$= \frac{1}{2^4}(N_x + N_y)$$

So, total 4 half lives are passed, so, age of rock is $4 \times 50 = 200$ years

72. **(c)** When ${}^{m}_{n}X$ emits one α-particle then its atomic mass decreases by 4 units and atomic number by 2. Therefore, the new nucleus becomes ${}^{m-4}_{n-2}Y$. But as it emits two β^- particles, its atomic number increases by 2. Thus the resulting nucleus is ${}^{m-4}_{n}X$.

NOTES No radioactive substance emits both α and β particles simultaneously. Also, γ rays are emitted after the emission of α or beta particle.

73. **(b)** Initially $P \to 4N_0$

$Q \to N_0$

Half life $T_P = 1$ min.

$T_Q = 2$ min.

Let after time t number of nuclei of P and Q are equal, that is

$$\frac{4N_0}{2^{t/1}} = \frac{N_0}{2^{t/2}}$$

$$\Rightarrow \frac{4N}{2^{t/1}} = \frac{1}{2^{t/2}}$$

$$\Rightarrow 2^{t/1} = 4.2^{t/2}$$

$$2^2.2^{t/2} = 2^{(2+t/2)}$$

$$\Rightarrow \frac{t}{1} = 2 + \frac{t}{2} \Rightarrow \frac{t}{2} = 2$$

$$\Rightarrow t = 4 \text{ min}$$

$$N_P = \frac{(4N_0)}{2^{4/1}} = \frac{N_0}{4}$$

at t = 4 min.

$$N_0 = \frac{N_0}{4} = \frac{N_0}{4}$$

or population of R

$$\left(4N_0 - \frac{N_0}{4}\right) + \left(N_0 - \frac{N_0}{4}\right)$$

$$= \frac{9N_0}{2}$$

74. (d) $N = N_0 e^{-\lambda t}$

Here, $t = 5$ minutes

$$\frac{N_0}{e} = N_0 \cdot e^{-5\lambda}$$

$\Rightarrow \quad 5\lambda = 1,$

$$\lambda = \frac{1}{5},$$

Now, $T_{1/2} = \dfrac{\ell n2}{\lambda} = 5\ \ell n2$

75. (d) Activity is given by

$$A = \frac{dN}{dt} = -\lambda N$$

Activity at time t_1 is

$A_1 = -\lambda N_1$

and activity at time t_2 is $A_2 = -1 N_2$

As $t_1 > t_2$, therefore, number of atoms remained after time t_1 is less than that remained after time t_2. That is, $N_1 < N_2$.

$\therefore$ number of nuclei decayed in $(t_1 - t_2)$

$$= N_2 - N_1 = \frac{(A_1 - A_2)}{\lambda}$$

NOTES

If a radioactive element A disintegrates to form another radioactive element B, which inturn disintegrates to form another element C,

Rate of disintegration of $A = \dfrac{dN_1}{dt} = -\lambda_1 N_1$

Rate of disintegration of $B = \dfrac{dN_2}{dt} = -\lambda_2 N_2$

Net rate of formation of B = Rate of disintegration of A – Rate of disiutegration of B = $\lambda_1 N_1 - \lambda_2 N_2$

76. (c) A radioactive substance when emits one alpha particles $\left({}_2^4He\right)$, its mass number reduces by 4 and charge no. reduces by 2 and after emission of two β-particles its charge no. increase by 2 thus the charge no. i.e. atomic number remains the same.

NOTES

Isotopes of an element have the same atomic number but different mass number.

77. (d) ${}_Z^AX \longrightarrow {}_{Z+1}^{A}Y : \beta, \ {}_{Z+1}^{A}Y \longrightarrow {}_{Z-1}^{A-4}B^* : \alpha$

${}_{Z-1}^{A-4}B^* \longrightarrow {}_{Z-1}^{A-4}B : \gamma \qquad (\beta, \alpha, \gamma)$

NOTES

Mass number and charge number of a nucleus remains unchanged during γ decay)

78. (c) Let the required time be t. Then

$N_1 = N_0 e^{-\lambda_1 t}$; $N_2 = N_0 e^{-\lambda_2 t}$

Where

N_1 = number of nuclei of X_1 after time t

N_2 = number of nuclei of X_2 after time t

N_0 = initial number of nuclei of X_1 and X_2 each.

Now, $\dfrac{N_1}{N_2} = \dfrac{N_0 e^{-5\lambda t}}{N_0 e^{-\lambda t}}$ Here $\dfrac{N_1}{N_2} = \dfrac{1}{e}$

$$\therefore \frac{1}{e} = \frac{e^{-5\lambda t}}{e^{-\lambda t}} \Rightarrow e^{-1} = e^{-4\lambda t} \Rightarrow 4\lambda t = 1$$

$$\therefore t = \frac{1}{4\lambda}$$

79. (c) $\lambda_A = 5\lambda$ and $\lambda_B = \lambda$

At $t = 0$, $(N_0)_A = (N_0)_B$

Given, $\dfrac{N_A}{N_B} = \left(\dfrac{1}{e}\right)^2$

According to radioactive decay,

$$\frac{N}{N_0} = e^{-\lambda t}$$

$$\therefore \frac{N_A}{(N_0)_A} = e^{-\lambda_A t} \qquad \text{..... (1)}$$

$$\frac{N_B}{(N_0)_B} = e^{-\lambda_B t} \qquad \text{..... (2)}$$

From (1) and (2),

$$\frac{N_A}{N_B} = e^{-(5\lambda - \lambda)t}$$

$$\Rightarrow \left(\frac{1}{e}\right)^2 = e^{-4\lambda t} = \left(\frac{1}{e}\right)^{4\lambda t}$$

$$\Rightarrow 4\lambda t = 2 \qquad \therefore t = \frac{1}{2\lambda}.$$

80. (a) In beta minus decay (β^-), a neutron is transformed into a proton, and an electron is emitted from the nucleus along with antineutrino.

$$n = p + e^- + \bar{\nu}$$

81. (d) Let at time t_1 & t_2, number of particles be N_1 & N_2. So,

$$R_1 = \frac{dN_1}{dt} = -\lambda N_1; \quad R_2 = \frac{dN_2}{dt} = -\lambda N_2$$

$$\frac{R_1}{R_2} = \frac{\lambda N_1}{\lambda N_2} = \frac{N_1}{N_1 e^{-\lambda(t_2 - t_1)}} = e^{\lambda(t_2 - t_1)}$$

$$R_1 = R_2 e^{\lambda(t_2 - t_1)} = R_2 e^{-\lambda(t_1 - t_2)}$$

82. (a) 100 g will become 25 g in two half lives, so, it is 3200 years.

83. (b) $_{-1}e^0$ represents a β-decay.

84. (b) Using the relation for mean life.

Given : $t = 2\tau = 2\left(\frac{1}{\lambda}\right) \qquad \left(\therefore \tau = \frac{1}{\lambda}\right)$

Then from $M = M_0 e^{-\lambda t} = 10e^{-\lambda \times \frac{2}{\lambda}}$

$= 10\left(\frac{1}{e}\right)^2 = 1.35\text{g}$

NOTES

The law of radioactive disintegration in terms of mass can be written as

$M = M_0\, e^{-\lambda t}$

Here, M_0 = mass of radioactive nuclei at time t = 0

M = mass of radioactive nuclei at time t

85. (d) $N = 4 \times 10^{16}\left(\frac{1}{2}\right)^{\frac{30}{10}} = \frac{1}{2} \times 10^{16}$

Atoms decayed $= 4 \times 10^{16} - \frac{1}{2} \times 10^{16}$

$= 3.5 \times 10^{16}$

86. (d) $_8O^{16} + {}_1H^2 \longrightarrow {}_2He^4 + {}_7N^{14}$

87. (a) $t = \frac{1}{\lambda}\log\frac{a}{a-x}$ when $t = T, x = \frac{a}{2}$

$T = \frac{1}{\lambda}\log\frac{a}{a-\frac{a}{2}} = \frac{1}{\lambda}\log_e 2 \Rightarrow \lambda = \frac{1}{T}\log_e 2$

88. (c) Atomic hydrogen is unstable and it has life period of a fraction of a second.

89. (d) We know that alpha particles are the nucleus of ionized helium atoms which contain two protons and two neutrons. These are emitted by the nuclei of certain radioactive substances. Streams of alpha particles, called α-rays, produce intense ionisation in gases through which they pass and are easily absorbed by matter.

90. (a) Emission of 1α particle led to decrease in atomic number by 2 while mass number by 4. On the other hand, emission of 2β particles increases atomic number by 2. Hence, overall emission of, 1α and 2β particles led to decrease in mass number by 4.

91. (c) $80 = 20 \times n_A \Rightarrow n_A = 4$

$80 = 40 \times n_B \Rightarrow n_B = 2$

$\frac{N_A}{N_B} = \frac{\left(\frac{1}{2}\right)^4}{\left(\frac{1}{2}\right)^2} = \frac{1}{2} \times \frac{1}{2} = \frac{1}{4}$

92. (c) $\frac{dN}{dt} = KN$

$9750 = KN_0$ .... (1)

$975 = KN$ .... (2)

Dividing (1) by (2)

$\frac{N}{N_0} = \frac{1}{10}$

$K = \frac{2.303}{t}\log\frac{N_0}{N} = \frac{2.303}{5}\log 10$

$= 0.4606 = 0.461$ per minute

93. (c) $_0n^1 \longrightarrow {}_1p^1 + {}_{-1}e^0 + X$

X must have zero charge and almost zero mass as electron is emitted. Hence X must be anti-neutrino.

94. (a) The penetrating power of radiation is directly proportional to the energy of its photon.

Energy of a photon $= \frac{hc}{\lambda} \propto \frac{1}{\lambda}$

∴ Penetrating power $\propto \frac{1}{\lambda}$

λ is minimum for γ-rays, so penetrating power is maximum of γ-rays.

95. (d) $^{200}X_{90} \longrightarrow {}^{168}Y_{80}$. We know that

$^{200}X_{90} \longrightarrow n_2He^4 + m_{-1}\beta^0 + {}^{168}Y_{80}$.

Therefore, in this process,

$200 = 4n + 168$ or $n = \frac{200-168}{4} = 8$.

Also, $90 = 2n - m + 80$

or, $m = 2n + 80 - 90 = (2 \times 8 + 80 - 90) = 6$.

Thus, respective number of α and β-particles will be 8 and 6.

96. (c) Half-life = 30 minutes; Rate of decrease (N) = 5 per second and total time = 2 hours = 120 minutes. Relation for initial and final count rate

$\frac{N}{N_0} = \left(\frac{1}{2}\right)^{\text{time/half-life}} = \left(\frac{1}{2}\right)^{120/30} = \left(\frac{1}{2}\right)^4 = \frac{1}{16}$.

Therefore, $N_0 = 16 \times N = 16 \times 5 = 80\ s^{-1}$.

97. (c) Increase of charge number by 1 indicates β emission. Decrease of mass number by 4 and charge number by 2 indicates α emission. No change of mass number and charge number indicates γ-emission.

98. **(a)** α-particle = $_2He^4$. It contains 2 p and 2 n. As some mass is converted into B.E., therefore, mass of α particle is slightly less than the sum of the masses of 2 *p* and 2 *n*.

99. **(d)** $\frac{N}{N_0} = \left(\frac{1}{2}\right)^{6400/1600} = \left(\frac{1}{2}\right)^4 = \frac{1}{16}$

100. **(b)** $_6C^{12} + {}_0n^1 \rightarrow {}_6C^{13} \rightarrow {}_7N^{13} + {}_{-1}\beta^0 + \text{Energy}$

101. **(d)** No. of half lives, $n = \frac{t}{T} = \frac{6400}{800} = 8$

$$\frac{N}{N_0} = \left(\frac{1}{2}\right)^8 = \frac{1}{256}$$

102. **(a)** From eqn. (ii), B has 2 units of charge more than C.

From eqn. (i), *A* loses 2 units of charge by emission of alpha particle. Hence, *A* and *C* are isotopes as their charge numbers are same.

103. **(c)** Curie is a unit of radioactivity.

104. **(d)** In two half lives, the activity becomes one fourth.

Activity on 1–8–91 was 2 micro–curie

∴ Activity before two months,

4 × 2 micro-curie = 8 micro curie

NOTES

Activity of substance decreases as number of undecayed nuclei decreases with time.

Also, Activity $\propto \frac{1}{\text{Half life}}$

105. **(d)** Two successive β decay increases the atomic number by 2. Therefore, (d) is correct.

28 Semiconductor Electronics : Materials, Devices and Simple Circuits

Trend Analysis with Important Topics & Sub-Topics

Topic Name	Sub-Topic	2021		2020		2019		2018		2017		2016	
		QNS.	LOD	QNS	LOD	QNS.	LOD	QNS.	LOD	QNS.	LOD	QNS.	LOD
Solids, Semiconductors and P-N Junction Diode	Depletion region in a p-n Junction	1	E	1	E								
	P-type semiconductor, p-n Junction diode					1	E	1	A	1	A	1	E
	Zener diode	1	E										
Junction Transistor	Structure of Transistor			1	E					1	A		
	Current amplification factor, base & Collector current, Voltage and power gain							1	A	1	A	1	A
Digital Electronics and Logic Gates	Logic gates and its combination	1	A	1	A			1	A	1	A	1	A
	Boolean operation					1	A						

LOD - Level of Difficulty | E - Easy | A - Average | D - Difficult | Qns - No. of Questions

Topic 1 Solids, Semiconductors and P-N Junction Diode

1. The electron concentration in an n-type semiconductor is the same as hole concentration in a p-type semiconductor. An external field (electric) is applied across each of them. Compare the currents in them. *[2021]*
   (a) No current will flow in p-type, current will only flow in n-type
   (b) Current in n-type = current in p-type
   (c) Current in p-type > current in n-type
   (d) Current in n-type > current in p-type

2. Consider the following statements (A) and (B) and identify the correct answer. *[2021]*
   (A) A zener diode is connected in reverse bias, when used as a voltage regulator.
   (B) The potential barrier of p-n junction lies between 0.1 V to 0.3 V.
   (a) (A) is incorrect but (B) is correct.
   (b) (A) and (B) both are correct.
   (c) (A) and (B) both are incorrect
   (d) (A) is correct and (B) is incorrect.

3. The increase in the width of the depletion region in a p-n junction diode is due to : *[2020]*
   (a) reverse bias only
   (b) both forward bias and reverse bias
   (c) increase in forward current
   (d) forward bias only

4. For a p-type semiconductor, which of the following statements is **true**? *[2019]*
   (a) Electrons are the majority carriers and trivalent atoms are the dopants.
   (b) Holes are the majority carriers and trivalent atoms are the dopants.
   (c) Holes are the majority carriers and pentavalent atoms are the dopants.
   (d) Electrons are the majority carriers and pentavalent atoms are the dopants.

5. An LED is constructed from a p-n junction diode using GaAsP. The energy gap is 1.9 eV. The wavelength of the light emitted will be equal to *[NEET Odisha 2019]*

(a) 654×10^{-11} m (b) 10.4×10^{-26} m
(c) 654 nm (d) 654 Å

6. In a p-n junction diode, change in temperature due to heating *[2018]*

(a) Affects only reverse resistance
(b) Affects only forward resistance
(c) Affects the overall V-I characteristics of p-n junction
(d) Does not affect resistance of p-n junction

7. Which one of the following represents forward bias diode ? *[2017, 2006, 2000]*

(a) 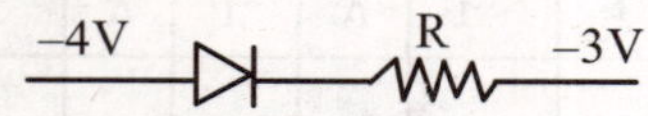

(b) 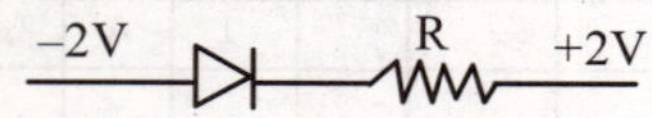

(c) 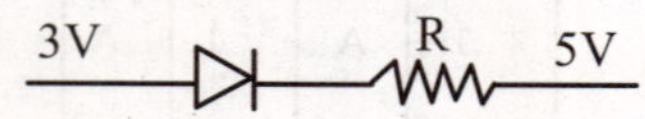

(d) 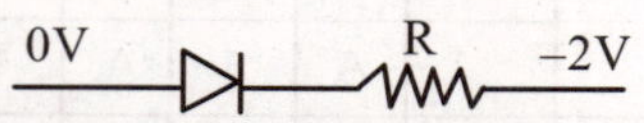

8. Consider the junction diode as ideal. The value of current flowing through AB is : *[2016]*

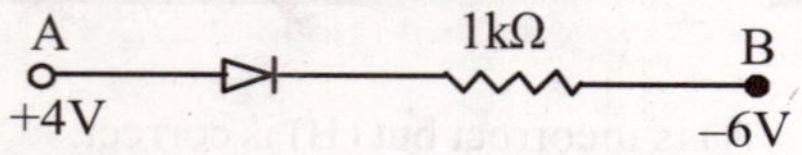

(a) 0 A (b) 10^{-2} A
(c) 10^{-1} A (d) 10^{-3} A

9. If in a p-n junction, a square input signal of 10 V is applied as shown, then the output across R_L will be *[2015]*

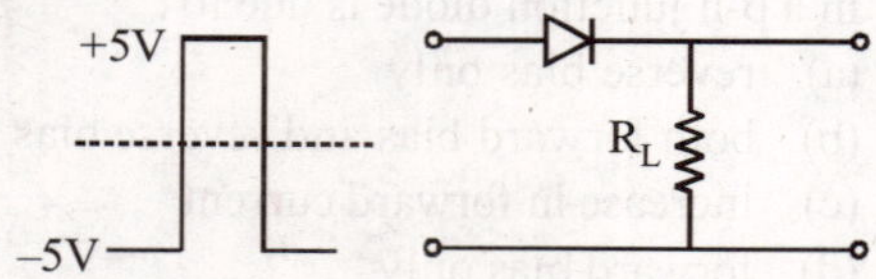

(a) 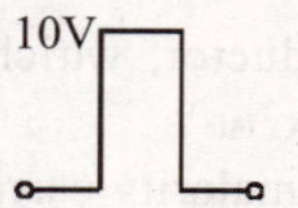

(b) 

(c) 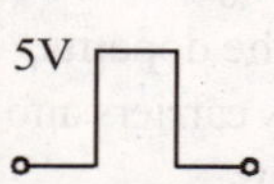

(d) 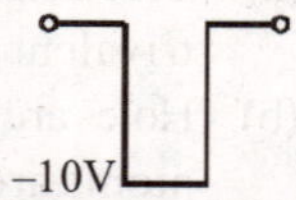

10. In the given figure, a diode D is connected to an external resistance R = 100 Ω and an e.m.f. of 3.5 V. If the barrier potential developed across the diode is 0.5 V, the current in the circuit will be: *[2015 RS]*

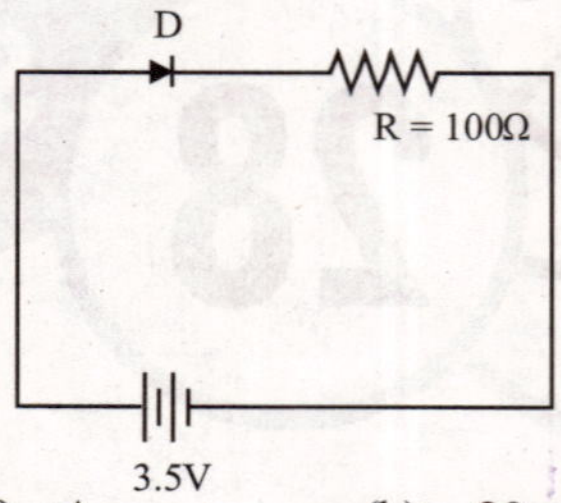

(a) 40 mA (b) 20 mA
(c) 35 mA (d) 30 mA

11. The given graph represents V - I characteristic for a semiconductor device.

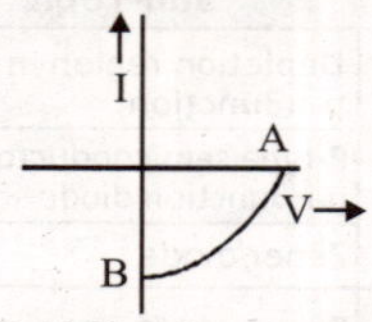

Which of the following statement is correct ? *[2014]*

(a) It is V - I characteristic for solar cell where, point A represents open circuit voltage and point B short circuit current.
(b) It is a for a solar cell and point A and B represent open circuit voltage and current, respectively.
(c) It is for a photodiode and points A and B represent open circuit voltage and current, respectively.
(d) It is for a LED and points A and B represent open circuit voltage and short circuit current, respectively.

12. The barrier potential of a p-n junction depends on: *[2014]*

(A) type of semi conductor material
(B) amount of doping
(C) temperature

Which one of the following is correct ?

(a) (A) and (B) only (b) (B) only
(c) (B) and (C) only (d) (A), (B) and (C)

13. In a n-type semiconductor, which of the following statement is true? *[2013]*

(a) Electrons are minority carriers and pentavalent atoms are dopants.
(b) Holes are minority carriers and pentavalent atoms are dopants.
(c) Holes are majority carriers and trivalent atoms are dopants.
(d) Electrons are majority carriers and trivalent atoms are dopants.

14. In an unbiased p-n junction, holes diffuse from the p-region to n-region because of *[NEET Kar. 2013]*
(a) the potential difference across the p-n junction
(b) the attraction of free electrons of n-region
(c) the higher hole concentration in p-region than that in n-region
(d) the higher concentration of electrons in the n-region than that in the p-region

15. Two ideal diodes are connected to a battery as shown in the circuit. The current supplied by the battery is : *[2012]*

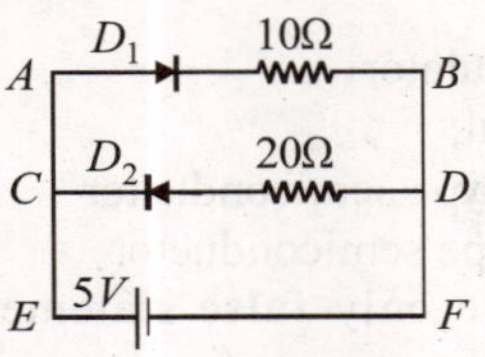

(a) 0.75 A (b) zero
(c) 0.25 A (d) 0.5 A

16. C and Si both have same lattice structure, having 4 bonding electrons in each. However, C is insulator whereas Si is intrinsic semiconductor. This is because : *[2012]*
(a) In case of C the valence band is not completely filled at absolute zero temperature.
(b) In case of C the conduction band is partly filled even at absolute zero temperature.
(c) The four bonding electrons in the case of C lie in the second orbit, whereas in the case of Si they lie in the third.
(d) The four bonding electrons in the case of C lie in the third orbit, whereas for Si they lie in the fourth orbit.

17. In forward biasing of the p–n junction *[2011]*
(a) the positive terminal of the battery is connected to p–side and the depletion region becomes thick
(b) the positive terminal of the battery is connected to n–side and the depletion region becomes thin
(c) the positive terminal of the battery is connected to n–side and the depletion region becomes thick
(d) the positive terminal of the battery is connected to p–side and the depletion region becomes thin

18. If a small amount of antimony is added to germanium crystal *[2011]*
(a) it becomes a p–type semiconductor
(b) the antimony becomes an acceptor atom
(c) there will be more free electrons than holes in the semiconductor
(d) its resistance is increased

19. In the following figure, the diodes which are forward biased, are *[2011M, 2004]*

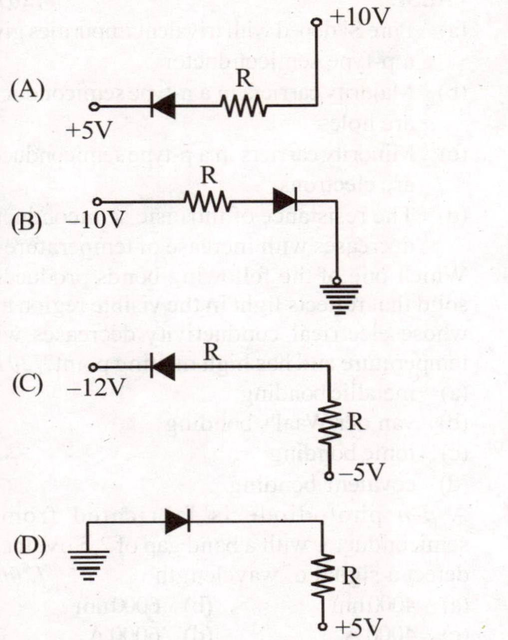

(a) (C) only (b) (C) and (A)
(c) (B) and (D) (d) (A), (B) and (D)

20. Pure Si at 500K has equal number of electron (n_e) and hole (n_h) concentrations of 1.5×10^{16} m^{-3}. Doping by indium increases n_h to 4.5×10^{22} m^{-3}. The doped semiconductor is of *[2011M]*
(a) n–type with electron concentration $n_e = 5 \times 10^{22}$ m^{-3}
(b) p–type with electron concentration $n_e = 2.5 \times 10^{10}$ m^{-3}
(c) n–type with electron concentration $n_e = 2.5 \times 10^{23}$ m^{-3}
(d) p–type having electron concentration $n_e = 5 \times 10^{9}$ m^{-3}

21. A zener diode, having breakdown voltage equal to 15V, is used in a voltage regulator circuit shown in figure. The current through the diode is *[2011M]*

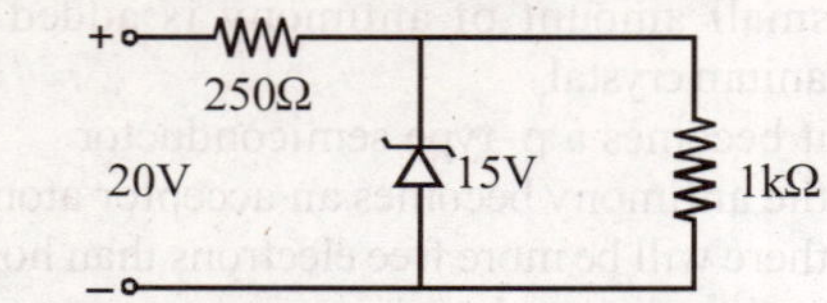

(a) 10 mA (b) 15 mA
(c) 20 mA (d) 5 mA

22. Which one of the following statement is FALSE ? *[2010]*
(a) Pure Si doped with trivalent impurities gives a p-type semiconductor
(b) Majority carriers in a n-type semiconductor are holes
(c) Minority carriers in a p-type semiconductor are electrons
(d) The resistance of intrinsic semiconductor decreases with increase of temperature

23. Which one of the following bonds produces a solid that reflects light in the visible region and whose electrical conductivity decreases with temperature and has high melting point? *[2010]*
(a) metallic bonding
(b) van der Waal's bonding
(c) ionic bonding
(d) covalent bonding

24. A *p-n* photodiode is fabricated from a semiconductor with a band gap of 2.5 eV. It can detect a signal of wavelength *[2009]*
(a) 4000 nm (b) 6000 nm
(c) 4000 Å (d) 6000 Å

25. Sodium has body centred packing. Distance between two nearest atoms is 3.7 Å. The lattice parameter is *[2009]*
(a) 4.3Å (b) 3.0 Å
(c) 8.6 Å (d) 6.8 Å

26. A p-n photodiode is made of a material with a band gap of 2.0 eV. The minimum frequency of the radiation that can be absorbed by the material is nearly *[2008]*
(a) 10×10^{14} Hz (b) 5×10^{14} Hz
(c) 1×10^{14} Hz (d) 20×10^{14} Hz

27. For a cubic crystal structure which one of the following relations indicating the cell characteristics is correct? *[2007]*
(a) $a \neq b \neq c$ and $\alpha = \beta = \gamma = 90°$
(b) $a = b = c$ and $\alpha \neq \beta \neq \gamma = 90°$
(c) $a = b = c$ and $\alpha = \beta = \gamma = 90°$
(d) $a \neq b \neq c$ and $\alpha \neq \beta$ and $\gamma \neq 90°$

28. In the energy band diagram of a material shown below, the open circles and filled circles denote holes and electrons respectively. The material is

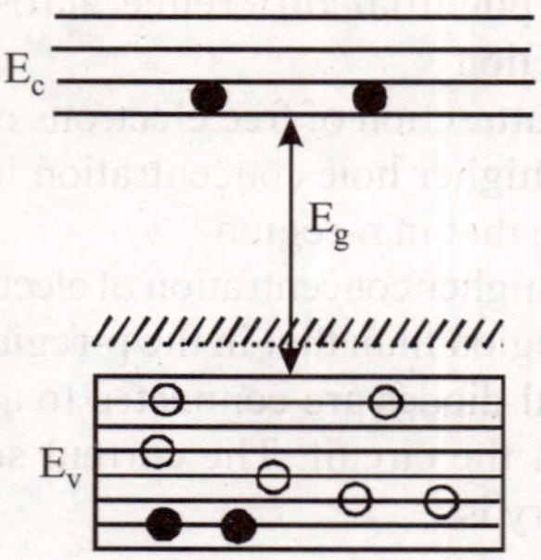

(a) an insulator *[2007]*
(b) a metal
(c) an *n*-type semiconductor
(d) a *p*-type semiconductor

29. Choose the only **false** statement from the following *[2005]*
(a) In conductors, the valence and conduction bands may overlap
(b) Substances with energy gap of the order of 10 eV are insulators
(c) The resistivity of a semiconductor increases with increase in temperature
(d) The conductivity of a semiconductor increases with increase in temperature

30. Carbon, silicon and germanium atoms have four valence electrons each. Their valence and conduction bands are separated by energy band gaps represented by $(E_g)_C$, $(E_g)_{Si}$ and $(E_g)_{Ge}$ respectively. Which one of the following relationships is true in their case? *[2005]*
(a) $(E_g)_{si} > (E_g)_{Ge} < (E_g)_C$
(b) $(E_g)_C < (E_g)_{Si}$
(c) $(E_g)_C = (E_g)_{Si}$
(d) $(E_g)_C < (E_g)_{Ge}$

31. Application of a forward bias to a *p–n* junction *[2005]*
(a) widens the depletion zone
(b) increases the potential difference across the depletion zone
(c) increases the number of donors on the n side
(d) increases the electric field in the depletion zone

32. Zener diode is used for *[2005]*
(a) amplification
(b) rectification
(c) stabilisation
(d) producing oscillations in an oscillator

33. In semiconductors, at room temperature [2004]
(a) the conduction band is completely empty
(b) the valence band is partially empty and the conduction band is partially filled
(c) the valence band is completely filled and the conduction band is partially filled
(d) the valence band is completely filled

34. The peak voltage in the output of a half-wave diode rectifier fed with a sinusoidal signal without filter is 10V. The d.c. component of the output voltage is [2004]
(a) $20/\pi$ V (b) $10/\sqrt{2}$ V
(c) $10/\pi$ V (d) 10V

35. In a *p-n* junction photocell, the value of the photo-electromotive force produced by monochromatic light is proportional to [2004]
(a) the voltage applied at the *p-n* junction
(b) the barrier voltage at the *p-n* junction
(c) the intensity of the light falling on the cell
(d) the frequency of the light falling on the cell

36. If a full wave rectifier circuit is operating from 50Hz mains, the fundamental frequency in the ripple will be [2003]
(a) 100 Hz (b) 25 Hz
(c) 50 Hz (d) 70.7 Hz

37. Barrier potential of a *p-n* junction diode does not depend on [2003]
(a) doping density (b) diode design
(c) temperature (d) forward bias

38. Reverse bias applied to a junction diode [2003]
(a) increases the minority carrier current
(b) lowers the potential barrier
(c) raises the potential barrier
(d) increases the majority carrier current

39. A d.c. battery of V volt is connected to a series combination of a resistor R and an ideal diode D as shown in the figure below. The potential difference across R will be [2002]

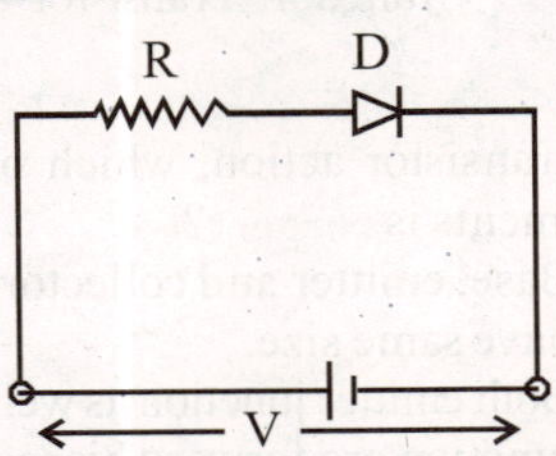

(a) 2V when diode is forward biased
(b) Zero when diode is forward biased
(c) V when diode is reverse biased
(d) V when diode is forward biased

40. In a *p-n* junction [2002]
(a) The potential of the *p* and *n*-sides becomes higher alternately
(b) The *p*-side is at higher electrical potential than the n side
(c) The *n*-side is at higher electrical potential than the *p*-side
(d) Both the *p* and *n*-sides are at the same potential

41. The intrinsic semiconductor becomes an insulator at [2001]
(a) 0ºC (b) 0 K
(c) 300 K (d) –100ºC

42. An alternating current can be converted into direct current by a [2001]
(a) transformer (b) dynamo
(c) motor (d) rectifier

43. In forward bias, the width of potential barrier in a p-n junction diode [1999]
(a) increases (b) decreases
(c) remains constant (d) first '1' then '2'

44. A depletion layer consists of [1999]
(a) electrons (b) protons
(c) mobile ions (d) immobile ions

45. Which of the following when added acts as an impurity into silicon produced n-type semi-conductor? [1999]
(a) P (b) Al
(c) B (d) Mg

46. In a junction diode, the holes are due to [1999]
(a) protons (b) extra electrons
(c) neutrons (d) missing electrons

47. A semi-conducting device is connected in a series circuit with a battery and a resistance. A current is found to pass through the circuit. If the polarity of the battery is reversed, the current drops to almost zero. The device may be [1998]
(a) a *p-n* junction
(b) an intrinsic semi-conductor
(c) a *p*-type semi-conductor
(d) an *n*-type semi-conductor

48. The cause of the potential barrier in a *p-n* diode is [1998]
(a) depletion of positive charges near the junction
(b) concentration of positive charges near the junction
(c) depletion of negative charges near the junction
(d) concentration of positive and negative charges near the junction

49. To obtain a *p*-type germanium semiconductor, it must be doped with [1997]
(a) arsenic (b) antimony
(c) indium (d) phosphorus

50. When arsenic is added as an impurity to silicon, the resulting material is [1996]
(a) *n*-type semiconductor
(b) *p*-type semiconductor
(c) *n*-type conductor
(d) insulator

51. Which of the following, when added as an impurity, into the silicon, produces *n*-type semiconductor? [1995]
(a) Phosphorous (b) Aluminium
(c) Magnesium (d) Both b and c

52. When a *p-n* junction diode is reverse biased the flow of current across the junction is mainly due to [1994]
(a) diffusion of charges
(b) drift of charges
(c) depends on the nature of material
(d) both drift and diffusion of charges

53. In the diagram, the input is across the terminals A and C and the output is across B and D. Then the output is [1994]

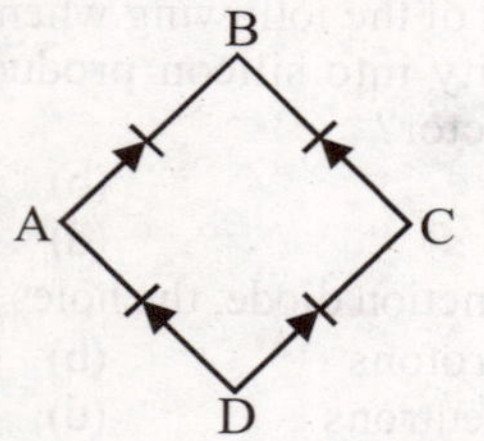

(a) zero (b) same as the input
(c) full wave rectifier (d) half wave rectifier

54. A piece of copper and other of germanium are cooled from the room temperature to 80 K, then [1993]
(a) resistance of each will increase
(b) resistance of copper will decrease
(c) resistance of copper will increase while that of germanium will decrease
(d) resistance of copper will decrease while that of germanium will increase

55. Diamond is very hard because [1993]
(a) it is a covalent solid
(b) it has large cohesive energy
(c) high melting point
(d) insoluble in all solvents

56. Which one of the following is the weakest kind of bonding in solids [1992]
(a) ionic (b) metallic
(c) Vander Waal's (d) covalent

57. For an electronic valve, the plate current I and plate voltage V in the space charge limited region are related as [1992]
(a) I is proportional to $V^{3/2}$
(b) I is proportional to $V^{2/3}$
(c) I is proportional to V
(d) I is proportional to V^2

58. The depletion layer in the *p-n* junction region is caused by [1991]
(a) drift of holes
(b) diffusion of charge carriers
(c) migration of impurity ions
(d) drift of electrons

59. When *n*-type semiconductor is heated [1989]
(a) number of electrons increases while that of holes decreases
(b) number of holes increases while that of electrons decreases
(c) number of electrons and holes remain same
(d) number of electrons and holes increases equally.

60. p-n junction is said to be forward biased, when [1988]
(a) the positive pole of the battery is joined to the *p*-semiconductor and negative pole to the *n*-semiconductor
(b) the positive pole of the battery is joined to the *n*-semiconductor and *p*-semiconductor joined to negative pole of the battery
(c) the positive pole of the battery is connected to *n*- semiconductor and *p*-semiconductor is connected to the positive pole of the battery
(d) a mechanical force is applied in the forward direction

61. At absolute zero, Si acts as [1988]
(a) non-metal (b) metal
(c) insulator (d) none of these

Topic 2 Junction Transistor

62. For transistor action, which of the following statements is correct? [2020]
(a) Base, emitter and collector regions should have same size.
(b) Both emitter junction as well as the collector junction are forward biased.
(c) The base region must be very thin and lightly doped.
(d) Base, emitter and collector regions should have same doping concentrations.

63. In the circuit shown in the figure, the input voltage V_i is 20 V, $V_{BE} = 0$ and $V_{CE} = 0$. The values of I_B, I_C and β are given by *[2018]*

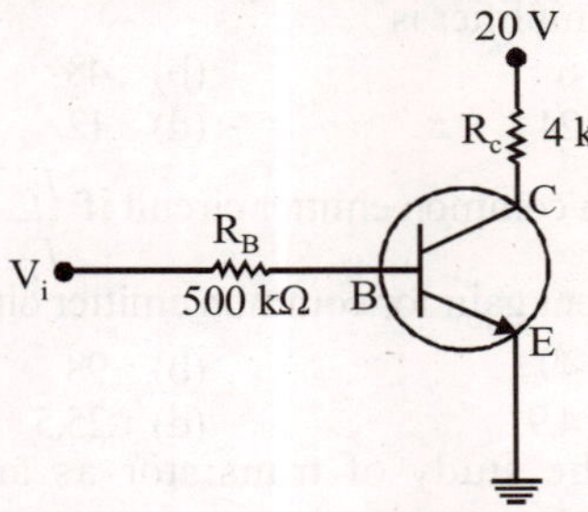

(a) $I_B = 40\ \mu A$, $I_C = 10\ mA$, $\beta = 250$
(b) $I_B = 25\ \mu A$, $I_C = 5\ mA$, $\beta = 200$
(c) $I_B = 40\ \mu A$, $I_C = 5\ mA$, $\beta = 125$
(d) $I_B = 20\ \mu A$, $I_C = 5\ mA$, $\beta = 250$

64. In a common emitter transistor amplifier the audio signal voltage across the collector is 3V. The resistance of collector is 3 kΩ. If current gain is 100 and the base resistance is 2 kΩ, the voltage and power gain of the amplifier is *[2017]*
(a) 15 and 200 (b) 150 and 15000
(c) 20 and 2000 (d) 200 and 1000

65. A npn transistor is connected in common emitter configuration in a given amplifier. A load resistance of 800 Ω is connected in the collector circuit and the voltage drop across it is 0.8 V. If the current amplification factor is 0.96 and the input resistance of the circuit is 192Ω, the voltage gain and the power gain of the amplifier will respectively be : *[2016]*
(a) 4, 3.84 (b) 3.69, 3.84
(c) 4, 4 (d) 4, 3.69

66. The input signal given to a CE amplifier having a voltage gain of 150 is $V_i = 2\cos\left(15t + \frac{\pi}{3}\right)$. The corresponding output signal will be : *[2015 RS]*
(a) $75\cos\left(15t + \frac{2\pi}{3}\right)$
(b) $2\cos\left(15t + \frac{5\pi}{6}\right)$
(c) $300\cos\left(15t + \frac{4\pi}{3}\right)$
(d) $300\cos\left(15t + \frac{\pi}{3}\right)$

67. In a common emitter (CE) amplifier having a voltage gain G, the transistor used has transconductance 0.03 mho and current gain 25. If the above transistor is replaced with another one with transconductance 0.02 mho and current gain 20, the voltage gain will be *2013]*
(a) 1.5G (b) $\frac{1}{3}$G (c) $\frac{5}{4}$G (d) $\frac{2}{3}$G

68. One way in which the operation of a n-p-n transistor differs from that of a p-n-p *[NEET Kar. 2013]*
(a) the emitter junction is reversed biased in n-p-n
(b) the emitter junction injects minority carriers into the base region of the p-n-p
(c) the emitter injects holes into the base of the p-n-p and electrons into the base region of n-p-n
(d) the emitter injects holes into the base of n-p-n

69. In a *CE* transistor amplifier, the audio signal voltage across the collector resistance of 2kΩ is 2V. If the base resistance is 1kΩ and the current amplification of the transistor is 100, the input signal voltage is : *[2012]*
(a) 0.1 V (b) 1.0 V
(c) 1 mV (d) 10 mV

70. Transfer characteristics [output voltage (V_0) vs input voltage (V_1)] for a base biased transistor in CE configuration is as shown in the figure. For using transistor as a switch, it is used : *[2012]*

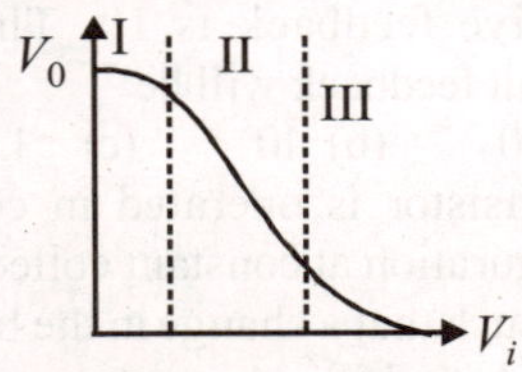

(a) in region III
(b) both in region (I) and (III)
(c) in region II
(d) in region (I)

71. The input resistance of a silicon transistor is 100 Ω. Base current is changed by 40 μA which results in a change in collector current by 2 mA. This transistor is used as a common emitter amplifier with a load resistance of 4 KΩ. The voltage gain of the amplifier is : *[2012M]*
(a) 2000 (b) 3000
(c) 4000 (d) 1000

72. A transistor is operated in common emitter configuration at $V_C = 2V$ such that a change in the base current from 100 μA to 300 μA produces a change in the collector current from 10mA to 20 mA. The current gain is [2011]
(a) 50 (b) 75 (c) 100 (d) 25

73. A common emitter amplifier has a voltage gain of 50, an input impedance of 100Ω and an output impedance of 200Ω. The power gain of the amplifier is [2010, 2007]
(a) 500 (b) 1000 (c) 1250 (d) 50

74. For transistor action : [2010]
(1) Base, emitter and collector regions should have similar size and doping concentrations.
(2) The base region must be very thin and lightly doped.
(3) The emitter-base junction is forward biased and base-collector junction is reverse based.
(4) Both the emitter-base junction as well as the base-collector junction are forward biased.
(a) (3), (4) (b) (4), (1)
(c) (1), (2) (d) (2), (3)

75. A transistor is operated in common-emitter configuration at $V_c = 2$ V such that a change in the base current from 100 μA to 200 μA produces a change in the collector current from 5 mA to 10 mA. The current gain is [2009]
(a) 100 (b) 150 (c) 50 (d) 75

76. The voltage gain of an amplifier with 9% negative feedback is 10. The voltage gain without feedback will be [2008]
(a) 90 (b) 10 (c) 1.25 (d) 100

77. A transistor is operated in common emitter configuration at constant collector voltage $V_c =$ 1.5V such that a change in the base current from 100 μA to 150 μA produces a change in the collector current from 5 mA to 10 mA. The current gain (β) is [2006]
(a) 75 (b) 100 (c) 50 (d) 67

78. An n-p-n transistor conducts when [2003]
(a) both collector and emitter are negative with respect to the base
(b) both collector and emitter are positive with respect to the base
(c) collector is positive and emitter is negative with respect to the base
(d) collector is positive and emitter is at same potential as the base

79. In the case of a common emitter transistor amplifier, the ratio of the collector current to the emitter current I_c/I_e is 0.96. The current gain of the amplifier is [2002]
(a) 6 (b) 48
(c) 24 (d) 12

80. For a common emitter circuit if $\frac{I_C}{I_E} = 0.98$ then current gain for common emitter circuit will be
(a) 49 (b) 98 [2001]
(c) 4.9 (d) 25.5

81. In the study of transistor as amplifier, if $\alpha = \frac{I_C}{I_E}$ and $\beta = \frac{I_C}{I_B}$, where, I_C, I_B and I_E are the collector, base and emitter currents, then [2000]
(a) $\beta = \frac{(1+\alpha)}{\alpha}$ (b) $\beta = \frac{(1-\alpha)}{\alpha}$
(c) $\beta = \frac{\alpha}{(1-\alpha)}$ (d) $\beta = \frac{\alpha}{(1+\alpha)}$

82. The transfer ratio β of a transistor is 50. The input resistance of the transistor when used in the common emitter configuration is 1 kΩ. The peak value of the collector A.C. current for an A.C. input voltage of 0.01 V peak is [1998]
(a) 100 μA (b) 0.01 mA
(c) 0.25 mA (d) 500 μA

83. The current gain for a transistor working as common-base amplifier is 0.96. If the emitter current is 7.2 mA, then the base current is
(a) 0.29 mA (b) 0.35 mA [1996]
(c) 0.39 mA (d) 0.43 mA

84. When an *n–p–n* transistor is used as an amplifier then [1996]
(a) the electrons flow from emitter to collector
(b) the holes flow from emitter to collector
(c) the electrons flow from collector to emitter
(d) the electrons flow from battery to emitter

85. An oscillator is nothing but an amplifer with [1994]
(a) positive feedback (b) negative feedback
(c) large gain (d) no feedback

86. The part of the transistor which is heavily doped to produce large number of majority carriers is
(a) emitter [1993]
(b) base
(c) collector
(d) any of the above depending upon the nature of transistor

87. For amplification by a triode, the signal to be amplified is given to *[1992]*
(a) the cathode (b) the grid
(c) the glass-envelope (d) the anode

88. To use a transistor as an amplifier *[1991]*
(a) the emitter base junction is forward biased and the base collector junction is reverse biased
(b) no bias voltage is required
(c) both junctions are forward biased
(d) both junctions are reverse biased.

89. In a common base amplifier the phase difference between the input signal voltage and the output voltage is *[1990]*
(a) 0 (b) $\frac{\pi}{4}$
(c) $\frac{\pi}{2}$ (d) π

90. Radiowaves of constant amplitude can be generated with *[1989]*
(a) FET (b) filter
(c) rectifier (d) oscillator

Topic 3 Digital Electronics and Logic Gates

91. For the given circuit, the input digital signals are applied at the terminals A, B and C. What would be the output at the terminal y? *[2021]*

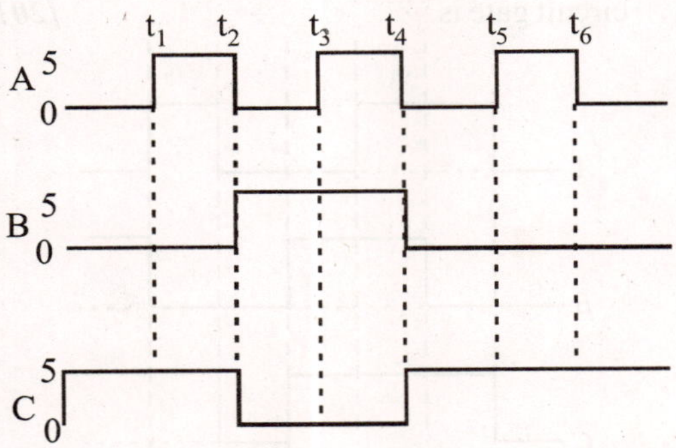

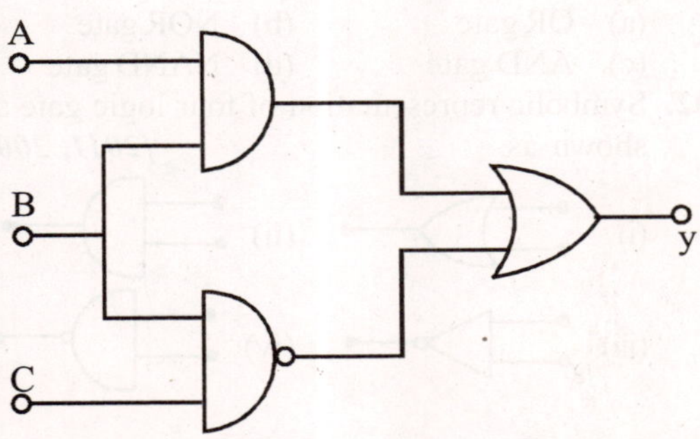

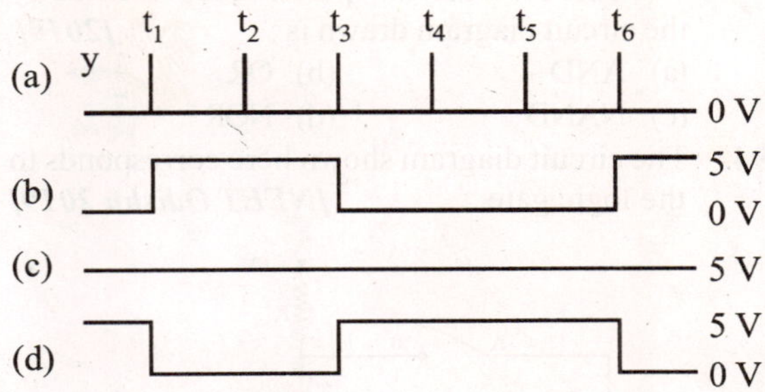

92. For the logic circuit shown, the truth table is: *[2020]*

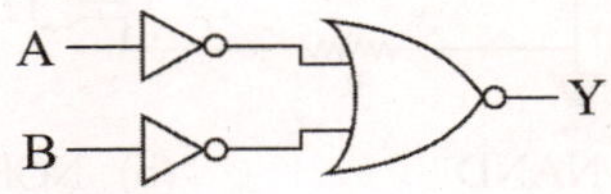

(a)

A	B	Y
0	0	0
0	1	1
1	0	1
1	1	1

(b)

A	B	Y
0	0	1
0	1	1
1	0	1
1	1	0

(c)

A	B	Y
0	0	1
0	1	0
1	0	0
1	1	0

(d)

A	B	Y
0	0	0
0	1	0
1	0	0
1	1	1

93.

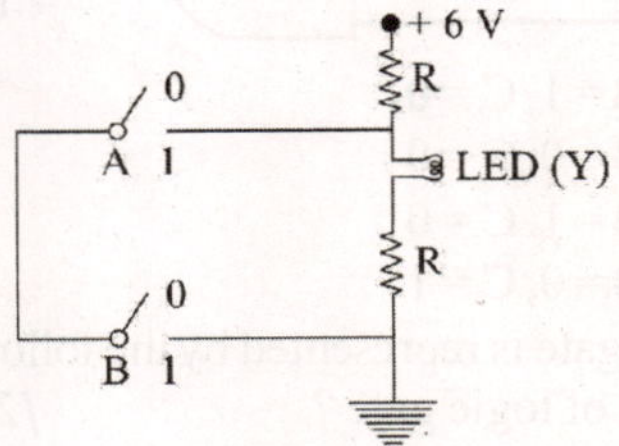

The **correct** Boolean operation represented by the circuit diagram drawn is : *[2019]*

(a) AND (b) OR
(c) NAND (d) NOR

94. The circuit diagram shown here corresponds to the logic gate, *[NEET Odisha 2019]*

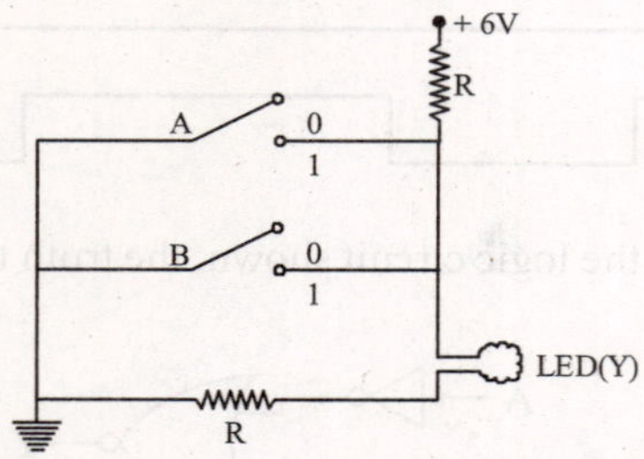

(a) NAND (b) NOR
(c) AND (d) OR

95. In the combination of the following gates the output Y can be written in terms of inputs A and B as *[2018]*

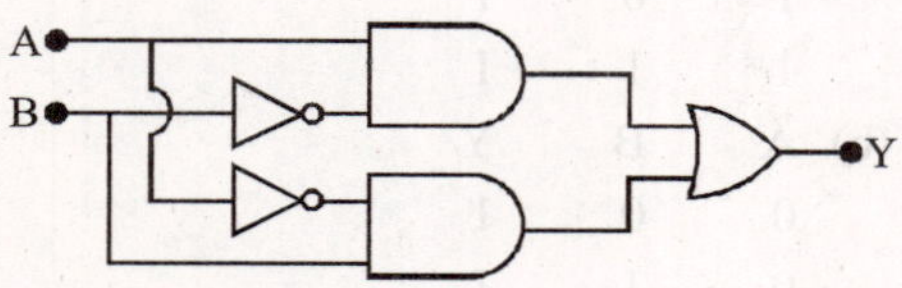

(a) $\overline{A\cdot B}$ (b) $A\cdot\overline{B}+\overline{A}\cdot B$
(c) $\overline{A+B}$ (d) $\overline{A\cdot B}+A\cdot B$

96. The given electrical network is equivalent to : *[2017, 2006, 2000]*

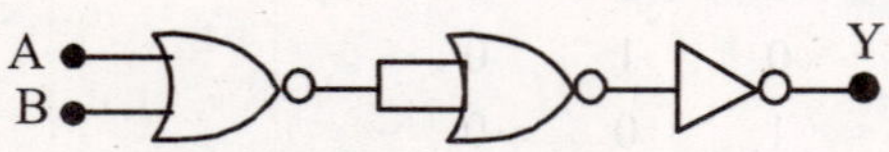

(a) OR gate (b) NOR gate
(c) NOT gate (d) AND gate

97. To get output 1 for the following circuit, the correct choice for the input is *[2016, 2012, 2010]*

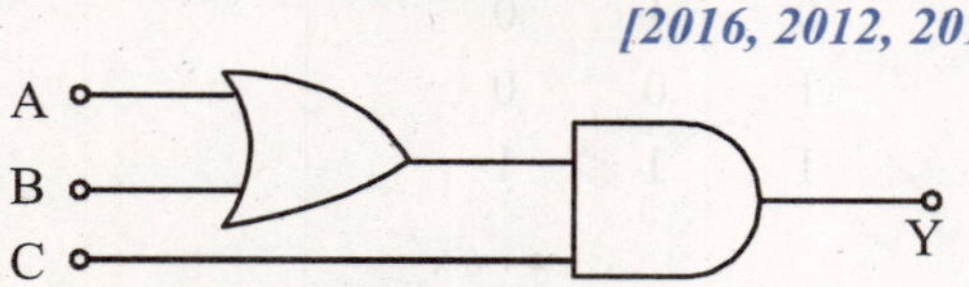

(a) A=0, B=1, C=0
(b) A=1, B=0, C=0
(c) A=1, B=1, C=0
(d) A=1, B=0, C=1

98. Which logic gate is represented by the following combination of logic gate ? *[2015]*

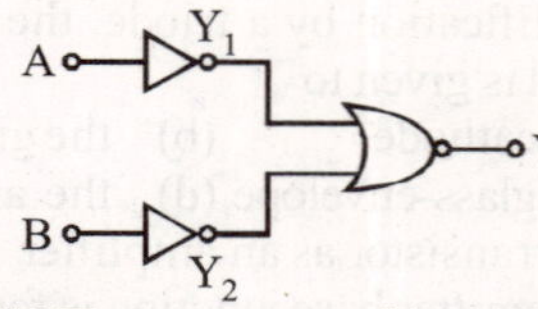

(a) NAND (b) AND
(c) NOR (d) OR

99. The output from a NAND gate is divided into two in parallel and fed to another NAND gate. The resulting gate is a *[NEET Kar. 2013]*

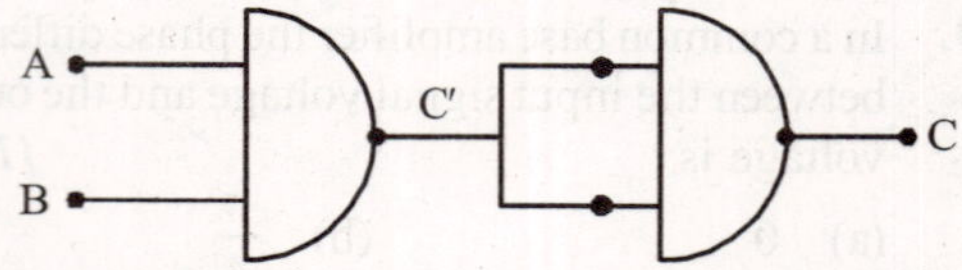

(a) NOT gate (b) AND gate
(c) NOR gate (d) OR gate

100. The output(X) of the logic circuit shown in figure will be *[2013]*

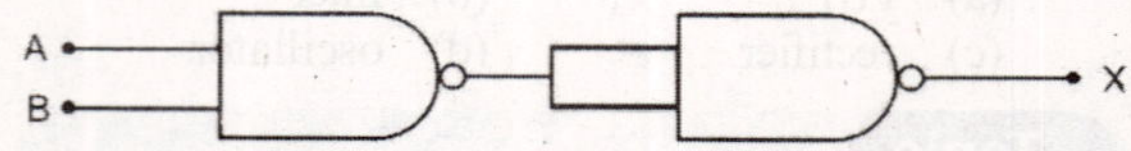

(a) $X=\overline{A.B}$ (b) X=A.B
(c) $X=\overline{A+B}$ (d) $X=\overline{\overline{A}}+\overline{\overline{B}}$

101. The figure shows a logic circuit with two inputs *A* and *B* and the output *C*. The voltage wave forms across *A*, *B* and *C* are as given. The logic circuit gate is : *[2012]*

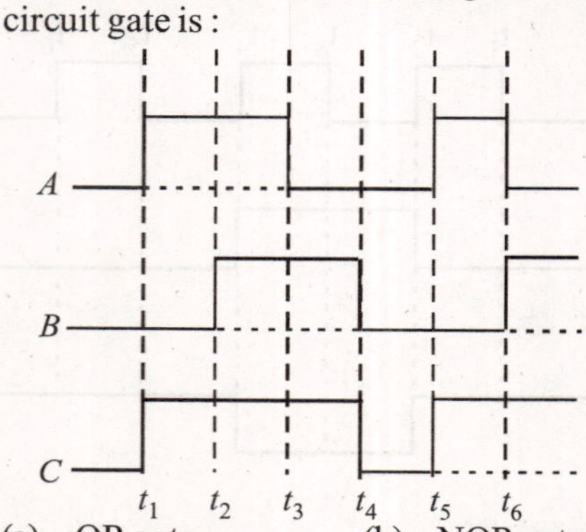

(a) OR gate (b) NOR gate
(c) AND gate (d) NAND gate

102. Symbolic representation of four logic gate are shown as *[2011, 2009]*

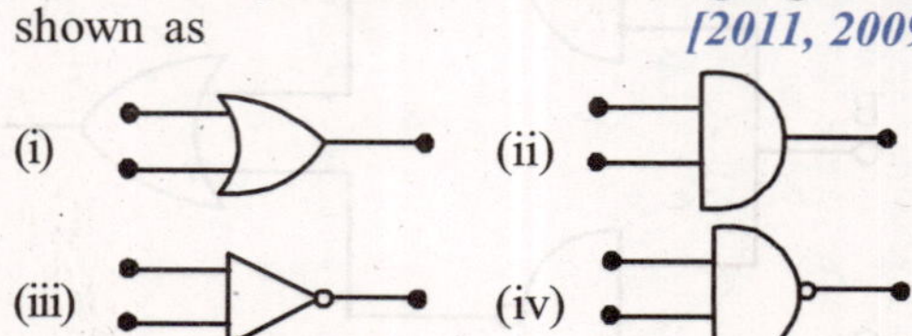

Pick out which ones are for AND, NAND and NOT gates, respectively

(a) (ii), (iii) and (iv) (b) (iii), (ii) and (i)
(c) (iii), (iii) and (iv) (d) (ii), (iv) and (iii)

103. The device that can act as a complete electronic circuit is *[2010]*

(a) junction diode (b) integrated circuit
(c) junction transistor (d) zener diode

104. The following Figure shows a logic gate circuit with two inputs *A* and *B* and the output *Y*. The voltage waveforms of A, B and Y are given :

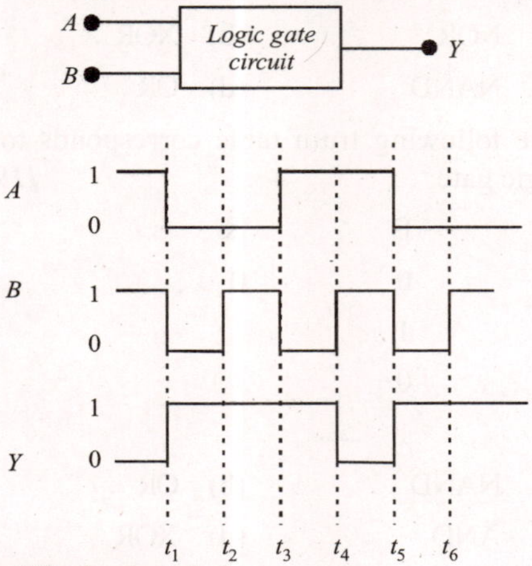

The logic gate is : *[2010]*

(a) NAND gate (b) NOR gate
(c) OR gate (d) AND gate

105. The circuit *[2008]*

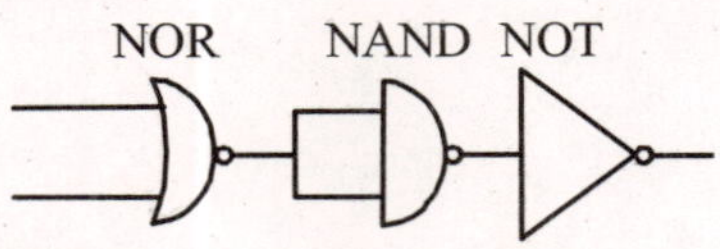

is equivalent to

(a) AND gate (b) NAND gate
(c) NOR gate (d) OR gate

106. In the following circuit, the output *Y* for all possible inputs *A* and *B* is expressed by the truth table *[2007]*

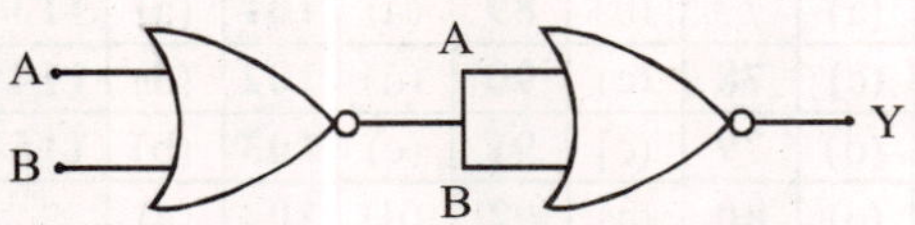

(a)

A	B	Y
0	1	1
0	1	1
1	0	1
1	1	0

(b)

A	B	Y
0	0	1
0	1	0
1	0	0
1	1	0

(c)

A	B	Y
0	0	0
0	1	1
1	0	1
1	1	1

(d)

A	B	Y
0	0	0
0	1	0
1	0	0
1	1	1

107. The following figure shows a logic gate circuit with two inputs *A* and *B* and the output *C*. The voltage waveforms of *A*, *B* and *C* are as shown below

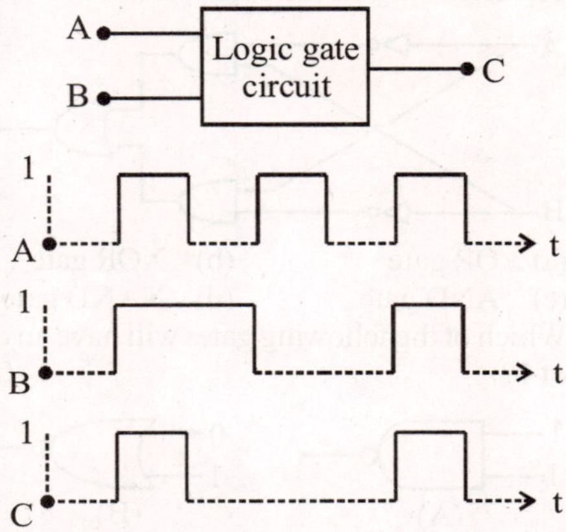

The logic circuit gate is *[2006]*

(a) NAND gate (b) NOR gate
(c) OR gate (d) AND gate

108. The output of OR gate is 1 *[2004]*

(a) if either input is zero
(b) if both inputs are zero
(c) if either or both inputs are 1
(d) only if both inputs are 1

109. Following diagram performs the logic function of *[2003]*

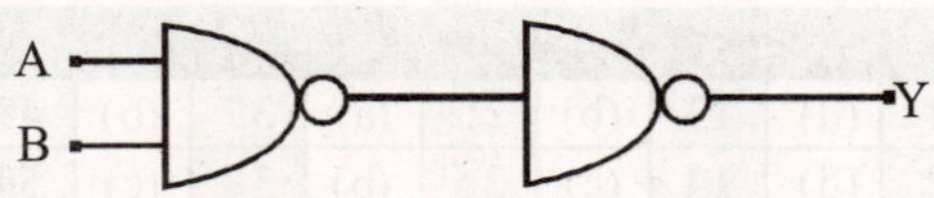

(a) XOR gate (b) AND gate
(c) NAND gate (d) OR gate

110. Which gate is represented by the following truth table ? *[2001, 1998, 1994]*

A	B	Y
0	0	1
1	0	1
0	1	1
1	1	0

(a) XOR (b) NOT
(c) NAND (d) AND

111. A gate has the following truth table

P	Q	R
1	1	1
1	0	0
0	1	0
0	0	0

The gate is *[2000]*

(a) AND (b) NOR
(c) OR (d) NAND

112. The following circuit represents *[1999]*

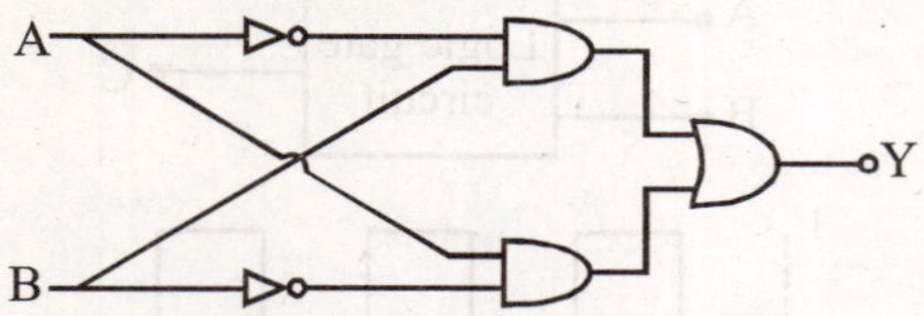

(a) OR gate (b) XOR gate
(c) AND gate (d) NAND gate

113. Which of the following gates will have an output of 1? *[1998]*

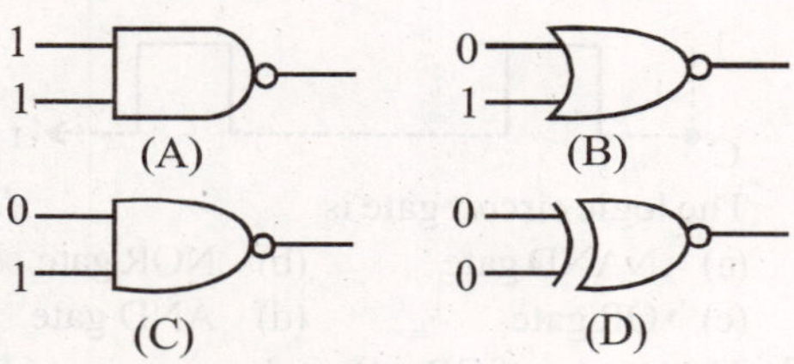

(a) D (b) A
(c) B (d) C

114. The following truth table belongs to which of the following four gates? *[1997]*

A	B	Y
1	1	0
1	0	0
0	1	0
0	0	1

(a) NOR (b) XOR
(c) NAND (d) OR

115. The following truth table corresponds to the logic gate *[1991]*

A	B	X
0	0	0
0	1	1
1	0	1
1	1	1

(a) NAND (b) OR
(c) AND (d) XOR

ANSWER KEY

1	(d)	13	(b)	25	(a)	37	(b)	49	(c)	61	(c)	73	(c)	85	(a)	97	(d)	109	(b)
2	(d)	14	(c)	26	(b)	38	(c)	50	(a)	62	(c)	74	(d)	86	(a)	98	(b)	110	(c)
3	(a)	15	(d)	27	(c)	39	(d)	51	(a)	63	(c)	75	(c)	87	(b)	99	(b)	111	(a)
4	(b)	16	(c)	28	(d)	40	(b)	52	(b)	64	(b)	76	(d)	88	(a)	100	(b)	112	(b)
5	(c)	17	(d)	29	(c)	41	(a)	53	(c)	65	(a)	77	(b)	89	(a)	101	(a)	113	(d)
6	(c)	18	(c)	30	(a)	42	(d)	54	(d)	66	(c)	78	(c)	90	(d)	102	(d)	114	(a)
7	(d)	15	(b)	31	(c)	43	(b)	55	(b)	67	(d)	79	(c)	91	(c)	103	(b)	115	(b)
8	(b)	20	(d)	32	(c)	44	(d)	56	(c)	68	(c)	80	(a)	92	(d)	104	(a)		
9	(c)	21	(d)	33	(c)	45	(a)	57	(a)	69	(d)	81	(c)	93	(c)	105	(c)		
10	(d)	22	(b)	34	(c)	46	(d)	58	(b)	70	(b)	82	(d)	94	(d)	106	(c)		
11	(a)	23	(a)	35	(c)	47	(a)	59	(d)	71	(a)	83	(a)	95	(b)	107	(d)		
12	(d)	24	(c)	36	(a)	48	(d)	60	(a)	72	(a)	84	(a)	96	(b)	108	(c)		

Hints & Solutions

1. **(d)** Current $\frac{I_n}{I_p} = \frac{\mu_e}{\mu_h} \because \mu_e > \mu_h$

$\therefore I_n > I_p$

Therefore current in n-type > Current in p-type

> **NOTES** The mobility of electron is 2-3 times the mobility of holes.

2. **(d)** • A zener diode is connected in reverse biased when used as voltage regulator as in reverse biased after breakdown, voltage becomes constant.
• Potential barrier of silicon diode is nearly 0.7V.

3. **(a)** Due to reverse biasing, the width of the depletion region increases in a *p-n* junction diode.

4. **(b)** In p-type semiconductor, trivalent impurities are added to intrinsic semiconductor, which creates holes which are majority charge carriers.

5. **(c)** $\lambda = \frac{12400}{E_g} Å = \frac{12400}{1.9} = 6526Å \simeq 654\text{ nm}$

6. **(c)** On heating, number of electron-hole pairs increases, so overall resistance of diode will change.
Hence forward biasing and reversed biasing both are changed.

7. **(d)** 

In forward bias, $V_1 > V_2$ i.e., in figure (d) p-side is at higher potential w.r.t. n-side of p-n junction diode.

8. **(b)** Since Diode is in forward bias, so the value of current flowing through AB

$$i = \frac{\Delta V}{R} = \frac{4-(-6)}{1\times 10^3} = \frac{10}{10^3} = 10^{-2}\text{A}$$

9. **(c)** Here P-N junction diode rectifies half of the ac wave i.e., acts as half wave rectifier. During + ve half cycle
Diode → forward biased output across will be

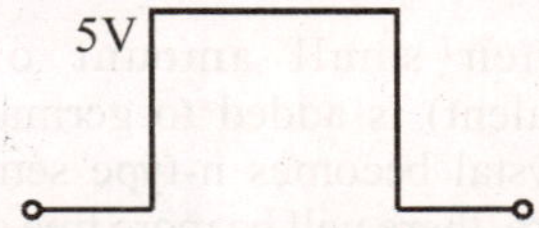

During –ve half cycle Diode → reverse biased output will not obtained.

10. **(d)** Current $I = \frac{V}{R} = \frac{(3.5-0.5)}{100}$ A

[$\because$ Barrier potential $V_B = 0.5$V]

$= \frac{3}{100} = 30$mA

> **NOTES** On an average the potential barrier in P-N junction is v 0.5 V and the width of depletion region 10^{-6} m
> Barrier electric field, $E = \frac{V}{d} = \frac{0.5}{10-6} = 5\times10^5$ V/m

11. **(a)** The given graph represents V-I characteristics of solar cell.

12. **(d)** The barrier potential of a p-n junction depends on amount of doping, type of semiconductor material and temperature.

13. **(b)** In a n-type semiconductor holes are minority carriers and pentavalent atoms are dopants.

14. **(c)** In p-region of p-n junction
holes concentration > electrons concentration and in n-region of p-n junction.
electrons concentration > holes concentration.

15. **(d)** Here D_1 is in forward bias and D_2 is in reverse bias so, D_1 will conduct and D_2 will not conduct. Thus, no current will flow through *DC*.

$I = \frac{V}{R} = \frac{5}{10} = \frac{1}{2}$ Amp.

16. **(c)** Electronic configuration of ^{6}C
$^6C = 1s^2, 2s^2\,2p^2$
The electronic configuration of ^{14}Si
$^{14}Si = 1s^2, 2s^2\,2p^6, 3s^2\,3p^2$
As they are away from Nucleus, so effect of nucleus is low for Si even for Sn and Pb are almost metalic.

17. **(d)** In forward biasing of the p-n junction, the positive terminal of the battery is connected to p-side and the negative terminal of the battery is connected to n-side. The depletion region becomes thin.

> **NOTES** An ideal junction diode when forward biased offers zero resistance. Voltage drop across such a junction diode is zero. In reverse bias diode offers infinite resistance and voltage drop across it is equal to voltage applied.

18. **(c)** When small amount of antimony (pentavalent) is added to germanium crystal then crystal becomes n-type semi conductor. Therefore, there will be more free electrons than holes in the semiconductor.

19. **(b)** Only in (A) and (C) diodes are forward biased as p-type should be at higher potential and n-type at lower potential.

20. **(d)** $n_i^2 = n_e n_h$

$(1.5 \times 10^{16})^2 = n_e (4.5 \times 10^{22})$

$\Rightarrow \quad n_e = 0.5 \times 10^{10}$

or $\quad n_e = 5 \times 10^9$

Given $\quad n_h = 4.5 \times 10^{22}$

$\Rightarrow n_h >> n_e$

$\therefore$ Semiconductor is p-type and $n_e = 5 \times 10^9 \text{ m}^{-3}$.

21. **(d)** Voltage across zener diode is constant.

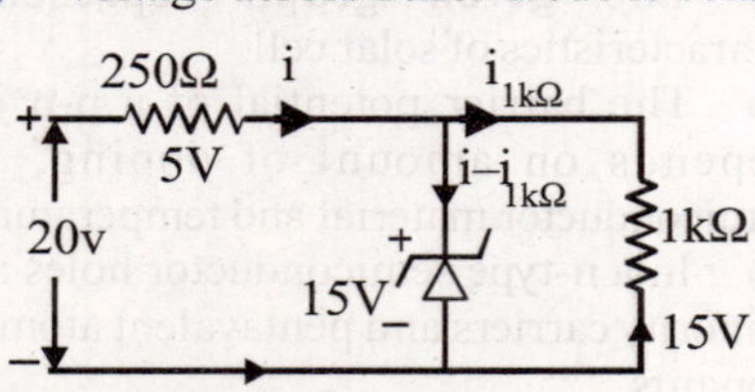

Current in 1kΩ resistor,

$$(i)_{1k\Omega} = \frac{15\text{volt}}{1k\Omega} = 15 \text{ mA}$$

Current in 250Ω resistor,

$$(i)_{250\Omega} = \frac{(20-15)\text{V}}{250\Omega} = \frac{5\text{V}}{250\Omega}$$

$$= \frac{20}{1000}\text{A} = 20 \text{ mA}$$

$$\therefore (i)_{\text{zener diode}} = (20-15) = 5\text{mA}.$$

22. **(b)** Majority carriers in an n-type semiconductor are electrons.

> **NOTES** The current due to minority carriers in the junction diode is independent of the applied voltage. It only depends upon the temperature of the diode.

23. **(a)** For a metal, conductivity decreases with increase in temperature.
Also, metal has high melting point.

24. **(c)** $\lambda_{max} = \frac{hc}{E}$

$$= \frac{6.6 \times 10^{-34} \times 3 \times 10^8}{2.5 \times 1.6 \times 10^{-19}} \simeq 5000 \text{ Å}$$

The wavelength detected by photodiode should be less than λ_{max}. Hence it can detect a signal of wavelength 4000Å.

25. **(a)** $d = \frac{\sqrt{3}}{2} a$

$$3.7 = \frac{\sqrt{3}}{2} a$$

$$a = \frac{2 \times 3.7}{\sqrt{3}} = 4.3\text{Å}$$

26. **(b)** $E_g = 2.0 \text{ eV} = 2 \times 1.6 \times 10^{-19} \text{ J}$

$E_g = h\nu$

$$\therefore \nu = \frac{E_g}{h} = \frac{2 \times 1.6 \times 10^{-19} \text{ J}}{6.62 \times 10^{-34} \text{ Js}}$$

$= 0.4833 \times 10^{15} \text{ s}^{-1} = 4.833 \times 10^{14} \text{ Hz} \simeq 5 \times 10^{14} \text{ Hz}$

27. **(c)** For a cubic crystal,
$a = b = c$ and $\alpha = \beta = \gamma = 90°$

28. **(d)** For a *p*-type semiconductor, the acceptor energy level, as shown in the diagram, is slightly above the top E_v of the valence band. With very small supply of energy an electron from the valence band can jump to the level E_A and ionise acceptor negatively.

> **NOTES** In p–type semiconductor-acceptor energy level lies just above the valence band. And in N-type semiconductor-donor energy level lies just below the conduction band.

29. **(c)** (a) is true as in case of conductors either the conduction & valence band overlap or conduction band is partially filled.

(b) is true as insulators have energy gap of the order of 5 to 10 eV.

(c) is false as resistivity (opposite of conductivity) decreases with increase in temperature.

(d) is true as with increase in temperature more and more electrons jump to the conduction band. So, conductivity increases.

30. **(a)** Due to strong electronegativity of carbon.

31. (c)

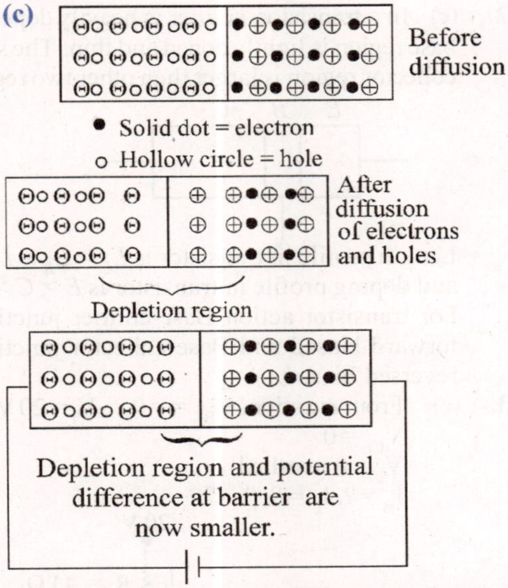

Number of donors is more because electrons from –ve terminal of the cell pushes (enters) the n side and decreases the number of uncompensated pentavalent ion due to which potential barrier is reduced. The neutralised pentavalent atom are again in position to donate electrons.

32. (c) At a certain reverse bias voltage, zener diode allows current to flow through it and hence maintains the voltage supplied to any load. It is used for stabilisation.

33. (c) In semiconductors, the conduction band is empty and the valence band is completely filled at 0 K. No electron from valence band can cross over to conduction band at 0K. At room temperature some electrons in the valence band jump over to the conduction band due to the small forbidden gap, i.e. 1 eV.

34. (c) $V = \frac{V_0}{\pi} = \frac{10}{\pi} V$

35. (c) Electromotive force depends upon intensity of light falling, it does not depend on frequency of barrier voltage.

36. (a) In case of full wave rectifier, Fundamental frequency = 2 × mains frequency = 2 × 50 = 100Hz.

37. (b) Barrier potential does not depend on diode design. Barrier potential depends upon temperature, doping density, and forward biasing.

NOTES Potential barrier is the potential difference created across the P-N junction due to the diffusion of electrons and holes.

38. (c) In reverse biasing, the conduction across the *p-n* junction does not take place due to majority carriers but takes place due to minority carriers if the voltage of external battery is large. The size of the depletion region increases thereby increasing the potential barrier.

39. (d) In forward biasing, the diode conducts. For ideal junction diode, the forward resistance is zero; therefore, entire applied voltage occurs across external resistance R i.e., there occurs no potential drop, so potential across R is V in forward biased.

40. (b) For conduction, *p-n* junction must be forward biased. For conduction *p*-side should be connected to higher potential and *n*-side to lower potential.

41. (a) At 0K, motion of free electrons stop. Hence conductivity becomes zero. Therefore, at 0K intrinsic semiconductor becomes insulator.

42. (d)

43. (b) We know that in forward bias of *p-n* junction diode, when positive terminal is connected to *p*-type diode, the repulsion of holes takes place which decreases the width of potential barrier by striking the combination of holes and electrons.

44. (d) Depletion layer is formed by diffusion of holes and electrons from *p*-type semiconductor to n-type semiconductor and vice-versa. Hence, donor and acceptor atom get positive and negative charge leading to formation of *p-n* junction. The donor and acceptor are immobile.

45. (a) *n*-type of silicon semiconductor is formed when impurity is mixed with pentavalent atom. Out of given choices only phosphorus is pentavalent.

46. (d) Holes are produced due to missing of electrons.

47. (a) In reverse bias, the current through a *p-n* junction is almost zero.

48. (d) During the formation of a junction diode, holes from *p*-region diffuse into *n*-region and electrons from n-region diffuse into *p*-region. In both cases, when an electron meets a hole, they cancel the effect of each other and as a result, a thin layer at the junction becomes free from any of charge carriers. This is called depletion layer. There is a potential gradient in the depletion

layer, negative on the p-side and positive on the n-side. The potential difference thus developed across the junction is called potential barrier.

49. **(c)** p-type germanium semiconductor is formed when it is doped with a trivalent impurity atom.

> **NOTES** Doping is the process of addition of impurity. When pure or intrinsie semiconductor is mixed with small amounts of certain specific impurities with valency different from that of the parent material the number of mobile electrons/holes transtically changes.

50. **(a)** Arsenic contains 5 electrons in its outermost shell. When Arsenic is mixed with silicon there is one electron extra in silicon crystal. Hence, such type of semi conductor is n-type semi conductor.

51. **(a)** Phosphorous (P) is pentavalent and silicon is tetravalent. Therefore, when silicon is doped with pentavalent impurity, it forms a n-type semiconductor.

52. **(b)** When p-n junction is reverse biased, the flow of current is due to drifting of minority charge carriers across the junction.

53. **(c)** The given circuit is a circuit of full wave rectifier.

54. **(d)** Copper is a conductor, so, its resistance decreases on decreasing temperature as thermal agitation decreases whereas germanium is semiconductor, therefore, on decreasing temperature resistance increases.

55. **(b)** Diamond is very hard due to large cohesive energy.

56. **(c)** Vander Waal's bonding is the weakest bonding in solids.

57. **(a)** According to Child's Law,

$I_a = KV_a^{3/2}$

Thus, $I \propto V^{3/2}$

58. **(b)** The depletion layer in the p-n junction region is caused by diffusion of charge carriers.

59. **(d)** Due to heating, when a free electron is produced then simultaneously a hole is also produced.

60. **(a)** For forward biasing of p-n junction, the positive terminal of external battery is to be connected to p-semiconductor and negative terminal of battery to the n-semiconductor.

61. **(c)** Semiconductors are insulators at room temperature.

> **NOTES** At absolute zero temperature (OK) conduction band of semiconductor is completely empty i.e., $\sigma = 0$. So semiconductor behaves as an insulator.

62. **(c)** In a transistor, emitter is heavily doped, the base region is lightly doped and thin. The size of collector region is larger than other two regions.

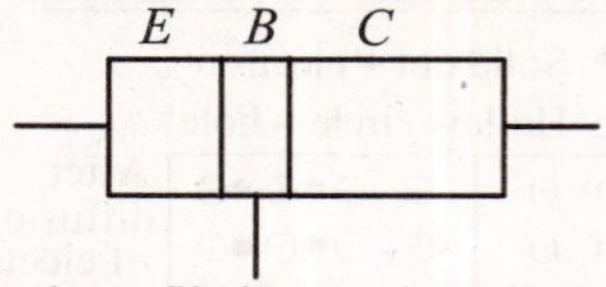

Length profile in transistor is $L_C > L_E > L_B$ and doping profile in transistor is $E > C > B$. For transistor action Base-emitter junction is forward biased and Base-collector junction is reversed biased.

63. **(c)** From question, $V_{BE} = = 0$, $V_i = 20$ V

$V_{CE} = 0$

$V_b = 0$ (earthed)

$I_B = ?, I_C = ?, \beta = ?$

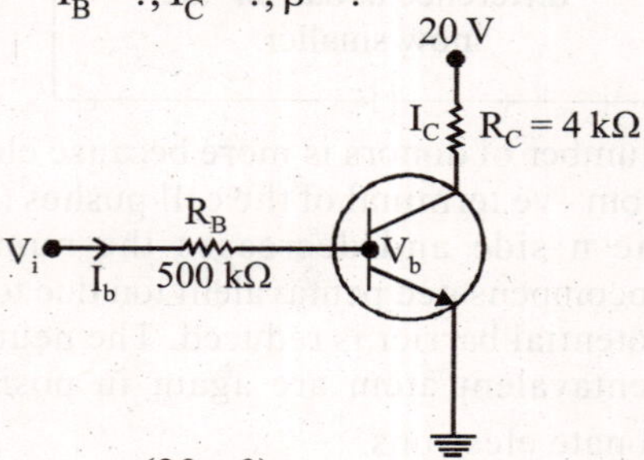

$$I_C = \frac{(20-0)}{4\times10^3} = 5\times10^{-3} = 5 \text{ mA}$$

$V_i = V_{BE} + I_B R_B$ or, $V_i = 0 + I_B R_B$

$\Rightarrow 20 = I_B \times 500 \times 10^3$

$$\Rightarrow I_B = \frac{20}{500\times10^3} = 40\ \mu\text{A}$$

$$\beta = \frac{I_C}{I_b} = \frac{25\times10^{-3}}{40\times10^{-6}} = 125$$

64. **(b)** Given, current gain $\beta = 100$, $R_c = 3$kΩ, $R_b = 2$kΩ

$$\text{Voltage gain } (A_v) = \beta\frac{R_c}{R_b} = 100\left(\frac{3}{2}\right) = 150$$

Power gain $= A_v\,\beta = 150\,(100) = 15000$

65. **(a)** Given: amplification factor $\alpha = 0.96$

load resistance, $R_L = 800\ \Omega$

input resistance, $R_i = 192\Omega$

$$\text{So, } \beta = \frac{\alpha}{1-\alpha} = \frac{0.96}{0.04} \Rightarrow \beta = 24$$

Voltage gain for common emitter configuration

$$A_v = \beta.\frac{R_L}{R_i} = 24\times\frac{800}{192} = 100$$

Power gain for common emitter configuration

$P_v = \beta A_v = 24 \times 100 = 2400$

Voltage gain for common base configuration

$$A_v = \alpha, \frac{R_L}{R_P} = 0.96 \times \frac{800}{192} = 4$$

Power gain for common base configuration

$P_v = A_v \alpha = 4 \times 0.96 = 3.84$

NOTES Transistor provides good power amplification when they are used in CE configuration. CC configuration amplifier is a power amplifier or current booster.

66. **(c)** **Given :** Voltage gain $A_V = 150$

$$V_i = 2\cos\left(15t + \frac{\pi}{3}\right); V_0 = ?$$

For CE transistor phase difference between input and output signal is $\pi = 180°$

Using formula, $A_V = \frac{V_0}{V_i}$

$\Rightarrow V_0 = A_V \times V_i$

$$= 150 \times 2\cos\left(15t + \frac{\pi}{3}\right)$$

$$\text{or } V_0 = 300\cos\left(15t + \frac{\pi}{3} + \pi\right)$$

$$V_0 = 300\cos\left(15t + \frac{4}{3}\pi\right)$$

67. **(d)** Voltage gain $A_v = \beta \frac{R_{out}}{R_{in}}$

$$\Rightarrow G = 25 \frac{R_{out}}{R_{in}} \quad \text{...(i)}$$

$$\text{Transconductance } g_m = \frac{\beta}{R_{in}}$$

$$\Rightarrow R_{in} = \frac{\beta}{g_m} = \frac{25}{0.03}$$

Putting this value of R_{in} in eqn. (i)

$$G = 25 \frac{R_{out}}{25} \times 0.03 \quad \text{...(ii)}$$

$$\therefore \quad G' = 20 \frac{R_{out}}{20} \times 0.02 \quad \text{...(iii)}$$

From eqs. (ii) and (iii)

Voltage gain of new transistor $G' = \frac{2}{3} G$

68. **(c)** In p-n-p transistor holes are injected into the base while electrons are injected into the base of n-p-n transistor. Emitter-base junction is forward biased.

69. (d) From the figure,

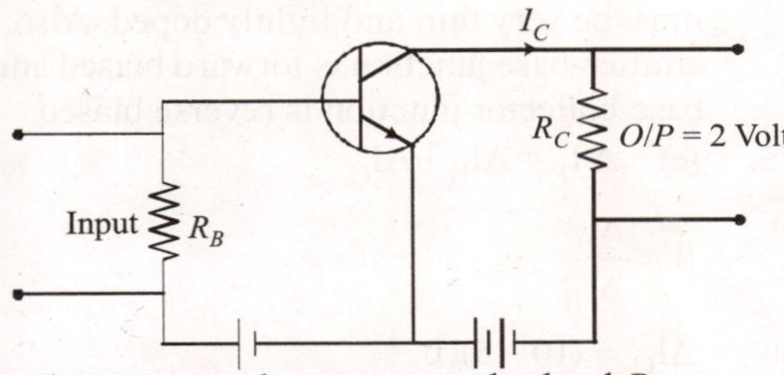

The output voltage, across the load R_C

$V_0 = I_C R_C = 2$

The collector current (I_C)

$$I_C = \frac{2}{2 \times 10^3} = 10^{-3} \text{ Amp}$$

Current gain (β)

$$(\beta) \text{ current gain} = \frac{I_C}{I_B} = 100$$

$$I_B = \frac{I_C}{100} = \frac{10^{-3}}{100} = 10^{-5} \text{ Amp}$$

Input voltage (V_i)

$V_i = R_B I_B = 1 \times 10^3 \times 10^{-5} = 10^{-2}$ Volt

$V_i = 10$ mV

70. (b) I $\rightarrow$ ON

III $\rightarrow$ OFF

In IInd state it is used as a amplifier it is active region.

71. (a) Voltage gain $(A_V) = \frac{V_{out}}{V_{in}} = \frac{I_{out}}{I_{in}} \times \frac{R_{out}}{R_{in}}$

$$A_V = \frac{2 \times 10^{-3}}{40 \times 10^{-6}} \times \frac{4 \times 10^3}{100} = 2 \times 100 = 2000$$

NOTES In a common-emitter amplifier, the voltage signal obtained across the collector and the emitter is 180° out of phase with the input voltage signal applied across the base and the emitter. This is indicated in the equation for A_V by its negative sign.

72. **(a)** The current gain

$$\beta = \frac{\Delta I_C}{\Delta I_B} = \frac{10\text{mA}}{200\mu\text{A}} = \frac{10 \times 10^3}{200} = 50$$

73. **(c)** Power gain = voltage gain × current gain

$$= V_G \cdot I_G = \frac{V_0}{V_i} \cdot \frac{I_0}{I_i}$$

$$= \frac{V_0^2}{V_i^2} \cdot \frac{R_i}{R_0} = 50 \times 50 \times \frac{100}{200}$$

$$= \frac{2500}{2} = 1250$$

74. **(d)** For transistor action, the base region must be very thin and lightly doped. Also, the emitter-base junction is forward biased and base-collector junction is reverse biased.

75. **(c)** $\Delta I_E = \Delta I_B + \Delta I_C$

$$\beta = \frac{\Delta I_C}{\Delta I_B}$$

$$\Delta I_C = (10-5)10^{-3} = 5\times10^{-3}\text{ A}$$

$$\Delta I_B = (200-100)10^{-6} = 100\times10^{-6}\text{ A}$$

$$\beta = \frac{5\times10^{-3}}{100\times10^{-6}} = \frac{5}{100}\times1000 = 50$$

76. **(d)** Negative feedback is applied to reduce the output voltage of an amplifier. If there is no negative feedback, the value of output voltage could be very high. In the options given, the maximum value of voltage gain is 100. Hence it is the correct option.

77. **(b)** $\Delta I_b = +50\mu A,\ \Delta I_c = 5\times10^{-3}\ A$

$$\beta = \frac{\Delta I_c}{\Delta I_b} = \frac{5\times10^{-3}}{50\times10^{-6}} = \frac{5\times1000}{50} = 100$$

78. **(c)** When the collector is positive and emitter is negative w.r.t. base, it causes the forward biasing for each junction, which causes conduction of current.

NOTES A transistor is mostly used in the active region of operation i.e., emitter base junction is forward biased and collector base junction is reverse biased.

79. **(c)** $\frac{I_c}{I_e} = 0.96$

$\Rightarrow\ I_c = 0.96 I_e$

But $I_e = I_c + I_b = 0.96 I_e + I_b$

$\Rightarrow\ I_b = 0.04 I_e$

$\therefore$ Current gain, $\beta = \frac{I_c}{I_b} = \frac{0.96 I_e}{0.04 I_e} = 24$

80. **(a)** We have,

$$\beta = \frac{\alpha}{1-\alpha} = \frac{0.98}{1-0.98} = \frac{0.98}{0.02} = 49$$

81. **(c)** As we know that $I_e = I_c + I_b$

Divide both side by I_e

$$\frac{I_e}{I_c} = 1 + \frac{I_b}{I_c} \Rightarrow \frac{1}{\alpha} = 1 + \frac{1}{\beta}$$

$$\beta = \frac{\alpha}{1-\alpha}$$

82. **(d)** $i_c = \beta\cdot\frac{V_s}{R_{in}} = 50\times\frac{0.01}{1000} = 500\times10^{-6}\text{ A}$

$= 500\ \mu\text{A}$

83. **(a)** Current gain $(\alpha) = 0.96$

$I_e = 7.2\text{ mA}$

$$\frac{I_c}{I_e} = \alpha = 0.96$$

$I_c = 0.96\times7.2\text{ mA} = 6.91\text{ mA}$

$I_e = I_c + I_b$

$\Rightarrow I_b = I_e - I_c = 7.2 - 6.91 = 0.29\text{ mA}$

84. **(a)** In an *n-p-n* transistor, the charge carriers, are free electrons in the transistor as well as in external circuit; these electrons flow from emitter to collector.

85. **(a)** A positive feed back from output to input in an amplifier provides oscillations of constant amplitude.

NOTES Due to the presence of some in herent electrical resistance amplitude of oscillation decreases. In order to obtain oscillations of constant amplitude, an arrangement for positive feedback from the output circuit to the input circuit so that the losses in the circuit can be compensated.

86. **(a)** The function of emitter is to supply the majority carriers. Emitter is heavily doped.

87. **(b)** The amplifying action of a triode is based on the fact that a small change in grid voltage causes a large change in plate current. The AC input signal which is to be amplified is superimposed on the grid potential.

88. **(a)** To use a transistor as an amplifier the emitter base junction is forward biased while the collector base junction is reverse biased.

89. **(a)** The phase difference between output voltage and input signal voltage in common base transistor or circuit is zero.

90. **(d)** Radiowaves of constant amplitude can be produced by using oscillator with proper feedback.

91. **(c)** Boolean expression for combination of logic gates

Output, $y = A.B + \overline{B.C}$

Time duration	Input Signals			Output Signal(y)		
	A	B	C	AB	$\overline{B.C}$	$y = A.B + \overline{B.C}$
$0 - t_1$	0	0	1	0	1	1
$t_1 - t_2$	1	0	1	0	1	1
$t_2 - t_3$	0	1	0	0	1	1
$t_3 - t_4$	1	1	0	1	1	1
$t_4 - t_5$	0	0	1	0	1	1
$t_5 - t_6$	1	0	1	0	1	1
$t_6 - t_7$	0	0	1	0	1	1

Output y is high (1) so output at the terminal y, $V_0 = 5V$

92. (d)

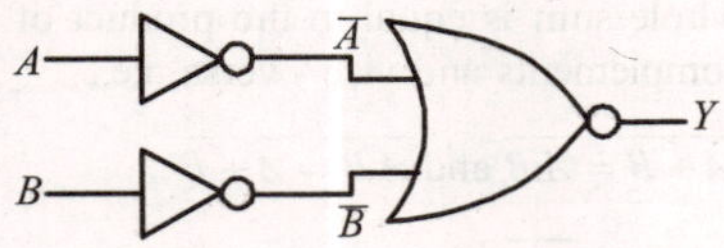

$Y = \overline{\overline{A} + \overline{B}} = \overline{\overline{A \cdot B}} = A \cdot B \Rightarrow$ AND Gate

Truth Table is :

A	B	Y
0	0	0
0	1	0
1	0	0
1	1	1

93. (c) From the given logic circuit, LED will glow, when voltage across LED is high.

Truth Table

A	B	Y
0	0	1
0	1	1
1	0	1
1	1	0

Boolean expression $Y = \overline{A.B}$

i.e.,circuit represents NAND gate.

94. (d) We assume that at any instant thickness of ice is x. And time taken to form additional thickness (*dx*) is *dt*.

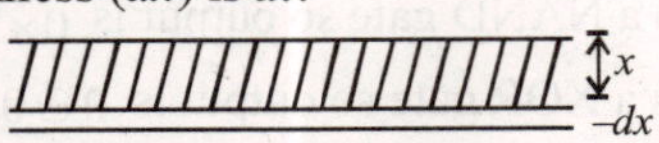

$$mL = \frac{KA[26-0]dt}{x} \Rightarrow (Adx)\rho L = \frac{KA(26)dt}{x}$$

$$\text{So, } \frac{dx}{dt} = \frac{26K}{x\rho L}$$

95. (b)

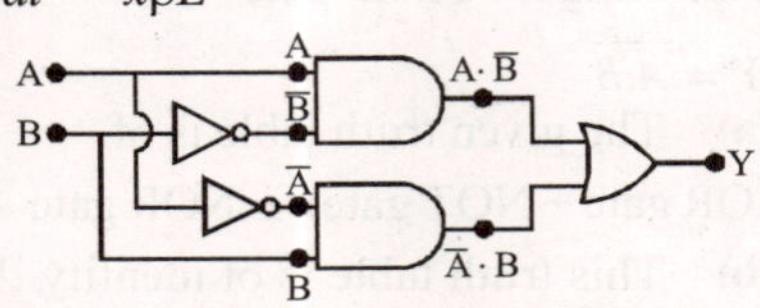

$Y = (A \cdot \overline{B} + \overline{A} \cdot B)$

96. (b)

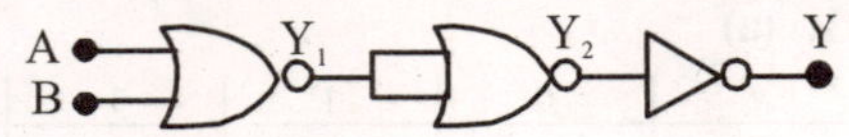

$Y_1 = \overline{A + B}$

$Y_2 = \overline{Y_1 + Y_1} = \overline{Y_1} = A + B$

$Y = \overline{Y_2} = \overline{A + B}$ i.e. NOR gate

97. (d) The Boolen expression for the given combination is

output $Y = (A + B) \cdot C$

Truth table

A	B	C	$Y = (A+B) \cdot C$
0	0	0	0
1	0	0	0
0	1	0	0
0	0	1	0
1	1	0	0
0	1	1	1
1	0	1	1
1	1	1	1

Hence, A = 1, B = 0, C = 1

98. (b) First two gates are NOT gates and the last gate is NOR gate.

Thus, $Y_1 = \overline{A}$, $Y_2 = \overline{B}$ and $Y = \overline{Y_1 + Y_2}$

The truth table corresponding to this is as follows:

A	B	$Y_1 = \overline{A}$	$Y_2 = \overline{B}$	$Y_1 + Y_2$	$Y = \overline{Y_1 + Y_2}$	A . B
0	0	1	1	0	0	0
0	1	1	0	1	0	0
1	0	0	1	1	0	0
1	1	0	0	0	1	1

Thus the combination of gate represents AND gate.

99. (b) $C' = \overline{A \cdot B} \Rightarrow C = \overline{\overline{A \cdot B}} = A \cdot B$

Hence the resultant gate is AND gate.

100. (b)

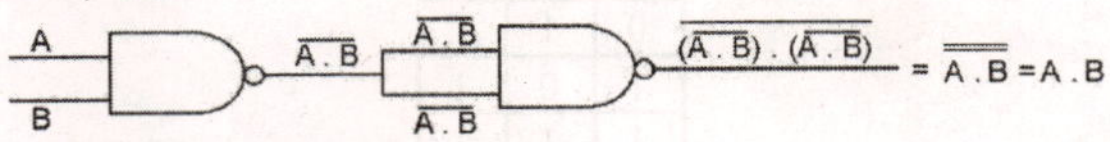

i.e., output X = A.B

Truth Table

A	B	X
0	0	0
1	0	0
0	1	0
1	1	1

101. (a)

A	0	1	1	0
B	0	0	1	1
C	1	1	1	1

OR gate

102. (d)

103. (b) Integrated circuit can act as a complete electronic circuit.

104. (a) From the given waveforms, the truth table is as follows.

A	B	Y
1	1	0
0	0	1
0	1	1
1	0	1

The above truth table is for NAND gate. Therefore, the logic gate is NAND gate.

105. (c) Let A and B be inputs and Y the output.

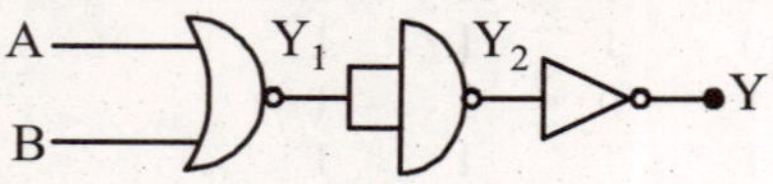

Then $Y_1 = \left(\overline{A+B}\right) = \bar{A}.\bar{B}$

(By De-Morgan's theorem)

$Y_2 = \overline{Y_1} = \left(\overline{\bar{A}.\bar{B}}\right) = \bar{\bar{A}} + \bar{\bar{B}} = A + B$

$\therefore\ Y = \overline{Y_2} = \left(\overline{A+B}\right)$

Hence, the given circuit is equivalent to a NOR gate.

> **NOTES** NOR and NAND gates is a universal gates as they can be used to perform the basic logic functions. AND, OR and NOT gate.

106. (c)

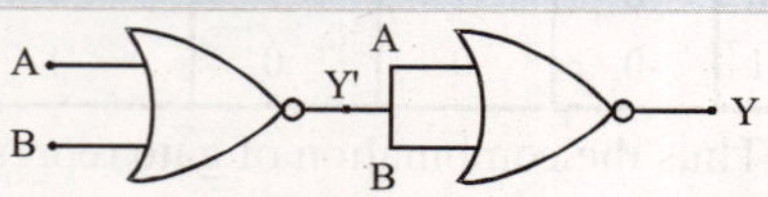

$Y' = \overline{A+B}$; $Y = \overline{\overline{A+B}} = A+B$

Therefore truth table :

A	B	Y
0	0	0
0	1	1
1	0	1
1	1	1

107. (d) On the basis of given graph, following table is possible.

A	B	C
0	0	0
1	1	1
0	1	0
1	0	0

It is the truth table of AND gate.

108. (c) Output will be one if A or B or both are one. $Y = A + B$

109. (b)

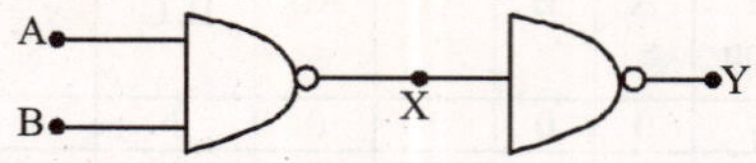

$X = \overline{A.B}$

$\therefore Y = \bar{X} = \overline{\overline{A.B}}$

$Y = A.\ B$ by Demorgan theorem

∴ This diagram performs the function of AND gate.

> **NOTES** De morgan's theorem - the complement of the whole sum is equal to the product of individual complements and vice - versa, i.e.,
>
> $\overline{A+B} = \bar{A}.\bar{B}$ and $\overline{A.B} = \bar{A} + \bar{B}$

110. (c) $Y = \overline{A.B}$

A	B	Y
0	0	1
0	1	1
1	0	1
1	1	0

Which is truth table of NAND gate.

111. (a) P, Q and R are related as $R = P.\ Q$ which is relation of AND gate.

112. (b) Output of upper AND gate $= A.\bar{B}$

Output of lower AND gate $= \bar{A}.\ B$

∴ Output of OR gate, $Y = A.\ \bar{B} + B.\ \bar{A}$

This is Boolean expression for XOR gate.

113. (d) (a) is a NAND gate so output is $\overline{1 \times 1} = \bar{1} = 0$

(b) is a NOR gate so output is $\overline{0+1} = \bar{1} = 0$

(c) is a NAND gate so output is $\overline{0 \times 1} = \bar{0} = 1$

(d) is a XOR gate so output is $0 \oplus 0 = 0$

Following is NAND Gate

$Y = \overline{A.B}$

114. (a) The given truth table is of (OR gate + NOT gate) ≡ NOR gate

115. (b) This truth table is of identity, $X = A + B$, hence OR gate.